Fundamentals of Weather and Climate

You have spread out the heavens like a tent
and built your homes on the waters above.
You use the clouds as your chariot
and ride on the wing of the wind.
You use the winds as your messengers
and flashes of lightning as your servants.

PSALM 104, vv. 2–4
Good News Bible

Fundamentals of Weather and Climate

SECOND EDITION

Robin McIlveen

OXFORD
UNIVERSITY PRESS

Great Clarendon Street, Oxford OX2 6DP

Oxford University Press is a department of the University of Oxford.
It furthers the University's objective of excellence in research, scholarship,
and education by publishing worldwide in

Oxford New York

Auckland Cape Town Dar es Salaam Hong Kong Karachi
Kuala Lumpur Madrid Melbourne Mexico City Nairobi
New Delhi Shanghai Taipei Toronto

With offices in

Argentina Austria Brazil Chile Czech Republic France Greece
Guatemala Hungary Italy Japan Poland Portugal Singapore
South Korea Switzerland Thailand Turkey Ukraine Vietnam

Oxford is a registered trade mark of Oxford University Press
in the UK and in certain other countries

Published in the United States
by Oxford University Press Inc., New York

First published in 1986 as *Basic Meteorology: A physical outline*
First edition published by Chapman and Hall 1992
Second edition published by Oxford University Press 2010

British Library Cataloguing in Publication Data
Data available

Library of Congress Cataloging in Publication Data
Data available

Typeset by MPS Limited, A Macmillan Company
Printed in Great Britain on acid-free paper by
CPI Antony Rowe
Chippenham, Wiltshire

ISBN 978–0–19–921542–3

10 9 8 7 5 6 4 3 2 1

Contents

5 Atmospheric thermodynamics

6 Cloud and precipitation

7 Atmospheric dynamics

8 Radiation, convection, and advection

9 The atmospheric engine

10 Surface and boundary layer

11 Smaller-scale weather systems

12 Large-scale weather systems in mid-latitudes

13 Large-scale weather systems in low latitudes

14 Climate and climate change

Preface to the second edition

It is 24 years since I wrote the preface to the ancestor of this book. The modest success of *Basic Meteorology and Fundamentals of Weather and Climate* (first edition) for the first 20 of those years has pleased me more than I can say, especially when expressed personally by readers. The basic premiss remains that there is a need for a book which covers the fundamentals of weather and climate in an intellectually rigorous but accessible format. This, together with much of the core text, remains in spirit, but has been completely rewritten in detail, and three major changes have been made.

- The treatment of climate and climate change has been greatly extended and moved to the end of the book. It is still necessarily very selective and discursive, but it includes a thorough outline of two important aspects of the current ice age which are often covered in mere outline or a computer-generated graph—the geometry of Earth's orbit and spin, and the effect of varying degrees of glaciation on our tilt and orbital seasons. In commenting on current rapidly changing perceptions of the scale and importance of climate change, I have tried to stay close to physical reality while steering a course between the apocalyptic and a bureaucratic blandness which seems to assume that it is all happening somewhere else.
- The treatment of atmospheric energy has been greatly extended and moved into the heart of the book. It covers energy, heat, and the thermodynamics of heat engines in a basic, painstaking way which recognizes that many students nowadays come to this core topic needing help to fill the gap between excellent thermodynamic texts which say almost nothing about the atmosphere, and advanced theoretical treatments which assume familiarity with fairly advanced meteorology, physics, fluid dynamics, and applied mathematics. The chapter on atmospheric dynamics has been extended by an introduction to conservation of angular momentum which tries to demystify the topic, just as I tried to demystify the Coriolis effect in the first edition of *Fundamentals*.
- The presentation has been restructured throughout to meet a complaint about the previous edition that it was readable rather than memorable from a students' point of view. Text is structured much more than before, and Boxes are scattered through the text to fill in theoretical, conceptual, or practical details, or discuss important issues which lie on the edge of the main narrative. And a series of short Numerical Notes repeatedly show how mechanisms being discussed in the narrative can be explored and checked numerically and related to familiar personal, local, or global experience.

As I near the end of my intellectual life (though with hopes of a little more to come) I become more and more convinced that I am most receptive to conceptual awakenings and understandings when I am puzzling over how to make simple calculations about important things. In that sense I part company from a trend in secondary and tertiary education to suppose that concepts are primary, and associated practicalities can look after themselves. The simple truth is that concepts and their numerical expression must remain in harness with each other, and with observation—the third element of the troika which has made such astonishing progress since it first began to gallop with Copernicus, Galileo, and Newton—and that this applies to our personal progress in science understanding, however humble or exalted, just as much as it obviously applies to the ongoing human science project.

As is usual and proper, I would like to mention some of the people near and far, obscure and famous, who I know have helped me to enjoy coming to terms, however limited, with the nature of things, especially of the physical atmosphere: the writer of Psalm 104 for showing me that to look attentively is to rejoice; my parents for giving me an education they never had themselves; my school physics and mathematics teachers for showing me how to get into physics by the three-fold way; Frank Ludlam and others in the golden autumn of Imperial College's brilliant little Meteorology Department for sharing their infectious and yet demanding love for their subject; Richard Feynman and James Clerk Maxwell for taking time in their busy lives to show how good technical teaching texts can be; a generation of staff and students in the Department of Environmental Sciences at Lancaster University for earthing me when I threatened to float off into the heavens; the staff at Oxford University Press for encouragement, help, and forbearance; and my wife Janice for helping and sorting me out when in the toils.

Robin McIlveen

Acknowledgements

I am grateful to the many people who have helped to provide the wide range of graphical material which is designed to be a significant feature of the book. The great majority of the line drawings were produced over 20 years ago by Jane Rushton of Lancaster University's Department of Environmental Science in response to my pencil sketches. Her work has been redrawn and extended by the staff of Oxford University Press for the present edition. Quite a number of the photographs of the previous edition have been replaced, and I am very grateful to Dewi Jackson of OUP for tracking down some excellent digital images. The following list acknowledges the valued supply of photographs, graphs and drawings, either freely or with explicit permission. Entries beginning with 'After', 'Data from', etc, acknowledge more indirect dependence on cited material. Unlisted photographs and line drawings are from my cameras and rough sketches.

Figure acknowledgements

Chapter 1

Figure 1.1 NASA–Goddard Space Flight Center, data from NOAA GOES.
Figure 1.2 NASA.
Figure 1.3 NOAA GOES Project Science Office.
Figure 1.7 After Gleick, J (1987) *Chaos*. Sphere, London; and Stewart, I. (1989), *Does God play dice?* Penguin, Harmondsworth.

Chapter 2

Figure 2.7 Larsson, P. and Beadle, M. (1985). *[88]*
Figure 2.8 National Weather Service, NOAA.
Figure 2.11 After Ludlam, FH. (1966). *The cyclone problem*. Inaugural lecture, Imperial College, London.
Figure 2.12b Online National Climatic Data Centre, NOAA.
Figure 2.13 NOAA/NWS, Reno, Nevada.
Figure 2.14 NEODAAS/University of Dundee, UK.
Figure 2.15 NEODAAS/University of Dundee, UK.

Chapter 3

Figure 3.4 NASA.
Figure 3.6 After Shanklin, JD. and Gardner, BG. (1989). British Antarctic Survey pamphlet, Cambridge, UK.
Figure 3.7 Robert A Rohde (robert@globalwarmingart.com).
Figure 3.8 After Bolin, B. (1971). *The carbon cycle*. WH Freeman, San Francisco.
Figure 3.9a After Siegenthaler, U. and Oeschger, H. (1987). *Tellus* 38B.
Figure 3.9b After Department of the Environment. (1989) *Global Climatic Change*. HMSO, London.
Figure 3.10 © Fotolia.
Figure 3.11 With kind permission from Ray Stern.
Figure 3.13 Martin Rushton, Bridge House Hotel, Grasmere.
Figure 3.14 NASA.
Figure 3.15 After Wayne, RP. (1985)—Bibliography.
Figure 3.16a © Nancy Nehning/iStockphoto.com.
Figure 3.16b NASA.

Chapter 4

Figure 4.1 Bill Butterfield, courtesy of KFYR-TV.
Figure 4.3 Plotted data from the Met Office, Exeter, UK.
Figure 4.5 NASA.
Figure 4.9 Plotted data from US Weather Bureau, US Department of Commerce, Washington, DC.
Figures 4.10, 4.11, 4.13 After Palmen, E. and Newton, CW. (1969). *[76]*

Chapter 5

Figure 5.2 After Warner, J. and Telford, J. (1967). *J. Atmos. Sci.* **24**: 374–82.
Figure 5.4 © Oscar Gutierrez/iStockphoto.com.
Figure 5.7 Plotted data from UK Met Office, Exeter, UK. Tephigram adapted from Met Office Metform 2810.
Figure 5.8 Tephigrams adapted from Met Office Metform 2810.

Chapter 6

Figure 6.6 Tephigram adapted from Met Office Metform 2810.
Figure 6.9 Robert Nutbrown.
Figure 6.11 © Alan Clements.
Figure 6.13 After Mason, BJ. (1962)—Bibliography.
Figure 6.14 Sergeant Al Roberts, Royal Air Force. © British Aerospace, Warton, Lancs, UK.
Figure 6.17 © Stephan Hoerold/iStockphoto.com.
Figure 6.22 After Wallace, JM and Hobbs, PV. (1977). *[17]*
Figure 6.23 © Deepdesert Dreamstime.com.
Figure 6.25 NASA/University of Alaska Fairbanks.
Figure 6.27 AEML, Department of Aviation Technology, Purdue University.
Figure 6.28 NOAA.
Figure 6.29 Goddard, JWF, STFC-Chilbolton Observatory, Chilbolton, Stockbridge, Hants, UK.

Chapter 7

Figure 7.3 Howcheng (from http://commons.wikimedia.org/wiki/File:Windblown_tree_Ka_Lae_Hawaii.jpg).
Figure 7.12a John Bowman, Dept of Environmental Science, University of Lancaster, UK.
Figure 7.23 After the Daily Weather Report, Met Office, Exeter, UK.
Figure 7.26 After Hide, R and Corby GA (eds). (1970). *The general circulation of the atmosphere*, Royal Meteorol. Soc., London.
Figure 7.28 NASA.

Chapter 8

Figure 8.1 After Gast, PR in Scorer, RS. (1978)—Bibliography.
Figure 8.4 After Garing, JS in Valley SL (ed). (1965). *Handbook of Geophysics and Space Environments*, McGraw-Hill, New York.
Figure 8.8 Data from London, J and Sasamori, T. (1971). *[37]*
Figure 8.9 After London, J and Sasamori, T. (1971). *[37]*
Figure 8.11 Data from Budyko, MI. (1966). *[66]*
Figure 8.15 Image created by www.theweatherchaser.com using data from the Japanese Meteorological Agency and the Australian Bureau of Meterology.
Figure 8.16 Data from London, J and Sasamori, T. (1971). *[37]*
Figure 8.18 After London, J and Sasamori, T. (1971). *[37]*
Figure 8.22 © Thomas Bradford/iStockphoto.com.
Figure 8.24 NASA.

Chapter 9

Figure 9.7 © iStockphoto.com.
Figure 9.8 After US Global Change Research Program. http://www.usgcrp.gov.

Chapter 10

Figure 10.9 After Oke, TR (1987). *[42]* and Munn, RE. (1966). *[71]*
Figure 10.11 After P Larsson, M Beadle, Dept Environmental Science University of Lancaster, UK.
Figure 10.12 Dr NE Holmes.
Figure 10.17 Data from Lettau, HH and Davidson, B (eds). (1957). *[44]*
Figure 10.19 After Oke, TR. (1987). *[42]*
Figure 10.20 Data from Lettau, HH and Davidson, B (eds). (1957). *[44]*
Figure 10.21 After Munn, RE. (1966). *[71]*
Figure 10.23 After Oke, TR. (1987). *[42]*
Figure 10.24 After Defant, A. in Barry, RG. (1981). *[21]*
Figure 10.25 © Dainis Demcs/iStockphoto.com.
Figure 10.26 After Simpson, JE, Mansfield, DA, Milford JR. (1977). *[73]*
Figure 10.30 After Monteith, JL. (1989). *[38]*
Figure 10.32 (Golden Gate, Sails) http://www.flickr.com/photos/trevino/191200266/.

Chapter 11

Figure 11.6 After Palmen, E and Newton, CW. (1969). *[76]*.
Figure 11.8 After Saunders, PM. (1961). *[74]*
Figure 11.11 After Ludlam, FH. (1980). *[29]*
Figure 11.14 After Ludlam, FH. (1966). *[75]*
Figure 11.15 Data from the Met Office, Exeter, UK. Tephigram adapted from Met Office Metform 2810.
Figure 11.16 © NOAA, Rockville, MD 20852, USA.
Figure 11.17 After Ludlam, FH. (1980). *[29]*
Figure 11.18 After Newton, CW. (1963) in *Meteorological Monograph* **27**: 33–58, and Palmen, E and Newton, CW. (1969). *[76]*
Figure 11.20 After Collinge, V, *et al.* (1990). *[77]*

Figure 11.21 After Palmen, E and Newton, CW. (1969) *[76]*, and Ludlam, FH. (1980). *[29]*
Figure 11.22 After Wallace, JM and Hobbs, PV. (1977). *[17]*
Figure 11.23a From http://www.flickr.com/photos/auntiep/440131048/.
Figure 11.23b R.J. Reynolds Tobacco Company Slide Set, Bugwood.org.
Figure 11.24 Courtesy of Gareth Berry.
Figure 11.25 NEODAAS/University of Dundee.
Figure 11.26 NASA.
Figure 11.27a After Browning, KA. (1971). *[78]*
Figure 11.27b After Matejka, TJ and Hobbs, PV. (1980). *[79]*
Figure 11.28 © NOAA, Rockville, Maryland 20852, USA.
Figure 11.29 Dr NE Holmes and JFR McIlveen. (1976). *[88]*
Figure 11.33 NEODAAS/University of Dundee.
Figure 11.34 Ludlam, F.H. (1980) Clouds and storms. Pennsylvania State University Press, Pennsylvania & London.

Chapter 12

Figure 12.1 After Fitz-Roy, R. (1863) in Figure 12.1.1 of Petterson, S. (1956). *[80]*
Figure 12.2 After Figure 12.1.4 of Petterson, S. (1956). *[80]*
Figure 12.4 After Figures 11.11.1 and 11.11.2 in Petterson, S. (1956). *[80]*
Figure 12.5 After Godske, CL, Bergeron, T, Bjerknes, J, and Bundgaard, RC. (1956). *[81]*
Figure 12.6 NEODAAS/University of Dundee.
Figure 12.7 NEODAAS/University of Dundee.
Figure 12.9 Hustvedt (http://commons.wikimedia.org/wiki/File:Moon_22_halo_colorado.jpg).
Figure 12.12 After New Zealand Meteorological Office, Wellington, New Zealand, courtesy Rod Stainer.
Figure 12.13 The Chief Engineer, Lancaster City Council.
Figure 12.16 After Carlson, TN. (1980). *[53]*
Figure 12.17 © Matthew Rambo/iStockphoto.com.
Figure 12.18 After Lundulph, PE. (1977). *[82]*
Figure 12.19 After Flohn, H. (1969). *[83]*
Figure 12.20 After Eady, ET. (1947). *[31]*
Figure 12.21 After Petterssen, S. (1956). *[80]*
Figure 12.23 After Figure 9.8 of Ludlam, FH (1980). *[29]*
Figure 12.26 After Willett, HC. (1944). *Descriptive meteorology*. Academic Press, New York.
Figure 12.27 After Stevenson, CM. (1967). *[84]*
Figure 12.28 After Figure 65 of Wickham, PG. (1970). *[13]*

Chapter 13

Figure 13.1 After the Daily Weather Report of the Met Office, Exeter, UK.
Figure 13.2 Simon Eugster.
Figure 13.3 Data from the Met Office, Exeter, UK.
Figure 13.7 After Ramage, CS. (1971) *[85]*, and Palmen, E and Newton CW. (1979). *[76]*
Figures 13.8–13.10 After Palmen, E and Newton CW. (1979). *[76]*
Figure 13.11 NASA.
Figure 13.12 After Palmen, E and Newton CW. (1979). *[76]*
Figure 13.13 NOAA.
Figure 13.14 After Palmen, E and Newton CW. (1979). *[76]*
Figure 13.15 NOAA.
Figure 13.17 After Palmen, E and Newton CW. (1979). *[76]*

Chapter 14

Figure 14.1 After Barry, RG and Chorley, RJ. (1989). *[87]*
Figure 14.2 Graph by Hanno based on data from J W Hurrell, Climate Analysis Section, National Centre for Atmospheric Research, Boulder, Colorado, USA.
Figure 14.3 After Moller, F. (1951). *Zeitschrift fur Meteorologie*, **1**: 1–7.
Figure 14.4 After Barry, RG and Chorley, RJ. (1989). *[87]*
Figure 14.5 Data from RR McKenzie, Hazelrigg Weather Station, Lancaster University, UK.
Figure 14.6 Historical Central England Temperature Data. Met Office, Exeter UK. **© Crown Copyright 2009.**
Figure 14.7 Mark A. Wilson (Department of Geology, The College of Wooster Ohio, USA).
Figure 14.8 After Lamb, HH. (1982). *[61]*
Figure 14.9 After Figure 5.19 in Lockwood, JG. (1979) —Bibliography.
Figure 14.13 After Figure 15.23 in Stanley, SM. (1993). *[90]*
Figure 14.15a From http://www.flickr.com/photos/gruban/137421445/.
Figure 14.15b http://en.wikipedia.org/wiki/File:Frost_Fair_of_1683.
Figure 14.16a The Chief Engineer, Lancaster City Council, UK.

List of tables

Useful information

Commonly used prefixes for powers of ten

Name	Symbol	
nano	n	10^{-9}
micro	μ	10^{-6}
milli	m	10^{-3}
centi	c	10^{-2}
kilo	k	10^{3}
mega	M	10^{6}
giga	G	10^{9}
tera	T	10^{12}
peta	P	10^{15}

Spherical properties (radius *R*)

Diameter	$2R$
Circumference	$2\pi R$
Maximum cross-section area	πR^2
Surface area	$4\pi R^2$
Surface area between latitudes 30°	$2\pi R^2$
Volume	$4/3\pi R^3$

SI units and common equivalents

Primary

Quantity	Name	Units	Equivalence
mass	kilogram	kg	
	tonne		10^3 kg

(*Continued*)

Quantity	Name	Units	Equivalence
length	metre	m	
	angstrom	Å	10^{-10} m
	nautical mile		1,852 m
time	second		
temperature	kelvin	K	
	Celsius	°C	Same unit size but T°C $\equiv (T + 273.2)$ K
quantity of pure element or compound	mole	Mol (formerly called gram molecule)	
angle	radian	rad	57.3°
	degree	°	0.017 45 rad (360° = 2π rad)
electrical current	ampere	A	

Some Secondary Units Used in Meteorology

frequency	hertz	Hz	one cycle per second
volume	cubic metre	m^3	
	litre	l	10^{-3} m^3
electric charge	coulomb	C or A s	
specific mass	kilogram per kilogram		
	gram per kilogram	g kg^{-1}	10^{-3}
	parts per million	ppm	10^{-6}
velocity	m per s	m s^{-1}	
	knot (nautical mile per hour)	kt	0.514 m s^{-1}
angular velocity		rad s^{-1}	57.3° s^{-1}
temperature gradient		K m^{-1} or °C m^{-1}	
velocity gradient, divergence, vorticity		s^{-1}	
density	kg per cubic m	kg m^{-3}	
acceleration	m per s per s	m s^{-2}	
momentum		kg m s^{-1}	
force	newton	N or kg m s^{-2}	
pressure	pascal	Pa or N m^{-2}	
	millibar	hPa or mbar	100 Pa
pressure gradient		Pa m^{-1}	
		hPA per 100 km	10^{-3} Pa m^{-1}

(Continued)

Quantity	Name	Units	Equivalence
angular momentum (or moment of momentum)		kg m^2 s^{-1}	
moment of force		N m	
energy	joule	J or N m	
	gram calorie	cal	4.18 J
power	watt	W or J s^{-1}	
energy flux density		W m^{-2}	
langley (cal per cm^2) per minute		Ly min^{-1}	697 W m^{-2} (radiation)
heat capacity		J K^{-1} or J °C^{-1}	
specific heat capacity		J kg^{-1} K^{-1}	
		cal g^{-1} °C^{-1}	4,180 J kg^{-1} K^{-1}

(See Box 7.6 for further treatment of units)

Universal constants

Speed of light *in vacuo*	c	3.00×10^8 m s^{-1}
Planck constant	h	6.626×10^{-34} J s
Avogadro number	N	6.022×10^{23} mol^{-1} (molecules per mol)
Universal gas constant	R^*	8.314 J K^{-1} mol^{-1}
Gravitational constant	G	6.672×10^{-11} N m^2 kg^{-2}
Stefan–Boltzmann constant	σ	5.670×10^{-8} W m^{-2} K^{-4}

Useful values in units normally used

Solar and Terrestrial	
Solar Diameter	1.39×10^6 km
Solar–terrestrial separation	150 ($\pm$ 5) $\times 10^6$ km
Effective blackbody temperature of solar surface	5,800 K
Solar constant	1,380 W m^{-2}
Sidereal day (period of terrestrial rotation)	86,164 s (23.934 hours)
Angular velocity of terrestrial rotation	7.292×10^{-5} rad s^{-1}
Coriolis parameter	1.0×10^{-4} rad s^{-1} (or s^{-1}) at latitude 43.3°
Solar day (mean period of apparent terrestrial rotation relative to the Sun)	86,400 s (24 hours)

(Continued)

Year (period of terrestrial orbit of Sun)	3.156×10^7 s (365.3 days)
Calendar year	3.154×10^7 s (365 days)
Angle between equatorial and ecliptic planes	23.4° (latitudes of tropics)
Effective blackbody temperature of planet Earth	255 K
Earth's planetary albedo in solar wavelengths	about 0.30
Terrestrial mean radius	6,370 km
Average gravitational acceleration at mean sea level	9.81 m s^{-2}
Terrestrial surface area	5.10×10^{14} m^2
Meridional length subtending latitude degree	111.2 km or 60 nautical miles
Elevation of highest surface above mean sea level (Mt Everest)	8,848 m
Depression of lowest surface below mean sea level (Dead Sea surface)	392 m

Atmospheric

Standard temperature	0 °C	Standard temperature and pressure (STP)
Standard atmospheric pressure	1,013.2 hPa	
Effective molecular weight of dry air	29.0	
Specific gas constant for dry air	287 J kg^{-1} K^{-1}	
Specific heat capacity for dry air at constant volume	717 J kg^{-1} K^{-1}	
Specific heat capacity for dry air at constant pressure	1,004 J kg^{-1} K^{-1}	
Density for dry air at STP	1.292 kg m^{-3}	
Decadal scale height of isothermal dry atmosphere at –20 °C	17.1 km	
Velocity of sound in dry air at 0 °C	331.5 m s^{-1}	
Thermal conductivity of dry air at 0 °C	2.43×10^{-2} W m^{-1} K^{-1}	
Dynamic viscosity of dry air at 15 °C	17.8×10^{-6} kg m^{-1} s^{-1}	
Saturated vapour at 0 °C		
Density	4.85×10^{-3} kg m^{-3}	
Pressure	6.11 hPa	
Specific latent heat of vaporization of water at 0 °C	2.50 MJ kg^{-1}	
Specific latent heat of fusion of ice at 0 °C	0.334 MJ kg^{-1}	
Density of water at 0 °C	1,000 kg m^{-3}	
Density of ice at 0 °C	917 kg m^{-3}	
Density of mercury at 0 °C	13, 595 kg m^{-3}	
Specific heat capacity of water at STP	4,187 J kg^{-1} K^{-1}	

Selection from International Civil Aviation Organization Standard Temperate Atmosphere

Height/m	Pressure/hPa	Temperature/K	Density/kg m^{-3}
0	1,013.3	288.2	1.23
5,000	540.5	255.7	0.736
10,000	265.0	223.2	0.414
15,000	121.1	216.7	0.195
20,000	55.3	216.7	0.0889

Typical magnitudes

Solar horizontal irradiance at Earth's surface	0–1,000 W m^{-2} low latitudes and mid-latitude summer 0–200 W m^{-2} mid-latitude winter
Average tropospheric wind speed	10 m s^{-1}
Wind speed in jet core	50 m s^{-1}
Height of jet core above sea level	9 km polar front jet stream 12 km subtropical jet stream
Height of tropopause above sea level	10 km high latitudes 15 km low latitudes
Height of freezing level above sea level	4 km low latitudes 0–3 km mid-latitudes
Depth of sub-cloud layer	1 km
Depth of precipitable water	30 mm
Average annual precipitation	1,000 mm
Specific humidity	10 g kg^{-1} low troposphere 1 g kg^{-1} mid-troposphere
Sea-level pressure gradient associated with gale force wind in mid-latitudes	5 hPa per 100 km
Equivalent tilt of isobaric surface	40 m per 100 km

Weather Systems

Type	*Breadth*	*Lifetime*	*Surface wind*	*Updraught*	*Rainfall*
cumulonimbus	10 km	1 hr	15 m s^{-1}	5 m s^{-1}	10 mm hr^{-1}
extratropical cyclone	1,000 km	1 week	15 m s^{-1}	5 cm s^{-1}	1 mm hr^{-1} (background)
hurricane	500 km	1 week	40 m s^{-1}	10 m s^{-1}	40 mm hr^{-1} (active annulus around the eye)

Introduction 1

1.1 The shallow film of air

We live out our lives under a constantly changing sky whose importance we dimly sense from early childhood. Familiarity and modern sheltered lifestyles can easily dull our awareness of it, even though weather is a common subject of casual conversation; however, an unusually colourful sunset or threatening storm can rekindle interest, admiration, and even fear. Unfortunately our small human scale and typical surface-based viewpoint can easily hamper efforts to comprehend the size, power, and complexity of the atmosphere. To set the scene for this book, we consider the atmosphere in brief outline, beginning with its very different horizontal and vertical scales.

Horizontally the atmosphere is enormously larger than we can see from any point on the Earth's surface. Since it completely covers the Earth, the *order of magnitude* (Box 1.1) of its horizontal extent is just that of the Earth's surface ~100,000 km. A higher viewpoint helps greatly, and photographs like Fig. 1.1 allow us to appreciate the vast panorama in ways that were impossible before the first meteorological satellites took cameras into orbit in the 1960s.

By comparison, the vertical extent of the atmosphere is very small, since the atmosphere is only a shallow film of air held to the Earth's surface by gravity (Fig. 1.2). Distances we could walk in a few hours in the horizontal are a very significant fraction of the vertical depth of the atmosphere. Our impression of the great height of a towering thundercloud, for example, arises more from a sense of the effort it would require to climb a comparable mountain, than from its actual height (~10 km). But although the atmosphere is so very shallow (3–4 orders of magnitude broader than the 10 km depth assumed) its influence on our living conditions is enormous, as will be apparent throughout this book.

Figure 1.1 The face of the Earth centred on the equator at about longitude 105° W as seen in visible light by geostationary satellite GOES-9 at 1800 UTC, July 11 1995. The satellite is hanging over the equator at a point just W of the dark curve of clear, cold water round N Peru and Equador. This is the Humboldt current whose sporadic failure is part of the huge El Niño event. The panorama shows a series of tropical disturbances in the E central Pacific, the Gulf of Mexico and beyond Florida. The large area of shallow low cloud in the Equatorial E Pacific indicates widespread subsiding air. The great swirls of cloud in higher latitudes are the extratropical cyclones. The influence of land on cloud is obvious over N and S America.

BOX 1.1 **Order of magnitude**

High numerical accuracy is vital for many scientific purposes but not for all; it is sometimes enough to know values to their nearest integral power of 10—their *order of magnitude*. The Earth's circumference is very nearly 40,000 km, which is *of order* 100,000 km (written ~100,000 km) since 40,000 is four times larger than 10,000 but only 2.5 times smaller than 100,000. The horizontal extent of the atmosphere is therefore 8 orders of magnitude larger than the human body (~1 m)—a difference which impedes human observation and conceptualization.

Orders of magnitude are also useful for quantifying measurables which vary greatly in magnitude. For example wind speeds range from about 0.1 m s^{-1} to nearly 100 m s^{-1} (i.e. across 3 orders of magnitude), and associated wind forces on people, buildings, etc. range over 6 orders of magnitude (i.e. a million-fold) since they tend to vary with the square of the wind speed (Box 9.1).

Unlike the oceans, the atmosphere has no definite upper surface: its pressure falls smoothly with increasing height, from values close to 10^3 hPa (one *atmosphere*) at the surface, to values 8 orders of magnitude (10^8 or 100 million times) smaller in the thin interplanetary gases a few hundred kilometres above the surface. Despite the lack of a top edge, the atmosphere's vertical distribution can be specified very usefully by its *decadal scale height* (Box 1.2) which lies between

BOX 1.2 Scale heights

The *decadal scale height* for atmosphere pressure is the height interval in which pressure varies 10-fold. Since pressures at mean sea level (MSL) and at 16, 32, and 48 km above MSL, are observed to be about 1000, 100, 10, and 1 hPa respectively, it follows that the decadal scale height for pressure lies close to 16 km through the first 50 km. Indeed such nearly pure *exponential decay* (Box 4.3) persists to over 90 km above MSL, where pressure is only about 0.001 hPa. Above this level the scale height increases significantly as the proportions of lighter gases rise relative to the heavier ones (Sections 3.1.1 and 3.2).

It follows from the arithmetic of exponential decay that the *binary* scale height for atmospheric pressure (the height interval in which atmospheric pressure halves or doubles) is almost exactly 0.33 of its decadal scale height, and is therefore close to 5 km throughout the first 90 km of the atmosphere, which means that atmospheric pressures at 5, 10, and 15 km above MSL are about 500, 250, and 125 hPa respectively.

Since pressure is effectively proportional to the mass of overlying atmosphere (Box 1.3), and the pressure 16 km above MSL is only 10% of the MSL value, it follows that about 90% of the mass of the atmosphere lies in that first 16 km. Similarly another 9% of its mass lies between 16 and 32 km above MSL, and 0.9% lies between 32 and 48 km. Assuming a corresponding binary scale height of 5 km, similar reasoning shows that the layers bounded by the first three binary scale heights above MSL contain 50%, 25%, and 12.5% of the mass of the atmosphere.

Figure 1.2 The limb of the Earth as seen by hand-held camera in relatively low orbit. Clouds (in the troposphere) are veiled by its haziness. The transition from haze to blue and deep black on the horizon is about 50% deeper than the Moon's disc, and contains 99% of the mass of the atmosphere.

16 and 17 km in the first 100 km above the Earth's surface, even though pressure falls by over 6 orders of magnitude through this layer. Since pressure is proportional to the weight (and therefore mass, Box 1.3) of overlying atmosphere, it follows from Box 1.2 that about 90% of the mass of the atmosphere is concentrated in the lowest 16 km—squeezed by the weight of overlying air into a layer whose depth is only 0.25% of the Earth's radius.

1.2 The stratified atmosphere

Since so much of the atmospheric mass is squeezed into such a shallow layer overlying the Earth's surface, the distributions of atmospheric pressure, temperature, humidity, etc. are all much more closely packed in the vertical than they are in the horizontal. For the same reason *isopleths* (lines or surfaces of equal quantity, such as *isobars* of pressure or *isotherms* of temperature) on vertical sections through the atmosphere (e.g. Fig. 4.9) are usually horizontally layered or *stratified* in appearance.

Corresponding to the stratification of atmospheric structure, there is an equally marked stratification of behaviour. Immediately overlying the Earth's surface is the *atmospheric boundary layer* (ABL, also called the *planetary boundary layer*)—a shallow, turbulent region, often literally torn between the land or sea surface below and the atmosphere above. Its depth is ~100 m, though it can vary three-fold either way with time and location. On the scale of Fig. 3.1 the ABL is hardly thicker than the printed horizontal axis marking zero height but it contains about 1% of the mass of the overlying atmosphere.

The rest of the atmosphere (about 99% of its total mass) is termed the *free atmosphere* because it is much less directly influenced by the underlying surface. The ABL, together with the first 10–15 km of the free atmosphere (the larger value applying in low latitudes), is collectively called the *troposphere* (from the Greek for 'turning') because it is full of weather-related turning and overturning. The active and cloudy troposphere is separated from the overlying relatively quiet and cloud-free *stratosphere* by the often sharply defined *tropopause*. Commercial aircraft cruising 12 km above sea level are usually in the high troposphere in low latitudes, whereas in middle and high latitudes they are usually in the low stratosphere. This book deals mainly with the troposphere because it is the site of most atmospheric activity which is related to surface weather.

Atmospheric stratification is already quite obvious on scales comparable with its binary scale height (about 5 km). In fact its activity is dominated by even larger, highly *flattened* disturbances to such an extent that, for ease of examination, vertical sections through the troposphere are drawn with vertical scales expanded by about two orders of magnitude (e.g. Fig. 4.9). On much smaller scales (~500 m and smaller), weather-related activity is much more *isotropic* (i.e. similar in scale, etc. in all directions). This is particularly the case in the ABL, which is heavily influenced by a hierarchy of turbulent *eddies* ranging in size from ~10 mm to ~100 m. In fact most of the troposphere is often slightly turbulent, as is obvious on many aircraft journeys, though less vigorously so than the ABL, as noticeable on take-off or final approach.

Figure 1.3 A low-latitude panorama from the geostationary satellite GOES 11 hanging over the E central Pacific. Deep clouds of the InterTropical Convergence zone form a broad line from the central Pacific to the horizon beyond the Caribbean, with shallow, low clouds (under gently subsiding air) widespread in the visible higher latitudes of both hemispheres. The cloud-free zone over the Humboldt current is even larger than in Fig 1.1, and there are large patches of rain cloud across the Amazon basin.

1.3 The disturbed atmosphere

The atmosphere, especially the troposphere, is in a state of continual commotion. *Cumulonimbus* clouds, taller than the mountains they can resemble, erupt and dissolve in an hour or so, as their updrafts, and the showers they sustain, grow and fade (Fig. 6.20). In middle latitudes, swirls of air (Fig. 1.1) as broad as small continents grow and decay in a few days, in their brief maturity coiling long bands of cloud (with swathes of rain and snow) round the areas of low surface pressure which give them their popular name (*depressions* or *lows*). Technically they are known as *extratropical cyclones*, to distinguish them from the *tropical cyclones* of low latitudes. The most energetic of the latter are popularly known as *hurricanes*, *typhoons*, or *cyclones* in different geographical regions (Section 13.5). Both extratropical and tropical cyclones contain and interact with activity on small scales: for example both contain regions populated by vigorous cumulonimbus, whose turbulent updrafts and downdrafts can make flying through them uncomfortable or even dangerous.

Superficially the atmosphere is a scene of huge chaos, but the very existence of characteristic types and scales implies underlying order. There is method in the apparent madness which *atmospheric science* aims to identify and describe in the most concise and comprehensive manner. Though by its Greek origins *meteorology* (the science of things aloft) could cover everything above the Earth's surface, in practice it has come to mean the study of the physical rather than the chemical nature of the lower atmosphere, especially the troposphere. And its meaning has narrowed even further in the last two centuries to focus on the physical aspects related to weather lasting from hours to days, leaving the term *climatology* to cover longer-term averages and types of weather-related activity. The time scales of meteorology and climatology overlap on the seasonal time scale (~100 days). In the last fifty years there has been an explosion of interest in *climate change*, which is the variability of climate on time scales from a few years to the age of the Earth—(Fig. 1.5 and Section 14.5).

Figure 1.4 Several swelling cumulus (congestus) showing flat bases, knobbly sides and tops, and the mountainous bulk typical of rapidly growing convective clouds. Notice the silver lining produced by strong forward scatter of sunlight, and the cloud shadow cast on the haze nearer the camera.

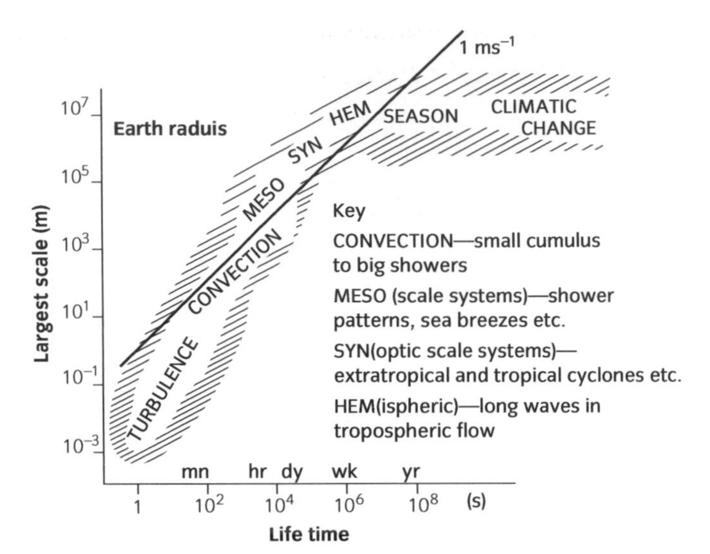

Figure 1.5 Types of weather-related atmospheric disturbance displayed according to lifetime and maximum space scale. Each type is mentioned in this chapter and discussed at length later in the book. Note that logarithmic scales (equal lengths along an axis corresponding to equal multiples of numerical value) cover wide ranges of value very compactly, but require careful interpolation.

Curiosity is the mainspring of meteorology, as it is of every branch of science, but so much atmospheric activity affects humankind so vitally, even disastrously, that meteorology has been developed since the mid-nineteenth century with the aim of routinely *forecasting* weather for as little as six hours ahead, especially as it affects surface conditions. In this period, forecasting has developed complex and specialized techniques practised routinely by nationally and internationally organized networks of observers and analysts (Section 2.8). In the last couple of decades, the challenge of forecasting climate change over years, decades, and longer, has begun to attract nearly comparable international effort (Section 14.8), particularly as we have become aware of past and likely future climate changes, and their potential importance for human and other life.

Though details of weather forecasting are considered only incidentally in this book, its bias towards types of tropospheric activity which strongly affect human activity is reflected in its coverage, which is why we largely ignore beautiful optical phenomena such as the rainbow, and are very selective in treating the vast range of chemical activity in the atmosphere, which has attracted much attention in recent decades under the heading of *atmospheric chemistry* (Sections 3.3–3.6 and *[1]*).

Much of the weather-producing activity of the atmosphere can be arranged in a spectrum according to space and time scales (Fig. 1.5). There is a range of over 10 orders of magnitude in space scale, and only slightly less in time scale. Space and time scales are related in such a way that big events last longer than small ones, the ratio of space to time being ~1 m s^{-1} for all the items depicted. This ratio is not so much a speed as a measure of the intensity or busyness of Earth's atmospheric activity, and its small value contrasts with the much larger values for the violently disturbed visible surface of the Sun, where *granulations* (violent solar convection) with horizontal diameters ~500 km appear and disappear within minutes. The list in Fig. 1.5 is necessarily incomplete; less important types are deliberately omitted for clarity, and no doubt other types remain as yet unidentified among the welter of observation.

1.4 The application of physical law

We believe that the observed structure of the atmosphere can be related to the operation and cooperation of natural physical and chemical laws, amongst which the forecasting bias of meteorology highlights physical laws in particular. Of these, the most important are the laws of statics and dynamics (complicated in expression but not in essence by the atmosphere's fluid nature), heat and electromagnetic radiation, and the properties of matter.

Consider for example the forces acting on a small volume of air, conventionally called an *air parcel*. In the lower atmosphere there are three such forces, arising respectively from pressure gradients, gravity, and friction. At a given time and place the *resultant* $\boldsymbol{F}$ of these forces is related to the acceleration $\boldsymbol{a}$ of the air parcel of mass M by a form of Newton's second law of motion, often called the *equation of motion* of the parcel,

$$\frac{\boldsymbol{F}}{M} = \boldsymbol{a} \qquad 1.1$$

where $\boldsymbol{F}$ and $\boldsymbol{a}$ are printed in bold to show that they are vectors rather than the scalar M (Box 1.3). In different atmospheric zones, air motion (represented here by its acceleration $\boldsymbol{a}$) can differ considerably in character, depending on the dominant forces acting. For example, in the ABL intense turbulence is associated with large transient frictional forces (Section 10.7), whereas in the free atmosphere friction is usually unimportant in comparison with gravitational and pressure gradient forces, and can often be ignored without serious error. Detailed applications of relations like Eqn 1.1 usually depend on judicious separation of primary and secondary factors, since only then are the relations mathematically simple enough to be analysed and solved (Section 7.9).

Note that Eqn 1.1 is insufficient on its own to describe fully any physical situation in the atmosphere. Formal expressions of other constraints, such as the conservation of mass and the laws of thermodynamics, must be combined with Eqn 1.1 before a well-defined solution is possible even in principle.

BOX 1.3 Physical quantities

These are measurable features of the atmosphere, such as the mass of an air parcel, its speed, velocity, acceleration, temperature, cloud content, electrical potential, etc. Some quantities have magnitude only, and are called *scalars*, whereas others have direction as well as magnitude, and are called *vectors*. In the present list, mass, speed, temperature, cloud content, and electrical potential are scalars, whereas velocity, acceleration, and all forces are vectors.

The magnitude of parcel *velocity* is just its *speed* (making an important technical distinction ignored in everyday speech) expressed in metres per second ($m\ s^{-1}$) in the SI system, and its horizontal direction is measured in relation to true North, and represented on a weather map by an arrow pointing in the direction of the air flow.

Two other quantities not distinguished in everyday speech are mass and weight. The *mass* of an air parcel is a scalar quantity measured by its reluctance to accelerate when acted on by a chosen force (as in Eqn 1.1) and expressed in kilograms (kg) in SI units. By contrast the *weight* of an air parcel is the force of the gravitational pull on it towards the centre of the Earth, and is therefore a vector quantity, with direction down the local vertical (Section 7.4.2), and magnitude expressed in newtons (N)—the SI unit of force. In meteorologically relevant conditions the mass of a body is conserved, whereas its weight varies with the slight variations in the strength of the surrounding gravitational field. For example, the weight of a certain mass of air falls by about 0.5% as it rises 15 km from sea level, since the Earth's gravitational pull falls in the same proportion (Table 7.1). Because this variation is quite small, the numerical distinction between mass and weight is significant only in the most accurate meteorological work, but the conceptual distinction is basic (Section 7.4.2).

Mass M, length L, time T, and temperature θ, are basic independent measurable quantities from which other quantities can be derived. For example, a volume is a length multiplied by breadth (another length) multiplied by height (yet another length). We say that the *dimensions* of volume are $[L]^3$. The dimensions of density and acceleration are respectively $[M]\,[L]^{-3}$ and $[L]\,[T]^{-2}$, consistent with their measurement units in SI or any other consistent system. All physically valid equations must be dimensionally consistent, since they would otherwise balance in some unit systems but not in others (Box 7.6) and lack the universal validity demanded of a law of nature.

BOX 1.4 Cause and effect

It seems natural to regard $\boldsymbol{F}$ and $\boldsymbol{a}$ in Eqn 1.1 as cause and effect respectively, because in homes, factories, and laboratories we often arrange forces to achieve desired effects; but it is important to realize that the basic laws identified by natural science describe relationship, rather than cause or effect—an important point missed in most popular and some scientific descriptions of physical behaviour. Actual events may well involve cause and effect (according to some of the many meanings of the terms), but such distinctions are notoriously difficult to make in the atmosphere, given its largely self-activating nature.

The temptation to identify cause and effect is strong because it satisfies our innate desire to explain things in terms of origins and consequences, but it is enhanced in meteorology by the nature of much weather-producing activity. Above the ABL, the dominant types of disturbance are cumulus and larger-scale weather systems (Fig. 1.5). In their combination of regularity, individuality, and transience, cumulus, thunderstorms, depressions, hurricanes, etc. resemble living organisms, and you will often find in forecasts and books (including this one) a tendency to treat them as such. This does not matter so long as it remains a descriptive device (like calling a particular hurricane Katrina), and so long as we are not tempted to think that we are truly distinguishing 'chicken and egg'. Thus a statement such as 'the weakening of the anticyclone over the British Isles will allow a belt of rain to move in from North West Ireland' is descriptively convenient, but must not be taken to imply will, purpose, or even priority in the interaction. In fact the weakening no more allows the advance than the advance causes the weakening—the two events are related (in fact quite subtly). Such relationships abound in meteorology, and long experience shows that it is positively unhelpful to distinguish priority in the endless flux of atmospheric activity. There is, however, a resounding exception to this rule, in the conspicuous role of the Sun as prime mover of all atmospheric activity (Sections 1.5, 8.1, and 8.2).

1.5 The Sun and the atmospheric engine

On any sunny day overland, it is obvious that the Sun has an important and direct influence on the atmosphere: its energy, about half of it in the form of visible sunlight, begins to warm the land surface soon after dawn, and with it the overlying air, which begins to convect, produces cumulus clouds if conditions allow (Fig. 1.4 and Section 5.10). Much of this book confirms and amplifies the

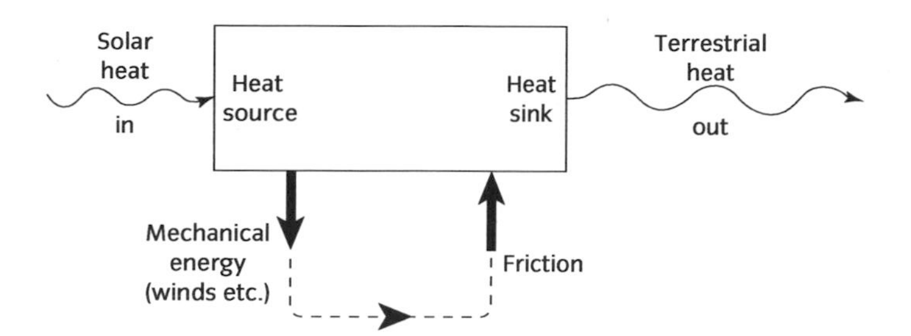

Figure 1.6 Schematic diagram of the atmospheric heat engine with heat source, sink, and flow of mechanical energy. This is subsequently transformed to heat by surface friction and internal friction in the atmosphere and oceans (Ch. 9).

initial impression that the atmosphere is quite literally solar powered. In this sense the Sun can be regarded as the prime mover of almost all atmospheric activity.

Sir Napier Shaw, an early twentieth century pioneer of modern meteorology, wrote that 'the weather is a series of incidents in the working of a vast natural engine'. Just as in an artificial engine, heat is taken into the atmospheric engine at a *heat source*, and exhausted from it at a cooler *heat sink* (Fig. 1.6), and *mechanical energy* is generated by temporary conversion of some of the throughput of heat. The heat source is located wherever solar energy is absorbed, mainly at the Earth's surface, and the heat sink is located wherever far infrared

BOX 1.5 Determinism and its limitations

Equations like Eqn 1.1 show that present conditions of a system are inescapably linked to all past and future conditions of the system. As Newton's legacy consolidated in the eighteenth century, it began to be assumed that the operation of such natural laws in any physical system 'determined' its future conditions. This claim was stated most ambitiously by the French natural philosopher and mathematician Laplace, who claimed in 1812 that perfect knowledge of the present state of everything in the universe, together with perfect understanding of all its laws of behaviour, would determine the future of the universe to the end of time. Such *Laplacian determinism* enthused and inspired the scientific community during the nineteenth century, even when investigating subjects, like the atomic nature of solids and fluids, vastly more complex than Laplace's ballistic model of the Sun and planets. And meteorologists inherited this confidence in the early twentieth century, as they began to apply fundamental physical laws to weather forecasting.

The first obvious breach of Laplacian determinism came in the 1920s, when quantum theory postulated, and observation confirmed, an inescapable uncertainty (named after Heisenberg) in our knowledge of the state of individual atomic and sub-atomic particles. All individual events on these small scales are blurred by an uncertainty which can be reduced only by considering the statistics of large numbers. Fortunately atoms are so small and numerous that this is exactly what happens in the much larger scales of activity which are studied in much of atmospheric science.

All larger-scale behaviour was still assumed to be deterministic in the Laplacian sense (and therefore perfectly forecastable in principle), until the development of electronic computers in the 1960s allowed numerical investigation of the full range of behaviour described by equations representing natural laws. This greatly enlarged the relatively small range of behaviour previously revealed by analytic (roughly 'algebraic') solutions of the equations. Quite unexpected results emerged separately from a wide range of physical, chemical, and biological studies. When brought together in the 1970s, these triggered one of those profound reviews of accepted wisdom which seem to agitate the scientific community every 50 years or so. Interestingly, the pioneering study was made by a meteorologist, Edward Lorenz *[2]*.

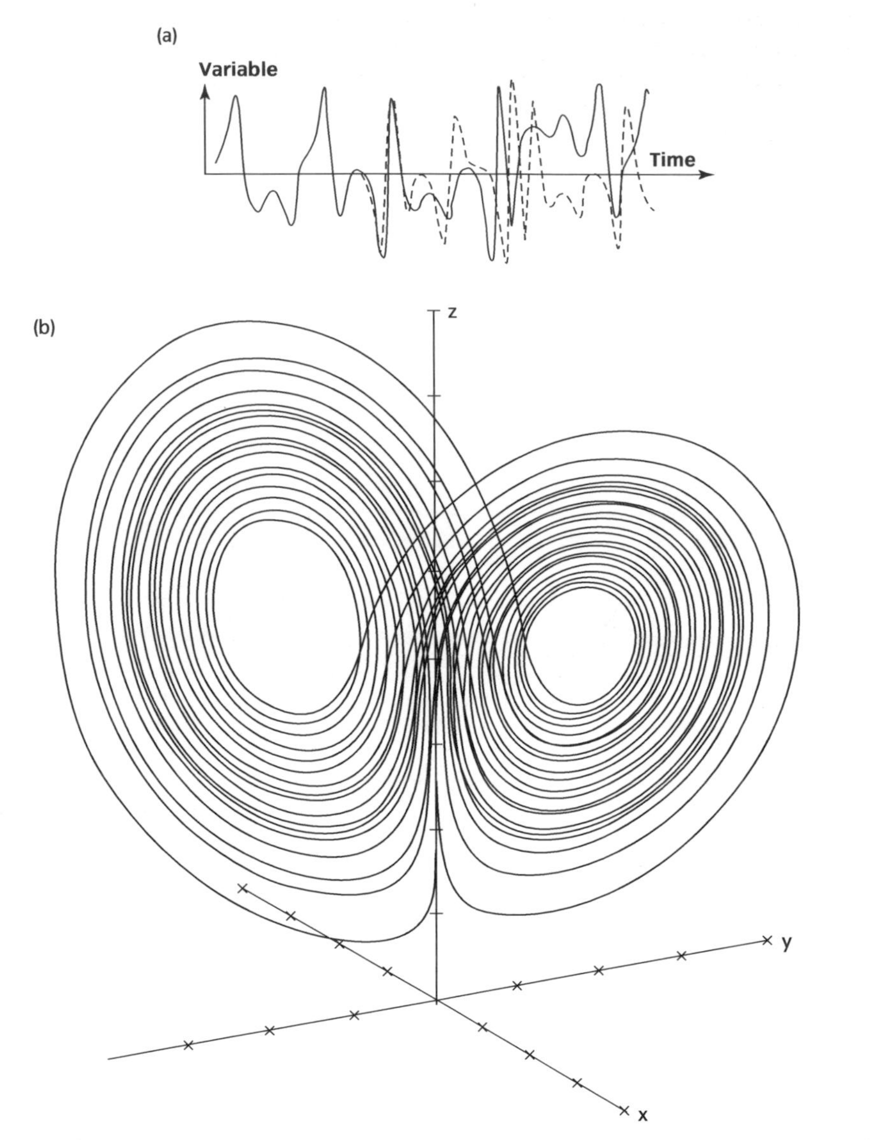

Figure 1.7 Portraits of chaos. (a) The butterfly effect: two runs of the same Lorenz atmospheric model, starting from minutely different initial conditions. Resemblance between between behaviour of a chosen variable (displacement about the vertical axis) virtually disappears about two thirds the way through the time period (the hoizontal axis) of the otherwise identical simulations. (b) Loose repetition with unexpected switches: the line representing the continuous sequence of states of the Lorenz convection model turns in upon itself as the convection cell rolls about a horizontal axis. However, the rolling changes slightly and progressively until it switches to slowly varying rotation in the opposite direction.

terrestrial radiation is emitted to space, mainly in the upper troposphere, but also at the Earth's surface under cloudless skies (Fig. 1.1). The mechanical energy produced appears in the form of the motion and vertical displacement of the air parcels and masses heated and cooled in this way (Section 9.7), and is quite quickly reduced to heat, mainly by friction.

Unlike artificial engines, the atmospheric engine is not constrained by a prefabricated structure of turbines, etc.: it generates and maintains its structure as it runs, including the hierarchy of disturbances associated with weather. The atmospheric engine also controls its intake of solar energy by producing cloud masses which can reflect (and therefore not absorb) a significant proportion of sunlight back to space (Fig. 1.1). Clouds also regulate the important output of terrestrial radiation to space (Section 8.2) from the heat sinks. All such regulation keeps the activity of the atmospheric engine within its typical range. Obviously an engine which continually repairs, regenerates, and regulates itself is more complex and subtle than an artificial engine, but there is no difference in the underlying principles at work, which include the *first* and *second laws of thermodynamics* (Chapters 5 and 9) in addition to the equation of motion (Eqn 1.1) and other physical and dynamical laws (Chapter 7). And underlying all, the atmospheric engine is kept going by the continual heat flow from the Sun, just as an artificial engine is kept going by its heat source.

1.6 Chaos in the atmosphere

In the 1960s Edward Lorenz used a primitive weather forecasting computer program to simulate months of weather by allowing it to run for very long periods from realistic initial conditions. He found by accident that his model atmosphere

BOX 1.6 The meaning of scientific explanation—a personal view

I believe that, by allowing key words to carry more meaning than they can logically bear, scientists can overstate what human science has achieved in increasing our understanding of the world. What are we entitled to mean when we say that we have scientifically 'explained' some aspect of the world, for example a thunderstorm? It is tempting to say that we have thereby explained why the thunderstorm is the way that it is. But if we do this we are presupposing that natural laws somehow explain and justify their own existence and can therefore be said to explain any event described in their terms. Since natural laws in fact justify themselves only in so far as they match observation, it is a circular argument to say that they can explain such observation. It is fairer to say simply that we have, by the application of natural law, described the thunderstorm more fully than ever before. The seeds of confusion between description and explanation were sown in the eighteenth century European Enlightenment, and their philosophically doubtful harvest reaped by Laplace, whose famous determinism (Box 1.5) effectively treats natural law as a sufficient cause for the behaviour it so impressively describes.

Consider a falling hailstone. For simplicity of statement, we say that it falls because it is acted on by gravity, but taken literally this carries a spurious implication of cause and effect. In fact what we really mean is that the hailstone falls towards the Earth in ways which are accurately described by the laws of dynamics, including gravitational attraction, atmospheric friction, and the various other laws which describe the complex growth of hailstones in clouds (Section 6.11). The identification and application of these laws is a triumph of human curiosity and intellect which wonderfully adds to previous descriptions of the ways of nature, thereby increasing, rather than reducing, the sum total of what is to be explained, in the most fundamental sense of the word. There is ample historical evidence (see below) that our scientific understanding of the nature of things is served better by persistent, humble questioning than by over-dogmatic answering which risks dimming our perception of the nature of things by overbearing preconception.

In 1862 William Thomson (later Lord Kelvin) made the first of many estimates of the age of the Earth by calculating its rate of cooling from an initially molten state to its present condition on the basis of the newly established laws of heat conduction and radiation. The answers were and remained around 30 million years, and geologists were confidently told that their much longer estimates (made on the messy, empirical basis of sedimentation rates, etc.) must be wrong. The completely unexpected discovery of radioactivity in 1892 revealed vast resources of nuclear energy which degrade to heat within the Earth, slowing its cooling and providing dating mechanisms which now show the Earth to be just over 4,500 million years old, and establishing the vast time scale for the slow and complex development of the Earth's surface, atmosphere, and biosphere to their present conditions (Sections 3.9 and 14.5). Kelvin's great reputation remains secure in very many ways (viz. the Kelvin scale of absolute temperature), but not on this account.

was so sensitive that minute differences in its assumed starting conditions (far smaller than could possibly be measured in the real atmosphere) would give rise to completely different simulated weather conditions right across the model globe only a few simulated months later. Lorenz rightly guessed that this was a property of the real atmosphere rather than his greatly simplified model. To illustrate this extreme sensitivity to initial conditions, he coined the graphic image of the 'butterfly effect', in which the flapping of a butterfly's wings in the Amazon would lead to a completely different pattern of the world's weather a few months later.

A little later Lorenz found similar extreme sensitivity in a primitive model of atmospheric convection (cumulus, etc.), and subsequent work by others has found it also in much simpler systems, such as the compound pendulum and the annual sequence of animal populations suffering growth and predation. In every case the sensitivity arises from nonlinear relations between parts of a system (i.e. relations more sensitive than simple direct or inverse proportionality), and can produce a type of irregular, effectively unpredictable behaviour called *deterministic chaos*, often simplified to *chaos*. It now seems that virtually every natural system has a capacity for such chaotic behaviour, and that the atmosphere in particular is probably at least sporadically *chaotic* on most of the scales depicted in Fig. 1.5.

Chaos is not random, despite superficial appearances, and its abbreviated title. The present state of any chaotic system still relates to past and future states according to natural laws, but the future behaviour of a chaotic system is predictable only in a limited sense, in both observational practice and mathematical principle. Sensitivity to initial conditions means that uncertainties arising from inaccurate observations grow more and more rapidly as the system evolves in time, and the same applies to any model simulation of the system. As a result, doubling the accuracy of measurement of initial conditions, even in the most perfect weather forecasting model, yields much less than a doubling of its useful forecasting period. Although current weather forecasting models will undoubtedly improve with increasing understanding of physical and chemical processes and computational skill, it is becoming thinkable that it may never be possible to forecast weather with current confidence and detail much more than a couple of weeks in advance (current practice manages about one week), confirming Lorenz's early doubts. Less specific seasonal forecasts may well become possible, as understanding and prediction of short-term climatic change improves (Sections 14.3 and 14.4), but they too will be limited, if climatic systems too are chaotic, as seems almost inevitable.

In addition to their extreme sensitivity to initial conditions, another characteristic property of chaotic systems is that they never repeat themselves exactly, although they may behave in a loosely cyclic fashion. This is just what we see in the atmosphere: broadly similar but never identical depressions form every few days in preferred parts of the middle latitudes, but at constantly varying intervals and locations; seasonal variations are broadly but never precisely the same; and no two cumulus clouds, turbulent eddies, and snowflakes are ever identical.

More subtly, it is known from the study of simple chaotic systems that aspects of typical behaviour are closely related to details of the nonlinear relations at work within them *[3]*. Although the chaotic nature of such a large and complex system as the atmosphere is only beginning to be examined in sufficient detail, it seems at least possible that typical atmospheric structure (for example the sizes and lifetimes of depressions) may be explicable to this extent some day.

Checklist of key ideas

You should now be familiar with the following ideas.

1. The atmosphere is shallow and nearly horizontally stratified in structure and behaviour.

2. Air pressure falls quickly and smoothly with increasing height, with a nearly uniform scale height.

3. Most weather activity is confined to the troposphere and is organized in meteorologically small and large (synoptic scale) weather systems.

4. Meteorology is the science of the physical behaviour of the lower atmosphere, especially those aspects relevant to weather forecasting.

5. The atmosphere obeys well-established quite simple physical and dynamical laws whose cooperation is often complex.

6. Atmospheric physical quantities are either scalar (having magnitude only) or vector (having magnitude and direction).

7. It is usually unhelpful to try to distinguish cause and effect in the atmosphere, apart from the Sun, which drives the atmospheric engine by input of solar radiant energy.

8. The variable aspects of weather and climate are dominated by deterministic chaos whose presence fundamentally limits predictability.

Problems

Outline answers to these problems can be found on the **Online Resource Centre**. Answers to odd numbered problems can be found under Student Resources, answers to even numbered problems under Lecturer Resources.

Level 1

1.1 Estimate the order of magnitude of the depth/breadth ratio of the troposphere.

1.2 What is the approximate atmospheric pressure two decadal scale heights above sea level?

1.3 Place the following in order of ascending height: troposphere, planetary boundary layer, stratosphere, free atmosphere, stratopause, tropopause. Deal with overlaps using brackets.

1.4 What is the ratio (in m s^{-1}) of the space and time scales of solar granulations?

1.5 Divide the following atmospheric quantities into vectors and scalars: wind velocity, temperature, gravitational force, wind speed, pressure (noting that experiment shows that pressure at any point in a fluid is independent of direction).

1.6 Identify the location of the effective heat source for the atmospheric engine: upper troposphere, tops of clouds, Earth's surface, stratosphere.

1.7 Figure 13.12 represents a vertical section through a severe tropical cyclone. If the vertical scale of the diagram were redrawn to be equal to the horizontal scale, what would be the length of the new vertical axis as a percentage of the horizontal?

Level 2

1.8 On a certain weather map point A has temperature 23°C, and point B temperature 15°C. If isotherms are drawn at intervals of 2°C including 16°C, at least how many isotherms must there be between A and B? Why 'at least'?

1.9 Horizontal accelerations of air in large weather systems are observed to be $\sim 10^{-4}$ m s^{-2}. What is the implied net force on an air parcel of mass 1 tonne? Note that Eqn 1.1 gives the answer in newtons if you enter M and $\boldsymbol{a}$ in SI units. Express the answer as a percentage of your weight, which is $9.8 \times M$, where M is now your mass in kg.

1.10 Estimate the volume of the air parcel in Exercise 1.9 near sea level (see Useful Information).

1.11 List four different meanings of 'cause' and 'because' in everyday English.

Level 3

1.12 Sketch an example of a temperature field for Exercise 1.8 with more than the minimum number of isotherms.

1.13 Actually the spuriously animistic statement in Box 1.4 about the anticyclone and the rain belt can be rewritten just as simply without implying animism, cause, or effect. Try it.

1.14 A Hollywood film director is threatening to make a disaster movie in which a rogue tornado ravages the countryside for several days. Give tactful advice.

1.15 What ultimately happens to the mechanical energy generated by the atmospheric engine? If you find the answer puzzling (as I did once), can you articulate your puzzlement? Very often articulation of misunderstanding helps understanding.

1.16 Improve the usual verbal definition of pressure to deal with the ambiguity riddling the end of Exercise 1.5.

1.17 In an introductory lecture on thunderstorms in 1962, Frank Ludlam asked us if we had heard the old story about thunderstorms coming up against the wind. I said that I had (from my mother who was terrified of thunder), gleefully supposing I was about to be given ammunition to demolish such an obviously silly notion in future. He told me that it often happened, and then explained how. As cocky beginners with little background in meteorology, Ludlam knew that we would have little idea of how wind flow can vary with height, and caused constructive embarrassment which I still remember. Identify my unwarranted assumption, skim Section 11.6.3 to see how wrong it was, and read again the last paragraph of Box 1.6.

Observations 2

2.1 The development of observation

Observational meteorology presumably began as our prehistoric ancestors learnt to scan the morning skies and weigh up sensations of warmth, wind, cloud, etc. to assess the weather's suitability for a day of hunting or gathering. Awareness must have increased with the onset of the agricultural revolution about 10,000 years ago, as the climate moved into its present interglacial (Section 14.5) and farmers became especially sensitive to weather at planting and harvesting. The growth of coastal and ocean voyaging in the last three millennia encouraged even greater sensitivity among sailors, whose accumulated experience, and need for reliable warning of heavy weather around coasts and harbours, eventually established modern observational meteorology and forecasting in the mid-nineteenth century. The development of aviation in the twentieth century produced a need for observations and forecasts of wind and weather conditions at airports and along flight paths in the upper atmosphere.

2.1.1 **Observational instruments**

Though we can personally sense a wide range of meteorological conditions, most bodily sensations are too indefinite to be reliable, even when focused by an anxious or enquiring mind. Our three-dimensional hunter's vision is very acute, but this advantage is limited in the absence of a conceptual framework for interpreting what we see. To unscramble the seeming confusion of atmospheric activity, our observations must be as unambiguous and definite as possible, and history shows that this requires the development of instrumental aids. Much of the observational information mentioned in this book was gleaned only after meteorological instruments began to appear in the seventeenth century, and you can judge their contribution to modern understanding by reading the erratic meteorological speculations made with the unaided senses by even the most perceptive observers of the ancient world *[4]*.

2.1.2 **Good observational practice**

Several general points arise when deciding how best to design and use observational instruments, and these are listed below and exemplified and developed in the rest of this chapter.

(i) *Field experience is crucial* For example, although temperature is obviously more reliably measured by thermometer than by bodily sensation of warmth, it is not immediately clear how best to expose a thermometer. As well as being in contact with the air, should the thermometer be exposed to sunlight, whose warming effect is so obviously important, or should it be shaded? In fact the value of consistent shading was not fully realized for nearly a century after thermometers were first used meteorologically in the mid-seventeenth century (Section 2.2), until extensive field experience proved that results from shaded thermometers are more consistent across ranges of location and time. Similarly, though measurement of atmospheric pressure by barometer began in the early seventeenth century, its extension to ships took a century, and its crucial role in weather mapping was not fully realized until the middle of the nineteenth century (Section 2.5). And the appearance of the meteorological satellite in the early 1960s was followed by two decades of vigorous trial before the present pattern of use was established (Section 2.11).

(ii) *All measurement is subject to limitation and error* However impressive the trappings of modern instrumentation, and the computerized logging and display of its output, instrumental data must never be accepted without critical appraisal. Every meteorological measurement is less than perfect to an extent which must be known and stated when it is recorded, and recalled and assessed when it is being interpreted. To avoid implying unwarranted accuracy, values should never be recorded or quoted to a higher resolution than these limitations allow.

(iii) *All instruments take a finite time to respond* to changes in what they measure, and this degrades their accuracy in ways which must be understood. Response time is often defined by observing the reaction of the measuring instrument to an abrupt change in the variable being measured (temperature in the case of a thermometer). The *lag* of the instrument is then defined as the time taken to register 63% of the abrupt change, a numerical value which arises naturally from the nearly *exponential* response of many instruments (Box 2.1). Lag is a measure of the sluggishness of an instrument's response, and therefore of the degree of time averaging and delay automatically imposed on what it measures. As a rule of thumb, an instrument cannot reliably measure variations lasting less than 10 times its lag as defined above.

(iv) *Instrumental networks* Imperfections in individual observations are compounded by problems of *representativity* and *consistency* when comparing data from widely separated places and times, as is inevitable in meteorology (Section 2.8). The need for consistency across national and international networks is obvious in principle, but in practice demands ruthless attention to every detail of instrumental design, maintenance, exposure and reading (e.g. Section 2.9.1). Problems of representativity are less obvious, but can be surprisingly intractable—the areal representativity of rain-gauge measurements of rainfall being disturbingly uncertain to this day (Section 2.6).

2.1.3 Basic surface measurements

The following sections on meteorological instruments and their data selectively exemplify all these points. Comprehensive descriptions are found in official observing manuals *[5,6]*. Previous comments about the time taken to come to terms with new data may apply to you personally, depending on your previous experience of observational meteorology. It may not be clear to you for some time just why certain observations and measurements are made the way they are, or are made at all. Persevere and reflect that current observational systems have been developed globally by many thousands of people over several centuries.

2.2 Temperature

Our bodies are sensitive to temperature in many competing ways. We feel hot or cold depending on whether our bodies are having to waste or conserve heat to maintain core temperature at about 37 °C, which depends on clothing, exposure to sunlight and terrestrial radiation, wind, humidity (which affects our ability to lose heat by sweating), state of health, and the time, size, and temperature of our latest meal, as well as on the temperature of the air around us *[7]*. Clearly a more focused and objective sensor is required.

The familiar alcohol-in-glass thermometer was developed in the mid-seventeenth century by Ferdinand II de Medici but although it was soon used to measure air temperature, over a century elapsed before the need for carefully standardized exposure was generally recognized. By then Fahrenheit had developed the mercury-in-glass thermometer and the temperature scale bearing his name. A thermometer, like the human body, is sensitive to both solar and terrestrial radiation, from which it must be shielded if it is to isolate the effects of air temperature. But shielding must allow sufficient ventilation for the air temperature inside the shield (i.e. in contact with the thermometer) to be nearly the same as outside.

The familiar *Stevenson screen* (Fig. 2.1) is a simple, robust compromise solution to the dilemma. Its white surfaces absorb little sunlight; and its thick wooden

(a)

(b)

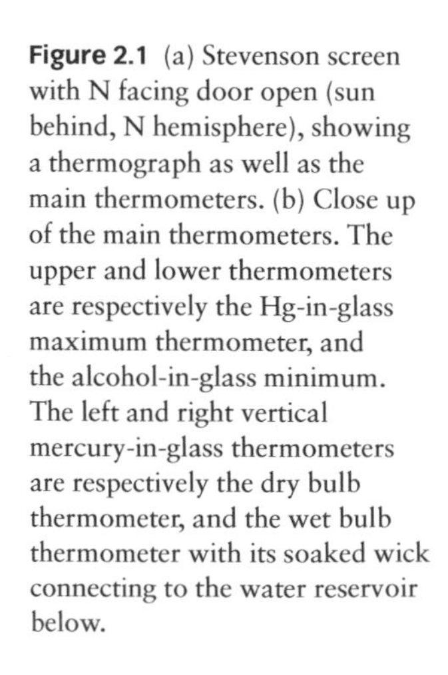

Figure 2.1 (a) Stevenson screen with N facing door open (sun behind, N hemisphere), showing a thermograph as well as the main thermometers. (b) Close up of the main thermometers. The upper and lower thermometers are respectively the Hg-in-glass maximum thermometer, and the alcohol-in-glass minimum. The left and right vertical mercury-in-glass thermometers are respectively the dry bulb thermometer, and the wet bulb thermometer with its soaked wick connecting to the water reservoir below.

walls insulate the interior from warming by residual solar absorption, and from warming or cooling by absorbing or emitting terrestrial radiation. It also shields thermometers from wetting by rain and consequent evaporative cooling. The access door is placed on the poleward side of the screen so that direct sunlight does not enter while thermometers are being read. The louvred walls and floor allow interior ventilation by available wind, but though this is adequate in many conditions, on still, sunny days it can allow air temperatures inside the screen to remain a degree or more above those outside. Ventilation can be improved by using a fan to force a draught of outside air through the screen interior, but the instrument then requires a power source, and is more expensive and vulnerable to breakdown—important points to consider in a large observation network which must be reliable in routine use.

2.2.1 Other thermometers

After nearly 400 years, the *mercury-in-glass thermometer* is still in widespread use, despite two serious disadvantages:

(i) it fails in low temperatures (i.e. at high enough latitudes and altitudes), since mercury freezes at about –40 °C; and

(ii) it cannot provide an automatic record of temperature. A manual record of the highest temperature reached since last reset is made by having a restriction between the mercury bulb and the graduated stem, which holds the mercury column at the highest temperature until it is shaken down to the current temperature.

Alcohol-in-glass thermometers can work to much lower temperatures, and are widely used in horizontally exposed *minimum thermometers* (Fig. 2.1), in which the alcohol meniscus pushes a small iron index along the inside of the thermometer bore as temperature falls, but leaves it at the minimum value until moved back by a magnet (after reading) to the meniscus at the current temperature. Note that when the *screen minimum* temperature falls below 0 °C there is said to be an *air frost*, regardless of whether or not frost is deposited nearby. Crude continuously recording thermometers (*thermographs*) have a temperature-sensitive *bimetallic strip* which moves a pen vertically across a chart on a drum turning horizontally on a vertical axis.

In recent decades thermometers have been developed to use the temperature sensitivity of a sensor's electrical resistance. The ease of storing, transmitting, and processing electrical data is encouraging the use of unmanned stations. The traditional mercury-in-glass thermometer used in near-surface measurements was designed to be accurate to within ±0.2 °C when individually calibrated against international standards, and read and maintained by trained staff, and this sets the standard for all replacement types.

Sluggishness of thermometer response (Box 2.1) increases with the heat capacity of the sensor (and so with sensor volume), and decreases with increasing rate of heat transfer between sensor and air (proportional to sensor surface area), so that the balance is proportional to sensor volume/area—i.e. to the diameter of a spherical sensor. The large mercury-in-glass thermometer bulbs used in Stevenson screens have lags of about a minute in a brisk wind, and longer in lighter winds. Though adequate for routine hourly near-surface measurements, this lag is much too long to resolve the rapid temperature fluctuations which accompany atmospheric turbulence (Section 10.7) and require much smaller thermometers, often using fine wires or beads as sensors.

2.2.2 Temperature scales

Temperatures are officially measured and recorded on the *Celsius* scale, still sometimes called the centigrade scale because its calibration points (the freezing and boiling points of pure water at standard atmospheric pressure) are 0 and 100 °C respectively, i.e. 100 degrees apart. In 1948 it was officially renamed after its inventor Celsius, partly because the older *Fahrenheit* scale was also centigrade

NUMERICAL NOTE 2.1 Correspondence between temperature scales

Since the modern standard calibration points of the Celsius and Fahrenheit scales are respectively (0 °C and 32 °F) and (100 °C and 212 °F), it follows that 100 Celsius degrees correspond to 180 Fahrenheit degrees, so that 9 Fahrenheit degrees correspond to 5 Celsius degrees.

In the temperate British climate, old climatology texts record that screen (i.e. near-surface) temperatures seldom lie outside 15 °F and 100 °F. To convert these to Celsius values, note that 100 °F is 68 Fahrenheit degrees above the freezing point of water (32 °F), and is therefore $68 \times 5/9 \approx 37.8$ °C above 0 °C. In the same way 15 °F is 17 Fahrenheit degrees below the freezing point of water, which is $17 \times 5/9$ Celsius degrees below 0 °C, i.e. −9.4 °C. The range of British surface temperatures on the Celsius scale is therefore −9.4 °C to 37.8 °C. The corresponding values in degrees kelvin are found by adding 273.2 to the Celsius temperatures, i.e. 263.8 to 311.0 K, though never quoted in this way in climate records.

BOX 2.1 Instrumental lag

All observational instruments respond sluggishly to changes in what they measure. In the following we use the behaviour of a sluggish thermometer to establish the widely used concepts of *lag* and *exponential response*.

Suppose that an accurate thermometer has been exposed to air of a certain temperature for so long that its reading θ is steady. If the air temperature suddenly rises to a new steady value T, Δ_0 higher than the original value, the thermometer reading θ will begin to rise towards T as heat flows into it from the warmer air (Fig. 2.2). According to Newton's law of cooling (and warming), the rate of warming of the thermometer is proportional to the temperature difference $\Delta = (T - \theta)$ between air and thermometer—being fastest at first and slowing progressively as Δ decays towards zero. Formally, the rate of change of Δ with time t is given by

$$\frac{d\Delta}{dt} = -k\Delta \qquad \text{B2.1a}$$

where the minus sign ensures that Δ decays rather than grows with time, and the value of the constant k describes the speed of response of the particular thermometer—larger k corresponding to faster response. The solution to Eqn B2.1a is shown in mathematics texts *[8]* to be

$$\Delta = \Delta_0\, e^{-kt} \qquad \text{B2.1b}$$

where t is the time elapsed after the sudden rise in air temperature, and e^{-kt} is the negative branch of the *exponential function* given in all mathematical tables and scientific calculators, whose shape is the *exponential decay* graphed on Fig. 2.2.

Consider four key points on the decay curve.

(i) Where $t = 0$, tables, etc. show that $e^0 = 1$, so that $\Delta = \Delta_0$, as it must. The temperature difference between air and thermometer has just jumped to its maximum value Δ_0.

(ii) At time $t = 1/k$, $k\,t = 1$, $e^{-1} = 0.3679$ and $\Delta \approx 0.37\,\Delta_0$. The thermometer has now warmed to the point where Δ has fallen to 37% of its initial value, i.e. has fallen by 63%. The time $t = 1/k$ is the *lag* of the thermometer and is denoted by λ. In a time interval equal to λ, the thermometer's misreading has dropped to 37% of its initial Δ_0.

(iii) At time $t = 2.3026/k$, $k\,t = 2.3026$, $e^{-2.3026} = 0.1$ and $\Delta = 0.1\ \Delta_0$. The thermometer misreading has now fallen to 10% of its initial value, i.e. has fallen by 90%.

(iv) Though the thermometer's misreading continues to fall, the shape of the decay curve ensures that it never quite reaches zero. For example at time 4 λ after the sudden rise in air temperature, the thermometer's misreading is still $(0.37)^4 \approx 0.019$, i.e. nearly 2% of its initial misreading.

Such numerical relationships are common to all exponential behaviour, however it arises (Box 4.3).

Response to oscillations

The steady air temperature in Fig. 2.2a is unrealistic: actual air temperatures vary continually on time scales from seconds to seasons and longer. Fig. 2.2b depicts the response of a sluggish thermometer (lag λ) to the simplest possible regular variation of air temperature—a sinusoidal oscillation with time period P and amplitude A_T. Suppose that the oscillation has started long before the segment shown in Fig. 2.2b; the thermometer reading has settled into a steady oscillation which misreads the pattern of true air temperature by (i) underestimating its amplitude, and (ii) delaying thermometer temperature peaks and troughs behind those of true air temperature. Both effects obviously result from the sluggish thermometer continually failing to 'catch up' with the varying air temperature.

The situation is explored by putting a sinusoidal pattern for T into Eqn B2.1a and solving for the thermometer reading θ. Standard analysis *[8]* shows that the amplitude A_θ of the thermometer oscillation is smaller than the amplitude A_T of the true air temperature oscillation according to

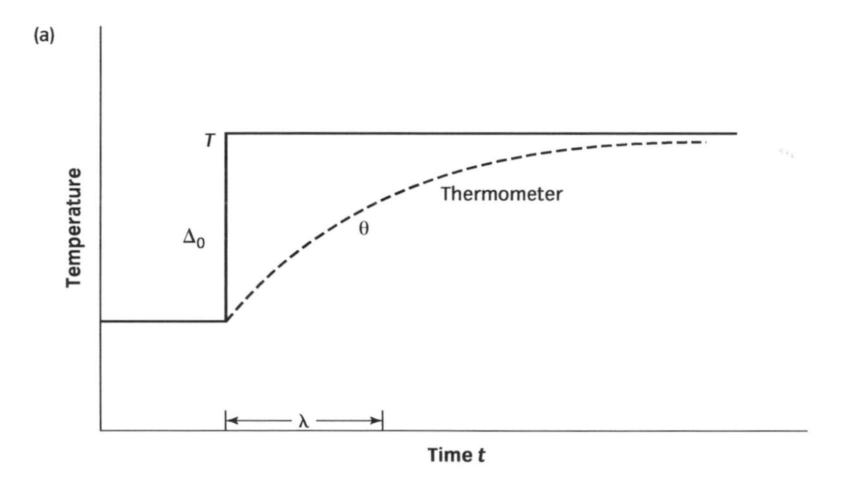

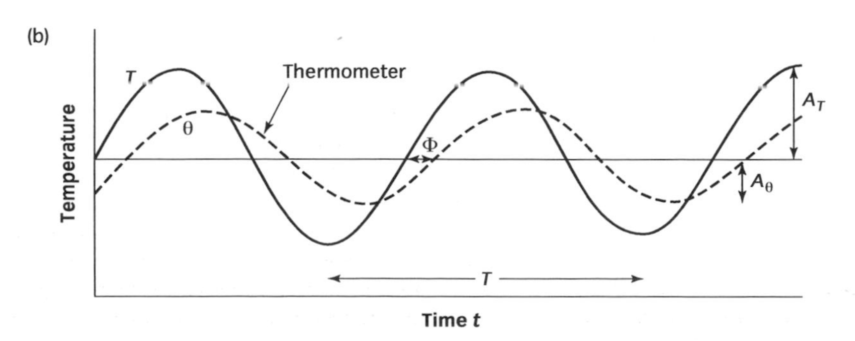

Figure 2.2 Thermometer response to (a) a sudden rise in ambient temperature, and (b) a regular sinusoidal oscillation of ambient temperature.

$$\frac{A_\theta}{A_T} = \left[1 + \left(\frac{2\pi\lambda}{P}\right)^2\right]^{-1/2} \qquad \text{B2.1c}$$

For example, if the thermometer is trying to following an oscillation of air temperature whose time period P (peak to adjacent peak, etc.—Fig. 2.2b) is equal to $2\pi\lambda$ (i.e. $\approx 6.3\lambda$),

$$A_\theta = \frac{A_T}{\sqrt{2}} \approx 0.71\, A_T$$

showing that the thermometer is under-reading the true temperature amplitude by nearly 30%. Analysis also shows that the amount Φ by which the thermometer curve lags the air temperature in this case is $P/8$—i.e. one eighth of one full cycle on Fig. 2.2b.

These results show that if a screen thermometer with a lag of 1 minute tries to follow an air temperature oscillation with period just over 6 minutes, it will underestimate the temperature range between peaks and troughs by 30% and be over 45 seconds late in their timing. Further analysis supports the rough rule that an instrument should not be relied on to follow oscillations with period shorter than 10 times its lag λ.

The above concepts, equations, and solutions apply to any instrument (not just thermometers) which responds exponentially to changes in the quantity it measures. Even when the response is not strictly exponential, (for example initially faster and then slower than the pure exponential decay shown in Fig. 2.2a), the instrumental lag is still often defined as the time for 63% response.

when first developed—its calibration points being 0 °F, the lowest reproducible laboratory temperature available in the early eighteenth century, and 100 °F, a slightly generous value for the average human core body temperature. On the Fahrenheit scale the freezing and boiling points of water are now exactly 32 and 212 °F. Volumes of trapped air or other *ideal gases* (Section 4.1) decrease steadily with falling temperature in ways which extrapolate to zero at about −270 °C. When further refined, the *Kelvin* scale takes −273.2 °C to be the lowest conceivable temperature (*absolute zero*), and uses the Celsius degree, so that the freezing and boiling points of water are 273.2 K and 373.2 K respectively.

2.3 Humidity

The amount of *water vapour* (the invisible gas of H_2O molecules) in air is measured by several types of *hygrometer*, and expressed in terms of several related *humidity measures*, outlined here and discussed more fully in Chapters 3–6.

The absolute amount of vapour is specified by *vapour pressure*, which is the contribution to the total pressure of moist air arising from its vapour content. *Vapour density* can be used for the same purpose and, for the same temperature, is directly proportional to vapour pressure since both water vapour and air behave as *ideal gases* (Section 4.1). *Relative humidity* is vapour pressure expressed as a percentage of *saturation vapour pressure*—the maximum vapour pressure which can be accommodated at the prevailing temperature without water condensing onto available surfaces (Section 6.2). Since saturation vapour pressure depends only on temperature (Fig. 5.5), simultaneous measurement of relative humidity and air temperature determines the actual vapour pressure (NN 2.2). A further measure of the humidity of moist air is the *dew-point temperature*, or simply *dew-point*—the temperature of a surface chilled just far enough to receive dew deposition from the adjacent warmer moist air.

NUMERICAL NOTE 2.2 Relative humidity and vapour pressure

On a certain cool, humid British summer day, near-surface temperature and relative humidity are 18.0 °C and 80% respectively. Figure 5.5 shows that the saturation vapour pressure at 18.0 °C is close to 20 hPa. Since relative humidity is 80%, the actual vapour pressure is close to $0.8 \times 20 = 16$ hPa—a typical value in temperate conditions, and about 1.6% ($= 100 \times 16/1{,}010$) of typical total atmospheric pressures at sea level.

2.3.1 Types of hygrometer

(i) *Relative humidity meters* Animal tissues, such as hair or skin, respond directly to relative humidity, and this is exploited in several simple instruments. In the common *hair hygrometer*, a bundle of human hair is kept under slight tension so that its decrease in length with increasing relative humidity is mechanically registered, usually on a dial. The length variation is sufficiently regular to allow calibration in a range of known relative humidities (determined by other methods) to produce a semi-quantitative instrument. Drifting of calibration with age and repeated humidity cycling limits use to registering changes rather than absolute values, but their simplicity and ability to produce a continuous graphical record make *hair hygrometers* and *hygrographs* useful for routine monitoring. *Skin hygrometers* use similar sensitivity of a small stretched piece of fine animal skin, and are more reliable than hair hygrometers. Hair and skin hygrometers have lags ~10 seconds at room temperature, but are much more sluggish at lower temperatures.

In recent decades, hygrometers have been developed to measure relative humidity through its effect on the electrical resistance of *hygroscopic* (water-absorbing) surfaces; these are more reliable than the hair and skin types, and can provide electrical output.

(ii) *Psychrometers* measure the humidity of unsaturated air by the evaporative cooling of a water-soaked cotton wick enclosing the thermometer bulb (the *wet bulb*). Their simplicity and reliability make them the most common instruments not requiring calibration. The simplest type consists of a *wet-bulb thermometer* mounted beside an otherwise identical *dry-bulb* thermometer (i.e. without wick) inside a Stevenson screen (Fig. 2.1). Natural ventilation allows water to evaporate from the wick of the wet bulb (kept soaked by osmosis along a fibre wick from an adjacent reservoir), cooling its *wet-bulb temperature* below the *dry-bulb temperature* by an amount which varies with the humidity and temperature of the air entering the screen. In very dry air, as found near the surfaces of warm deserts, this *wet bulb depression* can be very large (> 10°C), whereas it is zero in saturated air, as in fog. From simultaneous wet and dry-bulb temperatures, a book of *hygrometric tables [9]* or a special slide rule, gives the associated vapour pressure (or density), relative humidity, and dew-point temperature. When the Stevenson screen is poorly ventilated (on still days), the air inside it can be significantly moistened by evaporation from the wet bulb, and the humidity outside the screen therefore overestimated. This problem is reduced in the *aspirated psychrometer*, by mechanical ventilation (aspiration) of both bulbs. The wet bulb has much the same lag as the corresponding dry bulb. Electrical psychrometers with tiny wet and dry bulbs have much smaller lags and can detect rapid humidity fluctuations associated with atmospheric turbulence.

(iii) *Dew-point meters* have a polished metal surface which is chilled progressively below the temperature of the surrounding moist air, and whose surface temperature is measured at the moment when dew first mists the surface, this being the *dew-point* of the adjacent air. Until recently dew-point meters have been too elaborate and cumbersome for widespread use, though potentially accurate and needing no calibration. Relationships between vapour pressure, vapour density, relative humidity, wet-bulb temperature, dew-point, and other measures of humidity are detailed in Sections 5.4 and 5.5. The *dew-point depression* below the ambient air temperature is zero in saturated air, and roughly twice the wet-bulb depression in unsaturated air.

2.4 Wind

Horizontal *wind directions* are measured by *wind vane*, mechanically like those on church steeples, etc. but with an electromechanical system to allow remote and recordable reading. The standard meteorological vane has a lag of several seconds in normal winds but is more sluggish and ultimately unresponsive in very light winds, which are anyway very variable in direction (*fluky*). Much smaller and lighter vanes are needed to register the rapidly varying directions associated with atmospheric turbulence.

By meteorological convention, the direction of a horizontal wind is reported as the direction from which it is blowing, as in everyday speech (Fig. 2.3). This arose naturally from the age-old practice of sailors and farmers to look upwind to see approaching weather, and reflects the tendency (common but not universal—Section 11.6) for weather systems to move in much the same direction as surface winds. Though verbal and informal reports still use compass bearings, directions are recorded in degrees of *azimuth*, counted clockwise from zero at true North (NN 2.3). Because of incessant rapid variation in wind direction, routine readings of near-surface wind direction are averaged over a 10-minute interval and reported to the nearest 5° of azimuth.

Wind speed is measured by various types of *anemometer* developed in the late nineteenth century, but earlier procedures classified *wind strength* or *force* according to its effects on the sea surface and sailing ships. The famous *Beaufort scale* (Box 2.2) of wind force was soon extended to include wind effects on land, and 200 years later is still used widely to describe the strength of near-surface winds. Later measurement by anemometer has associated each

NUMERICAL NOTE 2.3 Wind directions by compass bearing and azimuth

By definition, northerly, easterly, southerly, and westerly winds (denoted N'ly, E'ly, S'ly, and W'ly) have azimuths 0°, 90°, 180°, and 270° respectively. As shown in Fig. 2.3, a NW'ly lies half way between a W'ly and a N'ly, and therefore has azimuth $270° + 45° = 315°$. A WNW'ly lies half way between a W'ly and a NW'ly and therefore has azimuth $315° - 22.5° = 292.5°$, whereas a NNW'ly has azimuth $315° + 22.5° = 337.5°$. The finest compass points (dotted in the sketch) differ by 11.25° azimuth, which is effectively 10°.

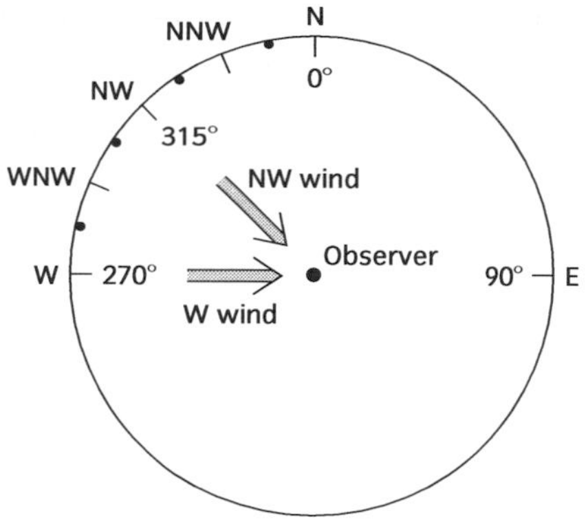

Figure 2.3 Horizontal wind directions on an azimuth circle, with sample degrees and compass bearings. The arrows represent airflow towards the observer/wind vane at the centre of the circle.

Beaufort *wind force* with a range of wind speeds (Box 2.2). Another maritime relic is the continued used of the *knot* (1 nautical mile per hour) as a unit of wind speed in parallel with the proper SI unit, the metre per second. The precise correspondence of 1 m s^{-1} = 1.944 kt can be rounded to 2 for many purposes. The knot is also still widely used for aircraft speeds.

BOX 2.2 **The Beaufort scale *[6]***

Table 2.1 Beaufort scale of wind force

Force	Specifications for use on land	Equivalent mean wind speed 10 m above ground	
0	**Calm:** smoke rises vertically	0 kt	0 m s^{-1}
1	**Light Air:** wind direction shown by smoke drift, but not by vanes.	2	0.8
2	**Light Breeze:** wind felt on face; leaves rustle; vanes move.	5	2.4
3	**Gentle Breeze:** leaves and small twigs moving; light flags lift.	9	4.3
4	**Moderate Breeze:** dust and loose paper lift; small branches move.	13	6.7
5	**Fresh Breeze:** small leafy trees sway; crested wavelets on lakes.	19	9.3
6	**Strong Breeze:** large branches sway; telegraph wires whistle; umbrellas difficult to use.	24	12.3
7	**Near Gale:** whole trees move; inconvenient to walk against.	30	15.5
8	**Gale:** small twigs break off; impedes all walking.	37	18.9
9	**Strong Gale:** slight structural damage.	44	22.6
10	**Storm:** seldom experienced on land; considerable structural damage; trees uprooted	52	26.4
11	**Violent Storm:** rarely experienced; widespread damage.	60	30.5
12	**Hurricane:** at sea, visibility is badly affected by driving foam and spray; sea surface completely white	$\geq$ 64	$\geq$32.7

Note that speeds in m s^{-1} are rounded to the nearest 0.1 m s^{-1}, while knots are rounded to the nearest knot.

By about 1800 CE there was obvious value in classifying wind strengths by their effects on the sea and the management of sailing ships. This was done in 1806 by Francis Beaufort, while surveying for the British Royal Navy, and his scale became the international standard for wind observations until 1946, and is still used in weather bulletins for shipping, and general forecasts of very windy weather (Table 2.1).

Beaufort also devised a concise code of *Beaufort letters* for surface weather description, which is still in use for official weather diaries. By predating the establishment of national observing networks by several decades, Beaufort is recognized as a pioneer of the drive to make meteorological observation as consistent and concise as possible.

Table 2.1 lists simplified descriptions of wind effects on fairly open land surfaces, together with equivalent ten-minute average wind speeds measured 10 m above the land surface—equivalents established by anemometry long after the scale was first devised. The current official name of each wind force category is given in bold. The full Beaufort scale for land and sea appears in official observing manuals *[6]*.

Comments

In any wind force, gusts (i.e. the highest wind speed in the 10-minute averaging period) are typically about 25% faster than the average wind speed, and usually last for a less than a minute. Average wind speeds at 2 m are about 80% of speeds at 10 m, depending on the *aerodynamic roughness* (Section 10.9.2) of the local land surface.

Some of the comments about frequencies of forces 10–12 apply best to the extratropical zones in which the Beaufort scale was first established. Such strong winds are more common in tropical coastlands visited by tropical storms and tropical cyclones. In extratropical zones, winds do occasionally reach hurricane force (even overland, as along parts of the South of England for a few hours on October 16th 1987), but that does not make the weather system a hurricane, despite media confusion. True hurricanes, etc. can occur only in tropical zones (Section 13.5), and can have maximum wind speeds over twice the 64 kt corresponding to the force 12 needed for official categorization (and naming) as a hurricane, typhoon, etc.

The common *cup anemometer* (Fig. 2.4) is insensitive to horizontal wind direction, unlike the *propeller anemometer* which has to be kept pointed into the wind by a trailing vane. The three cups rotate around a vertical axle as they catch the wind in their mouths on one side of the axle and push backwards into the wind on the other side. This drag asymmetry means that they respond more quickly to rising than to falling winds, and tend to overestimate average wind speeds in gusty conditions. With careful design the *rate* of cup rotation can be made directly proportional to wind speed over a wide speed range. Even after smoothing by instrumental lag, instantaneous wind speeds are too unsteady to be useful; instead the *run of wind* in a chosen time interval is found by converting the number of cup revolutions into an equivalent length of air passing the cups. Dividing this by the time interval (10 minutes, 12 hours, or 24 hours in the common types of observation) gives the average wind speed in this interval (NN 2.4). A 10-minute average, together with the maximum *gust* (itself smoothed somewhat by anemometer lag), is used to define wind speeds in routine hourly near-surface observations.

The lags of anemometers and wind vanes tend to decrease with increasing wind speed, complicating their response to gusts and lulls. It is therefore customary to

Figure 2.4 A standard British Meteorological Office cup anemometer. Its robust construction enables it to cope with stormy conditions with minimum maintenance. Revolutions are counted and displayed on the instrument (as here, 2 m above ground), or remotely by electrical signal when the anemometer is 10 m above ground.

NUMERICAL NOTE 2.4 Run of wind

A small cup anemometer held at height 2 m turns 18 times when walked quickly through 30 metres of still air. One turn therefore corresponds to 1.67 m of air moving past the cups—a simple method of calibration.

In one minute's exposure to a certain light wind the cup anemometer turned 115 times, corresponding to a run of wind of 192 metres, and a 1-minute average wind speed of 192/60 = 3.2 m s^{-1} or just over 6.2 knots. The corresponding wind speed at height 10 m is about 4 m s^{-1}, which corresponds to force 3 on the Beaufort scale.

express their lags in terms of the run of wind needed to produce 63% response to a step change of wind speed. For example, the standard near-surface anemometer shown in Fig. 2.4 requires about 15 m of air to pass its cups in order to register 63% of a sudden change of wind speed. At the lowest wind speed to which this anemometer responds (about 2 m s^{-1}), such a run corresponds to a lag in time of about 8 s, whereas the lag in response to gusts in a gale is roughly 1 s.

2.4.1 Other types of anemometer

The only anemometer not requiring prior calibration in a wind tunnel (at least in principle) is the *pitot tube*, which measures the pressure rise produced in air as it rams to a halt in the open mouth of a small tube (blocked at its other end by a pressure sensor) pointed into the wind by a vane. Since this pressure rise is proportional to the square of the wind speed *[10]*, the instrument is relatively insensitive to low wind speeds.

All routine measurements are of horizontal wind speed. Vertical wind components are measured in non-routine studies of turbulence, convection, etc. and are usually much weaker than horizontal winds. Near the surface, a calibrated propeller on a vertical axle can be used, turning one way in rising air, and the other way in sinking air. Higher above the surface, convective updrafts and downdrafts are estimated from their effects on the rates of rise or fall of aircraft flying across them. The very weak but persistent updrafts which maintain the great sheets of frontal cloud (nimbostratus) are much too small to be directly measurable, and have to be inferred from three-dimensional continuity of air flow (Section 7.15).

Very fast response anemometers use hot wires calibrated to give wind speed from its cooling effect, or the slowing or speeding of narrow beams of sound by head or tail winds.

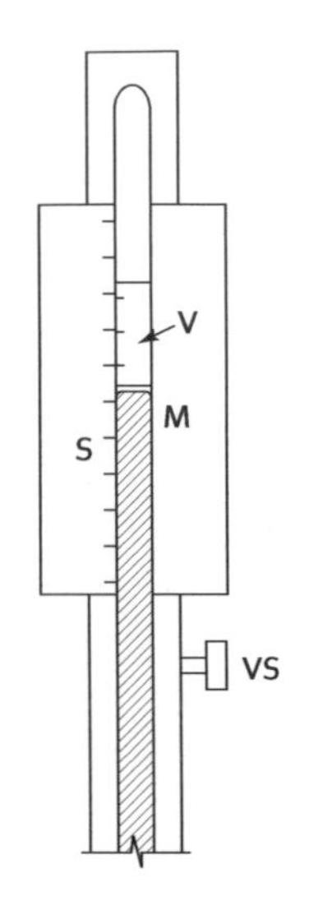

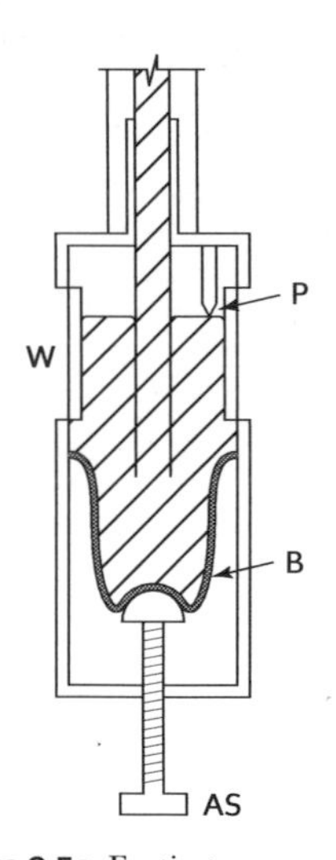

Figure 2.5a Fortin type mercury barometer. The mercury reservoir in the soft bag B is adjusted by the adjusting screw AS until the exposed mercury surface just touches the point P at the bottom of the fixed vertical scale S, visible through the window W. The vernier slide V is adjusted by VS until it appears to rest on top of the mercury meniscus M.

2.5 **Pressure**

The human body is largely insensitive to weather-related pressure changes below audible frequencies (below $\approx$ 20 Hz), with the result that the existence of much larger, slower changes was unsuspected before the development of the barometer in the mid-seventeenth century. The relation between atmospheric pressure changes and weather sequences was then soon noted, giving the barometer an important place among meteorological instruments which it has held ever since. Though the relation between pressure and weather is much more complex than suggested by messages printed on domestic barometers, it is very important in ways which are outlined in Chapters 7 and 12.

Because it does not require calibration, the basic instrument is still the *mercury-in-glass barometer*, working on the same principle as that developed by Torricelli in the 1640s. The atmosphere presses on an exposed reservoir of mercury to support a column of mercury in a vertical glass tube whose top end is sealed and evacuated (Fig. 2.5). The vertical height h of the top of the mercury column above the reservoir surface is related to the atmospheric pressure p pressing on the reservoir surface by the *barometric equation*

$$p = g\,\rho\,h \qquad 2.1$$

where g is the local *gravitational acceleration* and ρ is the density of mercury. Such barometers are heavy and fragile, but when read carefully and corrected for variations in g (Table 7.1) and mercury temperature (which affects ρ), they are accurate to better than 0.1 hPa. Marine instruments are mounted on gimbals to allow them to remain vertical on a heaving ship.

The *pascal* (1 newton per square metre) is the SI unit of pressure, but is inconveniently small for meteorology which uses the *hectopascal* (hPa), or 100 pascals, as the standard unit of atmospheric pressure. This is numerically identical with the *millibar* (mbar or mb) which was used for most of the twentieth century. Older official barometers, and domestic instruments of all ages, use millimetres or inches of mercury at 0 °C.

Figure 2.5b A barograph in its glass case. The aneroid capsule moves the pen arm up or down as the capsule respectively squeezes or expands. The line on this drum chart shows that pressure has been falling unsteadily in the last 36 hours.

NUMERICAL NOTE 2.5 Comparison of pressure units

The global average atmospheric pressure at mean sea level is about 1,013 hPa or mbar. We can rearrange Eqn 2.1 to calculate the equivalent barometer column height in millimetres of mercury. Inserting 9.81 m s^{-2} for *g* and 13,595 kg m^{-3} for the density of mercury (ρ), we use $h = p/(g\ \rho) = 1{,}013 \times 100\ /(9.81 \times 13{,}595) = 0.7596$ m or 759.6 mm. Note the 100 factor needed to convert 1,013 hPa to Pa so that all units are SI.

During the passage of a typical vigorous mid-latitude depression, sea level pressures often fall by about 40 hPa (for example from 1010 to 970 hPa) and recover again in the space of a few days. Use this calculation to show that a 40 hPa pressure change corresponds to 30 mm of mercury, a useful conversion ratio for domestic barometers.

Barometers can be operated indoors, because the time taken for indoor pressures to adjust to changes in outside pressures is only ~10 s, even in a draught-proof building, which is very short in comparison with the big, slow variations in atmospheric pressure associated with large-scale weather systems (NN 2.5). Small changes (~1 hPa) in atmospheric pressure can occur in tens of seconds in association with air movements in nearby thunderstorms, but these are smoothed away when large-scale pressure patterns are drawn on weather maps. In or near tornado funnels, atmospheric pressure may fall and recover by ~100 hPa in a minute or so, imposing damaging stresses on buildings which are too hermetically sealed to equalize internal and external pressures in time.

In recent decades, the more compact and robust *precision aneroid barometer* has largely replaced the mercury barometer in the meteorological surface network. A partly evacuated (*aneroid*) metal capsule with springy corrugations separating flat ends expands as ambient air pressure falls and vice versa, and these movements are measured and calibrated against a master mercury barometer.

The common domestic aneroid barometer uses the same principle with a cruder aneroid capsule mechanically linked to the familiar clockface display. Bearing friction and looseness can lead to errors of several millibars. Even so, this simple instrument can give a useful impression of the pressure variations associated with depressions and highs. Manually tapping such barometers can temporarily 'unstick' linkages and usefully show whether pressure is rising or falling.

A major shortcoming of all these barometers is their inability to give a continuous record. This is overcome in the traditional *barograph* (Fig. 2.5b), in which the movement of the aneroid capsule is taken mechanically to a pen writing on a graph fixed to a drum turning by clockwork. Linkage can still stick, but the pressure trends (*pressure tendencies*) graphically depicted are very useful for weather analysis and forecasting (Sections 12.3 and 13.5). Instruments using the pressure sensitivity of electrical properties of certain crystals are beginning to replace the aneroid barograph, and promise to become as accurate as basic mercury-in-glass instruments, which remain the calibrating standard. As always, electrical output simplifies data storage, display, and transmission.

Note that all barometers measure ambient atmospheric pressures, called *station pressures*. As discussed in Section 2.9 these need substantial correction to produce comparable values which are consistent across a network of stations sited at different altitudes.

2.6 Precipitation *[6]*

Though the term *precipitation* officially covers all the forms in which water and ice fall out of the atmosphere (the several types of snow and hail, as well as rain and drizzle), rain predominates heavily at the Earth's surface outside cold regions. Rain and drizzle are collected and measured directly in various types of *rain gauge*; and snow and hail are measured after thawing and expressed as rainfall equivalents. The quantity of rainfall during a typical 12 or 24 hour observation period is specified by the depth of water collected on a small horizontal area (~100 cm^2). Depths collected in these periods can range from zero to well over 100 mm, with most non-zero values lying in the range 1–20 mm.

The standard *manual gauge* (Fig. 2.6a) has a circular mouth with internal diameter 127 mm (5 inches), and a knife-edge rim to discard all rainwater falling outside the mouth. Rain is funnelled into a glass bottle to minimize evaporation, which can account for several millimetres daily from an exposed water surface, and then measured in a cylinder with graduations arranged to allow direct reading of rainfall depth (NN 2.6). *Tipping bucket gauges* (Fig. 2.6b) register electrically the times when a pair of small 'buckets' see-saw beneath the collecting funnel, alternately filling and tipping into reserve position while the other bucket fills and tips. Tip numbers in an observation period convert to rainfall totals in millimetres, and time intervals between consecutive tips convert to *rainfall rates* in millimetres per hour within the period.

Average depths of snow or hail lying on open ground are recorded at official sites, and are roughly an order of magnitude greater than equivalent rainfall when freshly fallen, though the factor decreases very considerably with subsequent compaction.

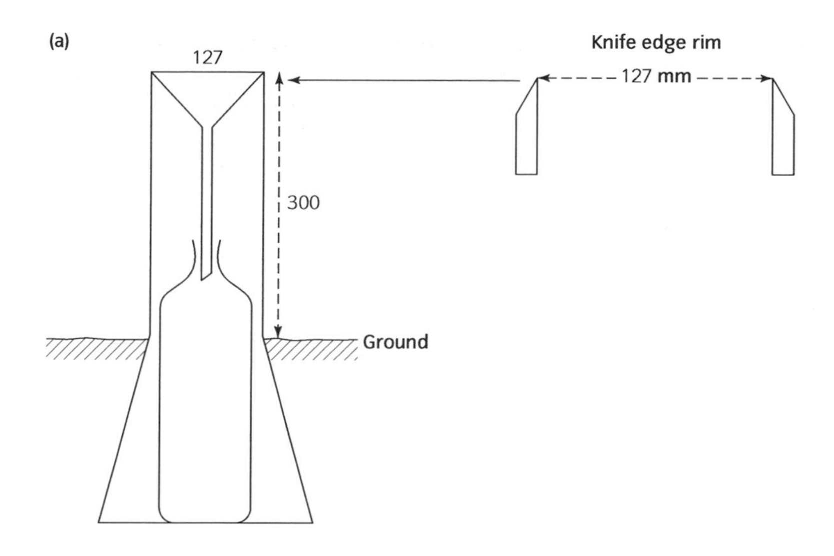

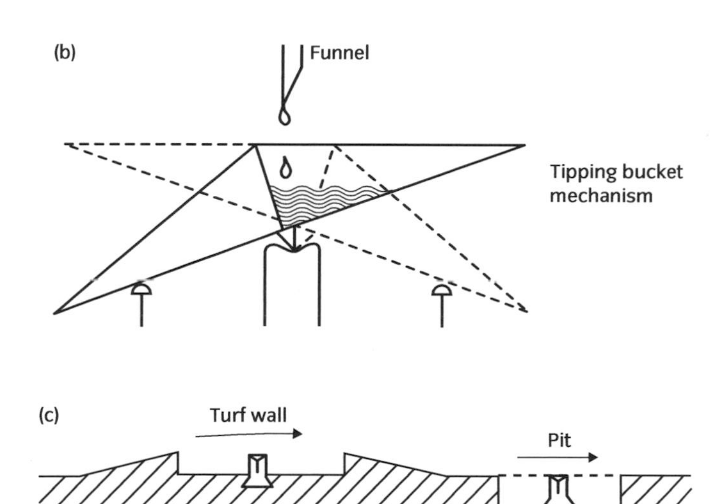

Figure 2.6 (a) A standard 127 mm diameter manual rain gauge, with a close up of the knife edge which minimizes systematic splash-in from droplets falling on the rim. (b) A vertical section through the mechanism of a tipping bucket type of recording rain gauge. When full, the collecting bucket see-saws from under the funnel spout and empties to the right, bringing the empty left bucket under the spout and sending an electrical pulse to a counter timer. (c) An idealized vertical section through rain gauges set inside a turf wall and in a pit, to minimize rduction of catch by airflow over the gauge mouths.

Sampling rate rather than instrumental lag determines a gauge's ability to resolve temporal variations of rainfall. Manual gauges are read once or twice per day, which is adequate for many purposes but obscures all timings and rate variations between successive readings. Such details require automatic gauges, which show that most of the rainfall accumulated in 24 hours (for example) typically falls in relatively short bursts of rain falling at much higher rates (*heavier* rain) than the implied average (Fig. 2.7). Rainfall accumulations are skewed in this way towards short, heavy bursts across all collection periods from minutes to years.

The great weakness of all these gauges is that their collection areas are mere points on a weather map. Their spatial representativity is therefore always in doubt: X mm of rain fell into gauge A in a certain storm, but what would have fallen into identical gauges if they had been in operation nearby? Research using

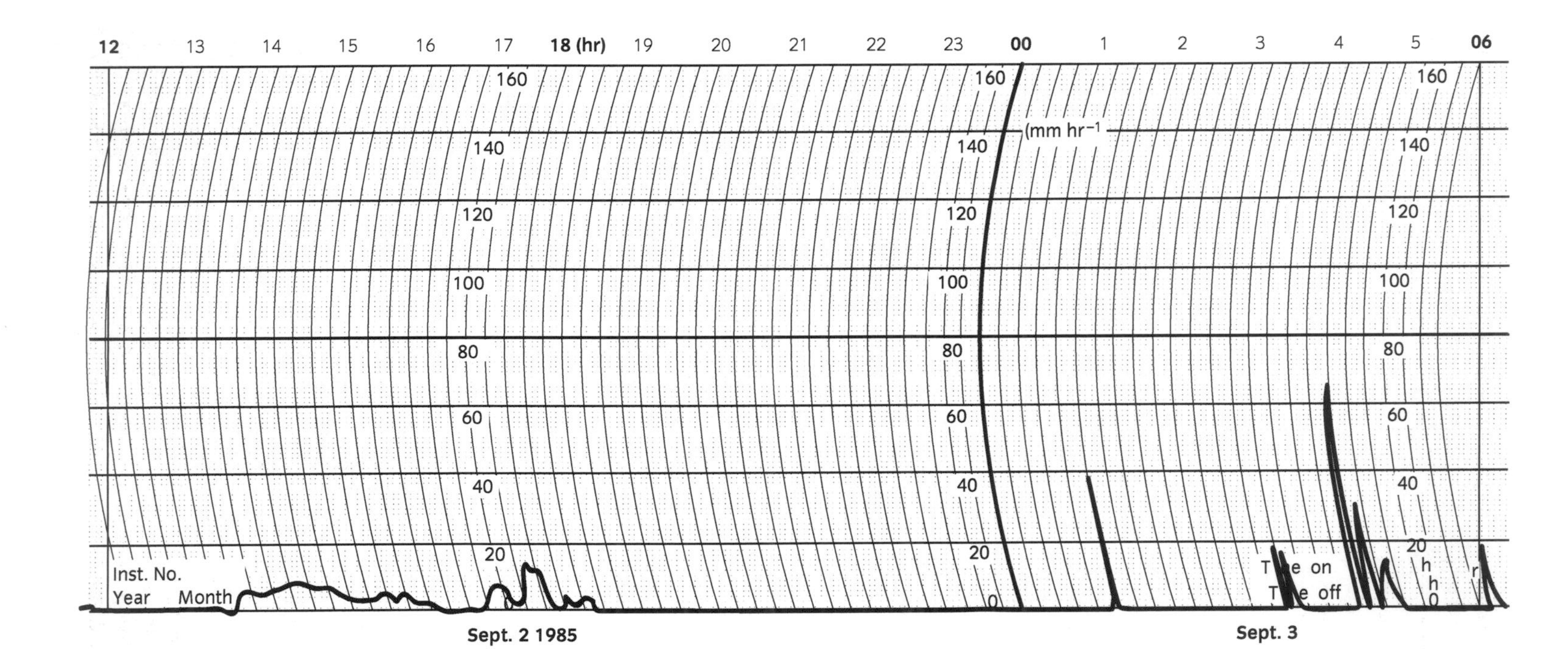

Figure 2.7 Rate of rainfall recorded on a Jardi type recording gauge, in which flow from a very large collector lifts a float and moves a pen, according to the rate of rainfall. The continuous rain (with embedded heavier bursts around 1800 Z reaching 15 mm hr-1) fell from a passing occluded front. The isolated spikes early next day correspond to showers in the following cooler NW'ly flow, with one peaking at over 60 mm hr-1.

special very dense networks of gauge show that rainfall can be strongly patterned in space—the patterning arising from gauge exposure and topographical effects, as well as precipitation patterns inherent in weather systems. Taking these in order:

(i) Gauge exposure. To avoid flooding or splash-in, the mouths of gauges used to be raised above the surrounding land surface, but by no more than about 30 cm to minimize airflow deformation and resulting loss of catch. To improve the compromise, flush-mounted gauges surrounded by a dummy nearly splash-free surface are now widely used (Fig. 2.6c). Exposure problems are very severe in ship-borne gauges, where the effects of sea spray and the ship's large bulk and motion are considerable. As a result the distribution of ocean rainfall is still much more uncertain than it is over land, and is the most inadequately observed major element of the world's weather.

(ii) Topographical influence. Rainfall can be strongly enhanced on the windward side of high ground (Section 10.11.4), and reduced in its lee, where there may be a *rain-shadow*. Since such patterns may affect estimates of overall rainfall, and yet be too detailed to be properly represented by any feasible large-scale gauge network, quite subtle techniques have been devised to interpolate between gauges as realistically as possible *[11]*.

(iii) Inherent rainfall patterns. In ascending scale, these range from the narrow stripes of wetted ground left by small individual showers, to the extensive swathes associated with fronts and other major weather systems. Showers are particularly difficult to deal with, since they can produce large variations between gauge totals (and hence in estimated area totals) depending on whether their rainshafts (width ~1–10 km) fall on gauges or miss them.

Since about 1950, *weather radar* has been used increasingly to observe the distribution and intensity of precipitation over wide areas (Fig. 2.8). The technique was developed after heavy precipitation was found to produce strong radar echoes in military radars. Echo strengths increase with the average size of raindrops, snowflakes, and hailstones, and with their numbers per unit volume of air. Since these are both directly related to the rate of precipitation, radars can be calibrated to estimate precipitation rates in ways which avoid many of the snags of conventional point gauges, although calibration varies depending on whether the precipitation is rain, hail, or snow.

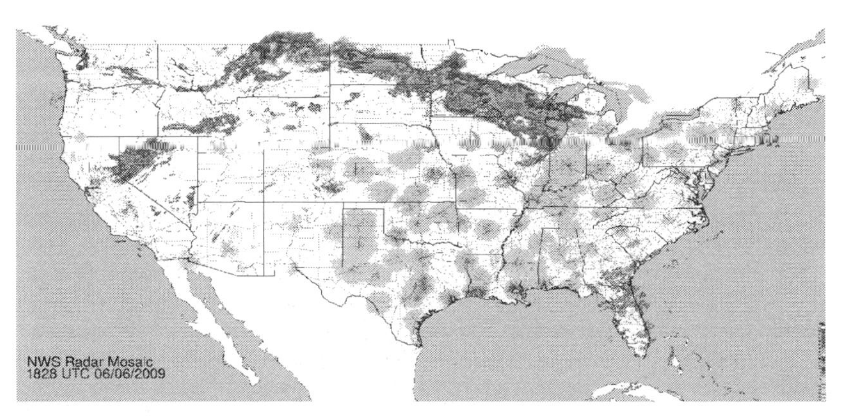

Figure 2.8 Composite picture from US weather radars for 1828 Z June 6 2009, showing a very large belt of rain extending W from Lake Michigan associated with an open frontal system separating cool air to the N from warmer air to the S. The extensive area of moderate rainfall in the frontal zone contrasts with the spotty pattern with higher maximum rainfall rates associated with outbreaks of heavy showers over Florida and Georgia. (See Plate 1).

NUMERICAL NOTE 2.6 Rain-gauge calibration

A typical standard British manual rain gauge has a circular mouth with internal diameter 127 mm, and a measuring cylinder with internal diameter 35 mm. In the automatic equivalent, the cylinder is replaced by a tipping bucket arrangement with bucket capacity 6.3 ml. By considering responses to a rainfall of 1 mm we can calibrate cylinder depth and bucket capacity in millimetres of rainfall.

If the gauge mouth is temporarily blocked just below its rim by a horizontal circular tray, the collected rain will stand 1 mm deep in the tray, before being funnelled into the collector. Rainfall volume is mouth area × depth = $\pi\,(127/2)^2 \times 1 = 12{,}668\ \text{mm}^3$ or about 12.7 ml for 1 mm rainfall, which suggests that the automatic version has been designed to register 0.5 mm rainfall per tip. The water depth in the cylinder of the manual version is found by noting that water volume = depth × area is conserved as rain passes from gauge mouth to cylinder, so that cylinder reading = rainfall depth × (gauge mouth diameter/cylinder diameter)^2 = 13.17 × rainfall depth ≈ 13.2 mm per mm rainfall. Cylinder gradation intervals of 2.63 mm (= 13.17/5) will correspond to rainfalls of 0.2 mm. The base of the cylinder is usually tapered, and its gradations widened, to ease the reading of small rainfalls.

2.7 Other surface observations

2.7.1 Bright sunshine

For nearly a century, periods of *bright sunshine* have been recorded by the *Campbell–Stokes sunshine recorder* (Fig. 2.9a), in which sufficiently bright sunshine focused by a glass ball is able to char a line across a time-marked curved cardboard strip as the Sun moves across the sky. The total length of the chars is measured manually at the end of each day, converted to hours of *bright sunshine*, and a fresh strip inserted. It has proved difficult to replace this excellent manual instrument by one with an electrical output without altering the criterion for bright sunshine.

2.7.2 Solar and terrestrial irradiance

Although not in routine use at all surface stations, various types of *solarimeter* have been devised and deployed in recent decades to measure the input of *direct sunlight* (direct from the Sun), and *diffuse sunlight* (scattered by the sky). The simplest types use a *thermopile* to measure the *irradiance* (the rate of radiant energy input to unit area—usually horizontal) by the warming of a blackened surface beneath a glass dome (Fig. 2.9b), as measured by calibrated electrical output. Unshaded, the solarimeter measures *global* solar irradiance (direct and diffuse sunlight); with minimum shading (the solarimeter just shaded by a distant disk) it measures diffuse solar irradiance only. The difference between the two gives the direct solar irradiance. A downward facing solarimeter measures the solar irradiance reflected upwards from the underlying surface. The net

(a)

(b)

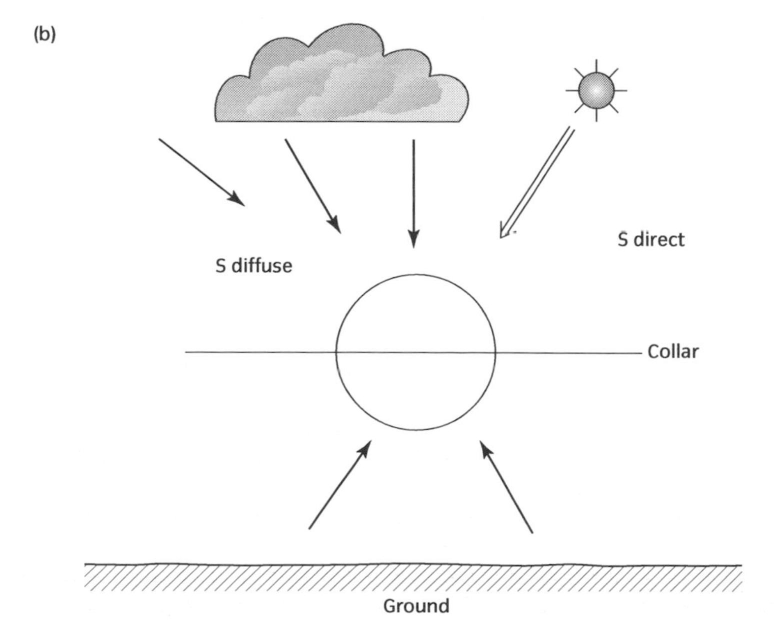

Figure 2.9 (a) A Campbell-Stokes sunshine recorder. The glass ball focusses the sunlight on the recording card on the polar side of the mounting, where the image of a really bright sun charts a trace from W to E (left to right) across the printed time scale. A new card is inserted daily. (b) A solarimeter seen edgeways on, with a white collar to separate the collection of sunlight coming down from the sky from the backscattered sunlight coming up from the sunlit surface about 1.5 metres below the instrument. Each translucent glass dome allows sunlight to fall on a blackened thermopile in the plane of the collar, which produces an electrical signal proportional to the incident energy flux.

solar input to the surface is given by the difference between global and reflected irradiances.

A similar thermopile under a translucent polythene dome measures the intensity of the direct and diffuse sunlight together with terrestrial radiation from the sky, since polythene is nearly transparent to both visible and far infrared radiation, whereas glass is opaque in the far infrared. An identical dome pointing downwards measures the solar and terrestrial radiation coming up from the underlying surface. The difference between the output from the two polythene domes is the net radiative input to the surface, which is very important in determining local climate (Section 8.6). The difference between the net radiative input and the net solar input give the net terrestrial input (usually negative, being an output) to the surface.

2.7.3 Surface and ground temperatures

Temperatures at and under the ground surface are important for road transport and agriculture. *Daily minimum* surface temperatures are especially significant when near 0 °C, and are usually measured by alcohol-in-glass minimum thermometers exposed without shielding just above short grass. This *grass minimum temperature* is often used as a measure of the degree of *ground frost*, whether or not frost is deposited. A similar thermometer exposed over a concrete slab is relevant to road icing. Thermometers are also buried at various shallow depths down to a metre under the soil surface, to measure the temperatures experienced by crop roots.

2.7.4 Visual observations

Several types of routine observation not mentioned so far are still made by the human eye, because they help define the current and future state of the atmosphere and yet are too varied and subjective to record instrumentally.

Amounts, types and heights of clouds are judged by eye, using internationally agreed categories which have been perfected over the last 150 years and enshrined in the *International Cloud Atlas* and derived handbooks *[12]*. For example, the high clouds shown in Fig. 2.10 are characteristic forerunners of a warm front, with its associated cloud, wind, and rain. Cloud types and names are listed in Box 2.3, and further examples appear in Chapters 11 to 13.

Present and past weather are reported by selecting one from a series of 99 descriptions arranged in order of increasing meteorological significance, only the most significant type being reported on any occasion. For example, all types of fog are less significant than rain, light rain is less significant than moderate rain, and the most significant weather type of all is a heavy hail shower with thunder (indicating extreme convective instability). *Past weather*, meaning weather in the period (often 6 hours) ending at the time of observation, is similarly described using a list of 10 descriptions. Present and past weather types are outlined in Box 2.4 and fully described and discussed in operational manuals for observers and forecasters *[13]*.

Visibility was assessed visually until recently—each station having a series of visible landmarks covering the set ranges from 10 m to 30 km, of which the most

Figure. 2.10 Plumes of cirrus advancing from the W ahead of a warm front. The plume tops are just catching the last rays of sunlight, but the ice crytals in each plume are sifting down into the shadow of dusk, probably darkened by shading by thicker frontal clouds 50 km or more to the W.

BOX 2.3 International cloud classification

Clouds are divided into distinguishable types or *genera*, each with a full Latin name describing its visual appearance, and an abbreviated form. Each *genus* is assigned to one of three bands of height above sea level (low, medium, and high), according to the height of its base, even though it may occupy higher height bands. For example, cumulonimbus are classed as low clouds, even though they usually extend to medium and high levels from their low level bases; and Nimbostratus often extend to high levels from its middle level bases. Official descriptions have been shortened in the following.

Low-level clouds (0–2 km above sea level)

Stratus (St)
Low, featureless, extensive cloud sheet, often just deep enough to produce drizzle or snow grains at the surface.

Stratocumulus (Sc)
Shallow cloud sheet, broken into more or less regular *cumuliform* lumps, giving drizzle or snow grains at the surface if cloud base not too high.

Cumulus (Cu)
Scattered clouds with knobbly hill-shaped upper parts and flat bases, usually reaching from a common base level to various heights in the field of view.

Cumulonimbus (Cb)
Large cumulus, usually showering and reaching to high levels from dark, low-level bases. Low and middle levels look like cumulus conglomerates, but high levels often have a *cirriform* (like cirrus—see below) *anvil*.

Medium-level clouds (2–4, 7, or 8 km respectively in polar, temperate, or tropical regions)

Altocumulus (Ac)
Shallow sheet broken into fairly regular, rounded patches or rolls, sometimes with *virga* (see below). Distinguished from lower Sc by smaller apparent size, and from *cirrocumulus* (see below) by the presence of shadows.

Altostratus (As)
Largely featureless, extensive sheet of watery-looking cloud, often thick enough to make the Sun appear as if shining through thick ground glass.

Nimbostratus (Ns)
Extensive sheet of cloud, usually precipitating, thick enough to blot out Sun, Moon, and stars, making days dark and nights pitch black.

High-level clouds (3–8 km in polar, 5–14 in temperate, and 6–18 in tropical regions)

Cirrus (Ci)
Detached, white, fibrous clouds, often with a silky sheen in direct sunlight.

Cirrocumulus (Cc)
Shallow patch or sheet, broken into more or less regular, unshadowed, and apparently small (being at high altitude) blobs or ripples, usually partly fibrous.

Cirrostratus (Cs)
Shallow extensive sheet of white cloud, thin enough to be largely translucent and produce haloes round the Sun and Moon. May look fibrous or smooth or both.

Genera can be amplified by adding a descriptive name, misleadingly called a *species*, as in the following selection.

Fractus (fra)
broken or jagged (of Cu and St);

Lenticularis (len)
elements lens-shaped (of Sc, Ac and Cc);

Humilis (hum)
of only slight vertical extent (of Cu);

Congestus (con)
growing markedly, often bulging like a cauliflower heart (of Cu);

Capillatus (cap)
distinctly fibrous upper parts (e.g. Cb with anvil).

The further addition of a list of *varieties*, such as *radiatus* (banded), *supplementary features* such as *virga* (precipitation streaks tapering from cloud base down to extinction above the surface), *accessory clouds* such as *pileous* (thin, smooth webs apparently stretched over the crowns of Cu con or Cb), and the newer category of *mother-cloud* (actively producing other cloud types) completes a scheme some of whose details suggest a forced marriage between the fluid complexity of clouds and biological taxonomy.

BOX 2.4 Present and past weather

The following Present Weather Table lists the categories used internationally by observers. Each category is given a code number, and at any observation time only the highest applicable number is reported. Descriptions have been simplified, since the full version is worded legalistically to avoid ambiguity.

Present Weather

00–19	*No precipitation at station at time of observation*
00	Cloud development unobserved during past hour
01	Clouds generally dissolving during past hour
02	Clouds generally unchanged during past hour
03	Clouds generally increasing during past hour
04	Visibility reduced by smoke
05	Haze
06	Widespread dust
07	Dust or sand raised by local winds, but not by dust storm or sandstorm
08	Dust or sand whirls seen in last hour, but no dust storm or sandstorm
09	Dust storm or sandstorm seen at or near station in past hour
10	Mist
11	Shallow patchy fog or ice fog
12	Shallow continuous fog or ice fog
13	Lightning but no thunder
14	Precipitation within sight not reaching surface
15	Precipitation reaching surface at a distance
16	Precipitation reaching surface near but not at station
17	Thunderstorm but no observed precipitation
18	Squalls seen at time of observation or during last hour
19	Funnel cloud(s) seen at time of observation or during last hour

20–29	*Precipitation, fog, or thunderstorm at station in past hour but not at time of observation. Note that freezing means freezing on impact with solid surfaces.*
21	Drizzle (not freezing) or snow grains, not in showers
22	Rain (not freezing), not in showers
23	Rain and snow or ice pellets, not in showers
24	Freezing drizzle or freezing rain
25	Rain showers
26	Rain and snow showers, or snow showers
27	Hail and rain showers, or hail showers
28	Fog or ice fog
29	Thunderstorm
30–39	*Dust storms, sandstorms, drifting or blowing snow*
40–49	*Fog or ice fog at time of observation*
50–59	*Drizzle at time of observation*
50	Not freezing, intermittent, slight
51	As 50 but continuous
52	Not freezing, intermittent, moderate
53	As 52 but continuous
54	Not freezing, intermittent, heavy
55	As 54 but continuous
56	Slight freezing drizzle
57	Moderate or heavy freezing drizzle
58	Slight drizzle and rain
59	Moderate or heavy drizzle and rain
60–69	*Rain at time of observation: as 50–59 but with drizzle replaced by rain, and rain (in 58, 59) replaced by snow*
70–79	*Solid precipitation, not in showers*
70	Snow flakes, intermittent at time of observation
71–75	Pattern as in 51–55
76	Ice prisms with or without fog
77	Snow grains with or without fog
78	Isolated starlike snow crystals with or without fog
79	Ice pellets
80–90	*Showery precipitation*
80	Rain showers, slight
81	Rain showers, moderate or heavy
82	Rain showers, violent
83	Rain and snow showers, slight
84	Rain and snow showers, moderate or heavy
85	Snow showers, slight
86	Snow showers, moderate or heavy
87	Slight showers of snow pellets, whether or not enclosed by ice, with or without showers of rain, or rain and snow
88	As 87 but moderate or heavy
89	Slight hail showers, with or without rain or rain and snow, without thunder
90	As 89 but moderate or heavy
91–94	*Current precipitation with thunderstorm in past hour*
91	Slight rain
92	Moderate or heavy rain
93	Slight snow, or rain and snow, or hail
94	Moderate or heavy snow, or rain and snow, or hail
95–99	*Current precipitation and thunderstorm*
95	Slight or moderate storm without hail but with rain and/or snow
96	Slight or moderate storm with hail
97	As 95 but heavy storm
98	Storm with sandstorm or dust storm
99	Heavy storm with hail

Past Weather

0	Cloud covering half or less of the sky throughout the period
1	Cloud covering more than half of the sky for part of the period
2	Cloud covering more than half of the sky throughout the period
3	Sandstorm, dust storm, or blowing snow
4	Visibility less than 1 km because of fog, ice fog, or thick haze
5	Drizzle
6	Rain
7	Snow or mixed rain and snow
8	Showers
9	Thunderstorm with or without precipitation

distant visible is reported at the time of observation. At automatic stations and many manned ones, this has been replaced by an instrument which measures the optical extinction of a beam of light folded by multiple reflections along a fairly short fixed horizontal path, and converts its output to the traditional visibility range. Visibility information is particularly important for shipping and aircraft, especially at ports and airports.

2.8 Observation networks

By the early nineteenth century it was clear that a great deal of weather in middle latitudes was organized in moving patterns with horizontal scales of 1000 km or more. However, while the highest speed of data transport was that of a galloping horse (later a train), analysis of faster movements and developments of weather patterns could only be retrospective (Fig. 2.11). The development of the electric telegraph by Morse and others in the 1840s transformed the situation: within 20 years, networks of meteorological observing stations connected by telegraph to a centre for analysing and forecasting were established in France, Britain, and the USA, and other industrial countries quickly followed suit, beginning the development of the national networks which now cover the land areas of the globe, coordinated from Geneva by the World Meteorological Organisation (WMO). The full observing, communicating, and analyzing and forecasting system is called World Weather Watch (WWW). In the following we outline the Global Observing System (GOS)—the observing part of WWW—under the separate headings of Surface, Upper Air, and Satellite networks. Consider first some general points.

(i) GOS is geared to record the present state of the atmosphere with just enough detail and accuracy to permit usefully accurate forecasting. Initially, when only the surface network existed, forecasting was mainly for coastal shipping, but it is now aimed at all sea and air transport, and land transport in hazardous conditions, as well as agriculture, industry, leisure, and the public at large. In vulnerable regions, special hazards such as hurricanes, tornadoes, snowfall, and drought receive special attention. In addition climate is now monitored globally to feed developing forecasts of climate change.

(ii) Just how much observational coverage and accuracy will meet these demands, is answerable only by experience. Our limited understanding of the great complexity of weather and climate might tempt us to go for all the data we can get, but limitations of finance and personnel mean that some observations can be made only if others are not, and so GOS has to be selective.

(iii) Optimization of GOS is never-ending, as you will see from WMO websites. As new observational techniques emerge, the balance of effort has to be reassessed. Automatic buoys are helping to fill gaps in the oceanic part of the Surface network, but are expensive to deploy and maintain. Satellites have taken some pressure off the Upper Air network, are highly cost-effective in terms of data quantity and coverage, but are individually very expensive. And weather radar networks resolve small and meso-scale precipitation patterns and support important short-term forecasts, including tornado warnings, but at very considerable cost.

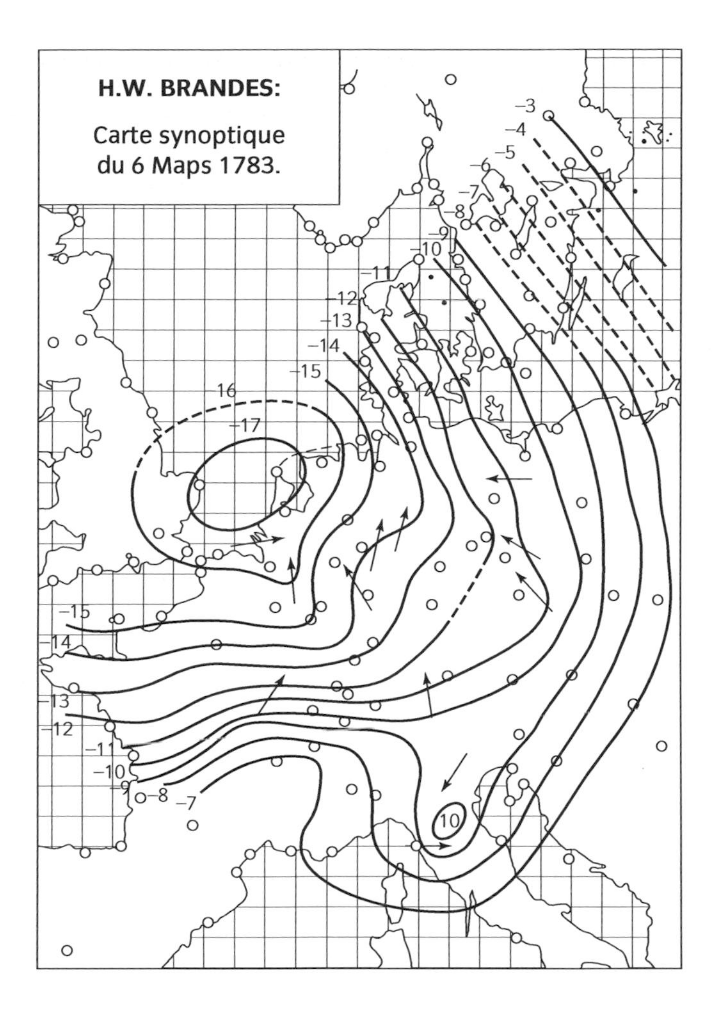

Figure 2.11 An early synoptic chart reconstructed by the pioneer Brandes some decades after the event. The isobars (of pressure difference from station averages to allow for the effects of variable station altitude) clearly reveal the SE quadrant of a depression centred between England and Holland, with cyclonic surface winds angled towards low pressure by friction (Section 7.13.4)

(iv) Increasing understanding of weather and climate can also shift priorities, easing some problems but highlighting demands for better or different data elsewhere. The increasing recognition of the importance of very large connected patterns of weather and climate such as El Niño (Section 14.3) has increased the need for highly accurate data on a global scale.

(v) Even when a good compromise network is agreed at any given time through WMO, this can be fully implemented by only the most wealthy nations: others have networks which are more or less inadequate, and the oceans are still very poorly covered. Weather radar is particularly restricted by cost. Patchiness of observational coverage, in the face of a global need for homogeneous coverage, is a persistent problem for GOS.

(vi) As already mentioned, standardization of observational practice is crucial, otherwise data from different stations can differ more on account of differences in observational practice than atmospheric conditions. Though obvious in principle, effective implementation requires surprising rigour in even the smallest matters, as exemplified below. Such good practice has been a preoccupation of WMO from the outset, and underpins every good observation, from the Stevenson screen to the satellite.

2.9 Surface network

Observations of wind, temperature, humidity, pressure, visibility, cloud, and present and past weather types are ideally made at each station of the global network every hour, just before the hour (Greenwich Mean Time or Universal Time, denoted by Z for Zulu), providing a vast amount of worldwide synoptic hourly data. In regions where such frequencies cannot be maintained, priority is given to observations at 0000, 0600, 1200, and 1800Z. The treatment of precipitation totals is different, since manual gauges are traditionally read at 0900 and 2100 hrs local time to capture the systematic differences between night and day overland precipitation. All instruments are designed to conform to international standards, and are operated by staff trained to use them correctly and consistently. The scale of weather phenomena captured by networks of surface and upper air stations making *synoptic* (simultaneous) observations has become known as the *synoptic scale*, and includes tropical and extratropical cyclones, anticyclones, fronts and jet streams and showery air masses.

2.9.1 Data standardization

As far as possible, observations are made at sites representative of the local terrain. Full details, even down to the length of grass in the fenced enclosure, are specified in handbooks such as *[6]*.

Pressure

Sites vary considerably in altitude, and since atmospheric pressure decreases by over 1 hPa (0.1%) for every 10 m rise near sea level, substantial corrections must be made to station pressures to prevent the effects of station altitude from swamping those arising from weather systems. Corrections are made by estimating the additional pressure which would have arisen had the atmosphere continued from station level down to mean sea level (the agreed datum level). The correction is calculated from a version of Eqn 2.1

$$\Delta p = g\,\rho\,h \qquad 2.2$$

where Δp is the pressure to be added to the station pressure to find pressure at MSL, g is the local gravitational acceleration, h is the height of the station above

NUMERICAL NOTE 2.7 Correction of station pressure

Imagine a vertical hole bored from station level down to MSL. Air pressure in the hole will increase downwards according to Eqn 2.2. Assuming g is 9.81 m s^{-2} (see Section 7.4.2), and ρ is 1.2 kg m^{-3}, MSL pressure at a station with altitude 123 m, will exceed station pressure by $\Delta p = 9.81 \times 1.2 \times 123 = 1{,}448$ Pa $\approx$ 14.5 hPa. Since air density varies significantly with air temperature and pressure (Section 4.5), in operational practice ρ is estimated from station level values using $p = \rho RT$ (Eqn 4.1). Corrections for high altitude stations allow for air density increasing downwards to MSL (Section 4.5).

MSL, and ρ is the air density of the virtual air column (NN 2.7), which is usually assumed to be the air density at the station surface. This is satisfactory for the many stations which lie within a few hundred metres of MSL, but needs modification at mountain stations, which may have to allow for a kilometre or more of virtual atmosphere.

Temperature

No attempt is made to correct air temperatures for station altitude, but the height of the Stevenson screen above the local ground surface must be standardized since diurnal temperature variations fall away sharply with increasing height above the ground surface, because heating and cooling are concentrated at the surface (Section 10.3 et seq.). Temperatures would be more consistent at 10 metres, but of course less representative of surface conditions. The compromise solution is to place all Stevenson screens in the network at the same height above the local surface, so that they are subject to similar daily cycles. The standard height of the thermometer bulbs inside the screen (often called *screen level*) is 1.5 m (with slight national variations), which is obviously related to human stature.

Wind speed

Wind speed increases with height above ground level, as well as with distance from upwind obstructions. The latter effect is minimized by choosing an open site, but the former requires arbitrary choice of measurement level, since wind speed increases (though more and more gradually) for several 100 m above the surface. The level chosen for both anemometer and wind vane is 10 m, which is about the highest level which can be reached using a small mast. At many secondary observation stations even a small mast is unavailable, and the anemometer is sited at 2 m and its data multiplied by a factor of 1.28*[6]* to provide an estimate of the 10 m wind speed.

2.9.2 Networks

The distribution of surface synoptic stations across the British Isles is shown in Fig. 2.12. The average separation is about 50 km, which is typical of industrial, densely populated countries. In poorer or more sparsely populated areas, networks are more open. There are about 4,000 stations like these worldwide, together with another 7,000 reporting less frequently. At sea there is now only a handful of dedicated weather ships, because of prohibitive cost. Instead about 3,000 merchant ships at any time are making observations on a voluntary basis. This is valuable, though data from merchant shipping are concentrated along popular shipping lanes and suffer from inevitable non-standard exposure. In addition there are nearly 1,000 instrumented drifting buoys, automatically reporting to collector satellites for subsequent downloading.

In addition to the standard synoptic network, Britain has a much denser network of voluntarily manned stations, many of them at schools, etc. where limited standard observations are made each day at 0900 Z (i.e. local morning). These data are not telegraphed to forecasting centres, but are collected and scrutinized on a monthly basis by the British Meteorological Office (the national forecasting service) where they serve to define the British climate and its variations on a much finer space scale than is possible with the synoptic network. Many other countries have similar non-standard systems.

Networks of *weather radars* are still largely confined to the wealthiest nations. The British Isles has a network of 11 sets, which is typical of coverage in western Europe and the United States. Each set continually monitors rainfall patterns within a range of over 100 km in non-mountainous terrain. Data are analysed, 'decluttered' of steady ground echo, amalgamated and displayed in real time by computer. Small images are displayed on public websites within about 15 minutes, because of their value as short-term rain predictors.

2.10 Upper air network

2.10.1 Radiosondes

Observations of wind, temperature, relative humidity, and pressure are made by *radiosondes*—free-flying balloons released from stations of the synoptic *upper air* network (Figs. 2.12a and 2.12b). The radiosondes are released at 0000 and 1200 Z daily, and climb at about 5 m s^{-1} until they burst randomly between 20 and 30 km above MSL, falling back to the surface by parachute drogue at much the same speed. Whilst ascending, temperatures, humidities, and pressures are measured in an instrumented package suspended well below the balloon and sent by radio to the launching station; and the sonde's position is monitored by automatic tracking radar.

The sonde has so much drag and so little inertia that it moves with the horizontal wind of each layer it rises through. Successive horizontal sonde positions (for example at 1 minute intervals) are analysed to yield a vertical profile of horizontal wind throughout its hour-long ascent. Data are analysed while the sonde is in flight, so that within minutes of bursting, ground station staff can transmit profiles of wind, temperature, and pressure from launch to the bursting level. Humidity data are usually ignored above 10 km because sensors are slow and unreliable in the very low temperatures there.

A radiosonde's flight is not strictly vertical, and can be a rather gradual slope in high winds. However, compared with the grossly flattened nature of all synoptic scale systems, even a slope of 1:10 is effectively vertical in relation to the very oblique structure of synoptic scale weather systems.

At 0600 and 1800 Z (i.e. midway between the full radiosonde ascents), balloons bearing only a radar reflector are flown and tracked to give profiles of wind alone. The six-hourly alternation of these *rawinsondes* with the full radiosondes provides sufficient resolution in time to define the structure of the synoptic-scale weather systems.

Network

Figure 2.12a shows the network of upper air stations in the British Isles, which again is typical of such networks in affluent countries only. The global total of upper air stations making the full set of observations is about 700, with another 200 making soundings only once per day. Upper air data from sea areas are limited to about 15 large commercial ships equipped with fully automatic

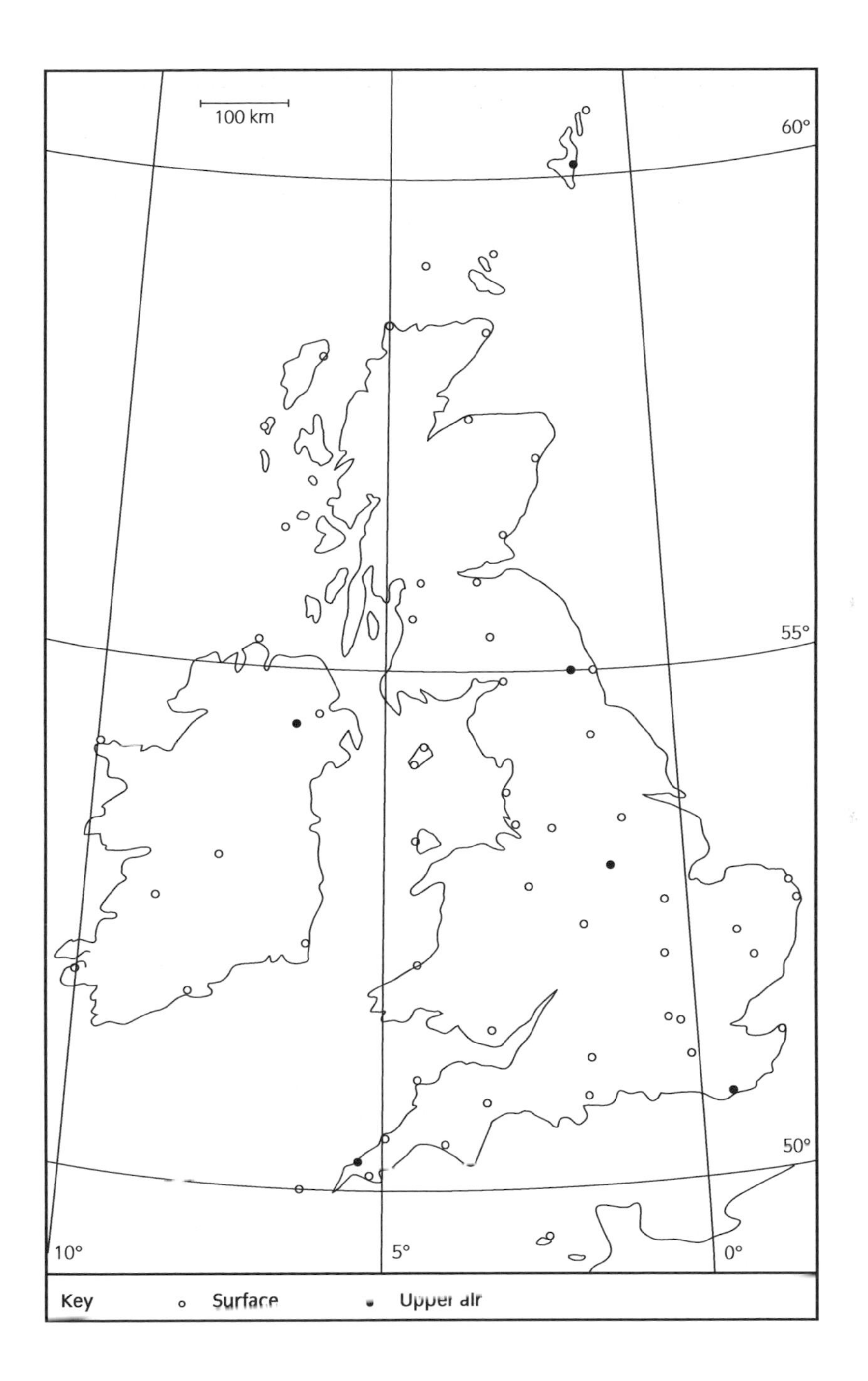

Figure 2.12a Map of British Isles surface (open circle) and upper air (filled circle) stations. Radiosonde are now mostly prepared, released, and tracked automatically, greatly reducing the costs of operation, and automation of the surface stations has encouraged a modest increase in numbers.

radiosonde systems sending data to collector satellites—so far largely confined to the North Atlantic. At its best the synoptic upper air network copes adequately with the synoptic scales of big weather systems, but becomes increasingly inadequate in the looser parts of the network, and everywhere fails to resolve mesoscale and small-scale structures (Fig. 1.5).

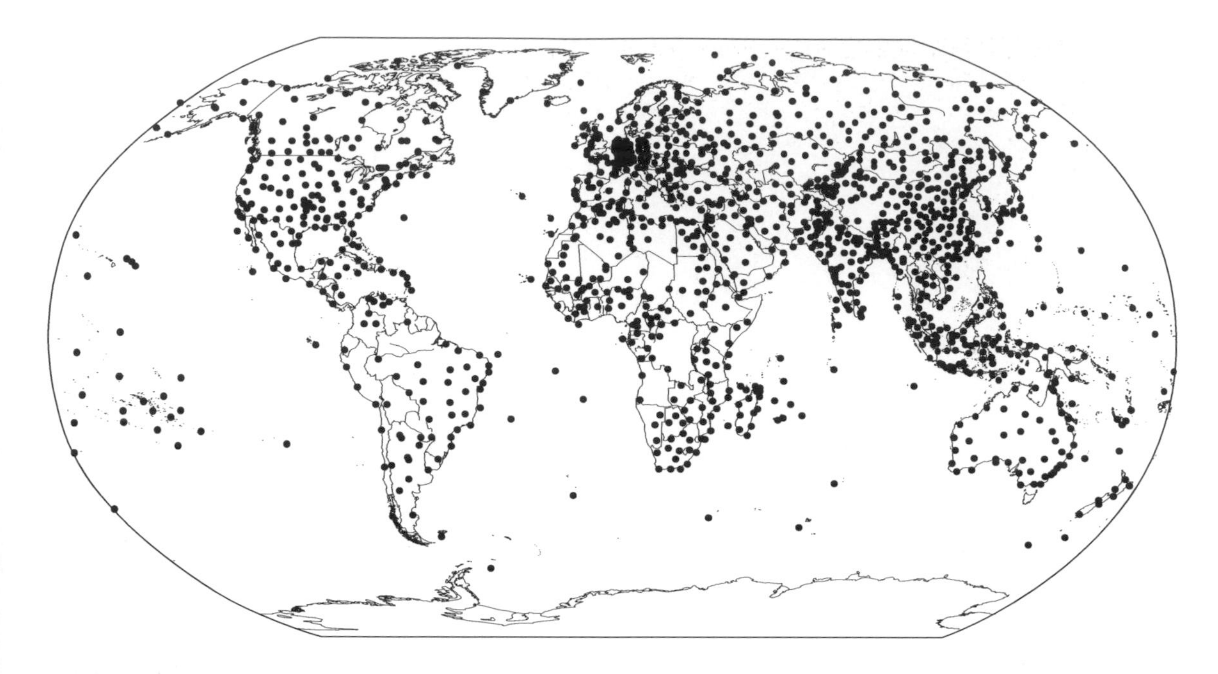

Figure 2.12b Upper air stations of the Global Observations System of World Weather Watch. Of the 900 worldwide, 600 fly sondes at 0000 and 1200 Z daily, and 200 fly once per day. There are about 15 equipped merchant ships, mostly in the Atlantic, and about 4,000 surface stations in the Regional Base Synoptic Networks.

2.10.2 Other data

In recent years many large civil aircraft have been fitted with automatic sensors which transmit pressure, temperature, and wind data in flight—mostly at cruising level (around the 250 hPa level), but with some from selected levels during ascent and descent. These provide valuable data, though concentrated heavily along popular flight paths in the northern hemisphere. Aircraft and ships provide a valuable though very inadequate sampling of the 70% of the atmosphere which overlies the oceans, but are especially sparse over the great oceans of the southern hemisphere.

Even after including ship and aircraft contributions, sampling intervals in space and time are much greater in the synoptic upper air network than in the much denser surface network. Fortunately weather systems are smoother and larger-scale aloft than they are near the surface, where local inhomogeneities generate significantly smaller and more transient details, so that the relatively coarse upper air network is satisfactory, at least in well-endowed areas. It is a major task of GOS to improve the existing upper air network, and/or devise complementary systems. Fortunately the satellite network has developed greatly in its five decades of existence.

2.11 The satellite network

The first meteorological satellites were launched in the early 1960s and attracted interest on three counts.

(i) Their panoramic views of the atmosphere directly revealed structures of large cloudy weather systems which had previously required painstaking assembly

Figure 2.13 Manual launch of a radiosonde. On a signal from the rest of the team (out of sight to the left), the operator will release the helium-filled balloon which will rise at about 5 m s^{-1}, lifting the instrument package (with its drogue parachute for later descent) when the 30 m line (invisible beyond the operator) pulls taut. The rising sonde can be tracked by radar, radio, or gps systems.

and analysis of synoptic data. Meteorologists in the 1960s were impressed to see the careful work of the pre-satellite generations so vividly confirmed in pictures such as Fig. 1.1.

(ii) The relatively high resolution of satellite images (a few kilometres compared with tens of kilometres in the synoptic surface network and even more in the upper air network) revealed a bewildering range of sub-synoptic scale structures in cloudy systems, many of which have still to be incorporated in standard observational models. For example, the quasi-geometric patterning of shower clouds in vigorously convecting oceanic zones (Fig. 2.14) was quite unknown before satellites were deployed, and has yet to be understood well enough to help forecasting.

(iii) Satellite observation can be spread relatively uniformly over the Earth's surface, helping fill the enormous gaps in the synoptic surface and upper air networks over oceans and low-latitude land areas.

2.11.1 Satellites and instruments

Meteorological satellites have multiplied and developed greatly since the 1960s, and the broad outline of a permanent network has emerged, although details are bound to change further as technology develops. Satellites are orbiting platforms for electromagnetic scanning of the atmospheric from above (*top-scanning*).

Figure 2.14 High resolution visible image from a polar orbiting satellite, showing a vast panorama of patterned shower clouds over the NE Atlantic. High latitudes are in very oblique sunlight, with dusk creeping in from the top right of the picture, but the stronger illumination at lower latitudes is picking out mesoscale structure on the top of the frontal cloud mass W of Ireland. The cool showery airstream S of Greenland is feeding cool air behind fronts off the upper right edge of the picture. There is a structured cyclonic swirl of cloud SW of Iceland.

Though most is passive, in that satellites receive radiation emitted or reflected from the atmosphere, some are beginning to use radar emitted and received by the satellite.

The field of view from a satellite is scanned line by line, either by a special television camera, or by spinning the satellite to sweep a narrow-field radiometer across the view, shifting it by one 'line' with each rotation. In each case data are sent in sequence to receiving stations on Earth for reconstitution of the whole image. The width of the scan line on the Earth's surface fixes the limit of resolution, details less than a few lines across being unresolved. Most satellites have a resolution of a few kilometres vertically beneath the satellite, widening with increasing obliqueness of view.

Cameras and radiometers are sensitive to one or more bands of visible or infrared wavelengths.

(i) If visible wavelengths are used, reflected sunlight picks out cloud vistas which are only slightly blurred versions of photographs taken manually by astronauts. A major disadvantage is the inevitable blindness on the Earth's night side.

(ii) Radiometers sensitive to wavelengths in the far infrared (i.e. $\lambda > 3$ μm) operate beyond the solar spectrum and respond instead to the terrestrial radiation emitted day and night by the Earth's surface and atmosphere. Since radiation intensity increases with the temperature of the emitting materials (Box 8.1), pictures can have a brightness pattern which corresponds to the temperature pattern in the panorama. Intensity can be displayed on a grey scale, or an arbitrary colour scale, usually with reds representing highest temperatures.

Because the cloud-free atmosphere is nearly opaque in some parts of the terrestrial spectrum and nearly transparent in others (Fig. 8.5), quite different pictures emerge in different wavelengths. In wavelengths in which cloudless air is transparent to terrestrial radiation, we get temperatures of the top surfaces of clouds, and uncovered ground or sea surfaces. Monochrome cloud vistas produced in this way resemble visible pictures if the grey scale is suitably chosen (Fig. 2.15). By contrast, pictures taken in wavelengths in which the air itself (actually its water vapour) is opaque to terrestrial radiation can show distributions of air temperatures in the cloud-free middle and upper troposphere, from which water vapour is radiating freely through the drier air above. A selection of wavelengths

Figure 2.15 Infrared image of frontal clouds over the NE Atlantic and NW Europe, taken from a polar orbiting satellite. Being independent of sunlight, the image is no darker in high latitudes. In fact it should be paler there, since the artificial grey scale has been arranged to make colder surfaces look paler. The very cold tops of the frontal cloud masses W of the British Isles and over N Europe and Scandinavia, contrast sharply with the warmer ground under the clear skies over Britain, and the even warmer (darker) adjacent seas. The picture was taken at about 0900 S on October 18 1979. The land surfaces must have darkened as they warmed under the rising sun, before they were covered by the occluding front advancing from the W.

covering a range of opaqueness can even be analysed to produce useful vertical temperature profiles in cloud-free air. These are potentially valuable additions to radiosonde measurements, and are under active development for that purpose.

(iii) Satellite-borne radar sets transmitting microwave radiation down into the atmosphere, and receiving reflections from a range of heights within it, will allow satellite soundings to reach inside clouds for the first time, with obvious potential for detecting rain areas on a grand horizontal scale.

2.11.2 Satellite orbits

Two types of satellite orbit are in use. *Sun-synchronous* orbits hold satellites at about 860 km (nearly one seventh of an Earth radius) above the surface, passing near the poles but making an angle to the meridians which keeps the orbit effectively fixed relative to the Sun (Fig. 2.16). At this altitude, satellites take about 102 minutes between successive passes over a given pole while their radiometers scan the helical swath of planet unrolling below. Any particular geographical location is overflown once every 12 hours at predictable clock times. Radiometer data can be received in real time by any suitably equipped station as it is overflown. In addition, data are accumulated over many orbits and downloaded to a master station to provide a complete data set.

In the *geosynchronous* or *geostationary* type of orbit, satellites are parked 35,780 km (over 5 Earth radii) above the equator, moving in the same sense as the rotating Earth. Since the orbital period at this radius is exactly one sidereal day, the satellite hangs vertically above a fixed point on the Equator (Fig. 2.16). The field of view at this range is nearly a full hemisphere, though the view of the polar, western, and eastern limbs is very oblique (see Fig. 1.1). A very high-resolution radiometer scans the whole view in about 20 minutes in solar and terrestrial wavelengths, sending data to ground-based stations and providing resolution which is nearly as good as that from the much lower level Sun-synchronous type.

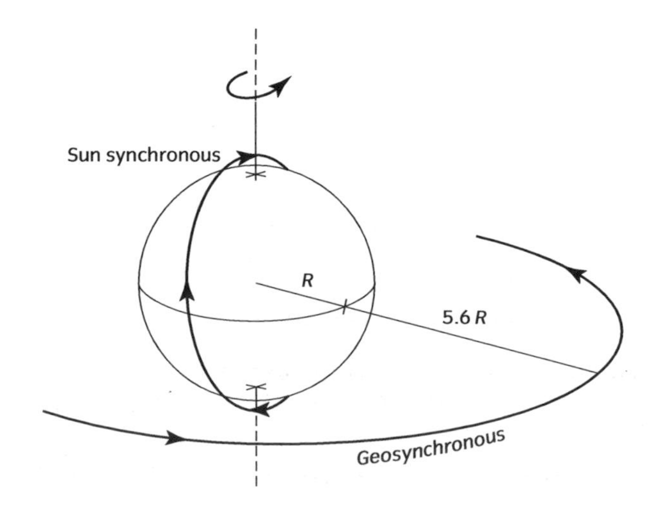

Figure. 2.16 Satellites in geosynchronous and sun-synchronous orbits. For ease of portrayal, the geosynchronous orbit has been drawn at a reduced scale relative to the Earth. Note that the retrograde sun-synchronous orbit misses the poles by about 10° latitude, whereas the geosynchronous satellite 'hangs' over a fixed position on the equator.

For GOS, global coverage is achieved by having a network of five geosynchronous satellites equally spaced along the equator. The bias of good coverage towards low latitudes is especially valuable because of the very poor synoptic network in those vast areas—half of the Earth's surface lies between latitudes 30° North and South. In particular it has transformed the detection and tracking of tropical cyclones, and greatly improved the vital forecasts of their damaging landfall.

2.12 Epilogue

A very large scale of individual and collective effort is needed to maintain World Weather Watch, even in its present incomplete state. Shortly before every major synoptic hour (i.e. 0000, 0600, 1200, 1800 Z) tens of thousands of observers all over the Earth are making their routine observations from the surface, and smaller numbers are launching and tracking radiosondes. Many more people are active around the clock in the associated telecommunication and data collection nets, and at the numerous forecasting offices. The completely international directorate and administration of the World Meteorological Organization coordinates the activities of the national networks and deals with matters ranging from standardizing equipment and procedure to planning and monitoring improvements to WWW.

After a century which has seen much of the optimism of the nineteenth century overwhelmed by confusion and strife, and a comprehensive approach to the human predicament frustrated by inflexible and unenlightened sectional interests, it is heart-warming to see what people can and will do when their efforts are focused by a common need. Of course the need is well-defined—to forecast the short-term behaviour of the lower atmosphere—and the atmosphere is inescapably global, mercifully independent of humankind's random subdivisions of the Earth's surface. Nevertheless the continuing achievement is encouraging, if only because the future of civilization in coping with serious anthropogenic climate change, and the social stresses this will surely generate, depends on such low-key, practical, and sincere cooperation, rather than on the strident rhetoric and divisive mayhem still so painfully evident in other aspects of human affairs.

Checklist of key ideas

You should now be familiar with the following ideas.

1. General principles of good observational practice.

2. The concept and measurement of instrumental lag.

3. Near-surface measurement of air temperature, humidity, wind direction, strength and speed, air pressure, and rainfall.

4. Rainfall observation and measurement by radar.

5. Observing and reporting sunshine, ground temperatures, clouds, and weather.

6. The Global Observing System of the WMO.

7. Ensuring representativity and consistency in the surface synoptic network.

8. Radiosondes and the upper air network.

9. Geosynchronous and geostationary orbits and the satellite network.

Problems

Outline answers to these problems can be found on the **Online Resource Centre**. Answers to odd numbered **problems** can be found under Student Resources, answers to even numbered problems under Lecturer Resources.

Level 1

2.1 A manufacturer seeks your advice on plans to build a transparent Stevenson screen. Advise him rationally and politely.

2.2 Why should a Stevenson screen have a floor?

2.3 The vapour content of a certain room is rising, though its air temperature and pressure are constant. Consider which of the following should be rising and which falling: vapour density, relative humidity, wet-bulb depression, dew-point depression.

2.4 What are the azimuths of the following winds: E, SE, S, and NNW (halfway between NW and N)?

2.5 Given that Hg density decreases with increasing temperature, if you read an Hg barometer at room temperature and use standard barometric tables (for Hg at 0 °C) without correcting for the barometer's actual (room) temperature, will you over or underestimate the atmospheric pressure?

2.6 Use Boxes 2.2, 2.3, and 2.4 to assess current wind strength, cloud types, and present and past weather at some convenient place and time.

2.7 Ideally how tall should the mast of a fully instrumented ocean buoy be? And if you have made the obvious choice, when might this be inadequate?

2.8 List all the ways of getting information about distributions of temperature, cloud, and precipitation through the troposphere.

Level 2

2.9 Suppose that a pitot tube anemometer gives an output of 1 unit in a wind speed of 5 m s^{-1}. What maximum output will enable it to register force 12 at 10 metres?

2.10 Careful work with a water-filled barometer shows that atmospheric pressure falls by about 1.2 cm of water for each 10 m height increase near sea level. Use realistic values of g and ρ in Eqn 2.1 to show that the corresponding pressure fall is about 1.18 hPa.

2.11 On a certain occasion the rain collected by a 127 mm (5 inch) diameter rain gauge filled the 25.4 mm (1 inch) diameter measuring cylinder to a depth of 300 mm. Find the true depth of rainfall, the volume of rainwater deposited on a catchment of area 500 km^2, and the rise in water level if half of this is collected in a 1 km square reservoir with vertical sides.

2.12 The surface of the Dead Sea lies 392 m below MSL. Assuming air density to be 1.2 kg m^{-3}, what correction must be applied to the station pressure at Dead Sea level to find the associated pressure at MSL?

2.13 When must a radiosonde be launched so that its flight to 20 km above launch level is completed by 1200 Z. Assuming that it is rising through a 20 m s^{-1} W'ly wind for the first half of its flight, while for the second half it is in an 80 m s^{-1} SW'ly jet, find the horizontal distance and direction of the final sonde position from its launching site. Such extreme translations trouble the Japanese weather service when the subtropical jet stream is blowing strongly overhead.

2.14 The lag of a moderately fast response thermometer is 3 s. How low is the thermometer reading10 s after an instantaneous 3 °C rise in ambient temperature?

2.15 Discuss the pros and cons of mounting your anemometer on a 2 m pole on the roof of a two-storey house, rather than on a 10 m pole near the house.

Level 3

2.16 In strong sunlight, sunshine often seems more intense to the human senses inside a greenhouse than in the open air. Actually the reverse must be the case. Explain.

2.17 Reconsider Problem 2.10 to show that water density is 833 times the local air density. Consider what would be observed if the same observations were made inside a tall Moon base filled with air at normal terrestrial surface pressure and temperature, assuming lunar g to be 0.17 times the terrestrial value.

2.18 In cool windy weather, large rain drops on the outside of a singly glazed window are often accompanied by misted patches on the inside of the window of a warm inhabited room. Explain.

2.19 In the absence of very dense-gauge networks, the spatial variation of rainfall round small objects such as hedges and houses can only be guessed. However, typical patterns of snowfall should provide some evidence. Consider factors which support and others which reduce the value of such evidence.

2.20 Section 2.2.1 suggests that small thermometers respond more rapidly because they have a smaller heat capacity. This argument is too simple, since small size must tend to slow response by limiting heat fluxes between thermometer and air. Argue the case more thoroughly for the simple case of a spherical thermometer to show that the speed of response is likely to be inversely proportional to bulb radius.

2.21 Station pressures on the occasion of Figure 2.11 were reduced to the plotted values by subtracting the station average in each case. Why was the modern correction procedure eventually preferred?

3 The constitution of the atmosphere

3.1 The well-stirred atmosphere

Up to nearly 100 km above sea level, the atmosphere consists largely of a mixture of gases whose main components are in remarkably uniform proportions, together with small and variable amounts of water vapour concentrated almost entirely in the troposphere (the first 10–15 km). We call the nearly uniform mixture *dry air* and use the term *moist air* for any mixture of dry air and water vapour. In addition to these gases, the atmosphere contains small quantities of condensed matter—liquid water and ice in the form of cloud and precipitation, and a population of even smaller solid and liquid particles known as *aerosol particles* (whose suspension in air is an *aerosol*), all heavily concentrated in the troposphere. Figure 3.1 presents the vertical distributions of these materials, together with other materials and processes to be mentioned shortly.

Over 99.9% of the mass of dry air consists of a mixture of molecular nitrogen (N_2), molecular oxygen (O_2) and atomic argon (Ar) in the proportions shown in Table 3.1—proportions which vary only minutely up to the 90 km level. The *partial density* of any of these constituents (for example the mass of N_2 per unit volume of air) falls enormously with ascent through this height range, but so does the overall density of the air (Section 4.5), with the result that the ratio of

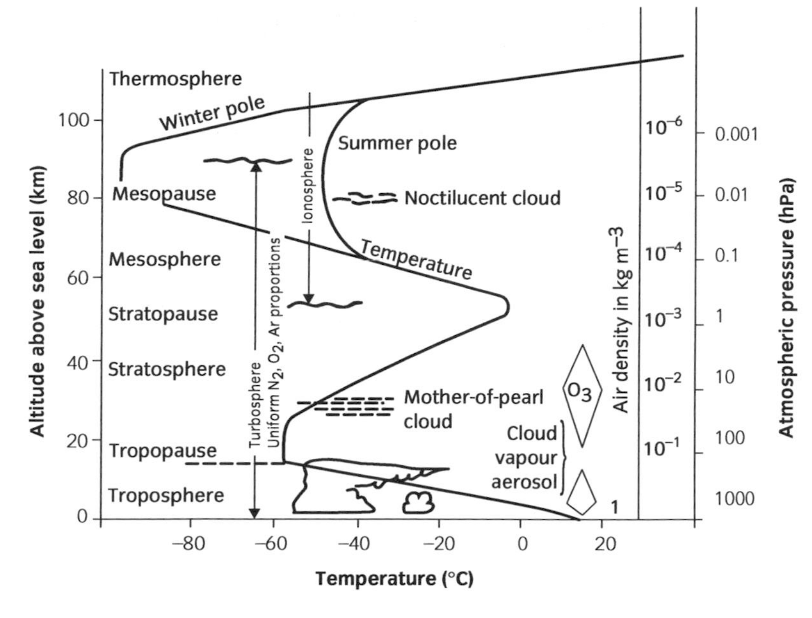

Figure 3.1 Schematic vertical distribution of atmospheric temperature, pressure, density, and atmospheric materials. Note that equal increments of height correspond well to equal multiple proportions of pressure and density, and that the large seasonal variation in the temperature of the high-latitude mesopause is induced by variations of direct input of solar hard UV.

Table 3.1 Composition of the turbosphere

Component gases	Specific mass	Molecular weight	Specific gas constant
Dry air			
N_2	0.755	28.02	297
O_2	0.231	32.00	260
Ar	0.013	39.95	208
CO_2	0.0005	44.01	189
Plus hundreds of trace gases			
Moist air			
Dry air	0.997	28.96	287
H_2O vapour	≈0.003 overall (~0.03 maximum)	18.02	461
Particles and droplets			
Cloud	~5 × 10^{-5} overall (~0.004 maximum)	Water and ice	
Precipitation	~3 × 10^{-6} overall (~0.004 maximum)	Water and ice	
Aerosol	10^{-8} overall (~10^{-7} maximum)	Various solids or concentrated solutions in water	

the two is very nearly uniform throughout. Calling this ratio (i.e. the mass of N_2 per unit mass of dry air) the *specific mass* of N_2, we must consider why it, and the specific masses of O_2 and Ar, are each so uniform throughout so much of the atmosphere.

In principle the reason is obvious: the atmosphere is continually mixed by its Sun-driven motion, and this mixing tends to maintain uniform distributions. But the atmosphere is also subject to selective processes which tend to degrade that uniformity. The uniform composition of the major components of dry air shows that mixing predominates in their cases, while the non-uniformity of water vapour and aerosol particles, etc. shows that selective processes predominate for them. Mixing is considered in more detail in the rest of this section, while selective processes occupy much of the rest of the chapter.

3.1.1 Turbulent and molecular diffusion in the atmosphere

In the atmosphere, turbulent diffusion (Box 3.1) predominates from the surface to over 90 km, and this deep layer is therefore called the *turbosphere* (Fig. 3.1). Above this there is a transition zone (the *turbopause*) where turbulent and molecular diffusion compete more equally, and above about 100 km is the *thermosphere*, where molecular diffusion predominates.

The distinction between turbosphere and thermosphere would matter much less than it does, if the equilibrium distributions for turbulent and molecular diffusion (Box 3.1) were identical, as they are on the domestic scale discussed in Box 3.1. However, the two equilibria can be very different in the free atmosphere because its vertical scale is much greater than the vertical scale of self-compression under gravity (Fig. 3.1). In fact since air density falls by over 10% for each kilometre height rise (NN 3.1), on larger height scales *convective* and *diffusive* equilibria (Box 3.1) begin to differ substantially.

Convective equilibrium

Despite the complexities of atmospheric turbulence (Section 10.7), the convective equilibrium it promotes in the compressible atmosphere is very simple on all vertical scales. Just as on the very small scale of a beaker of stirred water, the turbulence endemic throughout the turbosphere maintains a mixture in which each major component of dry air has a uniform characteristic specific mass (e.g. 0.755 for N_2). Such convective equilibrium is observed throughout the turbosphere, as shown in Table 3.1 and all but the top of Fig. 3.1, but it breaks down through and above the turbopause, as diffusive equilibrium takes over.

Diffusive equilibrium

Unlike convective equilibrium, diffusive equilibrium is significantly altered as we go from the laboratory scale to vertical scales of a kilometre or more. Consider molecular diffusion at work in such a deep air column. In the absence of turbulence, any one gas of the mixture (argon for example) will diffuse towards an equilibrium distribution unchanged by further molecular diffusion. The presence of nitrogen, oxygen, and other molecules may slow the approach to equilibrium, but will not otherwise alter it, so that we can ignore the rest and focus entirely on the equilibrium distribution of argon (in this case).

As shown in Section 4.4, the equilibrium profile (nearly pure exponential decay) is steeper for relatively heavy atoms like Ar (atomic weight 40) than it is for lighter molecules like O_2 (molecular weight 32). Vertical distributions of different gases at diffusive equilibrium will therefore differ according to their

BOX 3.1 Turbulent and molecular diffusion in liquids and gases (Fig. 3.2)

In laboratory scale liquids, mixing promotes uniformity in initially heterogeneous mixtures: an uneven distribution of ink in water becomes more uniform as soon as stirring begins, and quickly approaches an apparently uniform distribution of ink throughout the water. Since it is unaltered by further mixing, uniformity is by definition the *convective equilibrium* distribution for this mixing. Whether driven by stirring rod, convection or shaking, such mixing proceeds towards convective equilibrium by *turbulent* (or *eddy) diffusion* of the ink throughout the water. In such mixing substantial volumes (*eddies*) of more or less inky water form, move and mix out of recognizable existence, intermingling myriads of water and ink molecules in the process.

Even without turbulent mixing, uniformity would eventually result from the intermingling of ink and water by *molecular diffusion*. On the laboratory scale *diffusive equilibrium* would be reached in days rather than the seconds taken to reach convective equilibrium, because molecular diffusion is limited by the very slow *random walk* of individual water and ink molecules among their neighbours in the course of thermal molecular motion. Though the processes are very different, the equilibria reached are identical. In fact molecular diffusion is always at work, and maintains the uniform mixture of ink after turbulent stirring has ceased, unless there is a selective process at work like ink coagulation and settling.

The diffusive behaviour of gases is less easily observed than that of liquids, but quite simple laboratory demonstrations show it to be very similar, though very much faster because gas molecules move ballistically at about the speed of sound between collisions. At normal gas densities any turbulent mixing present will still dominate molecular diffusion, as it always does in liquids; but if gas densities are low enough, molecular diffusion will dominate, because molecules move much longer distances between consecutive collisions.

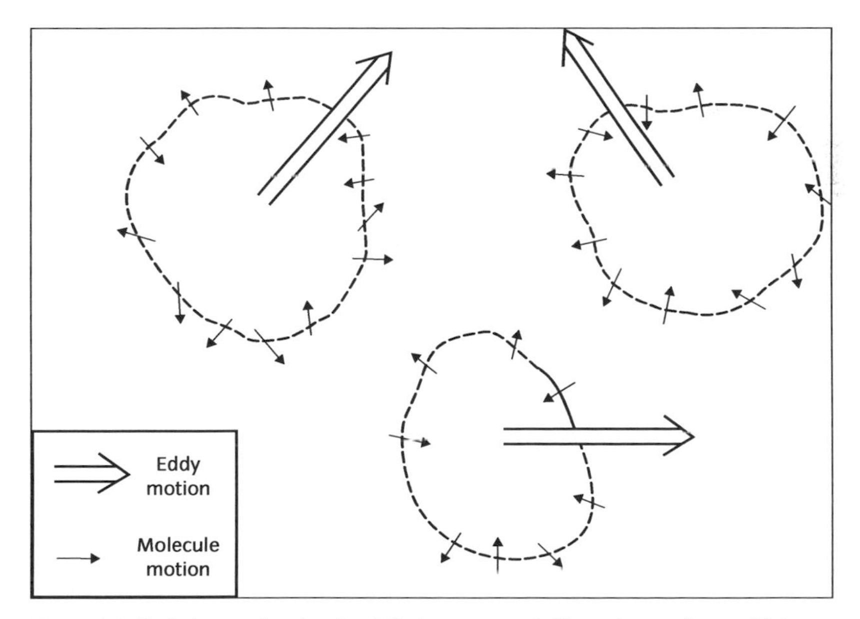

Figure 3.2 Turbulent and molecular diffusion compared. Three air parcels are eddying amongst each other and others not shown. On a much smaller scale individual molecules are entering and leaving each parcel. If the parcels and their motion are actual scale (corresponding to the meteorological turbulent microscale), the molecular arrows should be invisibly short (between collisions) and so numerous as to turn the page black.

NUMERICAL NOTE 3.1 Density lapse in an air column

Air pressure (we consider density later) falls exponentially with increasing height, with a binary scale height of about 5 km throughout the turbosphere (Box 1.2), which means that pressure falls from p_1 to p_2 as height increases from z_1 to $z_2 = z_1 + \Delta z$ (in km) according to $p_2/p_1 = (2)^{-\Delta z/5}$. (Check that $p_2 = p_1/2$ when $\Delta z = 5$ km.) For $\Delta z = 1$ km this gives $p_2/p_1 = 2^{-0.2} = 0.87$, showing that pressure falls by 13%, as air density would too if the air column had uniform temperature. Since in fact air temperatures can fall by up to 10 °C in 1 km (Section 4.2), there is an offsetting density increase of 3.7% (10 K in 273 K) per km. Combined with the pressure effect, there is an overall density decrease of over 9% as height increases by 1 km. In fact density variations throughout the depth of the turbosphere largely follow the nearly exponential decay of pressure because proportional temperature variations on the absolute scale are very much smaller than the huge proportional variations in pressure (Section 4.5).

scale heights—the scale height for the Ar profile being about 32/40 (=80%) of the scale height for O_2, since 32/40 is the inverse of the ratio of their molecular weights in uniform temperatures. If Ar and O_2 were equally abundant (by mass) in the middle heights of a complete air column (Fig. 3.3), Ar would predominate at the column base while O_2 would predominate at its top, and the proportion of Ar relative to O_2 would fall continually with increasing height. In the real atmosphere O_2 is 20 times as abundant as Ar overall, but the ratio of Ar to O_2 would still fall with increasing height, and have the shape shown in Fig. 3.3, though with much lower absolute values.

Corresponding curves can be drawn for the absolute and relative amounts of O_2, N_2, Ar, and other gases in diffusive equilibrium in the deep atmosphere, and all show exponential profiles teased apart by the selective effect of gravity, with heavier atoms enriched towards the base of the column, and lighter atoms enriched aloft. In principle such separation must occur in laboratory-scale gas

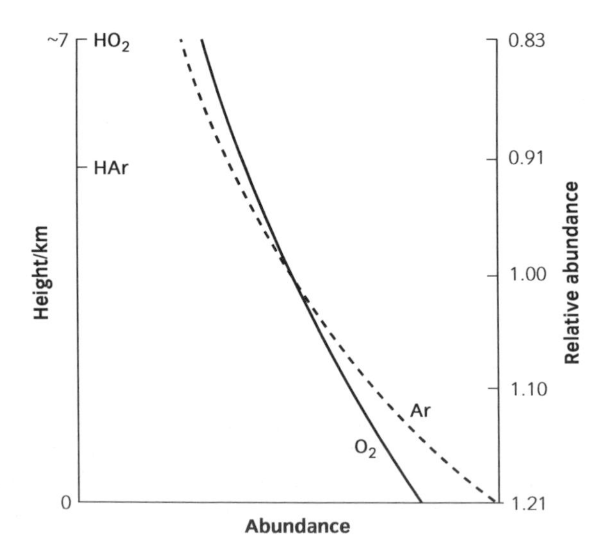

Figure 3.3 Vertical profiles of the relative abundance of atmospheric Argon and molecular Oxygen in diffusive equilibrium through one exponential scale height of air, with the relative abundance of Ar multiplied by 20 for clarity. In the first 100 km above sea level turbulent diffusion overwhelms this molecular diffusion and makes the profiles identical apart from the factor of 20.

volumes too, but vertical scales (~1 m) are so small compared with typical scale heights (~1 km) that the gravitational separation is unobservably small.

Convective and diffusive regimes

Below the turbopause, atmospheric mixing (by weather systems, convection, and turbulence) continually overwhelms the vertical separation of heavier and lighter components of air by molecular diffusion. Diffusive separation is still happening, but larger-scale mixing is moving air parcels and masses up and down much faster than its constituent molecules are being stratified by molecular diffusion. Convective equilibrium therefore prevails throughout the turbosphere, and the relative proportions remain remarkably uniform, as apparent in Table 3.1 and Fig. 3.3. Above the turbopause, however, diffusive equilibrium prevails.

3.2 Selective processes near and above the turbopause

Above about 100 km, diffusive equilibrium under gravity stratifies constituent gases according to molecular weight as just described. However, several other selective processes can significantly modify the expected purely gravity-driven diffusive separation.

(i) Although lighter gases should predominate at great heights, the concentrations of hydrogen and helium (the two lightest molecules) have fallen far below their cosmic abundances by preferential leakage to space throughout Earth's long history—their molecular speeds being higher than those of heavier molecules at any given temperature according to the kinetic model of gases.

(ii) As height increases there is increasing *photodissociation* of molecules into their constituent atoms, and this begins to affect the composition of air above the 80 km level. For example, above about 120 km the reaction

$$O_2 + \text{photon (solar UV)} \rightarrow 2O$$

maintains more than half of all oxygen in the atomic form (O), strongly absorbing solar ultraviolet with wavelengths between 0.1 and 0.2 μm, and maintaining the relative warmth of the *thermosphere* (Fig. 3.1). Since atomic oxygen (O) acts like a gas with half the molecular weight of O_2, the overall density of oxygen is reduced further by subsequent upward diffusion of the lighter O atoms towards diffusive equilibrium under gravity.

(iii) *Photoionization* also increases with altitude. For example the reaction

$$O + \text{photon (solar UV)} \rightarrow O^+ + \text{free electron}$$

maintains a population of ionized oxygen atoms (O^+) and free electrons, absorbs solar UV with wavelengths less than about 0.1 μm, and warms air over a wide range of heights. The free electrons ensure that the upper turbosphere is the base of the *ionosphere*, by keeping electrical conductivity high

Figure 3.4 Launch of a high-altitude research balloon. The instrument package contains radiometers etc. to measure water vapour and trace gases in the stratosphere and low mesosphere and hangs about 20 metres below the base of the slack balloon enevelope. As the balloon ascends, the bubble of helium expands against falling outside air pressure, fully filling the spherical envelope at flotation altitude (about 50 km according to mission). Afterwards the drogue parachute returns the package to the surface.

enough to reflect and channel long radio waves emitted from man-made radio transmitters on the Earth's surface, and allowing radio propagation around the spherical Earth. The specific masses of all types of ions increase sharply with height above the turbopause, and at altitudes greater than about 150 km can seriously affect the gross movement of air in the presence of the Earth's magnetic field. The atmosphere at these and higher levels is a diffuse *plasma* with properties very different from those of the electrically nearly neutral lower atmosphere *[14]*.

3.3 Selective processes in dry air

Consider the major gases of dry air (Table 3.1) in increasing order of their chemical activity in the surface and turbosphere.

3.3.1 Argon

Since this is chemically inert, its only short-term activity is the physical one of being continually mixed throughout the turbosphere. The uniformity of its specific mass there is a measure of the predominance of convective over diffusive equilibrium, despitė having the heaviest molecule of the major atmospheric gases. In the very long term the picture is less simple. The present abundance of argon consists almost entirely of the isotope ^{40}Ar, which has accumulated by very slow radioactive decay of potassium (^{40}K) buried in the outer shell of the Earth. This ^{40}Ar leaks slowly to the Earth's surface (Section 3.9) and then into

NUMERICAL NOTE 3.2 The build up of atmospheric Ar

The half-life for radioactive decay of ^{40}K is 1.3×10^9 years, meaning that the amount of undecayed ^{40}K in a certain mass of rock falls to 50% in one half life, to 25% in two, to 12.5% in three, and so on. For simplicity taking the age of the Earth to be 3.5 such half lives (i.e. 4.55 billion years), the mass of undecayed ^{40}K falls to $(0.5)^{3.5} \approx 9\%$ during this enormous time period. Currently therefore, about 91% of the original ^{40}K has decayed to gaseous ^{40}Ar and become liable to leak into the atmosphere, compared with just 50% when the Earth was only 1.3 billion years old (one half-life for the decay of ^{40}K). Assuming negligible permanent trapping of ^{40}Ar within the solid Earth, the amount of ^{40}Ar in the atmosphere has very nearly doubled ($91/50 \approx 1.8$) between when the Earth was 1.3 billion years old and now (about 4.55 billion years). Further increase will be small however long the Earth survives.

Computational tip: $0.5^{3.5} = \sqrt{0.5} \times 0.5^3$ i.e. $0.7071 \times 0.125 \approx 0.089$ i.e. 8.9%.

the atmosphere, where it is mixed throughout the turbosphere almost instantaneously on these huge time scales, maintaining a distribution which is uniform at any time but grows gradually throughout Earth's history (NN 3.2).

3.3.2 Nitrogen

In its molecular form (N_2), nitrogen accounts for very nearly 75% of the mass of the atmosphere and virtually all of its nitrogen. Although not very active chemically in terrestrial conditions, it is far from inert. Atmospheric N_2 is *fixed* (chemically combined with other substances) and removed from the atmosphere in several natural and artificial ways. It is also returned to the atmosphere by processes dominated by *denitrification*.

The current total fixation rate is believed to be about 290 megatonnes (Mt) per annum, made up as follows *[15]*:

- fixation by biological micro-organisms, mainly on land 240
- similar fixation via artificial fertilizers 20
- fixation by lightning and wildfires 10
- fixation by agricultural burning 20

The current total rate of return to the atmosphere is estimated to be about 295 Mt per annum, made up as follows:

- denitrification from land biomass 275
- denitrification from land via nitrous oxide (N_2O) 10
- net marine denitrification 10

The overall picture is near equilibrium dominated by fixation and denitrification to and from land, with a small part by net denitrification in the oceans. Rounding and equalizing the totals we can say that about 300 Mt of N_2 are extracted from the atmosphere each year and similar amounts returned. Comparing this with the total amount of nitrogen in the atmosphere, which is very accurately known to be 3.95×10^9 Mt (NN 3.3), we see that individual N_2 molecules must spend about 13 million years ($3.95 \times 10^9/300$) in the atmosphere between entering it and next leaving. This is the *residence time* for nitrogen in the atmospheric branch of the nitrogen cycle, found by dividing the mass of the reservoir by the rate of influx or efflux (Box 3.2). This time is so long that the mixing of N_2 throughout the turbosphere is almost instantaneous by comparison. As in the case of argon we therefore expect the specific mass of N_2 to be very nearly uniform throughout the turbosphere, as is the case.

Uncertainties about magnitudes of components in the N_2 cycle suggest that in fact there may be significant global imbalance. If a slight excess of injection over removal of atmospheric N_2 were to be maintained at 10 Mt per yr for 400 million years (less than 10% of the life of the Earth so far), the mass of atmospheric nitrogen would double to about 8×10^9 Mt, nearly doubling the total atmospheric mass, and with it the surface pressure. Such extrapolation assumes that all other physical, chemical, and biological processes involving nitrogen remain static, which is completely unrealistic in a terrestrial system we know to be riddled by rapid self-adjustment. Though our understanding of the nitrogen cycle is incomplete, it is already clear that human activity is significantly affecting the system. Indeed there are already strong localized effects, with nitrogen-rich runoff from intensively fertilized agricultural land depleting oxygen in rivers and lakes to produce disruptive algal blooms, and increasing oxides of nitrogen (~10^{-3}% of N_2 in abundance) significantly distorting natural air chemistry *[15]*.

3.3.3 Oxygen

Oxygen is chemically and biologically the most reactive of the major atmospheric gases—so reactive that it may seem surprising that unreacted molecular oxygen (*free* O_2) should make up nearly one quarter of the mass of the atmosphere.

BOX 3.2 Residence time

Consider atmospheric N_2 as an example of a reservoir maintained in dynamic equilibrium by balanced production and loss. Suppose M is the mass of atmospheric N_2, and F is its mass rate of production (denitrification), which is also its rate of loss (fixation). If labelled individual molecules of N_2 entered and left the atmosphere steadily in the same chronological order, each would spend a time T in the atmosphere given by $T = M/F$. T is called the *residence time* of N_2 in the atmosphere. Of course molecules do not enter and leave in such an orderly manner, so T is an average period of residence, with some molecules staying in the atmosphere for much shorter times and others staying much longer.

The concept of residence time is useful in all systems in or near dynamic equilibrium. The quantity M is usually a mass or something proportional to mass, but it can be any conservative quantity, e.g. energy in Section 8.3, or equivalent water volume in the hydrologic cycle (Section 3.8). Quantities M and F may be expressed in any consistent units, and the residence time T will emerge in whatever time units are used for F, e.g. years if F is expressed in throughput per year.

If during a cycle the quantity M passes through successive reservoirs, without accumulation or deficit, the flux F must remain the same along the sequence, and so must the ratio of capacity to residence time, so that $M_1/T_1 = M_2/T_2 = M_3/T_3$ showing that residence times T_i decrease with reservoir capacity M_i. This chain rule applies nicely to the vapour, cloud, precipitation sequence in the atmospheric branch of the hydrologic cycle (Section 3.8).

Residence time is still a useful measure in systems which are significantly out of balance, as many cycles now are because of human interference. If the mass efflux F_e is larger than its influx F_i by a steady amount $\Delta F = F_e - F_i$, then the atmospheric reservoir will empty in time $M/\Delta F$ which is much longer than M/F_e when the imbalance is only slight (i.e. $\Delta F << F_e$ or F_i). It follows that changes in M during one residence time are small, and that residence time retains its physical meaning despite the imbalance.

In fact, virtually all inorganic terrestrial materials are already fully oxidized, so that the reactivity of O_2 is very largely with living or recently dead organisms.

Atmospheric O_2 is consumed by the many types of *respiration* by which living organisms produce energy internally by highly regulated oxidation of food, and by *decomposition*, in which bacteria act similarly on the corpses of dead organisms. Detailed estimates *[15]* suggest that these processes (RD for brevity) collectively consume oxygen at a rate of 190,000 Mt per annum. This very nearly balances the release of O_2 as a by-product of *photosynthesis* (PS), the reverse process by which green plants on land, and *phytoplankton* in water, build their tissues from carbon dioxide and water, using energy from sunlight. RD and PS can be represented by Eqn 3.1—PS driving from left to right by input of solar radiant energy, and RD working from right to left, releasing heat in the process.

$$CO_2 + H_2O + \text{photon (solar visible)} <-> [CH_2O] + O_2 \qquad 3.1$$

The term $[CH_2O]$ represents the sugars and more complex carbohydrates which are produced by photosynthesis and consumed by respiration.

O_2 is also consumed by *combustion* of all sorts (CB), ranging from natural wildfires to the coal, oil, and gas-fired furnaces of human industry, but even after recent sharp increases in domestic and industrial combustion, the current total annual oxygen consumption by CB is only about 1/12 of the consumption by RD. The residence time of O_2 in the atmosphere is still dominated by the near balance between PS and RD.

Estimated values of oxygen reservoir and throughput imply a residence time of just over 6000 years for oxygen in the atmosphere (NN 3.3). Though much shorter than the residence time for N_2 (consistent with oxygen's much greater activity), this is still very long compared with the speed of mixing through the atmosphere. We should therefore expect the specific mass of O_2 to be very uniform throughout the turbosphere, which is almost always the case. A sad exception proves the rule: in the firebombing of the city of Dresden in 1945, pockets of people who would otherwise have survived, died from oxygen starvation as the surrounding fires temporarily lowered its specific mass below life-sustaining levels.

Because of its chemical reactivity, the present large amounts of atmospheric O_2 cannot have survived the Earth's convulsive origin, just as O_2 is not found in gases vented by volcanoes. Almost all atmospheric O_2 is believed to have arisen from a slight excess of PS over RD, which has persisted throughout much of the long development of life on Earth, but has been especially marked in the explosive development of life in the last few hundred million years (Section 3.9.2). This excess has been maintained by the burial of organic debris before full decomposition has been able to complete the PS RD cycle (Eqn 3.1).

Using different oxygen isotopes as tracers, laboratory investigation shows that, of the two free oxygen atoms produced in each photosynthetic reaction (on the right-hand side of Eqn 3.1), one comes from water on the left-hand side and the other comes from the CO_2. There are very large reservoirs of water and CO_2 on Earth, both in the oceans (which contain over 50 times more dissolved carbon dioxide than the atmosphere). Using these reservoirs and the power of sunlight, photosynthesizing life has built up the reservoir of free oxygen on which an important part of the biosphere (including ourselves) now depends.

Humankind's frantic combustion of fossil fuels cannot last more than a few hundred years, even in the case of the most abundant reserves (coal, oil, and gas). But even so, this sudden completion of the decomposition of myriads of ancient plants,

NUMERICAL NOTE 3.3 Mass and residence time for atmospheric nitrogen and oxygen

N_2
Because hydrostatic equilibrium is so perfect in the atmosphere (Box 4.2), the mass of air resting on unit horizontal surface area is p_s/g, where p_s is surface pressure, and g is gravitational acceleration. Using global average MSL pressure 1,013 hPa for p_s and 9.8 m s^{-2} for g, the atmospheric mass is $1{,}013 \times 100/9.8 \approx 10{,}340$ kg m^{-2}, i.e. 10.34 tonnes m^{-2}. From Table 3.1 the proportion of this mass arising from the presence of N_2 in the air column is $0.755 \times 0.997 = 0.753$, so that the column contains about 7.78 tonnes of N_2. Multiplying this by the surface area of the Earth (5.1×10^{14} m^2, assuming radius R is 6,370 km and spherical surface area $4\pi R^2$), we find the total mass of atmospheric N_2 to be 3.97×10^{15} tonnes, i.e. $\approx 4.0 \times 10^9$ Mt.

Applying Box 3.2, the residence time T for N_2 in the atmosphere is given by $T = M/F$, where M is the total atmospheric mass of N_2 ($\approx 4 \times 10^9$ Mt), and F is its rate of throughput (about 300 Mt yr^{-1} according to Section 3.3). We have $T = 4.0 \times 10^9/300 \approx 13$ million years.

O_2
According to Table 3.1, the mass of atmospheric O_2 must be in proportion to the mass of N_2 in the ratio $0.231/0.755 = 0.306$, which gives 1.21×10^9 Mt O_2. Since the rates of gain and loss by PS and RD respectively are each about 190,000 Mt per yr, the residence time for O_2 molecules in the atmosphere is $1.21 \times 10^9/190{,}000 \approx 6{,}400$ yr.

will significantly affect parts of Earth's surface system. The oxygen freed over aeons in this way by the action of ancient sunlight is now being suddenly reclaimed from the atmosphere, with consequences which are poorly understood, except that they will be felt much more immediately in the small reservoir of free CO_2 in the atmosphere and oceans, than in the much larger reservoir of O_2 (Section 3.5).

3.4 Ozone *[16]*

A minute fraction of total atmospheric oxygen is kept in the form of ozone (O_3) by a series of photochemical reactions (the *Chapman reactions*) in the stratosphere involving *hard solar UV* (i.e. wavelengths less than 0.2 μm), and *soft solar UV* (wavelengths between 0.2 and 0.3 μm).

$$\begin{aligned}
&\text{(i)}\ O_2 + \text{photon (hard solar UV)} \rightarrow 2O \\
&\text{(ii)}\ O_2 + O + M \rightarrow O_3 + M \\
&\text{(iii)}\ O_3 + \text{photon (soft solar UV)} \rightarrow O_2 + O \\
&\text{(iv)}\ O + O_3 \rightarrow 2\,O_2
\end{aligned} \qquad 3.2$$

Reaction (i) is increasingly strong above 100 km, where it helps maintain the warmth of the thermosphere (Section 3.2.2). It weakens progressively with decreasing height, but still maintains a small but important proportion of atomic oxygen (O) in the upper and middle stratosphere (50–30 km) before fading away in the lower stratosphere.

In reaction (ii), the atomic oxygen combines with molecular oxygen (O_2) to form ozone. The third body M of the triple collision can be any other gas molecule, since its role is simply to take away surplus energy which would otherwise cause the newly formed ozone molecule to fly apart immediately after the collision which formed it.

In reaction (iii) this ozone is *photodissociated* by soft solar UV (as distinct from the hard UV in reaction (i)) which is strongly absorbed in the process. The atomic oxygen produced combines with ozone to re-form molecular oxygen by reaction (iv) and so complete the chain of reactions.

These reactions collectively maintain a maximum *number density* of O_3 (number of O_3 molecules per unit volume) between 20 and 40 km above sea level (Fig. 3.5). The process operates by reactions (i) and (iv) balancing to maintain a small but important proportion of oxygen in the form of *odd oxygen* (O and/or O_3) as distinct from *even* common molecular oxygen (O_2). Reactions (ii) and (iii) then maintain a dynamic equilibrium between O_3 and O, which favours O_3 in the middle stratosphere because the altitude is high enough to have significant O (which diminishes downwards) and yet low enough (and hence dense enough) for the triple collision to be reasonably frequent (Fig. 3.5).

The O_3 maximum maintained in this way is only a minute fraction (maximum specific mass ~10^{-5}) of the small atmospheric mass at these levels. And yet it provides vital protection for life on Earth, because the photons of soft UV which it strongly absorbs would otherwise reach the Earth's surface in such numbers as to destroy exposed living tissue, even though each soft photon is individually less damaging than the much scarcer photons of hard UV (Fig. 8.1). All living things not shielded by at least several metres of water, or its equivalent, would quickly die of sunburn, as their weak molecular bonds were ruptured by the deluge of soft UV. The development of this frail but vital filter through Earth history is obviously related to the growth of atmospheric O_2, and through this to the development of the biosphere which it protects, so that an understanding of these relationships is central to current attempts to trace the origin and development of life on Earth (Section 3.9).

More generally, the maintenance of the O_3 maximum in the stratosphere exemplifies the presence of selective reactions acting quickly enough to compete with moderately vigorous mixing. In the case of ozone the competing effects are so nicely balanced that ozone would be more strongly concentrated in the low-latitude stratosphere, especially in winter, if meridional air currents did not continually carry ozone poleward from there. The O_3 distribution observed in the mid-twentieth century is believed to have been typical of the period since atmospheric O_2 reached its current abundance at least 300 million years ago *[16]*, but the precarious balance between selection and mixing has since been seriously disturbed by human activity since (Box 3.3).

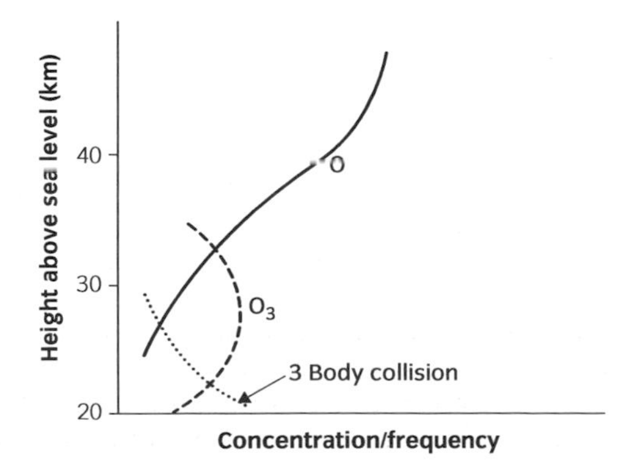

Figure 3.5 Schematic vertical profiles of ozone concentration and mechanisms maintaining its maximum in the lower middle stratosphere.

BOX 3.3 The ozone hole

In the mid-1980s, observers in Antarctica were surprised to observe very sharp temporary falls in stratospheric ozone in the early local spring (September to December—Fig. 3.6). Further investigation showed that these had been recorded elsewhere in Antarctica for a decade previously, but had been automatically disregarded as being unrealistically low. Subsequent observations have shown seasonal falls increasing in Antarctica and spreading to other high southern latitudes, and high northern latitudes too, together with an accumulating loss in global ozone levels *[16]*.

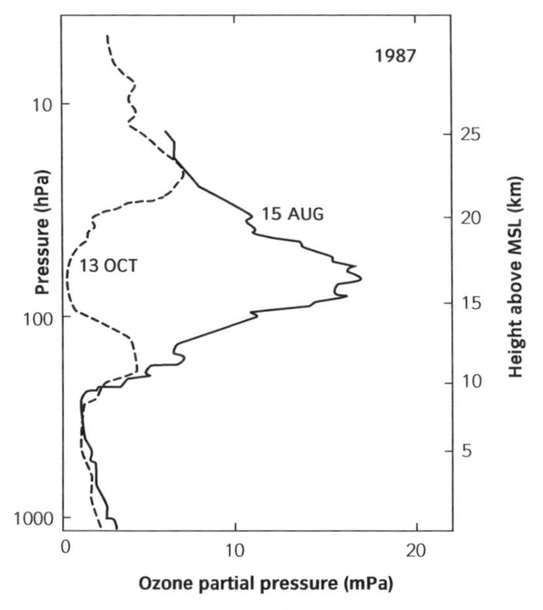

Figure 3.6 Observed profiles of O_3 abundance above Halley Base (75°S, 28° W), Antarctica in 1987, during heavy springtime depletion (October) and after summer recovery (August). These observations were made only a few years after the ozone hole was first discovered; overall levels are still falling in 2009 by the accumulation of incomplete annual recoveries since.

Although the chemistry was hard to unravel initially, it is now accepted that this ozone 'hole' is being produced by residues from domestic and industrial *chlorofluorocarbons* (CFCs) used since the 1950s in refrigerators and spray propellants. These gases were released mainly in the industrial northern hemisphere, but spread by convection and air currents throughout the troposphere and into the stratosphere. Although chosen for their chemical inactivity in the lower troposphere in which they were used, it is now known that CFCs can become chemically very active in conditions which prevail in parts of the stratosphere.

The release of chlorine from CFCs is encouraged in the winter Antarctic stratosphere, as thin but extensive ice-cloud forms in the westerly circumpolar vortex, cooling in relative isolation from the rest of the atmosphere. Chemical reactions on the surfaces of the myriad tiny ice crystals release relatively large quantities of *molecular chlorine* (Cl_2) which photodissociates to the very active *atomic chlorine* (Cl) when sunlight returns in spring. Local ozone is then consumed at several per cent per day by a series of catalytic reactions. The destruction ceases when the cold circumpolar vortex breaks down in early summer and intrusions of air from lower latitudes bring heat and ozone from lower latitudes.

In view of its importance for life on Earth (Section 3.4), there is great concern over any reduction of the Earth's stratospheric ozone shield. Although worst over Antarctica, the seasonal weakening of the shield is now significant over many populated regions, with consequent damaging increases in soft UV at sea level. And the global loss is accumulating year by year as post-spring recovery of ozone fails to make good successive spring losses.

Fortunately the discovery of the 'ozone hole' has triggered international efforts to stop the manufacture of CFCs, and has played a major role in moving environmental concerns up the political agenda nationally and internationally. Production of CFCs has been largely ended by international agreements, reached quite quickly after the problem and its origin were identified. However CFCs are so long-lived in most of the atmosphere, on account of the chemical inactivity in the troposphere which made them attractive in the first place, that the legacy of past production is still reaching the polar stratospheres and fuelling annual bursts of ozone depletion. Current measurements suggest that accumulating depletion will bottom out in the next few years, but a forecast of full recovery by 2070 is still speculative. If the ozone shield recovers its natural strength by then, significant depletion will have lasted for over a century, with a widespread human cost in extra skin cancers and eye cataracts caused by exposure to enhanced soft UV. Since there is no conceivable way of discriminating between damage by 'natural' and enhanced UV, the human toll is being buried in private grief and impersonal mortality statistics, obscuring its importance in public perception, which is that the problem has been solved. In fact the ozone hole stands as an object lesson in the need for extreme caution in releasing radically new substances or organisms into the global environment, where ignorance of risk can easily be misinterpreted as evidence of safety.

3.5 Carbon dioxide *[15]*

Though *carbon dioxide* (CO_2) is only a trace component of the atmosphere (specific mass 0.0005) it interacts very vigorously with the biosphere, being produced by combustion (CB), and respiration and decay (RD), and consumed by photosynthesis (PS), as shown in Eqn 3.1. Since these interactions exactly complement the production and consumption of O_2, the mass fluxes of CO_2 into and out of the atmosphere exceed the fluxes of O_2 in precisely the ratio of their molecular weights (44/32), which means that they are about 260,000 Mt per yr (NN 3.4). However, the mass of CO_2 in the atmosphere is so much smaller than the mass of O_2 (Table 3.1 and NN 3.4) that the resulting residence time of CO_2 in the atmosphere between successive involvements with the biosphere is only about 10 years (NN 3.4).

A nearly comparable exchange of CO_2 takes place between the atmosphere and the *photic zone* (the sunlit surface layer) of the oceans, which contain nearly as much dissolved CO_2 as the atmosphere in a few tens of metres of water. If the exchange rate exactly equalled that between atmosphere and land biosphere, the total atmospheric inputs and outputs of CO_2 would be doubled and the residence time of CO_2 in the atmosphere would be halved to 5 years. This time is so short that we should expect to observe variations of the specific mass of atmospheric CO_2, as local productions and consumptions temporarily exceed the smoothing effects of atmospheric mixing.

Diurnal and seasonal variations

Photosynthesis can proceed so rapidly in sunlit vegetation on land, especially when dense and well watered, that CO_2 levels in the immediate vicinity can fall 20% below the twenty-four-hour average. At night, respiration may raise CO_2 levels as far above the average. Such diurnal variations are enhanced further within dense stands of vegetation by the reduction in ventilating air motion there.

Seasonal cycles of PS and RD produce large seasonal variations in CO_2 levels in the low troposphere, as shown by the Hawaiian observations which have been

NUMERICAL NOTE 3.4 Carbon dioxide capacity and residence time

From Table 3.1, the mass of atmospheric CO_2 is only about 0.0005 (0.05%) of the mass of the atmosphere. Proceeding as in NN 3.3, the mass of CO_2 in an air column resting on one horizontal square metre at sea level is nearly 5.2 kg, and the global total is a little over 2.6×10^{15} kg, i.e. 2.6×10^6 Mt CO_2.

The rate of throughput of CO_2 can be calculated from the rate of throughput of O_2 by noting from Eqn 3.1 that for every O_2 molecule (molecular weight 32) leaving the atmosphere, a CO_2 molecule (molecular weight 44) enters, and vice versa. The CO_2 mass flux is therefore numerically equal to the O_2 mass flux (190,000 Mt per year according to Section 3.3) multiplied by 44/32, which comes to just over 260,000 Mt per year.

As usual, we find the residence time for CO_2 in the atmosphere by dividing its capacity by the rate of throughput, i.e. $2.6 \times 10^6/(2.6 \times 10^5) = 10$ yr.

the global benchmark since 1957 (Fig. 3.7). In spring and early summer the burst of PS in land plants in middle latitudes consumes more CO_2 each day than is released by nocturnal RD, driving a progressive net reduction until autumn, when daily RD begins to exceed PS and drive a progressive net increase until spring. The effect is particularly pronounced in the N hemisphere because of the concentration of land masses there, with their continental (heavily seasonal) climates and huge masses of seasonal vegetation.

The seasonal oscillation observed in Antarctica is weaker and six months out of step with Hawaii, showing that tropospheric mixing is able to spread the seasonal rhythm from vegetated lower S latitudes, but not swamp it with the much larger signal from the N hemisphere, especially since transport is relatively poor across the equatorial region, as suggested graphically by the Hadley circulation (Section 4.7.3) and confirmed by many studies.

The vertical distribution of the seasonal oscillation of CO_2 amounts (Fig. 3.8) is consistent with the vertical structure of atmospheric mixing suggested in Section 1.2. The oscillation is largest in the atmospheric boundary layer, whose extremely well-mixed volume extends only a few hundred metres above the surface (Fig. 10.1), where all sources and sinks of CO_2 are localized. The seasonal oscillation is somewhat smaller throughout the rest of the troposphere, which extends to heights of between 10 and 13 km in the middle and high latitudes depicted in Fig. 3.8, and is efficiently mixed by convection and larger scale ascent and descent. However, only a small fraction of the seasonal oscillation survives in the stratosphere because the intervening tropopause acts as a weakly permeable lid on the troposphere.

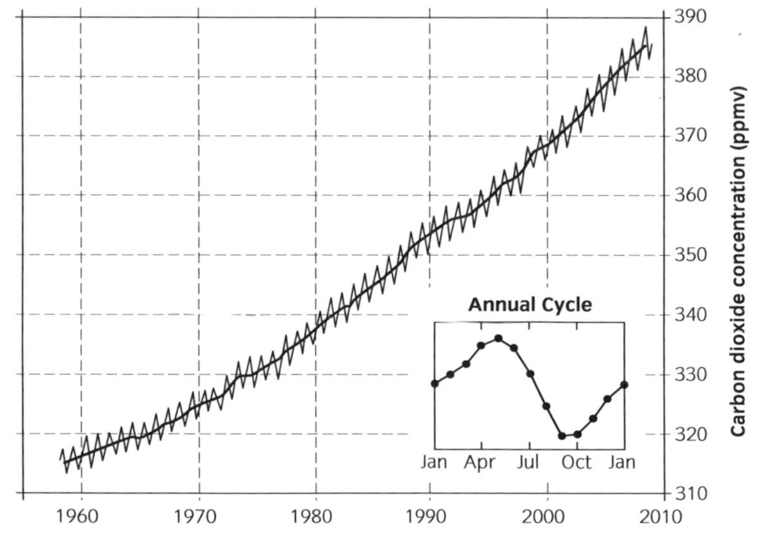

Figure 3.7 The famous 'Keeling curve' of atmsopheric abundance of CO_2 from Mauna Loa, Hawaii, showing increasingly rapid rise with global industrialization of a rising world population (population doubling in the time frame). Many of the wobbles in the smoothed curve correspond to periods of world economic instability, the curve for 2009 to 2011 will presumably show a relatively sharp dip. The annual cycle shows depletion from May to October as photosynthesis in the N hemisphere consumes atmospheric CO_2, and recovery from October to May by organic respiration and decay. Recovery exceeds consumption by the net industrial input during the year, less the still considerable proportion being stored by solution in the oceans.

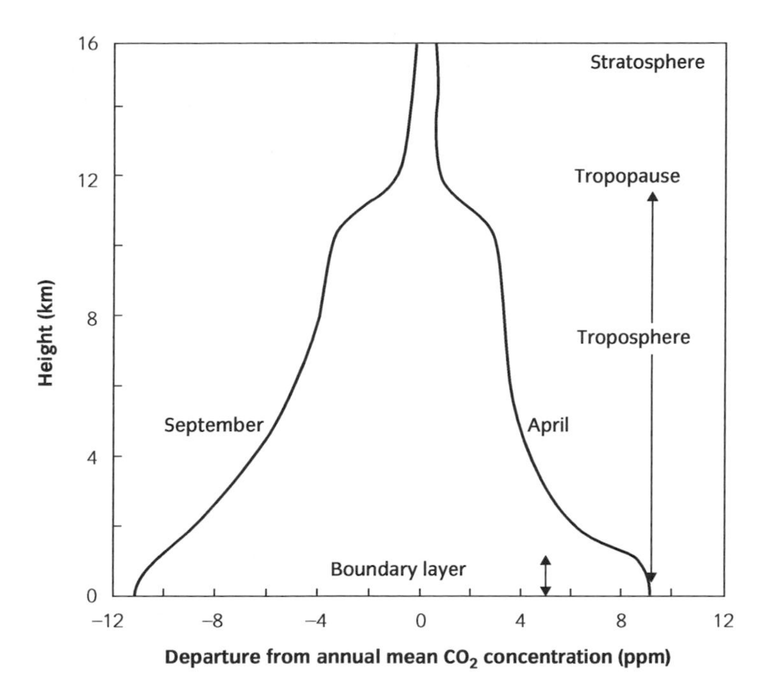

Figure 3.8 Schematic vertical profiles of autumn and spring variations in CO_2 in the troposphere and low stratosphere N of latitude 30°.

Longer-term variations

The steady rise in the seasonally averaged CO_2 apparent in Fig. 3.7 is being driven by the increasingly rapid combustion of fossil fuels which began with the British industrial revolution and is now nearly global. Rapid though it is, this *anthropogenic* increase in CO_2 is much slower than its speed of mixing through the atmosphere, with the result that it is observed as clearly in the most remote locations as it is in the most industrial, unlike the seasonal variations discussed earlier. The rapid dissolution of CO_2 into the photic layer of the oceans temporarily buffers the rise in atmospheric CO_2, so that the anthropogenic production is believed to be considerably greater than the total implied by Fig. 3.7, which is about 440,000 Mt (over 17%) between 1960 and 2000.

BOX 3.4 Anthropogenic CO_2

Figure 3.9 sets the recent anthropogenic rise in atmospheric CO_2 in the context of the large natural variations associated with climate change over the last 160,000 years. These are dominated by increased oceanic absorption of CO_2 as ocean temperatures fell into the depths of the glacial epoch, followed by release back into the atmosphere as the oceans warmed into the current interglacial. In the 300 years since the beginnings of modern industry, atmospheric concentrations have risen by over 35%, and are now well above those reached in the previous, somewhat warmer interglacial, 120,000 years ago. In fact CO_2 levels are now probably higher than at any time since the global climate slid into the current ice age several million years ago (Section 14.5). Such relatively long-term variations involve interactions with the large reservoir of CO_2 in the deep oceans (~50 times the atmospheric reservoir) which are much too slow to influence year-to-year activity, and also with the huge carbonate reservoir in ocean sediments, which contains the CO_2 equivalent of over 50,000 atmospheres, both of which are poorly understood as yet *[15]*.

On the long time scale of Fig. 3.9, the current rate of rise of atmospheric CO_2 (about 10% per decade) is too fast to be resolved by the thickness of the printed line. Even so, the observed rise in the atmosphere represents only part of the anthropogenic CO_2 accumulated so far, the rest being dissolved in ocean surface layers or taken up by land biomass. These could continue to buffer industrial CO_2 production if they too were not being altered by human activity—by global deforestation, and reduction of the CO_2 capacity of the ocean reservoirs through global warming.

All this represents a large and increasingly rapid shift away from the natural CO_2 economy, and there is great concern that it may be inducing significant changes in global climate, because CO_2 is such an important greenhouse gas (Sections 8.5 and 14.9.1). As with ozone depletion, but to a larger and increasing extent, this is attracting national and international attention, as it should. Reducing anthropogenic production of CO_2 is much more difficult than reducing CFCs, which played only a marginal role in human affairs. Global industrialization currently depends on continual, increasing consumption of fossil fuels and consequent release of CO_2. With concerns about safety and waste disposal still limiting power production by nuclear fission, and feasible power production by nuclear fusion still beyond the horizon despite half a century of development, there is no easy way out of

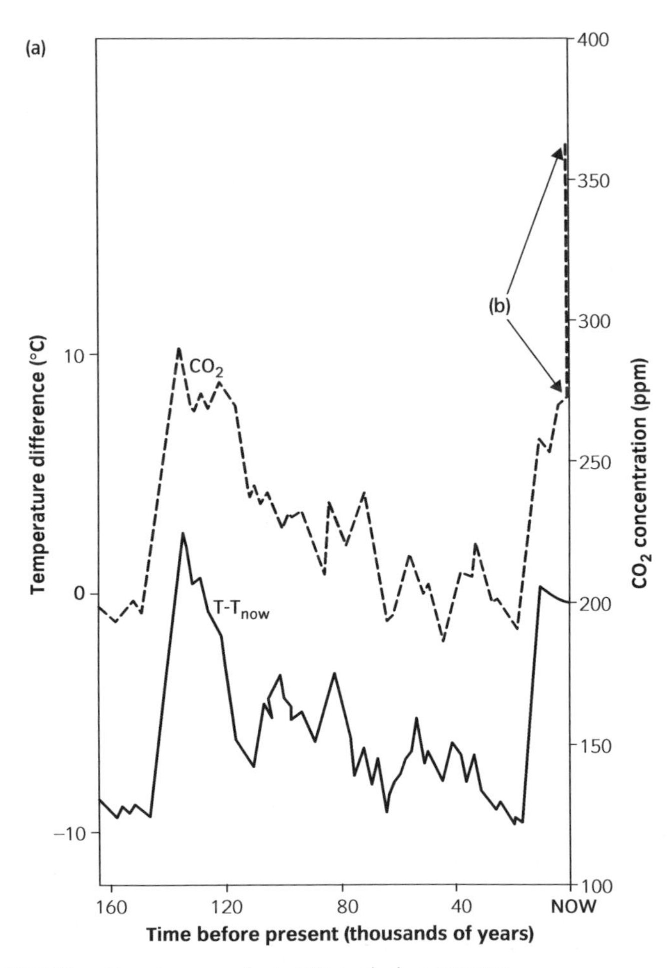

Figure 3.9a Variations in atmospheric CO_2 in the last 160,000 years. High correlation with global mean surface temperatures deduced from ^{18}O abundance (all from ice cores from Vostok (Antarctica)). After the last glacial minimum, CO_2 levels approached values prevalent in the previous interglacial (c. 130,000 yr BP).

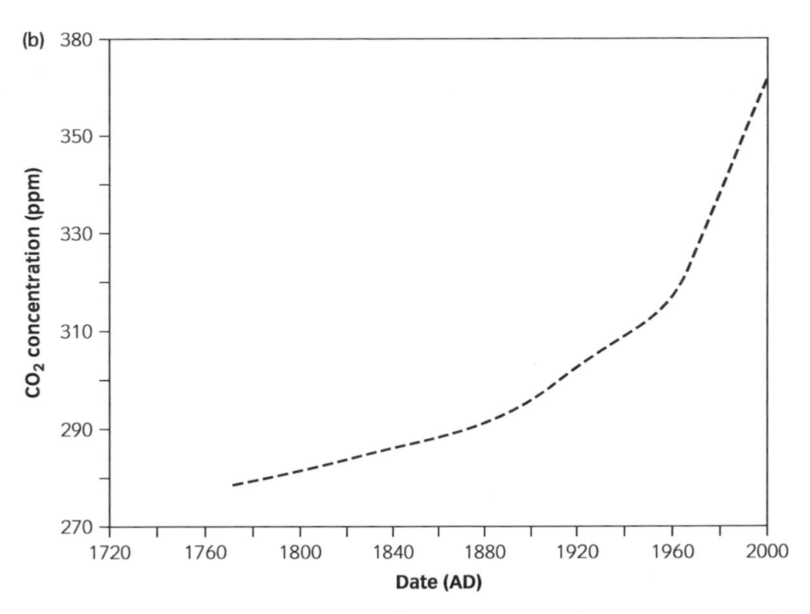

Figure 3.9b Recent Antarctic CO_2 data and direct measurements from Hawaii (since 1957) show a large and accelerating rise since the onset of modern industrialization in the early 1700s. On the time scale of Fig 3.9a this anthropogenic spike is too thin to be printed. Values in 2009 stood at over 385 ppmv, probably higher than at any time since the onset of the present ice age 3.5 Myr BP.

Note the unfortunate omission of v from ppmv on the CO_2 scales of these two diagrams; fortunately the numerical values make it clear that this is the 'volumetric' ratio rather than the larger specific mass.

the impasse. We need to work urgently on many fronts: encourage energy conservation, reduce CO_2 emissions from power stations, transport, and daily living, develop sustainable energy resources, and probably make limited temporary use of nuclear fission to bridge the gap before these changes begin to work—all matters of public debate as I write in 2009. And over all this hangs the need to revisit and be prepared to rethink the national and global economic presumptions underlying recent industrial and social development.

Two practical strategies are relevant to the focus of the present box.

(i) There is considerable potential for repeatedly planting, harvesting, and replanting trees and other vegetation specifically for combustion, since CO_2 extracted from the atmosphere during growth is simply returned during combustion, without overall increase. In terms of Eqn 3.1, such combustion short-circuits some of the respiration and all of the decay phase (right to left process), so that we can use some of the energy captured earlier from sunlight during photosynthesis (left to right). For example, Brazil has pioneered the production of a petrol substitute by fermenting sugar. However, problems can arise from changing land use on a global scale, especially if food production suffers as wealthy nations offer to pay more for their mobility than poor nations can afford for their nutrition.

(ii) Techniques are being developed for extracting CO_2 before or after the combustion of fossil fuels. If extracted before combustion, the combustible residue can be hydrogen, whose combustion produces water vapour which adds harmlessly to the hydrologic cycle which self-regulates on a time scale of days (Section 3.8). If extracted after combustion, the damaging CO_2 combustion product has to be removed from flue gases. In each case the CO_2 must be *sequestered* (hidden away) in permanent secure stores, or converted to carbonates—the former raising the possibility of long-term leakage, a problem which also continues to limit enthusiasm for nuclear fission as a major power source.

Longer-term variations

The steady rise in the seasonally averaged CO_2 apparent in Fig. 3.7 is being driven by the increasingly rapid combustion of fossil fuels which began with the British industrial revolution and is now nearly global. Rapid though it is, this *anthropogenic* increase in CO_2 is much slower than its speed of mixing through the atmosphere, with the result that it is observed as clearly in the most remote locations as it is in the most industrial, unlike the seasonal variations discussed earlier. The rapid dissolution of CO_2 into the photic layer of the oceans temporarily buffers the rise in atmospheric CO_2, so that the anthropogenic production is believed to be considerably greater than the total implied by Fig. 3.7, which is about 440,000 Mt (over 17%) between 1960 and 2000.

3.6 Sulphur dioxide *[15]*

Sulphur dioxide (SO_2) is one of the most biologically destructive of the common atmospheric trace gases. Background concentrations of $\sim 10^{-4}$ ppmv (i.e. specific mass $\sim 10^{-10}$, Box 4.1) are maintained by natural oxidation of reduced sulphur compounds such as hydrogen sulphide (H_2S) and dimethyl sulphide $(CH_3)_2S$, which are emitted by living and decaying organisms. However, artificial combustion of sulphur-rich coal and some oils, and smelting of sulphide ores, now add greatly to local and total production of SO_2.

Irritation and damage to plant and animal tissues is caused when concentrations are raised significantly above the natural background, though details are still being established. In light winds, poor ventilation allows local concentrations to rise sharply around strong industrial sources, producing total destruction of vegetation in extreme cases. Even after implementing stringent air pollution controls in recent decades, concentrations of about 1 ppmv (10,000 times background) are still recorded in some modern cities, so that anthropogenic SO_2 is one of the biggest causes of serious lung damage after tobacco smoking. The situation is worse in countries where governments are unwilling or unable to meet the considerable cost of controlling pollution. The most serious pollution episodes occur in calm or nearly calm conditions, especially where the effluent is trapped beneath the convectively stable atmospheric layer only a few hundred metres above the surface source which is typical of anticyclonic weather systems (Sections 10.12 and 12.5). This acts as a lid to surface-based convection, so that the pollution accumulates in a shallow surface-based layer whose air mass may be only 10% of the full troposphere which would otherwise be available for mixing and dilution.

Continuing oxidation of SO_2 tends to produce SO_3 which is hydrated to H_2SO_4 (sulphuric acid) in cloud droplets, giving much greater acidity than corresponds to the pH value of about 5.6 maintained by natural solution of atmospheric CO_2. In this way the vast urban areas spawned in the twentieth century in the industrial heartlands of Europe, North America, China, and Japan now maintain broad downwind swathes of seriously acidified rain and snow, with biological damage to rivers, lakes, soils, and vegetation (especially trees—Fig. 3.10). The mechanisms by which such *acid rain* does damage are still being analysed, and it seems likely that oxides of nitrogen and other substances and factors are also involved.

Figure 3.10 Trees devastated by industrial pollution, probably mostly by acid rain. Grassy vegetation in the foreground is still vigorous, as are some trees, but others are dying back from their crowns, and several are clearly dead. (See Plate 2).

In the context of the present chapter, the point is once again that all such pollution occurs when local production and deposition outruns the atmosphere's dispersive capacity, and this can occur because of heavy production, heavy deposition, or reduced dispersion, or any combination of these.

3.7 Aerosol *[17]*

So far we have considered purely gaseous mixtures. However, the turbosphere, especially the troposphere, also contains particles and droplets of solid and liquid material, many of which play important roles despite their very small specific masses (Table 3.1). We consider the smallest of these here (aerosol particles), and defer the larger elements (cloud and precipitation) to the next section.

Very large numbers of solid and liquid *aerosol particles* are effectively suspended in the lower atmosphere, especially the low troposphere, because they are so small that their fall speeds in still air are much slower than typical up and down movements of air. They range in size from aggregates of only a few hundred molecules ($\sim 10^{-3}$ μm across), to particles about 10^4 times larger (~ 10 μm). The latter are comparable in size with true cloud droplets but have much larger proportions of solids and/or solutes in water. Consider aerosol particles and droplets in order of increasing size.

(i) **Aitken nuclei** Particles with radii less than 0.1 μm are called *Aitken nuclei* because they *nucleate* artificial cloud formation in a cloud chamber—a method used by the Scottish pioneer John Aitken to collect and count them.

They dominate the total aerosol particle population numerically, but because of their small individual masses, make up only a fifth or so of the total aerosol particulate mass.

Aitken nuclei are produced by natural and man-made combustion, by gas-to-particle conversion of gases produced by biological metabolism and decay, and by volcanoes. Since most enter the atmosphere at or near land surfaces, their concentrations are particularly high in continental air, with number densities often exceeding 10^6 per litre. Specific masses reduce somewhat above the atmospheric boundary layer, but remain largely uniform throughout the rest of the troposphere. Above the tropopause amounts are smaller, with a fairly uniform distribution consistent with their arrival as micro-meteorites from space.

Gravitational settling is not directly effective in removing Aitken nuclei, since their fall speeds ($<$ 10 cm per day, Fig. 6.10), are slower than their speeds of random *Brownian motion* through 'still' air, and are completely negligible compared with even the weakest updraft. Once carried away from its source however, an Aitken population reduces quite rapidly by Brownian collision and coagulation of Aitken nuclei to form particles large enough to settle out under gravity. Aitken nuclei are also removed by being dragged by water vapour diffusing quickly onto the surfaces of swelling cloud droplets. Depletion is especially marked in maritime air masses, since re-injection from the surface is much smaller than overland.

(ii) **Large nuclei** Particles with radii between 0.1 and 1 μm are called *large nuclei*. Though usually 10 or more times less numerous than the Aitken nuclei, they comprise nearly half the total particulate mass. Fall speeds are still much smaller than updrafts, but there is a marked decrease in population above typical cloud base levels in the low troposphere, which is associated with their removal by *rain-out*. This happens as large nuclei efficiently nucleate cloud droplet formation in even the most slightly supersaturated air (Section 6.9), and are removed to the surface by subsequent rainfall from the cloud.

Large nuclei are produced directly in the same ways as Aitken nuclei, as well as by coagulation of Aitken nuclei, and by injection of small salt particles, left after evaporation of the fine spray produced by bursting bubbles on the surfaces of breaking sea waves. Though more numerous in continental than maritime air, the difference is smaller than in the case of Aitken nuclei. Dense populations of large nuclei scatter sunlight very effectively, producing haze which can reduce horizontal visibility and dim the low Sun. This is especially marked in anticyclones, whose convectively stable 'lids' confine and concentrate the haze in the first few hundred metres above the local surface (Fig. 3.11 and Section 13.2.2).

(iii) **Giant nuclei** Particles with radii larger than 1 μm are called *giant nuclei*. Though they comprise nearly half of the total aerosol particulate mass, giant nuclei are even less numerous than large nuclei, their numbers rarely exceeding 10^3 per litre even in the very low troposphere.

Giant nuclei are produced and lost in the same ways as large nuclei, except that they are too massive to be produced by Brownian coagulation. They are also produced when wind raises dust from arid lands. As with large nuclei, giant nuclei very effectively nucleate cloud droplet formation and are removed by rain-out. If caught below a raining cloud base, they are also efficiently

Figure 3.11 A thick industrial haze almost completely obscuring the urban area which is maintaining it. The polluted air is being trapped laterally by surrounding hills and vertically by a convectively stable layer probably compounded from local nocturnal cooling and an anticyclonic subsidence inversion. On such occasions the haze top is visually sharpest when seen from the same elevation. The camera on this occasion seems to have been a little below that level on the hill trapping the polluted air on this side. (See Plate 3).

washed out to the underlying surface (dramatically reducing low-level haze) as they are collected by the rain drops falling through the aerosol layer, being too massive to swerve out of the way of the drops, as smaller aerosol particles do. Fall speeds of giant nuclei are large enough (Fig. 6.10) to ensure that their population reduces quite quickly by gravitational settling when updrafts are weak—an effect which increases sharply with particle size and maintains a much steeper lapse of abundance with increasing height than is the case for large nuclei.

3.7.1 Total aerosol

The total mass of aerosol particles suspended in a column of air resting on a horizontal square metre of the Earth's surface amounts to only a few tenths of a gram—about 100 million times less than the mass of the supporting air column—and yet they can have important effects, because of their large numbers and aggregate surface areas. Acting as physical platforms they play a vital role in cloud formation, and potentially large aggregated cross-section area along a line of sight gives them an important role in reducing visibility (NN 3.5). Aerosol particles injected into the stratosphere in periods of enhanced volcanism can produce measurable global cooling by deflecting away incoming solar radiation (Section 14.7.4).

As in the case of ozone (Section 3.4), it is remarkable how minority constituents of the atmosphere can have an importance out of all proportion to their abundance. And the distribution of aerosol exemplifies once again the balance between the homogenizing tendency of atmospheric mixing, and differentiation by selective processes—of which the most obvious here is gravitational settling.

NUMERICAL NOTE 3.5 Surface cross-section and visibility

Suppose an aerosol contains 10^5 large nuclei per litre, i.e. 10^8 per cubic metre. If each nucleus has radius 0.6 μm, its cross-section area $\pi R^2 \approx 10^{-12}$ m^2, so that the total cross-section per cubic metre of aerosol is $\approx 10^{-4}$ m^2. A 1 km long 'tube' of such air will therefore have a particulate cross-section of 0.1 m^2 per unit tube cross-section. If none of these particles obscure each other in line of sight, 10% of the visual path along the tube will be blocked by particulates, causing significant loss of visibility. Lateral scattering from such particles in a sunbeam makes it visible as a haze to observers from the side.

3.8 Water

From previous sections it might seem that all the major atmospheric constituents have essentially uniform specific masses, and that significant variations occur only in trace materials such as ozone and aerosols. There is an extremely important exception to this tendency: the *water substance* (water as vapour, cloud, or precipitation) is very unevenly distributed through the atmosphere, being almost entirely confined to the troposphere, and heavily concentrated in the warm, low troposphere (Fig. 3.1). For example, the *specific humidity* (the specific mass of water vapour) of the atmosphere is only about 0.3% overall, whereas it often exceeds 3% in the warmest parts of the troposphere. Aviation experience confirms that cloud is almost entirely confined below the tropopause.

About 97% of the mass of the *hydrosphere* (the collective term for all the water substance on and above the Earth's solid surface) is in the oceans, which would cover the entire Earth's surface to a depth of about 2.8 km if evenly distributed. Of the remaining 3%, about 75% is in the ice caps Antarctica and Greenland, 25% is in lakes and groundwater, and only 0.03% is in the atmosphere, almost all of it in the form of vapour. In fact the total vapour content of the atmosphere would produce a global rainfall of only 3 cm if it were suddenly and completely precipitated, this being the *precipitable water content* of the atmosphere (Section 5.4). Comparing this with the equivalent depth of the oceans confirms that only about 0.001% of the total hydrosphere is in the atmosphere ($3 \times 10^{-2}/2{,}800 \approx 1 \times 10^{-5}$).

The hydrosphere is believed to have been formed by outgassing of steam from volcanoes during the early life of the Earth, the vapour condensing on the cool exterior surface to form the oceans (Section 3.9). The total mass of the current hydrosphere is believed to be effectively constant on all meteorologically relevant time scales. However, within the hydrosphere, a very rapid exchange takes place as the water substance continually moves through the *hydrologic cycle* (Fig. 3.12)—evaporating from oceans and moist land, existing briefly (Chapters 5 and 6) as vapour in moist air, condensing to form cloud, and then precipitating back to the surface, to be available for recycling either immediately or after involvement in groundwater, rivers, lakes, and oceans. In the ice ages which have dominated 10–15% of Earth's history, including the last few million years, polar regions have been cold enough to support permanent ice caps, fed by accumulating snow and drained by glaciers. These have

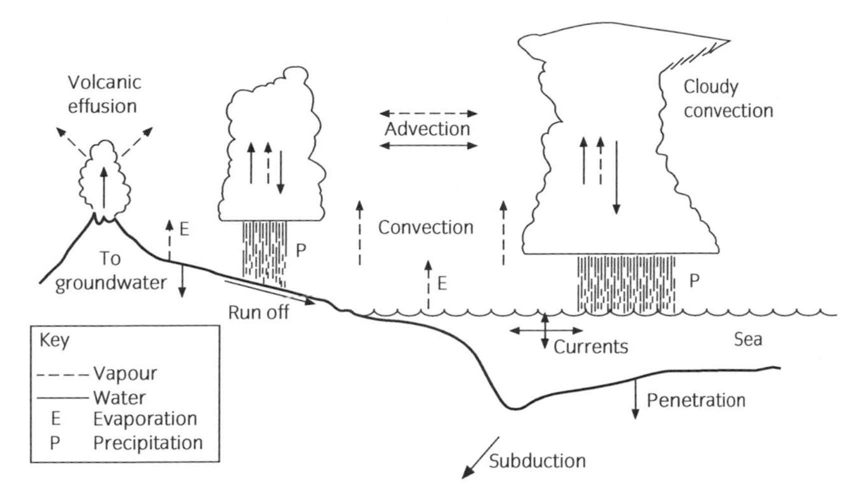

Figure 3.12 Schematic outline of the hydrologic cycle emphasizing the atmospheric branch hydrologic cycle. Note that though the vertical and horizontal proportions of the diagram are roughly realistic, the mass of water in the oceans is about 100,000 times the mass of water substance (mostly vapour) in the atmosphere.

diverted as much as 5% of the hydrosphere from the oceans (over twice the current amount, since we are in a warm phase of an ice age), burying extensive land surfaces under ice and exposing coastal shallows by lowering sea levels. However, ice caps grow or shrink over such meteorologically long time periods (hundreds and thousands of years) that they produce only very small imbalances of the global hydrologic cycle *[18]*, which we can treat as a globally steady state for meteorological purposes.

The water substance moves very quickly through the atmospheric branch of the hydrologic cycle. Globally, the annual precipitation is about one metre rainfall equivalent, most of it in the form of rain, rather than snow or hail. Since the precipitable water content of the atmosphere is so much smaller than the annual throughput of water vapour (3 cm compared with 1 m) the residence time for water as vapour in the atmosphere is only 11 days (NN 3.6) and the balance between precipitation and evaporation must be very close on a global scale. By contrast the residence time for water in the oceans is 2,800 years, on account of the vast amount of water (NN 3.6). The atmospheric and marine branches of the hydrologic cycle are highly asymmetrical in this respect, with a typical water molecule moving through the atmosphere for a week or two after evaporation, before returning for another few thousand years in the oceans.

Water begins its journey along the atmospheric branch of the hydrologic cycle when it evaporates from the sea, lakes, rivers, or moist land (including vegetation), or when it *sublimes* from unmelted ice. All such conversion to vapour is maintained by solar heating, either directly, as when moist land surfaces dry in direct sunlight, or less directly, as when a drying wind (air warmed by the Sun elsewhere) does the same. As discussed in Chapters 8 and 10, evaporation uses a substantial part of the solar input to the Earth's surface, which would otherwise be considerably warmer by day and colder by night, as observed in arid deserts.

Water vapour moves with horizontal and vertical winds, but since it remains highly concentrated in the warm, low troposphere, selective processes must be working quickly to offset the homogenizing effects of rapid mixing. These processes are the cooling, saturation, and cloud formation which take place very quickly in rising air, and the equally rapid precipitation which returns water and ice to the surface from all clouds more than about a kilometre deep.

NUMERICAL NOTE 3.6 Residence times for water

The residence time T_a for water in the atmosphere is given by the amount of water in the atmospheric reservoir divided by the rate of output (rainfall), or input (evaporation). Expressing each in terms of an equivalent depth of water in metres, we use 3 cm for global precipitable water content and 1 m for global annual rainfall to find $T_a = 3 \times 10^{-2}/1 = 1/300$ years, i.e. 11 days. Note depth units must be made consistent and that T_a emerges in years because the rainfall rate is expressed in metres per year.

The residence time for water in the oceans similarly comes to 2,800 years if we use the same rainfall rate and 2,800 m for the depth of an equivalent world ocean. In fact only 70% of the Earth is covered by oceans, and trying to see how this might alter the oceanic residence time touches on a number of important issues, including apportionment of global rainfall between land and oceans, which is still not accurately known.

Figure 3.13 The spectre of the Brocken. The figure and cameraman are on the slopes of Blencathra in the English Lake district, looking W'ward into a cloud-filled valley on a wintry morning. The shadow of the cameraman has a halo centred on his camera, arising from back-reflection from the nearly uniform population of cloud droplets catching the morning sun. The spectator sees the same effect centred on his eyes in his own shadow, but this is not visible to the camera. The dramatic title arises from a fanciful interpretation of observations made famous on the slopes of The Brocken in the Harz mountains of Germany, where such conditions are common. (See Plate 4).

This behaviour is considered in some detail in Chapters 5 and 6, but a simple outline now will serve to complete our summary of the constitution of the atmosphere.

3.8.1 Cloud and precipitation

Mechanisms

As mentioned in Section 3.7, many aerosol particles and droplets act as cloud condensation nuclei. Indeed cloud droplets form and grow on them so quickly, even in the most marginally supersaturated air (Section 6.8), that the maximum vapour content of rising cloudy air is effectively held at the saturation value appropriate to its temperature. This has two important and related consequences.

(i) In the presence of the sharp fall of temperature with increasing height imposed by expansive cooling of rising air (Section 5.6), and the very sharp fall of saturation vapour content with temperature, the *saturation limit* concentrates the great bulk of water vapour in the warm, low troposphere. In fact the saturation vapour content is so low in the very cold high troposphere (−40 to −60 °C) that this region acts as a nearly impermeable barrier to the upward leakage of vapour from the troposphere, keeping the stratosphere and higher layers very dry, even where they are relatively warm.

(ii) As moist air rises from the surface layers in convection, etc. it is brought to saturation less than a kilometre above all but the driest surfaces, which maintains the observed concentration of cloud in the low troposphere in the following way. And as cloudy air ascends above cloud base, air rising in its wake maintains that base, so that the cloud column or layer deepens above it. When cloud depths increase to more than a few hundred metres, some cloud droplets or crystals are able to grow large enough to fall through the cloudy rising air into the sub-cloud layer, and will reach the surface as precipitation if they do not evaporate en route. The efficiency of precipitation depends on complex microscale and cloud-scale mechanisms (Sections 6.11 and 11.5), but generally increases strongly with the vertical extent of the precipitating cloud. Deep frontal and shower clouds (kilometres deep) can return an appreciable fraction of their cloud water and ice to the surface within an hour of their first appearance.

In these ways water vapour, cloud, and precipitation are kept heavily concentrated in the lower troposphere, despite the vigorous vertical motion which would otherwise tend to keep their specific masses more uniform, as it does the specific masses of other tropospheric materials.

Cloud distribution

Note that the horizontal distributions of precipitation and cloud are very different. Though at any time clouds cover about half of the Earth's surface, most are so shallow that they precipitate little if at all before evaporating as they lose buoyancy and/or mix with drier surroundings. At any instant only the much smaller fraction of the surface lying below substantial clouds (i.e. those extending into the middle and upper troposphere) receives precipitation at the moderate and heavy rates which dominate rainfall (Fig. 2.7). Horizontally extensive sheets and veils of shallow cloud at all levels produce virtually no precipitation to the surface, although they are important for their ability to veil sunlight, and to close, at least partly, the atmospheric window to terrestrial radiative emission to space (Section 8.3).

A few exceptions prove the general rule that there is almost no cloud above the tropopause.

(i) The strongest updrafts in heavy thundershowers can overshoot the tropopause a little way before falling back by negative buoyancy and spreading out as *anvils* just beneath the tropopause (Section 11.4).

(ii) *Nacreous* or *mother-of-pearl* clouds occasionally appear in the middle stratosphere (about 20–30 km above sea level), and are believed to be ice clouds forming in the crests of waves of air flowing over hills far below (Section 11.8). The formation of diffuse forms of such clouds in the very cold polar winter stratosphere plays an important role in ozone depletion (Box 3.3).

Figure 3.14 Skeins of bluish-white noctilucent cloud photographed long after sunset but still in the sunlight about 80 km above the local high-latitude surface. This puts them in the mesopause, where air pressures are about 100,000 times smaller than surface pressures and temperatures between –50 and –100 °C according to season. (See Plate 5).

(iii) In middle and high latitudes *noctilucent clouds* (Fig. 3.14) are occasionally observed well after dusk (hence the name 'night shining') at altitudes of about 80 km, but their constitution is not well established.

Though cloud is extremely widespread in the troposphere, its typical transience, especially obvious in the cumulus family, suggests that residence times of the atmospheric water substance in the form of cloud and precipitation are much shorter than the 11 days found for water vapour. It follows from the chain rule for residence times (Box 3.2) that at any instant most of the atmospheric water substance is in the form of vapour, consistent with the values labelled 'overall' in Table 3.1.

Although the specific mass of atmospheric water in all its forms is very small, it is so heavily involved in several important aspects of atmospheric behaviour, that it is a major constituent in effect. Many of its roles mentioned in this section will be detailed later, but it must be clear already that the confinement of the water substance to a relatively shallow layer overlying the Earth's surface has very important consequences, the most obvious of which is the localization of weather as we know it in the same shallow layer.

3.9 The evolution of the atmosphere *[16]*

3.9.1 Origins

The Earth is believed to have formed in the aftermath of a supernova (the catastrophically explosive death of a big star) which left a *nebula* (cloud) of debris from which the solar system emerged by gravitational aggregation about 4.6 Gyr BP (4,600 million years before the present). It is likely that the composition of the proto-Earth's primordial atmosphere was consistent with that of the parent nebula, but that this was swept away by the violent activity of the young solar system, including the cataclysmic collision which left us with the Moon and a heavily modified young Earth. Certainly the present proportion of ^{36}Ar (not the abundant ^{40}Ar mentioned in Section 3.3) is about 10^6 times less than the proportion found in the solar system as a whole, and there seems to be no other way of losing this very heavy inert gas. It follows that our present atmosphere has evolved from a secondary atmosphere produced by outgassing from the young Earth's hot interior, rather than from the preceding proto-Earth.

Current evidence *[15]* suggests that initial fast outgassing of water vapour, hydrogen, carbon monoxide and dioxide, hydrogen chloride, and molecular nitrogen (N_2) was largely completed by about 4.4 Gyr BP (i.e. in about 200 My after Earth's formation). The water vapour probably condensed quite quickly to form deep *archaean* oceans, and the very light hydrogen molecules escaped from the Earth's gravitational field (Section 3.2), leaving the atmosphere dominated by CO_2 and N_2. The large amounts of atmospheric CO_2 (probably hundreds of times present values) helped keep the Earth's surface somewhat warmer than at present, even though the young Sun's heat output was about 30% below current values, by maintaining a *greenhouse effect* very much larger than today's (Section 8.5).

This *palaeoatmosphere* was still very different from today's atmosphere, in which carbon dioxide is reduced to an important trace, and molecular oxygen (O_2) is the most abundant gas after N_2, and these differences must be explained by the intervening evolution of the Earth's atmosphere. The range of current models of that evolution shows how uncertain we are of all but the broadest outline, but the following summary represents what may be an emerging consensus.

3.9.2 Evolution

It seems certain that much of the early CO_2 was removed by *chemical weathering* of exposed or submarine rock, or by *hydrothermal pumping* of sea water through the fissures of submarine mid-ocean ridges, both of which left it chemically locked in a variety of carbonates in the Earth's crust and upper mantle. Such removal of CO_2 may even have prevented a runaway *greenhouse effect,* like that suffered by our neighbouring planet Venus, as the young Sun slowly increased its heat output—a catastrophic heating which would have ended the prospects for life on Earth as it did on Venus. Instead, it seems that by about 3.8 Gyr BP, Earth's atmospheric CO_2 had been reduced to an important minor component, which, together with water vapour and cloud, was able to maintain surface temperatures by a modest greenhouse effect, much as continues today (Section 8.5). There is no fossil evidence of life at this stage, and it is known that in the absence of photosynthetic production, atmospheric O_2 levels must have been limited to ~10^{-9} present values, maintained by *photolysis* of H_2O and CO_2 by solar UV. The subsequent large increase in atmospheric O_2 is closely related to the evolution of life on Earth (Fig. 3.15).

From about 3.5 Gyr BP there is increasingly unambiguous fossil evidence of single-celled bacteria capable of feeble *anaerobic* photosynthesis (i.e. photosynthesis in the absence of oxygen—Section 3.3.3). By 2.7 Gyr BP large populations of *cyanobacteria* (formerly known as blue green algae—Fig. 3.16) were growing in colonies which have left large pillow-shaped *stromatolites* in shallow tropical seas, driving net production of atmospheric O_2. Such early life was necessarily aquatic, since water several metres deep was needed to filter out the damaging UV from the sunlight (Section 3.4). Oxygen levels in the atmosphere probably rose only very slowly, but as they did, so did the much smaller concentrations of ozone, maintained as outlined in Section 3.4, which in turn increased the protective UV filter, making progressively shallower waters hospitable for primitive life.

When atmospheric O_2 reached about 1% of present values, a new and much more efficient type of *aerobic* photosynthesizing organism developed, raising oxygen levels and further widening habitats. There is fossil evidence of such organisms

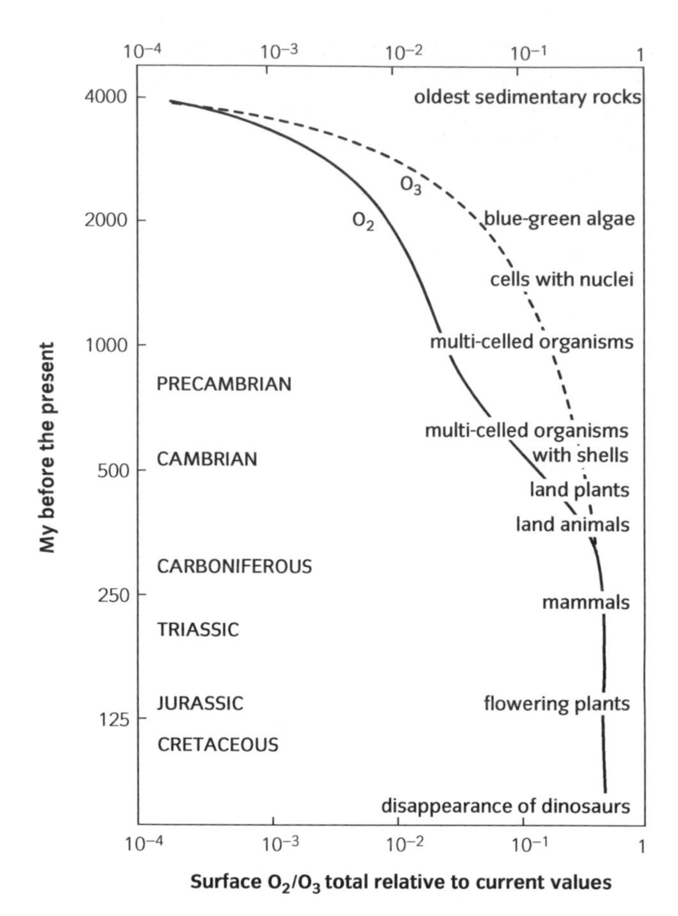

Figure 3.15 Probably co-evolution of atmospheric O_2, O_3 and life on Earth. The non-linear time scale focuses on the explosion of life in the last 10% of the life of the planet. On the scale of the diagram, the anthropogenic spike should be in the bottom 25 millionths of a millimetre.

from about 1.2 Gyr BP, distinguished by the appearance of a nucleus in each single cell. Even more efficient multicelled organisms followed, so that by about 0.6 Gyr (600 Myr) BP there were marine organisms with shells, whose known requirements for dissolved oxygen show that atmospheric O_2 levels had reached at least 10% of present values. As they died they formed fossil-rich beds of calcium carbonate in the shallow seas, further raising levels of O_2 and O_3 to the point where organisms no longer needed a shield of water against solar UV. Whether or not it followed immediately, the result was the vast Cambrian explosion of life, and its colonization of land surfaces. Widespread intense photosynthesis then raised O_2 levels towards present levels and eventually, in the late Carboniferous period (320–290 Myr BP), laid down the beds of hydrocarbon debris which are currently being extracted and burnt in humankind's brief frenzy of industrial combustion. And all this time ^{40}K was decaying radioactively to ^{40}Ar to produce the third major component of the present atmosphere, albeit a biologically and chemically inactive one.

This brief survey shows how far the present atmosphere has evolved from its ancient origins, how a variable greenhouse effect has maintained surface temperatures in the narrow range needed to allow the accelerating evolution of life, and how that evolving life has in turn built up the reservoir of free oxygen on which it now depends. Such obvious self-regulation has recently encouraged the view that the terrestrial biosphere should be regarded as a coherent

(a)

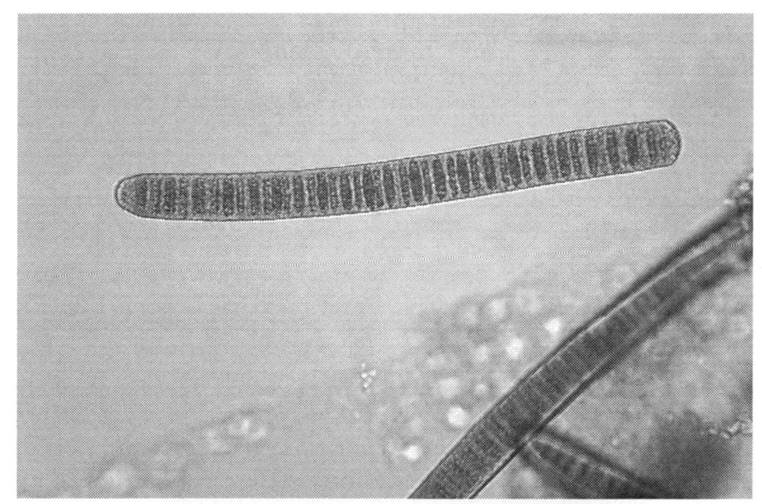

(b)

Figure 3.16 The pedigree of life is long with humble, resilient roots. (a) Cyanobacteria or blue-green algae have been flourishing for about 3,000 million years on and in every part of Earth's surface layers. These ones are linear colonies of single cells, and their green appearance suggests that they can photosynthesize. Others are single cells. Many help fix atmospheric N_2. (b) Living stromatolites—metre-sized colonies often dominated by cyanobacteria. The oldest fossil stromatolites are about 2,800 million years old. Cyanobacteria are believed to have played an important role in the conversion of life to oxygen dependence (Fig 2.15).

superorgansim, *Gaia [19]*, whose intrinsic feedback mechanisms enable it to regulate its environment to its own advantage. Whatever the merits of such a view, it is impossible not to be impressed by (and grateful for!) the continued cooperation of the Earth's biosphere and atmosphere. It would be a sad irony if the only part of the biosphere currently capable of being consciously aware of such cooperation should choose to ignore its obvious lessons.

Checklist of key ideas

You should now be familiar with the following ideas.

1. Turbulent and molecular diffusion in water and air.

2. Convective and diffusive equilibria under gravity in and above the turbosphere.

3. Concept, measurement, and physical meaning of residence time.

4. Selective physical, chemical, biological, and industrial processes affecting constituents of dry air (argon, nitrogen, oxygen, carbon dioxide, and sulphur dioxide).

5. Human destruction of ozone and production of carbon dioxide.

6. Aerosol particles and droplets and their input and removal.

7. Water substance below, in, and above the troposphere, the hydrologic cycle (evaporation, vapour, condensation, cloud, precipitation).

8. Co-evolution of biosphere and atmosphere in outline.

Problems

Outline answers to these problems can be found on the **Online Resource Centre.** Answers to odd numbered problems can be found under Student Resources, answers to even numbered problems under Lecturer Resources.

Level 1

3.1 In non-technical terms describe the mixing processes and forces competing to maintain the distributions of atmospheric gases.

3.2 List the order of component layering from the top down in diffusive equilibrium of an atmosphere of moist air (taking only the first three components of dry air), in which there is some photodissociation of molecular oxygen and water vapour.

3.3 Briefly define the following terms: denitrification, residence time, respiration, photosynthesis.

3.4 List the effects producing temporal variations of the specific mass of CO_2 at Mauna Loa (Fig. 3.7).

3.5 Why does the number density of large nuclei fall rapidly with increasing height in the vicinity of cloud-base level.

3.6 Starting from evaporation, sketch and label a likely sequence of events experienced by a water molecule which is eventually precipitated back to the surface. Add branches describing the further history as the water molecule (a) falls into the sea, and (b) falls onto the land.

3.7 An exciting and very wet experiment in school showed that atmospheric pressure can maintain a 10 m column in a water barometer. What is the total pressure at the bottom of a 2 m deep swimming pool, and on the floor of a 4 km deep oceanic basin?

Level 2

3.8 How in principle should convective and diffusive equilibria differ in the atmosphere of an orbiting space station (i.e. in the absence of apparent gravity)?

3.9 Assuming a reasonable value of 4 litres for accessible air capacity of a pair of human lungs, find the maximum mass of oxygen in them assuming an air density of 1.2 kg m^{-3}. Why is the actual value likely to be a little lower?

3.10 Suppose that the maximum specific mass of O_3 is 10^{-5} at 32 km above sea level. Use a reasonable value for air density there (two decadal scale heights above the surface), find the density and number density of O_3 in quantities per litre.

3.11 The reservoir of carbon in the living and dead terrestrial biosphere is believed to be about 2,800 units (ignoring fossil reservoirs like coal and oil), and the annual exchange of CO_2 with the atmosphere is about 260 units, where a unit is 1 Gt (10^3 Mt or 10^{12} kg). Find the implied residence time for carbon in the terrestrial biosphere.

3.12 In a region where air density is 1 kg m^{-3}, find the specific mass of an aerosol dominated by giant nuclei of radius 1 μm, density 2,200 kg m^{-3}, and number density 10^3 per litre. Compare with Table 3.1.

3.13 Assuming realistically that 97% of the mass of Earth's water substance is in ocean basins covering 70% of the Earth to a depth of 4 km, and assuming that three-quarters of the remaining 3% of water substance is currently in the ice caps of Antarctica and Greenland, find the rise in mean sea level if these were to melt into the existing ocean basins.

Level 3

3.14 On the laboratory scale, convective equilibrium always tends to encourage uniformity of distribution. Why then does it maintain the sharp vertical lapse of air density observed in the atmosphere?

3.15 Why is the altitude of maximum specific mass of ozone (or any other atmospheric component with a level of maximum concentration) greater than the altitude of maximum ozone density?

3.16 Assuming that all atmospheric water vapour eventually precipitates by falling 3 km at 3 m s^{-1} without evaporation, and assuming a reasonable value for the residence time for atmospheric water vapour, use the chain rule of Box 3.2 to estimate the mass of falling precipitation as a percentage of the mass of water vapour at any instant. Articulate and critically assess any assumptions made in the estimation.

3.17 Use Fig. 3.7 to estimate the typical seasonal rise (and fall) of atmospheric CO_2 as a multiple of the mean annual rate of change between 1980 and 2000. Converting the latter to Gt per year (from the global mass of atmospheric CO_2, NN3.4) express the former in global Gt per year, and consider why this is only a fifth of the values for the exchanges with the biosphere quoted in Section 3.5.

3.18 Write down an expression for the balance between water gain by land (by excess of precipitation over evaporation) and water loss by the oceans (by excess of evaporation over precipitation) and rearrange your expression to show that when expressed per unit surface areas of land and ocean, the excess must be 2.33 times the deficit.

4 The state of the atmosphere

4.1 The state of the air

The mixture of gases called air accounts for almost all of the mass of the atmosphere (Table 4.1), whose *state* (physical condition) and motion are outlined in the present chapter, beginning with its state.

The temperature, pressure and density of a small parcel of air relate to each other through the *equation of state* for an ideal gas, whose meteorological form (Box 4.1) is

$$p = \rho R T \qquad 4.1$$

Air in the parcel presses outwards on its surroundings (usually other air parcels) with a force p per unit area of parcel surface, which is the *air pressure* throughout the parcel if it is small. This air pressure is the parcel-scale result of continual bombardment by many of the huge number of air molecules within the parcel (about 2×10^{25} per kg). When expressed on the *absolute* or *Kelvin* scale, the *air temperature* T is a measure of the kinetic energy of the bombarding air molecules in their incessant thermal motion *[20]*. The *air density* ρ is the mass of air per unit parcel volume, and R is the *specific gas constant* for air, whose value for dry air is nearly 287 J K^{-1} kg^{-1} (Box 4.1), when pressure, absolute temperature, and density are expressed in SI units.

4.1.1 Density

Using this value for R, Eqn 4.1 relates simultaneously measurable values of pressure, temperature, and density for a parcel of dry air, and is often used to determine the remaining measurable when the other two are known. In practice, air pressure and temperature are often used to determine air density (NN 4.1), since direct measurement of air density (by weighing a rigid container of known volume before and after evacuation of air) is slow and clumsy in comparison with simultaneous measurement of pressure and temperature. Such measurements and calculations show that air densities in the low troposphere are about 1 kg m^{-3}, falling roughly with pressure throughout the turbosphere, for example falling by two thirds between the surface and upper troposphere (NN 4.1). Equation 4.1 is also used algebraically to put equations into a more convenient form, for example when deriving the pressure–height relationship in Section 4.3—again often replacing air density by the associated pressure and temperature.

Air in the troposphere always contains some water vapour, whose relatively low molecular weight (18 compared with the effective value of 29 for dry air—Box 4.1) reduces the density of moist air a little below the density dry air would have at the same temperature and pressure. Although the effect is small it can significantly increase the very small buoyancies which drive convection (NN 4.2 and Box 11.1).

Though atmospheric densities are at most only one thousandth of the density of water, their physical effects are very significant, as shown statically by the compression of air near sea level under the weight of overlying air, and dynamically by the effects of strong winds on land and sea surfaces, trees, buildings, and people (Fig. 4.1).

Figure 4.1 A brief storm has tipped over a stoutly built but insecurely founded barn. Wind pressure on walls can be surprisingly large, as shown by occasional spectacular damage to tents, caravans (trailers), high-sided trucks, parked aircraft, bridges, cooling towers, etc. in every part of the world. Presumably this was more securely tethered next time. Such pressures drove maritime transport until only 160 years ago.

NUMERICAL NOTE 4.1 Air density and the equation of state

To find air density near sea level, rearrange Eqn 4.1 in the form $\rho = p/(R\,T)$ and substitute typical values for air pressure (1,010 hPa $= 1.01 \times 10^5$ Pa), temperature (15 °C = 288.2 K), and specific gas constant (287 J kg^{-1} K^{-1}) to find $\rho \approx 1.2$ kg m^{-3}. Do the same for conditions typical of the upper troposphere (p = 300 hPa and $T = -50$ °C (223.2 K)) to find $\rho \approx 0.47$ kg m^{-3}, i.e. about 40% of the sea level value. Note that although air density falls with pressure, lower temperatures aloft offset this effect a little.

BOX 4.1 The equation of state for air

Ideal gases

In terrestrial conditions, the major atmospheric gases, and many minor ones too, closely obey the *ideal gas laws* discovered by Boyle, Charles, and Gay-Lussac. These laws combine in the *equation of state* to relate pressure p, volume Vol, and absolute temperature T of a given amount of any *ideal gas* by

$$p\,Vol = n\,R^*\,T \qquad \text{B4.1a}$$

where n is the number of *moles* of the gas in the volume (its mass in grams divided by its molecular weight M), and R^* is the *universal gas constant*, so-called because it has the same value (8.314 J K^{-1} mol^{-1}) for any ideal gas. According to the atomic nature of matter 1 mole of any pure molecular substance (the mass in grams equal to its molecular weight) contains 6.02214×10^{23} molecules. The zero of the absolute temperature scale was first determined from ideal gases, like air and its constituents, as the temperature at which the volume of a given mass of gas would fall to zero at atmospheric pressure, extrapolating from laboratory conditions. Later studies have confirmed that this *absolute zero* (0 K) lies at −273.2 °C, even though it can never be fully reached.

If the mass of ideal gas is m kg, the number of moles n is by definition $10^3\,m/M$, where 10^3 converts kilograms to grams. Using this in Eqn B4.1a and rearranging, we find the *meteorological form* of the equation of state

$$p = 10^3 \frac{m\,R^*\,T}{M\,Vol} = \frac{m}{Vol}\frac{10^3\,R^*}{M}\,T = \rho\,R\,T \qquad \text{B4.1b}$$

where ρ is the gas density (m/Vol) in kg m^{-3} and R ($=10^3\,R^*/M$) is the *specific gas constant* of the particular ideal gas in J K^{-1} kg^{-1}. Unlike the universal gas constant R^*, the numerical value of R is unique to each particular gas, since it varies inversely with its molecular weight M.

Values of the specific gas constants for the major atmospheric gases are given in Table 4.1.

Air as a mixture of ideal gases

Equation B4.1b applies to each component of dry air. To extend it to the dry air mixture, we use *Dalton's law of partial pressures*, which states that each component of a mixture of ideal gases obeys Eqn B4.1b, regardless of the presence of other gases. The total pressure p of the dry air mixture is therefore the sum of the *partial pressures* (the pressure each would exert in the absence of the others), each of which is given by Eqn B4.1b for that component, so that

$$p = p_n + p_o + p_a = [(\rho_n\,R_n) + (\rho_o\,R_o) + (\rho_a\,R_a)]\,T \qquad \text{B4.1c}$$

where p_n, etc. are the partial pressures of nitrogen, oxygen, and argon (ignoring carbon dioxide and rarer components for simplicity), R_n, etc. are the specific gas constants for nitrogen, etc. and the *partial densities* of nitrogen ρ_n, etc. are the masses of each component divided by the total volume of the parcel of dry air. Temperature T is common if the mixture is in thermal equilibrium, as we always assume. Since the overall density ρ of the mixture is simply the sum of the partial densities, we can multiply the right-hand side of Eqn B4.1c top and bottom by ρ and rearrange to recover the form of Eqn B4.1b and justify Eqn 4.1.

Table 4.1 Composition of the turbosphere

Component gases	Specific mass	Molecular weight	Specific gas constant
Dry air			
N_2	0.755	28.02	297
O_2	0.231	32.00	260
Ar	0.013	39.95	208
CO_2	0.0005	44.01	189
Plus hundreds of trace gases			
Moist air			
Dry air	0.997	28.96	287
H_2O vapour	$\simeq$ 0.003 overall ($\sim$0.03 maximum)	18.02	461
Particles and droplets			
Cloud	$\sim 5 \times 10^{-5}$ overall ($\sim$0.004 maximum)	Water and ice	
Precipitation	$\sim 3 \times 10^{-6}$ overall ($\sim$0.004 maximum)	Water and ice	
Aerosol	10^{-8} overall ($\sim 10^{-7}$ maximum)	Various solids or concentrated solutions in water	

$$p = \rho\,[(x_n\,R_n) + (x_o\,R_o) + (x_a\,R_a)]\,T = \rho\,R\,T \qquad \text{B4.1d}$$

where $x_n = \rho_n/\rho$ is the *specific mass* of nitrogen in the air—the ratio of its partial density (or mass) to the total density (or mass) of the air—and x_o and x_a are the specific masses of oxygen and argon.

Note that $R = [(x_n\,R_n) + (x_o\,R_o) + (x_a\,R_a)]$ is the specific gas constant for the mixture, whose value is determined by the types and proportions of the component gases. Inserting values for the x_s and R_s from Table 4.1 we find that the specific gas constant for dry air is 287 J K^{-1} kg^{-1}, which means that dry air behaves as if it were a single ideal gas with specific gas constant 287 J K^{-1} kg^{-1}. Inserted in $R = 10^3\,R^*/M$, this value implies that dry air behaves like a single ideal gas with molecular weight $M = 29.0$.

If we include water vapour in the mixture (specific mass 3% or more in warm moist air), its lower molecular weight (18) ensures that moist air is slightly less dense than dry air at the same temperature and pressure (NN 4.2). The effect is analysed in detail in the Appendix to Box 11.1.

Proportions by mass and by volume *[1]*

In the meteorological form of the equation of state (Eqns B4.1b and 4.1), the simplest measure of the proportion of a particular gas in a mixture is its specific mass (e.g. mass of N_2 per unit mass of dry air). Because chemical reactions involve whole numbers of molecules, chemists prefer ratios of molecular numbers, and therefore use Eqn B4.1a. Since numerical values for these two proportions can differ considerably in a given mixture, it is important to be able to relate them numerically and to know which is being quoted at any time.

Rearranging Eqns B4.1a and B4.1b shows that the molecular ratio v_n (for N_2 for example) is related to its specific mass x_n by

$$v_n = \frac{M}{M_n}\,x_n \qquad \text{B4.1e}$$

where M and M_n are the molecular weights of dry air and N_2 respectively. A similar relation applies to the other components of dry air, and to any ideal gas in any mixture of ideal gases.

In the case of N_2, the two measures of its proportion in dry air differ only slightly (0.782 by molecules compared to 0.755 by mass), because the molecular weights for dry air M and nitrogen M_n are nearly equal. The difference is much larger for gases with molecular weights very different from that of dry air. For example the specific mass of CO_2 (molecular weight 44) in dry air is about 5.0×10^{-4} (Table 4.1), or 500 ppm (parts per million), whereas its molecular proportion is (29/44) times this value, i.e. only 330 ppm. In practice the molecular ratio is nearly always written as 330 ppmv, where the final v means 'by volume'. The use of 'by volume' rather than 'by molecule' arises because reinterpretation of the equations of state for CO_2 and air shows that the molecular proportion of CO_2 in air is equal to the proportion of the total volume of an air parcel which would be occupied by its CO_2 content at the parcel pressure and temperature.

NUMERICAL NOTE 4.2 The density of moist air

The specific gas constant R_v for pure water vapour, molecular weight $M_v = 18$, is given by $R_v = 10^3 R^*/M_v = 10^3 \times 8.314/18 = 462 \text{ J K}^{-1} \text{ kg}^{-1}$. If a certain parcel of moist air has 3% specific mass water vapour, and 97% dry air, the specific gas constant for the moist air is $R_m = x_v R_v + x_d R_d = 0.03 \times 462 + 0.97 \times 287 = 292.3$, which means that at a certain pressure p and temperature T, the ratio of the densities of moist and dry air $\rho_m/\rho_d = (p/R_m T)/(p/R_d T) = R_d/R_m = 0.982$ for this vapour content. Moist air is less dense than dry air at the same pressure and temperature, in this case by 1.8%.

4.2 The vertical profile of temperature

Figure 4.2 is the very simplified picture of the vertical temperature profile up to 100 km above sea level over most of the Earth, first shown as Fig. 3.1. The dashed lines represent the much lower polar winter temperatures found in the troposphere and very high turbosphere in the seasonal absence of sunlight. We consider the major features of Fig. 4.2 and how they relate to the physical mechanisms at work in the atmosphere.

4.2.1 Troposphere

The first 10–15 km above sea level stands out as a region in which temperature *lapses* (falls) sharply and persistently with increasing height. This is the *troposphere*, whose characteristically large *temperature lapse* rate is directly related to the widespread occurrence of cloudless and cloudy convection. Observed lapse rates vary considerably with time and geographical location, even when averaged from surface to tropopause, but often lie fairly close to 6°C km^{-1}. For example, the lapse rate on the occasion of the radiosonde flight displayed in Fig. 4.3 is 6.5 °C km^{-1} when averaged through the troposphere. On this basis we might expect air temperatures on a 1,000 m peak to be around 6–7°C cooler than at adjacent sea level, as is often observed, although a number of competing effects complicate the issue in fuller treatments *[21]*.

Consider how the steep temperature lapse rate in the troposphere is maintained by vertical air movements associated with convection. As an air parcel rises into regions of lower ambient air pressure, it expands and cools as its air molecules collide with its expanding 'walls' (i.e. with the molecules of surrounding parcels making up those walls), losing some of their kinetic energy in the process. When there are

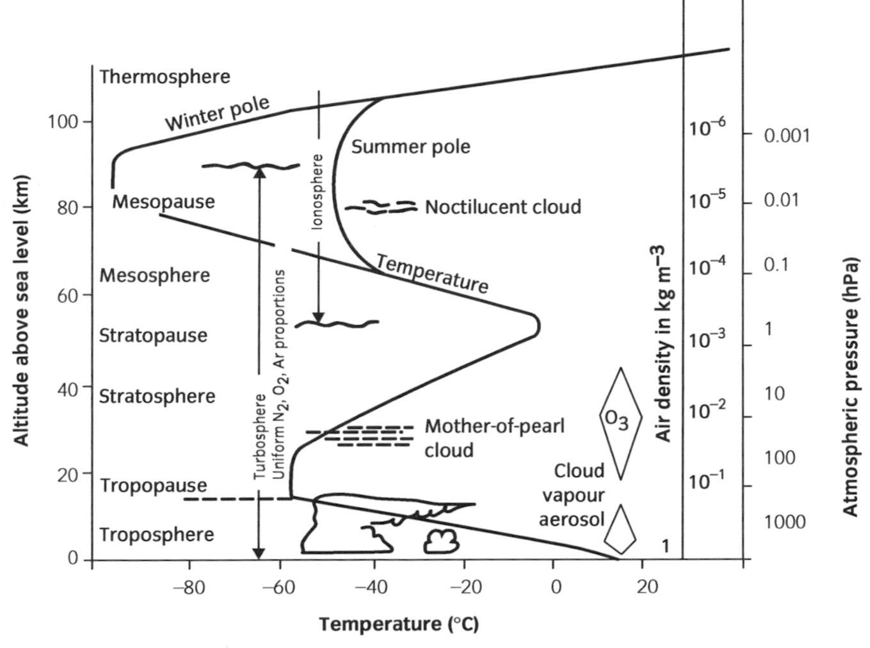

Figure 4.2 Schematic vertical distribution of atmospheric temperature, pressure, density, and atmospheric materials.

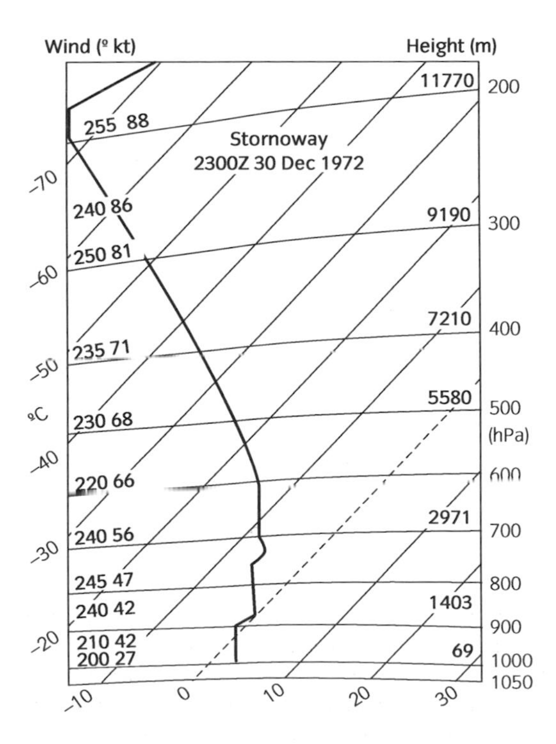

Figure 4.3 A typical winter sounding in an Atlantic depression, measured by routine radiosonde from a synoptic upper-air station in NW Scotland. It is displayed on a tephigram (Section 5.8.2) with temperature, height, wind azimuth, and speed plotted or listed against pressure. The tropopause is sharply defined a little above the 200 hPa level on this occasion, surmounted by a temperature inversion in the base of the stratosphere.

no exchanges of heat between the parcel and its immediate or distant environment (by conduction or radiation for example), and the process is sufficiently slow and near equilibrium to be *reversible*, it is said to be *adiabatic [22]*, and the cooling by such *adiabatic expansion* is given by thermodynamic theory (Sections 5.6 and 5.7). Sinking parcels undergo a corresponding warming by *adiabatic compression*.

Since convection is widespread and vigorous in the troposphere, we should expect there to be a convective equilibrium temperature profile with characteristic lapse rate(s), just as there is a convective equilibrium distribution of specific masses of the constituents of dry air. This is confirmed by observation and theory (Sections 5.9 and 5.10), which show that in cloudless air there is a single well-defined temperature lapse rate of 9.8°C per km (≈10°C km^{-1}), whereas in cloudy air the lapse rate ranges continuously from around 5°C km^{-1} in the warm low troposphere, to values very near the cloudless limit in the cold air found everywhere in the high troposphere.

On the very cloudy occasion of Fig. 4.3, the usual cloud-free zone overlying the surface was only about 250 m deep—too shallow to be clearly resolved on a height scale covering 12 km. However, the deep cloud layer filling the rest of the troposphere maintained lapse rates which increased with increasing height (i.e. decreasing temperature), in at least rough agreement with the above convective equilibrium model for cloudy air.

Two apparently minor features hint at complications ignored in this simple model.

(i) Although convective air movements in the troposphere are nearly adiabatic, they are not exactly so, because there is usually some non-adiabatic (*diabatic*) heating or cooling. Air parcels gain or lose some heat by direct interaction with solar and terrestrial radiation, and can do this at rates which produce significant temperature changes over periods of a day or more (Sections 8.6 and 8.9). Although this has little effect on air rising quickly in a shower cloud (where ascent to the tropopause can be completed in under an hour), vertical motion in the atmosphere is often much slower and more persistent than this, giving time for significant diabatic warming or cooling.

(ii) Two *temperature inversions* (in which temperature rises rather than falls with increasing height) appear as steps in the lower troposphere temperature profile in Fig. 4.3. Although apparently small, they are capable of stopping all but the most vigorous convection as if they were resilient flexible lids (Box 11.2); indeed the inversion at the 900 hPa level would have stopped lower convection from reaching higher levels. In fact the deep, extensive layers of precipitating cloud filling the rest of the troposphere on this occasion were not produced by convection, but by air rising in a vigorous *front* moving NE'wards across the British Isles. In such fronts, large sheets of cloud are maintained by ascent which is much slower, more persistent, and more extensive than ascent in shower clouds (Section 12.4). Ascent is so slow that air parcels usually take a day or more to rise from the low to the high troposphere, during which time they may move 1000 or more kilometres horizontally in the strong winds typical of frontal regions, which include jet streams in the upper troposphere. In Fig. 4.3 a weak jet maximum of 86 knots was reported at the 250 hPa level (the high troposphere). In fact air in fronts ascends on very long, very slightly sloping paths (gradients ~ 1:100), and not in the short, nearly vertical paths of either radiosondes or updrafts in showery convection (Fig. 4.4).

It follows that the temperature profile recorded by a radiosonde rising through a front (Fig. 4.4), is a nearly vertical section through a very slightly tilted dynamic 'sandwich' of air streams, each layer of which may be traced back to a geographically

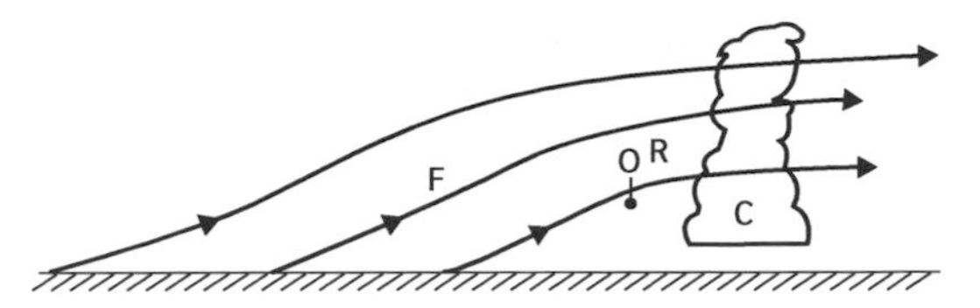

Figure 4.4 A radiosonde (R) rises at 5 m s^{-1} through the slightly tilted airflows of a large-scale atmospheric weather system such as the front (F) of an extratropical cyclone. Though its path is angled by wind shear, it is nearly vertical by comparison with the nearly horizontal airflows, and is essentially parallel with cumulus convection (C). Slopes like (F) are typically 1:100 while slopes in (F) and (C) are usually greater than 1:5.

Figure 4.5 Photo of convective overshoot. A hand held photograph of a vigorous cumulonimbus over Mali (central NW Africa) showing an anvil spreading asymmetrically from the main updrafts near its right-hand edge. Assuming the anvil to be at least 12 km above the local surface, the updraft overshoot into the stratosphere is between 0.5 and 1 km. Another updraft is reaching anvil level at its right hand edge, so that the centre of activity of the composite storm is moving at a different speed (and possibly direction) from its constituent cumulonimbus, as often happens in convective storms (Ch 11.6). Notice the many little cumulus in comparison with the few congestus and the maybe two cumulonimbus. (See Plate 6).

distinct and distant surface source region. In these circumstances, we must not expect to be able interpret all details of the observed temperature profile in terms of processes, such as showery convection, which are confined to the profile itself.

4.2.2 Tropopause and stratosphere

The characteristically steep temperature lapse rates of the troposphere come to an end at the *tropopause*, often abruptly as in Fig. 4.3, where a deep temperature inversion underlies the nearly isothermal layer which makes up much of the lower stratosphere. This marked difference between troposphere and stratospheric temperature profiles is clearly associated with the relative weakness and scarcity of vertical air motion in the stratosphere: most convective updrafts in the troposphere die at and just above the tropopause, and only the most violent are capable of overshooting by more than a kilometre (Fig. 4.5 and Box 11.2). Since frontal uplift ends there too, almost all cloud is confined below the tropopause. In fact temperature profiles in the lower stratosphere are largely determined by diabatic rather than adiabatic processes, i.e. by solar heating, and net heating or cooling by terrestrial radiation, rather than by vertical mixing.

The upper stratosphere is even more strongly dominated by the diabatic effects of radiation; its deep temperature inversion extends to the *stratopause* (Fig. 4.2), where air temperatures approach those of the Earth's surface 50 km below, maintained by absorption of soft solar UV by ozone (Section 3.3). Such warmth contrasts strongly with the abysmal cold which would prevail if adiabatic convection reached these levels from the Earth's surface. In fact temperatures in the upper stratosphere are maintained by a combination of radiative warming and weak, nearly horizontal movements of air from distant regions. For example, in middle and high latitudes temperatures can fall considerably in winter, as solar warming ends but cooling by net infrared emission continues (each by interaction with O_3—Section 8.6). However, especially in the N hemisphere, this cooling is often offset by sporadic *sudden warmings* of the lower stratosphere, believed to be triggered by large-scale activity in the underlying troposphere.

4.2.3 Mesosphere and above

Above the stratopause, temperatures lapse through the *mesosphere* to minimum values at the *mesopause* (Fig. 4.2) which can fall to −100 °C in high-latitude winters—among the lowest temperatures found anywhere in the planet. Though some of this lapse is assisted by weak vertical air motion, the typical average lapse rates of about 3 °C km^{-1} are much smaller than the 9.8 °C km^{-1} to be expected in cloud-free convective equilibrium (Section 5.10), again showing the importance of diabatic effects. In summer, continuous local warming by photodissociation and photoionization of air molecules (Section 3.1) raises mesopause temperatures to −30 °C. Above the mesopause, temperatures increase sharply with height in the *thermosphere* as increasing absorption of the harder components of solar UV swamps the weak opposing convective effect (Section 3.1).

4.2.4 Summary

The variations of temperature through the turbosphere are impressive on the Celsius scale, especially when compared with the narrow temperature tolerances of most living organisms, including ourselves, but it is important to note that they are relatively small on the Kelvin scale. In fact Fig. 4.2 shows that most atmospheric temperatures up to the 100 km level lie within 50 ° of 250 K, demonstrating the power of the Sun in keeping terrestrial temperatures far above the depths plumbed in the outer parts of the solar system and beyond, and keeping surface temperatures conditions suitable for the evolution of life.

4.3 The vertical profile of pressure

The vertical profiles of atmospheric pressure and temperature differ in two obvious ways: pressure falls consistently with height, whereas temperature rises and falls over two irregular cycles up to 100 km (Fig. 4.2). Pressure falls over a million-fold in this height range, whereas the range of temperature variations is about 20% on either side of 250 K. Moreover, this huge pressure variation is very regularly distributed in the vertical: pressures divide by about 10 for every height increase of 16 km. Formally we can say that the *decadal scale height* for pressure lies within 2 km of 16 km throughout the turbosphere (Box 1.2), as shown by the close correspondence between the linear height and logarithmic pressure scales of Fig. 4.2. More detailed observation reveals only quite small deviations from this simple pressure–height relation.

Why should the vertical distribution of atmospheric pressure be so smooth and so regular, and what does this tell us about processes at work in the atmosphere?

4.3.1 Smoothness

Atmospheric pressure varies very smoothly in the vertical; horizontal pressure variations are even more smoothly distributed (Section 4.6); and barographs show that pressures vary smoothly with time at fixed locations. Such observations show that some process is smoothing atmospheric pressure much more quickly than it is disturbed by atmospheric commotion. To recognize this process, consider what happens when a lightning flash causes a sudden substantial pressure change in a small part of the atmosphere.

When lightning flashes in or near a thundercloud *[23]*, the air along its narrow path expands explosively with the sudden conversion of electrical energy into heat (Section 6.12). An audible pressure shock wave spreads out at about the speed of sound. One second after the brief flash, the short loud shock wave has moved outwards over 300 metres, leaving most of the initial pressure excess distributed (and diluted) throughout a cylindrical volume about 7 orders of magnitude greater than that of the initial disturbance (NN 4.3). Ten seconds later, the now much quieter shock wave has reached out another 3 kilometres, and the pressure excess behind it has fallen another 2 orders of magnitude to almost inaudible levels. Note that at any particular location the audible shock from any small length of lightning flash is actually very brief, the familiar long rumble of thunder being maintained by the sequential arrival of short shocks from more and more distant parts of the kilometres-long lightning stroke (Section 6.12).

The spreading and dissipation of thunder shows how atmospheric pressure is spread and smoothed at about the speed of sound. The much smaller and less abrupt pressure excesses and deficits continually produced by other atmospheric disturbances are spread and smoothed just as quickly as thunder by a sea of mostly inaudible waves of sound and infrasound, maintaining the observed smooth distribution of atmospheric pressure.

NUMERICAL NOTE 4.3 Redistribution of pressure effects of lightning

Air is heated to incandescence in the few milliseconds duration of the lightning stroke in a tube of air of radius ~10 cm (0.1 m) centred on the stroke *[23]*. In 1 second the boundary shock wave travels outwards about 330 m from the stroke, increasing the bounded cylindrical volume by a factor of $(330/0.1)^2 \approx 10^7$, and to the same extent diluting the pressure excess throughout the volume. In a further 10 seconds the volume enclosed by the shock wave increases further by a factor of 100, to 10^9 times the volume initially disturbed.

4.3.2 Regularity

The smoothness explained above is not the same thing as uniformity. It is simply the absence of significant departure from the pressure distribution to be expected when air is fully adjusted to prevailing forces, by far the greatest of which is

BOX 4.2 Hydrostatic equilibrium

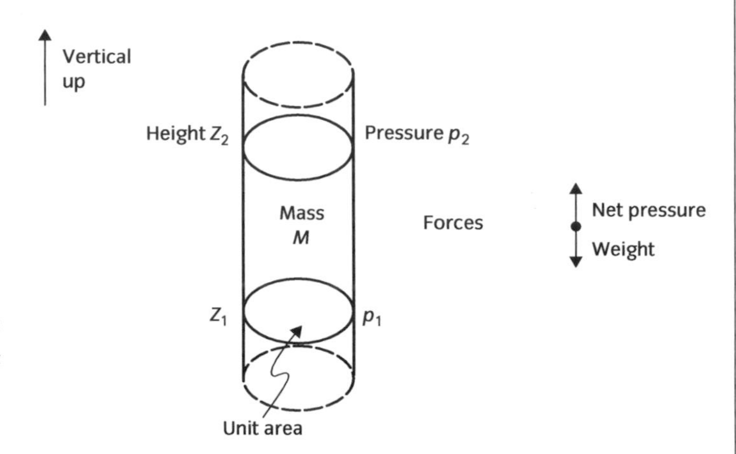

Figure 4.6 Forces on a vertical column of air in hydrostatic equilibrium. The weight of segment M is balanced by the excess of the pressure force up on its base over the pressure force down on its top.

Figure. 4.6 shows a vertical air column with unit horizontal cross-section area, containing mass M of air between heights z_1 and z_2. Since the pressure p_1 at height z_1 acts equally in all directions, we know that it must be pushing upwards on the base of the cylindrical air parcel with a force p_1 (pressure being numerically equal to force on unit horizontal area). In the same way, a force p_2 is pushing down on the top of the air parcel, so that the net upward force on the parcel is $p_1 - p_2$.

Since the air parcel has mass M, its weight is $M\,g$, where g is the local gravitational force on unit mass (its *gravitational acceleration* if it were to fall completely freely). If the cylindrical parcel and its surroundings are in hydrostatic equilibrium, we find the simplest form of the *hydrostatic equation* by equating weight and net upward pressure force

$$p_1 - p_2 = M\,g \qquad \text{B4.2a}$$

which confirms that p_1 must be greater than p_2, i.e. that pressure must lapse with increasing height.

Since the air parcel volume is numerically equal to $(z_2 - z_1)$ (again because the air column has unit horizontal area), its mass M is $\rho\,(z_2 - z_1)$, where ρ is the average density of the air parcel. After insertion and rearrangement, Eqn B4.2a generates a second form of the hydrostatic equation

$$\frac{p_2 - p_1}{z_2 - z_1} = \frac{\Delta p}{\Delta z} = -\rho\,g \qquad \text{B4.2b}$$

where the left-hand side is the vertical pressure gradient averaged through the vertical depth $(z_2 - z_1)$ of the air parcel. Equation B4.2b shows that this pressure gradient is directly proportional to the average parcel density, and the negative sign confirms that pressure lapses with increasing height.

A third form of the hydrostatic equation is found by letting the air parcel become very thin vertically, so that vertical averaging disappears. The left-hand side is now the vertical pressure gradient at height z, and ρ is the air density there.

$$\frac{\partial p}{\partial z} = -g\,\rho \qquad \text{B4.2c}$$

The left-hand side of Eqn B4.2c is written in *partial differential* form because it describes the variation of pressure with height alone, i.e. at a fixed horizontal location and chosen time. Pressure changes with varying horizontal location at a fixed time (as seen on synoptic weather maps), or with varying time at a fixed location (as seen on barograph records), are important in many ways, but not in the balance considered here.

gravity. Assuming air to be stationary, we can examine the static equilibrium of compressible air at rest under its own weight, looking for an explanation for the marked regularity of vertical distribution apparent in Fig. 4.2. Though in reality the atmosphere is never completely static, we will see later (Sections 7.9 and 7.14) that deviations from static equilibrium are very slight in most conditions.

In a static atmosphere, the upward and downward forces on any air parcel must balance, since otherwise there would be vertical acceleration. Gravitational attraction between the parcel and the Earth produces a downward force on the whole body of the parcel which we call its weight (Fig. 4.6). To balance this, the upward air pressure force on the lower surface of the parcel must be greater than the downward pressure force on its upper surface by just the required amount. Such *hydrostatic equilibrium* is qualitatively consistent with the observed height profile of pressure, but we can examine more closely to show that it is quantitatively consistent too (Eqns B4.2a, b and c in Box 4.2).

4.4 The hydrostatic atmosphere

(i) Equation B4.2a (Box 4.2) states that the air pressure at any height in the atmosphere is numerically equal to the weight of air in the overlying column resting on unit horizontal area there. If we consider an air column reaching from the Earth's surface to the top of the atmosphere, the pressure at the column base is the surface pressure p_s, the pressure at the column top is effectively zero, and we have

$$p_s = M g \qquad 4.2$$

where M is the mass of atmosphere resting on unit horizontal area of the Earth's surface, and we have ignored the small upward lapse in gravitational g (about 0.3% for every 10 km rise in the turbosphere) by using a single average value.

NN 4.4 shows that a typical sea-level pressure of 1,010 hPa corresponds to just over 10 tonnes of air resting on each horizontal square metre. However, even the flimsiest table does not collapse under this weight, because the nearly perfect fluidity of the air ensures that virtually the same air pressure is pressing upwards on its underside, i.e. that pressure is *isotropic* (magnitude independent of direction) in every small volume of air.

The pressure and height scales of Fig. 4.2, show that air pressure divides by about 10 for every 16 km increase in height. It follows that the mass of overlying air behaves in just the same way (i.e. has the same decadal scale height). From this and Fig. 4.2, we see that 90% of the mass of the atmosphere lies in and just above the troposphere, that a further 9% lies in the lower half (by height) of the stratosphere, while the upper half of the stratosphere contains only a further 0.9%. Altogether therefore, the troposphere and stratosphere account for 99.9% of the mass of the atmosphere, although confined within a layer reaching only 50 km above sea level—less than 1% of an Earth radius.

NUMERICAL NOTE 4.4 Air column masses

Rearrange Eqn 4.6 to form $M = p_s/g$, and substitute $p_s = 1{,}010$ hPa (i.e. 101,000 Pa) and $g = 9.8$ m s^{-2}, to find $M = 10{,}306$ kg m^{-2}—showing that about 10.3 tonnes of air rests on 1 square metre. On parts of the Tibetan plateau (over 5 km above MSL), surface pressures can be as low as 550 hPa, showing that the overlying air column contains just over 5 tonnes per square metre. In the centre of a deep depression (sea level pressure 950 hPa), there is nearly 10% less mass of air in the overlying atmosphere than in the centre of a high-pressure zone where sea level pressure is 1,040 hPa.

(ii) The second form of the hydrostatic equation (Eqn 4.2b of Box 4.2) shows that the lapse of pressure in a shallow horizontal layer is directly proportional to the average air density in the layer.

$$\frac{\Delta p}{\Delta z} = -\rho g \qquad 4.3$$

Near sea level, air densities are typically about 1.2 kg m^{-3}, and it follows from Eqn 4.3 that pressure falls by about 1.2 hPa for every 10-metre rise in pressure (NN 4.5). If air pressure continued falling at this rate, the pressure–height graph would be a straight line reaching zero pressure about 8.4 km above lea level (NN 4.5 and Fig. 4.7), where the atmosphere would have a definite top surface, like a very low-density ocean. This does not happen because air density falls with increasing height, and therefore so does the rate of fall of pressure, according to Eqn B4.2b. The profile of pressure against height is a smooth curve, steep at first, but becoming gradually less steep with increasing height (Fig. 4.7). There is therefore no definite height at which air pressure becomes zero: it fades more and more gradually into the near (but not complete) emptiness of space.

In the sea, by contrast, water is so incompressible that its density is nearly the same everywhere, so that the profile of pressure against depth is a nearly straight line, which extrapolates to great pressures at great depths (NN 4.5), and to zero (or rather to local atmospheric pressure) at a well-defined top surface (sea level), whose mean level (MSL) provides the datum for almost all height measurements in meteorology.

NUMERICAL NOTE 4.5 Pressure lapse in a shallow layer

We find the pressure lapse ($p_1 - p_2$) in a vertical 10 m near sea level, by rearranging Eqn 4.3 in the form $\Delta p = \rho g \Delta z$ and substituting 10 m for Δz, and a typical value for air density ρ (1.22 kg m^{-3} from NN 4.1), to find $\Delta p = 1.22 \times 9.81 \times 10 \approx 120$ Pa (i.e. 1.2 hPa). If pressure were to continue falling at this rate from a typical value of 1,010 hPa at sea level, it would reach zero at a height above sea level given by $1{,}010/1.2 \approx 842$ dam (decametres) = 8.42 km. Note that the calculation gave the height directly in decametres, because we used 10 m slices of air.

Water density being close to 1,000 kg m^{-3} everywhere in the sea, pressure increases uniformly by about $1{,}000 \times 9.8 \times 10 = 98{,}000$ Pa (980 hPa, i.e. nearly 1 atmosphere) for each 10 m increase in water depth.

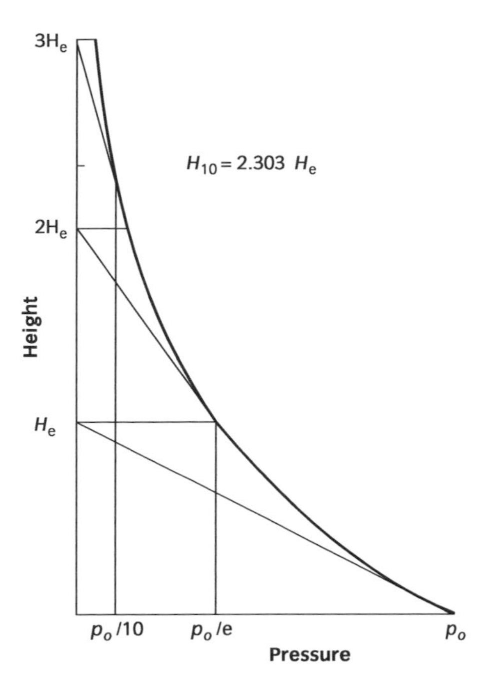

Figure 4.7 Pressure (or density) profile in an isothermal atmosphere in hydrostatic equilibrium. A tangent to the curve at any level reaches zero pressure one exponential scale height (H_e) above that level. The tangent from the surface would represent the pressure profile in an ocean of incompressible air with density equal to sea level values (about 1.2 kg m^{-3}). H_e is about 7,900 m for air at 270 K—more or less as the temperature is higher or lower. One decadal scale height is shown for comparison.

(iii) Finally we use the third version of the hydrostatic equation (Eqn B4.2c of Box 4.2), and replace the air density ρ using the equation of state for air (Eqn 4.1) to produce

$$\frac{\partial p}{\partial z} = -\rho\, g = -\frac{p\, g}{RT} = -\frac{p}{H_e} \tag{4.4}$$

where $H_e = R\,T/g$ is the *exponential scale height* for the distribution of pressure p with height z. When air temperature is uniform, H_e is constant along the profile (again ignoring the very small variations in g) and the rate of decrease of pressure with increasing height ($\partial p/\partial z$) is directly proportional to pressure itself. As shown in Box 4.3 the pressure–height profile is then the pure *exponential decay* curve sketched in Fig. 4.7 and described by

$$\frac{p_2}{p_1} = e^{(z_2 - z_1)/H_e} \tag{4.5}$$

According to this, as height increases by H_e (i.e. $z_2 - z_1 = H_e$) pressure falls from p_1 to $p_2 = e^{-1}\, p_1 = 0.3679\, p_1$.

The pure exponential pressure–height profile (i.e. uniform H_e since uniform T), has two important properties.

(a) The tangent to any point on the pressure profile reaches zero pressure at a height H_e above that point (Fig. 4.7).

(b) Standard integration (Box 9.5) shows that the centre of mass of the atmosphere lies at height H_e above the surface. Since H_e is proportional to absolute temperature T, the gravitational potential energy of an air column increases and decreases with T (Section 9.5).

BOX 4.3 Profile of air pressure

In hydrostatic equilibrium the vertical gradient of atmospheric pressure ($\partial p/\partial z$) varies according to

$$\frac{\partial p}{\partial z} = -\frac{p}{H_e} \qquad \text{B4.3a}$$

where $H_e = R\,T/g$ is the *exponential scale height* for the decay of pressure with height z. If H_e is constant throughout a range of height, standard integration *[8]* accumulates the infinite number of infinitely small lapses of p between heights z_1 and z_2 to find that

$$\frac{p_2}{p_1} = e^{-(z_2 - z_1)/H_e} \qquad \text{B4.3b}$$

where e is 2.718281..., the natural base for the exponential function e^x (and for the natural logarithms which are its mathematical inverse) given in all mathematical tables and calculators.

In a height interval $z_2 - z_1 = H_e$, B4.3b shows that $p_2/p_1 = e^{-1}$, which mathematical tables, etc. evaluate as 0.3678 (to four places of decimals), showing that pressure has lapsed from p_1 at height z_1 to $p_2 \approx 37\%\, p_1$ at z_2. Since the exponential curve has the same shape everywhere along its length, a further identical height interval from z_2 to z_3 will find pressure $p_3 = 0.3678\, p_2 = (0.3678)^2\, p_1 = 0.1353\, p_1$, and this continues ad infinitum along the decay curve. An equivalent statement is that p_1 falls to $p_1/2.718$ as height increases by H_e, and so on.

Equations like B4.3b describe many natural phenomena. With pressure replaced by temperature or radioactivity, and height by time, it describes Newton's law of cooling (Box 2.1) or radioactive decay. In every application, judicious scaling and choice of axes produces a graph which will fit exactly the pressure curve in Fig. 4.7. In meteorology, unlike most other applications, it is natural to plot the dependent variable (height) on the vertical axis, and pressure (the physical property) on the horizontal axis, as in Fig. 4.7.

The logarithmic equivalent

The logarithmic relation corresponding to Eqn B4.3b is

$$\ln\frac{p_2}{p_1} = -\frac{z_2 - z_1}{H_e} \qquad \text{B4.3c}$$

where ln is the standard abbreviation for logarithm to the base e. By rearranging as

$$\frac{z_2 - z_1}{H_e} = -\ln\frac{p_2}{p_1} \qquad \text{B4.3d}$$

we can find the height interval in which pressure p_2 falls to any chosen fraction of p_1. For example, $p_2 = p_1/10$ occurs when $(z_2 - z_1)/H_e = -\ln(0.1) = 2.303$, meaning that pressure falls to 1/10th of any chosen initial value in a height interval $H_{10} = 2.303\, H_e$, called the *decadal scale height* for pressure. In the same way, we find that $p_2 = p_1/2$ occurs in a height interval $H_2 = 0.693\, H_e$, which is the *binary scale height*—the height interval in which pressure falls to one half of its starting value.

According to Eqn B4.3c, a pure exponential curve of pressure against height corresponds to an absolutely straight line relationship between height and the logarithm of pressure. Figure 4.8 shows that when log pressure is plotted against height for the occasion of Fig. 4.3, the actual data lie close to the straight line corresponding to the exponential ideal, with a little curvature arising from the strong temperature lapse through the troposphere.

Thickness

Equation B4.3d defines the *thickness* (the vertical depth) of a layer of air bounded by pressures p_1 and p_2, and therefore containing a certain mass of air per unit horizontal area. Thickness is much used in meteorological theory and practice, and is always calculated assuming air to be an ideal gas in hydrostatic balance. Rearranging Eqn B4.3d we find

$$z_2 - z_1 = \frac{RT}{g}\ln\frac{p_1}{p_2} \qquad \text{B4.3e}$$

where T is the layer temperature. If T varies with z (or p), the value which satisfies B4.3a is the layer average weighted by ln p. Since T is the only variable on the right-hand side once p_1 and p_2 have been chosen, we see that the lateral (i.e. nearly horizontal) thickness gradient is proportional to the lateral gradient of layer mean temperature T.

When $\Delta p = p_1 - p_2$ is much smaller than p

$$z_2 - z_1 \simeq \frac{RT}{g}\frac{\Delta p}{p} \qquad \text{B4.3f}$$

relating the height thickness $\Delta z = z_2 - z_1$ to the proportional pressure thickness $\Delta p/p$.

NUMERICAL NOTE 4.6 Scale heights and thickness

Use $H_e = R\,T/g$ to find scale heights. Through the turbosphere, taking $T = 250$ K, $g = 9.66$ m s^{-2} (the value 50 km above MSL) and $R = 287$ J kg^{-1} K^{-1}, to find $H_e = 7{,}428$ m ≈ 7.4 km. A 20% increase in T and 1.5% increase in g will increase H_e by about 18.5%, because H_e is directly proportional to T and inversely proportional to g. In the same way, a 20% decrease in T and 1.5% decrease in g will decrease H_e by about 18.5%. Centred on 7.4 km, these percentages produce a range of from 6.0 to 8.8 km for H_e. Note the substantial allowance for the g variation through the turbosphere; the variation in g the troposphere is much smaller and is often ignored in meteorology.

Using Eqn B4.3e, and assuming a layer mean temperature of 273 K (i.e. an exponential scale height of 7,995 m if $g = 9.8$ m s^{-2}), we find that a layer bounded by pressures 900 and 800 hPa has thickness $= 7{,}995 \times \ln(900/800) = 7{,}995 \times 0.1178 \approx 942$ m. If the layer warms from 273 to 274 K, the thickness increases by 1 in 273—i.e. 3.5 m—as can be double checked by working through the calculation again with $T = 274$ K.

4.4.1 Values of scale heights in the atmosphere

Throughout the turbosphere, Fig. 4.2 shows that absolute air temperatures T lie mostly within 50 K (i.e. within 20%) of 250 K. In addition gravitational g lies within 1.5% of 9.66 m s^{-2}. It follows that values of H_e in the turbosphere lie mostly between 8.8 and 6.0 km (NN 4.6), with largest values in warmest air at low levels, and smallest in coldest air at high levels. Realistic pressure profiles therefore lie close to the exponential ideal for a temperature of 250 K, with slightly greater curvatures (smaller H_e values) at lower temperatures, and smaller curvatures (larger H_e values) at higher temperatures.

In the troposphere near MSL absolute temperatures T mostly lie within 15 K of 288 K, and $g \approx 9.8$ m s^{-2}, so that H_e values centre on 8.4 km, which is the value for the depth of the equivalent incompressible atmosphere found in (ii) of the previous subsection and sketched in Fig. 4.7.

As described in Box 4.3, the exponential scale height H_e is the most fundamental of several related scale heights. More practical scale heights are

(i) the *binary scale height* $H_2 = 0.693\,H_e$, in which pressure falls to one half, and

(ii) the *decadal scale height*, $H_{10} = 2.303\,H_e$, in which pressure falls to one tenth.

Corresponding to a typical 7.3 km for mid-tropospheric H_e, the binary and decadal scale heights are 5.1 and 16.8 km respectively.

Using H_{10} and starting with a pressure 1,000 hPa at sea level, pressure 100 hPa is found about 16.8 km above sea level, pressure 10 hPa is found at about 34 km, and so on up to about 100 km where air pressure is about 10^{-3} hPa or 0.1 Pa, in good agreement with the height and pressure scales on Fig. 4.2, allowing for the temperature sensitivity of scale height.

A value of 5.0 km for H_2 is in only fair agreement with Fig. 4.3, where the 500 hPa surface lies nearly 5.51 km above sea level. The discrepancy arises from the relative warmth of the lower troposphere on that occasion, where the temperature at the 750 hPa level (the middle of the lower troposphere) was about −6 °C, i.e. 267 K. The value of H_2 for $T = 267$ K is 5.42 km, much closer to the reported height. A thorough pressure-weighted average temperature removes the remaining 90 m discrepancy.

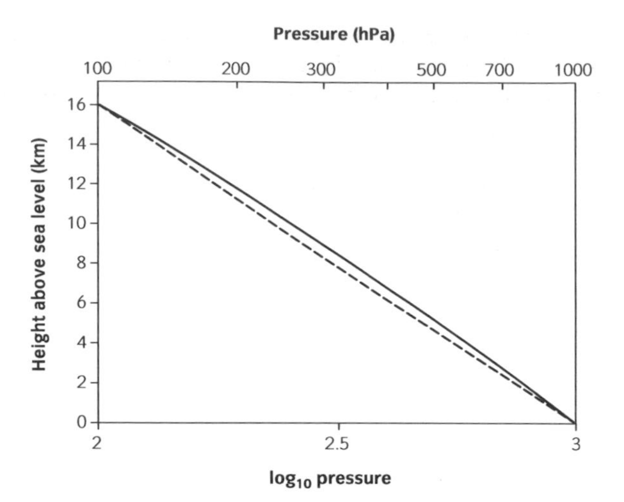

Figure 4.8 Logarithm of pressure (to the base 10 here, but the rule is general) from Fig 4.3 to show its near linearity with height, explaining the simple regularity of Figs 4.2 and 3.1.

In this section, we have confirmed that the vertical profile of atmospheric pressure through the whole turbosphere is determined essentially by the material constitution of the atmosphere, by its temperature profile, and by the prevalence of a very close approximation to hydrostatic equilibrium—the last despite the ceaseless commotion which keeps it so well mixed.

4.5 The vertical profile of density

Since air is an ideal gas (Eqn 4.1), the vertical distribution of air density follows from the distributions of pressure and temperature—dominated by the pressure profile and adjusted according to the relatively small temperature variations on the absolute scale. Vertical profiles of air density and pressure are therefore very similar—the density profile showing a nearly exponential decay with increasing height and a decadal scale height which is never very far from 16 km throughout the turbosphere. This has some important consequences.

(i) Air densities in the normal human environment (mainly within a few hundred metres of sea level), are the highest in the atmosphere, with values of about 1.2 kg^{-3}. Since the summit of Mount Everest is in the upper troposphere, where air densities are about 60% lower (NN 4.1), with each normal breath a climber there inhales only 40% of the mass of air inhaled at sea level, with oxygen reduced in the same proportion because of the uniformity of proportions. Such a reduced intake cannot support severe exertion by even the fittest climber for more than a few days, with the result that almost all use oxygen enriched air to reach the summit.

Air-breathing aircraft engines, such as the common turbojet, are designed to operate best at about 12 km, to take advantage of the reduced airframe drag there, but are unable to operate efficiently at much higher altitudes because of oxygen starvation—even though the drag would be smaller.

(ii) At much greater altitudes air is completely useless for human breathing and turbojets, and as its density falls to the extremely low values found in the upper stratosphere and mesosphere, it becomes ineffective in conducting and convecting heat to or from immersed bodies. A research balloon floating at an altitude of 50 km above sea level becomes extremely cold at night despite being immersed in the relatively warm air there (Fig. 4.2), because it cools much faster by emitting far-infrared radiation than it warms by the feeble conductive and convective contact with the ambient air, whose density is three orders of magnitude below surface values. However, even at 80 km altitude, where air density is five orders of magnitude smaller than at sea level, very fast-moving bodies like meteorites and returning space vehicles can be vaporized by frictional heating with the air unless the re-entry is very carefully managed.

(iii) Chemical interactions between molecules (Section 3.4), and molecular interactions with electromagnetic radiation, decrease with partial densities of molecular species, and therefore usually with overall air density. For example, solar wavelengths which are almost completely absorbed before reaching sea level may be relatively abundant only a few kilometres higher up, because so much absorption occurs in the relatively dense surface layers of the atmosphere. This is the case for ultraviolet wavelengths, which is why people on mountain holidays can suffer severe sunburn if they do not use heavy sunscreens. And the familiar blue of the cloudless daylit sky deepens rapidly towards black with increasing altitude, the blue being produced by preferential lateral and backscattering of the blue end of the visible solar spectrum by air molecules, so that the blueness fades as molecular number densities fall. For the same reason the blue atmospheric layer on the Earth's horizon is seen from satellites to be confined to a very shallow zone overlying the Earth's surface (Fig. 1.2).

4.6 Distributions from equator to poles

So far we have examined vertical distributions of temperature and pressure, etc. When we begin to examine horizontal distributions of these and other meteorological variables, especially in seasonal or annual averages, we enter the conventional domain of *climatology*. Vertical distributions are also very important climatologically, since they are closely related to prevailing weather types, but are often ignored in superficial accounts of climatological *air masses*, which are treated as vast 'solid' blocks of air confined by vertical walls, rather than fluid multi-layered nearly horizontal sandwiches with different layers moving at different speeds and directions. We examine horizontal and vertical distributions of meteorological variables, and their relationship with the state and activity of the atmosphere, at many points throughout the rest of this book, but begin now with a review of distributions between the equator and poles.

4.6.1 Pressure

Because of the vertical compression of the atmosphere, distributions of temperature, pressure, etc. would be crushed to illegibility if they were not stretched one-hundred-fold in the vertical, as in Fig. 4.9, etc. The largest and most persistent horizontal gradients are those in *meridional* directions (i.e. equator to pole), as expected given the strong meridional gradient of solar input to Earth. Significant horizontal gradients occur also in *zonal* directions (E–W), but they tend to be

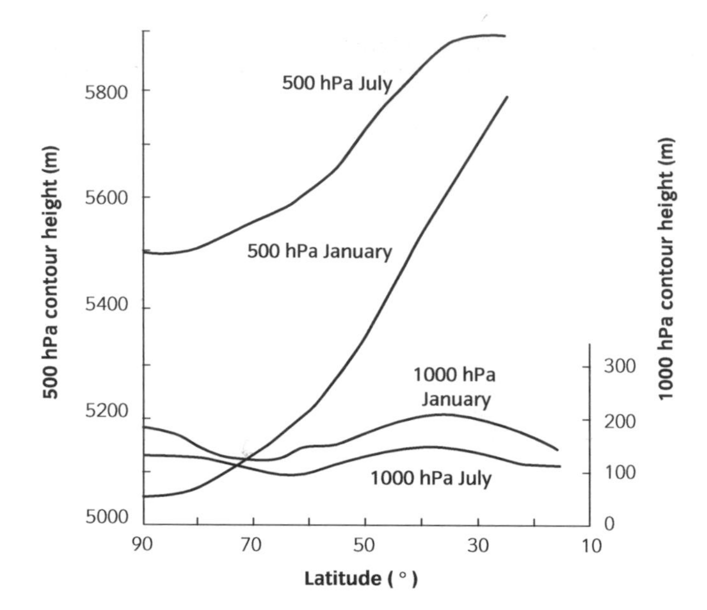

Figure 4.9 Monthly and zonally averaged height contours of the 1,000 and 500 hPa surfaces in the N hemsiphere in January and July. The vertical height scales are chosen to enhance contour gradients rather than absolute heights.

less pronounced, since they are mostly associated with land–sea contrasts, and must disappear in the complete zonal averaging used to produce a global meridional picture like Fig. 4.9.

Near sea level

Consider first the meridional distributions of monthly mean pressures at the base and middle of the troposphere (Fig. 4.9). Data are from the N hemisphere only, in this case, since observations are still less comprehensive in the mainly oceanic S hemisphere. However, we should expect the latter to be broadly similar after allowing for the six-month shift in seasons. In Fig. 4.9 the distribution of pressure at the base of the troposphere is represented by the heights of the 1,000 hPa isobar above mean sea level (MSL). Zonal averages of this *isobaric* surface lie between 100 and 200 m above MSL, which corresponds to MSL pressures of between 1,008 and 1,016 hPa, after allowing for the lapse of about 1.2 hPa per 10 m typical of the low troposphere (NN 4.5). For comparison, the global annual average pressure at MSL is 1,013.2 hPa. It is standard meteorological practice to represent the horizontal distribution of pressure by means of height contours of isobaric surfaces, partly for ease of display, but also because it simplifies the expression for the horizontal pressure gradient force (Box 7.7).

In Fig. 4.9 the elevation of the 1,000 hPa surface varies little with latitude, apart from a slight downward slope towards low latitudes, a broad shallow plateau in the subtropics, and an equally broad shallow depression in middle latitudes. In fact these apparently slight pressure differences are associated with quite dramatic differences in weather and climate between the calm, dry subtropics and the disturbed and wet middle latitudes, and arise in complex ways from the activity of the dominant weather systems in the zones (Chapters 12 and 13). On a coarser view the contour variations are so small that the 1,000 hPa surface is nearly horizontal from low to high latitudes, i.e. that the average pressure at MSL is nearly uniform, and that there is little variation between winter and summer.

Mid-troposphere

The pressure distribution in the middle troposphere is represented by the zonal average 500 hPa isobar on Fig. 4.9, which lies between 5 and 6 km above MSL and differs greatly from the 1,000 hPa surface in both meridional profile and seasonal variation. The 500 hPa surface slopes downwards consistently from low to high latitudes, and the slope is much greater in winter, because of the substantial fall in contour height from summer to winter in middle and high latitudes. Actually all these changes are large only in a relative sense: falls of 600 m in 60 degrees of latitude (over 6,000 km), and 400 m in six months at a given location, are revealed only by a meticulously operated radiosonde network making due allowance for small variations of g with latitude and height (Section 7.4.2).

Overall, Fig. 4.9 shows that the altitude of the 500 hPa surface falls by about 15% between low and high latitudes in winter, and by about half this amount in summer, whereas the 1,000 hPa surface stays practically level and constant. It follows that the vertical depth of the lower half of the atmosphere (by pressure and therefore effectively by mass) varies by the same amount as the height of the 500 hPa level, and so therefore must its volume per unit horizontal area and its average density. Figure 4.10 shows the obvious reason for such behaviour: the mean temperature of this layer falls consistently and significantly from low to high latitudes, the fall being steepest in winter because high latitudes are much cooler then, whereas low latitudes are only slightly cooler.

The relationship between layer depth (*thickness*) and layer mean temperature is clear in the expression for any scale height, but the binary scale height H_2 applies directly in the present case, since the bounding isobaric surfaces in Fig. 4.9 differ in pressure by a factor of two. According to Box 4.3

$$H_2 = 0.693\frac{RT}{g}$$

which has values of about 5 km for T about 250 K (−23 °C). The meridional gradients of thickness and layer mean temperature are therefore in direct proportion, as are their seasonal variations.

In view of the relatively steep and extensive horizontal pressure gradients in the middle troposphere in Fig. 4.9, it may seem surprising that sea-level pressures are so nearly uniform. Since the total mass of the atmosphere is so evenly distributed (as seen by the 1,000 hPa contours), the significant localized concentrations and depletions in its lower layers must be closely balanced by compensating distributions in higher layers, as we see below.

4.6.2 Temperature

Figure 4.10 shows the zonally averaged meridional distribution of air temperature from the surface to the 100 hPa level across both hemispheres in January and July. Since it is bounded by pressures of 100 and just over 1000 hPa, the layer contains just over 90% of the weight of the atmosphere, together with most of its weather. As in many meteorological diagrams, the vertical dimension is shown using horizontal isobars and a vertical logarithmic pressure scale, which

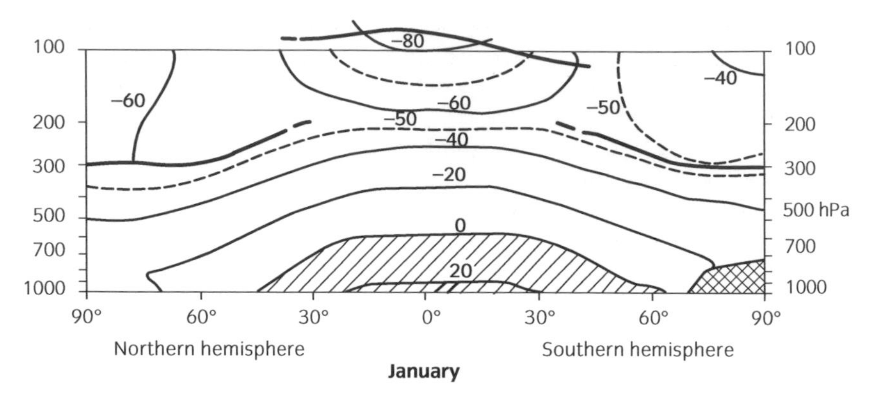

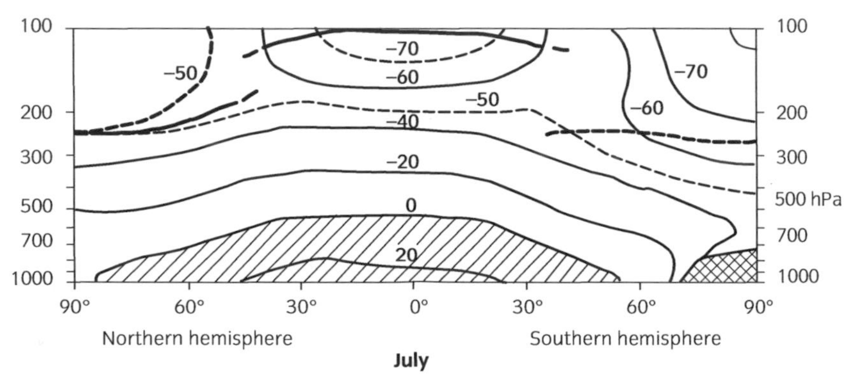

Figure 4.10 Monthly and zonally averaged isotherms in January and July. Typical troposphere locations are shown by thick lines, and areas above 0° C are shaded. Note the consistent vertical temperature lapse throughout the troposphere, the concentration of tropospheric baroclinity in midlatitudes especially in winter, and the reversal of normal baroclinity about the 200 hPa level in summer.

is directly proportional to height according to Eqn B4.3c. Such diagrams cannot show the important variations in the heights of isobaric surfaces above MSL discussed above, but that is not a serious problem now, as we focus on the much larger height variations of the *isotherms* in Fig. 4.10.

From the locations of the tropopause in Fig. 4.10, we see that the diagram includes almost all of the deep equatorial troposphere, whereas in middle and high latitudes it includes a substantial part of the low stratosphere as well, since the troposphere is much shallower there. The troposphere is deepest in low latitudes because of the deep, vigorous convection maintained by strong heating from the high-angled Sun. But although surface temperatures in low latitudes are amongst the highest anywhere on Earth, the great depth of the local troposphere, together with its steep lapse of temperature, means that temperatures at the 100 hPa level can be lower in low latitudes than in high latitudes. The effect is most marked in summer (July in the N hemisphere, January in the S hemisphere), when the long hours of daylight warm the low stratosphere in middle and high latitudes. The poleward lapse of temperature, which from our surface-bound experience we tend to regard as universal, reverses above the 200 hPa level in summer, and shows an equatorward temperature lapse of over 30 °C at 100 hPa. In winter, the low stratosphere cools sharply during the long polar nights (especially over Antarctica) so that temperatures at the 100 hPa level are nearly uniform, which is still very different from the strong poleward temperature lapse familiar in surface layers in those latitudes.

Recalling the relation between thickness and layer mean temperature, it is clear that the poleward downslope apparent on the 500 hPa surface (Fig. 4.9) must persist and increase up to somewhere between the 300 and 200 hPa levels. Above this the increasing depth of relatively warm low stratosphere in middle- and high-latitude summers progressively reduces the isobaric slope, reversing it by the 100 hPa level.

Comparing Figs. 4.9 and 4.10, we see that both isobars and isotherms tilt downwards towards the poles from about latitude 30°, but that isotherms slope more steeply than isobars, often by a factor of 10. It follows that isotherms and isobars must intersect throughout the region, and that intersections are most numerous in middle latitudes in winter. An atmosphere with intersecting isotherms and isobars is called *baroclinic*, whereas it is called *barotropic* when isotherms and isobars are parallel. The troposphere depicted in Figs. 4.9 and 4.10 is nearly barotropic between the tropics, whereas it is definitely baroclinic at higher latitudes, especially in the winter hemispheres. As discussed in Chapters 12, 13 and elsewhere, the distinction between near *barotropy* in low latitudes and strong *baroclinity* in middle latitudes underlies the quite different dynamic regimes in these regions, and associated differences in weather an climate.

4.7 The general circulation

In many geographical locations, even casual observation suggests that surface winds have a preferred direction which either persists throughout the year or varies seasonally. In middle latitudes, there is a pronounced W'ly bias, in that winds with a W'ly component are more frequent and/or stronger than winds from other directions. And winds in large areas between the tropical and equatorial zones have an E'ly bias. These casual impressions are confirmed by instrumental measurement: when flow at a particular location is averaged over time periods much longer than the longest-lived weather systems (usually over a season or a year), a systematic residual speed and direction usually survives. The global distribution of such residuals defines the *general circulation* in the season or year. In this section we examine *zonal* averages of air flow (data averaged round a hemisphere in a narrow latitude range) averaged over decades.

Consider the general circulation in the troposphere and just above it. Because only horizontal winds are measured routinely, the circulation is everywhere horizontal. That cannot be the whole picture, however, since vertical motion is enforced as air flows over uneven land, and implied wherever continuing horizontal flows would produce persistent concentration or depletion of air (as in Fig. 4.12a). True air flow is intrinsically three-dimensional, both in the general circulation and in all the weather systems which maintain and disturb it, even though vertical speeds are often at least two orders of magnitude smaller than horizontal wind speeds, and are usually too small to be measured directly—another consequence of the flattening of the atmosphere by gravity.

Many important features of mean horizontal flow are apparent on vertical meridional sections similar to those already used to display pressure and temperature. In Fig. 4.11 the meridional cross-sections are curved over a quadrant of cross-section, and expanded vertically by a factor of 250 relative to the underlying Earth quadrants. Vertical elevation is represented by a linear pressure scale in decibars (100 hPa), which expands the geometrical height scale at low levels and compresses it aloft, thereby highlighting the troposphere, especially its lower half, at the expense of higher layers.

4.7.1 Westerly components

Consider zonal components of mean flow, i.e. W'ly and E'ly components calculated by simple geometry (NN 4.7) from wind speeds and directions measured at many locations and heights, averaged in zones round each hemisphere, averaged over seasons or years, and further averaged over decades (Fig. 4.11).

In middle and high latitudes, W'ly components predominate throughout the troposphere. At the surface, this fits with day-to-day experience, on the oceans and W continental margins especially. And we see that W'lies are even more extensive and stronger in the upper troposphere, with maximum speeds between the 200 and 300 hPa levels. In winter the maximum lies a little poleward of the tropics, and is especially strong, with mean W'ly components exceeding 40 m s^{-1} (nearly 80 knots) in the N hemisphere. Speeds are even higher in the S winter hemisphere (not shown). In the N summer hemisphere, the maximum is weaker and lies further poleward—a seasonal displacement which is much smaller in the S hemisphere (not shown). In both seasons and hemispheres upper tropospheric W'lies die away more abruptly on the equatorward side of the maximum than

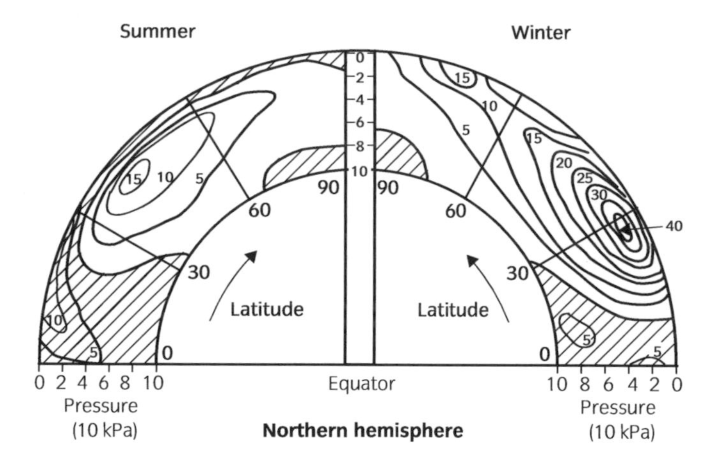

Figure 4.11 Seasonally and zonally averaged zonal wind speeds in the N hemisphere in winter and summer, with isotachs labelled in m s^{-1} and regions with E'ly components shaded—the rest being W'ly. The local verticals represent pressure labelled in multiples of 10 kPa.

NUMERICAL NOTE 4.7 Winds and components

It helps to sketch the geometry of winds and azimuths in each case before following the calculation.

If a certain mean wind is 30 m s^{-1} SSW'ly (i.e. azimuth 202.5° clockwise from N—Section 2.4), its W'ly component is given by 30 cos θ, where θ is the angle between the wind and due W (270° − 202.5° = 67.5°).

Since cos (67.5°) = 0.383, the zonal (W'ly) component is 11.5 m s^{-1}. The meridional (S'ly) component is 30 cos (22.5°) = 30 × 0.924 = 27.7 m s^{-1}.

If the exercise is repeated for a mean NE'ly Trade Wind of 6 m s^{-1}, the zonal E'ly components and meridional N'ly components are each 4.2 m s^{-1}.

on the polar side. Each hemisphere contains a vast mass of troposphere moving from W to E at speeds which, even after seasonal and zonal averaging, correspond to gale force in summer, and far more in winter. Each belt of zonal flow is termed a *circumpolar vortex*, reflecting its continuity around the hemisphere and truly enormous scale.

In middle and high latitudes the strongest mean winds occur in the upper parts of the baroclinic zones outlined in Section 4.6, and the circumpolar vortices are rotating from W to E, like the solid Earth but faster. These features are dynamically related in important ways which we begin to examine in Box 4.4, anticipating later discussion of the dynamics, energetics, structure, and behaviour of large-scale weather systems (Chapters 7, 8, 12, and 13).

Above the high troposphere, mean W'ly wind speeds fall away quite sharply with increasing height, at least in the summer hemisphere, where mean winds become E'ly everywhere above the lower middle stratosphere (visually diminished by vertical compression in Fig. 4.11).

In the winter hemisphere, mean winds remain W'ly at all levels poleward of the tropics, and increase again with height above a minimum at the same level as the tropical wind maximum. As in the mid-latitude troposphere, in middle and high latitudes there is a clear association between baroclinity and zonal flow in the low stratosphere: W'ly flow increases with height through a layer with a poleward temperature lapse, and decreases and even reverses as the temperature gradient reverses.

4.7.2 Easterly components

In Fig. 4.11 E'ly components are apparent in both the low and high troposphere at low latitudes, and in the low troposphere at high latitudes. Though locally significant, the high-latitude E'lies cover only about 5% of the total surface area of the Earth, as compared with nearly 50% covered by the low-latitude E'lies (half the surface area of a sphere lies between latitudes 30°).

The surface and low troposphere E'lies of low latitudes are the zonal components of the *Trade Winds* (or *Trades*), so-called because their day-to-day persistence was relied upon in the days of commercial sailing ships. Unlike the W'lies of high latitudes and altitudes, the Trades are not true zonal winds even when averaged, since they have a systematic equatorward bias described in the next subsection. The marked weakening of the zonal component of the Trades in summer is confined to the N hemisphere and is a consequence of the large-scale distortion of air flow associated with the SW'ly summer *monsoon* of southern Asia (Section 13.2). As mentioned in the previous subsection, the low-latitude summer E'lies in the high troposphere are associated with the reversed meridional temperature gradient there.

4.7.3 Meridional components

When wind data are analysed for meridional components of flow, and averaged over latitude zones, seasons, and decades, we find the patterns of meridional circulation shown in Fig. 4.13. These and the zonal patterns of Fig. 4.11 can be related and visualized with the help of the schematic flows of Fig. 4.12 a and b. In Fig. 4.12a we see the NE and SE Trades converging towards the Equator with comparable equatorward and W'ward components of motion. Figure 4.13

shows that mean equatorward flow in the very low troposphere exceeds 2 m s^{-1} throughout the year, and 3 m s^{-1} in winter, and is accompanied by poleward flow in the high troposphere, which is also stronger in winter.

As shown schematically in Fig. 4.12b, these equatorward and poleward (meridional) flows are the horizontal branches of two hemispheric scale meridional circulations (one on each side of the equator though biased a little towards the N hemisphere), which is named the *Hadley circulation*, after George Hadley, whose discussion in CE 1735 was one of the earliest successful dynamic descriptions of atmospheric motion.

The Hadley circulation

The horizontal flows in each *Hadley cell* are completed by a rising branch in the rainy equatorial zone, and a descending branch in the arid subtropical zones of each hemisphere (Figs. 4.12b and 8.21).

The near-equatorial zone in which the opposing trade winds converge is called the *Inter-tropical Convergence Zone* (ITCZ—Fig. 4.12a and b), and its convergence feeds the updrafts (~1 m s^{-1}) of the vast populations of warm shower clouds which dominate its climate. However, since these updrafts are embedded in much larger volumes of gently sinking air (Section 11.5), the speed of net upflow averaged across the rising branch of the Hadley circulation is ~1 cm s^{-1}, which is much too small to be measured directly.

Similarly, the diverging pattern of near-surface flow in the subtropics is fed by continual descent associated with the zones of high sea-level pressure called the *subtropical high-pressure zones,* or *subtropical highs* (STHs), which virtually encircle the Earth in the subtropics of each hemisphere (Section 13.2). The absence of cloud and rain in the STHs indicates the persistence of sinking air there, though the descent is so slow (a few cm s^{-1} at most) that it is too small to measure directly, and is termed *subsidence* rather than downflow.

Note again that equatorward motion in the low troposphere is associated with W'ward motion. In the high troposphere too, meridional motion is accompanied by zonal motion, but here poleward motion is associated with E'ward motion which is so much faster, particularly in the poleward half of each Hadley cell, that air moves in poleward spirals which encircle the Earth at least once between their equatorial source and subtropical sink.

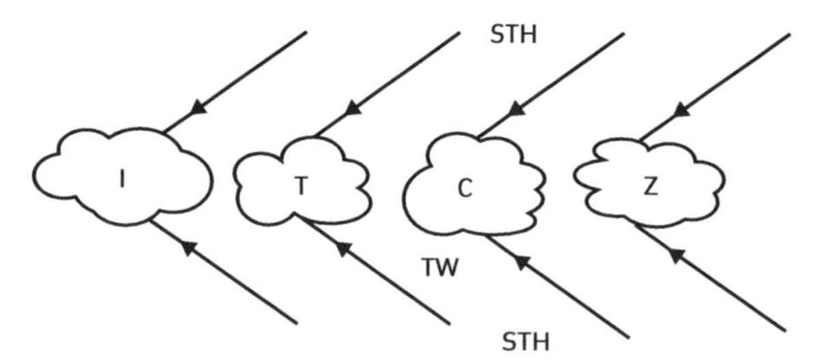

Figure 4.12a Schematic plan view of low-tropospheric airflow in the Hadley circulation. The Trade Winds (TW) blow from the subtropical high pressure (STH) zones to the intertropical convergence zone (ITCZ), where they feed the cloudy updrafts of tropical weather systems. The symmetrical flow shown is found when the ITCZ lies near the equator; when its does not, the cross-equatorial flow is deflected to become roughly SW'ly in the N hemisphere (NW'ly in the S hemisphere).

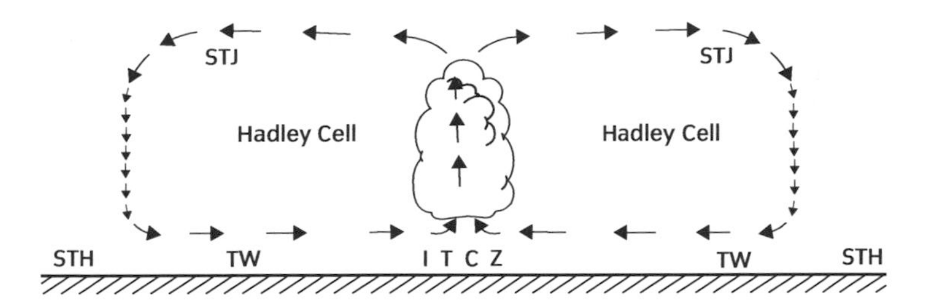

Figure 4.12b A schematic meridional vertical cross-section through the Hadley circulation. The low level flow is the elevation for the plan of Fig 4.12a. The rest shows the poleward flow from the vigorous anvils of convection in the ITCZ to widespread gentle subsidence in the STHs.

Circulation and disturbance

Throughout the rest of the troposphere, mean meridional components of motion are very small. There is some evidence in Fig. 4.13 of a relatively weak meridional circulation poleward of the Hadley cell but in the opposite sense. Though this *Ferrel cell* plays a significant role in global energetics, in these and higher latitudes it is swamped by the sporadic, intense disturbances associated with middle-latitude weather systems, such as extratropical cyclones (Chapter 12). Even the relevance of the Hadley circulation for day-to-day patterns of air flow and weather conditions is easily exaggerated. Here too the presence of rain or sunshine, gentle or strong winds, relates more obviously to the distribution and activity of a range of transient weather systems, each of which may be associated with large deviations of local

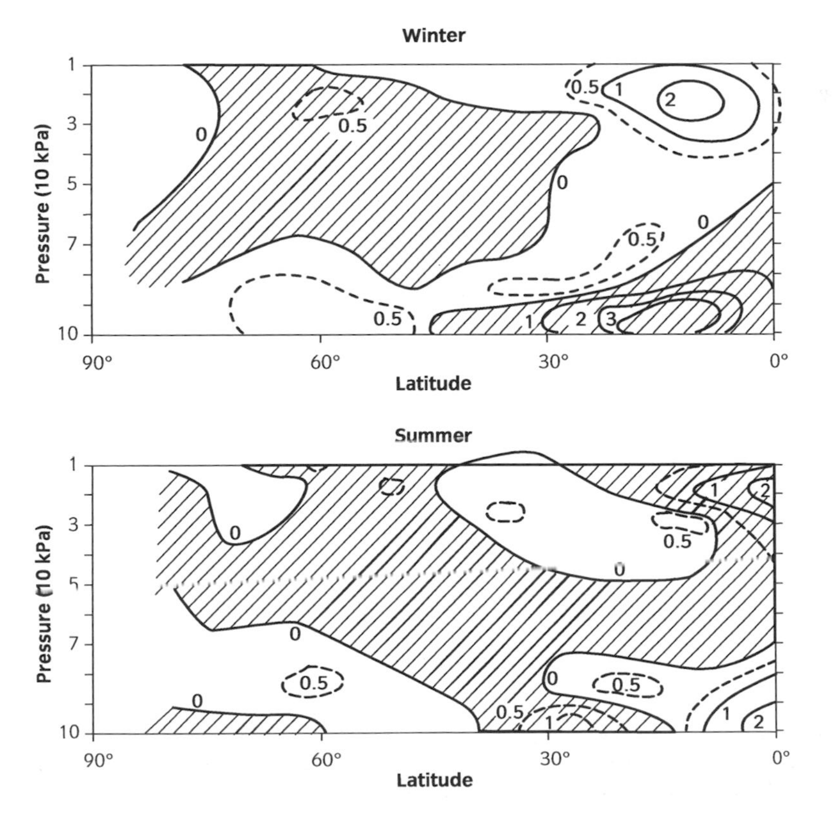

Figure 4.13 Seasonally and zonally averaged meridional wind speeds in the N hemisphere in winter and summer. Axes and labelling are as in Fig 4.11, with shaded areas distinghishing N'ly from S'ly components. Notice that speeds are much lower than in Fig 4.11, and that patterning is strongest in the region dominated by the Hadley circulation.

conditions from the mean (Chapter 13). In fact the Hadley circulation is important mainly because it maintains the trades, feeds and constrains the embedded weather systems of the trades and ITCZ, and feeds the subsidence of the STHs.

BOX 4.4 Elementary kinematics and dynamics of rotation

Angular velocity

Consider a point moving at a steady circumferential speed V round a circle of radius R (Fig. 4.14a). If it takes time T to complete one circumference (length $2\pi R$), the speed V of the point is $2\pi R/T$. By definition of *radian* angular measure (Fig. 4.14a), the angle subtended at the centre by one circumference is 2π radians, so that we can regard $2\pi/T$ as the *angular velocity* Ω of the orbiting point—its rate of rotation expressed in radians turned per unit time. The expression for circumferential speed V therefore has two equivalent forms.

$$V=\frac{2\pi R}{T}=\Omega R$$

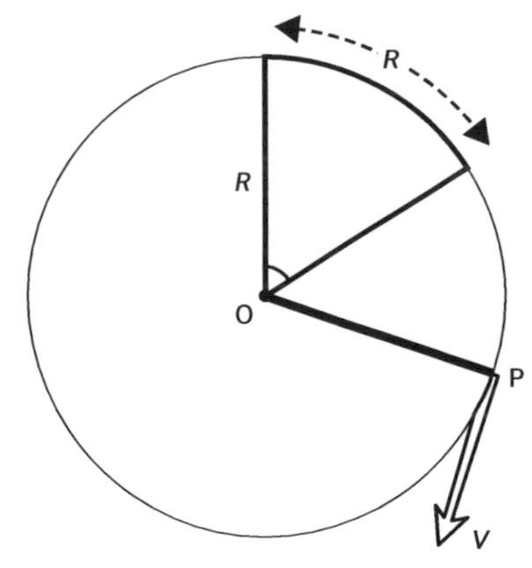

Figure 4.14a Circular angles and motion. One radian is the angle subtended at the centre of a circle by an arc of one radius. The angle subtended by its circuference is $2\pi R/R = 2\pi$, so that a radian is $360/(2\pi) \approx 57.3°$. If P moves steadily round the circle at speed V, it completes the circle in time $2\pi R/V$, so that its angular velocity in radians per unit time is $2\pi/(2\pi R/V) = V/R$.

Now consider a N hemisphere quadrant section of the Earth, rotating with angular velocity Ω about the polar axis (Fig. 4.14b). Simple arithmetic (NN 4.8) shows that the angular velocity of the Earth is 7.29×10^{-5} rad s^{-1} as it spins on its axis relative to the stars, and that the absolute E'ward speed of a point P fixed on the equator at sea level is 464 m s^{-1} (903 knots). At latitude 30° the E'ward speed at sea level has dropped to 402 m s^{-1} because P is swinging at the same angular rate Ω around a radius which is now only R cos 30°, i.e. 0.866 R. If air slides freely poleward from an equatorial source fixed to the Earth's surface there, it flows over regions where the underlying surface is moving progressively more slowly E'ward, and we might expect the air to flow progressively faster E'ward relative to the underlying surface, as it maintains the strong E'ward speed of its equatorial source. In the case of air moving from the equator to latitude 30°, this argument suggests that there should be a W'ly wind of 62 m s^{-1} (121 kt) at latitude 30° (NN 4.8).

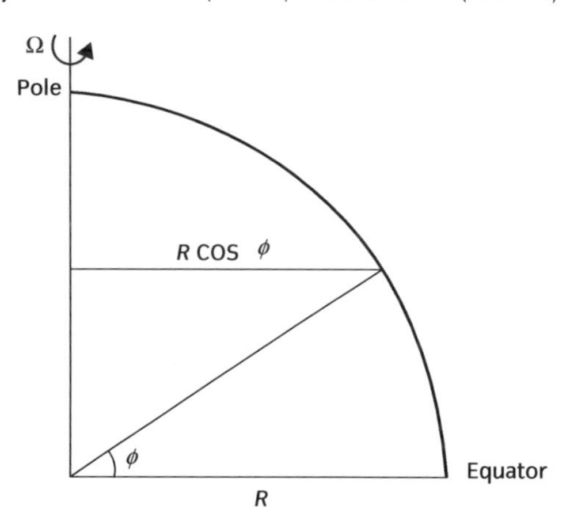

Figure. 4.14b A quadrant of the Earth showing how the radius of rotation of a point on its surface varies with its latitude ϕ. All fixed points on and in the solid Earth have the same angular velocity Ω, so that tangential speeds of fixed surface points vary as ΩR, i.e. $\Omega R \cos\phi$.

In fact even this large effect considerably underestimates the effect of the Earth's rotation.

Angular momentum

In the previous paragraph we have assumed that the poleward sliding airstream conserves its initial tangential momentum. This ignores important kinematic and dynamic consequences of the Earth's rotation and spherical shape whose more thorough analysis

(Section 7.16) shows that, in the absence of any twisting force (*torque*) about the Earth's rotational axis, what is actually conserved is the *angular momentum* of the airstream about Earth's axis of spin.

The angular momentum of a body whirling about an axis is the product of its tangential momentum ($m\ V$) and its perpendicular distance d from the axis, i.e. ($m\ V\ d$). As d reduces V increases and vice versa, so that their product ($d\ V$) remains constant, unless there is a force with a turning moment (*torque*) acting to increase or reduce its angular momentum about the axis of spin. The increase in tangential speed seems mysterious, until you realize that there must be a force acting towards the axis, and that this does work on the body as it moves in or out (Box 7.13).

Applying this to the case of air moving poleward over the Earth's surface (Fig. 4.14b), we see that air flowing East at wind speed U relative to a fixed point P at latitude ϕ (itself moving East at speed $\Omega\ R \cos \phi$ as the Earth spins) has angular momentum

$$m\ [U + \Omega\ R \cos \phi]\ R \cos \phi \qquad \text{B4.4a}$$

and it is this (rather than E'ward momentum $m\ [U + \Omega\ R \cos \phi]$) which is conserved as the airstream slides freely (i.e. without torque) to higher or lower latitudes.

If the air starts from rest relative to the Earth's surface at the equator ($U = 0$ at $\phi = 0$, $\cos \phi = 1$), its initial angular momentum is $m\ \Omega\ R^2$. Conservation of this requires that at latitude ϕ, the eastward speed U of the airstream over the Earth's surface must satisfy

$$m\ (U + \Omega\ R \cos \phi)\ R \cos \phi = m\ \Omega\ R^2$$

which rearranges to

$$U = \Omega R \left(\frac{1}{\cos \phi} - \cos \phi \right) \qquad \text{B4.4b}$$

where $\Omega\ R$ is the E'ward speed of the Earth's surface at the equator.

Section 7.16 contains a fuller treatment and discussion.

NUMERICAL NOTE 4.8 Angular velocity and circumferential speed

Since the Earth turns through $2\ \pi$ radians in 23 hr 56 min 4 s (=86,164 s), its angular velocity Ω is $2\ \pi/86{,}164 = 7.29 \times 10^{-5}$ rad s^{-1}. Taking the Earth to be a perfect sphere of radius 6,370 km, the E'ward speed of any point fixed on the equator at MSL is $\Omega\ R$, i.e. $7.29 \times 10^{-5} \times 6{,}370{,}000 = 464$ m s^{-1}. At latitude 30° the radius of rotation about the Earth's axis has fallen to 0.866 (= cos 30°) of the equatorial value, reducing the E'ward speed of a fixed surface point to 402 m s^{-1}. At latitude 60°, the surface E'ward speed has fallen to 232 m s^{-1}. And the latitude ϕ at which the surface E'ward speed falls to 20 m s^{-1} is given by $\cos \phi = 20/464 = 0.043$, i.e. $\phi \approx 87.5°$.

Air at rest on the equator is actually moving E'ward at 464 m s^{-1}. If it maintains this absolute speed as the airstream slides to latitude 30°, it will appear there as a W'ly wind of $464 - 402 = 62$ m s^{-1}. However, using Eqn B4.4b to ensure conservation of angular momentum (rather than E'ward tangential momentum), the W'ly wind speed at latitude 30° is given by $U = \Omega\ R\ (1/\cos \phi - \cos \phi) = 464 \times (1/0.866 - 0.866) = 134$ m s^{-1}. Adding this to the E'ward speed of Earth's surface there shows that the absolute E'ward speed of the air at latitude 30° is now $402 + 134 = 536$ m s^{-1}.

4.7.4 Zonal and meridional components

In the Hadley circulation (Figs. 4.12 and 4.13) the poleward flow in the upper troposphere is full of strong W'lies, whereas the shallow equatorward flow in the low troposphere is full of the weaker but persistent E'ly components of the trades. These are prime examples of a universal tendency for unexpected lateral motion to appear whenever air or water flows persistently over the surface of the rotating Earth. Though superficially surprising, such effects are inevitable whenever we observe the moving masses (air or water) from *reference frames* embedded in the surface of the rotating Earth, as we do for obvious convenience (Box 4.4). According to NN 4.8 the absolute E'ward speed of the Earth's surface (and embedded observational reference frames) is 464 m s^{-1} at the equator,

402 m s^{-1} at latitude 30, 232 m s^{-1} at latitude 60, and only falls below 20 m s^{-1} at latitudes greater than 87.5 °.

Consider air initially at rest relative to the Earth's surface at the equator, for example rising in the core of one of the great cumulonimbus of the ITCZ. If it then moves poleward to latitude 30° in the high-tropospheric branch of the Hadley cell, without frictional or other dynamic connection with the underlying surface, we might suppose that it would maintain its original E'ward speed and be observed there as a W'ly wind of 62 m s^{-1} (464–402). In fact in these conditions air constrained to orbit the rotating Earth (rather than fly off into space) will maintain its *angular momentum* (absolute E'ward speed × distance from the Earth's spin axis—Box 4.4) rather than its linear momentum, with the result that the W'ly wind observed at latitude 30° is 134 m s^{-1}. This means that its absolute E'ward speed is even faster than it was at the equator (NN 4.8), which seems counter-intuitive, given our assumption of no dynamic connection. The situation is examined further in Section 7.16 (and dynamics texts *[24]*) where it appears that forces acting directly towards the axis of spin produce the speed increase, and these are exactly the forces which maintain the poleward motion of the air. The situation is directly analogous to the spinning figure skater, who increases her rate of spin by pulling in her extended arms. Leaving further consideration to Section 7.16, consider the actual consequences for air moving poleward in the Hadley circulation.

In fact air seldom moves so far poleward on such a grand scale. The forces driving the air poleward in the Hadley circulation are insignificant beyond latitude 30°, for complex reasons associated with the location of convection in equatorial regions and subsidence in the subtropics, which are often much less than 30° of latitude apart. And even though air in the upper troposphere is at least 10 km above the surface, it is restrained by types of friction associated with undulating air flow high above mountains, and scattered cumulonimbus injecting slow-moving air from the low troposphere. As a result, though the tendency to conserve angular momentum maintains the strong W'lies in the upper troposphere at the polar limits of the Hadley circulation, observed speeds never reach values associated with perfect conservation.

The tendency to conserve angular momentum also applies to air moving equatorward from rest at higher latitudes, which must tend to encourage weaker E'ward flows at lower latitudes, just as the spinning skater slows by extending her arms again. Air spreading from the bases of the subtropical high-pressure zones usually has little zonal motion relative to the local Earth's surface, having lost its earlier W'ly momentum relative to the local surface during slow subsidence through the high-pressure system; it is therefore nearly enough at rest relative to the Earth's surface at about latitude 30°. If it now conserves angular momentum as it flows equatorward it will develop E'ly winds as it goes. However, unlike the much freer air flow in the high troposphere, the flow in the very low troposphere is kept in check by intense turbulent friction with the underlying surface, so that meridional and zonal speeds remain comparable. The modest E'ly components of the NE and SE Trades are the result.

4.8 Instantaneous and average patterns

The traditional distinction between weather and climate is between short-term or current atmospheric conditions (weather), and average or typical conditions (climate). Though intuitively reasonable, making the distinction raises difficulties in practice.

(i) The old assumption that climate is a steady background for rapidly varying weather has been deeply eroded by mounting evidence that climate is itself changeable on every time scale we can observe (Section 14.5). There are in fact no clear gaps in the activity spectrum linking the shorter periods we associate with weather, the longer periods we associate with climate, and the even longer periods we associate with climate change: instead we observe a continuous spectrum of variations from the briefest turbulent flutter to climatic epochs lasting hundreds of millions of years. We defer further discussion of climate and climate change to Chapter 14.

(ii) The assumption of a steady climatic background could still be useful if the disturbances were small and symmetrical about the mean, which is often the case with the small, short-lived turbulent fluctuations endemic in the atmospheric boundary layer. But it is of little help when considering relationships between means and deviations associated with convection, depressions, and even larger scale events. In such cases, mean conditions are littered with the footprints of many important types of disturbance.

To enlarge on point (ii) in particular, and highlight some important aspects of tropospheric disturbances, we re-examine several features of the mean conditions outlined in the previous three sections.

4.8.1 Mid-latitudes

The most striking feature of the temperature distribution in Fig. 4.10 is the strongly baroclinic zone in middle latitudes. On any instantaneous meridional section, however, this baroclinity is much less smoothly distributed than it is on a seasonal, zonal mean. At any instant, baroclinity is usually concentrated in one or two narrow *fronts* separating broad, nearly barotropic *air masses* (i.e. with relatively small internal horizontal temperature gradients). In fact a well-marked front is always associated with an *extratropical cyclone*, of which it forms a vital part. The behaviour of extratropical cyclones and their fronts is complex, though some aspects have become familiar to the general public through weather maps on television and newspapers. A fuller description is given in Section 12, where it is shown that a sharply defined front can persist for several days in the form of an intensely baroclinic zone $\sim$2000 km long and 200 km broad, having almost any orientation, and moving at speeds $\sim$10 m s^{-1}, usually with an E'ward component, and often curving or even coiling round the cyclonic vortex.

The transience, localization, motion, and distortion of fronts spreads and greatly weakens their intense baroclinity on zonal and seasonal averages, so that the broad baroclinic zone of middle and high latitudes (Fig. 4.10) is rarely observed in instantaneous synoptic observations. The average picture is the heavily smoothed residue of a litter of intense localized disturbances which erupted sporadically over the range of latitudes and longitudes in the averaging period.

Similar remarks apply to the meridional distribution of mean surface pressure apparent in Fig. 4.9. Inhabitants of middle and high latitudes are familiar with the incessant variations of surface pressure associated with the development and movement of extratropical cyclones (depressions) and anticyclones (high pressures). Any one depression appears on an instantaneous meridonal section as a sharp, deep dip in the 1,000 hPa isobaric profile—typically ~400 m deep (corresponding to a pressure deficit of about 40 hPa at MSL) and concentrated within 10° of latitude. The simultaneous development and decay of several extratropical cyclones and anticyclones around each latitude zone is further smoothed by seasonal averaging, to give the broad, shallow depression evident in Fig. 4.9.

4.8.2 Low latitudes

Here the situation is somewhat different. The high average pressures apparent in the subtropics in Fig. 4.9 are much more representative of instantaneous meridional profiles than is the case at higher latitudes, because they result from the presence of unusually stable and persistent weather systems—the very large *subtropical highs* (STHs) which effectively girdle the Earth at these latitudes.

The pattern of mean W'ly winds in Fig. 4.11 is also recognizably related to patterns on instantaneous meridional sections. The core of maximum wind speeds at about the 200 hPa level in the subtropics is associated with the frequent presence of the powerful *subtropical jet stream* (STJ), whose strong W'ly flow is maintained by the conservation of angular momentum discussed in Sections 4.7 and 7.16. Though both the intensity and latitude of this powerful flow vary with time and longitude, the variations are sufficiently small for the mean pattern to be only moderately weakened and spread compared with instantaneous local patterns.

The STJ is more intense in winter months, and moves poleward and weakens in the summer, in association with variations in the Hadley circulation (Fig. 4.13). The STJ was the last of the major features of tropospheric flow to be clearly identified, being recognized only in the early 1940s, when military aircraft flying W'ward at high altitudes near Japan reported very strong headwinds, sometimes strong enough to bring slower aircraft almost to a standstill relative to the surface. The STJ can be especially intense in that region because of strong confluence in the upper troposphere of air flows divided upstream by the Himalayan and Tibetan massifs.

4.8.3 Mid-latitudes again

The poleward extension of the W'ly wind maximum into middle latitudes in the upper troposphere arises from the presence of the *polar front jet stream* (PFJ). Like the fronts with which it is intimately connected by the thermal wind relation (Section 7.12), the PFJ is disjointed and sporadic, and so narrow and highly mobile that the mean section gives no useful impression of its appearance on instantaneous local sections. In these the PFJ appears as a relatively narrow stream of fast-moving air (almost always with a W'ly component, though orientation varies widely) centred on the 300 hPa level and closely paralleling its associated fronts (Fig. 12.10). We will see (Section 7.12) that the W'ly wind maximum in mid-latitudes is dynamically related to the baroclinic zones in those regions, and that the relationship applies both to average and instantaneous sections.

Note that the STJ and PFJ appear in the height zones used by modern civil aircraft. To minimize fuel wastage in flying against such strong air flows, flights are individually routed to avoid forecast headwinds either by lateral or vertical displacement. Aircraft flying with a jet stream can take advantage of the following wind, though particularly strong vertical wind shears below and above the jet cores are avoided because they can contain uncomfortable or even dangerous turbulence (Section 11.8). Only supersonic civil aircraft, and some military aircraft, routinely fly well above the influence of jet streams.

Checklist of key ideas

You should now be familiar with the following ideas.

1. Absolute temperature, pressure, and density of air, the meteorological form of the equation of state.

2. Equation of state for single and mixed ideal gases and air, partial densities, partial pressures, specific masses, molecular (volumetric) proportions.

3. Vertical profile of temperature to 100 km above sea level, temperature lapses and inversions and their physical significance in troposphere, stratosphere, mesosphere, and above.

4. Maintenance of smooth, nearly exponential vertical profile of pressure, concept of hydrostatic equilibrium.

5. Theory and application of hydrostatic equilibrium to air pressure (exponential, binary, and decadal scale heights and thickness of isothermal layers) and density.

6. Vertical distribution average air pressure and temperature and its variation with latitude and season, barotropy and baroclinity.

7. General circulation of the atmosphere in low, middle, and high latitudes, zonal and meridional flow, circumpolar vortex, jet streams, the Hadley circulation and associated climatic patterns.

8. Concepts of angular velocity and angular momentum, conservation of angular momentum, and the subtropical jet stream.

9. Relation between instantaneous and seasonally averaged distributions of pressure, baroclinity, and weather systems.

Problems

Outline answers to these problems can be found on the **Online Resource Centre**. Answers to odd numbered problems, can be found under Student Resources, answers to even numbered problems under Lecturer Resources.

Level 1

4.1 If the pressure of a rising air parcel halves while its temperature remains constant, what happens to its density? Since parcel temperature always falls during ascent, will your first answer be an over- or underestimate?

4.2 A certain very large, high-altitude research balloon measures 100 m from the top of its envelope to the instrument package. Using Fig. 4.2 as typical, describe how the temperatures of the package and top will compare as it ascends from the surface to the stratopause.

4.3 State the conditions for perfect exponential decay of atmospheric pressure with increasing height.

4.4 Deep anticyclones have a characteristically warm core throughout all but the highest parts of the troposphere. By considering the distribution of layer thickness, outline the expected shape of the 1,000, 500, and 300 hPa surfaces on a vertical section through the middle of such a system.

4.5 Sketch isotherms on a vertical section through two adjacent air masses with uniform surface pressure. One air mass is warm and barotropic, and the other becomes progressively colder away from the vertical boundary between the masses. Sketch the set-up using two isotherms and two isobars (the given and another) and labelling warm and cool at the surface and high and low pressure on an upper horizontal.

4.6 Using Figs. 4.10 and 4.11, etc. as evidence, brief a party of inexperienced mountaineers about likely conditions on the summit of Everest.

4.7 In mid-latitudes especially, mean conditions should not be considered to be the steady background which is disturbed by occasional weather systems. Justify this statement from a typical barograph trace.

Level 2

4.8 Use the equation of state for dry air to calculate the air density at the summit of Everest on an occasion when pressure and temperature there were 313 hPa and −38.2 °C respectively. If the temperature then falls by 10 °C isobarically (possible though unlikely), estimate the % change in air density without repeating the calculation.

4.9 In a certain isothermal atmosphere with exponential scale height 7 km, find the height intervals between the following standard radiosonde pressure levels: 1,000, 700, 500, 300 hPa.

4.10 Use data from Fig. 4.9 to estimate the mean temperature of the 1,000–500 hPa layer at latitudes 30° and 70° in winter, assuming an isothermal atmosphere in each case. Use the expression for binary scale height.

4.11 Inspect Figs 4.11 and 4.13 and estimate the time taken for air to travel (a) around the Earth at constant latitude in the vicinity of the core of the subtropical jet stream, and (b) between latitudes 30° and 10° in the winter Trades.

4.12 Estimate the different travel times for aircraft shuttling with an air speed of 250 m s^{-1} between airports A and B which are 1,000 km apart, when there is a jet stream flowing from A to B at 60 m s^{-1} at cruising altitude. Ignore ascents and descents to and from cruising level.

4.13 Express a typical polar front jet stream as a fraction of the tangential speed of a fixed point on Earth's surface at latitude 50°. Find the tangential speed and centripetal acceleration of a similar point at the equator and express the latter as a fraction of g. Treat Earth as a sphere.

4.14 Using the method of Box 4.1, show that the specific gas constant for an atmosphere of carbon dioxide is 189 J kg^{-1} K^{-1}, and hence find the density of Martian surface 'air', assuming pressure 10 hPa and temperature 220 K, and compare with a typical Earth value.

Level 3

4.15 Given that the specific heat capacity of CO_2 at constant pressure is about 900 J kg^{-1} K^{-1}, and that near-surface g is 3.8 m s^{-2}, find the 'dry' adiabatic lapse rate of the Martian atmosphere. Using this and other Mars data from Problem 4.14, estimate the decadal scale height of the Martian atmosphere in two stages, first using surface temperature, and then using your first estimate of temperature in the middle layer.

4.16 Consider possible changes in low latitudes in Fig. 4.10 if deep convection there were to be considerably reduced. And similarly consider how the appearance of Fig. 4.11 might differ in low latitudes if the Earth did not rotate. In fact convection is likely to increase in some part of low latitudes in the absence of rotation. Where and why?

4.17 Supposing that the trade winds start from rest at latitude 25° and move equatorwards conserving their original angular momentum, what W'ward wind speed will they have at 5°? Repeat the calculation for air rising from rest at latitude 40° to a polar front jet core at 60°. Compare with typical observations and discuss.

Atmospheric thermodynamics

5

5.1 Introduction

Heat enters and leaves the atmosphere in many different ways, warming, cooling, or otherwise changing the state of the air. For example, air often warms by absorbing solar radiation, and cools by emitting more infrared terrestrial radiation than it absorbs (Section 8.6). Sometimes there is cooling or warming without obvious output or input of heat, as when rising air expands and chills, or sinking air is compressed and warms. On these occasions heat is converted into mechanical energy as air expands against its surroundings; or mechanical energy is converted into heat as air is compressed by its surroundings. Expansional cooling of rising air can be significantly reduced by cloud formation, as water vapour condenses on swelling cloud droplets and releases the heat latent as vapour since evaporation.

In describing these processes quantitatively, we use concepts, terms, and techniques developed in the science of *thermodynamics*, a subject which has grown in the last two centuries from ad hoc speculations about the nature of temperature and heat to a cornerstone of physical science *[***20 and 25***]*. In this chapter we apply basic thermodynamics to atmospheric behaviour in just sufficient detail to clarify the physical processes at work and establish useful relationships. The nature of heat as a form of energy, and a fuller outline of the roles of energy in the atmosphere, are explored in Chapter 9.

5.2 The first law of thermodynamics

5.2.1 Basic form

To describe the thermal behaviour of a typical air parcel, we need to allow for its thermal response to the injection or removal of heat, and to expansion or compression. This is done by using a form of the *first law of thermodynamics*, relating a very small heat input to the resulting warming and expansion of unit mass of air (Eqn 5.1). The process is assumed to be *reversible*—staying so near equilibrium as it proceeds in some direction as to be put into reverse by a minute change of circumstances. Though never strictly reversible, many weather-related processes are sufficiently so for the ideal to be a useful approximation—for example adiabatic warming of sinking air. Other atmospheric processes are not even approximately reversible—such as the violent heating and expansion of air in and around a lightning flash.

The first law of thermodynamics is usually written in the form

$$dQ = C_v \, dT + p \, dVol \qquad 5.1$$

where dQ is a very small heat input (from contact with Sun-warmed land for example), dT is the associated very small rise in parcel air temperature, and $dVol$ is the very small increase in the parcel volume Vol (Fig. 5.1). Negative values describe heat outputs, temperature falls and volume decreases. Much larger (finite) changes can be described by accumulating (integrating) very many consecutive very small (infinitesimal) changes.

The first term on the right-hand side of Eqn 5.1 represents the amount of the heat input which is used to warm the air parcel, i.e. to increase its *internal energy*—the kinetic (and rotational and vibrational) energies of its randomly moving air molecules, of which absolute temperature T is the parcel scale measure (Box 5.1). This term is directly proportional to the temperature rise dT, the constant of proportionality C_v being the *specific heat capacity* of air *at constant volume*, whose name is justified by setting $dVol$ to zero in Eqn 5.1. Laboratory measurements show that C_v for dry air is 717 J kg^{-1} K^{-1} across the range of conditions found in the turbosphere. With this value the first two terms of Eqn 5.1 imply that it takes 717 joules of heat energy to warm 1 kg of any tropospheric air by 1 Kelvin or Celsius degree, provided the air volume is kept constant.

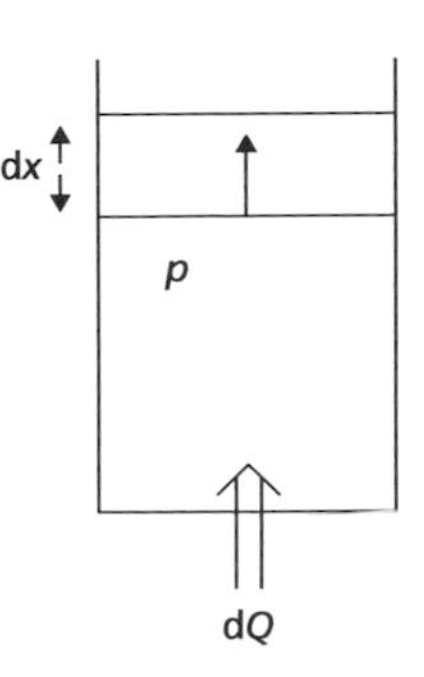

Figure 5.1 A little heat dQ enters air trapped at constant pressure p in a cylinder, warms the air and pushes the light frictionless piston outwards through a little distance dx, doing work $p\,A\,dx$ ($= p\,dVol$), where A is the area of the piston face and $dVol$ is the increase in volume of the trapped air.

The second term on the right-hand side of Eqn 5.1 comes into play when the parcel volume does NOT remain constant, and it represents the amount of heat input which is used as the parcel does mechanical *work* by expanding against the surrounding air. As shown in Fig. 5.1 this work is given by the product of the parcel air pressure and the increase in parcel volume. If the parcel contracts rather than expands, the surroundings do work on the parcel, raising its temperature and in effect adding to the input of heat dQ, rather than offsetting it. It has been known since Newton's time that work done quantifies a change in *mechanical energy* (i.e. *kinetic* and/or *potential energy*). In the present case it quantifies a transformation from heat energy to whatever mechanical energy is produced by the expansion of the air parcel—for example gravitational potential energy when warmed air expands and slightly lifts itself and the overlying atmosphere against the downward pull of gravity (Section 5.3 and Section 9.6). In the SI system the unit of work (and hence energy) is the *newton metre* (N m), which is called the *joule* (J).

Equation 5.1 shows that the input of heat energy to the air parcel is shared between increasing its internal energy and doing work on its surroundings. The overall balance of the equation is an example of the *principle of conservation of energy*, which is believed to hold in all circumstances, and whose consequences for the atmosphere are considered more fully in Chapter 9. Any particular example of the principle, such as Eqn 5.1, is of course valid only in so far as it corresponds to the actual process being described. Fortunately the very simple scheme of Eqn 5.1 accounts very well for several important atmospheric processes.

5.2.2 Meteorological form

The involvement of $dVol$ in Eqn 5.1 is not well suited to describing a largely unconfined gas like air, whose parcels are more often concepts than observable realities. Replacing Vol by the reciprocal of air density ($Vol = 1/\rho$ for unit mass of air) and using the equation of state (Eqn 4.1) for air as an ideal gas, we can find (Box 5.1) the meteorological form of Eqn 5.1,

$$dQ = C_p\,dT - \frac{dp}{\rho} \qquad 5.2$$

which relates the heat input dQ to the rise of temperature at constant pressure and the increase of air parcel pressure dp. According to Box 5.1, the *specific heat capacity* of dry air *at constant pressure* (C_p) is larger than C_v by R, the specific gas constant for dry air.

$$C_p = C_v + R$$

From the values for C_v and R (Table 4.1, etc.), and from direct laboratory measurements, we find that C_p stays very close to 1,004 J kg^{-1} K^{-1} throughout the turbosphere, which we can often round to 1,000 J kg^{-1} K^{-1}.

There is no mystery in C_p being 40% larger than C_v. For the same heat input dQ, the temperature rise at constant volume (Eqn 5.1 with zero dVol) will be 40% larger than the temperature rise at constant pressure (Eqn 5.2 with zero dp) because, at constant volume, all the heat input warms the air, whereas at constant pressure, 40% of the heat input is used in expanding the air against its surroundings (NN 5.1). As we see in the following section and elsewhere, C_p applies to a very wide range of atmospheric circumstances.

Note that values of C_v, R, and C_p differ slightly with the addition of water vapour to dry air. However, the saturation limit (Fig. 5.5) in the atmosphere is low enough to ensure that dry air values can be used for most meteorological purposes, regardless of humidity.

NUMERICAL NOTE 5.1 Warming air at constant volume and at constant pressure

Suppose 1,000 J of heat energy is added to a 1 kg parcel of dry air. If this happens at constant pressure we can use $dT = dQ/C_p$ (Eqn 5.2 with $dp = 0$) to find $dT = 1{,}000/1{,}004 \approx 1.0$°C (or K). If the warming happens at constant volume, we use $dT = dQ/C_v$ (Eqn 5.1 with $dVol = 0$) to find $dT = 1{,}000/717 \approx 1.4$°C. Since 1.4°C warming (at constant volume) uses all of the 1,000 J input, 1.0°C warming (at constant pressure) uses only 1,000 × 1/1.4 = 714 J, so that the remaining 286 J of the heat input is used to expand the air rather than raise its temperature, numerically confirming $C_p = C_v + R$ to within small rounding errors.

BOX 5.1 Deriving the meteorological form of the first law of thermodynamics

Since Eqn 5.1 is written for unit mass of air, parcel volume $Vol = 1/\rho$, and we have

$$dQ = C_v dT + d\left(\frac{1}{\rho}\right) \qquad \text{B5.1a}$$

We can use standard differentiation *[8]* and the equation of state for air ($p = \rho R T$) to rewrite

$$p\,d\left(\frac{1}{\rho}\right) = d\left(\frac{p}{\rho}\right) - \frac{dp}{\rho} = d(RT) - \frac{dp}{\rho} = R dT - \frac{dp}{\rho}$$

and use this to replace the last term of Eqn B5.1a to get

$$dQ = C_v dT + R dT - \frac{dp}{\rho} \qquad \text{B5.1b}$$

Although terms 2 and 3 on the right-hand side of Eqn B5.1b together represent the expansion term (p dVol) of Eqn 5.1, terms 1 and 2 on the right-hand group naturally to form

$$dQ = (C_v + R)\, dT - \frac{dp}{\rho}$$

Since only ($C_v + R$) dT on the right-hand side remains when pressure is constant ($dp = 0$), we see that that ($C_v + R$) is C_p, the specific heat capacity of air at constant pressure, so that

$$dQ = C_p\, dT - \frac{dp}{\rho} \qquad \text{B5.1c}$$

When dealing with an air parcel of mass m rather than 1, it is easy to confirm that

$$dQ = mC_p dT - m\frac{dp}{\rho} \qquad \text{B5.1d}$$

5.3 Isobaric heating and cooling

Atmospheric heating and cooling often occurs while air pressure is constant or nearly so. For example, the daytime warming of the atmospheric boundary layer, and its cooling by night, occur at pressures determined hydrostatically by the weight of the overlying atmosphere, which often changes very little during the few hours of warming or cooling. The equation describing such *isobaric* heating or cooling of unit mass of air is simply Eqn 5.2 with zero dp, i.e.

$$dQ = C_p \, dT$$

and the more general expression for mass m follows similarly from Eqn B5.3d

$$dQ = m \, C_p \, dT \qquad 5.3$$

5.3.1 Boundary layer warming

On sunny mornings overland, screen temperatures typically rise at a couple of degrees per hour for several hours as the Sun-warmed surface injects heat into the overlying air at nearly constant pressure. Making measurements above the surface layer is difficult, but has been done in special studies.

In the Australian study shown in Fig. 5.2, the shallow surface-based convecting layer deepened from about 400 m to nearly 750 m in 75 minutes, warming from about 23.2 °C to about 25.0 °C in the process. Deepening was limited because the warming layer was trapped under a strong temperature inversion maintained by anticyclonic subsidence of overlying air (Section 12.5). This enhanced the degree of warming since the temperature rise of the layer is inversely proportional to the mass (depth) of air being heated (Eqn B5.3).

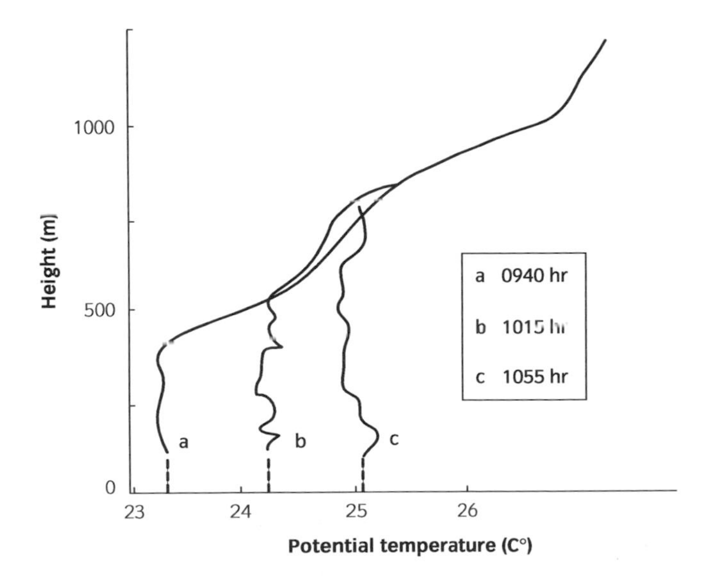

Figure 5.2 Morning warming and deepening of a convecting layer over extensive grassland in Australia. Potential temperature is used to reduce the well-mixed unsaturated temperature-height profile to a vertical (Sections 5.6 & 5.11), which shows as a mean through the wriggly lines up to the rising base of the stable layer. The wriggles are an artifact of measuring temperature by horizontal traverses by aircraft, separated by longish time intervals, rather than using shoals of radiosondes. The overlying stable layer is maintained by anticyclonic subsidence (Section 13.2).

Temperature profiles are speculative below 125 m in Fig. 5.2 because the measuring aircraft could not safely fly lower. However, the dashed interpolations are realistic since buoyant thermals erupting from just above the heated surface keep the convecting boundary layer well mixed and maintain a nearly isothermal layer which warms en bloc as time passes, as confirmed by other studies using tall instrumented towers.

We can simplify the situation depicted in Fig. 5.2 by saying that a 500 m deep layer of air was warmed by 1.8 °C in 75 minutes. Application of Eqn 5.3 shows that such warming at constant pressure requires an input of 1.08 MJ (i.e. $m\ C_p\ \Delta T$) of heat energy from each horizontal square metre of underlying Sun-warmed ground surface, which corresponds to an average rate of 240 W m^{-2} over the 75 minute warming period (NN 5.2). Of this 1.08 MJ m^{-2} total , 0.77 MJ m^{-2} (i.e. $m\ C_v\ \Delta T$) was used to warm the layer as observed by thermometers, while the remaining 0.31 MJ m^{-2} (i.e. $m\ R\ \Delta T$) was used as the expanding layer lifted the weight of the overlying atmosphere. A simple calculation (NN 5.2) shows that the warming layer must have thickened by just over 3 metres to do this amount of work against gravity, which is far too small to be observed directly, being swamped by the much larger deepening by convective mixing apparent in Fig. 5.2.

Note that in such calculations we do not have to consider the complex turbulent realities of the mixing of heat through the warming layer, because energy is conserved for the layer and for each and every constituent parcel.

NUMERICAL NOTE 5.2 Warming boundary layer

Apply Eqn 5.3 to a 500 m deep vertical column of air resting on 1 m^2 of horizontal ground surface, to relate a small heat input ΔQ to a small rise ΔT in column air temperature. ΔT is used instead of dT as a symbolic reminder that the change is much more than infinitesimal. Assuming air density 1.2 kg m^{-3}, the 500 m column (volume 500 m^3 per horizontal square metre) contains 600 kg of air, which is the value for m. Using 1,004 J kg^{-1} K^{-1} for C_p and 1.8 °C for ΔT, $dQ = 600 \times 1{,}004 \times 1.8 \approx 1.08$ MJ m^{-2}. Averaged over the 75-minute warming period, the rate of heat injection is $1{,}080{,}000/(75 \times 60) = 240$ W m^{-2}. Of this total input, $m\ C_v\ \Delta T = 600 \times 717 \times 1.8 \approx 0.77$ MJ m^{-2} is used in warming the air, while the rest ($m\ R\ \Delta T = 600 \times 287 \times 1.8 \approx 0.31$ MJ m^{-2}) is used in doing work by vertical expansion against gravity.

Assuming that the mass M of atmosphere overlying the warming layer was 10,000 kg m^{-2}, this could be lifted a distance Δh by the thickening warming layer such that $M\ g\ \Delta h = 0.31$ MJ m^{-2}, which gives $\Delta h = 0.31 \times 10^6/(10{,}000 \times 9.8) \approx 3.2$ m. In fact about 3% of the 0.31 MJ is used to raise the centre of gravity of the thickening layer itself (Section 9.6.3).

Solar energy arriving at 1,100 W m^{-2} from a solar elevation of 40°, means $1{,}100 \times \sin 40° = 707$ W m^{-2} falls on a horizontal surface (Box 8.2). A dark grass surface might absorb 45% of this ($\approx$ 318 W m^{-2}) and reflect the rest, so that 320 W m^{-2} is a reasonable estimate for the input of solar energy to the surface (Section 10.3) in Fig. 5.2, suggesting that on this occasion about 80 W m^{-2} (= 320 − 240) of the solar input was not being used to warm and inflate the air.

A rough estimate (NN 5.2) suggests that the rate of input of solar heat into the land surface on the occasion of Fig. 5.2 was about 320 W m^{-2}—considerably larger than the value just estimated from the observed warming of the air. This apparent discrepancy is not surprising, since on this occasion significant amounts of solar heat must have been used to evaporate dew and soil moisture from the grassy land surface (NN 5.2), and smaller amounts to warm the ground itself (Section 10.5). Such vapour is carried aloft by rising air parcels, which are systematically moister than sinking ones, driving a substantial upward flux of *latent heat* in addition to the more obvious flux of *sensible heat* measured (sensed) by thermometers as in Fig. 5.2. The latent heat flux invisibly connects the heat used to drive surface evaporation with the same amount of heat released during cloudy condensation or dew formation (Fig. 5.3), either of which can occur well away from the evaporating source. Cloudy condensation is much greater than dew deposition (Fig. 5.4) in all but the most arid zones, where the latent heat flux is very small anyway.

Since vapour, evaporation, and condensation play very important roles in atmospheric thermodynamics, we need to re-examine how water vapour is observed and quantified.

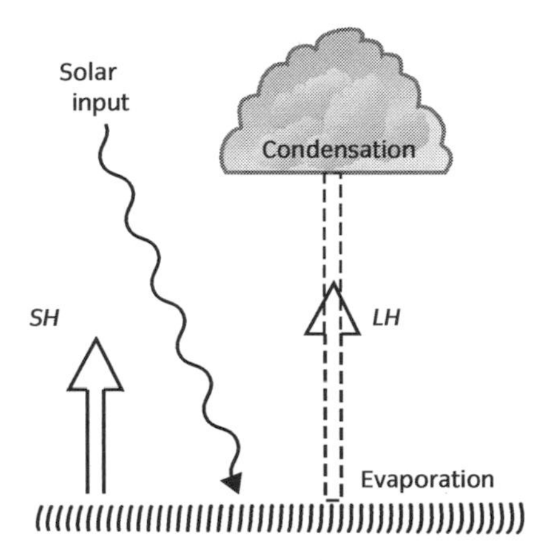

Figure 5.3 Schematic diagram showing the flux of latent heat (latent in the vapour flux) from surface evaporation to lifting condensation level. The solar heat used to evaporate the water is unavailable for surface warming and the flux of sensible heat it drives.

Figure 5.4 Dew on a spider's web on a humid morning. The air is slightly supersaturated and the very tight curvatures of the web make it act like a mesh of condensation nuclei. It is even possible that the spider's secretions encourage condensation by a Raoult effect (Section 6.7). (See Plate 7).

5.4 Measures of water vapour

5.4.1 Vapour pressure

Absolute vapour content is measured by *vapour pressure e* or *vapour density* ρ_v. Since water vapour behaves as an ideal gas, its equation of state (Eqn 4.2) relates vapour pressure e to vapour density and absolute temperature T through

$$e = \rho_v R_v T \qquad 5.4$$

where R_v is the specific gas constant for water vapour, value 462 J kg^{-1} K^{-1} according to Table 4.1 and the text. Note that e is the partial pressure of water vapour (Box 4.1), and ρ_v is its partial density. Values of e and ρ_v are ~10 hPa and 10^{-2} kg m^{-3} respectively in the low troposphere, falling by 2–3 orders of magnitude to the tropopause, essentially because the saturation limit falls sharply with temperature (Fig. 6.2).

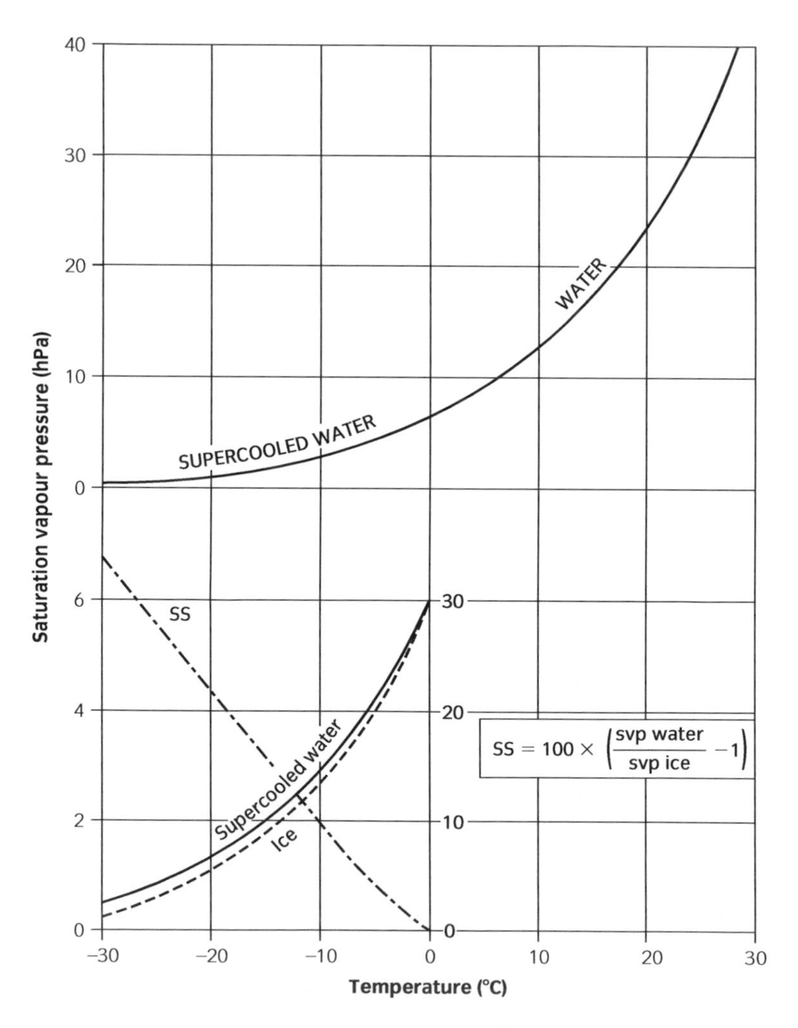

Figure 5.5 Variation of saturation vapour pressure with temperature. The curve is a nearly pure exponential, but below 0°C saturation values over supercooled water exceed those over ice, as shown by the separation of the full and dashed lines. The chain line in the lower part of the diagram shows the supersaturation (SS in %) relative to ice for a vapour which is just saturated relative to supercooled water. You can find coresponding vapour densities by using the equation of state for vapour (Section 4.1).

5.4.2 Specific humidity

To specify the relative rather than the absolute abundance of water vapour, we can use the *specific humidity q*—the proportion by mass of vapour in a parcel of moist air (the specific mass of vapour in air)—defined by

$$q = \frac{m_v}{m} = \frac{\rho_v}{\rho} \qquad 5.5$$

where $m = m_v + m_d$ is the mass of vapour and dry air in the parcel, and the final form of Eqn 5.5 follows by using mass = density × volume and cancelling the common parcel volume.

Replacing each density by the equations of state for water vapour (e.g. Eqn 5.4) and moist air, and cancelling the common absolute temperature T, we find

$$q = \frac{e\,R}{p\,R_v}$$

where e and R_v are as defined for Eqn 5.4, p is the total air pressure, and R is the specific gas constant for moist air, usually assumed to be the same as the gas constant for dry air, although the ~ 1% error can matter occasionally. According to Eqn B4.1b, the ratio of the gas constants R/R_v is simply the inverse ratio of their effective molecular weights (i.e. 18/29 = 0.622), often represented by ε.

$$q \simeq \frac{\varepsilon\, e}{p} \qquad 5.6$$

Typical q values in the low troposphere are $\sim 10^{-2}$, usually written 10 g kg^{-1}; and values in the high troposphere are one or two orders of magnitude smaller, again because of the saturation limit. According to Table 4.1, the q value for the whole atmosphere is 3 g kg^{-1}.

Note that the proportion of vapour to the mass of *dry* air (the *humidity mixing ratio r*) is used instead of q for some purposes.

$$r = \frac{m_v}{m_d} = \frac{q}{(1-q)}$$

Since the possible ~1% difference between them seldom matters, we use q in this book.

5.4.3 Relative humidity

Laboratory measurement shows that the vapour pressure and density of a *saturated vapour* (a vapour in thermal equilibrium with a plane surface of pure water—Section 6.2) depend only on the vapour temperature, which is also the air temperature, since all its components are in thermal equilibrium. Figure 5.5 shows that saturation values fall sharply with falling temperature, in nearly pure exponential decay. Since any particular temperature has a definite *saturation vapour pressure* (*svp*)—for example 17 hPa at 15 °C—we can always express the actual vapour content of a sub-saturated vapour as a percentage of the saturation value at its temperature. This is the *relative humidity* (*RH*) of the air (strictly of the vapour, but the usage is universal)

$$RH = 100\,\frac{e}{e_s} = 100\,\frac{\rho_v}{\rho_{vs}} \qquad 5.7$$

where saturation values are subscripted s. The last version in Eqn 5.7 follows by using the equation of state for vapour and cancelling the common temperature T.

Values of RH range from 100% (saturation) to about 10% in the troposphere, and exceed 50% near all but the driest surfaces. In growing clouds they may exceed 100%, but only by such small margins that saturation is still the effective upper limit (Section 6.8). At temperatures well below 0 °C, we must distinguish saturation with respect to ice from saturation with respect to supercooled water, because the two have substantially different values and supercooled water abounds in clouds (Fig. 5.5 and Section 6.2).

5.4.4 Dew-point

The *dew-point temperature* of an air parcel is the temperature at which its air just reaches saturation when cooled isobarically without addition or removal of water vapour. Since the proportion of vapour in air (its specific humidity q) is unaltered by cooling, at least until condensation begins, and since the total air pressure is fixed at the ambient value, vapour pressure is unaltered by isobaric chilling. The *saturation vapour pressure* (SVP) at the dew-point T_d is therefore the same as the initial vapour pressure at temperature T, which is written

$$e(T) = e_s(T_d) \qquad 5.8$$

Measurement of T_d, together with tables of SVP versus temperature, finds the vapour pressure of the sample air, and hence all other measures of humidity. For example, the relative humidity RH is given by

$$RH = 100\,\frac{e_s(T_d)}{e_s(T)} \qquad 5.9$$

When it is not zero, the *dew-point depression* ($T - T_d$) is roughly twice as large as the *wet-bulb depression* ($T - T_w$), because saturation at T_d is achieved purely by chilling, whereas saturation at T_w is reached sooner by simultaneous chilling and moistening. Because the thermodynamics of the process are complex, no simple relation exists between numerical values of T, T_w and T_d, and dedicated thermodynamic tables *[9]*, calculators, or diagrams (Section 5.8) have to be used.

5.4.5 Thermodynamic wet-bulb temperature

The *thermodynamic wet-bulb temperature* of a certain air parcel is defined to be the lowest temperature to which it can be cooled by evaporating water into it, at constant pressure and in thermal isolation from its surroundings (i.e. isobarically and adiabatically). As water evaporates, the parcel is chilled and its vapour content increased until saturation is reached at the thermodynamic wet-bulb temperature T_w. A simple *psychrometric equation* can be derived (Box 5.2) relating the initial specific humidity of the air $q(T)$ to the depression ($T - T_w$) of the thermodynamic wet-bulb temperature below the initial dry-bulb temperature T, and the saturation specific humidity at the wet-bulb temperature $q_s(T_w)$.

$$q(T) = q_s(T_w) - \frac{C_p}{L}(T - T_w) \qquad 5.10$$

where $q(T)$ and $q_s(T_w)$ are used in pure ratios (rather than g kg^{-1}), $q(T)$ is the required specific humidity (at temperature T), T and T_w are observed, and $q_s(T_w)$ can be read from tables for saturation at T_w. The calculation is built into special *hygrometric tables*, which give $q(T)$ directly for any observed T and $(T - T_w)$.

BOX 5.2 The psychrometric equation

Consider an air parcel of mass m being cooled from its initial temperature T to its thermodynamic wet-bulb temperature T_w by evaporating a mass m_v of water into it. Since there is no other heat source or sink, the latent heat required for evaporation ($L\,m_v$) must come from cooling the air parcel, and since the cooling is isobaric we can use Eqn 5.3 to find

$$L\,m_v = m\,C_p\,(T - T_w)$$

which rearranges to

$$\frac{m_v}{m} = \frac{C_p}{L}(T - T_w) \qquad \text{B5.2a}$$

where L is the *specific latent heat of vaporization* of water—the quantity of heat needed to evaporate unit mass of water, always close to 2.5 MJ kg^{-1} in terrestrial conditions—and we ignore the heat capacity of the imaginary water reservoir, because its mass is so much smaller than the mass of air in realistic conditions. For the same reason we can treat m as constant throughout the process (ignoring the addition of m_v). Then, because the denominator m is constant, m_v/m can be regarded as the increase in the specific humidity of the air as it goes from its initial $q(T)$ to its final $q_s(T_w)$:

$$m_v/m = q_s(T_w) - q(T)$$

Incorporate this in Eqn B5.2a and rearrange to find the *psychrometric equation*:

$$q(T) = q_s(T_w) - \frac{C_p}{L}(T - T_w) \qquad \text{B5.2b}$$

Note that the relatively small heat capacity of the vapour has been ignored throughout. Replacing the q terms in Eqn B5.2b by Eqn 5.6 or 5.7 allows either the vapour density or vapour pressure of the air to be calculated, using tabulated or graphed (e.g. Fig. 6.2) values of $e_s(T_w)$ or $\rho_{sv}(T_w)$.

NUMERICAL NOTE 5.3 Humidity calculations

On a certain occasion, screen level temperature is 20 °C, thermodynamic wet-bulb temperature T_w is 15 °C, and air pressure is 1,013 hPa. Given that saturation vapour pressures at 20, 15, and 11.6 °C are 23.3, 17.0, and 13.7 hPa, we can find all possible humidity measures as follows.

By the psychrometric equation (5.10) $q(20) = q_s(15) - [10^3/2.5 \times 10^6](20 - 15)$. To find $q_s(15)$ for this equation, use Eqn 5.6 to find $q_s(15) = 0.622\, e_s(15)/p = 0.622 \times (17.0/1{,}013) = 0.0104$ (i.e. 10.4 g kg^{-1}). Inserting this in the psychrometric equation $q(20) = 0.0104 - 0.0020 = 0.0084$, i.e. 8.4 g kg^{-1} is the specific humidity of the air. Use Eqn 5.6 again to find the corresponding vapour pressure $e = qp/0.622 = 0.0084 \times (1{,}013/0.622) = 13.7$ hPa. Since this is the *SVP* at 11.6 °C, this temperature must be the dew-point T_d. Finally we find the relative humidity from Eqn 5.9 $RH = 100 \times 13.7/23.3 \approx 59\%$. We have found $q = 8.4$ g kg^{-1}, $T_d = 11.6$ °C, and $RH \approx 59\%$. Note that T_w lies roughly half way between T and T_d.

5.4.6 Precipitable water content

The total amount of vapour in a vertical column extending through all or part of the atmosphere can be expressed as the depth of rainfall which would result if it fell out as rain—the *precipitable water content*, normally expressed in millimetres. A column of air with pressure 'thickness' Δp has total mass $m = \Delta p/g$ per unit horizontal area (Section 4.4). If the air has specific humidity q (expressed as a pure ratio rather than in g kg^{-1}) then the column contains mass m_v of water vapour per unit horizontal area (kg m^{-2} in SI), where

$$m_v = q\,m = q\,\frac{\Delta p}{g} \qquad 5.11$$

Since 1 mm of rainfall lying on 1 square metre contains 1 kg of rainwater, the value of m_v in kg m^{-2} is numerically equal to the depth of precipitable water in millimetres. The precipitable water content of an actual air mass can be calculated layer by layer in this way from radiosonde profiles of humidity and temperature against pressure (NN 5.4).

Values of precipitable water content are ~10 mm in the bottom 100 hPa layer of air, falling rapidly at higher altitudes because of saturation limiting, so that total tropospheric values are ~30 mm, more in lower latitudes, less in high. The global atmospheric q value of 0.003 (Table 4.1) and mean surface pressure ≈1,000 hPa corresponds to a precipitable water content of nearly 31 mm (NN 5.4).

NUMERICAL NOTE 5.4 Precipitable water content

Suppose air is observed to have temperature 15 °C, *RH* 75%, and pressure 1,010 hPa. Since *SVP* at 15 °C is 17 hPa (tables), vapour pressure $e = 0.75 \times 17.0 = 12.8$ hPa, and specific humidity $q = \varepsilon\, e/p = 0.622 \times 12.8/1{,}010 = 7.9 \times 10^{-3}$ (i.e. 7.9 g kg^{-1}). If this q persisted through a layer of air of pressure thickness 100 hPa (containing $\Delta p/g = 1{,}020$ kg of air resting on 1 m^2) the mass of precipitable water in the layer would be $1{,}020 \times 7.9 \times 10^{-3} = 8.1$ kg, corresponding to a rainfall of 8.1 mm, which is the precipitable water content of the layer.

At −20 °C, *SVP* relative to ice is 1 hPa. If *RH* is 75% relative to ice, and air pressure is 550 hPa, then a corresponding calculation gives $q = 0.622 \times 0.75/550 \approx 8.5 \times 10^{-4}$ (i.e. 0.84 g kg^{-1}). Though geometrically about twice as thick, a 100 hPa 'thick' layer at this level would contain the same mass of air as the 100 hPa layer near sea level, but have a precipitable water content of only $1{,}020 \times 0.84 \times 10^{-3} = 0.86$ kg, i.e. 0.86 mm, about one tenth of the value in the low troposphere.

The total precipitable water content of a typical atmospheric column is $q\,p_s/g$, where q is the overall specific humidity as a pure ratio (0.003 in Table 4.1), and p_s is the mean surface pressure ≈1,000 hPa (less than the MSL value because of land elevation). The precipitable water content $= 0.003 \times 1{,}000 \times 100/9.8 = 30.6$ kg m^{-2} or mm rainfall depth.

Note that q values are used here as pure ratios, despite being quoted in g kg^{-1}.

5.5 Measuring water vapour

5.5.1 Relative humidity meter

Since animal hair and skin shrinks in direct response to rising *RH* (Section 2.3), hair and skin hygrometers have been used in relative humidity meters for over a century. Indeed the *hair hygrometer* is nearly universal in homes and other places where only semi-quantitative information is wanted. Higher-quality versions have been used more seriously in *hair hygrographs*, and the gold-beater's *skin hygrometer* was for long used in British radiosondes, but calibration has always been a problem with the whole range of hair and skin hygrometers. Similar problems still limit the use of modern electrical methods of measuring *RH*, so that reliable *RH* values are often determined in other ways.

5.5.2 Wet-bulb psychrometer

In a wet-bulb thermometer (Section 2.3) net evaporation continues from the soaked wick enveloping the thermometer bulb (Fig. 5.6), unless the air being measured is already saturated. Evaporation keeps the temperature of the wick, bulb, and film of surrounding moistened air, at the wet-bulb temperature T_w, which is $(T - T_w)$ below the air temperature T registered by an adjacent dry-bulb thermometer. Some ventilation (aspiration) is needed to ensure that evaporation from the wet bulb does not build a thick cocoon of saturated air around the wet bulb.

Complicating practicalities are avoided by using the thermodynamic wet-bulb temperature analysed in Section 5.4.5 and Box 5.2—a 'one-shot' model of the continuous process used in practical instruments. By a happy accident of opposing corrections, the wet-bulb temperature registered by a real, suitably aspirated wet-bulb thermometer (air draft 4 ms $^{-1}$ or more) approximates very closely to the ideal thermodynamic wet-bulb temperature T_w. The adjacent dry-bulb temperature T in the psychrometer corresponds to the initial air temperature T in the thermodynamic version. However, wet-bulb thermometers relying on natural rather than forced ventilation, in a Stevenson screen, for example, can differ enough from the thermodynamic T_w to need a slightly modified form of Eqn 5.10 for accurate calculation of water content, and therefore a slightly different set of hygrometric tables.

5.5.3 Dew-point meter

In the classic dew-point meter (Section 2.3), a polished metal surface is exposed to the moist air to be measured, and observed visually as it is cooled to the point (the dew-point), where it first becomes misted by dew. The surface temperature is an accurate measure of the dew-point if the cooling is slow and the temperature is read at the first signs of misting. Though needing no calibration, practicalities have in the past made the dew-point meter too cumbersome to be considered for

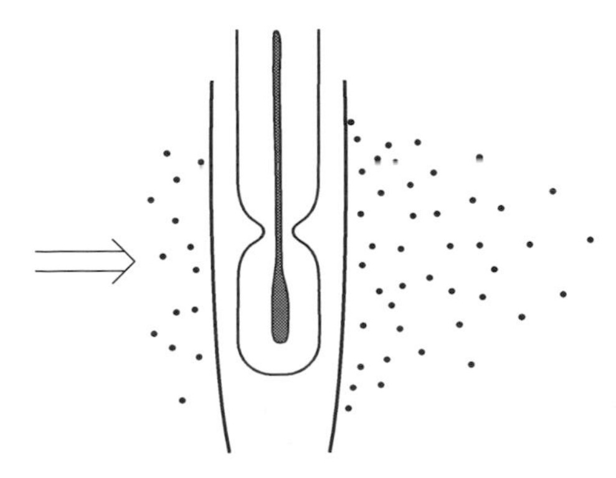

Figure 5.6 Evaporation from the soaked wick of a wet-bulb thermometer into an unsaturated airflow.

routine measurements, which instead use the wet- and dry-bulb psychrometer. In recent decades dew-point meters have become cheaper, smaller, and automated (detecting dew formation electrically rather than visually) and are becoming widespread in industry.

5.6 Dry adiabatic reference process

We have seen (Section 4.2) that air can rise or fall so rapidly in convection, etc. that temperature changes much more by expansion or compression than it does by heat inputs and outputs—by absorption or emission of radiation, or heat conduction. We can say that parcel motion in such weather systems and processes is nearly *adiathermal* (in thermal isolation) despite quite large changes of temperature. And since their expansions and compressions are usually quite slow (explosive expansion in a lightning flash is an obvious exception), such gentle parcel coolings and warmings are approximately *reversible* in the thermodynamic sense (Section 9.3, *[20]*). A process which is both adiathermal and reversible is said to be *adiabatic*, and since adiabatic processes are important benchmarks throughout thermodynamics, it is very useful in meteorology to define two ideal adiabatic *reference processes* for comparison with reality.

The simpler of the two is the *dry adiabatic reference process*, in which there is neither condensation nor evaporation of water within the moving, thermally isolated air parcel. The title 'dry' is universal but misleading, since the air can be very humid, provided it remains cloud free. In cloudy air by contrast, water may condense or evaporate, heating or cooling the air by liberation or absorption of latent heat. Since almost all weather-related cloud forms or evaporates in air held very close to saturation (Section 6.2), it is useful to define a *saturated adiabatic reference process* (Section 5.7) to compare with cloudy processes.

5.6.1 Dry adiabatic process

In the dry adiabatic reference process, changes in temperature and pressure are related by Eqn 5.2 with zero dQ, i.e. by the balance of the two terms on the right-hand side.

$$C_p \, dT = \frac{dp}{\rho} \tag{5.12}$$

The behaviour described by Eqn 5.12 is qualitatively consistent with common experience: compression (positive dp) is associated with warming (positive dT), and decompression is associated with cooling. For quantitative comparison with real atmospheric behaviour, air density ρ is replaced using the equation of state (Eqn 4.2), to give

$$\frac{dT}{T} = \frac{R}{C_p}\frac{dp}{p} \tag{5.13}$$

where T is the parcel's absolute temperature, and p is its total pressure. Equation 5.13 simply states that small proportional changes in absolute temperature are directly proportional to small proportional changes in parcel pressure, and that the constant of proportionality is R/C_p, whose value for dry air is close to 0.286 throughout the well-mixed turbosphere. For example, a 3% drop in air pressure (about 30 hPa in the low troposphere—corresponding to a rise of about 250 m) is associated with a fall of just under 1% in absolute air temperature (about 2.6 °C in 300 K) (NN 5.5). Values of R/C_p for realistically moist air differ so little that the dry value can be used regardless of vapour content.

When percentage changes in p and T are no longer small, Eqn 5.13 must be integrated between initial and final conditions to find *Poisson's equation.*

$$\frac{T_1}{T_2} = \left(\frac{p_1}{p_2}\right)^{R/C_p} \qquad 5.14$$

which relates distinctly different initial and final values, subscripted 1 and 2 respectively.

Consider for example, a fictional but realistic deep layer over a strongly sunlit subtropical desert like the Sahara. In such arid conditions, air can rise dry adiabatically (i.e. without reaching saturation and forming cloud) for 5 km or more above the hot surface, its pressure falling from 1,000 to 500 hPa in the process. If an air parcel starts from the 1,000 hPa level with a temperature of 32 °C (305.2 K), Eqn 5.14 shows (NN 5.5) that it will reach the upper level (500 hPa) with a temperature of −22.9 °C (250.3 K)—having cooled by 54.9 °C or K by *dry adiabatic* decompression.

NUMERICAL NOTE 5.5 Dry adiabatic cooling

(i) *Small pressure drop* Use Eqn 5.13 and substitute 0.286 for R/C_p, −30 hPa for Δp and 1,000 hPa for p, to find $\Delta T/T = -0.286 \times 30/1{,}000 = -8.6 \times 10^{-3}$, a fall of nearly 0.9%, corresponding to a fall of 2.6 K in 300 K.

(ii) *Sahara case* Invert and rearrange Poisson's equation (Eqn 5.14) to form $T_2 = T_1 \left(\frac{p_2}{p_1}\right)^{R/C_p}$

and insert 305.2 K (32.0 °C) for the initial temperature T_1, 1,000 hPa for the initial pressure p_1, and 500 hPa for the final pressure p_2. Inserting 0.286 for (R/C_p), the numerical equation becomes

$T_2 = 305.2 \times (500/1{,}000)^{0.286} = 305.2 \times 0.8202 = 250.3$ K, i.e. −22.9 °C, a temperature fall of very nearly 55 K or °C. Note that p_1 and p_2 can be expressed in any units (e.g. mm Hg) since they appear as a dimensionless ratio. T_1 and T_2 must be expressed on an absolute scale, and the kelvin scale is the only one in use. Alternatively, take the log of rearranged Poisson to any convenient base

$$\log T_2 = \log T_1 + (R/C_p) \log (p_2/p_1)$$

and insert numerical values for T_1, etc. to find the same result.

5.6.2 Dry adiabats and potential temperature

The initial, final, and all intermediate states of an air parcel rising or falling by dry adiabatic process are said to be on the same *dry adiabat*—the infinite sequence of states, defined by Eqn 5.14 and at least one pressure–temperature pair (305.2 K and 1,000 hPa in the Sahara case). Dry adiabats are specified in thermodynamic tables such as *[9]*, and plotted on several types of *thermodynamic diagram* (Section 5.8). In the *tephigram* in Fig. 5.7, the Sahara dry adiabat appears as a straight, bold, diagonal line running from its origin on the 1,000 hPa isobar to the 500 hPa isobar (i.e. the mid-troposphere), where convective ascent is supposed to end on this occasion. The Sahara dry adiabat is parallel to a family of standard dry adiabats which form part of the grid of lines of the tephigram.

It is useful to label each dry adiabat with a numerical value which is the same everywhere along it, and is therefore conserved during that particular dry adiabatic process. The obvious convention is to label each dry adiabat by its actual temperature as it crosses the 1,000 hPa isobar. This temperature is called the *potential temperature* of the air at any point along that particular dry adiabat, and it corresponds physically to the temperature which would be reached by an air parcel taken dry adiabatically from its initial temperature and pressure to 1,000 hPa. From this definition and Eqn 5.14, the potential temperature θ of air with absolute temperature T and pressure p (expressed in hPa) is given by

$$\theta = T\left(\frac{1{,}000}{p}\right)^{R/C_p} \qquad 5.15$$

Note that potential temperatures are greater than actual temperatures at all pressures less than 1,000 hPa, the numerical differences being already quite large in the middle troposphere, as shown in the Sahara case, whose potential temperature is exactly 32 °C everywhere along the temperature profile, but whose actual temperature at 500 hPa is nearly 55 ° colder (NN 5.5).

5.7 Saturated adiabatic reference process

5.7.1 In outline

Except in subtropical deserts like the Sahara, air in the low troposphere is usually so moist that it chills to saturation after rising quite modest distances above the local surface—often less than 1 km (Fig. 1.4), and sometimes only a few 100 m. If such air continues rising above the saturation level, cloud droplet formation and growth (Sections 6.3 to 6.5) holds the air very close to saturation while excess vapour is condensed and large amounts of latent heat are evolved within the cloud-filled air. The relevant reference process is now the *saturated adiabatic reference process*, in which the chilling moist air between the cloud droplets is held precisely at saturation by rapid condensation of excess vapour onto the droplets, and the latent heat evolved is retained within the cloudy air parcel, partly offsetting its decompressional cooling. The process reverses in descending, warming air, as evaporation from supposedly available cloud water maintains saturation, and evaporative cooling partly offsets compressional warming. For simplicity the reference process ignores the heat capacity of cloud water after its condensation, and also simply assumes that 'by magic' there is always enough to evaporate and saturate descending air. To this extent the reference process is not perfectly adiabatic, but the discrepancy is very small, and the simplification is very useful, since otherwise the range of possible initial cloud water amounts would seriously complicate the analysis to little purpose.

Below 0 °C, saturation is assumed to be relative to supercooled water rather than ice, because cloud droplets remain *supercooled* down to about −30 °C (Section 6.11)—unlike bulk water which freezes at 0 °C. At lower temperatures, freezing liberates some more latent heat (latent heat of *fusion*), but the effects are usually small given the small amounts of uncondensed vapour at such low temperatures. The wholesale freezing of supercooled cloud droplets can be important in the formation of cumulonimbus anvils (Section 11.4).

5.7.2 In detail

Consider a saturated parcel rising a short distance in a saturated adiabatic process. It cools slightly by adiabatic expansion and becomes slightly supersaturated, since saturation vapour density falls with falling temperature (Fig. 6.2). The excess vapour is immediately condensed as cloud, leaving the remaining vapour exactly at saturation. If the small change in saturation-specific humidity is dq_s, then the amount of latent heat given to unit mass of surrounding moist air is $-L\, dq_s$, where L is the specific latent heat of vaporization, and the minus signs ensure that a negative dq_s (a fall in q_s) gives positive heat to the air. The first law of thermodynamics (Eqn 5.2) now relates the liberation of latent heat ($dQ = -L\, dq_s$) to the changes of parcel air temperature and pressure:

$$dQ = -L\, dq_s = C_p\, dT - \frac{dp}{\rho}$$

$$C_p\, dT = \frac{dp}{\rho} - L\, dq_s \qquad 5.16$$

The first and second terms of Eqn 5.16 (i.e. on either side of the equals sign) describe the dry adiabatic cooling of air with falling pressure (and warming with rising pressure). In saturated ascent the third term is positive because dq_s is negative (q_s falling with falling temperature), partly offsetting the dry adiabatic cooling, and showing that cloud formation reduces decompressional cooling below the dry adiabatic value. By following the same argument for saturated *descent* we see that cloud evaporation reduces compressional warming, as expected.

The saturated adiabatic process is most directly compared with the dry adiabatic process when expressed in terms of changes in potential temperature θ, which by definition are zero in a dry adiabatic process. By rearranging the combination of Eqns 5.2 and 5.15 as shown in Box 5.3 we find

$$\frac{d\theta}{\theta} = -\frac{L}{C_p}\frac{dq_s}{T} \qquad 5.17$$

which confirms that condensation (negative dq_s) increases potential temperature, whereas evaporation reduces it.

An important aspect of the saturated adiabatic process follows from the exponential increase of vapour density (and therefore of saturation specific humidity) with temperature (Fig. 6.2). Since the slope increases sharply with increasing temperature, it follows that emissions or absorptions of latent heat are much larger at high temperatures than they are at low temperatures, where there is very little vapour left to condense. The contrast between the saturated and dry adiabatic processes is therefore much more marked at high temperature than it is at low.

There are *saturated adiabats* and *equivalent potential temperatures* for the saturated adiabatic reference process, just as there are dry adiabatics and potential temperatures for the dry adabatic reference process, but these are easier to visualize after outlining graphical ways of displaying and analysing upper air data.

BOX 5.3 Potential temperature and the saturated adiabatic process

Taking the log of Eqn 5.15 (defining potential temperature θ), and differentiating to relate very small changes $d\theta$ in θ and dT in temperature T, we find

$$\frac{d\theta}{\theta} = \frac{dT}{T} - \frac{R}{C_p}\frac{dp}{p} \qquad \text{B5.3a}$$

Taking Eqn 5.2 (the meteorological form of the first law of thermodynamics), rearranging it using $p = \rho R T$ (the equation of state for air), and dividing across by $(C_p T)$

$$\frac{dQ}{C_p T} = \frac{dT}{T} - \frac{R}{C_p}\frac{dp}{p} \qquad \text{B5.3b}$$

Comparing Eqns B5.3a and b we see that

$$\frac{d\theta}{\theta} = \frac{dQ}{C_p T} \qquad \text{B5.3c}$$

which shows that the very small potential temperature increase $d\theta$ for an air parcel is proportional to the very small heat input dQ divided by the absolute temperature T at which this input occurs.

In the saturated adiabatic process, the heat injected by condensation during a short adiabatic ascent is given by $-L\,dq_s$ (see Eqn 5.16 and discussion). When substituted for dQ in Eqn B5.3c, minor rearrangement gives

$$\frac{d\theta}{\theta} = -\frac{L}{C_p}\frac{dq_s}{T} \qquad \text{B5.3d}$$

which is Eqn 5.17. Since the latent heat is evolved or absorbed within the parcel, the parcel does not exchange heat with its surroundings and the process remains adiabatic.

5.8 Aerological diagrams

5.8.1 Outline

From the beginnings of thermodynamics, special graphs (*thermodynamic diagrams)* have been used to describe and analyse sequences of thermodynamic states of materials, especially gases. In the early twentieth century several were developed for meteorological analysis of upper air data from radiosondes, etc. Each point on such an *aerological diagram* corresponds to a particular air temperature and pressure, and humidity can be represented as well. Axes are chosen to highlight graphically meteorologically important activity, especially the dry and saturated adiabatic reference processes. The *tephigram* used in Britain and Canada, was devised by Napier Shaw, who was head of the British Meteorological Office, and one of a number of meteorologists in several countries who helped consolidate the scientific framework of meteorology about a century ago.

As shown in Box 5.4 the axes of the tephigram are temperature T and the logarithm of potential temperature θ, both labelled in °C. Since log θ is a direct measure of the important thermodynamic measure called *entropy* and represented by Greek phi (ϕ), the axes of the diagram are T and ϕ—hence 'tephigram'. It can be shown that areas on a tephigram are proportional to energy, and can be used, for example, to estimate the buoyant energy available in convective situations (Section 11.4). Other aerological diagrams are the popular *skew T log p* diagram, and the older, less convenient *Stuve* diagram *[26]*.

In the following, tephigrams are detailed and used to display and analyse radiosonde data. Many points apply with little alteration to the skew T log p diagram.

5.8.2 Tephigram

A tephigram is outlined and used in Fig. 5.7, and its construction is detailed in Box 5.4. The T axis slopes down page-right, so that isotherms of T slope up page-right (labelled in °C page bottom and left). The log θ axis slopes up page-right (i.e. parallel to T isotherms), so that isopleths of potential temperature θ (dry adiabats) slope up page-left, their separation closing progressively up page-right, according to the logarithmic relationship. Isobars are slightly curved, nearly horizontal lines labelled page-right, separating upwards logarithmically so that up-page distances are nearly linear in height (as in Fig. 3.1). Heights of standard isobars above mean sea level (which is usually just underneath the 1,000 hPa isobar) are reported for each individual sounding and written in metres on the isobars, page-right. *Saturation adiabats* are curved long-dash lines making angles of up to 45° or so with θ isopleths (*dry adiabats*) at the foot of the page (i.e. in the warm low troposphere), but curving asymptotically towards dry adiabats in the high, cold troposphere. Isopleths of saturation-specific humidity are nearly straight light-dashed lines nearly parallel to the isotherms.

5.8.3 Temperature profiles

Subtropical In outline

The synthetic Sahara sounding (Section 5.6) is the thick straight line running parallel to the θ isopleths from bottom-right to centre page of Fig. 5.7. Its potential temperature can be seen to be about 32 °C by extrapolating down-right to intersect the 1,000 hPa isobar to the right of the 30 °C isotherm, just visible far bottom-right. Its upper end is on the 500 hPa isobar, at an actual temperature of about −23 °C, as seen by interpolating between adjacent T isotherms. The tephigram plot determines conditions at all other points along the connecting dry adiabat: for example the temperature at 600 hPa is very close to −10 °C, as can be confirmed by calculation using the method of NN 5.5.

Mid-latitude In detail

Consider actual radiosonde data from a station in Northern Ireland plotted on Fig. 5.7. The segmented bold line running up the middle of the page connects a series of temperatures and pressures selected from the radiosonde transmission to minimize data bulk. The points (at the joints between straight segments) are a tiny proportion of the data sent by the rising radiosonde, but are selected so that straight line connections on the tephigram always lie within 1 °C of the detailed temperature/pressure record.

In the bottom 5 hPa or so (roughly the first 40 m above the radiosonde station), temperature falls more steeply with decreasing pressure (increasing height) than

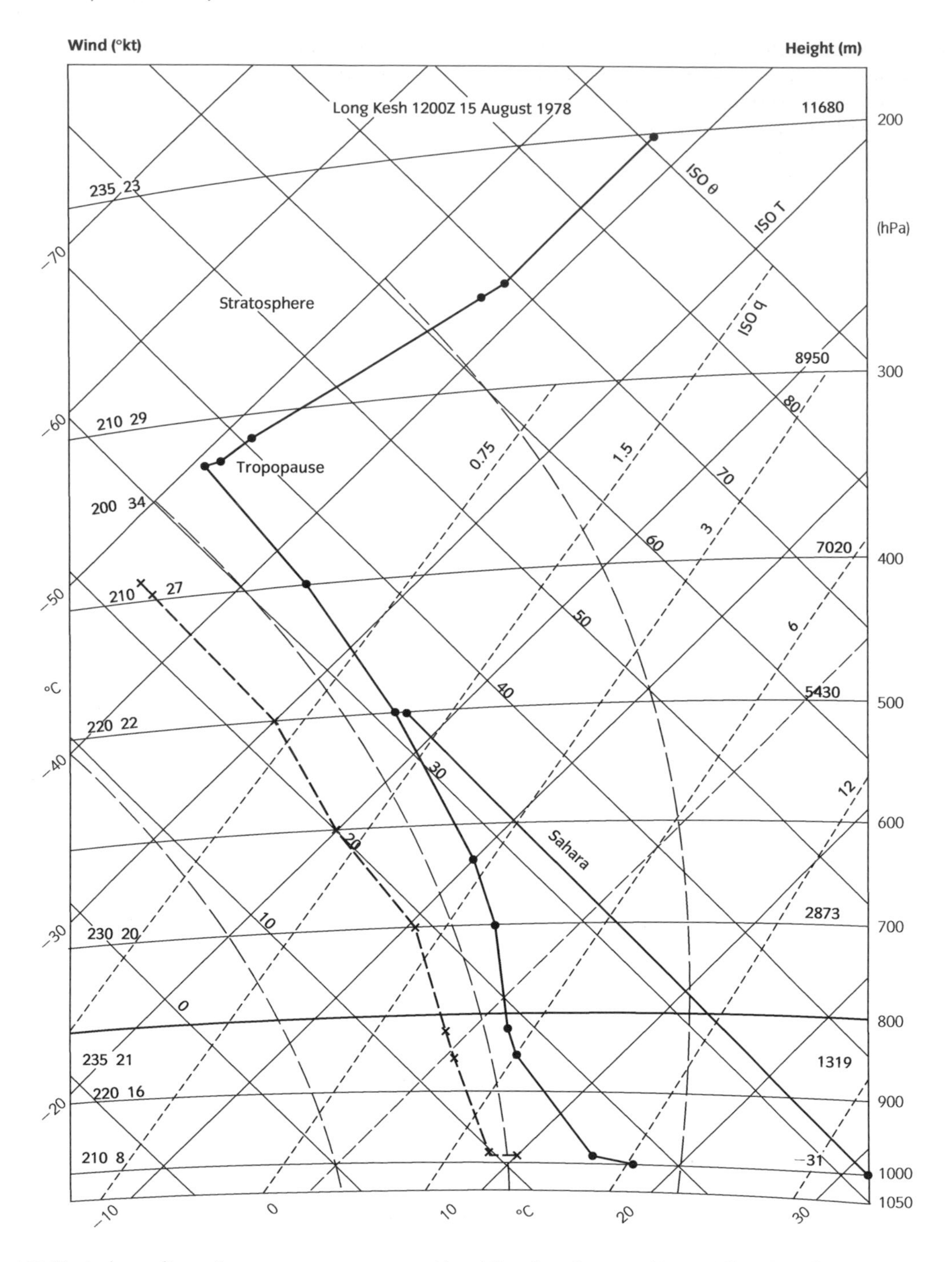

Figure 5.7 Vertical soundings of pressure, temperature, and humidity plotted on a tephigram (Ch. 5.8) with temperature and dew points represented by continuous and dashed lines respectively. The deep sounding is typical of many situations in middle latitudes, with a nearly dry adiabatic surface layer surmounted by a more stable layer which becomes nearly dry adiabatic before it becomes abruptly very stable at the tropopause. The warm dry adiabat labelled Sahara, represents the fictional but realistic Sahara example discussed in Ch. 5.7.

in a dry adiabat, indicating strong convective instability over the Sun-warmed land surface (Section 5.10). Above this there is a layer about 1,700 metres deep in which temperature nearly parallels the dry adiabats, strongly suggesting the presence of vigorous unsaturated ('dry') convection.

Above the bend in the temperature profile at the 810 hPa level (probably the base level of shower clouds widespread at the time of the radiosonde flight), temperature falls considerably more slowly with pressure, closely paralleling the nearest saturated adiabat (long-dash curve). It continues like this into the high troposphere, at first cooling much more slowly than the dry adiabats, and then steepening and gradually approaching the slope of the dry adiabats in the high troposphere (above the 400 hPa level, over 7 km above MSL).

At 315 hPa (about 8.5 km above MSL), the *tropopause* appears as a very sharp transition between the top of the convecting troposphere and the base of the *stratosphere*, which is nearly isothermal and therefore convectively very stable (Section 5.10). There is a *temperature inversion* above the tropopause, with temperatures rising 7 °C in just over 3 km. This temperature rise continues unsteadily through the middle and upper stratosphere, reaching 0 °C or so at the stratopause (Fig. 4.2) about 50 km above sea level and far above the top of Fig. 5.7. Such convective stability prohibits convection and inhibits the larger scales of vertical motion, both of which are rife in the troposphere. Apart from occasional powerful updrafts from tropospheric cumulonimbus fountaining a kilometre or so above the tropopause, before falling back (Section 11.4), vertical motion in the stratosphere is very much weaker and less important than in the troposphere. Because weather systems are so closely confined to the troposphere, radiosonde balloons are designed to burst below 25 km, and data is analysed only up to the meteorologically useful limit—about 200 hPa in middle latitudes, and 100 hPa in low latitudes.

Notice how the tephigram displays radiosonde data in ways which graphically highlight the presence of nearly dry and saturated adiabatic convection, cloud base, and other meteorologically important features, including the tropopause.

5.8.4 Humidity profile

The bold dashed line to the left of the troposphere's temperature profile in Fig. 5.7 is the dew-point temperature profile. For example, at 850 hPa air temperature T is about +4 °C, whereas the dew-point T_d is just over 0 °C, as seen by sliding page-left on the isobar while still reading the temperature scale. This dew-point depression ($T - T_d$) of 4 °C is a measure of the sub-saturation of the air (Section 6.5) at 850 hPa. Following the T and T_d profiles upwards from the surface (just above the 1,000 hPa isobar in Fig. 5.7), we see the two apparently trying to converge at the kink at about 815 hPa, suggesting that this was the base of local convective clouds. If the radiosonde had risen through one of these clouds, we should expect the T and T_d profiles to merge at cloud base, and remain coincident to the cloud top, probably at the tropopause on this occasion, to judge from the observed temperature profile. Two factors often intervene to prevent such neat agreement with expectation, and probably did so this time.

(i) By definition, shower clouds are scattered, often widely, so that a radiosonde is more likely than not to rise through the clear air between clouds, missing the fully saturated air above cloud base level.

(ii) Despite careful calibration, etc. radiosonde humidity sensors are much less reliable than radiosonde thermometers and barometers, and often show sub-saturation in conditions otherwise known to be saturated. The situation worsens in the low temperatures of the upper troposphere.

On this occasion of Fig. 5.7 the mere tendency of the T and T_d profiles to converge, together with the transition of the temperature profile from nearly dry adiabatic to nearly saturated adiabatic, would be regarded as quite good evidence for the presence of dry convection below cloud base (the kink at 815 hPa), and saturated convection above cloud base.

5.8.5 Wind and height profile

Selected wind azimuths (Section 2.4) and speeds (in knots) from the standard pressure levels (850, 700, 500 hPa, etc.) are entered on the isobars at which they were measured, with extra readings to cover important features appearing at non-standard pressures. On this occasion there was a local wind speed maximum just below cloud base, and an overall maximum of 34 knots from 200° azimuth (20° West of S) just below the tropopause. In addition the surface wind (from the 10 m anemometer at the launching site) is entered on the surface pressure isobar.

Heights in *geopotential metres* (actual metres adjusted for slight variations of gravity with location and height) above mean sea level are entered on the standard isobars. The negative height of the 1,000 hPa isobar shows that on this occasion it was 31 m below MSL (by extrapolation), and therefore 75 m below the station (44 m above MSL). The 500 hPa surface, the conventional middle of the troposphere (though well above it geometrically in this particularly shallow troposphere) was 5,430 m above MSL.

BOX 5.4 An outline tephigram

The following prescription outlines a tephigram with temperature (T) range –60 to +60 °C, potential temperature (θ) range –60 to +100 °C, and pressure (p) range 1,000–125 hPa.

(i) *Axes and isotherms* Convert extreme θ values to degrees K and find their logarithms to the base 10. Scaling to fit the page, draw the log θ axis up the vertical and the T axis to the right in the horizontal, and draw and label isotherms (constant T) and potential isotherms (constant θ), in 20 ° steps from 0 °C.

(ii) *Isobars* Draw and label the 1,000 hPa isobar as the locus of points where $\theta = T$. Take $\log_{10}$ of Eqn 5.15

$$\log \theta = \log T + (R/C_p) \log (1000/p)$$

and use it to draw and label the 500, 250, and 125 hPa isobars as follows. Since $\log \theta = \log T$ represents the 1,000 hPa isobar already plotted, the 500 hPa isobar is displaced up the log θ axis from the 1,000 hPa isobar by a distance (scaled for your page) equivalent to $(R/C_p) \log (1000/p)$, i.e. $0.286 \times \log (1{,}000/500) = 0.086$. An identical displacement from the 500 hPa isobar gives 250 hPa and so on. Other isobars follow from other choices of p. Draw and label these isobars.

(iii) *Isopleths of saturation-specific humidity* Convert q_s values of 10, 5 and 2.5 g kg^{-1} to dimensionless ratios (10^{-2}, etc.), and use these in $q = \varepsilon\, e/p$ (Eqn 5.6) to find the corresponding saturation vapour pressures (e_s) on the 1,000 hPa isobar. Values should agree closely with some of the following

e_s/hPa	16	8	4	2	1
T/°C	14	3.7	−5.6	−14.7	−24

and the corresponding temperatures can be used to locate the intersection of the 1,000 hPa isobar with the three q_s isopleths (10, 5 and 2.5 g kg^{-1}). Repeat for the 500, 250, and 125 hPa isobars, and connect the intersections to form the q_s isopleths, labelling them in g per kg values.

(iv) *Saturated adiabatics* Draw and label the saturated adiabat which passes through the intersection of the 10 g kg^{-1} q_s isopleth with 1,000 hPa, using the following useful three-step approximation to the continuous release of latent heat in rising air.

(a) Warm air by isobaric release of latent heat by condensing 5 g kg^{-1} of vapour. Equation 5.16, $C_p\, dT = -L\, dq_s + dp/\rho$, with zero dp shows that dT is almost exactly 2.5 °C as q_s falls by 1 g kg^{-1} and pro rata. Warm by 12.5 °C (i.e. 5 × 2.5) isobarically from the starting intersection (10 g kg^{-1} with 1,000 hPa), and then cool dry adiabatically to saturation by moving along a dry adiabat to the 5 g kg^{-1} q_s isopleth. Since the vapour content of the air has fallen to 5 g kg^{-1}, you have now reached the intersection of the chosen saturated adiabat with the 5 g kg^{-1} isopleth.

(b) Repeat the process of step (a) by starting at its end point (q_s = 5 g kg^{-1}) and condensing half of this to reach the intersection of the chosen saturated adiabat with q_s 2.5 g kg^{-1}.

(c) Find the asymptotic dry adiabat at the cold end of the saturated adiabat (i.e. the *equivalent potential temperature* θ_e—see Section 5.9) by using $\theta_e - \theta_w = 2.5\, q_s$ (Eqn 5.21) and noting that θ_w is the value of θ at the 1,000 hPa isobar—the starting point of the saturated adiabat now being outlined.

Connect the intersections and the asymptote by a smooth curve representing the chosen saturated adiabat (θ_w 13.7 °C). The model could be improved by using more (smaller) pairs of isobaric and dry adiabatic steps, but is surprisingly good as it stands.

(v) *Completion* Note that some parts of your tephigram represent conditions not found in the real atmosphere: sea level pressures hardly ever exceed 1,050 hPa; dry adiabats seldom exceed the potential temperatures of 45 °C found low over hot deserts. Deep saturated convection seldom rises from surface conditions with q values greater than 25 g kg^{-1}; and since this value saturates air at just over 28 °C at 1,000 hPa, it follows from section (ivc) above that the maximum equivalent potential temperature in the troposphere is just over 90 °C, realized around the 125 hPa level. Sketch a rough saturated adiabat to indicate the limiting condition.

Hatch the areas excluded by all these limits, and rotate your page 45 degrees clockwise, to orientate isobars left to right across the page, and compare with Fig. 5.7.

5.8.6 Equivalent and wet-bulb potential temperatures

It is useful to have a single temperature to label each saturated adiabat, like the potential temperature which labels each dry adiabat. Figure 5.8 shows two ways in which this can be done.

(i) At high altitudes (i.e. low pressures), temperatures and vapour contents are so low, even at saturation, that any particular saturated adiabat is asymptotic to a particular dry adiabat, whose potential temperature is called the *equivalent potential temperature* θ_e of that saturated adiabat. In physical terms θ_e is the temperature which would be reached by a parcel after saturated adiabatic ascent to nearly zero pressure and dry adiabatic descent from there to 1,000 hPa (Fig. 5.8).

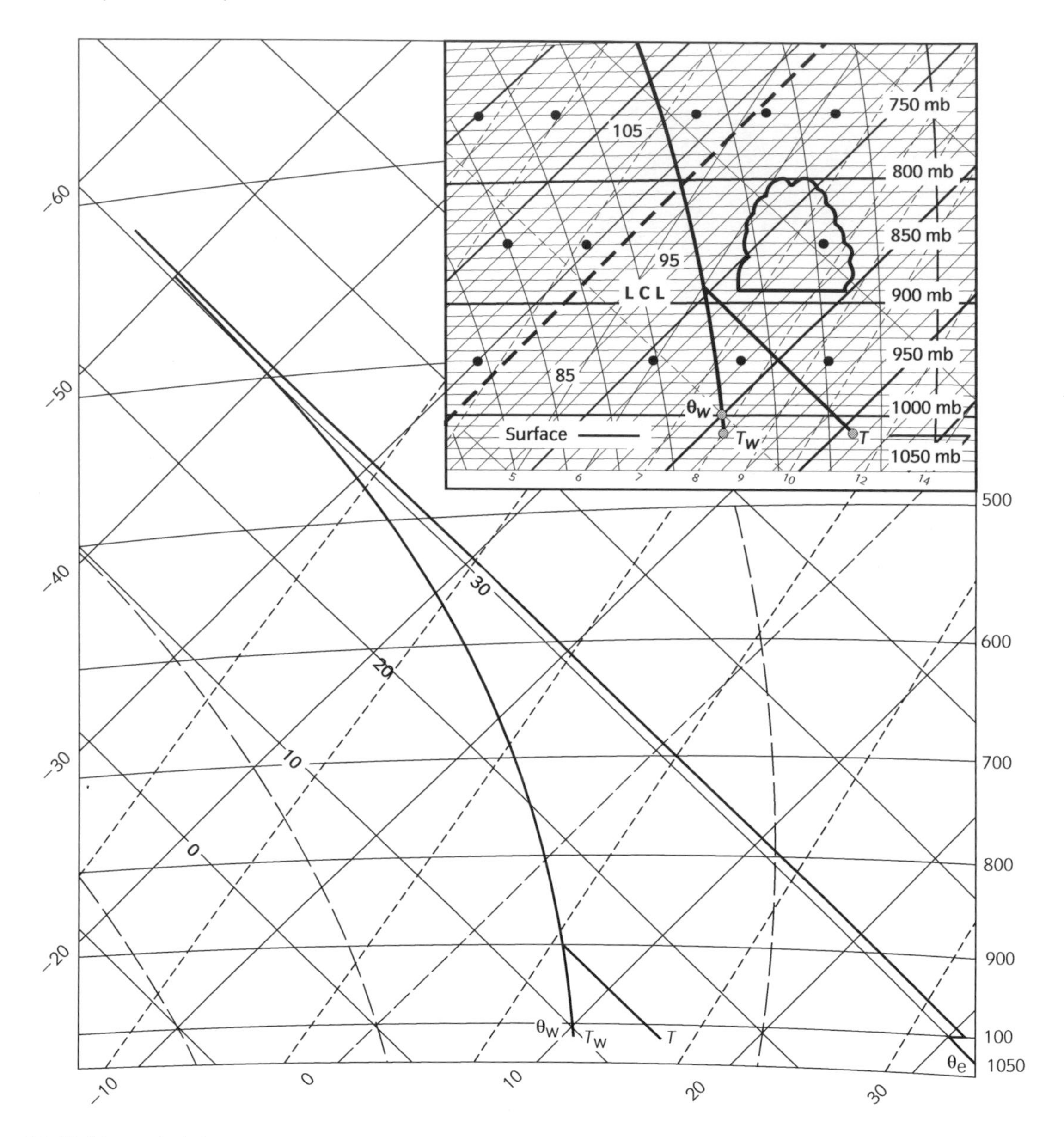

Figure 5.8 Tephigram depicting wet-bulb potential temperature and equivalent potential temperature. *Inset*: Enlargement of the base of the main diagram to depict lifting condensation level and Normand's theorem.

(ii) An alternative approach is to label each saturated adiabat by the actual temperature at which it intersects the 1,000 mbar isobar—as is done for dry adiabats. We will write this temperature as θ_w, and examine its relationship with θ_e.

The equivalent potential temperature θ_e is higher than θ_w (Fig. 5.8) because vapour condenses and releases latent heat within the air parcel as it rises from the 1,000 hPa level to the low pressures of the upper troposphere. By 200 hPa (about 12 km above MSL), strong adiabatic cooling and the falling saturation limit have condensed virtually all of the parcel's vapour. The magnitude of $(\theta_e - \theta_w)$ can be found very accurately from thermodynamic tables or diagrams, and can be estimated very usefully from

$$\theta_e - \theta_w = 2.5\, q_s \qquad 5.18$$

which is derived from Eqn 5.17 by integrating and simplifying (Box 5.5). Here q_s is the specific humidity (expressed in grams of water vapour per kilogram of moist air) of air which is saturated at the 1,000 hPa level. A typical value for q_s in temperate latitudes is 10 g kg^{-1}, which corresponds to a value of 25 °C for $(\theta_e - \theta_w)$. This agrees to within a fraction of a degree with the value given by thermodynamic tables and diagrams, and is well within the accuracy of reading even the large format version of Fig. 5.7 which is used for routine plotting of radiosonde data.

The physical meaning of θ_w is obvious in the case of saturated air: it is the temperature reached by saturated adiabatic descent to 1,000 hPa. However, there is a relevance to unsaturated surface conditions which is shown on the inset in Fig. 5.8. Suppose that an unsaturated air parcel with temperature T has wet-bulb temperature T_w and pressure p. T_w is less than T by an amount which is a measure of the sub-saturation of the air, and is plotted on the isobar passing through T in Fig. 5.8 because it is reached by isobaric evaporative cooling from T (Section 5.4). According to a thermodynamic theorem name after the meteorologist Normand, the dry adiabat through T, and the saturated adiabat through T_w intersect at the *lifting condensation level* (LCL) of the unsaturated air (Fig. 5.8)—the level at which air would become saturated if lifted dry adiabatically from its original state near the surface. Since *Normand's theorem* applies at all levels below the LCL, it follows that the saturated adiabat traces out the actual wet-bulb temperature of the adiabatically rising or falling parcel of unsaturated air. If this trace is followed down to the 1,000 hPa level, the temperature reached is known as the *wet-bulb potential temperature* of the air parcel (written θ_w)—the wet-bulb temperature which the air would have if taken adiabatically to the 1,000 hPa level—saturated adiabatically above the LCL, dry adiabatically below it. Normand's theorem is completed in Section 6.5 by involving the dew-point.

5.8.7 Conservation of θ_w

We have established that the wet-bulb potential temperature θ_w is conserved in a wide range of thermodynamic processes. In dry adiabatic warming or cooling Normand's theorem applies as described and θ_w is conserved. In saturated adiabatic warming or cooling, θ_w is conserved by definition. Adiabatic evaporation of water at constant pressure is part of the wet-bulb thermometer process (Section 5.5), which by definition does not alter the wet-bulb temperature of the air, and therefore does not alter its wet-bulb potential temperature. In fact θ_w is the most widely conservative thermodynamic marker used in meteorology. To illustrate this, consider an adiabatic process which is usually at work beneath the bases of showering clouds.

Chilling by sub-cloud rain

Suppose that a cumulonimbus cloud, whose base is at the LCL in Fig. 5.8, begins to produce thick, fine rain below cloud base, the rain consisting of a dense population of smallish droplets evaporating freely into the slightly unsaturated air as they fall. Each droplet chills itself and its surroundings like a wet bulb, so that the net effect is to cool the rain-filled air to the wet-bulb temperature of the originally slightly unsaturated air. The chilled air may then continue to sink, desaturating by descent and resaturating by further rain droplet evaporation, tracing a saturated adiabat downwards from cloud base. If this continues to the 1,000 hPa level, the saturated rain-filled air arrives with an air temperature T equal to θ_w. This sort of process accounts for sudden downdrafts of cool, moist air observed near the surface just at or before the onset of a heavy shower. They are often markedly cooler ($\sim$10 °C) than the previous surface air, which will tend to have a temperature given by the dry-adiabat through LCL, being the base of the air stirred by dry adiabatic convection up to cloud base. This warmer air is displaced upwards or sideways by the chilled shower-filled air, which often spreads horizontally ahead of the shaft of precipitation, giving a few minutes warning of the imminent shower.

If evaporation from rain drops does not keep pace with desaturation by compressive warming of the downdraft (large rain drops fall more quickly and evaporate less efficiently than small ones—Section 6.11), the downdraft will arrive with the same θ_w as before, but unsaturated and warmer than in the first case.

The equivalent potential temperature θ_e is obviously conserved in any process conserving θ_w, since it is simply an alternative labelling system for saturated adiabatic processes. But though clearly relevant to the life cycle of air ascending in very deep convection, θ_e lacks the direct identification with conditions near the surface (the1,000 hPa level, strictly) which is the special advantage of θ_w.

5.9 Adiabatic temperature lapse rates

So far we have considered variations of air parcel temperature with pressure in the adiabatic reference processes (Sections 5.6 and 5.7, Eqns 5.12–5.14). When pressure changes arise almost entirely from height changes, as is usually the case, we can find the corresponding changes of temperature with height by making use of the nearly perfectly hydrostatic relation between pressure and height (Section 4.4). We will do this thoroughly for the dry adiabatic reference process, before extending it descriptively to its saturated equivalent.

BOX 5.5 Dry adiabatic lapse rates of temperature and potential temperature

Temperature

The dry adiabatic variation of air parcel temperature T with pressure p is given by Eqn 5.12.

$$dT = \frac{1}{C_p}\frac{dp}{\rho} \qquad \text{B5.5a}$$

When the pressure change dp arises from vertical motion dz through an atmosphere with ambient vertical pressure gradient $\partial p/\partial z$, the parcel pressure and height changes are related by

$$dp = \frac{\partial p}{\partial z}\,dz$$

If the ambient air has density ρ' and is in hydrostatic equilibrium, Eqn 4.6 gives

$$\frac{\partial p}{\partial z} = -g\rho'$$

and the change dT of parcel temperature T with height change dz in dry adiabatic ascent or descent is found by combining these three equations to produce

$$\frac{dT}{dz} = -\left(\frac{\rho'}{\rho}\right)\frac{g}{C_p} \qquad \text{B5.5b}$$

Since air rises and falls very slowly compared with the speed of sound (except near lightning), moving parcels always have the same pressure as their horizontal surroundings, so that their equations of state ensure that

$$\frac{\rho'}{\rho} = \frac{T}{T'}$$

where T and T' are the absolute temperatures of the parcel and ambient air respectively. Equation B5.5b then becomes

$$\frac{dT}{dz} = -\left(\frac{T}{T'}\right)\frac{g}{C_p} \qquad \text{B5.5c}$$

which simplifies to

$$\frac{dT}{dz} = -\frac{g}{C_p} \qquad \text{B5.5d}$$

when parcel and surroundings have the same temperature. We say that the *dry adiabatic lapse rate* of T is (g/C_p), whose magnitude in the troposphere is just under 10^{-2} °C m^{-1}.

Potential temperature

By standard manipulation of Poisson's equation, we can relate small changes in potential temperature θ to small changes in temperature T and pressure p

$$\frac{d\theta}{\theta} = \frac{dT}{T} + \frac{R}{C_p}\frac{dp}{p} \qquad \text{B5.5e}$$

If the pressure change dp arises from vertical movement dz through surroundings in hydrostatic equilibrium and the temperature difference between parcel and surroundings is very small, repeating the above analysis changes the last term of B 5.5e to $(g/C_p)\,dz/T$.

After rearrangement we find an expression for the rate of change of θ with height for a parcel whose rate of change of temperature with height is dT/dz

$$\frac{d\theta}{dz} = \frac{\theta}{T}\left[\frac{dT}{dz} + \frac{g}{C_p}\right] \qquad \text{B5.5f}$$

When the parcel is undergoing a dry adiabatic ascent or descent $dT/dz = -g/C_p$ and $[\,] = 0$, confirming that the dry adiabatic lapse rate of potential temperature is zero, as it must be by definition of θ.

Note that in an isothermal process ($dT/dz = 0$)

$$\frac{d\theta}{dz} = \frac{\theta}{T}\frac{g}{C_p}$$

which approximates to g/C_p in the low troposphere, since $\theta \approx T$ on the absolute scale.

5.9.1 Dry adiabatic lapse rate

When an air parcel of density ρ moves dry adiabatically and vertically through ambient air which is in hydrostatic equilibrium and has local air density ρ', we find (Box 5.5) that parcel temperature T varies with parcel height z according to

$$\frac{dT}{dz} = -\left(\frac{T}{T'}\right)\frac{g}{C_p} \qquad 5.19$$

The appearance of both parcel and ambient air temperatures in Eqn 5.19 threatens complication. However, in all but the most vigorous types of convection, buoyant air parcels are observed to be only slightly warmer than their surroundings, 1 °C being typical. The corresponding ratio of absolute temperatures is about 1.004. Even in the most powerful cumulonimbus updrafts are seldom more than 5 °C warmer than their surroundings, which corresponds to a temperature ratio of 1.02. It follows that in realistic conditions (T/T') in Eqn 5.19 is almost always very close to unity, so that Eqn 5.19 closely approximates to

$$\frac{dT}{dz} = -\frac{g}{C_p} \qquad 5.20$$

The minus sign confirms that temperature falls (*lapses*) as height increases. The magnitude of the right-hand side of Eqn 5.20 is effectively constant throughout the troposphere since g variations are less than about 0.5% and C_p is kept highly uniform by mixing (Section 3.1). Inserting values for g and C_p, the magnitude of the vertical temperature gradient is almost exactly 0.098 °C m^{-1}, or 0.98 °C per 100 m, often rounded to 1 °C per 100 m, or 10 °C per km. This is the *dry adiabatic lapse rate*, often denoted by Γ_d, so that Eqn 5.20 becomes

$$\frac{dT}{dz} = -\Gamma_d \qquad 5.21$$

According to this, if an air parcel were to rise dry adiabatically through the full depth of the troposphere, it would cool by about 100° in the relatively shallow troposphere of high latitudes (about 10 km thick) and by about 150° in the deeper tropical troposphere. Such large temperature lapses are not observed in the real atmosphere because saturation always intervenes to reduce the temperature falls by liberation of latent heat (Section 5.7).

Note that the dry adiabatic lapse rate expressed in terms of potential temperature θ is very simply

$$\frac{d\theta}{dz} = 0$$

both by definition and by repeating the argument which led to Eqn 5.20 (Box 5.5). As shown in Fig. 5.2 and discussed further in Section 5.11, these profiles of T and θ are maintained by the dry adiabatic mixing and hydrostatic equilibrium which dominates cloud-free convection.

5.9.2 Saturated adiabatic lapse rates

In the saturated adiabatic reference process, continual release of latent heat by cloud formation in rising air reduces the temperature lapse rate below the dry adiabatic value, which corresponds to an increase of potential temperature with height.

According to Eqn 5.17, when the saturation-specific humidity of the remaining vapour falls a little (negative dq_s) by condensing into cloud, the potential temperature of the cloudy parcel rises according to

$$d\theta = -\left(\frac{L\theta}{C_p T}\right) dq_s \qquad 5.22$$

In the warm lower troposphere, the steep fall of the saturation limit with falling temperature produces large rises of potential temperature, reducing typical lapse rates well below the dry adiabatic value (≈1 °C per 100 m). In fact saturated adiabatic temperature lapse rates may be as low as half the dry adiabatic value (i.e. 0.5 °C per 100 m) in the warmest parts of the troposphere. As temperatures fall, the falling saturation limit ensures that there is less and less vapour to condense, so that the saturated adiabatic lapse rate increases towards the dry adiabatic value, becoming indistinguishable in the cold high troposphere.

5.10 Convective stability

In previous sections we have analysed the sequences of states experienced by individual air parcels undergoing dry or saturated adiabatic vertical motion. In assessing their relevance to the real atmosphere, it is important to notice that radiosondes, or the satellite-borne radiometers which may become viable alternatives, do *not* observe sequences experienced by a moving parcel: instead they observe vertical profiles of temperature, etc. either instantaneously (as by satellite or instrumented towers), or in relatively short transits (as by radiosonde or instrumented aircraft). Such observed profiles are snapshots of the environment of our theoretical parcels (slightly blurred in the case of radiosonde and aircraft data) taken from what we can call the *environmental viewpoint*. The distinction between the *parcel viewpoint* of adiabatic theory and the environmental viewpoint of observation is so basic that we might not expect them to have any simple relationship, and yet we have already seen good agreement between the two. For example, the dry adiabatic temperature lapse rate (Eqn 5.17) derived from the parcel viewpoint is observed (from the environmental viewpoint) to be extremely widespread in the low and high troposphere. Why should this be?

5.10.1 Dry adiabatic motion

Suppose that an air parcel is embedded in air with a certain observed vertical temperature gradient $\partial T/\partial z$. The *partial differential* $\partial T/\partial z$ represents the rate of variation (of temperature) with height alone—isolated from any variations with horizontal position or time—and therefore corresponds to an instantaneous snapshot of the parcel's vertical temperature environment at a fixed horizontal location. If the air parcel is moving vertically and dry adiabatically through its hydrostatic environment, its sequence of parcel temperatures traces out a temperature profile whose vertical temperature gradient is by Eqn 5.18, which is written

$$\frac{dT}{dz} = -\Gamma_d$$

where Γ_d is the dry adiabatic lapse rate discussed in Section 5.9. The difference in symbolism between $\partial T/\partial z$ and dT/dz corresponds precisely to the difference between the environmental and parcel viewpoints—the first being a snapshot of the local vertical environment, the second being the variation with height extracted from the history of a moving parcel.

Parcel and layer stabilities

As sketched in Fig. 5.9, the buoyant behaviour of the parcel is determined by the difference between the *environmental temperature gradient* $\partial T/\partial z$ and the parcel temperature gradient dT/dz, which is assumed to be $-\Gamma_d$.

(i) If the environmental temperature lapse rate is *superadiabatic* (temperature lapsing more quickly with height than the dry adiabatic—Fig. 5.9a),

$$\frac{\partial T}{\partial z} < -\Gamma_d \quad \text{or} \quad \frac{\partial \theta}{\partial z} < 0$$

upward displacement of an air parcel relative to its immediate environment (in the course of turbulent or convective motion, etc.) will make it slightly warmer and therefore less dense than its surroundings. The resulting buoyancy (Box 7.10) will then drive the parcel further upwards from its original position, generating more buoyancy, and so on. If instead the parcel moves downwards from its original position through the same superadiabatic environment, it becomes cooler and therefore denser than its environment, and the resulting negative buoyancy drives it further down. Since such *positive feedback* will amplify the slightest vertical displacement of the parcel, we see that it can remain in its initial set-up only if it is completely undisturbed. It is therefore in an *unstable* state in much the same way as a ball-bearing perched on the top of a smooth dome. And since every parcel making up this superadiabatic temperature profile is unstable in just the same way, we see that the whole layer is *convectively* (or *statically*) *unstable*.

(ii) If the environmental temperature lapse rate is sub-*adiabatic* (temperature lapsing more slowly with height than the dry adiabatic—Fig. 5.9b),

$$\frac{\partial T}{\partial z} > -\Gamma_d \quad \text{or} \quad \frac{\partial \theta}{\partial z} > 0$$

upward dry adiabatic displacement of an air parcel will make it colder and denser than its surroundings, while downward displacement will make it warmer and less dense, in each case generating buoyancy which opposes the initial displacement. Such *negative feedback* keeps the parcel in a stable state by returning it to its original level, like a ball-bearing in a bowl. And since every parcel in the sub-adiabatic layer is in the same situation, we see that the whole layer is *convectively* (or *statically*) *stable*.

(iii) In the same way we can show that a layer of air with an environmental lapse rate exactly equal to the dry adiabatic, i.e.

$$\frac{\partial T}{\partial z} = -\Gamma_d \quad \text{or} \quad \frac{\partial \theta}{\partial z} = 0$$

is in a state of *neutral convective* (or *static*) *stability*, with respect to dry adiabatic vertical displacement, meaning that any constituent parcel will tend to remain in its new position after vertical displacement, rather than move further away or back to its original position, like a ball-bearing on a flat horizontal plate.

5.10.2 Dry and saturated adiabatic convection and conditional instability

So far we have considered only the dry adiabatic parcel process. We can use the same kind of analysis to find the criteria for the convective stability of a layer whose constituent air parcels are following a saturated adiabatic process. Assuming that there is always enough water to keep descending air saturated, we can proceed exactly as before. Figure 5.9c is the equivalent of Fig. 5.9a except that the straight dry adiabat is replaced by the slightly curved saturated adiabat, whose lapse rates (Γ_s) are always less than the dry adiabatic value—quite significantly so in the low troposphere.

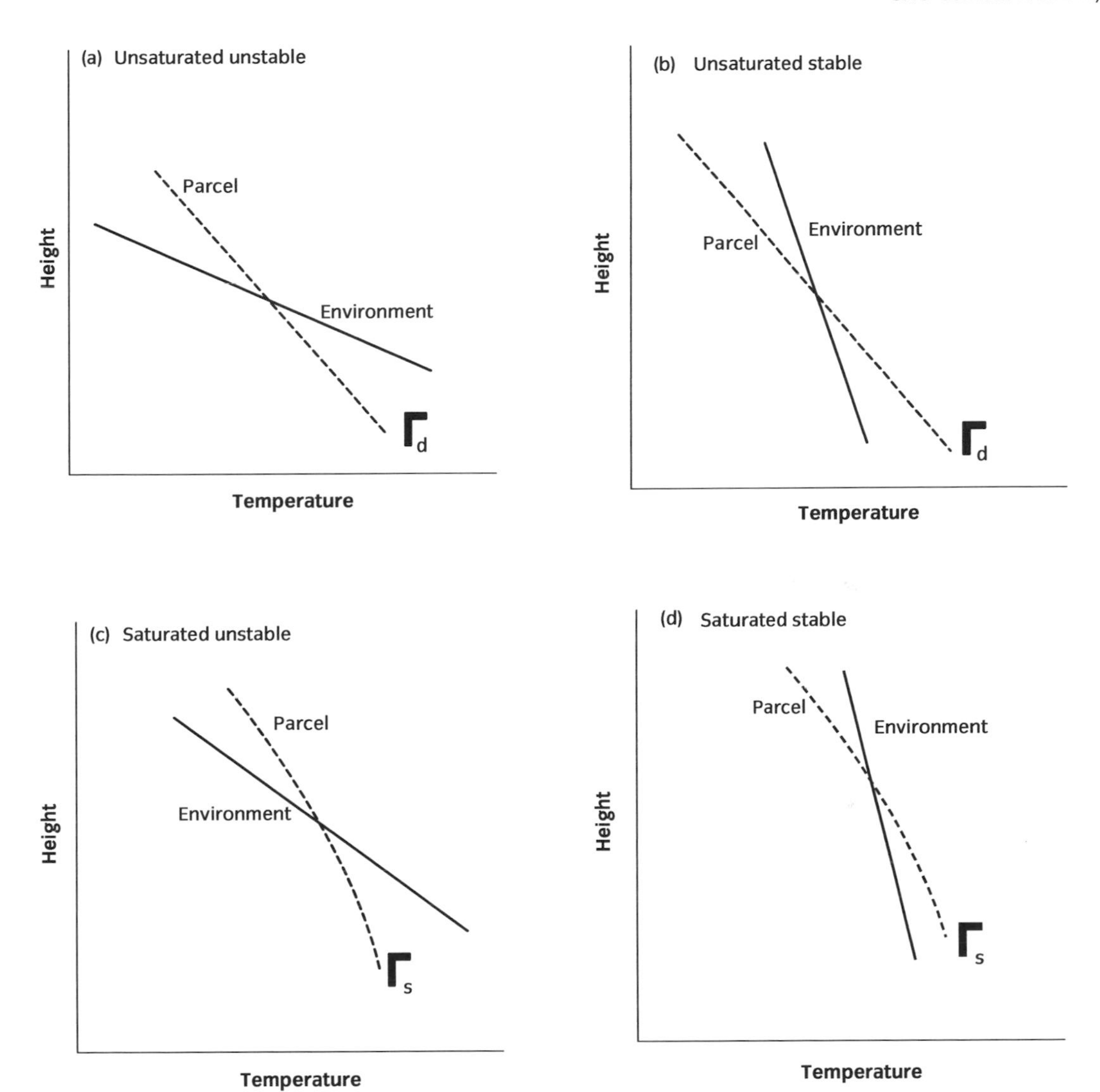

Figure 5.9 Parts (a) & (b) Parcel and environmental temperature profiles representing adiabatic convective stability and instability in an unsaturated atmospheric layer. Parts (c) & (d) represent the same stability regimes for saturated adiabatic convection in a saturated layer.

The criteria for convective instability, neutral stability, and stability are assembled in Table 5.1 for both dry and saturated adiabatic parcel processes. In each case, neutral convective stability occurs when the environmental temperature lapse rate exactly equals the lapse rate defined by the appropriate adiabatic parcel process, with instability or stability arising when the environmental lapse rate is respectively greater or less than the parcel lapse rate. Because two quite different parcel processes are possible (dry or saturated), there is a range of intermediate environmental lapse rates which may be either convectively stable or unstable depending on which parcel process is at work. Such a layer of air is said to be *conditionally unstable*, the condition being whether the parcel process is unsaturated or

Table 5.1 Criteria for convective stability

Environmental temperature gradient		Adiabatic parcel process		
in T	in θ	unsaturated		saturated
$\frac{\partial T}{\partial z} < -\Gamma_d$	$\frac{\partial \theta}{\partial z} < 0$	U		U
$\frac{\partial T}{\partial z} = -\Gamma_d$	$\frac{\partial \theta}{\partial z} = 0$	N		U
$-\Gamma_d < \frac{\partial T}{\partial z} < -\Gamma_s$	$0 < \frac{\partial \theta}{\partial z} < \left(\frac{d\theta}{\partial z}\right)_s$	S	conditional instability	U
$\frac{\partial T}{\partial z} = -\Gamma_s$	$\frac{\partial \theta}{\partial z} = \left(\frac{d\theta}{dz}\right)_s$	S		N
$\frac{\partial T}{\partial z} > -\Gamma_s$	$\frac{\partial \theta}{\partial z} > \left(\frac{d\theta}{dz}\right)_s$	S		S

Key U unstable N neutrally stable S stable

saturated, effectively whether the layer is cloudless or cloudy. *Conditional instability* can play an especially important role in the middle and low troposphere, where the difference between Γ_d and Γ_s can be large. It often arises when an unsaturated layer with a lapse rate significantly greater than the saturated adiabatic value appropriate to its temperature, is raised to saturation by slow ascent of the whole layer in a large-scale weather system (Section 11.6). Before the layer saturates it is convectively stable and quite quiet, but as it saturates and fills with cloud it becomes very unstable, generating vigorous convection within itself which may destabilize overlying layers if conditions are right, producing deep violent convection.

5.10.3 Real observations

Patterns like Fig. 5.9 can be identified on real data plotted on tephigrams, and analysed for the presence or absence of convection. For example, the shallow segment at the base of the sounding in Fig. 5.7 is strongly super dry-adiabatic and therefore convectively unstable—typical of shallow layers immediately overlying warmed land surfaces (Sections 5.3 and 8.9). The overlying layer, nearly reaching the 800 hPa layer in Fig. 5.7, is slightly but definitely sub-dry-adiabatic, and therefore convectively stable if unsaturated (as dew-point depressions suggest), but convectively unstable if saturated (cloud-filled). The rest of the troposphere has a temperature profile close to the saturated adiabatic, which suggests neutral convective stability if the air is saturated. In fact the considerable sub-saturation implied by the dew-point depression in this layer may be misleading, for reasons mentioned in Section 5.8.4 and discussed in *[27]*. The low stratosphere above the very sharp tropopause at 320 hPa in Fig. 5.7 is quite typical in containing a deep temperature inversion which is extremely stable for both saturated and unsaturated convection.

5.11 The maintenance of near-neutral stability

We have seen in principle how a layer becomes convectively unstable as soon as its environmental temperature lapse rate exceeds the appropriate parcel temperature lapse rate. Solar heating of the surface, together with cooling of the upper troposphere by net loss of terrestrial radiation (Section 8.6), continually push tropospheric environmental lapse rates toward these limits—the dry adiabatic lapse rate if the air is unsaturated, as is normal in the very low troposphere, and the saturated adiabatic lapse rate in rest of the troposphere. With that in mind we now examine how substantial layers of air respond to the appearance of convective instability within them.

Once convection is under way, according to the general law of mixing (Section 3.1) it will tend to establish an environmental temperature profile which is unaltered by further convective mixing of that sort. In the unsaturated low troposphere this is the dry adiabatic lapse rate, as widely observed. But in the patchily saturated middle and high troposphere, the situation is complicated because rising air tends to follow a saturated process, whereas sinking air does so only in the rather special case when precipitation is evaporating into it (Section 5.8.7); otherwise it tends to follow the dry process. In addition there is vigorous mixing between saturated clouds and their unsaturated environments. The complex situation is described further in Section 11.5, but is poorly understood.

5.11.1 Low troposphere

Although nearly dry adiabatic environmental lapse rates are widely observed in the surface-based layer, they are not universal. In high-latitude winters, strongly sub-adiabatic lapse rates, including quite strong temperature inversions, are maintained by seasonal surface cooling, especially over land and ice, because their effective heat capacities are much smaller than those of unfrozen seas and lakes (Section 10.4). In lower latitudes overland, weaker but still distinctly sub-adiabatic environmental lapse rates develop during cloudless nights, to be replaced by dry adiabatic lapse rates during the following morning in a repeating diurnal cycle which we now examine descriptively. The corresponding cycle over large water surfaces is almost imperceptible, because it is so heavily smoothed by their much larger heat capacities.

Figure 5.10 represents a typical environmental temperature profile observed just above a land surface in middle or low latitudes towards the end of a cloudless night. The ground and overlying air has cooled during the night by net emission of terrestrial radiation (*TR*), producing a convectively stable layer which may extend many tens of metres above the surface, and even contain a temperature inversion, as shown. This stability suppresses convection and indeed all vertical components of turbulent motion (Section 10.8) in the layer. And yet within a few hours of the dawn return of solar surface warming, the cool, quiet, stable layer has been replaced by a warming, deepening, convecting layer, whose dry adiabatic temperature profile is often maintained through the remaining daylight hours, until nocturnal cooling begins to stabilize the air from the surface upwards, re-establishing the stable layer to be destroyed by the next day's heating, and so on. Consider the stages of the cycle.

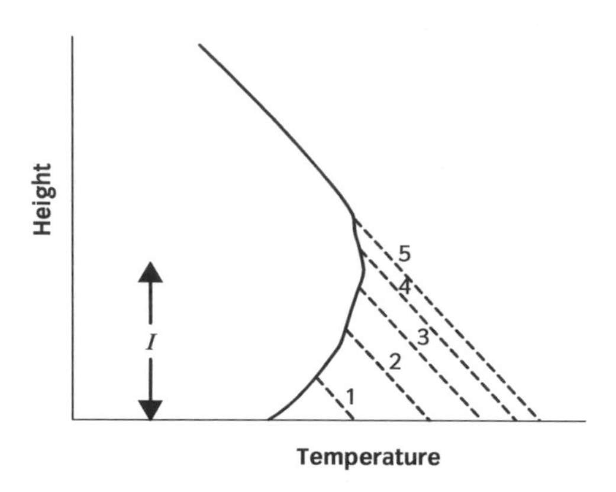

Figure 5.10 A convectively stable layer formed by cooling of the surface, with a temperature inversion throughout layer I. Numbered dashed lines represent successive stages in the evolution of the temperature profile during subsequent solar heating. In effect this is an idealized version of Fig 5.2, using temperature rather than potential temperature, and surmounted by a dry adiabatic layer (left from yesterday's convection) rather than a stable layer.

Morning

When the Sun comes up, the surface is warmed by absorption of solar radiation and begins to share this input with the immediately overlying air, mainly by conduction, but partly also by *TR* exchange (Section 8.6). As the surface air warms, the dry adiabatic lapse rate is exceeded in a very shallow surface-based layer, which begins to convect. Once started, convection pumps heat much more quickly into the overlying air than unaided conduction and *TR* exchange could have done, warming it and increasing the depth of the convecting layer (Fig. 5.10). Three processes are at work.

(i) The incessant mixing of the air parcels as they rise and fall in turbulent convection holds the vertical temperature profile of the convecting layer very close to the dry adiabatic temperature profile traced out by each moving parcel before it is mixed out of existence. This is the general law of mixing in action (Section 3.1), tending to produce a distribution (of temperature) unaffected by further mixing—the dry adiabatic lapse rate of the parcel process. It is true that rising parcels are consistently warmer than sinking ones (which is why they rise) but their small size and brief lifetimes ensure that temperature differences are small at any instant.

(ii) The gradual warming of the convecting layer (Figs. 5.2 and 5.10) is maintained because on average the air parcels rising into the layer are slightly warmer than the layer itself. As these warmer parcels mix out of recognizable existence, their slight temperature excesses are given to their surroundings, which therefore warm by many small stages while the environmental lapse rate remains close to the dry adiabatic value. In terms of Fig. 5.10, the temperature profile creeps bodily to the right as time passes, without changing slope.

(iii) The convecting layer eats progressively into the overlying stable temperature profile established by the previous night's cooling, increasing the depth of the convecting layer and incidentally bringing down some of the heat deposited there by previous days' heatings. The rate of warming of the convecting layer will tend to slow as its depth (and mass and heat capacity) increases, though this is offset by the increasing rate of heat input to the surface with the rising Sun. The net result is that air near the surface warms quickly at first, but then more and more slowly, though the slowing is restrained until late morning by the rising Sun. The rate of warming will become very small indeed if the deepening convecting layer connects up with the upper part of the nearly dry adiabatic layer left over from the previous day's convection, as often happens in the late morning, just before cumulus clouds appear. When clouds appear, convection in and above the cloud layer is invigorated by the release of latent heat by condensing vapour, and this can even re-invigorate convection in the sub-cloud layer if the conditions are right.

Afternoon and night

Although potential solar input begins to decline after local noon, it usually stays strong enough to maintain convection through the sub-cloud layer, and with it the dry adiabatic environmental lapse rate. Of course extensive cloud formation will reduce solar input to the surface below its clear air potential, at least patchily, but unless there is some large-scale change, like the arrival of a frontal cloud mass, solar surface warming will usually continue to drive convection and slowly warm the convective layer as the afternoon wears on.

Some little time before sunset, however, the rate of loss of heat by net *TR* from the ground (Fig. 8.6) begins to exceed the declining net input from the Sun, so that the temperature lapse rate quickly becomes sub-adiabatic very close to the surface. Convection there dies at once, and with it all convective heat flux from the surface. Air sufficiently far above the surface may still be stirred by residual convection and turbulence, but even if the stirring dies, the layer tends to retain a nearly dry adiabatic lapse rate because it is largely disconnected thermally from the cooling surface. Once formed, the new ground-based stable layer tends to deepen gradually through the night, reproducing by next dawn another version of the temperature profile with which the previous day began.

5.11.2 Additional points

When the solar input to a land surface is strong enough, the very small-scale convection which dominates the lowest metres of air is unable to pump sensible heat aloft as fast as it is arriving from the Sun. In this event the surface and immediately overlying air become warmer, and the resulting distinctly superadiabatic lapse rate drives the convection more and more vigorously, until the convected flux of sensible heat matches the input of solar heat (Sections 10.6–10.9). The resulting *superadiabatic* lapse rate can be quite marked in the lowest few metres of air, but falls so sharply with increasing height that, more than 10 metres or so above all but the most intensely heated surfaces, the dry adiabatic lapse rate prevails to within a few per cent to the top of the convecting layer, or to cloud base if lower.

Quite vigorous stirring is often observed in the absence of solar input to the surface—for example at night and over substantial depths of water. Provided that the surface is not being strongly cooled by net *TR* loss, a turbulent tumbling motion is often maintained by moderate or strong wind shears near the surface, as sketched in Fig. 5.10. This is termed *mechanical convection* to distinguish it from the more obvious *thermal convection* discussed throughout this section (Fig. 5.11). Over many ocean surfaces the sub-cloud layer is more or less continually stirred by a combination of mechanical and thermal convection. It is the vertical movement of stirred air, whether driven by wind shear or surface heating, together with the dry adiabatic parcel process, which holds the environmental lapse rates in the sub-cloud layer so close to the dry adiabatic value for so much of the time.

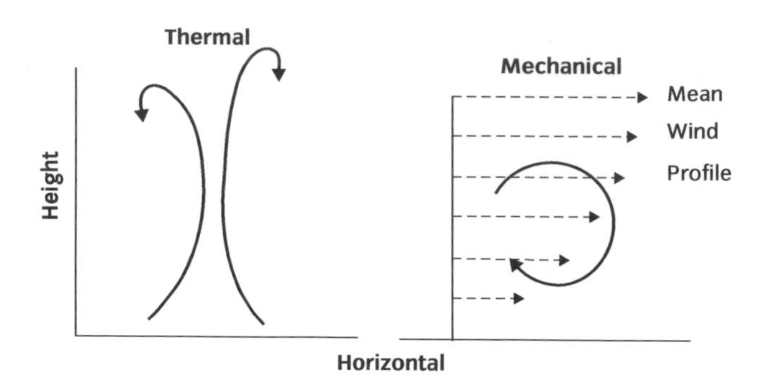

Figure 5.11 Thermal and mechanical origins of turbulence: thermal driven by buoyancy generated over a warm surface, and mechanical driven by the wind shear over the surface.

Checklist of key ideas

You should now be familiar with the following ideas.

1 The first law of thermodynamics in meteorological form, internal energy and external work, specific heat capacities of air at constant volume and constant pressure.

2 Ideal and actual isobaric heating and cooling of the atmosphere (including the atmospheric boundary layer and the importance of latent heat).

3 Measures of water vapour content (vapour pressure, specific humidity, relative humidity, dew-point, thermodynamic wet-bulb temperature, precipitable water content) and their interconnections.

4 Practical measurement of water vapour content.

5 The concept and application of the dry adiabatic reference process, dry adiabats and potential temperature.

6 The concept and application of the saturated adiabatic reference process, and relation with the dry adiabatic process.

7 Outline of the tephigram (including making a simple version) as an example of an aerological diagram, with application to real radiosonde data and associated weather.

8 The concept and application of equivalent and wet-bulb potential temperature.

9 Theory of dry and saturated adiabatic lapse rates, adiabatic motion and convective stability, instability, neutral stability, and conditional instability.

10 The maintenance of near neutral stability in the real atmosphere, with application to the diurnal cycle in the low troposphere overland.

Problems

Outline answers to these problems can be found on the Online Resource Centre. Answers to odd numbered problems can be found under Student Resources, answers to even numbered problems under Lecturer Resources.

Level 1

5.1 By applying the first law of thermodynamics qualitatively to a parcel of air, firstly in a rigid container, and secondly in a slack plastic bag, argue why the specific heat at constant pressure (C_p) should be larger than C_v.

5.2 Examine the following list and decide (with reasons) which processes are likely to be approximately isobaric and which are not: daytime warming of the full depth of the atmosphere; air in a shower cloud updraught; air flowing horizontally through the core of a jet stream; air spiralling horizontally into the core of a tornado; the atmospheric boundary layer during nocturnal cooling.

5.3 The vapour content of surface air tends to rise when there is evaporation from a land surface. Name and explain what surface activity tends to make it fall, and extend these to ocean and ice or snow surfaces. Briefly describe what happens in these last two cases.

5.4 Air in the upper branches of the Hadley circulation may take 10 days to move from the equatorial to the subtropical zones. Consider the applicability of the adiabatic assumption to this process.

5.5 If meteorology had been pioneered on the Tibetan Plateau (about 4 km above mean sea level), the reference level for potential temperature might well have been chosen to be 600 rather than 1000 hPa. In this case, consider the following pressure levels and decide in each case whether actual temperatures would be greater, equal to, or less than 'Tibetan potential temperatures': 1000, 700, 600, 300, 100 hPa.

5.6 In what circumstances are the following vertical temperature gradients unrealistic and realistic. Well above the surface: (a) 1 °C lapse in 200 m; (b) 1 °C in 80 m; (c) rise of 2 °C in 300 m. Close to the surface: (d) 2 °C lapse in 50 m; (e) 2 °C rise in 50 m.

5.7 On a certain nearly saturated temperature profile at 500 hPa, place the following in order of ascending numerical value: air temperature, wet-bulb potential temperature, equivalent potential temperature, dew-point, potential temperature, wet-bulb temperature.

Level 2

5.8 On a certain cloudless night, an infrared radiometer showed that the net rate of loss of radiant energy from a ground surface averaged 50 W m^{-2} for 8 hours. If the heat lost is drawn from the first 30 hPa of overlying air, find the resulting fall of air temperature, assuming it to be uniformly distributed through the layer. What actually happens usually?

5.9 The saturation vapour pressures at 15 and 10 °C are respectively 17.0 and 12.3 hPa. Find the associated saturation specific humidities when the atmospheric pressure is 1,000 hPa. Find the relative humidity of air which has a temperature of 15 °C and a dew-point of 10 °C.

5.10 On a certain summer's day in mid-latitudes the air temperature at screen level is observed to be 25 °C. If there is no significantly superadiabatic layer near the surface, what is the minimum height of the freezing level above the surface? Use a tephigram (Fig. 5.7) to estimate a more realistic value in mid-latitudes.

5.11 Given that certain air at 1,000 hPa pressure and 15 °C temperature has a wet-bulb temperature of 10 °C, find its specific humidity and relative humidity, assuming that the measured wet-bulb temperature is identical to the thermodynamic wet-bulb temperature, and taking the saturation specific humidity of air at 10 °C to be 7.65 g kg^{-1}, and using the saturation vapour pressure from Fig. 5.5.

5.12 In a real mishap which was fortunately not fatal, an airliner depressurized suddenly by losing a cargo door at altitude. If the internal and external pressures were 750 and 400 hPa just before depressurization, and the internal air temperature was 22 °C, estimate the internal temperature just after depressurization, assuming dry adiabatic cooling. After the crisis was over, several passengers remarked on the sudden chill.

5.13 In middle latitudes, air at the 300 mbar level is often observed to have temperatures as high as –37 °C in the vicinity of jet-stream cores. Assuming the air to be saturated, calculate its potential temperature, and use Fig. 5.7 to find its wet-bulb and equivalent potential temperatures.

Level 3

5.14 Recalculate the temperature fall in problem 5.8 if the heat loss is shared with the air layer and a 20 cm deep layer of ground. Assume ground density 2,000 kg m^{-3} and specific heat capacity 2,000 J kg^{-1} K^{-1}. In which direction would the answer be changed if the cooling produced fog?

5.15 In the event described in problem 5.12, passengers reported a temporarily dense fog in the cabin just after depressurization. Assuming realistically that the cabin wet-bulb temperature was 17 °C before the event, use Fig. 5.7 to make a more realistic estimate of the cabin temperature just after depressurization. Why was the fog only temporary?

5.16 Discuss the realism of dry adiabatic ascent from the surface to the jet core in Problem 5.13. If instead there was saturated adiabatic ascent, what global location for the surface source is suggested by the data?

5.17 Consider a situation in which nocturnal surface cooling under clear skies establishes a convectively stable layer in the first 300 m of the atmosphere, with a temperature inversion in the first 100 m. Sketch this on a graph of temperature against height, and on an equivalent graph of potential temperature versus height. If an approaching depression now partially reverses the cooling by bringing in warm cloudy overcast, and rising winds encourage mechanical convection through the lowest 200 m, sketch the altered temperature profile on the same two diagrams.

5.18 Estimate the precipitable water content of the Long Kesh sounding (Fig. 5.7), and compare it with values from saturated cold and warm atmospheres (0 °C and 20 °C saturated adiabats). Work in 100 hPa slabs and assume the humidity profile to be accurate.

Cloud and precipitation

6

6.1 Introduction

The troposphere is full of cloud: at any instant about half of the Earth's surface is overlain by cloud ranging in depth from a few tens of metres to more than 10 kilometres. Cloud is extremely variable in both time and space, and convective clouds are especially ephemeral. Within half an hour of its first appearance as a little cloudlet, a cumulonimbus may fill a thousand cubic kilometres of troposphere, and yet have vanished within a further hour, leaving no trace of its powerful maturity beyond a carpet of soaked ground and cool surface air, and decaying tufts of anvil cirrus in the high troposphere. Lifetimes of individual cloud droplets and crystals are even shorter than this. If you watch a big cumulus for a few minutes, you will see that it is in ceaseless commotion, with bumps and filaments of cloud emerging from the main mass and evaporating within a few tens of seconds. Observed rates of precipitation and estimates of cloud water mass suggest that even in cloud interiors, many individual droplets and crystals exist for only a few tens of minutes. The great swaths of altostratus and nimbostratus associated with mid-latitude fronts are apparently much more persistent, often lasting a week or more on weather maps and satellite panoramas, but again observation shows that air flows through them so quickly that any particular air parcel will remain within the cloud mass for only a day or two.

Clearly cloud formation and dissipation is rapid and widespread in the troposphere. In the present chapter we will examine these mechanisms in some detail, and find that they are surprisingly complex.

6.2 Saturation

The concept and physical reality of *saturation* is central to any description of the formation, maintenance, and dissipation of cloud in the atmosphere.

6.2.1 Concept

Figure 6.1a depicts a flux of water molecules evaporating from a plane surface of pure water into an overlying vapour, and an opposing flux of water molecules

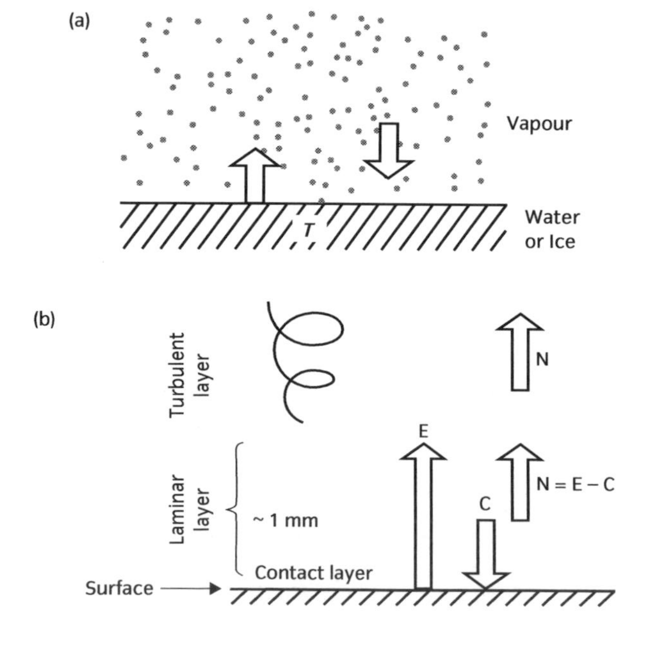

Figure 6.1 (a) A saturated vapour overlying a plane surface of pure ice or water. (b) Low level boundary layers: contact layer, remainder of laminar boundary layer, and the base of the turbulent boundary layer. The contact layer is saturated at the surface temperature, but is losing vapour by diffusion through the remainder of the laminar boundary layer, whose upper boundary is shedding vapour into the usually much drier base of the turbulent boundary. As a result the net upward vapour flux in the turbulent boundary layer is fed by net evaporation from the surface through the saturated contact layer.

from the vapour condensing onto the water. At any given common temperature of surface and vapour, there is a definite vapour density ρ_{sv} (the *saturation vapour density*) at which the fluxes balance, so that surface and vapour are in dynamic equilibrium. The *saturated vapour* has a corresponding *saturation vapour pressure* (*SVP*), usually written as e_s, which is given by the equation of state for water vapour, and whose dependence on temperature T has nearly the same shape as ρ_{sv} in the relatively small range of atmospheric Ts on the absolute scale.

$$e_s = \rho_{sv} R_v T$$

Figure 6.2 shows that *SVP* for plane pure water follows a nearly pure exponential growth with temperature across this temperature range, growing from 6.4 hPa at 0 °C to over 40 hPa at 30 °C. The rise of *SVP* with temperature corresponds to the increasing ease with which water molecules escape from the surface as its temperature and molecular agitation increases. The nearly exponential profile

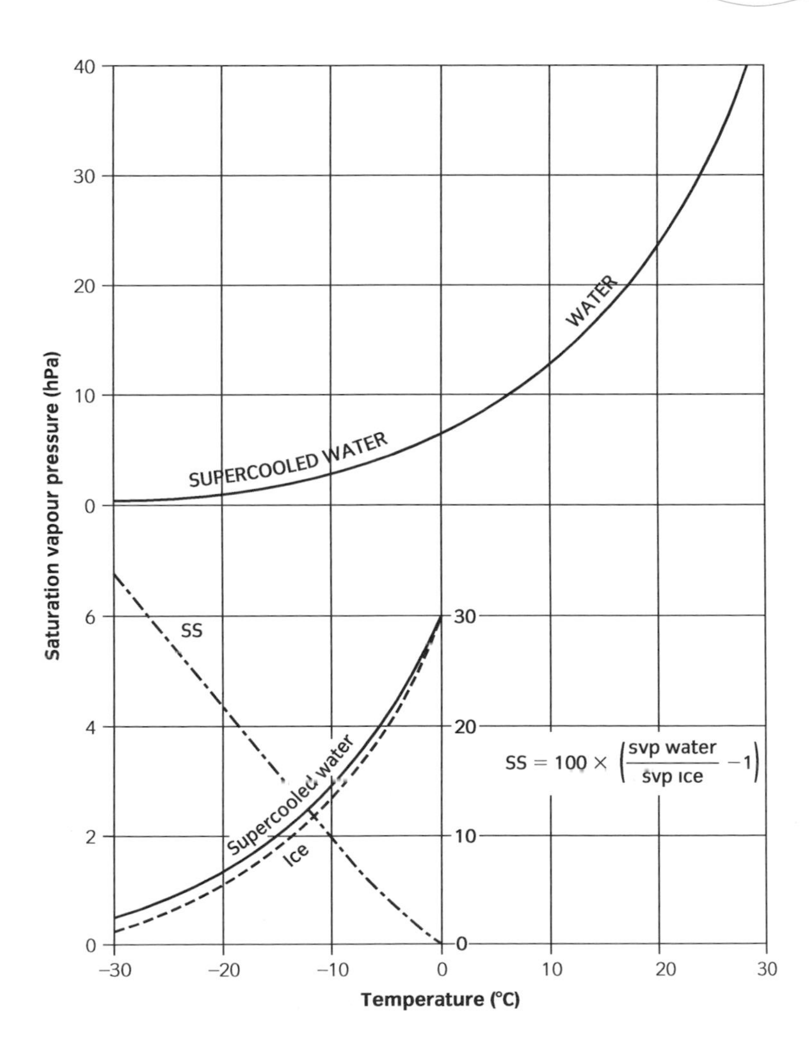

Figure 6.2 Variation of saturation vapour pressure with temperature. The curve is a nearly pure exponential, but below 0 °C saturation values over supercooled water exceed those over ice, as shown by the separation of the full and dashed lines. The chain line in the lower part of the diagram shows the supersaturation (SS in %) relative to ice for a vapour which is just saturated relative to supercooled water. You can find corresponding vapour densities by using the equation of state for vapour (Section 4.1).

continues below °C for a *supercooled* water surface (i.e. which has remained liquid despite being below its freezing point), falling to a fraction of 1 hPa at −30 °C, the lowest temperature at which supercooled water cloud is commonplace in the atmosphere.

Figure 6.2 also includes SVPs for water overlying a plane surface of pure ice—i.e. where water molecules *subliming* from the ice (evaporating directly without melting) are in dynamic equilibrium with vapour molecules condensing onto it. Saturation vapour densities and pressures over an ice surface are less than those over a supercooled water surface at the same temperature, because evaporation from ice is inhibited by its stronger intermolecular bonding. As a result, a vapour which is saturated with respect to supercooled water, at a certain temperature, is distinctly *supersaturated* with respect to ice at that temperature (i.e. its relative humidity RH with respect to ice is greater than 100%). Figure 6.2 shows that the supersaturation SS ($= RH - 100$) with respect to ice rises almost linearly with decreasing temperature below 0 °C, and exceeds 6% below about −26 °C, a significant result which plays an important role in the development of clouds containing both ice and supercooled water (Section 6.11.3).

Note that in all this it is the vapour and not the dry air which is saturated, despite conventional usage, since the presence of dry air has no effect on vapour pressure, according to Dalton's law of partial pressures (Section 4.1). We use the term 'saturated air' on the understanding that it means moist air whose vapour is saturated.

6.2.2 Reality (Fig. 6.1b)

In terrestrial conditions, air in contact with liquid or solid surfaces has a shallow laminar boundary layer (LBL), of order 1 mm thick, in which transport of vapour and other gases, and heat and momentum, are all dominated by molecular diffusion rather than the turbulence which prevails in the atmosphere at large (Fig. 6.1b, Sections 3.1 and 10.7). Heat and vapour diffuse so freely between a water or ice surface and its LBL that

(i) the surface and *contact layer* (the very shallow face of the LBL next to the surface) have the same temperature, and

(ii) the contact layer is kept saturated at this common temperature.

In terrestrial conditions this continues to be the case even when air in the rest of the LBL and the atmosphere beyond may have different temperatures and vapour pressures.

If air in the LBL beyond the saturated contact layer has a lower vapour pressure, vapour will diffuse outwards from the contact layer, but any would-be vapour deficit in the contact layer itself will be made good immediately by an excess of evaporation over condensation at the surface, so that the contact layer of the LBL remains effectively saturated even though vapour is passing through it from the surface to the larger air beyond. In this way a shallow pool of water can evaporate on a windy day, even though the air in contact with the water remains effectively saturated; and a patch of snow or ice can sublime away through a contact layer which remains effectively saturated with respect to ice. The same process maintains the chilled, saturated contact layer round a wet-bulb thermometer, while the warmer less humid flow of ambient carries away the evaporated water (Section 2.3).

Note that saturation has been defined in terms of plane surfaces of pure water or ice. Similar dynamic equilibria of vapour fluxes apply over the curved surfaces of impure water or ice typical of atmospheric aerosol and cloud droplets, though equilibrium vapour pressures and densities for tight curvatures (small droplets or crystals) can differ significantly from those over plane pure surfaces. And equilibria can vary with the amount and type of impurity in the water or ice. Such differences play an important role in cloud formation, development, and dissipation, and will be examined in some detail in following sections. For simplicity we reserve the term 'saturation' for equilibria with plane surfaces of pure water or ice.

6.3 Sub-cloud layer

Clouds in the lower troposphere are almost always separated from the Earth's surface by a cloud-free layer whose depth may range from less than a hundred metres to several kilometres, depending on weather conditions. Indeed the visible presence of cloud above a substantial cloud-free layer overlying the surface is so normal (Fig. 6.3) that we might suppose its explanation to be trivial. In fact a full explanation involves a complex web of meteorological factors, as suggested by the following descriptive outline.

Seventy per cent of the Earth's surface is water, and much of the rest is moist ground or ice. According to the previous section, the air in contact with all these wet, moist, or icy surfaces is saturated at surface temperatures. We might therefore expect vapour to diffuse from this saturated layer to build and maintain a much deeper layer of saturated air covering most of the Earth's surface. And given the great efficiency of the cloud-forming processes to be described presently (Sections 6.7 and 6.8), we should expect much of the Earth to be shrouded in fog. In fact very little fog is observed, which suggests that there are mechanisms at work removing water from the atmosphere quickly enough to prevent the accumulation of fog. These are the precipitation processes, which ensure that some cloud droplets and crystals are always growing big enough to fall back to

Figure 6.3 Smallish cumulus over the Scottish Highlands. The nearly uniform cloud base was just above the 1,200 m summit of Ben Lawers behind the camera, essentially unaffected by the uneven terrain, which includes Loch Tay deeply out of sight in the foreground.

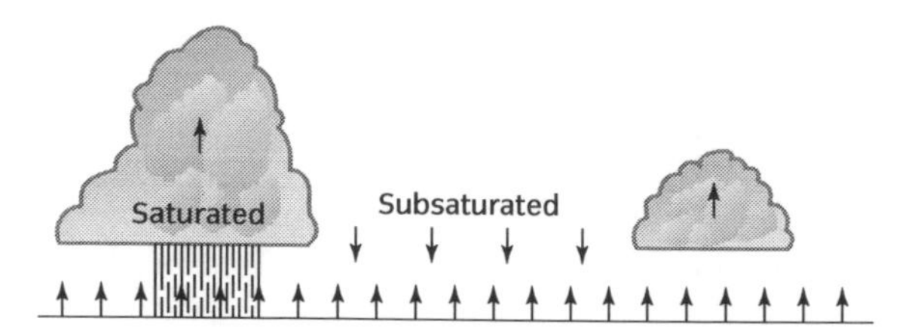

Figure 6.4 Vertical pathways of vapour, cloud, and precipitation in the atmosphere.

the surface under gravity. As a preliminary to later detail (Section 6.11), we present a brief outline of precipitation processes in the rest of this section.

The least efficient precipitation process is observed in deep fog layers, which continually lose part of their cloud water through a fine drizzle of droplets intermediate in size between cloud droplets and normal drizzle droplets; these are barely perceptible except when they touch and chill our faces or speckle car windscreens. The loss of fog water in this way is usually insufficient to clear the fog, though some fog clearance schemes use its artificial encouragement.

More efficient types of precipitation occur in strong or persistent updrafts and the much deeper layers they maintain. In these, rates of precipitation onto the underlying surface area can greatly exceed the rate of evaporation from that surface. The situation is sketched in Fig. 6.4: only a fraction of the Earth's surface is covered by efficiently precipitating cloud, but this cloud and its precipitation are maintained by evaporation from virtually the whole Earth surface. Between the cloudy updrafts there is sinking air, desaturated by dry adabatic descent after prior precipitation (Section 5.7). As this air approaches the surface it maintains a layer of sub-saturated and therefore cloud-free air before merging with the saturated contact layer clinging to the surface. Without knowing the details, we can sense that the typical depth of sub-cloud layers and the fraction of the global surface overlain by precipitating cloud, are in some way controlled by the efficiency of the precipitation processes at work in clouds. If these were all as feeble as those in fog, then our initial expectations of a largely fog-bound planet would be realized.

According to Fig. 6.4, the sub-cloud layer is fed by sub-saturated vapour from above and saturated vapour from beneath. In and just above the very shallow saturated contact layer, turbulent mixing is inhibited by viscosity in the LBL, and by the very small inefficient eddies in the base of the turbulent boundary layer (Sections 10.6 and 10.7), so that there is a strong vertical lapse of vapour content, as sketched in Fig. 6.5. This transition zone is seldom more than a metre deep, and underlies a deep layer, occupying almost all of the sub-cloud layer, in which turbulence maintains a nearly uniform distribution of water vapour. This well-mixed layer plays such a basic role in the formation of cloud in the low troposphere that it warrants detailed consideration in the following sections.

6.4 Well-mixed sub-cloud layer

Consider the vertical distribution of water vapour in a well-mixed layer extending upwards from a shallow transition zone (Fig. 6.5) immediately overlying an

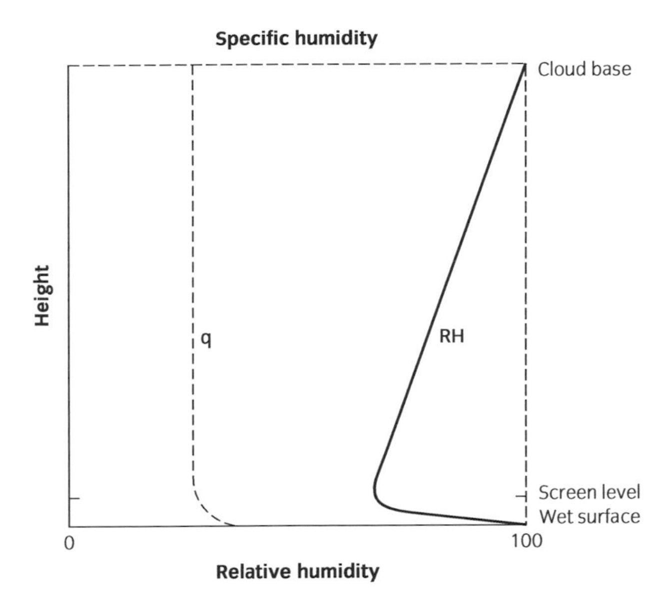

Figure 6.5 Idealized profiles of specific humidity q and relative humidity RH in a well-mixed subcloud layer over a moist surface.

evaporating surface. Ignoring relatively small volumes dampened by evaporation from falling precipitation, there are no sources or sinks of vapour between the lower and upper limits of the well-mixed layer (transition zone and cloud base respectively), so that continual efficient mixing of vapour and air will maintain a uniform specific humidity q throughout the well-mixed layer

$$q = \text{massof parcel vapour / mass of parcel moist air} = \frac{\rho_v}{\rho} \qquad 6.1$$

where ρ_v and ρ are the densities of vapour and moist air (Section 5.4). Since air density lapses with increasing height, vapour density must lapse in proportion, to maintain uniform q. In a typical 1 km depth of the sub-cloud layer, both air pressure and air density fall by about 10%, so that vapour pressure and density must do the same.

The stirring of the sub-cloud layer also establishes an equilibrium vertical distribution of temperature, which is the dry adiabatic profile (Section 5.9). This can be converted into a profile of saturation vapour density ρ_{sv} by using Fig. 6.2 or thermodynamic tables. For example, the temperature lapse of 10 °C expected in a kilometre depth of air is associated with a lapse of ρ_{sv} from 12.8 to 6.8 g m^{-3}, assuming that air temperature is 15 °C at the base of the well-mixed layer (the surface). In warmer layers the lapse of ρ_{sv} is even larger because of the increasing steepness of the slope of Fig. 6.2 at higher temperatures.

We see now that the lapse of saturation vapour density ρ_{sv} in the well-mixed layer greatly exceeds the 10% lapse of actual vapour density ρ_v. Since relative humidity RH is defined to be

$$RH = 100 \frac{\rho_v}{\rho_{sv}} \qquad 6.2$$

it follows that RH must increase with increasing height from a minimum at the base of the well-mixed layer (X), as shown in Fig. 6.5. This increase means that at some level, higher or lower depending on the degree of sub-saturation of the air at X, the value of RH reaches 100%, and since this corresponds to saturation, it follows that any further ascent by an air parcel will produce supersaturation and trigger cloud formation by condensation of excess vapour onto cloud droplets. The height of first saturation therefore corresponds to cloud base—the top of the sub-cloud layer. In fact, to the extent that the atmosphere behaves like our adiabatic and well-mixed model, cloud base coincides with the lifting condensation level (LCL) defined in Section 5.8.

Note that we have now explained something which might seem too commonplace to require explanation—that saturation and cloud develop in rising rather than sinking air. If, for example, the variation of ρ_{sv} with temperature had been very much smaller than shown in Fig. 6.2, so that the vertical lapse of ρ_{sv} in the well-mixed layer was less than the lapse of actual vapour density, then air would desaturate with ascent rather than descent, and there would be no familiar cloud base or cloud or precipitation. And without precipitation, Earth would be as fog-bound as naively assumed at the beginning of Section 6.3, with enormous consequences for the weather and climate of Earth's surface.

6.5 Dew-point profile

More detailed humidity profiles in the sub-cloud layer are shown best by dew-point profiles. Since air at the base of the well-mixed layer (Fig. 6.6) is significantly sub-saturated, its dew-point T_d is well below its temperature T. If follows from the definition of dew-point (Section 5.4) that its value at any particular height (or pressure) in the sub-cloud layer is found on Fig. 6.6 by sliding isobarically (i.e. at constant pressure) to the left until saturation is reached, i.e. until the

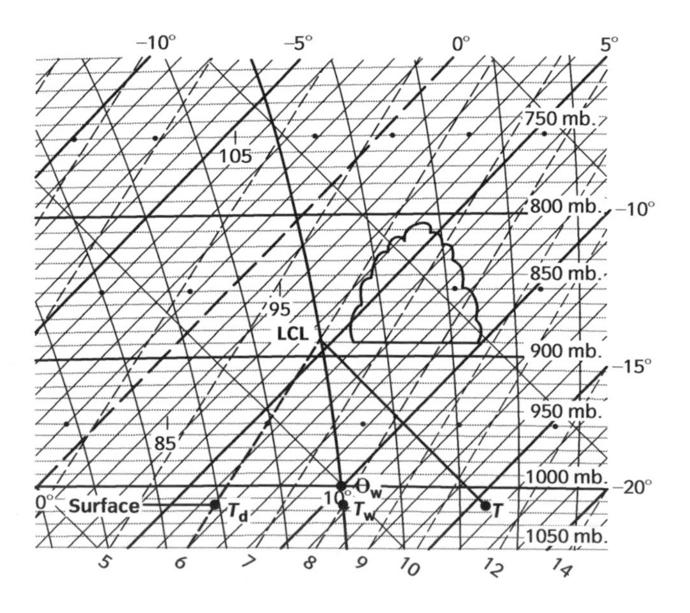

Figure 6.6 Tephigram of the lower troposphere showing idealized profiles of temperature T, wet-bulb temperature T_w and dew point T_d in a well-mixed, dry adiabatic sub-cloud layer, with lifting condensation level LCL. This is the completed version of Fig 5.8. Average temperatures and pressures in the subcloud layer are such that 1 hPa pressure lapse corresponds to about 9 m in elevation.

saturation specific humidity q_s (whose isopleths are the faint dashed lines obliquely crossing the isotherms) is equal to the unchanging specific humidity q of the air. The temperature at which this happens is the local dew-point T_d.

Now according to our simple model of the well-mixed layer, q is the same at all levels up to the lifting condensation level (LCL). It follows that the T_d profile in Fig. 6.6 must lie along a line of constant q_s ($= q$). The detailed thermodynamics underlying the tephigram ensure that q_s isopleths are nearly straight lines running almost perpendicularly across the dry adiabats, so that the T and T_d profiles in the well-mixed layer are nearly straight lines which converge sharply and meet at the LCL (where $T = T_d$).

Note that we can now complete the full expression of Normand's theorem (Section 5.11 and Fig. 6.6) by stating that the dry adiabat through the dry-bulb temperature T, the saturated adiabat through the wet-bulb temperature T_w, and the saturation specific humidity through the dew-point T_d, all meet at the lifting condensation level LCL. Normand's theorem applies to T, T_w, and T_d at any level in the well-mixed sub-cloud layer.

BOX 6.1 Dew-point lapse rate

According to more advanced texts *[17]* the variation of saturation vapour pressure e_s with absolute temperature T is given by the *Clausius–Clapeyron equation*

$$\frac{de_s}{dT} = \frac{L}{T(\alpha_{sv} - \alpha_w)}$$

where α_{sv} and α_w are the *specific volumes* (volume per unit mass or $1/\rho$) of saturated vapour and water respectively. We can ignore α_w because it is very much smaller than α_{sv}. Using the equation of state for water vapour ($\alpha_{sv} = R_v T/e_s$), we get

$$\frac{de_s}{dT} = \frac{L\ e_s}{R_v\ T^2} \quad \text{B6.1a}$$

where R_v is the specific gas constant for water vapour. A solution of this equation (for e_s as a function of T) fits the curve of Fig. 6.2 very closely.

Since the vapour pressure e of an unsaturated parcel is by definition equal to the saturation vapour pressure e_s at the dew-point T_d of the parcel, we can reinterpret Eqn B6.1a as a relationship between vapour pressure and dew-point.

$$\frac{de}{dT_d} = \frac{L\ e_s}{R_v\ T_d^2}$$

A simple expression for specific humidity q (Eqn 5.6) rearranges to show that vapour pressure e and total atmospheric pressure p are directly proportional

$$e = \frac{q\ p}{\varepsilon}$$

so that for constant q, $de = (q/\varepsilon)\ dp$, and

$$\frac{dp}{dT_d} = \frac{L\ p}{R_v\ T_d^2}$$

Assuming hydrostatic equilibrium (Eqn 4.6) to relate dp to dz, and the equation of state for air (Eqn 4.2) as usual, this can be rearranged to express the variation of dew-point with height

$$\frac{dT_d}{dz} = -\frac{g\ R_v}{L\ R}\frac{T_d^2}{T} \quad \text{B6.1b}$$

Since realistic T_d and T differ only slightly on the absolute scale this becomes

$$\frac{dT_d}{dz} \simeq -\frac{g\ R_v\ T_d}{L\ R} \quad \text{B6.1c}$$

At a typical dew-point of 283 K in the low troposphere, the right-hand side has magnitude 1.8×10^{-3} K m^{-1}, which means that dew-point lapses by 0.18 K per 100 m rise through the well-mixed layer, as can be checked from the crossing of isotherms and iso q_s in Fig. 6.6. Since T lapses by 0.98 K per 100 m (the dry adiabatic lapse rate), we see that T and T_d converge by 0.8 K per 100 m rise in a typical well-mixed layer in the low troposphere.

It is shown in Box 6.1 that in conditions typical of the low troposphere, vertical profiles of T and T_d profiles converge at such a rate that dew-point depression $T - T_d$ falls by 0.8 °C for each 100 m rise through the well-mixed sub-cloud layer. It follows that if the dew-point depression $T - T_d$ at the base of the layer well-mixed layer is D_o °C, then the height of the LCL above the base is 125 D_o metres. This rule of thumb, or its slightly more accurate equivalent on a tephigram, is often used to calculate cloud base height from Stevenson screen measurements of T and T_w (converted to T and T_d). In fact, heights calculated in this way often underestimate true cloud base heights by 10% or more, suggesting that our model of the well-stirred layer needs improvement. It seems likely that at 1.5 m above the ground surface, screen level may be a little too low to be fully representative of the base of the well-mixed layer, leading to an under-measurement of D_o and an underestimation of cloud base height.

The simple model discussed in this and preceding sections applies only to well-stirred layers, i.e. layers in which air parcels are continually moving up and down in thermal or mechanical convection or both. Though such stirring is widespread in the sub-cloud layer, it is not universal, being either modified or absent in the convectively stable layers which usually develop in the first 100 m over a land surface on cloudless nights. Falling temperatures and rising relative humidities are widespread near the surface as nocturnal cooling progresses, and saturation and fog may result, but no simple relations apply to their vertical profiles to compare with those in well-mixed layers.

6.6 Condensation observed

You can watch cloud forming in rising air most easily when weather conditions are encouraging fair weather cumulus (Fig. 6.3). Within a few tens of seconds of an updraft reaching its lifting condensation level (LCL), the slight haziness typical of the top of the sub-cloud layer becomes patched and veined with thickening cloud filaments which quickly merge to form a small cumulus. The nearly horizontal base, representing the LCL, and the cauliflower-shaped upper parts, are familiar from casual everyday observation (Fig. 6.7), but the actual speed of formation and development of such clouds are easily missed by the sporadic skyward glances encouraged by modern living. A few minutes spent watching the sky on a bright morning, just as cumulus are beginning to appear, will correct the misleading impression of the static drift of cloud left by casual observation: they are much more dynamic and transient than Wordsworth's famous but untypically inaccurate image.

I wandered lonely as a cloud that floats on high o-er vales and hills*

* Note that Wordsworth may have been referring to a solitary lenticular cloud, formed at the crest of a wave of air flowing high above one of the many small mountains in the English Lake District (Section 11.8). However, the problem with his image is then that 'floats' suggests motion with the flow, whereas such waves of air are typically locked onto the underlying mountain, like water over a boulder in a shallow river. The loneliness of such a cloud certainly can be visually striking, and that was the image he wanted.

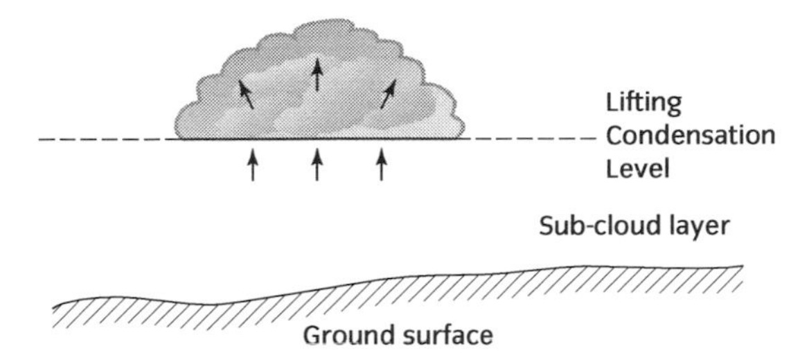

Figure 6.7 Outline of a typical small cumulus forming above the lifting condensation level (LCL). The buoyant air is rising from the ground, but the uniform horizontal LCL reflects the well-mixed sub-cloud layer rather than the uneven surface.

BOX 6.2 A simple cloud chamber

Find a large glass bottle (about 5 litres capacity is ideal), place a small quantity of water inside and fit the mouth with a bung pierced by a narrow glass tube tipped by a length of rubber tubing soft enough to form an air-tight seal under finger pressure (Fig. 6.8).

Illuminate the interior of the bottle with an intense, narrow beam of light. A useful extra is a water manometer tall enough to measure a pressure difference of 30 hPa (i.e. a 30 cm vertical difference in water level—NN 6.1).

After shaking the bottle to ensure that its air is saturated at the glass temperature, blow into the rubber tube and maintain the excess pressure for about 30 seconds by pinching. Although in principle the blowing must introduce some warm moist air, back-pressure keeps this so small that the main effect of blowing is to raise the interior air pressure, and hence the air temperature by compressive warming. A little further shaking while the tube is pinched ensures that the air resaturates at the higher temperature (making good the temporary desaturation produced by warming). Release the rubber tube to allow the excess interior pressure to fall to zero, chilling the saturated air by decompression and nearly instantaneously filling the bottle with an apparently thin cloud. Actually this would be a dense fog on the scale of a bathroom and larger, since the droplet number density would limit visibility to a few metres (Fig. 6.9).

This sequence shows that cloud can form very quickly in air which calculation shows to be only slightly supersaturated. If, for example, pressure falls by 30 hPa on depressurization, according to NN 6.1 the dry adiabatic fall of temperature must be about 2.5 °C.

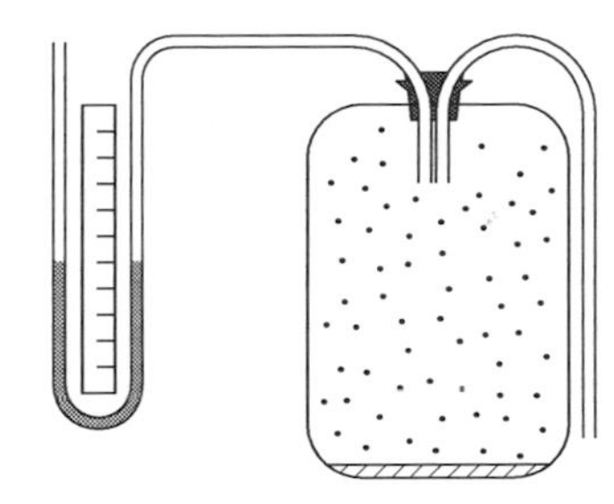

Figure 6.8 A simple cloud chamber showing bottle, inlet, and venting tube (hanging right), and water manometer.

Since cloud formation makes it a saturated rather than a dry adiabatic process (Section 5.7), the temperature fall is about half this, say 1.3 °C. According to Fig. 6.2, air saturated at the initial temperature (i.e. just before depressurization) should be about 7% supersaturated just after depressurization. More sophisticated observation and theory show that cloud droplets can grow so quickly at the expense of the supersaturating vapour during depressurization, that they hold the supersaturation to 1% or less. Temporary supersaturations are even lower in atmospheric cloud formation and growth, because depressurization is much slower than in the cloud chamber.

If you look closely into the cloud bottle just after depressurization, you will see a moving grain of bright pinpoints in the strongly illuminated cloud. Though this is all that can be seen of the individual cloud droplets without magnification, we can estimate their size very roughly by gauging their fall speed after initial swirling has died away. The droplets are then seen to

Figure 6.9 A thin (visibility about 300 m) winter morning fog in England, probably produced by inland advection of warm, maritime air and overnight ratiative chilling. Light from distant objects is scattered out of the line of sight and replaced by light scattered from intervening droplets.

be sinking at a rate of millimetres per second. A well-established relationship between droplet size and fall speed (Fig. 6.10) shows that droplets falling at such speeds have diameters of a few micrometres ($\mu m = 10^{-3}$ mm), which compare reasonably with droplet radii observed in atmospheric cloud, where special measurements by aircraft indicate average radii of about 5 μm.

Cleaning the chamber air

Further work with the cloud chamber reveals an interesting complication. Several minutes after depressurization, almost all cloud droplets have sunk to the bottom of the bottle. If it is again pressurized, resaturated, and depressurized, the resulting cloud is noticeably thinner than before. Further repetitions of the cycle produce less and less cloud even though the degree of pressurization and subsequent cooling is maintained, but the trend is dramatically reversed when a small amount of smoke, from a freshly snuffed match, for example, is introduced into the bottle neck. The merest wisp is enough to produce a dense cloud at the next depressurization, whose removal will require further cycles of pressurization, depressurization, and settling.

This complication shows that cloud forms much more readily in smoky air than it does in clean air. In fact, as mentioned in Section 3.7, smokes are just a part of the considerable range of atmospheric aerosol particles which have been shown (by work with cloud chambers) to act as *condensation nuclei*, encouraging cloud droplet formation at low supersaturations. Aitken and Coulier used cloud chambers in the late nineteenth century to count the particles in atmospheric aerosol by forming cloud on them and then allowing them to settle onto a counting plate on the chamber floor. In so doing they isolated the most effective condensation nuclei. In industrial regions the connection between smoke and cloud formation is sometimes quite obvious: Fig. 6.11 shows a cumulus developing where smoke from a coal-fired power station is rising to the lifting condensation level. To examine how condensation nuclei encourage cloud formation, we must look closely at the vapour exchange between small droplets and their immediate surroundings.

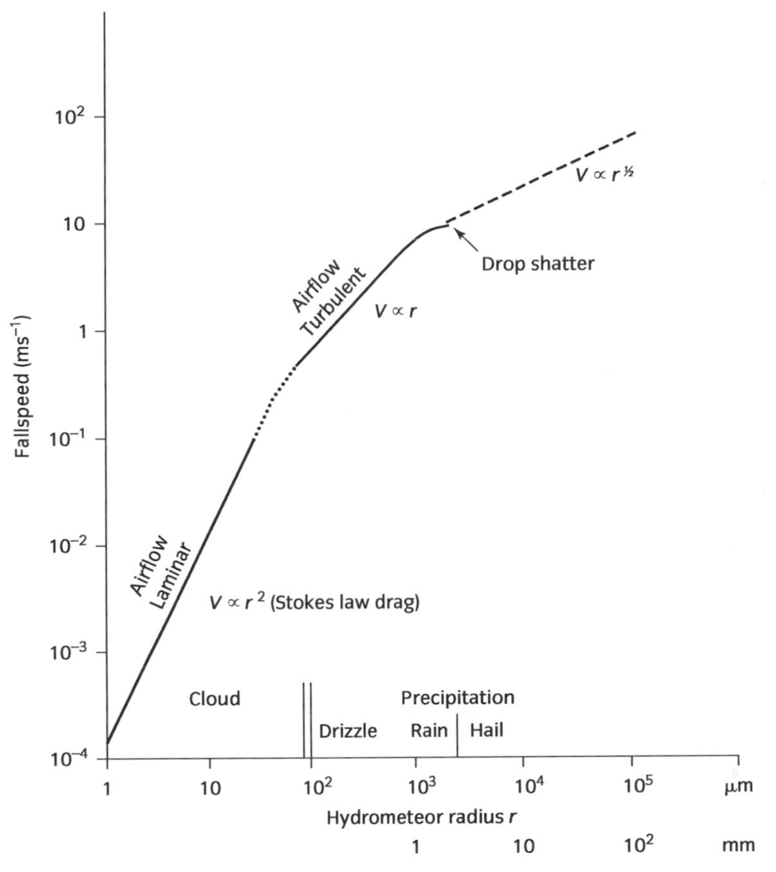

Figure 6.10 Graph of terminal velocity fallspeed versus radius for spherical bodies ranging in size from small droplets to giant hail. Speeds shown apply to the low troposphere; at higher altitudes speeds are somewhat larger on account of the lower drag. (Unpublished survey by the author, 1968)

Figure 6.11 Cumulus forming from a local water source. Water vapour in combustion products from the chimneys of an Arizonal power station condenses to form pale smoke, evaporates by mixing with surrounding drier air, and then recondenses at a fairly well-defined lifting condensation level. This is well below the natural cloud base visible in distant cumulus, and is probably lowered too by evaporation from the cooling grids surrounding the power plant.

NUMERICAL NOTE 6.1 Supersaturation in a cloud bottle

A height difference of Δz cm (i.e. $\Delta z/100$ m) in the arms of a water manometer corresponds to a pressure difference of $\Delta p = g\,\rho\,\Delta z/100$ Pa according to the barometer equation (Eqn 2.1), where gravitational g is $\approx$ 10 m s^{-2}, and ρ is the density of water (1,000 kg m^{-3}). Inserting these values we find $\Delta p = 100\,\Delta z$ Pa which is Δz hPa, showing that a water manometer reading in centimetres is numerically equal to a pressure difference in hectopascals.

According to Eqn 5.12, dry adiabatic decompression of air by Δp cools air by $\Delta T = \Delta p/(\rho\,C_p)$ where ρ is air density (about 1.2 kg m^{-3} near MSL) and C_p is the specific heat capacity of air at constant pressure (1,000 J kg^{-1} K^{-1}). For $\Delta p = 30$ hPa (3,000 Pa), the temperature drop is $3{,}000/(1.2 \times 1{,}000) = 2.5$ K or °C. For the same pressure fall, saturated adiabatic cooling is about half this because of the release of latent heat by cloud formation.

Careful measurement by sensors mounted on aircraft shows that air in most cloudy interiors is almost exactly saturated with respect to liquid water, having relative humidities of between 100 and 101% (i.e. *supersaturations* of between 0 and 1%). This, and other aspects of behaviour which play important roles in the cloudy atmosphere, can be observed in a simple cloud chamber which can be set up and used as described in Box 6.2, or simply read as a realistic description of cloud formation.

6.7 Condensation modelled *[28]*

We define saturation to be the state of water vapour in dynamic equilibrium with an exposed plane surface of pure water or ice. However, plane surfaces are not relevant to atmospheric cloud formation, since condensation occurs on very small aerosol particles and water droplets (radii $\sim$ 0.1–10 μm). Equilibrium between such droplet (or crystal) surfaces and vapour is significantly affected by their tight surface curvatures, and by the fact that the droplets often contain quite strong water solutions of salts. The latter condition applies because many aerosol droplets originate as particles of *hygroscopic* (water-attracting) material, such as sodium chloride, which form droplets of concentrated solution as soon as condensation begins.

6.7.1 Curvature

Consider the effect of tight surface curvature on the equilibrium between water (or ice) and ambient vapour. An escaping water molecule which is part of a convex surface of small radius of curvature r is less tightly bound to that surface than it would be to a plane surface in otherwise identical conditions, because there are fewer attracting neighbouring molecules within any short range (Fig. 6.12) of the escaper. Water therefore evaporates more easily, and the equilibrium vapour density ρ_{vr} is higher than the equivalent value for a plane surface in the same conditions, which is by definition the saturation vapour density ρ_{vs}. This is confirmed by detailed observation and theory, which show that

$$\frac{\rho_{vr}}{\rho_{vs}} = \exp\left[\frac{A}{rT}\right] \qquad 6.3$$

where A is a constant for the droplet liquid (water in the present case) and T is the absolute temperature of the droplet and its immediate surroundings. The effect is named after *Thomson* (later Lord Kelvin) who first derived Eqn 6.3.

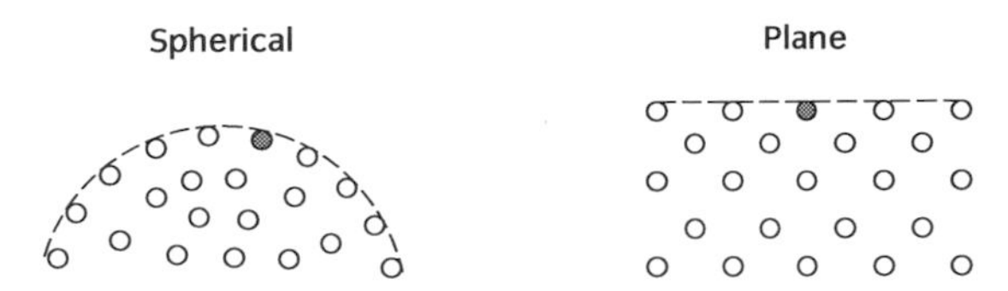

Figure 6.12 Molecular environment of a molecule at the tightly curved surface of a small liquid or solid body, in comparison with a plane surface.

According to Eqn 6.3 the ratio ρ_{vr}/ρ_{vs} increases as r decreases, becoming infinitely large as r vanishes, apparently making it impossible for a droplet to form! However, the finite size of molecules places a lower limit on r, and this is increased further by the production of clusters of water molecules by random multiple collisions in a sufficiently dense vapour. When these effects are included, together with realistic values for A and T, it can be shown that such *homogeneous nucleation* by multiple collisions becomes significant only when ρ_{vr}/ρ_{vs} exceeds a value of about four, i.e. when ambient relative humidities (RH) are about 400%, where 100% represents saturation for a plane surface. At lower RH, homogeneous nucleation is ineffective because embryonic droplets form infrequently and evaporate again almost immediately because the ambient vapour density ρ_v is less than the equilibrium value ρ_{vr} for the minute droplet. It follows that we should not expect clouds to form in perfectly clean air until it has a supersaturation (RH – 100) of about 300%. Such large values are not observed in the natural atmosphere, because condensation becomes widespread on its impurities at very much lower supersaturations.

Consider an air parcel containing particles much larger than the embryonic droplets formed by homogeneous nucleation. If they are water droplets, or at least coated by a film of water, according to Eqn 6.3 the parcel vapour density needed to exceed their equilibrium vapour density, and hence initiate condensation onto their surfaces, is much smaller than the value required for effective homogeneous nucleation. In fact the exponential decline of equilibrium vapour density with increasing particle radius is so steep that a supersaturation of only 1% is enough to enable a particle of radius 0.15 μm to grow by condensation. Such a low value confirms the practical importance of condensation nuclei in promoting cloud formation: without them, supersaturations would indeed have to rise to about 300% before cloud could begin to appear. In the simple cloud-chamber experiment, the repeated scavenging of condensation nuclei by cloud formation and settling produced air so clean that further cloud formation was seriously inhibited until smoke was introduced. In the real atmosphere the sources of aerosol mentioned in Section 3.7 ensure that there is almost always an adequate supply of condensation nuclei.

6.7.2 Dissolved impurity and curvature

The role of condensation nuclei in encouraging cloud formation is enhanced by the fact that many of them consist of hygroscopic solids which form droplets of quite concentrated solution in moist air. According to *Raoult's law* the presence of solute reduces the equilibrium vapour density over plane or curved water surfaces below the value for pure water—in fact by an amount which increases with solute concentration. If we consider a mass m of solute dissolved in a droplet of radius r, then the *Raoult effect* is represented by adding a negative term of form $B\ m/r^3$ to the right-hand side of Eqn 6.3, where the value of B varies with the type of solute and we assume that water is always the solvent. If we focus on conditions

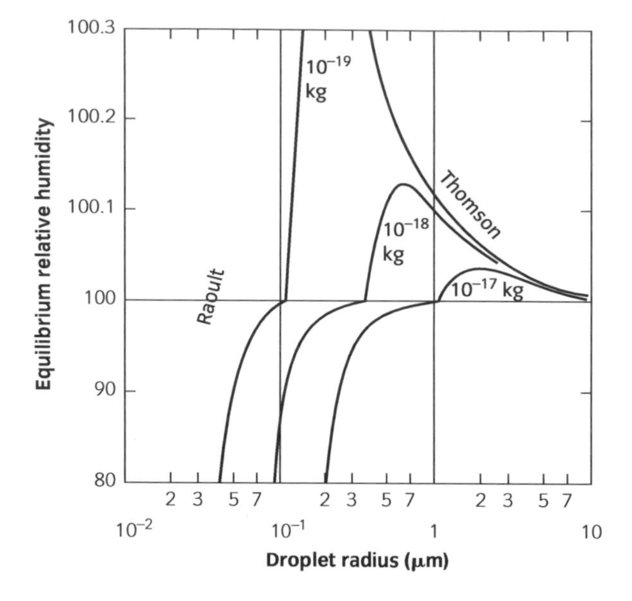

Figure 6.13 Kohler curves for droplets containing various masses of NaCl. The scale of the RH axis expands above 100% to clarify the structure there, introducing kinks in the curves which would otherwise be smooth. Note that 10^{-18} kg corresponds to 10^{-9} μg.

when equilibrium and saturation vapour densities are nearly equal, then the right-hand side of Eqn 6.3 approximates to $1 + A/(r\,T)$, and the Thomson and Raoult effects combine to form

$$\frac{\rho_{vr}}{\rho_{vs}} \simeq 1 + \frac{A}{rT} - \frac{Bm}{r^3} \qquad 6.4$$

Note that the dependence of the Raoult effect on solute concentration makes its negative effect large for small droplets, opposing the Thomson effect and encouraging droplet growth.

Figure 6.13 represents the combined Thomson and Raoult effects for several values of m typical of sodium chloride in atmospheric aerosol. Each *Köhler curve* (named after a Swedish pioneer meteorologist) represents the variation of *equilibrium relative humidity* ($100\ \rho_{vr}/\rho_{vs}$) with droplet radius r for a different fixed mass m of sodium chloride. When the salty droplet is small enough the salt concentration is so high that the Raoult effect dominates and equilibrium *RH* is reduced below the saturation value (100%) despite the opposing Thomson effect. This makes it possible for small droplets of sodium chloride solution (and other solutions too—all Köhler curves have the same general shape) to be in equilibrium with a sub-saturated vapour. At much larger radii the Raoult effect is negligible and the Thomson effect is small too, reducing the equilibrium supersaturation towards zero as the radius increases. Between these extremes there is a critical droplet radius r_c for each particular mass of solute at which the equilibrium supersaturation is a maximum. The existence of such a maximum vitally affects the behaviour of growing droplets, as outlined in the next section.

6.8 Cloud droplet growth

Consider a hygroscopic aerosol particle suspended in air whose relative humidity is rising because of adiabatic expansion and cooling—usually but not necessarily because the parcel is rising. At some ambient *RH* well below 100% the particle becomes a tiny droplet of concentrated solution whose equilibrium RH is equal to the ambient value. In Fig. 6.13 the droplet growth with rise (or fall) of ambient *RH* can now be traced by following the Köhler curve for its solute mass and type. As ambient *RH* rises the droplet grows by diffusion and condensation of vapour, its equilibrium *RH* rising to match the ambient value so rapidly that the difference between the two is negligible in the slowish rates of change typical of atmospheric updrafts. For droplets smaller than the critical droplet radius r_c (i.e. to the left of the peak of its Köhler curve in Fig. 6.13), droplet and vapour are in a state of neutrally stable equilibrium, since any little change in ambient *RH* produces rapid growth (or shrinkage) to a new equilibrium size. In the realistic case of a droplet containing 10^{-9} μg of NaCl, the equilibrium radius increases from 0.1 μm at 87% ambient *RH* to 0.2 μm at 99%, increasing the droplet volume and mass eight-fold.

Neutrally stable equilibrium continues until the droplet reaches its critical radius (at the peak of its Köhler curve), which occurs when the ambient vapour is slightly supersaturated ($SS = RH - 100 = 0.13\%$ in the case of 10^{-9} μg NaCl). If ambient *RH* rises further, the droplet enters a phase of runaway growth because the equilibrium *RH* falls from its peak value and can no longer match the ambient value. The droplet now grows at a rate which is limited only by the speed at which vapour can diffuse to the swelling droplet and latent heat can diffuse away (which otherwise warms the droplet, increases evaporation and slows the growth rate). Despite the typically small excess of ambient over equilibrium vapour densities which prevail in clouds, droplet radius grows 10- or 20-fold beyond its critical radius in a very short time, which accounts for the almost instantaneous appearance of micron-scale cloud droplets in the simple cloud-chamber experiment, and even in moist air flow around round fast-moving aircraft (Fig. 6.14).

If ambient vapour pressure remains constant at some value which is greater than the droplet's equilibrium value, droplet growth continues indefinitely, but at a slower and slower pace as measured by droplet radius, because the very rapid increase of droplet mass with radius outruns the ability of vapour to diffuse sufficiently rapidly down the gradient of vapour density (Box 6.3).

6.8.1 Activation

When a growing droplet exceeds its critical radius, the condensation nucleus is said to be *activated*—switched from a state of neutral stability to one of almost explosively unstable growth from an embryo to a full cloud droplet.

There is no stability for droplets greater than their critical radius (to the right of the Köhler peak). If ambient *RH* is greater than the equilibrium value for the droplet size, the droplet grows without obvious limit, as described. If ambient *RH* falls below the equilibrium value for the droplet size, for example because the droplet leaves the cloud and is immersed in unsaturated

Figure 6.14 Net rotation of airflow round the wings (clockwise looking astern from the near wing tip) of the Tornado spins off into trailing vortices whose core pressures are low enough to produce saturation and hence cloud. In more humid conditions the cloud may envelope all upper surfaces of the airframe. Assuming an air speed of 150 m s^{-1} and a cloud formation fetch of <1 m, cloud is forming in <7 ms, much faster than in any cloudy weather system, except possibly a natural tornado.

air, it shrinks by evaporation, faster and faster as it moves left along its Köhler curve, and its equilibrium RH climbs up and over the peak. In fact the droplet will shrink very rapidly until ambient and equilibrium RH coincide on the left-hand side of the Köhler peak, where neutrally stable equilibrium is re-established at a very small droplet size. You can see this very rapid evaporation when the cloud chamber is recompressed (and warmed) and the previous cloud vanishes, and as air leaves the low-pressure zones around a fast moving aircraft (Fig. 6.14). This instant vanishing contrasts with the persistence of *condensation trails* left by high-flying aircraft, which are formed by vapour (much of it from the engines themselves) condensing on condensation nuclei left in their exhaust gases.

Since its Köhler curve has no peak, a non-hygroscopic condensation nucleus has no neutrally stable growth regime corresponding to a hygroscopically nucleated droplet below its critical size. Instead it remains essentially dry until the ambient RH exceeds the value corresponding to its radius of curvature. Vapour then begins to condense and the droplet grows rapidly by uncontrolled condensation, like an activated hygroscopic nucleus. If ambient RH falls below the droplet's equilibrium RH, the droplet will shrink by uncontrolled evaporation until it is dry.

6.9 Cloud development

Previous sections have outlined mechanisms at work on individual droplets during their formation and growth. We now bring these together to describe the formation of a small cloud of such droplets just above a well-mixed cloud-free layer. Though this clearly relates to the formation of fair-weather cumulus (Fig. 6.3), the same synthesis applies to air rising into the base of an established cumulus congestus or cumulonimbus. Indeed with different time and space scales and geometry, it applies to all atmospheric cloud formation, from very slow, static cooling in fog, to rapid smooth uplift in lenticular cloud (Fig. 11.32), and from slow persistent ascent in fronts, to rapid horizontal decompressional cooling in air spiralling into tornado funnels.

6.9.1 Spectrum of condensation nuclei *[28]*

We begin with air rising towards the top of the sub-cloud layer, its temperature falling and its *RH* rising by dry adiabatic expansion. The population of condensation nuclei in the humidifying air is very important. Numbers and sizes in particular vary widely in the atmosphere, but as a rough guide one litre of air from the low troposphere at an inland site may contain 10^7 *Aitken nuclei* (radii between 5×10^{-3} and 0.1 mm), 10^5 *large nuclei* (radii between 0.1 and 1.0 mm) and 10^3 *giant nuclei* (radii exceeding 1.0 mm). No nuclei smaller than the largest Aitkens become activated in natural clouds, because maximum supersaturations are held sufficiently close to zero by the pre-emptive growth of droplets on larger nuclei (Section 6.9). The numbers of *large* nuclei can therefore be taken as representative of the total numbers of nuclei available for activation in the rising air. Over land these are believed to consist mainly of sulphates and sulphuric acid (derived from the combustion product SO_2), fine soil particles, and sea salt. Numbers are larger closer to localized sources such as active volcanoes and wild and man-made fires and furnaces. Over the oceans numbers can be down by a factor of 10 or more, and the proportion of sea salt can be much larger.

When the relative humidity of the rising air exceeds about 80%, many hygroscopic nuclei begin to swell by condensation, in equilibrium with the rising ambient *RH*. As this exceeds 90% such growth can produce a *wet haze* which reduces visibility quite noticeably, especially in horizontal lines of sight. Observers on mountains or in aircraft ascending or descending through the low troposphere often see a marked haziness extending tens or even hundreds of metres below the base level of ambient cumulus. In heavily industrial regions this elevated haziness may become visible even from the oblique viewpoints of ground-level observers. From Fig. 6.13 and equivalent data it transpires that the radii of haze droplets forming on all nuclei smaller than giants are less than 1 μm.

6.9.2 Selective activation

While the rising air remains unsaturated, haze droplets remain in equilibrium with the ambient *RH*, swelling as it rises, but not yet activated. But almost as soon as the air begins to supersaturate, some droplets exceed their critical radius, become activated and begin to grow very rapidly. At first only the most favoured are activated—those forming on giant hygroscopic nuclei—but as the supersaturation of the rising air increases, increasing numbers of less favoured droplets are activated, as are the largest non-hygroscopic nuclei. In a relatively short time so many droplets are growing so rapidly that their consumption of vapour slows, checks, and then reverses the rise of supersaturation. Once ambient supersaturation starts to fall, no further droplets are activated, and those still struggling towards activation begin to shrink again, trapped to the left of their Köhler peaks on Fig. 6.13. However, the equilibrium vapour pressure close to the surfaces of the activated droplets is still slightly below the ambient supersaturation (imagine their Köhler curves extended well to the right of the critical radius) and they continue their rapid growth for as long as the ambient air remains significantly supersaturated, which means in effect for as long as the rising air parcel remains within the body of the cloud.

The manner of activation of a cloud of droplets has three important consequences.

(i) The transition from wet haze to cloud proper tends to be quite rapid, since most of the droplets are activated at supersaturations below 1%. The activated droplets swell to micron scale in a few seconds, becoming very much more effective at scattering sunlight and producing the dramatic visual transformation apparent in the atmosphere and in cloud chambers. Together with the well-mixed nature of the sub-cloud layer this transformation also accounts for the familiar sharpness and uniformity of cumulus cloud-bases (Fig. 6.3).

(ii) The number of cloud droplets formed is equal to the number of droplets and particles activated, which in turn is determined by the aerosol population and the maximum supersaturation reached in the first stages of cloud formation. Although some cloud droplets may coagulate subsequently, to a large extent a cloud's droplet numbers are established by the conditions in which it is first formed. In the case of cumulus these are the conditions prevailing in the first few tens of metres above cloud base.

(iii) With the numbers of cloud droplets established, their ultimate size is determined by the amount of vapour available for condensation. For example, consider the case of a litre of typical low tropospheric air which contains 10^5 activated condensation nuclei and 10 mg of vapour (corresponding to a specific humidity of about 10 g per kg). If half the available vapour is condensed, as might well happen if the air ascends through an appreciable fraction of the troposphere, NN 6.2 shows that each cloud droplet will grow to a radius of about 23 μm. If all the vapour is condensed the droplets will grow to 29 μm. In fact growth to such a size by diffusion and condensation is relatively slow—only the initial stages are very fast, as droplets activate and race to radii of a few microns—and is usually by-passed by mechanisms we consider later (Section 6.11) to explain the observed speed with which clouds precipitate.

6.9.3 Deactivation

Deactivation of cloudy air occurs most obviously at the edges of all types of cumulus clouds. The cauliflower-like upper parts are in a state of rapid flux as parcels of cloudy air erupt from the interior and are turbulently mixed with the clear, sub-saturated ambient air. As the cloudy parcels become sub-saturated the cloud droplets shrink very rapidly below their critical radii. In fact the dryness of the ambient air usually takes them so far below their critical size that

NUMERICAL NOTE 6.2 Cloud droplet size

If 5 mg of water vapour is condensed onto 10^5 activated nuclei as cloud forms in 1 litre of moist air, and each spherical droplet grows to an identical radius r (volume $(4/3)\,\pi r^3$), then conservation of mass of water substance requires that

$$5 \times 10^{-6} = 10^5\,\rho\, 4\,\pi\, r^3/3$$

where ρ is the density of liquid water ($= 10^3$ kg m^{-3}) and the factor 10^{-6} on the left-hand side converts mg to kg. Rearrangement give $r^3 = 15 \times 10^{-2}/(10^5 \times 10^3 \times 4 \times \pi) \approx 1.2 \times 10^{-14}$ m^3 so that $r \approx 2.3 \times 10^{-5}$ m or 23 μm. If 10 mg of vapour is condensed, the radius of each droplet grows by the cube root of 2 (≈ 1.26) to 29 μm.

Figure 6.15 A decaying cumulus, showing brighter upper parts, where cloud number densities are still high, and duller ring-shaped lower parts, where cloud is thinning by detrainment and evaporation.

there is little or no wet haze to soften the edge of the cloud. On the sunlit side of the cumulus, the sharp contrast between the clear ambient air and the strong backscattering of sunlight by the cloud gives rise to its familiar brilliant, solid appearance (Fig. 11.7), while on the shaded side, strong forward scattering from the edge zone forms the 'silver lining' round the edge of the dark cloud body—dark because so much light has failed to get through, having been backscattered at the far side (Fig. 1.4).

Deactivation can occur gradually throughout a large cloud volume as the updrafts within weaken and die. As supersaturation gradually falls to zero by mixing with drier air, the population of cloud droplets is thinned by selective deactivation of its smaller droplets, so that the cloud loses its brilliant, sharp-edged appearance, and becomes fuzzy and dull before disappearing (Fig. 6.15).

6.10 Precipitation observed *[17]*

So far we have concentrated on the mechanisms which form and maintain tropospheric cloud. In the long run these are precisely balanced by the precipitation processes which return the condensed water and ice to the surface. Though there is widespread evaporation of cloud, and of precipitation falling through unsaturated air, this merely represents a redistribution of the water substance within the atmosphere. It is the ultimate return to the Earth's surface which is important, and this depends very largely on precipitation to the surface, since dew and frost formation are relatively unimportant.

6.10.1 Water

Over most of the Earth the water substance is observed to fall to the surface in the variety of forms summarized in Table 6.1, and selectively depicted in Fig. 6.16. Water droplets reach the surface with diameters ranging from about 100 μm (0.1 mm) to about 3 mm. The smallest of these are felt rather than seen in fog. Normal drizzle (diameters about 0.2 mm) reaches the surface from low cloud such as stratus, which can be considered to be fog layers lifted clear of the surface by stirring and slight desaturation of the surface layers. Raindrops cover the rest of the size range and account for the great mass of precipitation reaching the surface in all but the coldest conditions. Fine rain, verging on drizzle, is observed only when cloud bases are low, whereas the growth of larger raindrops (diameters > 2 mm) require several kilometres depth of cloud above cloud base—as found in many frontal and shower clouds. The largest raindrops fall from the heavy shower clouds which are endemic in all latitudes, but especially spectacular in equatorial regions.

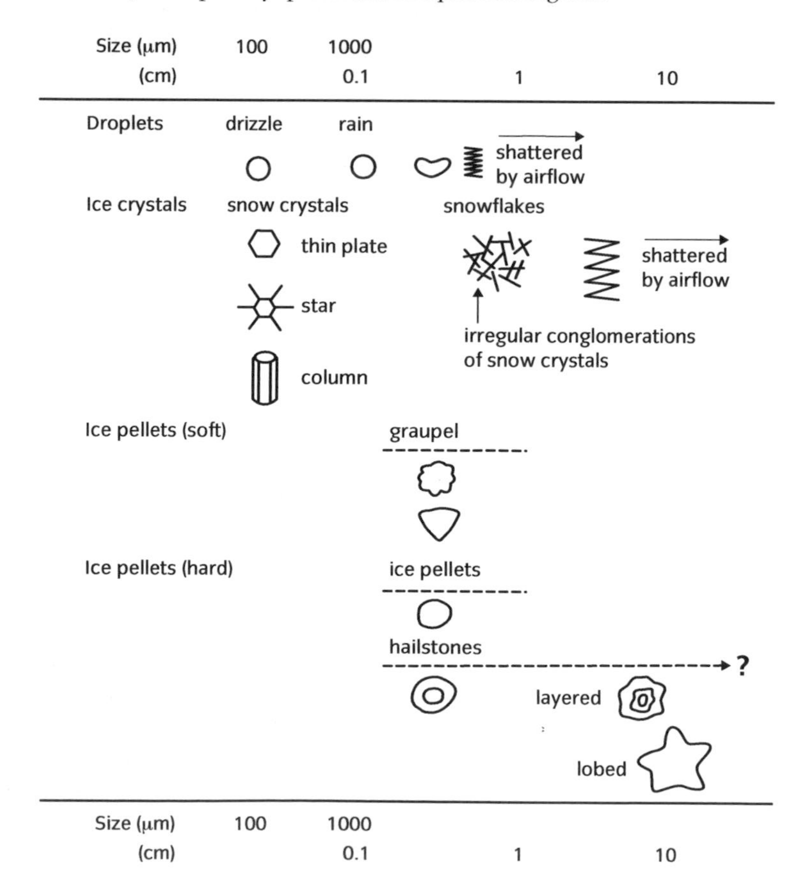

Figure 6.16 A selection of precipitation particles and drops. Note the hexagonal shape of all the ice crystals, reflecting the basic lattice structure of ice. Snow crystals grow by developing dendrites on the six corners, and snowflakes by the snagging and accumulation of crystals.

Table 6.1 Properties of selected precipitation droplets and particles

	Name	Diameter (mm)	Shape	Fall speed (m s^{-1})
Water				
	Drizzle	0.2	Sphere	0.8
	Rain	0.5	Sphere	4.0
	Rain	5.0	Unstable cap shape	10.0
Ice				
	Snow crystals	0.2 to 5.0	{Prism, plate} {Star, needle}	0.3 0.7
	Snowflakes	1.0 to	Irregular aggregates of from 2 to 100s	0.5 to
		20.0	of crystals (often stars)	1.0
	Graupel	0.5	Conical	0.5 to
	(soft hail)	5.0		2.5
	Hail	3.0	Roughly spherical	8.0
	Giant hail	20.0	Spherical with knobs	20.0

6.10.2 Ice

Ice is precipitated in an even greater range of sizes than water, and in innumerable shapes (Fig. 6.16). Six-pointed *crystals* of much the same diameter as water droplets fall to the surface from stratiform clouds when the low-level air is below 0 °C and the air aloft is colder still, as is usual. When the low-level air is not far below 0 °C, snow crystals aggregate by collision to form snow flakes which in calm conditions may be centimetres across, though much thinner and open textured. If unusually, a cold low layer is overlain by air above 0 °C, *freezing rain* may result as raindrops, supercooled by falling through the cold layer, freeze on impact with the surface.

Snow crystals and flakes can also fall from shower clouds when the low-level air is cold enough. However, the precipitating cores of larger showers often contain some hail, which is produced in a very different way (next section). In mid-latitudes in winter the hail particles are often millimetric cones of white, low-density ice, known as *graupel* or soft hail the whiteness and low density arising because air is trapped between the constituent ice grains of the particles. Over land in summer in mid-latitudes, larger hail may fall, which sometimes include layers of clear high-density ice alternating concentrically with layers of white ice to produce an onion-like appearance (Fig. 6.16). There seems to be no clear upper limit to the sizes of such hailstones, and radii of 3 cm are regularly reported in the most vigorous showers, often with a knobbly exterior.

6.10.3 Size

The minimum sizes of precipitation droplets and particles are readily explained. As soon as a droplet or particle falls out of its parent cloud into sub-saturated air it begins to evaporate, and it can be shown (Box 6.5) that the sensitivity of both evaporation rate and fall speed to droplet size combine to make the distance fallen for total evaporation extremely sensitive to size on exit from cloud base. In fact the simplified model in Box 6.5 suggests that the distance fallen increases with the fourth power of the radius, and that a droplet must have an initial radius of at least 100 mm to survive a fall of a few hundred metres in air sub-saturated by only a few per cent. Cloud droplets, being 10 times smaller in radius, survive for only a few centimetres, which is consistent with the observed sharpness of cloud edge already discussed. Large raindrops and hail can fall kilometres with little percentage shrinkage.

It follows that drizzle-sized droplets and crystals are the smallest which can survive falling through even quite shallow sub-cloud layers. When cloud bases are higher, the drizzle droplets evaporate in the first few hundred metres below cloud base, so that droplets reaching the surface are usually at least millimetric in size—the normal sizes for rain and snow. Solar backlighting of precipitation from a solitary cloud occasionally shows up the progressive evaporation of smaller droplets or flakes with increasing distance below cloud base, producing a downward tapering veil called *virga* (Fig. 6.17).

From the quoted sizes of precipitation droplets and particles, it is clear that all are very much larger than typical cloud droplets. In terms of radius, drizzle and raindrops are 10 and 100 times larger respectively, and in terms of volume these ratios become 10^3 and 10^6. These volume ratios indicate the enormous growth which has to be explained by any mechanism for precipitation: for example a typical raindrop has about one million times the water content of a typical cloud droplet.

Figure 6.17 Virga trailing from a surprisingly shallow cumulus. The rapidly evaporating precipitation is silhouetted against the evening sky. The suspiciously sharp lower cut-off suggests that this could be snow, becoming invisible below its melting level, but of course continuing to evaporate, and almost certainly not reaching the ground.

BOX 6.3 Droplet growth by diffusion and condensation

The mass flux F of vapour diffusing towards a spherical droplet is given by *Fick's law*:

$$F = 4\pi n^2 D \frac{d\rho_v}{dn} \qquad \text{B6.3a}$$

where D is the diffusion coefficient for water vapour in air and $d\rho_v/dn$ is the radial gradient of vapour density ρ_v at radius n from the centre of the droplet (Fig. 6.18). In a steady state, F is independent of n, and we can integrate the Eqn B6.3a to find a relation between F and the vapour density difference $\Delta\rho_v$ between the droplet surface ($n = r$) and the distant environment ($n = \infty$)

$$F\int_r^\infty \frac{1}{n^2}\,dn = 4\pi D \int_{\rho_{vr}}^{\rho_{v\infty}} d\rho_v$$

to find

$$F = 4\,\pi\, r\, D\,\Delta\rho_v \qquad \text{B6.3b}$$

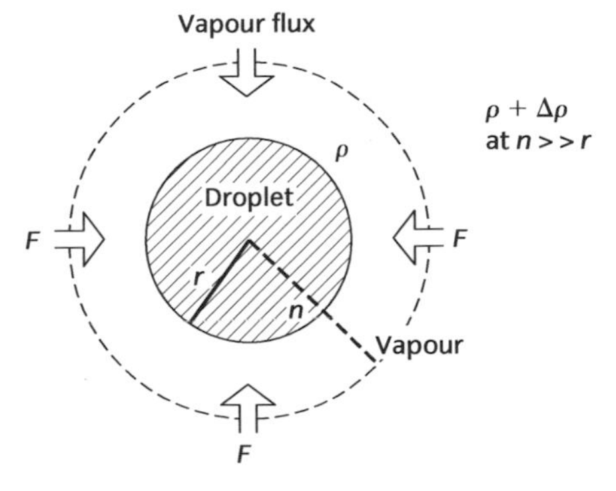

Figure 6.18 Vapour diffusing radially inwards towards the surface of a spherical droplet from a higher ambient vapour density.

As expected we see that F is positive if $\Delta\rho_v$ is positive also, i.e. if the flux is towards the droplet if the distant ambient vapour density exceeds the vapour density at the droplet surface. And the analysis shows that F is directly proportion to $\Delta\rho_v$. This vapour flux condenses at the droplet surface and adds to its mass.

$$F = \rho_w \frac{d}{dt}\left(\frac{4}{3}\pi r^3\right) = 4\pi\rho_w r^2 \frac{dr}{dt} \qquad \text{B6.3c}$$

where ρ_w is the density of the water droplet (or ice if it is a crystal). Equating the right-hand sides of Eqns B6.3b and 6.3c, we find

$$r\frac{dr}{dt} = D\frac{\Delta\rho_v}{\rho_w} \qquad \text{B6.3d}$$

Notice that to the extent that the vapour density difference $\Delta\rho_v$ remains constant, the rate of growth of droplet radius is inversely proportional to the radius—the droplet grows quickly at first and then more and more slowly.

Growth times

We can integrate Eqn 6.3d to find the time taken by a droplet to grow from one radius to another. First we interpret the vapour density difference $\Delta\rho_v$ in terms of the distant ambient vapour density ρ_{va} and the saturation vapour density ρ_{vs} at the surface of the droplet (which we assume is already too large to have any Thomson or Raoult effects).

$$\Delta\rho_v = \rho_{va} - \rho_{vs}$$

$$\Delta\rho_v = \frac{\rho_{vs}}{100}\left[100\frac{\rho_{va}}{\rho_{vs}} - 100\right] = \frac{\rho_{vs}}{100} SS \qquad \text{B6.3e}$$

where SS is the supersaturation (RH – 100) of the ambient air. We are supposing droplet temperature is identical to ambient air temperature, ignoring heating by continual release of latent heat at the droplet surface which is included in fuller treatments. Using Eqns B6.3d in B6.3e we find

$$r\frac{dr}{dt} = \frac{SS}{X} \qquad \text{B6.3f}$$

where $X = 100\,\rho_w/(D\,\rho_{vs})$, which has value 4.5×10^{11} s m^{-2} when saturation vapour density is 10 g m^{-3}, corresponding to a temperature of 11 °C. We integrate Eqn B6.3f to find the time t to grow from radius r_1 to r_2,

$$t = \frac{X}{2SS}\left(r_2^2 - r_1^2\right) \qquad \text{B6.3g}$$

Starting from radius 0.5 μm, the times taken to grow to larger sizes are as follows at 11 °C and SS 0.2%.

r	1	2	4	8	16	32 μm
t	0.9	4.3	18	72	291	1,163 s
				(1.2	4.8	19.4 min)

These results apply equally to the shrinkage of a droplet by evaporation, defining SS to be the sub-saturation (100 – RH) and reading the table from right to left.

6.11 Precipitation modelled

In previous sections we have followed the development of cloud droplets from their hazy origins to the phase of rapid growth by condensation which follows activation. Once clear of the complex curvature around the critical size (Fig. 6.13), droplet growth by condensation is controlled by the supersaturation of the ambient vapour (the equilibrium vapour density at the droplet surface being almost exactly equal to the saturation value for a plane, pure surface). As outlined in Box 6.3, the rate of growth of droplet radius by diffusion and condensation in the presence of a constant supersaturation is inversely proportional to droplet radius. Although the model ignores complications introduced by droplet warming by release of latent heat, the predicted lengths of time for growth agree reasonably well with more elaborate models *[28]*. These all show that whereas a droplet may grow to a radius of 5 μm in about 20 s in the presence of 0.2% supersaturation, it would take another 20 minutes to grow to a radius of 30 μm, and several days to reach the size of a small raindrop. This contrasts with observation of the real atmosphere which shows that a cloud can produce a shower of millimetric raindrops only half an hour after the cloud's first appearance, and that almost any cloud will precipitate if it lasts more than a few tens of minutes and is at least a few hundred metres deep from base to top. Clearly such familiar and important behaviour cannot be explained by diffusion and condensation alone.

Note that it is impossible for all the activated droplets in a typical cloud to grow to the size of raindrops: this would require a million times more vapour than is available in the cloud-forming air and would produce an unrealistic apocalyptic deluge. It seems that observed precipitation must depend on the favoured growth of a small fraction of the total population of cloud droplets.

6.11.1 Selective growth

A realistic cloud must contain a spectrum of droplet sizes. During initial formation, giant hygroscopic nuclei will become activated earlier than less favoured nuclei, so that there will be a size spectrum in the population of rapidly growing droplets. Now the *terminal velocities* (the equilibrium fall speeds at which particle weight is balanced by drag) of droplets in this size range increase with the square of the droplet radius, as outlined in Box 6.4 and depicted in Fig. 6.10, so that in the developing cloud the larger droplets may be falling up to 10 times faster than the smaller ones. This produces collisions, which encourage the growth of the larger, faster-falling droplets by coalescence with smaller droplets. This is called growth by *collision and coalescence*, to distinguish it from the growth by diffusion and condensation considered earlier. A very important property of growth by collision and coalescence emerges if we consider the volume of air swept out by a falling collector drop of radius r and terminal velocity V (Fig. 6.19). The volume swept per second is proportional to $r^2 V$, which is proportional to r^4, given the r^2 dependence of V already noted. Assuming that the mass of smaller droplets collected is proportional to the volume swept, then it follows that the rate of growth of the radius of the collecting drop dr/dt increases as r^2 (Box 6.6). Such accelerating growth contrasts with the decelerating growth produced by diffusion and condensation.

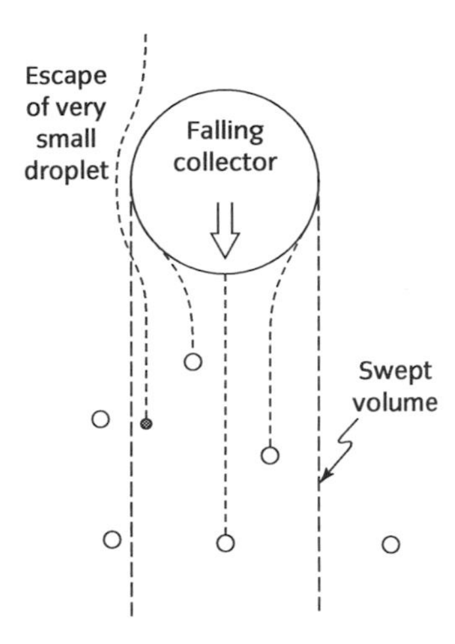

Figure 6.19 A large cloud droplet or a precipitation droplet is falling through smaller neighbours and collecting some of those within the swept path (envelope dashed). Dotted lines are paths of small bodies relative to the falling collector. One has escaped collection.

Observations in the atmosphere and special large cloud chambers, together with computer simulation of the growth of a model cloud population, confirm that collision and coalescence is effective in atmospheric clouds. However,

BOX 6.4 Terminal velocities of falling drops

When a spherical droplet of radius r and density ρ_w is immersed in air of density ρ, by Archimedes' principle, it suffers a net downward force given by its weight less the weight of the displaced air. Because air is so much less dense than water, *the downward force* is effectively the droplet weight W.

$$w = \frac{4}{3} \pi r^3 g \rho_w$$

If the droplet starts to fall from rest, it accelerates downwards, initially at g, but then more and more slowly as air resistance (upward drag F) builds up with increasing speed, until W and F balance and the droplet falls steadily at its *terminal velocity* V_t. Since W is proportional to droplet volume, while F tends to be proportional to droplet surface area, V_t increases with the droplet radius in ways which depend on just how the drag force F varies with V_t.

(i) When the droplet is very small, its terminal velocity V_t is dominated by the viscosity ν of the air. Theory and observation agree that

$$F = 6 \pi \rho \nu V_t r$$

so that at terminal velocity (where $W = F$) we have

$$V_t = Y r^2 \qquad \text{B6.4a}$$

where $Y = 2 g \rho_w/(9 \rho \nu)$.

This *Stokes law* regime of fall speeds (Fig. 6.10) applies to spheres of water or ice with radii up to about 30 μm, which includes all cloud and aerosol. The proportionality of V_t to r^2 means that it increases rapidly with size. In conditions typical of the low troposphere, the term Y has value 1.2×10^8 s^{-1} m^{-1}, which fits the bottom left of the curve of Fig. 6.10.

(ii) For droplets with radii between about 30 and 1000 μm (i.e. large cloud droplets to rain drops), the air flow around the falling drop is turbulent and drag increases more sharply with droplet size than in the Stokes regime, being proportional to the product $V_t r^2$ rather than $V_t r$. It follows from balancing W and F that V_t is directly proportional to droplet radius r. In the low troposphere the expression in SI units is

$$V_t = 8{,}000\, r \qquad \text{B6.4b}$$

which fits the middle section of Fig. 6.10.

(iii) For larger bodies still (large raindrops and virtually all hail) the air flow is so strongly turbulent that the drag is proportional to the product $V_t^2 r^2$, which gives V_t proportional to $\sqrt{r}$. In SI units we have

$$V_t = 250\sqrt{r} \qquad \text{B6.4c}$$

Turbulence in the flow round the drop becomes so great that rain drops shatter soon after reaching radii of about 2 mm. No such limit applies to hailstones, which are observed to grow to diameters of as much as 10 cm on rare occasions (Section 11.6).

BOX 6.5 Evaporation of falling drops

Consider a droplet falling at its terminal velocity V_t. In a very small time interval dt it will fall a very small distance dz (inverting the vertical coordinate for convenience) given by d$z = V_t$ dt. If the droplet is evaporating as it falls, its radius r is changing at rate dr/dt, which will be negative for the shrinking droplet. We can associate the change dr in droplet radius with the time taken dt = dr/(dr/dt), and hence with the distance fallen

$$dz = \left[\frac{V_t}{\frac{dr}{dt}}\right] dr \qquad \text{B6.5a}$$

By integrating Eqn 6.5a we can accumulate all the tiny falls dz and radius changes dr to find the overall distance z fallen while the droplet shrinks from radius r to zero.

$$z = -\int_0^r \frac{V_t}{\frac{dr}{dt}}\, dr \qquad \text{B6.5b}$$

In Boxes 6.3 and 6.4 we have expressions for V_t and dr/dt to insert into Eqn B6.5b to complete the calculation.

For a droplet with radius r less than about 30 μm, Eqn B6.3f determines dr/dt, with SS interpreted as the degree of sub-saturation (100 – RH) of the environment of the evaporating drop, and Eqn B6.4a determines V_t.

$$z = \int_0^r \frac{XY}{SS} r^3\, dr \qquad \text{B6.5c}$$

If SS is uniform along the path of the falling droplet, standard integration gives

$$z = \frac{XY}{4\, SS} r^4 \qquad \text{B6.5d}$$

Using the realistic values for X and Y quoted in Boxes 6.3 and 6.4, and assuming an ambient sub-saturation of 5% (i.e. RH 95%), the distances fallen by droplets evaporating completely from chosen initial radii are tabled according to Eqn B6.5d.

Table Distances fallen for complete droplet evaporation at RH 95%.

Initial radius	r	1	10	100	1000 μm
Distance fallen	z	2.7 μm	2.7 cm	(270 m	2700) km
				[240 m	240] km

The r^4 relationship is valid for droplets of radius up to 30 μm (Box 6.4), giving a 10,000-fold increase in distance fallen for each ten-fold increase in initial droplet radius. Beyond the fall of nearly 2.2 m for a 30 mm droplet (i.e. between the central columns of the table), the air flow round the falling droplet becomes increasingly turbulent, reducing V_t and enhancing dr/dt. A simplified approach combining the purely diffusive result for dr/dt (Eqn B6.3f) with the $V_t = 8{,}000\, r$ (Eqn B6.4b) and ignoring significant wet-bulb chilling of the droplet produces the [ ] values in the table, which agree reasonably with more thorough analyses *[29]* and reduce the ( ) values slightly for drizzle, and grossly for rain. The resulting set of z values shows that cloud droplets ($r \sim 10$ μm) fall only centimetres before vanishing, drizzle droplets ($r \sim 100$ μm) fall only a couple of hundred metres, but that raindrops ($r \sim$ mm) could fall many times the depth of the troposphere before evaporating completely.

BOX 6.6 Growth by collision and coalescence

Consider a collector droplet of radius r, falling at speed V_t and therefore sweeping air volume at the rate $\pi r^2 V_t$ (Fig. 6.19). If there is a population of much smaller droplets with specific mass m, their mass per unit air volume is $m\,\rho$, where ρ is the aggregate density of the cloud-filled air. If the smaller droplets have negligible terminal velocity, and all in the swept volume are collected by the falling collector droplet, then the rate of collection of mass is

$$R = \pi r^2 V_t m \rho \qquad \text{B6.6a}$$

As in Box 6.3 this produces a rate of increase of collector mass

$$R = \rho_w \frac{d}{dt}\left(\frac{4}{3}\pi r^3\right) \qquad \text{B6.6b}$$

where ρ_w is the density of the collected water or ice. Equating the expressions B6.6a and B6.6b, we find

$$\frac{dr}{dt} = \frac{m\,\rho\,V_t}{4\,\rho_w} \qquad \text{B6.6c}$$

(i) According to Box 6.3, for collector radius up to 30 μm we have $V_t = Y r^2$, so that

$$\frac{dr}{dt} = \frac{r^2}{A} \qquad \text{B6.6d}$$

where $A = 4\,\rho_w/(m\,\rho\,Y) = 3.3 \times 10^{-2}$ m s if m is 1 g kg^{-1}, and other values are as in Box 6.3. The time taken for the collector to grow from radius r_1 to r_2 is given by the integral of Eqn B6.6d

$$t = A\left(\frac{1}{r_1} - \frac{1}{r_2}\right) \qquad \text{B6.6e}$$

With $m = 1$ g kg^{-1}, it takes nearly 14 minutes for the collector radius to grow from 20 to 40 μm, and this time halves as m doubles and so on. Growth by collision and coalescence can therefore be faster than growth by diffusion and condensation in this important size regime provided there is a good supply of droplets suitable for efficient collection, i.e. neither too small nor too large.

(ii) Collector droplets with radii between 30 and 1000 μm have fall speeds given by Eqn B6.4b. When substituted in Eqn B6.6c we find

$$\frac{dr}{dt} = \frac{r}{B}$$

where $B = 4\,\rho_w/(m\,\rho\,8 \times 10^3) = 500$ s for $m =$ 1 g kg^{-1} and previously quoted values. The time taken for the collector to grow from radius r_1 to r_2 is given by

$$t = B \ln\left(\frac{r_2}{r_1}\right) \qquad \text{B6.6f}$$

With $m = 1$ g kg^{-1}, growth from 30 to 300 μm (a crucial transition from large cloud to small precipitation) takes 19 minutes, and this time halves as m doubles and so on, showing that growth by collision and coalescence can account for observed short time intervals when conditions are favourable.

the mechanism is a good deal more complex than has been suggested so far. For example, very small drops tend to flow round the would-be collector drop and so avoid collision. And even if they do collide, not all colliding droplets coalesce—significant numbers may bounce apart again. There are efficiencies of both collision and coalescence which are observed to vary with the sizes of both the collecting and collected droplets in quite complex ways *[28]*. Observation by sophisticated and painstaking laboratory work agrees with theory in showing that droplets with radii less than 20 mm are inefficient as collectors, and that droplets with radii more than 80% and less than 20% of the radius of the collector are not efficiently collected.

6.11.2 Warm, cold, and mixed clouds

We see that precipitation develops in two stages. First, diffusion and condensation produces a population of cloud droplets which includes some with radii exceeding 20 μm—though it is still a little difficult to envisage how the largest of these are produced in the shortest times observed. Subsequently these large droplets grow by collision and coalescence with their smaller neighbours, at a rate which increases with size, to produce millimetre-sized drops in a few tens of minutes. This two-stage process is believed to explain precipitation from all relatively warm clouds, i.e. those which do not contain temperatures below about –10 °C. Since much of the troposphere is colder than –10 °C, some of it very much colder, and clouds are at best only marginally warmer than the ambient air, a great deal of cloud is cold in this special sense. In low latitudes only the upper parts of tall shower clouds are cold, but in middle and high latitudes all clouds are cold, except for cloud bases in warm conditions. Can the prevalence of cold cloud have any bearing on the development of cloud, and especially precipitation?

Everyday experience suggests that water freezes as soon as the water temperature falls to 0 °C, or just a little below this if the water is contaminated by salt or some other impurity. However, it is observed that large amounts of atmospheric cloud persist in the form of supercooled droplets down to –30 °C or so. As in the case of equilibrium between vapour and liquid or ice, tiny volumes of water substance behave very differently from much larger volumes.

Ice nucleation [28]

Careful work in laboratory and atmosphere shows that water freezes by a process of *nucleation*—the crystal lattice of ice spreading rapidly through the water volume from a very small triggering nucleus of molecules. At about –40 °C the thermal movement of molecules in supercooled water is so sluggish that clumps of them settle into the lattice formation by chance and then freeze the entire volume—a process known as *homogeneous nucleation*, resembling homogeneous nucleation of condensation. At much higher temperatures, homogeneous nucleation is so rare in the small volumes of typical cloud droplets that their freezing usually requires the intervention of a tiny quantity of a substance which initiates freezing because its lattice structure is similar to that of ice. These are known as *freezing nuclei* and their action is termed *heterogeneous nucleation*.

A freezing nucleus may be present within a supercooled droplet from its formation, or it may make contact with its outer surface, but in either case, as the droplet cools, the nucleus becomes increasingly likely to initiate freezing in any given observation period. A complication is that most experimental work on ice clouds concentrates on the resulting ice crystals, even though some of these may be produced by the action of *sublimation nuclei* triggering the formation of an ice crystal directly from the vapour. Sublimation nuclei are probably much less important than freezing nuclei, but for simplicity we will refer to all nuclei encouraging the production of ice crystals in clouds as *ice nuclei*.

It is conventional to ascribe the increasing likelihood of heterogeneous nucleation with decreasing temperature to an increasing concentration of ice nuclei, even though it is their effectiveness which is increasing—their numbers presumably remain constant in the absence of a local source. Although there are still discrepancies between observations, it is certain that concentrations of ice nuclei are very much smaller than those of condensation nuclei, being about one per litre at –10 °C.

The net result of all ice nucleation processes is that ice crystal populations in clouds increase with decreasing temperature. Above –10 °C there is almost no ice; between –10 and –20 °C there is an increasingly significant minority of ice crystals which may convert to a majority between –20 and –30 °C; and below –30 °C cloud is predominantly ice. Because of homogeneous nucleation, no supercooled water can persist below –40 °C.

In these circumstances, it is clear that ice crystal concentrations in clouds increase with height, that clouds in the middle troposphere in middle latitudes contain both ice and supercooled water (called *mixed clouds*), and that ice clouds predominate in the high troposphere (Fig. 6.20). This last statement corresponds to simple observation: ice cloud has a characteristically fibrous appearance (hence the term *cirrus*—Latin for hair) which arises because of its slowness to evaporate in the presence of unsaturated air, and which can be seen frequently in the high troposphere (Fig. 2.10). The tops of vigorously growing shower clouds maintain the sharp-edged cauliflower appearance of water clouds until they reach the upper troposphere, when their edges can be seen to soften and diffuse into a fibrous anvil as the cloud of supercooled water

Figure 6.20 Ice and water cloud in showers. A cumulonimbus off the left side of the picture is spreading its fibrous (icy) anvil to the right. The knobbly underhang is water cloud and is too rough to be called mammatus. The distant congestus will soon glaciate and lose its crisp cauliflower (water cloud) appearance.

glaciates (transforms to ice cloud) quickly in the very low temperatures there. The Norwegian *Tor Bergeron* (a good forename for a student of cumulonimbus), observing the development of cumulus in visually clear westerly air flows approaching the coasts of Norway from the Atlantic, noted in the 1930s that clouds began to precipitate very shortly after their tops glaciated, and proposed the following connecting mechanism.

6.11.3 The Bergeron–Findeisen mechanism *[17]*

In principle it is apparent in Fig. 6.2 that air which is saturated with respect to water is considerably supersaturated with respect to ice, at least when the temperature is significantly below 0 °C; in fact at –15 °C the supersaturation is over 15%. If an ice crystal is produced in a cloud which consists mainly of supercooled water, the vapour will be grossly supersaturated relative to the ice crystal, which will therefore grow very much more rapidly by diffusion and sublimation than its supercooled neighbours—tens of times faster according to the ratio of the relevant supersaturations. Indeed as more ice crystals grow more and more rapidly, they may reduce the ambient vapour below vapour equilibrium relative to super-cooled water and drive wholesale transfer of water substance from droplets to crystals. This powerfully selective mechanism has been named the *Bergeron–Findeisen* mechanism after Bergeron (as above) and the German *Walter Findeisen* who pioneered the use of very large cloud chambers to study this and other cloudy processes. It is believed to be important in the mixed clouds present in cumulonimbus and nimbostratus, and therefore an important factor in precipitation in all latitudes.

In practice crystals growing by diffusion and sublimation preserve the hexagonal shape of the ice crystal lattice, with variations on the hexagonal theme (Fig. 6.16) arising in different conditions. Terminal velocities are often considerably less than those of droplets of the same mass, so that it is at least possible that some fine crystals falling out of fairly shallow but cold clouds may have time to grow by this mechanism alone, which could explain some drizzle from cold stratocumulus, where the crystals fall into air warmer than 0 °C and melt before reaching the surface.

It is much more likely, however, that the Bergeron–Findeisen mechanism operates in conjunction with the collision and coalescence mechanism, though the latter is complicated in mixed clouds by the geometry of the collector crystal. An ice crystal falling through smaller supercooled water droplets grows by *accretion*, the droplets freezing on impact with what amounts to a giant freezing nucleus to produce a rapidly growing hailstone. If the droplets are only slightly supercooled they may form a layer of water which freezes subsequently to form clear, dense ice. If the droplets are heavily supercooled they freeze instantaneously and trap air pockets to form white, spongy, low-density *graupel*. Hail containing alternate concentric shells (rings in section) of white and clear ice has probably oscillated between higher and lower temperature before finally falling out of the parent cloud: air flow in very vigorous showers is known to be capable of allowing a substantial vertical oscillation of the growing hail (Section 11.6).

If a large ice crystal is falling through smaller ice crystals, it tends to grow by *aggregation*, the smaller crystals colliding with and sticking to the collector

Figure 6.21 A rainbow segment beneath melted snow. The heavy snow shower is melting to give rain in the lowest few hundreds of metres. The sun behind the camera produces a rainbow by internal reflection in the nearly spherical drops, but produces unstructured backscatter from the snow.

crystal. Not all collisions lead to aggregation however: it is observed that significant aggregation occurs only when temperatures are above −10 °C, presumably because ice surfaces are coated with a slightly sticky liquid film. Aggregation of snow crystals (where diffusion and sublimation has produced pronounced hexagonal spikes—Fig. 6.16) is enhanced by snagging or interlocking of adjacent spikes or *dendrites*, producing very large snowflakes in calm conditions.

Lastly, note that because so much of the troposphere is so cold, even in equatorial regions, a great deal of precipitation begins in the form of ice crystals and develops into snow crystals and flakes. If these fall into air which is warmer than about 2 °C they melt and arrive at the surface as rain. In middle latitudes almost all rain is melted snow (Fig. 6.21), even in the height of summer. Only the larger hailstones from summer thunder showers reach the surface largely unmelted because of their large mass, high fall speeds and small surface to volume ratio. In winter the melting level for snow often fluctuates between sea level and a kilometre above, as the synoptic scale weather pattern changes, giving the variable snow cover on hills and mountains and the occasional blanketing of lowlands so typical of maritime climates such as Washington State's and Britain's, and so testing of the local weather forecaster's skill.

6.12 Atmospheric electricity

Showers are often associated with thunder and lightning, and violent showers may contain nearly continuous thunder and lightning which, together with their torrential rain, large hail, and damaging gusts of wind, earns them the formal title *severe local storms*. Thunder and lightning also occur in some fronts, especially vigorous cold fronts, and in rather special weather systems (variously called *squall lines* or *line squalls* in different regions) which are intermediate between fronts and lines of severe storms (Section 11.7.4). All intense tropical cyclones produce vigorous thunder and lightning. At any instant several thousand thunderstorms are active in the troposphere, concentrated in low latitudes and over daylit land.

Despite the usual word order, in 'thunder and lightning' lightning always comes first, and is always accompanied by thunder, whether audible to an observer or not. The loudness of nearby thunder and the vividness of lightning, often accentuated by the brooding darkness of the clouds in which they occur, have impressed people from earliest times and teased human powers of explanation. Gods like Zeus and Thor have been invoked, and the latter has left his mark in the very name thunder, and the technical term *anvil* for the shape of the glaciated upper parts of mature cumulonimbus—thunder being supposed to be the sound of Thor beating on his gigantic anvil. Lucretius in 55 BC *[4]* supposed that thunder was the noise of great clouds crashing together, recognizing that large clouds are always involved. In AD 1752 Benjamin Franklin and d'Alibard proved independently that thunderstorms have an electrical nature, by the highly dangerous practice (fatal in d'Alibard's case) of flying kites nearby on conducting tethers. Franklin in particular was able to show that the electrical charges involved were sometimes positive and sometimes negative. In the 1920s C.T.R. Wilson established the typical distribution of charge within a thunderstorm (Fig. 6.22) by analysing measurements of electrical field strength made on the ground at a range of distances from the storm. In the following decade this work was confirmed and amplified by G. Simpson and others who flew electric field strength meters into thunderstorms on balloons. Meanwhile very fast cameras had been developed to examine the complex

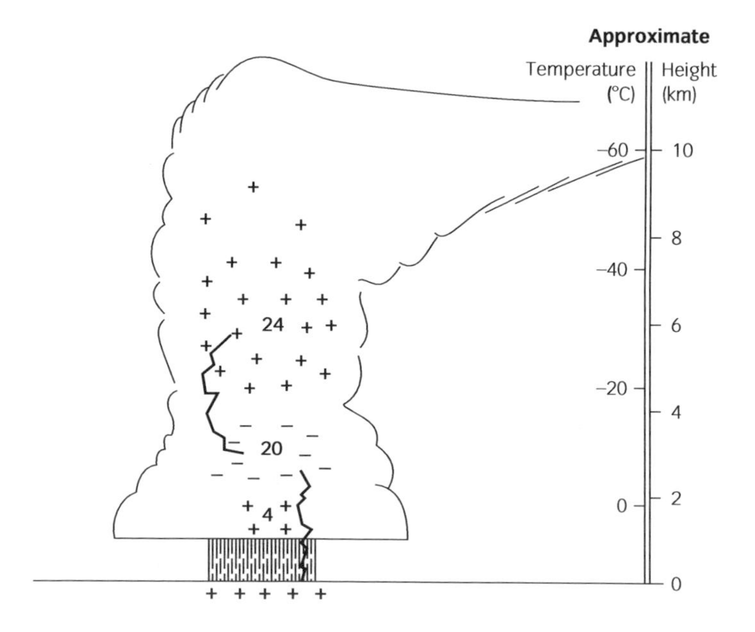

Figure 6.22 Typical distribution of electrical charge within an active thunderstorm in middle latitudes, as deduced by remote observation of changes in electrical field strength accompanying lightning strokes. Quantities of separated charge are given in coulombs.

Figure 6.23 Lightning under the cloud base of an evening thunderstorm. The flash probably extends upwards for at least the same distance again inside the cloud, which later at night would be seen to light up from within. The flash is on the edge of a shaft of heavy precipitation which is visibly spreading out as its associated cool downdraft reaches the surface. The incomplete right fork is possibly reaching the surface within the heavy rain. (See Plate 8).

structure of lightning—the giant sparks jumping within clouds, and from cloud to ground.

The typical distribution of electric charge shown in Fig. 6.22 implies that positive and negative charges are being separated vertically within the thundercloud. Lightning within the cloud temporarily neutralizes the separated charges by connecting them through an electrically conducting channel, but observations show that further charge separation requires only 20 s or so. In recent years, extensive work in the laboratory and by instrumented aircraft in and near thunderstorms has been directed at uncovering the charge-separating mechanisms at work. Results are still not completely conclusive *[23]*, but suggest that several distinct mechanisms are important, with the common feature that small particles or droplets tend to become positively charged while larger particles become negatively charged. The greater fall speeds of the larger particles (many of them precipitation elements) allows gravity to drive the required charge separation. In one proposed mechanism, a small cold ice crystal collides with a hailstone warmed by release of latent heat from the freezing of impacting supercooled droplets. During their brief collision the very mobile positive ions migrate towards the colder end of the crystal (i.e. away from the hailstone) leaving the warmer end and the hailstone negatively charged (Fig. 6.24). On separation the stone carries its negative charge downwards and the crystal its positive charge upwards. The charge separation at each collision is very small, but there are so many collisions in the cubic kilometres of active precipitating cloud that the net result is at least potentially consistent with observed charging rates.

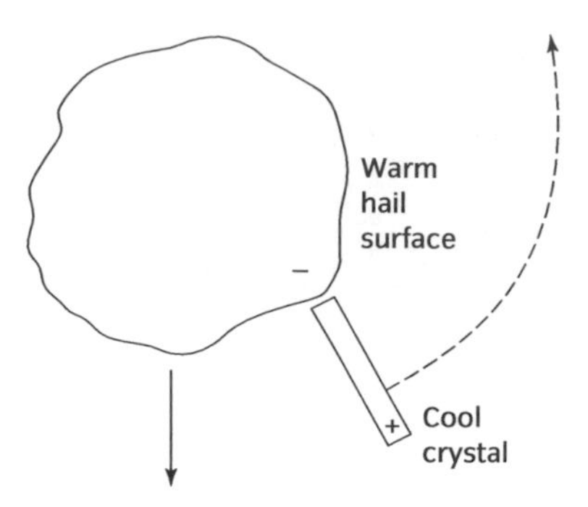

Figure 6.24 Charge separation during collision between a rising ice crystal and a falling warmer hailstone.

6.12.1 Lightning

Lightning is simply a giant spark produced when the electrical insulation of a mixture of air, cloud, and precipitation breaks down, which is believed to occur when the local electric field strength reaches about 1 MV m^{-1}. From basic electrostatics the field strength at the edge of a 2 km radius sphere containing 24 C (coulombs) of electrical charge is only about one tenth of this critical value, so that it has to be supposed that the critical value is being reached because charge is being concentrated dynamically within the active volume. However it begins, a lightning spark is then able to propagate quickly across regions where the electric field is considerably weaker than the critical value.

Lighting almost always originates within cloud, but may then either remain within the cloud, connect within a neighbouring cloud (*cloud-to-cloud* lightning), or connect to the local surface (*cloud-to-surfac*e lightning). Because it is more readily observable, cloud-to-surface (especially land) lightning has been studied much more than the other types. Observation suggests that many ground strikes begin in the strong field between the small positive charge just below the melting level and the large negative charge higher up. Once begun, a so-called *stepped leader* makes its way from the negative charge towards the surface, its jagged steps outlining the familiar forked lightning shape. As it nears the surface the electric fields there, already often strong enough to support a flickering *corona discharge* around sharp points such as the twigs of trees, cause a spark to jump to meet the stepped leader. The main surge of free electrons towards the Earth then occurs in the established ionized channel, but in such a way that the surge propagates upwards in a *return stroke* which is the main part of the lightning flash. The current in a channel a few centimetres across may reach 10,000 amps briefly, heating the air to the familiar incandescent flash. The explosive expansion of the air sends out shock waves which are heard as thunder.

As the negatively charged region of the cloud may be only partly discharged by this process, there are often several return strokes at intervals of about 1/20th of a second, each preceded by a downward *dart leader which* reactivates the ionized channel. The typical succession of several brilliant return strokes is just perceptible to the eye as the flicker of a lightning flash occurring near the edge of vision, where movement and change are most apparent, for physiological reasons. After full discharge, the recharging processes prepare for another flash a few tens of seconds later, which almost always occurs in a fresh location, since the ionized channel has long since dispersed and the storm is always unsteady and usually travelling. Storms producing much more frequent flashes to ground must have several separate charging zones at work simultaneously. Although their location makes detailed comparison difficult, technical study suggests that cloud-to-cloud and internal discharges differ somewhat from cloud-to-ground discharges, for example in being longer lasting and less intrinsically brilliant *[23]*.

In everyday speech it is customary to distinguish between fork lightning and sheet lightning as if they were quite different types of lightning, sheet lightning being apparently silent. In fact most if not all sheet lightning is simply the loom of a lightning fork obscured inside or behind an intervening cloud, and too distant to be audible (see below).

Other types of lightning have been distinguished since about 1990. Very powerful bolts of *positive lightning* can occasionally shoot sideways and downwards from storm tops (including anvils) to ground, possibly accounting for anecdotes of 'bolts from the blue' and rare cases of severe aircraft damage. Relatively diffuse *blue jets* are observed reaching tens of kilometres upwards from storms to the base of the ionosphere, and even more diffuse, usually reddish *sprites* (Fig. 6.25) are observed in the ionosphere. These last two types appear to be upward extensions of the electrical activity of the tropospheric storm, and there is intense speculation about their mechanism *[30]*.

6.12.2 Thunder

Thunder is the noise of shock waves rippling out from the explosive expansion of air in the incandescent lightning flash. Since they spread at the speed of sound compared with the effectively instantaneous transmission of the light of the visible flash, a distant flash produces audible thunder after a time delay which is directly proportional to the distance from the nearest point of the flash, the delay

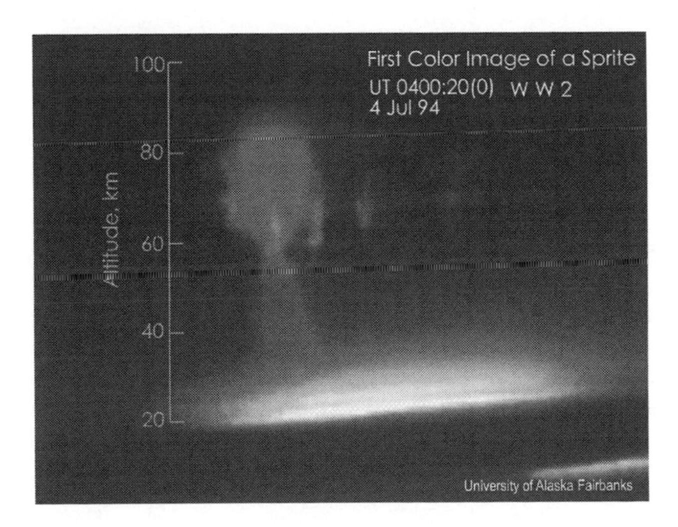

Figure 6.25 Sprite lightning. This one reaches over 70 km above its parent thunderstorm, whose ordinary lightning appears as a brilliant blur. The flickering dance of these tall ghostly figures gives them their attractive name.

being about 3 seconds per kilometre, according to the speed of sound. Thunder is seldom audible more than about 10 km away, but lightning is clearly visible for many tens of kilometres at night if there are no intervening clouds. Such silent flashes, bright but indistinct because of range, are popularly called *heat lightning*, because they are often seen on hot summer nights in middle latitudes, in addition to the term sheet lightning mentioned above.

The prolonged booming peals of thunder are rarely caused, as popularly supposed, by echoes from clouds or mountains. Since lightning flashes can be several kilometres long, a ground strike near the observer may rumble for 10 s or more as the shock waves from the more distant parts of the flash follow the hissing, tearing bellow of the nearest part of the flash. And the forked path produces booming concentrations in the long rumble as sound energy coming from lengths of the lightning zigzag perpendicular to the line of sight is compressed into shorter time intervals on arrival at the ear.

6.12.3 Electrical cycle and energy

Thunderstorms play an important role in the atmospheric electrical cycle. The cloud-to-surface flashes are continually pumping electrons Earthwards at a rate of several thousands of amperes world-wide, and maintaining the upper troposphere and higher atmosphere at an electrical potential of about half a million volts above earth potential. Individual storms produce potentials 10 or more times higher than this in their upper parts, but much of this is wasted by inter and intra cloud lightning and other electrical leakage. The remainder maintains the *fair-weather* electric field as depicted in Fig. 6.26, the field strength being a maximum near the surface where it is about 100 V m^{-1}. In this field the small concentration of ions produced by cosmic radiation and natural radioactivity in the surface and atmosphere maintains a gentle, persistent current through the air which world-wide balances the concentrated spasmodic currents driven by thunderstorms.

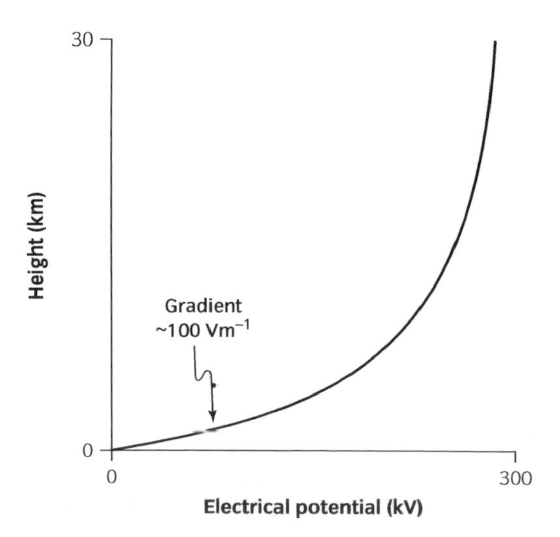

Figure 6.26 Vertical profile of electrical potential in fair weather. The gradient of profile (potential gradient) is the strength of the vertical component of the electrical field.

The dissipation of energy in lightning flashes is spectacular because they are so brief and localized. However, when averaged throughout the volume and lifetime of a thunderstorm they represent only a small fraction of the storm's total energy budget, by far the largest components of which are the buoyant potential energy of the air which is about to rise or fall, and the kinetic energy of the rising and falling air (Section 9.7). In effect thunder and lightning are spectacular sideshows in the wings of the main drama, in which millions of tonnes of air, water, and ice are in great commotion. The noises which Lucretius thought were cloudy collisions are more properly the relatively slight sounds of clouds tearing themselves apart under gravity and buoyancy.

6.13 Practical applications

6.13.1 Cloud seeding *[28]*

Speculation about the role of the Bergeron–Findeisen process in initiating precipitation led to extensive work in artificial stimulation of precipitation beginning in the 1940s and continuing to the present day. The ability of ice nuclei to trigger precipitation from suitable clouds, and the belief that such nuclei might be in short supply in some conditions, led people to believe that the artificial introduction of ice nuclei might enhance precipitation, which could be very valuable for agricultural production in marginally arid areas. Some early spectacular claims for the effectiveness of such so-called *cloud seeding* encouraged widespread work, but although it can occasionally change the physical appearance of clouds quite dramatically, years of exhaustive investigation have shown that the effects on precipitation are quite slight, probably not increasing it by more than 20% above the level expected before intervention. Because precipitation is naturally so variable, the detection of such small increases requires prolonged experimental study and very careful statistical analysis to distinguish between real effects and happy accident.

Since stimulated precipitation could be economically very valuable, work still continues in some countries, sustained more by hope than objective results. Silver iodide (AgI) has been used quite widely as a cloud seeder, because laboratory studies have shown it to be one of the most efficient *freezing nuclei*, beginning to be effective at –4 °C. Myriads of tiny crystals of AgI (whose crystal structure is similar to ice) are produced from small burners which can be flown through clouds or sited on the ground to feed updrafts rising into clouds. Seeding must be carefully controlled: if a cloud is swamped by freezing nuclei the effect might be to produce larger numbers of smaller ice crystals than would have occurred without seeding, which is the opposite of the selective growth required of all precipitation mechanisms.

6.13.2 Aircraft icing

As soon as aircraft began to fly in supercooled water clouds, the icing-up of exposed surfaces, especially the leading edges of wings, began to be a dangerous

Figure 6.27 A small civil aircraft shortly after heavy icing, presumably while descending to land. The de-icing boot on the wing leading edge has done its job well at this aerodynamically crucial place, but the rest of the wing looks bady iced.

hazard. Aircraft surfaces cooled to ambient sub-freezing temperatures can act like giant hailstones and collect masses of ice by the impact-freezing of supercooled cloud droplets and freezing rain (Fig. 6.27). In the worst conditions these can accumulate in a few minutes to the point where their weight, and their distortion of the lifting aerofoil shape of the wings, can destroy the plane's ability to fly. Many aircraft have been lost in this way and others have survived only by the last-minute melting and loosening of the burden of ice as they fell out of control into the warm low troposphere. Although flying lower or higher (where the collection efficiency of the more numerous ice crystals is very low) does reduce the problem, almost all modern aircraft have systems for dealing with icing, including heating or even flexing the leading edges of wings and other crucial parts.

6.13.3 Radar

Radars emitting wavelengths of about 10 cm receive strong echoes from precipitation particles and droplets, which can be used to measure precipitation over wide areas (Section 2.6). Radar has also been used to investigate the invisible and fairly inaccessible interiors of cumulonimbus (shower clouds), nimbostratus (precipitating frontal clouds), and the active inner zones of severe tropical cyclones (Fig. 13.12). Indeed some of the major features of severe local storms (Section 11.6) have been clarified with the help of radar, including types which scan vertically to provide a vertical section of echo (Fig. 6.28) rather than the more common plan view. Many large civil aircraft now carry small forward-pointing

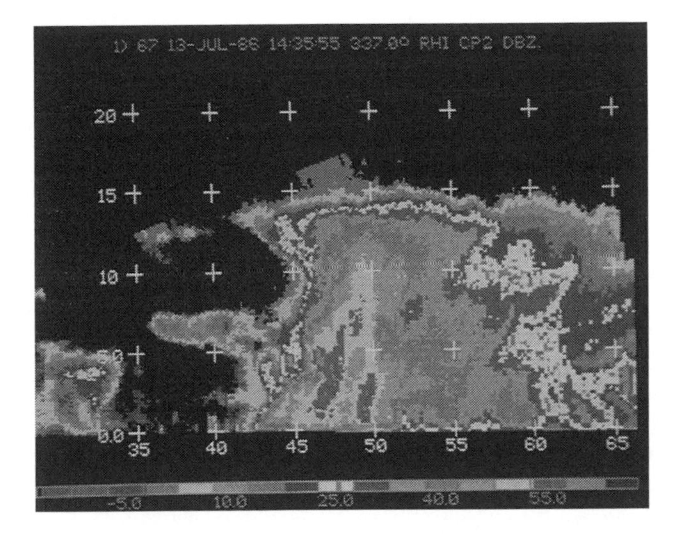

Figure 6.28 Radar echo from a tall cumulonimbus (assuming the vertical units are in in km) showing vertical shafts of heavy precipitation and a very clearly marked top where updrafts run out of buoyancy against the tropopause, and some limited overshoot. The vertically scanning radar has been programmed to produce a display of echo strength (brightness) against range and height (RHI display). (See Plate 9).

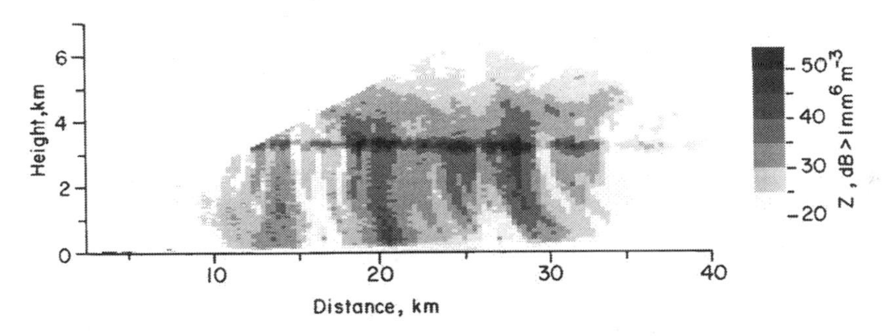

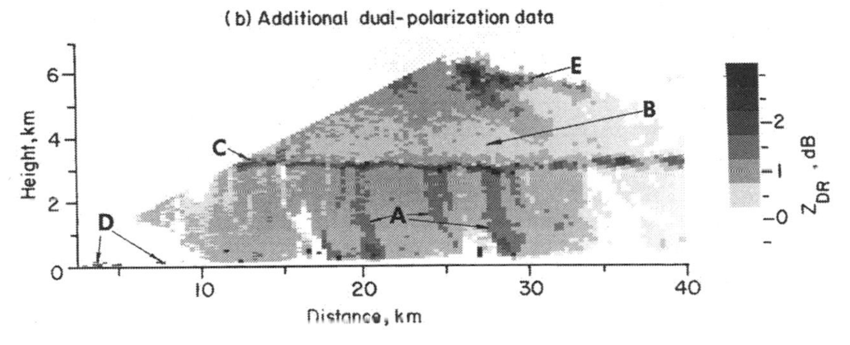

Figure 6.29 Radar echo strength from a section of frontal cloud over southern England, showing the horizontal melting or 'bright band' of enhanced echo at altitude 3.2 km separating the snow above from the rain beneath. The upright stripes of stronger echo show small-scale cells of heavier rain which are tilted by the strong wind shear in the lowest 1.5 km.

radars which enable them to detect and avoid regions of heavy precipitation, mainly because of their associated unpleasantly and even dangerously turbulent downdrafts and updrafts. Nimbostratus often maintains a horizontal band of strongly enhanced echo where snowflakes become coated with water as they

Checklist of key ideas 198

melt, reflecting radar waves more efficiently than either the dry snow and ice above or the rain beneath. Such *melting bands* (Fig. 6.29) confirm that most surface rain in middle and high latitudes is formed by the melting of snow falling from higher levels.

6.13.4 Lightning conductors

In a classic example of the connection between pure research and useful application, Benjamin Franklin followed his dangerous confirmation of the electrical nature of lightning by proposing that buildings could be protected from lightning damage by connecting a stout electrical conductor from just above the highest part to a grounded metal stake. This he supposed would offer the lightning an easy route to Earth—so easy that little damaging heat would be generated *en route*. (Buildings and trees are damaged by the heat generated as the current surges through their natural resistance, explosively boiling interior moisture or sap.) His supposition was amply confirmed by experience, and damage to church spires and other exposed buildings was dramatically reduced after protection. However, there was resistance from traditionalists and cynics, and decades were to pass before the lightning conductor became universally accepted.

Sailing ships had long suffered damage and crew casualties, arising from lightning strikes to masts, spars, and rigging. Injury and death occurred because crew were often aloft at the time of the lightning strike, adjusting sails and spars in response to the squalls and shifts of wind associated with thunderstorms. The British Royal Navy then maintained a large number of sailing ships in all seas and weather and suffered accordingly. However, the Admiralty did not take kindly to advice from an outsider, especially after the American revolution cast Franklin in the role of rebel statesman, and so British sailors continued dying in significant numbers. Eventually as a compromise the Admiralty fitted temporary lightning conductors, to be placed aloft when danger threatened, but otherwise to be coiled away below. After a period in which more British sailors were killed hauling lightning conductors aloft, permanent copper strips were attached to the masts and damage and injury was reduced far below previous levels.

Every year a few thousand people worldwide are killed or injured through inadvertently becoming lightning conductors. Reports often note that the victim was struck on a metal cap badge or something similar, and this indeed is often the case. However, it is probably quite wrong to suppose that the metal object contributed to the accident or that its absence might have avoided it. As the stepped leader reaches down to within about 100 m of the surface the electric fields there increase dramatically, especially in the vicinity of any fairly sharp upward projection from the surface. An upright human form will localize the upward spark and the subsequent return stroke. If there is a sharp protruding badge or golf club no doubt this will be selected, but the badge only focuses a stroke already connecting to the person's body. The only sensible precaution to take when caught in the open by a thunderstorm is to get down on the ground and stay there until lightning strikes are at least a mile off. Standing under a solitary tree is no protection, because the tree is a prime target and a strike on it will easily produce lethal fields and currents within many metres of trunk and branches, as shown by the deaths of cattle sheltering against the sting of hail and heavy rain. Nor is it sensible to suppose that only a few people are struck by lightning; only a few people are caught in the open by thunderstorms, but a significant fraction of them are struck.

Checklist of key ideas

You should now be familiar with the following ideas.

1. The concept of a saturated vapour and how this places an upper limit on the vapour density close to a water or ice surface.

2. Maintenance of the global sub-cloud layer by vertical mixing, cloud production, and precipitation.

3. Vertical profiles of specific and relative humidity and dew-point in a well-mixed sub-cloud layer and the relation between screen-level observations and the lifting condensation level.

4. Observations of cloud in the atmosphere and a simple cloud chamber; the observed relation between size and terminal velocity of cloud and precipitation elements.

5. The effects of droplet curvature and impurity on surface-vapour equilibrium and the role of condensation nuclei in cloud formation.

6. Stable and unstable droplet growth and decay related to Köhler curves and the concept of activation.

7. Individual droplet growth and decay, and observed spectra of condensation nuclei, related to observed growth and decay of atmospheric haze and clouds.

8. Observed precipitation sizes, shapes, fall speeds, speed of production in cloud and speed of evaporation outside cloud.

9. Growth of precipitation in cloud modelled by diffusion and condensation, collision and coalescence; warm, cold, and mixed clouds and the Bergeron–Findeisen process.

10. Observed formation of drizzle, rain, snow, and hail related to the models.

11. The global cycle of atmospheric electricity and the role of thunderstorms; a summary of observations of lightning and thunder.

12. Brief outlines of cloud seeding, aircraft icing, radar observation of precipitation, and lightning safety.

Problems

Outline answers to these problems can be found on the **Online Resource Centre**. Answers to odd numbered problems, can be found under Student Resources, answers to even numbered problems under Lecturer Resources.

Level 1

6.1 State which of the following affects the density of a vapour in equilibrium with a water surface and briefly mention how: dissolved water impurity, air pressure, air purity, water temperature, wind speed, water surface shape?

6.2 Define a saturated vapour.

6.3 A friend gazes at rain beating on the window and exclaims in exasperation that if only there could be less rain there would be more sunshine. Outline the inconsistency.

6.4 Rain leaves a hilly road equally wet at two adjacent locations separated by 500 m in altitude. If the ensuing sunshine and wind plays equally on each, give two reasons why you would expect the lower level to dry faster.

6.5 List the deductions about cloud formation which can be drawn from the simple cloud-chamber experiments described in the text.

6.6 Given the large number densities of aerosol, why are the number densities of cloud droplets so comparatively small?

6.7 Explain why, when the upper parts of a mixed (ice and water) cloud are dissolving, the last vestiges are always ice cloud.

6.8 Although rain drops with radii larger than about 2 mm are not observed (Box 6.4), much larger drops are observed beneath trees in wet weather. Explain.

6.9 It is a common fallacy to suppose that sheet lightning is a special type of silent lightning. Explain what actually happens.

6.10 Figure 6.29 shows a melting band on a range–height radar display. Consider how it might appear on the more normal plan–position display (such as Fig. 2.8), in which the radar beam continually scans around at a small angle above the horizontal. Remember the curvature of the Earth's surface.

Level 2

6.11 Use Fig. 6.2 to estimate vapour densities and pressures at the surface of an ice cap in summer (–10 °C) and an equatorial ocean in winter (25 °C). Why might a similar estimate for the Sahara be very misleading?

6.12 In a certain subtropical desert, dew is observed to form each night when the surface temperature falls to 5 °C. Assuming constant atmospheric pressure and vapour content, estimate the relative humidity when the surface temperature reaches 50 °C during the heat of the day. Use Fig. 6.2 and the fact that the svp at 50 °C is 123 hPa. Assuming that the air temperature at screen level is then 45 °C and that this is at the base of the well-mixed layer, use the simple rule of thumb for lapse of dew-point depression to find the implied height of cloud base above the surface.

6.13 Show from the approximate form of the relationship between equilibrium and saturation vapour densities (Eqn 6.4) that on any particular Köhler curve the radius r at which these densities are equal is given by $r^2 = (T\,B\,m)/A$. Given that $A = 3.16 \times 10^{-7}$ m K for water at 0 °C, and that for NaCl $B = 1.47 \times 10^{-4}$ m^3 kg^{-1}, find r values corresponding to salt particle masses of 10^{-18} and 10^{-16} kg.

6.14 Use Eqn B6.3g (the last equation of Box 6.3) to calculate the time taken for a cloud droplet of radius 5 μm to evaporate to 0.5 μm in an environment with ambient relative humidity 90% when droplet and environment have temperature 11 °C. Repeat the process at –20 °C, given that saturation vapour density is then 1.07×10^{-3} kg m^{-3}, and note the relevance of the result to the persistence of ice cloud in the upper troposphere.

6.15 Find the time taken for a droplet to grow from radius 15 to 500 μm by collision and coalescence in the low troposphere with cloud water of specific mass 3 g kg^{-1}. Allow for the change in terminal velocity regime by calculating in two stages: the first to cover growth to 30 μm and the second to cover further growth. How much longer is needed to grow to radius 1 mm?

6.16 Lightning strikes the surface 200 m away from an observer. Assuming that the flash extends to a maximum height 2 km vertically above the observer, and has a nearly horizontal section 1 km above the observer, describe in detail the observed peal of thunder, given that the speed of sound in air is 330 m s^{-1}.

Level 3

6.17 Consider the realistic development of a layer of dew 1 mm thick in 1 hour to find the implied flux of condensing water in molecules per second per square metre. Compare this with the number of impacts (X) (per second per unit area) of molecules of water vapour according to the following expression derived from the kinetic theory of gases *[20]*: $X = 10^3\, A\, e/(2\sqrt{3}\, R^*\, T\, M)$ where A is Avogadro's number, R^* is the universal gas constant, e is the vapour pressure, M and T are the molecular weight and absolute temperature of the vapour. Notice that the result shows that the deviation from perfect dynamic equilibrium is very small.

6.18 By analysing the shape of the Köhler curve described by Eqn 6.4, show that the droplet radius r_m for maximum equilibrium vapour is $\sqrt{3}\, R$, where R is the droplet radius for equal equilibrium and saturation vapour densities (i.e. the answer to Problem 6.13). Show further that associated maximum supersaturation SS_m is given by $200\, A/(3\, r_m\, T)$, where terms are as defined in

Problem 6.13 and R values are 0.62 and 6.2 μm. Calculate r_m and SS_m for the salt particles cited in Problem 6.13 (i.e. R values 0.36 and 3.6 μm).

6.19 Discuss the role of droplet and turbulent eddy scale processes in maintaining the sharp, regular bases of cumulus clouds, and their sharp irregular tops and sides.

6.20 Consider the growth by collision and coalescence of a droplet falling through cloud with cloud water specific mass m, and show that droplet growth dr is proportional to distance fallen dz (assuming no updraft in the cloud) according to

$$dr = \{m\,\rho/(4\,\rho_w)\}\,dz$$

where ρ and ρ_w are the density of air and water. Hence find the depth of very wet cloud ($m = 4$ g kg^{-1}, typical of warm, low clouds over hills in wet weather) which could produce an increase of 1 mm in droplet radius. Such rapid growth partly accounts for the familiar increase of precipitation over hills.

7 Atmospheric dynamics

7.1 Introduction

In this chapter we relate the movements of air parcels to the forces which act on them, using *Newton's second law of motion*, the law underpinning all *dynamics* (Section 1.4). The approach is necessarily more mathematical than in most other chapters: symbols and equations are used freely to provide concise and unambiguous descriptions of behaviour which is often too complex for purely verbal description; equations are rearranged and developed to reveal physical meanings which might otherwise remain hidden, and some are solved by algebraic or numerical means, though complete solutions are rare. It is important in all this to remember that mathematical representation can help us understand the dynamics of the real atmosphere only when we cultivate the habit of looking at every term, equation, magnitude, graph, etc. to grasp the physical behaviour that it represents, and assess its relevance to what we know of the real atmosphere. Realistic magnitudes of terms are estimated, partly for their intrinsic interest and the sense of physical reality they impart, and partly to distinguish dominant terms from others that can be ignored at least initially—all of which can ease interpretation and solution. Graphs and diagrams can be particularly helpful, and are worth examining and annotating freely to make them your own.

7.2 Equation of motion *[24]*

According to Newton's second law of motion, when a force F acts on a body of mass m (in our case usually a parcel of air) moving with velocity $\mathbf{V}$, the rate of change with time of the body's *momentum* ($m\,V$) is equal in both magnitude and direction to F. We write

$$F = \frac{\mathrm{d}}{\mathrm{d}t}(m\,\mathbf{V}) \qquad 7.1$$

where $\mathrm{d}(m\,V)/\mathrm{d}t$ represents the rate of change of momentum with time, i.e. the very small change $\mathrm{d}(m\,V)$ of momentum divided by the very small time interval $\mathrm{d}t$ in which the change occurs.

Since momentum is the product of mass and velocity, it can change with changing mass or changing velocity or both, as we see by expanding the right-hand side of Eqn 7.1.

$$F = \frac{\mathrm{d}}{\mathrm{d}t}(m\,\mathbf{V}) = m\frac{\mathrm{d}\mathbf{V}}{\mathrm{d}t} + \mathbf{V}\frac{\mathrm{d}m}{\mathrm{d}t} \qquad 7.2$$

In most cases we will meet, mass is constant, so that $\mathrm{d}m/\mathrm{d}t$ is zero and Eqn 7.2 becomes

$$\boldsymbol{F} = m\frac{\mathrm{d}\boldsymbol{V}}{\mathrm{d}t} \quad \text{or} \quad \boldsymbol{F} = m\boldsymbol{a} \quad \text{or} \quad \boldsymbol{a} = \frac{\boldsymbol{F}}{m} \qquad 7.3$$

where $\mathrm{d}\boldsymbol{V}/\mathrm{d}t$ is the rate of change of parcel velocity with time—its *acceleration* $\boldsymbol{a}$. As usual, $\boldsymbol{F}$, $\boldsymbol{V}$, and $\boldsymbol{a}$ are printed in bold to show that they are *vector* quantities (i.e. having direction as well as magnitude), as distinct from *scalar* quantities, like mass m and time t (and temperature), which have magnitude only. The three forms of Eqn 7.3 are completely equivalent, and the last is the *equation of motion* introduced in Section 1.4.

7.2.1 Units and dimensions

Equation 7.3 states that when a force acts on an air parcel, the parcel's acceleration is proportional to the force and inversely proportional to the parcel's mass—its inertia. In using the equation in observation or calculation we must ensure that all its terms are expressed in consistent units, such as SI, if necessary by conversion from practical units if these are not SI (NN 5.1 and Useful Information).

The third form of Eqn 7.3 shows that force per unit mass ($\boldsymbol{F}/m$) is equivalent to an acceleration, whose SI units are m s^{-2}. Such equivalence appears in any consistent set of units (not just SI) and is an example of the *dimensional consistency* to be expected of all valid equations relating measurable physical quantities. Stated by Joseph Fourier in the early 1800s and justified by vast experience since, dimensional consistency is a very useful constraint for checking units and physical significance, and the validity of uncertain equations (Boxes 1.3 and 7.6).

7.2.2 Direction, axes, and components

The second and third forms of Eqn 7.3 state that resultant force $\boldsymbol{F}$ and acceleration $\boldsymbol{a}$ have exactly the same direction, as in the verbal statement of Newton's second law which began this section. However, though $\boldsymbol{a}$ and $\boldsymbol{F}$ have the same direction (as does $\mathrm{d}\boldsymbol{V}$, which has the same direction as $\boldsymbol{a}$, since $\mathrm{d}t$ is scalar), this need not be the direction of velocity $\boldsymbol{V}$. In fact $\boldsymbol{a}$ may be at any angle to $\boldsymbol{V}$: perpendicular to it, as in circular motion, or opposite to it, as in linear deceleration, both of which are common in the atmosphere (Section 7.13), or indeed in any other direction. And if the air parcel is acted on by two or more forces acting in different directions, as is very common, each may differ from the direction of the resultant force. So despite the overall directional simplicity of Eqn 7.3, its component terms are often directionally complex, demanding care in drawing, analysing, and understanding.

To analyse the position and motion of an air parcel, we use a framework of perpendicular axes. The conventional *meteorological frame* (Fig. 7.1) has an origin O at a convenient fixed point on the Earth's surface, an x axis pointing horizontally East, a y axis pointing horizontally North, and a z axis pointing vertically upwards. The slight curvature of the x and y axes can be ignored unless the coverage becomes global.

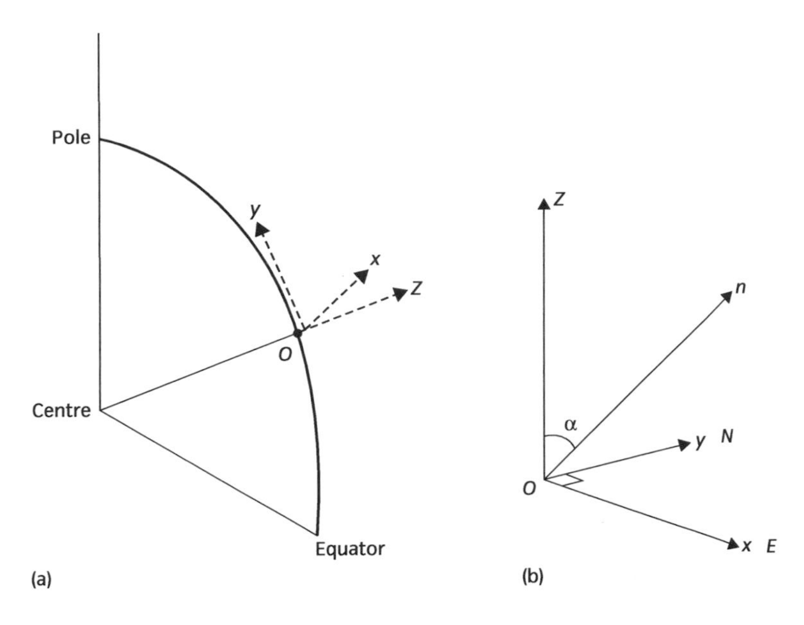

Figure 7.1 (a) Northern hemisphere quadrant showing conventional meteorological axes with origin O on the surface. (b) Axes x (N), y (E), and z (up) related to a generalized axis n.

An air parcel has position (x, y, z), velocity components (u, v, w), acceleration components $(\mathrm{d}u/\mathrm{d}t, \mathrm{d}v/\mathrm{d}t, \mathrm{d}w/\mathrm{d}t)$, and is acted on by a net force with components (F_x, F_y, F_z). The vector equation 7.3 can be written as three scalar equations (independently valid because mutually perpendicular), each of which gathers the components for that axis (the x axis for example) of every term of the vector equation.

$$\frac{\mathrm{d}u}{\mathrm{d}t} = \frac{F_x}{m} \qquad \frac{\mathrm{d}v}{\mathrm{d}t} = \frac{F_y}{m} \qquad \frac{\mathrm{d}w}{\mathrm{d}t} = \frac{F_z}{m} \tag{7.4}$$

For example, a preliminary analysis of convection might begin with the vertical component of Eqn 7.4

$$\frac{\mathrm{d}w}{\mathrm{d}t} = BUF - g - FRF_z \tag{7.5}$$

where the upward acceleration $\mathrm{d}w/\mathrm{d}t$ of a buoyant air parcel is driven by the buoyant force BUF, opposed by gravity g, and restrained by the vertical component of friction FRF_z, each of these being expressed as force per unit parcel mass. The simplest term is g, which has magnitude very close to 9.8 m s^{-2} throughout the lower troposphere. Even casual observation of buoyant accelerations in cloudy convection suggests that vertical accelerations $(\mathrm{d}w/\mathrm{d}t)$ are several orders of magnitude smaller than g, and that convective parcel dynamics must therefore be dominated by the near balance

$$BUF - g - FRF_z \approx 0$$

Though a more thorough study of convection will involve the two horizontal components as well (Section 7.14), the simplicity and power of the scalar component approach is obvious.

7.2.3 Reference frame

Newton's laws of motion describe dynamic behaviour as observed from an unaccelerated (*inertial*) reference frame. However, the meteorological reference frame outlined in Fig. 7.1 is far from inertial, since it is constantly accelerating towards the polar axis as Earth spins. This acceleration must be included explicitly in the equation of motion, which otherwise would be unbalanced by its absence, and this has important meteorological consequences outlined in Box 7.4 and Section 7.6.

7.3 Solving and forecasting

7.3.1 Solving

Equation 7.3 is called the equation of motion because it can provide an observably valid description of the motion of an air parcel, or any other body. We say that this potential is fully realized only if we can *solve* the equation, but solving can have several different meanings which we need to clarify.

For example, Eqn 7.5 describes the vertical acceleration of a buoyant air parcel.

In a loose sense, we begin to solve this equation as soon as we start to analyse its physical implications for a given atmospheric situation, especially if it simplifies in the process, and you see this process in action already in Eqn 7.5. Indeed if simplified further (by eliminating friction and vertical acceleration for example), the simplified residual equation leads back to the very useful concept of hydrostatic equilibrium (Section 4.4) which links atmospheric statics and dynamics.

Solving equations in this loose sense makes up much of the present chapter, indeed much of the theoretical content of this book, and is a vital part of the process of scientific investigation. However, the word 'solution' has more specific meanings which we must outline, even if we cannot use them much in this book.

Consider Eqn 7.5 in a situation in which the terms on its right-hand side cancel to zero. The air parcel has zero vertical acceleration, so that w = constant will be a solution, since its substitution in $\mathrm{d}w/\mathrm{d}t$ gives zero on the left-hand side and balances the equation. The behaviour of the air parcel can now be described by $w\ (= \mathrm{d}z/\mathrm{d}t)$ = constant, so that we can write

$$\frac{\mathrm{d}z}{\mathrm{d}t} = C$$

where z is the parcel height above MSL and C is its constant upward speed. A solution to *this* equation is (since its substitution gives C on its left-hand side)

$$z = C\,t + D \qquad 7.6$$

and is the ultimate solution of Eqn 7.5 with exactly balanced forces—the parcel is climbing steadily at speed C from a height D above MSL at zero time. If C and

D are known (for example by observation) we can specify the height of the parcel at any later time t, and can say that we have *fully solved* Eqn 7.5 by extracting from it a full description of the parcel's vertical motion.

Real atmospheric behaviour is never so simple; indeed frictional and other interactions between adjacent air parcels mean that full equations are not soluble (in the full sense) by any of the many mathematical relationships and procedures devised in recent centuries. Perceptive simplification (to ease the mathematics while maintaining relevance to the real atmosphere) has produced a few heroic solutions relevant to the early stages in the growth of depressions and anticyclones *[31]*, but in the main the equation of motion for the atmosphere does not have *analytic* solutions (algebraic solutions such as Eqn 7.6), and this fact obviously inhibits attempts to set meteorology and weather forecasting on a firm mathematical footing.

7.3.2 Numerical forecasting

The development of the electronic computer from the 1940s onwards has opened up the possibility of finding *numerical* solutions to the full equation of motion, and this process now lies at the heart of all detailed forecasting of weather and climate.

In principle we put current observations into the equation of motion (or whatever other equation is in use) and find the little changes which it implies for a little *time step* dt forward into the future, and repeat this process progressively through the forecasting period. For example, start again with a version of Eqn 7.5 for a parcel of mass m.

$$\mathrm{d}w = m\, F_z\, \mathrm{d}t$$

If the height, upward speed, and net upward force are respectively z, w, and F_z at the start of the time step dt, then at its end, the parcel's upward speed will be

$$w + \mathrm{d}w \quad \text{i.e.} \quad w + m\, F_z\, \mathrm{d}t$$

and the parcel's height will be

$$z + \mathrm{d}z \quad \text{i.e.} \quad z + \left(w + \frac{\mathrm{d}w}{2}\right)\mathrm{d}t = z + \left(w + \frac{m\, F_z}{2}\right)\mathrm{d}t$$

If we do the same for all the other parcels in the atmosphere, and work out how they are interacting (by bumping, rubbing, and squeezing), we can predict their positions, motions, and states at the end of the time step dt. These are then used as the starting condition for another time step forwards, and so on, time step after time step, to produce a forecast extending into the future. Equally we could do this for a past situation, and produce a forecast from that earlier time to compare with the observed outcome and test the accuracy of the particular *numerical model* being used.

Obviously the process is tedious and complex: current forecasting models cover the entire Earth with a three-dimensional grid of fixed points where the state and motion of the air are observed and repeatedly calculated (an *Eulerian* approach which is a simpler alternative to the above *Lagrangian* approach of following parcels as they move). In current practice, grid points are about 30 km apart in the horizontal, and 1 km apart in the vertical, and temperature, pressure, humidity, wind speed, and wind direction are measured and predicted at each point in time steps of a few minutes in order to produce forecasts for multiples of 6 hours ahead, out to 6 days in the future. To do this in time to run the whole process again 6 hours later needs the biggest available computers and teams of expert handlers, as well as others continually checking the forecasts against subsequent actual weather, and yet more people working to improve computation schemes and adapt them to continuing advances in observation techniques, and computer software and hardware.

The details of such forecasting procedures are far too specific and complex to cover further in this book. Some details appear in websites and advanced texts, but many are buried in technical reports by the WMO and national forecasting services *[32]*. It is impressive to see how much potential is realized in meteorology, as in many other branches of physical science, by Newton's second law of motion. The story of the falling apple may be apocryphal, but whether its impact was virtual or real in 1665, its repercussions are enormous today.

7.4 Forces *[33]*

The equation of motion relates air parcel movement to the action of one or more forces acting on the parcel. Three types of force are at work in the atmosphere: pressure forces arising from parcels pressing on each other, gravitational forces, and frictional forces arising from parcels rubbing in relative motion. The first and last are *contact forces*, since they depend on contact between the air parcel and its neighbours, or the Earth's surface. Gravity is a *body force*, applying to every part of the air parcel because every molecule in the parcel is being pulled by Earth's gravitational field.

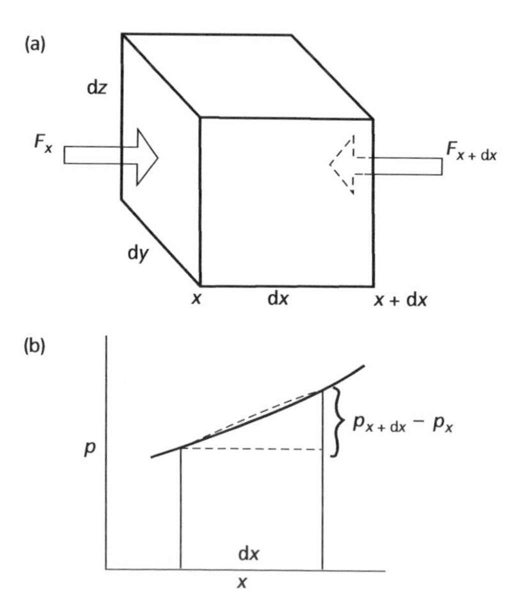

Figure 7.2 (a) Net x component of force on a cubical air parcel embedded in a pressure gradient. (b) Pressure difference and pressure gradient in the x direction.

BOX 7.1 Pressure gradient force

Consider a very small block-shaped air parcel in a positive pressure gradient $\partial p/\partial x$ along the x axis (Fig. 7.2a and b). Because of this gradient, the ambient pressure p_{x+dx} pressing in the negative x direction on the parcel face at position $x + dx$ is slightly greater than the pressure p_x pressing in the positive x direction on the parcel face at x. The excess pressure $p_{x+dx} - p_x$ pressing in the negative x direction can be expressed by

$$p_{x+dx} - p_x = \frac{\partial p}{\partial x} dx$$

where dx is the very small x dimension of the parcel (Fig. 7.2b). If the other two parcel dimensions are dy and dz, the faces at x and $x + dx$ each have area (dy dz) and the net pressure gradient force on the block in the positive x direction is

$$(p_{x+dx} - p_x)\, dy\, dx = -\frac{\partial p}{\partial x} dx\, dy\, dz = -\frac{\partial p}{\partial x} dvol$$

where $dvol$ (= dx dy dz) is the very small volume of the parcel, and the minus sign shows that the net force acts down the pressure gradient (e.g. towards negative x if pressure is increasing with x). If the parcel air has density ρ, its mass is $\rho\, dvol$, and the pressure gradient force per unit parcel mass along the x axis (PGF_x) is given by

$$PGF_x = -\frac{1}{\rho\, dvol}\frac{\partial p}{\partial x} dvol = -\frac{1}{\rho}\frac{\partial p}{\partial x}$$

In the same way we find PGF components $-(1/\rho)\, \partial p/\partial y$ and $-(1/\rho)\, \partial p/\partial z$ in the y and z directions, and $-(1/\rho)\, \partial p/\partial n$ in the general n direction.

7.4.1 Pressure forces

Surrounding air presses inwards on all the walls of an air parcel, so that the net force is zero unless there is a pressure gradient across the parcel, in which case a *pressure gradient force* (PGF) acts on the parcel, pushing down the pressure gradient, from higher to lower pressure (which assumes that there is a smooth pressure field running through all air parcels, as discussed in Section 4.3). The pressure gradient (and associated PGF) could lie along one of the meteorological axes (Fig. 7.1), but is much more likely to be angled between them, as shown by the general axis n (Fig. 7.2), which points towards higher pressure along the line of maximum pressure gradient.

By considering the pressure forces on a block-shaped parcel of air density ρ (Fig. 7.2 and Box 7.1), we find that the PGF per unit mass of air parcel is given by

$$PGF = -\frac{1}{\rho}\frac{\partial p}{\partial n} \qquad 7.7$$

where $\partial p/\partial n$ is the rate of increase of pressure with increasing distance along the axis n, and the minus sign ensures that PGF acts back down the pressure gradient. The *partial derivative* $\partial p/\partial n$ is used to represent the pressure gradient because it describes pressure variation with n alone—ignoring pressure variations in other directions, or with time.

The PGF components along the conventional x, y, and z axes are found in the usual vector way: for example, if α is the angle between the n axis and the vertical (the z axis—Fig. 7.2), then the vertical PGF per unit mass is given by

$$-\frac{1}{\rho}\frac{\partial p}{\partial z} = -\frac{1}{\rho}\frac{\partial p}{\partial n}\cos\alpha$$

NUMERICAL NOTE 7.1 Pressure gradient, PGF, and acceleration

Consider a mid-latitude depression in which pressure falls N'ward by 6 hPa per horizontal 100 km in the low troposphere, where air density is 1.2 kg m^{-3}. The pressure gradient in SI units is 600 Pa per 100 km, which is $600/(100 \times 1000) = 6 \times 10^{-3}$ Pa m^{-1}, so that the horizontal *PGF* per unit mass is 5×10^{-3} m s^{-2}. Note that this has the units of acceleration as required by the equation of motion (Eqn 7.3).

If an air parcel accelerates for an hour at this rate from rest, its N'ward speed increases by acceleration × time = $5 \times 10^{-3} \times 60 \times 60 = 18$ m s^{-1}—about Beaufort force 8 if occurring at a height of 10 m. In calculating this we have ignored the slowing effect of friction (strong near the surface), and the substantial turning of the Earth which in an hour will 'throw' the parcel significantly E'ward (to the right of its initial motion—Section 7.7) in the N hemisphere.

The observed atmospheric pressure gradient is almost always so nearly vertical (α so small that cos α is very nearly unity) that the vertical pressure gradient $\partial p/\partial z$ is very much larger than the horizontal pressure gradients $\partial p/\partial x$ and $\partial p/\partial y$ (Section 4.3). For example, horizontal pressure gradients associated with extratropical cyclones rarely exceed 5 hPa per 100 km, whereas 5 hPa in 50 m is typical of vertical pressure gradients in the low troposphere. *PGF* components are unequal to exactly the same extent, since the air density is common. However, the independence of the vertical and horizontal components of the equation of motion ensures that the relatively very small horizontal *PGF*s can still be very important for horizontal motion, provided the other dominant terms are similarly small, as is the case (Section 7.9). Only on scales below ~10 m (in turbulent swirls and eddies) does the great dominance of vertical *PGF*s (and nearly balancing *g*) fade away in the equation of motion.

7.4.2 Gravitational force

The gravitational attraction between an air parcel and the Earth accounts for the downward force on the parcel which we call its *weight*, and an equal and opposite force on the planet Earth which is unobservable in practice. Measurement by calibrated weighing machines (but not balances) at rest in the Earth-bound reference frame (Fig. 7.1) shows that the weight of any body of mass m is given by $m\ g$, where g has an average value of about 9.81 m s^{-2} at sea level and varies with latitude ϕ and altitude z as shown in Table 7.1. The factor g is known as the *gravitational acceleration* because it is identical to the acceleration of bodies falling freely in a vacuum under their own weight in an Earth-bound reference frame. The gravitational force *GRF* per unit mass on an air parcel is therefore $-g$, the minus sign indicating that it acts towards negative z (downwards).

Inserting this in Eqn 7.5 shows that such an air parcel would fall down, accelerating at 9.81 m s^{-2} if it were sealed in a plastic bag and released in a vacuum, i.e. dropping like a stone. Of course nothing like this happens in the real atmosphere, because the parcel is largely sustained by the net upward forces from surrounding air parcels, which maintain the *PGF* component arising from the vertical lapse of ambient air pressure. And any motion through its fellow parcels is restrained by friction. However, *PGF* and frictional forces have no effect on the gravitational force, since it is a body force independent of the presence of other bodies, except

Table 7.1 Apparent gravitational acceleration

Altitude in km	Latitude 0°	45°	90°
0	9.780	9.806	9.832
20	9.719	9.745	9.770
40	9.538	9.684	9.709

through gravitational attraction between them, which is minute in the case of air (or even mountains).

As shown in Table 7.1 the magnitude of *GRF* varies slightly with altitude and latitude, because weight (and therefore free-fall acceleration) are measured in the standard reference frame fixed to the Earth's surface (Fig. 7.1), which is continually accelerating towards the Earth's axis of spin. As discussed later (Section 7.6) this frame's acceleration combines with true gravitational acceleration to produce the slightly variable *apparent* g shown in Table 7.1. These variations are too small to matter for many (but not all) meteorological purposes—a very convenient outcome which arises from gross self-adjustment by the bulk of the spinning Earth over geological time.

7.4.3 Frictional forces *[10]*

Friction arises whenever bodies (liquids and gases as well as solids) in contact move or try to move relative to each other, and it always restrains the relative motion. In the atmosphere it plays a crucial role in limiting velocity differences within the air, and between the air and surface. Just as the frictional force between a sliding block and a bench top is parallel to the sliding motion, so the frictional force *FRF* on an air parcel is parallel to the sliding movements between it and its neighbours or a solid or liquid surface (Fig. 7.3). Such relative motion is concentrated in *wind shears*—regions where wind speed and/or direction vary vertically or horizontally across the axis of air flow. The strongest and most persistent wind shears are the vertical shears of horizontal wind that are endemic in the *atmospheric boundary layer* (ABL)—the lowest few hundred metres of the atmosphere (Chapters 1 and 10). Figure 7.4 depicts a typical profile of average wind speed U in the ABL, with shear $(\partial U/\partial z)$ strongest near the surface and diminishing

Figure 7.3 A windswept tree in an exposed location. Though the biological mechanism must be quite subtle, the message is obvious.

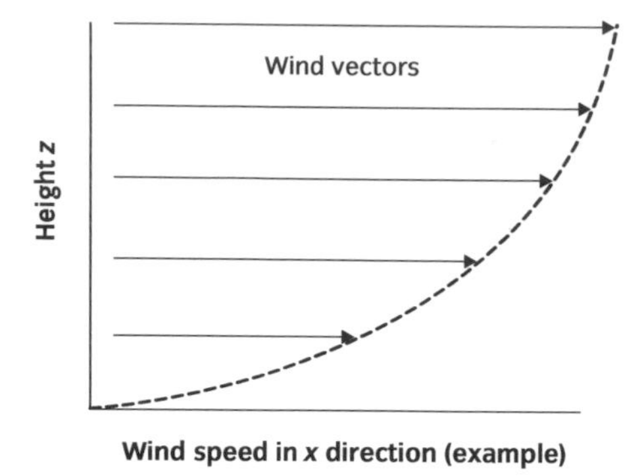

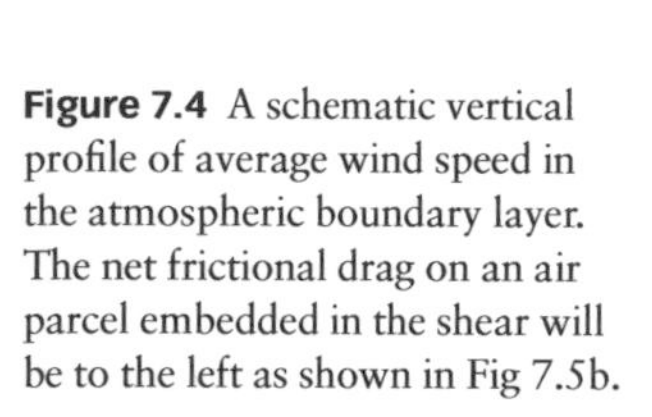
Figure 7.4 A schematic vertical profile of average wind speed in the atmospheric boundary layer. The net frictional drag on an air parcel embedded in the shear will be to the left as shown in Fig 7.5b.

upwards. (The x direction here is arbitrary; winds can come from any direction in the real ABL.) Significant vertical shears are also found above and below the cores of jet streams, and on a much smaller scale there are intense, short-lived, randomly orientated shears in and around every turbulent eddy.

In terms of dynamics, friction between parcels moving at different speeds exchanges momentum between the parcels, with drag forces acting until the shear is removed. Since force is measured by the rate of change of momentum it tends to produce (Eqn 7.4), frictional drag represents a flow of momentum across an imaginary surface in the sheared layer which will exert a measurable drag force on the air, etc. it reaches, and an equal and opposite force on the air, etc. it leaves. Momentum flows across wind shears because air itself is transferred, and this happens in two very different ways in the atmosphere—by diffusion of air molecules and by *turbulent diffusion* of eddying air parcels.

Molecular diffusion

If air flow is very smooth, as it is within millimetres of a smooth solid surface, it is called *laminar*, because it has the appearance of thin plates (lamina), sliding over one another (Fig. 7.5a). As outlined in Box 7.2 the *viscous shearing stress* T (the tangential viscous drag on unit area of one side of a plate-shaped parcel) is given by

$$T = \mu \frac{\partial u}{\partial z} \qquad 7.8$$

where μ is the *dynamic* (or *Newtonian*) *coefficient of viscosity* of air, whose value in atmospheric conditions is close to 18×10^{-6} Pa s in SI units.

- In the uniform shear of Fig. 7.5a, the opposite faces of a plate-shaped parcel will experience equal and opposite drags, so that the parcel suffers zero net frictional force *FRF*, even though it is being continually sheared. The x-ward momentum of the air flow is being passed from parcel to parcel across the shear and hence to the surface, which feels a force T per unit surface area in the x direction.
- If the shear is not uniform (if the profile of flow speed is not a straight line) Fig. 7.5b and Eqn 7.8 show that each air parcel will experience a net force in the positive or negative x direction, because the viscous forward drag on its more

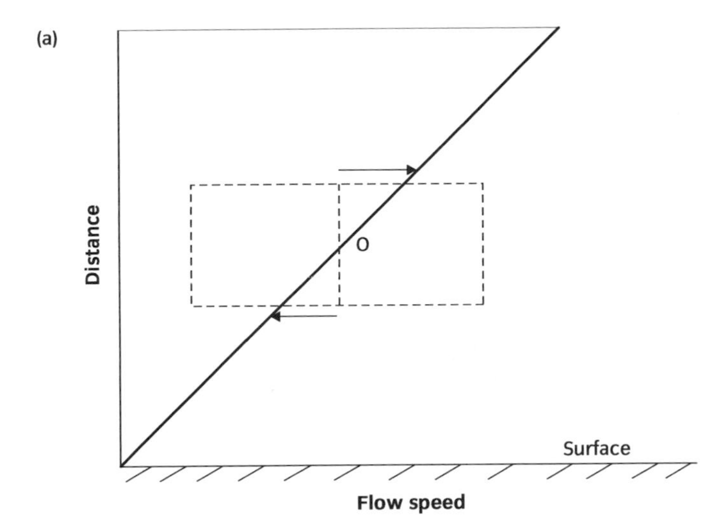

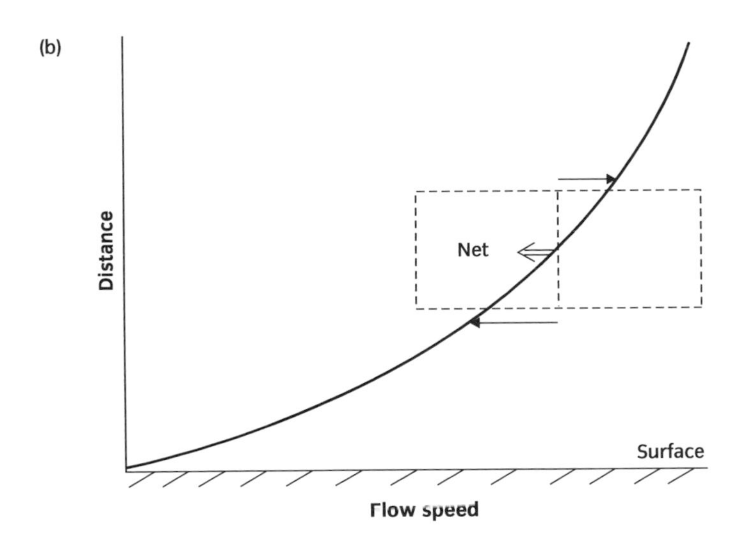

Figure 7.5 (a) Balancing frictional forces on a block-shaped parcel in a uniform shear. (b) Unbalanced frictional forces on a block-shaped parcel in a non-uniform shear. Provided the shear stress on each face is proportional to the shear there, the net force will be against the flow, allowing the parcel to feel the drag of the nearby surface although not in direct contact.

strongly sheared surface (larger $\partial u/\partial z$) is no longer balanced by the backward drag on the less sheared surface. As shown in Box 7.2, the net viscous force FRF_x in the x direction per unit mass of air parcel is given by

$$FRF_x = \frac{\mu}{\rho}\frac{\partial}{\partial z}\left(\frac{\partial u}{\partial z}\right) = \nu\frac{\partial^2 u}{\partial z^2} \qquad 7.9$$

where ν is the *kinematic coefficient of viscosity* of air (μ/ρ), which varies mainly with temperature and is about 15×10^{-6} m^2 s^{-1} in the low troposphere, where air density is about 1.2 kg m^{-3}. The term $\partial^2 u/\partial z^2$ is the shear of the wind shear (the curvature of the wind profile), and is negative in the example shown in Fig. 7.5b, since $\partial u/\partial z$ is decreasing with increasing z. According to Eqn 7.9 each parcel in this flow experiences a net viscous drag in the negative x direction, as is particularly obvious for a parcel in the fastest moving layer, since it is dragged by slower air both above and below. For the same reason a parcel located in a speed minimum is dragged forwards, showing that viscosity acts to smooth out irregularities in shear, working towards an equilibrium flow profile with uniform shear—a fundamental role of viscosity.

Turbulent diffusion

On scales much greater than a millimetre (in fact from centimetres to hundreds of metres), air flow in the troposphere and stratosphere is turbulent rather than laminar, with parcels jostling chaotically like giant transient 'fluffy' molecules.

BOX 7.2 Viscous forces

Straight flow profiles

Newton sheared thin layers of common liquids between parallel plates (Fig. 7.5a), one fixed and one moving, and established that the drag per unit plate area (the *shearing stress T*) needed to maintain a steady shear $\partial u/\partial z$ across the liquid layer is given by

$$T = \mu \frac{\partial u}{\partial z} \qquad \text{B7.2a}$$

where the *dynamic coefficient of viscosity* μ of the liquid has a well-defined value for a given liquid, pressure and temperature. For practical reasons, air was examined much later than water, etc. when its μ value in typical conditions was found to be $\approx 18 \times 10^{-6}$ Pa s. The vertical axis z is used with the atmospheric boundary layer in mind, though at these millimetric scales it could as easily be the x or y axis. Such experimental work usually produces straight line flow profiles between the laboratory plates.

Curved flow profiles

Figure 7.5b depicts a flow profile more typical of flow in the atmosphere, with shear (and hence T) varying across the layer. Consider the frictional forces acting on a thin slice of air of unit horizontal area bounded by heights z and $z + \mathrm{d}z$. At height $z + \mathrm{d}z$ the faster air above drags the top surface of the slice in the positive x direction with a force numerically equal to the tangential shearing stress $T_{z+\mathrm{d}z}$. At level z the slower air below drags the bottom surface of the same slice in the negative x direction with a force T_z. The small stress difference $\mathrm{d}T = T_{z+\mathrm{d}z} - T_z$ can be written in terms of the gradient of stress across the narrow gap dz, because all variations are gradual on scales much bigger than the molecular scale.

$$\mathrm{d}T = T_{z+\mathrm{d}z} - T_z = \frac{\partial T}{\partial z}\,\mathrm{d}z$$

Because the slice has unit horizontal area, dT is the net frictional drag force on the slice in the positive x direction, and the little slice volume d*vol* is numerically equal to dz. Since the little mass dm of the thin slice is ρ dz (i.e. ρ d*vol*), the net frictional drag force per unit mass is given by

$$FRF_x = \frac{\mathrm{d}T}{\mathrm{d}m} = \frac{1}{\rho\,\mathrm{d}z}\frac{\partial T}{\partial z}\,\mathrm{d}z = \frac{1}{\rho}\frac{\partial T}{\partial z} \qquad \text{B7.2b}$$

The sign of the balance confirms that FRF_x is in the x direction when x-ward stress T increases with z, and in the negative x direction when T decreases with increasing z.

Turbulence diffuses momentum across wind shears on these and larger scales very much more rapidly than does the very short-range molecular diffusion, so that turbulent flow is dominated by *eddy viscosity*, even though molecular viscosity is still at work throughout the turbulent air, feeding on the localized shears generated round the smaller whirling eddies (Section 10.7).

We might expect there to be a turbulent equivalent of Eqn 7.8 linking *eddy stress* T to average wind shear $\partial U/\partial z$ (the shear of U averaged to remove turbulent gusts and lulls) via a *dynamic coefficient of eddy viscosity* K:

$$T = K\frac{\partial U}{\partial Z} \qquad 7.10$$

In fact turbulent motion is so much more complex than molecular motion that K has not been clearly described by any statistical or mechanistic description of turbulence developed so far, even after 150 years of close study. Although this severely limits the usefulness of eddy viscosity in fundamental studies, its simplicity and conceptual similarity with μ keeps it in widespread empirical use (Section 10.9).

Values of K derived by applying Eqn 7.10 to field measurements of wind drag and shear, range across several orders of magnitude, but more than a few centimetres above a sea or land surface they are usually at least four orders of magnitude larger than viscous μ values (being at least 0.1 Pa s^{-1}). And K values tend to increase further with height (Section 10.9), as suggested by the typical wind profile in the ABL (Fig. 7.4), whose shear $\partial U/\partial z$ decreases with increasing height as eddy viscosity increases with eddy size and dynamic effectiveness in reducing wind shear.

7.5 Relative and absolute acceleration

The distinction between real and apparent gravity in Section 7.4 highlights the importance of the acceleration ***FA*** of the conventional meteorological reference frame (Fig 7.1) as it rotates with the Earth (its *non-inertial* behaviour). If a parcel acceleration $\boldsymbol{a}$ is measured relative to the conventional frame, its absolute acceleration is $\boldsymbol{a} + \boldsymbol{FA}$ and the proper statement of the equation of motion is

$$\boldsymbol{a} + \boldsymbol{FA} = \frac{\boldsymbol{F}}{m} \quad \text{or} \quad \boldsymbol{a} = \frac{\boldsymbol{F}}{m} - \boldsymbol{FA} \qquad 7.11$$

If a non zero ***FA*** is omitted from Eqn 7.11, the right-hand side of the second version shows that a mysterious force will seem to act in the opposite direction to ***FA***, just as the upward acceleration of an elevator feels like extra body weight to its passengers, or the *centripetal* (inward) acceleration of a hobby horse on a fairground carousel feels like a *centrifugal* (outward) body force to its rider. The presence of significant ***FA*** has great theoretical and practical importance for large-scale distributions and movements of air and water, as outlined in the following simple and general models.

7.5.1 Simple model

Consider the motion of a toy train moving on a rotating domestic turntable, and compare what is seen by an observer rotating with the turntable, with what is seen by an observer standing in the playroom. We ignore the relatively very slow rotation of the playroom with the Earth.

The toy train T is running at a steady speed V round a circular track of radius R fixed on the turntable, which is turning with angular velocity Ω round O at the centre of the track (Fig. 7.6) in the same direction as the train's motion. Since T is moving at speed V relative to the track, and any point on the track is moving at speed ΩR relative to the playroom (Box 4.4), by simple addition of velocities the train is moving round O at radius R at speed $V + \Omega R$ relative to the playroom. Using Eqn B7.3, the centripetal acceleration of T in Fig. 7.6 is

(i) $\dfrac{V^2}{R}$ relative to the turning track and turntable;

(ii) $\dfrac{1}{R}\left(\Omega R+V\right)^2$ relative to the playroom.

The difference between (ii) and (i) is the part of the absolute acceleration which we miss when we make uncorrected observations from the rotating turntable instead of from the non-rotating playroom. Multiplying out (ii) and subtracting (i) we find that the missing terms are

$$\underset{(a)}{\Omega^2 R} + \underset{(b)}{2\,\Omega V} \qquad 7.12$$

Term (a) Simple centripetal

$\Omega^2 R$ is the acceleration of any fixed point on the rotating track relative to the room. It increases with the square of the rotation rate Ω and with the distance R from the axis of rotation O, and is always directed towards O (*centripetal*). According to Eqn 7.11 and text, its presence will be sensed in the turntable frame

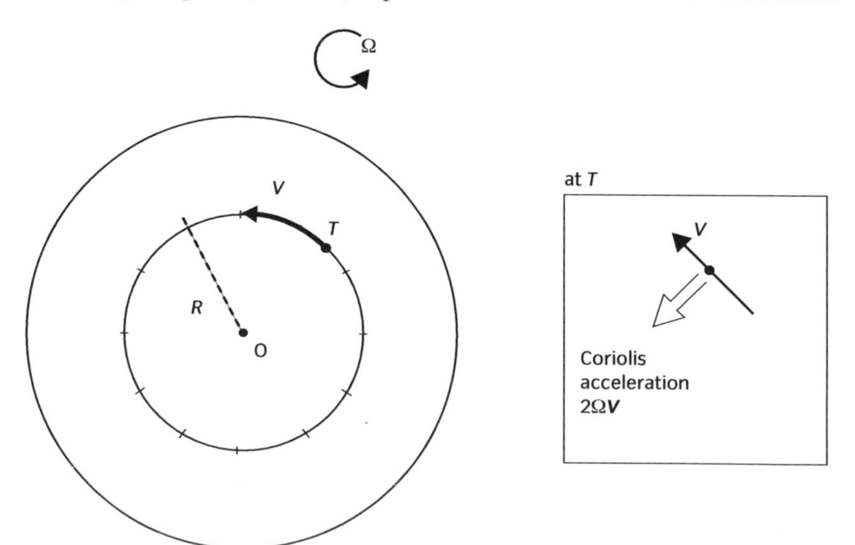

Figure 7.6 Toy train T travelling round a circular track on a turntable rotating in the same direction. The inset shows the magnitude and direction of the Coriolis acceleration at T.

as a *centrifugal* body force on the train with magnitude $\Omega^2 R$ per unit train mass, with obvious relevance to the acceleration of the standard reference frame fixed to the surface of the rotating Earth (Section 7.6).

Term (b) Coriolis

$2\ \Omega\ V$ combines the turntable rotation Ω and the train velocity V relative to track and turntable, but *does not involve* R. It is called the *Coriolis acceleration*, after the French engineer who highlighted its dynamical importance in the early nineteenth century. Though less familiar than term (a), the Coriolis acceleration is clearly a part of the absolute acceleration of a moving body (the train) which becomes distinctive when we choose to observe it from a rotating frame (the turntable) in which the body is *not* at rest. Since this is exactly what we do when we observe the moving atmosphere from the rotating meteorological reference frame, we should expect the Coriolis effect to be relevant to the observed flow of Earth flow and ocean currents, which is very much the case as we will see. Again according to Eqn 7.11 and text, the Coriolis acceleration will be sensed in the turntable frame as a body force ($2\ \Omega\ V$ per unit mass) acting in the opposite direction.

BOX 7.3 **Centripetal acceleration**

We have seen in Box 4.4 that a point P describing a circle of radius R with steady angular velocity Ω, is continually moving with a circumferential speed $V = \Omega\ R$, provided Ω is expressed in radians per unit time. Figure 7.7a depicts this by showing the position of the swinging position vector $\boldsymbol{R}$ at some instant, together with P's simultaneous tangential velocity vector $\boldsymbol{V}$.

In Fig. 7.7b the acceleration vector $\boldsymbol{A}$ is given by swinging $\boldsymbol{V}$ in just the same way as $\boldsymbol{V}$ is given by swinging $\boldsymbol{R}$. Redraw the swinging velocity vector in Fig. 7.7b to find the acceleration vector at the instant shown in Fig. 7.7a, and notice that $\boldsymbol{A}$ is directed towards O (centripetal). Find A (the magnitude of $\boldsymbol{A}$) by seeing that the same 'swinging geometry' which gives $V = \Omega\ R$ in Fig. 7.7a, must give $A = \Omega\ V$ in Fig. 7.7b, so that

$$A = \Omega V = \Omega^2 R \quad \text{or} \quad A = \frac{V^2}{R} \qquad \text{B7.3a}$$

using $\Omega = V/R$. These are equivalent expressions for the centripetal acceleration of a point or body moving steadily round a circular path. The version of Eqn B7.3a involving Ω applies naturally to the rotation of a solid platform like the Earth (Section 7.6).

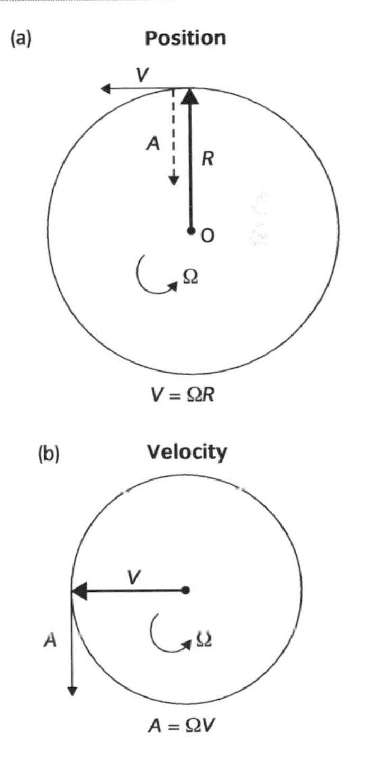

Figure 7.7 Position, velocity, and acceleration of a body moving steadily round a circle. The position diagram (a) describes path and velocity in terms of a swinging radius vector, whereas the velocity diagram (b) describes velocity and acceleration in terms of a swinging velocity vector.

Note that the Coriolis acceleration is not necessarily centripetal. If the train in Fig. 7.6 is reversed and run counter to the turntable's rotation, the equivalent of 7.12 becomes

$$\Omega^2 R - 2\,\Omega\,V$$

which shows that the Coriolis acceleration is opposing the centripetal acceleration of the track and is therefore directed *away* from O in the turntable frame. If the train moves in other directions on the turntable, the Coriolis acceleration has other directions, as shown in the following general model.

7.5.2 General model

The special track layout in the simple model produces a very simple expression for the Coriolis acceleration. To see if this is general, we need to consider all other positions and orientations of the rail track relative to the turntable pivot O. This is done in Box 7.4 and Fig. 7.8, where the train is moving with steady speed V along a straight track placed at random on the turntable. This arrangement is now completely general, apart from the choice of a straight track and consequent absence of the curvature term, which anyway is the same for both frames of reference, as seen in Eqn 7.12.

By inspecting Fig. 7.8 we see that the train is affected by the turntable rotation in two distinct ways: through the centripetal acceleration of any point on the track as it orbits O, *and* through the lateral twisting of the track as the train drives along it. As in the simple model (and Box 7.3), the first is given by an $\Omega^2 R$ term. However, the twisting effect is very different, being independent of R, since the rate of twisting is the same everywhere on the turntable, and increasing with both the rate of twisting Ω and the speed V of the train along the twisting track. In fact this is the Coriolis acceleration, which the geometry and kinematics of Box 7.4 and Fig. 7.8 confirm has magnitude $2\,\Omega\,V$, and direction perpendicular to the train velocity relative to the rotating frame (on the side it is twisting towards), and is confined to the plane of rotation, all of which is consistent with the simple model.

As always, a hidden acceleration in one direction (perpendicularly to the left of the train motion in Fig. 7.8) is equivalently experienced as a body force in the opposite direction—respectively *Coriolis acceleration* and *Coriolis force* (per unit mass). Because the basic centripetal effect ($\Omega^2 R$) is effectively absorbed in the bulging Earth (Section 7.6), the apparently marginal Coriolis effect is very important for large-scale movements of air and water, as outlined throughout much of this chapter.

Though centrifugal and Coriolis forces are often described as apparent rather than real forces, it is important to remember that there is nothing imaginary about their dynamical effects. The toy train will derail outwards if the turntable's rotation Ω is increased to the point where $2\,\Omega\,V$ makes it overbalance, and vast tonnages of air and water are continually moving laterally over the Earth's surface in near balance with local Coriolis terms (Section 7.10). And beware of glib distinctions between apparent and real forces; a thorough debate about whether to ascribe such effects to forces or accelerations leads to depths explored by Mach and Einstein. Cultivate a pragmatic physical and geometrical grasp of what is going on and expect to be puzzled occasionally. The Coriolis effect is very neatly but enigmatically described by vector algebra *[34]*.

BOX 7.4 Coriolis kinematics

To check that the outcome from the simple model (Fig. 7.6) is generally true, we now consider the train T moving steadily along a segment of straight track placed at random on the turntable (Fig. 7.8). As the turntable rotates with angular velocity Ω, the segment midpoint X orbits about the turntable axis O (not shown) with the usual centripetal acceleration $\Omega^2 R$. In addition the track segment *twists* with the turntable at the same rate Ω, moving the train laterally with the Coriolis acceleration, which we now evaluate.

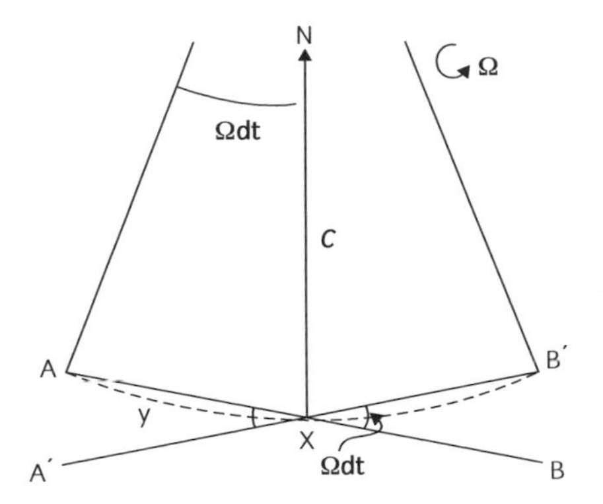

Figure 7.8 As the body moves along the rotating straight line AB, the rotation takes B to B′ and the body describes the arc AXB′ in non-rotating space. The distance to the rotation axis O does not appear in the expression for the lateral (Coriolis) acceleration of the body moving along the arc; it appears in the simple centripetal acceleration of the fixed point X towards O which is to be added to the Coriolis effect.

In a little time interval dt the train travels at speed V from point A to point B on the track, while track and turntable twist by the little angle $d\phi = \Omega\, dt$; as a result the train moves along the curved line AXB′ relative to the non-rotating playroom. The geometry of Fig. 7.8 shows that AXB′ is the arc of a circle of radius C, whose radii from A and B′ meet at N at an angle $2\, d\phi$. Since arc AXB′ has length $V\, dt$, since $V\, dt/C = 2\, d\phi$ (by radian measure), and since $d\phi = \Omega\, dt$, we have

$$C = \frac{V\, dt}{2\, d\phi} = \frac{V\, dt}{2\,\Omega\, dt} = \frac{V}{2\,\Omega}$$

The Coriolis acceleration towards N is the centripetal acceleration of T travelling at speed V on the arc of a circle of radius C, which is V^2/C by the standard expression (Eqn B7.3a). As a final result we have

$$\text{Coriolis acceleration} = \frac{V^2}{C} = 2\,\Omega\, V \qquad \text{B7.4a}$$

which is identical to the Coriolis term in the simple model. Viewed from an observer turning with the turntable, the train experiences a lateral body force per unit mass in the opposite direction to the acceleration observed by someone fixed in the room. Sensitive gauges measuring the sideways pressure between wheel flanges and rails would confirm the magnitude and direction of the effect—it is not imaginary!

7.6 Earths' rotation and apparent g

If the Earth were a perfect sphere with concentric mass distribution, the true gravitational force on a body on its surface would everywhere act towards the Earth's centre with the same strength. However, when we measure gravity from a point fixed on the rotating Earth's surface (the meteorological reference frame—Fig. 7.1), true g is offset by a centrifugal force (or partly lost in the centripetal acceleration) at all latitudes except the poles, so that the measured *apparent* g is less than true g, and directed slightly poleward of the Earth's centre, except at the equator and Poles (Fig. 7.9).

The discrepancy between true and apparent g is greatest at the equator because the centripetal effect ($\Omega^2 R$, Box 7.3) is greatest there, where it would

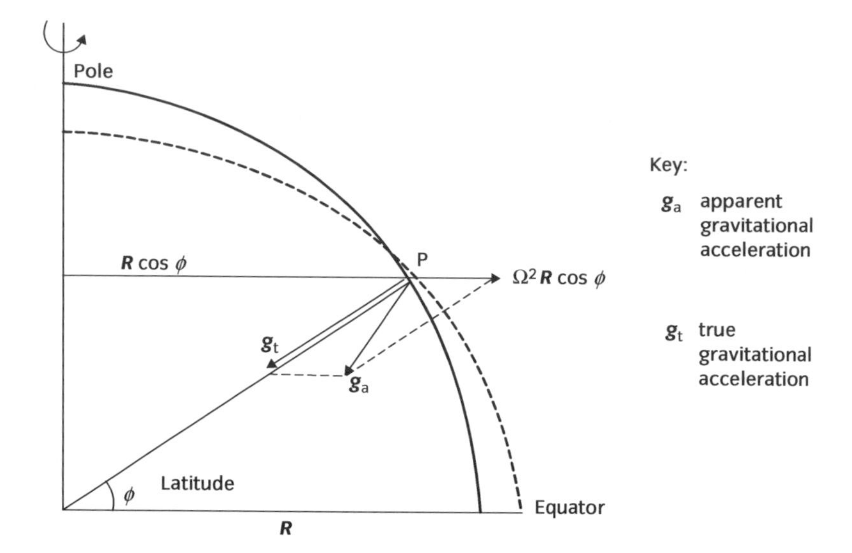

Figure 7.9 True and apparent accelerations (g_t and g_a) on the surface of a spherical Earth (solid quadrant). The dashed quadrant represents the actual Earth, though its oblateness has been greatly exaggerated for clarity.

be 0.034 m s^{-2} for a perfectly spherical Earth (Box 7.5). The observed equatorial excess 0.052 m s^{-2} in Table 7.1 is larger than this because the Earth is not a perfect sphere, as discussed below.

Even more importantly, on a perfectly spherical Earth, the resultant angling of apparent *g* away from the local vertical would maintain a component acting along the local surface (i.e. horizontally) towards the equator with a value peaking at 0.017 m s^{-2} (i.e. about 0.17% *g*) at latitude 45° (Box 7.5).

The consequences of having such an equatorward component of apparent *g* would be huge, with the atmosphere and oceans sliding towards the equator and accumulating there at the expense of higher latitudes. No such displacements are observed, because the 'solid' Earth itself is fluid on geological time scales and has bulged at the equator in response to the centrifugal effect of its own rotation, in essentially the same way as we have just envisaged for the atmosphere and oceans on a rigid spherical Earth. As a result the Earth has bulged to an equilibrium shape in which apparent *g* has zero component parallel to the Earth's surface, so that Mean Sea Level is the global horizontal datum surface. The equilibrium is complicated in detail but not in essence by the presence of concentric zones of increasing density towards the Earth's centre, and by the distortion of the gravitational field of the deformed Earth, with the result that its equatorial radius exceeds its polar radius by about 21 km and apparent *g* varies with latitude as shown in Table 7.1.

As a result of this adjustment, the potentially huge centrifugal effects of the Earth's rotation on the atmosphere and oceans have been almost completely absorbed by the relatively slight equatorial bulging of the Earth. The bulge itself (20 km in over 6,000) has no meteorological significance, and the small meridional and vertical variations of apparent *g* are significant only in the most accurate representations of large-scale fields of atmospheric pressure.

BOX 7.5 Centrifugal effects on a sphere

Assuming Earth to be a perfect sphere of radius R turning with angular velocity Ω (Fig. 7.9), the centripetal acceleration X of a point P on its surface at latitude ϕ is

$$X = \Omega^2 R \cos \phi$$

which has a maximum value of $\Omega^2 R$ at the equator, and is zero at the Poles. With $\Omega = 7.29\ 10^{-5}$ rad s^{-1} and $R \approx 6{,}350$ km, the equatorial value ($\Omega^2 R$) is 0.034 m s^{-2}, which is about 0.35% of 9.81 m s^{-2}.

At latitude ϕ this acceleration (directed towards the polar axis) makes an angle $(90 - \phi)$ with the local poleward horizontal, giving a horizontal poleward component of centripetal acceleration

$$X \sin \phi = \Omega^2 R \cos \phi \sin \phi = \frac{1}{2} \Omega^2 R \sin 2\phi$$

This horizontal component is zero at equator and Poles, and reaches a maximum value of $1/2\ \Omega^2 R$ at $2\phi = 90°$ (latitude 45°), which is about 0.17% g.

7.7 Coriolis on a sphere

The centrifugal bulge of the oblate Earth outlined in the last section accommodates by far the largest effects of the Earth's rotation, and leaves the static atmosphere and oceans dynamically unaffected by that rotation provided we treat mean sea level as our horizontal datum, as we do. But once the air and sea begin to move relative to the spinning Earth, the smaller but still significant Coriolis effect comes into play, with terms proportional to wind and current speeds. To examine the resulting atmospheric (and oceanic) dynamics we need to find the Coriolis terms in three dimensions at every point on and above the spherical globe.

Figure 7.10a represents the Coriolis effect for a body moving across the surface of a turntable rotating counter-clockwise when viewed from above, as analysed in Section 7.5. As we see below, the Coriolis effect is complicated in detail but not in principle by the fact that the Earth is effectively a rotating sphere rather than a flat turntable. In Fig. 7.10b it is apparent that at any point P at latitude ϕ in the N hemisphere, Earth's angular velocity Ω can be represented by two perpendicular turntables—one turning with angular velocity ($\Omega \sin \phi$) about the z axis (the local vertical through P), and the other turning at ($\Omega \cos \phi$) about the local y axis (pointing North through P). We can therefore apply the situation of Fig. 7.10a separately to each of the turntables in Fig. 7.10b to find all the Coriolis terms for the x, y, and z components of the equation of motion.

The two turntables are shown separately in Fig. 7.11, looking at P down the z and y axes, with directions of rotation as for the N hemisphere. Taking the component wind speeds u, v, and w in turn, we find the components of the Coriolis acceleration in the standard x, y, and z directions (E'ward, N'ward, and upward).

$$\begin{array}{ll} x & -2\,\Omega\, v \sin \phi + 2\,\Omega\, w \cos \phi \\ y & 2\,\Omega\, u \sin \phi \\ z & -2\,\Omega\, u \cos \phi \end{array} \qquad 7.13$$

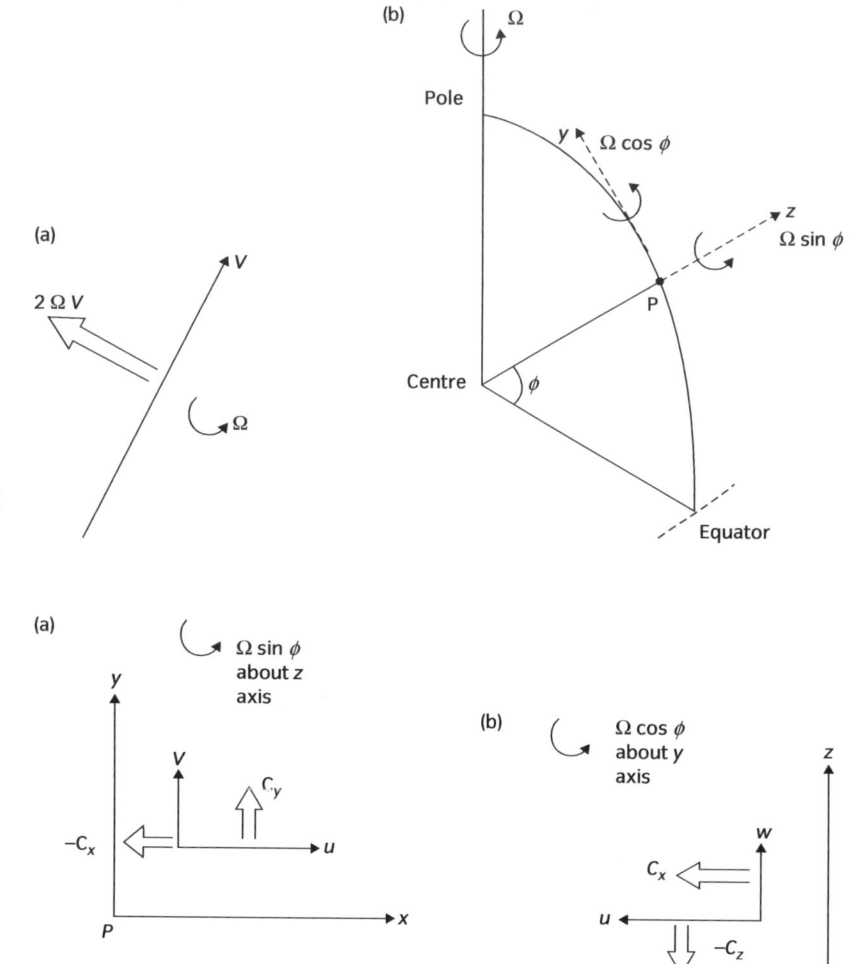

Figure 7.10 (a) Coriolis acceleration for relative motion on a counterclockwise rotating turntable. (b) Components of Earth's rotation at position P at latitude ϕ, about the local vertical axis z, and about the local N'ward axis y.

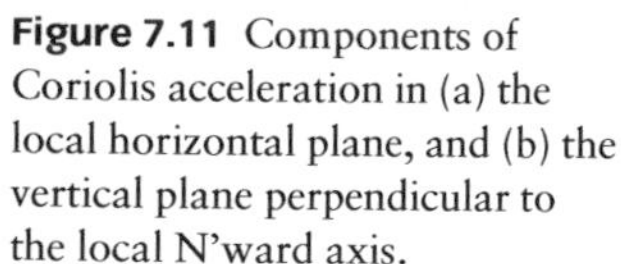

Figure 7.11 Components of Coriolis acceleration in (a) the local horizontal plane, and (b) the vertical plane perpendicular to the local N'ward axis.

According to these, a pure North wind (negative v, and zero u and w) has only one Coriolis component—a horizontal acceleration towards the East (the x direction). An updraft (positive w, and zero u and v) also has a single Coriolis acceleration towards the East. However, a West wind (positive u, and zero v and w) has Coriolis accelerations N'ward and vertically downward. Air flow with components in all three directions therefore has Coriolis components in all three directions, each of which is a product of relative velocity, a component of the Earth's angular velocity, and the tell-tale factor 2.

In most realistic meteorological situations, these Coriolis terms are not equally important, and some are often negligible, but discrimination is possible only when all other terms of each component of the equation of motion have been collected and compared.

7.8 The full equation of motion

We can now collect all the terms of an air parcel's equation of motion (Eqn 7.2) introduced in previous sections. They are written symbolically, as a balance between accelerations (*A* terms) and forces per unit parcel mass (*F* terms), and are in bold to show that they are vectors.

$$\boldsymbol{RA} + \boldsymbol{FA} = \boldsymbol{PGF} + \boldsymbol{GRF} + \boldsymbol{FRF} \qquad 7.14$$

The relative acceleration ***RA*** of an air parcel is measured relative to the standard meteorological reference frame (Fig. 7.1). The term arising from the frame acceleration ***FA*** is simply the Coriolis acceleration ***CA*** (Eqn B7.4a), since the large fixed component of centripetal frame acceleration has been offset by the bulge of the 'fluid' Earth and its residue absorbed into the apparent gravitational force ***GRF*** (Section 7.6). Conventionally ***CA*** is moved to the right-hand side of Eqn 7.14 and thereby treated as a force, so that

$$\boldsymbol{RA} = -\boldsymbol{CA} + \boldsymbol{PGF} + \boldsymbol{GRF} + \boldsymbol{FRF} \qquad 7.15$$

where the remaining terms on the right-hand side are the pressure gradient force ***PGF***, acting from higher to lower pressure, the gravitational force ***GRF*** (omitting 'apparent' from now on) acting down the local vertical, and the frictional forces ***FRF*** smoothing air flow everywhere and slowing it near the surface.

The vector form of Eqn 7.15 is very compact, but inconvenient for detailed analysis or solution, so we break the equation of motion into its *x*, *y*, and *z* components, producing three scalar equations which can be regarded as independent even though they may have terms in common. We can write out these component equations by collecting the terms detailed in previous sections. Conventionally we write the frictional components as negative, because they usually tend to oppose air motion.

x component $$\frac{du}{dt} = 2\,\Omega \sin\phi\, v - 2\,\Omega \cos\phi\, w - \frac{1}{\rho}\frac{\partial p}{\partial x} - FRF_x \qquad 7.16a$$

y component $$\frac{dv}{dt} = -\,2\,\Omega \sin\phi\, u - \frac{1}{\rho}\frac{\partial p}{\partial y} - FRF_y \qquad 7.16b$$

z component $$\frac{dw}{dt} = 2\,\Omega \cos\phi\, u - \frac{1}{\rho}\frac{\partial p}{\partial z} - g_a - FRF_z \qquad 7.16c$$

The complete analytical solution (Section 7.3) of these equations, together with any other constraints which may apply, is far beyond even the very considerable powers of modern mathematics. Indeed the friction terms are not well defined by current understanding of turbulence, and have to be estimated semi-empirically (Sections 7.3 and 10.9). Even so, we can gain useful insights into some of the types of motion described by Eqns 7.16 by removing terms known to be relatively unimportant. A first step in this process is to examine, by a process called *scale analysis*, the relative magnitudes of the various terms in each component by substituting observed values. Since the relative sizes of these values can vary enormously with the time and space scales of types of atmospheric motion, we need to distinguish between such scales and deal with them separately. We choose the two scales of organized air motion which between them dominate weather worldwide (Fig. 1.5): synoptic scale systems (Sections 7.9–11) and small-scale systems (Section 7.14). Their consideration and analyses of atmospheric rotation occupies most of the rest of this chapter.

7.9 Synoptic scale motion

7.9.1 Scale analysis

Synoptic scale weather systems such as extratropical cyclones and anticyclones dominate the troposphere in middle latitudes, and their tropical counterparts are nearly as important in low latitudes. Typical values of length and time scales and other values are as follows, expressed in SI units rounded to integral powers of 10.

horizontal scale	L	1,000 km	10^6 m
vertical scale	H	10 km	10^4 m
time scale	t	1 day	10^5 s
horizontal pressure change	Δp	10 hPa	10^3 Pa
vertical pressure change	p	1,000 hPa	10^5 Pa
air density	ρ		1 kg m^{-3}
Earth's angular velocity	Ω		10^{-4} rad s^{-1}
gravity	g		10 m s^{-2}

L is the horizontal distance in which there is usually a substantial change in any observable property of the system, such as pressure or wind. H and t are defined in the same way for vertical and time scales. Δp is a typical horizontal pressure variation in such systems, whether from centre to edge at an instant or at a fixed point as the system passes. The vertical pressure change is dominated by the huge pressure lapse with height and is comparable with surface pressure p because large systems usually fill most of the depth of the atmosphere as measured by mass and pressure.

Values of other measurable properties and terms in the component equations of motion follow from the basic list.

Note that near the equator a Coriolis term involving sin ϕ will be at least an order of magnitude smaller than given above, as will a term in cos ϕ near the Poles.

horizontal wind speed	$U \sim \dfrac{L}{t}$	10 m s^{-1}
vertical wind speed	$W \sim \dfrac{H}{t}$	10^{-1} m s^{-1}
horizontal acceleration	$\dfrac{U}{t}\left(\text{or } \dfrac{U^2}{L}\right)$	10^{-4} m s^{-1}
vertical acceleration	$\dfrac{w}{t}\left(\text{or } \dfrac{w^2}{H}\right)$	10^{-6} m s^{-1}
main Coriolis acceleration	$\Omega\, U$	10^{-3} m s^{-2}
minor Coriolis acceleration	$\Omega\, w$	10^{-5} m s^{-2}
horizontal pressure gradient	$\dfrac{\Delta p}{L}$	$10^{-3} \text{ Pa m}^{-1}$

Ignoring friction terms for the moment, we can assign an order of magnitude to each term of Eqns 7.16. The units are m s^{-2} throughout (i.e. acceleration or force/mass)

$$\begin{array}{lllll} \dfrac{du}{dt} = & 2\,\Omega \sin\phi\, v & 2\,\Omega \cos\phi\, w & & -\dfrac{1}{\rho}\dfrac{\partial p}{\partial x} \\ 10^{-4} & 10^{-3} & 10^{-5} & & 10^{-3} \end{array} \qquad 7.17a$$

$$\begin{array}{lllll} \dfrac{dv}{dt} = & -2\,\Omega \sin\phi\, u & & & -\dfrac{1}{\rho}\dfrac{\partial p}{\partial y} \\ 10^{-4} & 10^{-3} & & & 10^{-3} \end{array} \qquad 7.17b$$

$$\begin{array}{lllll} \dfrac{dw}{dt} = & 2\,\Omega \cos\phi\, u & & -\,g & -\dfrac{1}{\rho}\dfrac{\partial p}{\partial z} \\ 10^{-6} & 10^{-5} & & 10 & 10 \end{array} \qquad 7.17c$$

Note that these component equations are strictly true for a tangent plane at any point on the Earth's surface. If the horizontal axes are curved to fit the surface (as must be done in large-scale dynamic analysis) then curvature terms are needed to account for the centripetal accelerations implied by motion on curved axes. These are usually relatively small, provided lines of latitude are not used as axes in very high latitudes!

Comments

(1) Because large-scale uplift w is so much slower than horizontal wind speeds u and v, the second Coriolis term in Eqn 7.17a is two orders of magnitude smaller

than the first, and can be ignored. This leaves the two horizontal component equations (7.17a and b) looking much more symmetrical.

(2) The relative acceleration terms (du/dt and dv/dt) in Eqns 7.17a and b are smaller than the Coriolis and pressure-gradients terms, but only by an order of magnitude, suggesting that they may be unimportant usually but not always. When they are unimportant, the symmetry of the Coriolis terms ensures that a relatively simple *geostrophic balance* prevails, as outlined in the next section.

(3) The vertical component equation (7.17c) is completely dominated by the gravitational and vertical pressure gradient terms, whose balance corresponds to the hydrostatic equilibrium discussed in Section 4.4. But note the absurdity of concluding that vertical accelerations (dw/dt) are generally unimportant; if they were truly zero there would be no updrafts, no deep cloud and no precipitation larger than drizzle!

Reinstating the friction terms omitted from Eqns 7.17:

(4) If friction arises solely from molecular viscosity, the equation of motion will include terms involving the curvature of the vertical profile of horizontal wind (like $\nu\, \partial^2 u/\partial z^2$) which are especially important at the base of the atmospheric boundary layer (Section 7.4). To estimate their magnitudes by scale analysis, note that $\partial^2 u/\partial z^2$ is the gradient of a velocity gradient—a velocity divided by a length, and by the same length again. If the length is taken to be 300 m (the typical depth h of the ABL), then since $\nu \sim 10^{-5}$ m^2 s^{-1}, terms $\nu\, \partial^2 u/\partial z^2$, $\nu\, \partial^2 v/\partial z^2$ are $\sim \nu\, U/h^2 \sim 10^{-9}$ m s^{-2}. These are so very much smaller than the other terms in Eqns 7.17a and b that we can say that viscosity is dynamically unimportant even in the heavily sheared ABL. But beware of dismissing viscosity as irrelevant more generally: the energetics of the atmosphere and oceans are profoundly affected by its presence (Section 9.8).

(5) If we use the much larger kinematic coefficient of eddy viscosity K/ρ (~ 1 m^2 s^{-1} or even more) in place of ν, the friction term $(K/\rho)\, U/h^2 \sim 10^{-4}$ m^{-2} is only one order of magnitude smaller than the Coriolis and pressure gradient terms, suggesting that turbulent friction may be able to distort the geostrophic balance, as happens systematically in the atmospheric boundary layer (Section 7.13).

7.9.2 Laminar and turbulent flow *[10]*

By very general observation, comparison between relative accelerations (du/dt, etc.) and viscous terms in the equation of motion are found to distinguish between flows which are significantly controlled by viscosity and those which are not. The comparison is expressed by the ratio of the acceleration and viscous terms, which is called the *Reynolds number* (Re) after the pioneer fluid dynamicist Osborne Reynolds. Using the same length scale L in both the acceleration term (U^2/L) and the viscous friction term ($\nu\, U/L^2$), their ratio Re is

$$Re = \frac{UL}{\nu} \qquad 7.18$$

In an enormously wide range of flows, from water in millimetric bore pipes to planetary atmospheres, it is observed that when Re is less than $\sim 10^3$ the flow

(a)

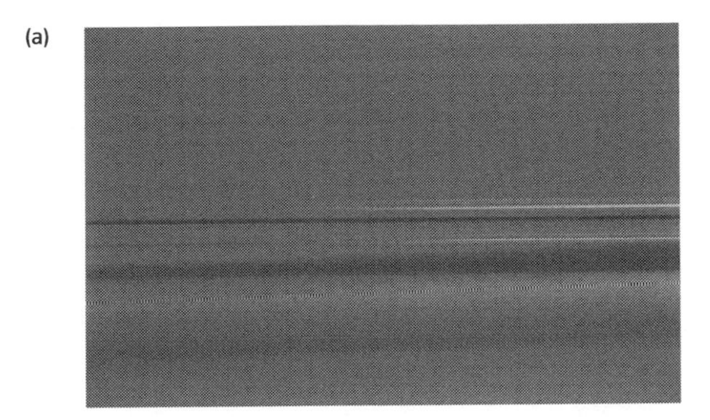

(b)

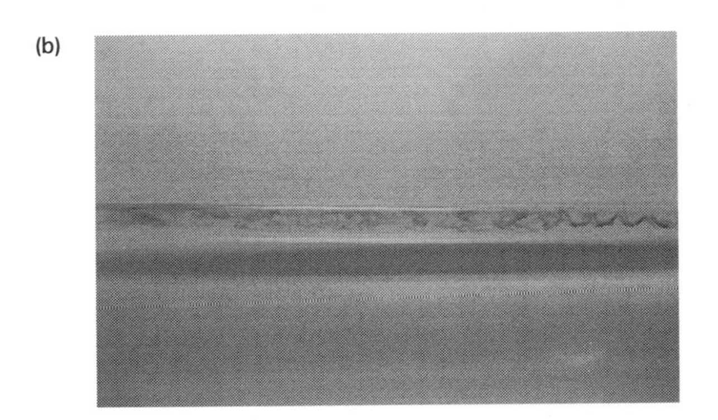

Figure 7.12 Water flow in a glass pipe with dye injected centrally at the right-hand end. In laminar flow (a), the dye plume's downstream motion and spreading are imperceptible In faster, turbulent flow (b), the plume diffuses quickly to the pipe walls while swirling downstream (to the left).

is always very smooth (*laminar*) because it is dominated by viscosity. However, when Re is larger than this critical value, the flow is usually *turbulent*, in the sense that it is continually breaking into transient, apparently random, eddies—Fig. 7.12.

Inserting synoptic scale values for U and L in expression 7.18, we find $Re \sim 10^{12}$, which is consistent with turbulence being widespread in the troposphere—endemic in the ABL as we see and feel in daily life, severe in convection, and quite widespread in the shear zones above and below jet streams, to the occasional discomfort and worse of aircraft cruising there. Working backwards from the critical value of 10^3 for Re, and assuming $U \sim 1$ m s^{-1}, we see that laminar flow should prevail in systems of scale $L < 1$ cm, which is consistent with the presence of the shallow *laminar boundary layer* covering the Earth's surface (Section 10.6).

The Reynolds number is one of several *dimensionless numbers* used to describe the importance of terms on the right-hand side of the equation of motion in comparison with the relative accelerations on its left-hand side. Since every term in the equation of motion has the dimensions of acceleration, the ratio of any two terms is dimensionless, so that its numerical value in any particular case or condition is independent of the system of units used (SI nowadays, but many others in the past), giving dimensionless number their very general validity and utility (Box 7.6).

BOX 7.6 Using dimensions

As mentioned in Box 1.3, all measurable physical quantities, however complex, can be regarded as combinations of a few independent physical quantities, such as mass M, length L, time T, and temperature θ. We can use the fact that all valid equations relating physical quantities must be dimensionally consistent (Section 7.2) to find the dimensions of a composite physical quantity, to help understand its physical significance, to check the validity of a proposed equation, or to constrain the form of an incompletely defined equation.

Dimensions of important quantities

Angle is dimensionless since radian measure = arc/radius, length has dimensions $[L]$, area $[L^2]$, volume $[L^3]$, density $[M\ L^{-3}]$, Velocity $[L\ T^{-1}]$, acceleration $[L\ T^{-2}]$, force (= mass × accel) $[M\ L\ T^{-2}]$, pressure or shear stress (force/area) $[M\ L^{-1}\ T^{-2}]$, energy (= work done by force) $[M\ L^2\ T^{-2}]$, specific heat capacity (energy/mass/temperature rise) $[L^2\ T^{-2}\ \theta^{-1}]$.

Checking dimensions

(i) The exponential scale height of an isothermal atmosphere in hydrostatic equilibrium is given by $H = R\ T/g$. The dimensions of the specific gas constant R follow from the equation of state for the ideal gas

$$[R] = [p/(\rho\ T)] = [M\ L^{-1}\ T^{-2}]\ [M\ L^{-3}]^{-1}\ [\theta]^{-1} = [L^2\ T^{-2}\ \theta^{-1}]$$

showing that $[H] = [L^2\ T^{-2}\ \theta^{-1}]\ [\theta]\ [L\ T^{-2}]^{-1} = [L]$ a length, as it must be, as a height.

(ii) Pressure gradient force per unit mass, $PGF = (1/\rho)\ (\partial p/\partial n)$

$$[PGF] = [M\ L^{-3}]^{-1}\ [M\ L^{-1}\ T^{-2}]\ [L]^{-1} = [L\ T^{-2}]$$

an acceleration, as it must be to be a term in the equation of motion.

(iii) Reynolds number $U\ L/\nu$. From $\nu = \mu/\rho$, $\mathbf{T} = \mu\ \partial u/\partial z$, and $\mathbf{T}$ = force/area, we have $\nu = \mathbf{T}/(\rho\ \partial u/\partial z)$, $[\nu] = [M\ L\ T^{-2}\ L^{-2}]/\{[M\ L^{-3}]\ [L\ T^{-1}\ L^{-1}]\} = [L^2\ T^{-1}]$. But these are the dimensions of $U\ L$, so that $Re = U\ L/\nu$ is dimensionless, as we know it must be as the ratio of two terms in the equation of motion.

Checking an equation

If you wrongly remember the expression for the exponential scale height to be $H = R\ T\ g$, its dimensional inconsistency will show it to be invalid. You could correct the expression for H by trial and error, or more methodically, as follows. Suppose you were not sure of the powers of R and g in the expression for H

$$H = R^a\ T\ g^b \quad [H] = [L^2\ T^{-2}\ \theta^{-1}]^a\ [\theta]\ [L\ T^{-2}]^b$$

Do not confuse the symbol for time T with the symbol θ for temperature. Clearly a = 1 for the θs to cancel in the first and second [ ] groups on the right-hand side, and b = –a for the Ts to cancel in the first and last [ ] groups. Hence we recover $H = R\ T/g$. More elaborate cases lead to simultaneous equations in the unknown powers.

Limitations

(i) Suppose that in the previous subsection you incorrectly remembered $H = C_v\ T/g$. You would find it dimensionally consistent, but that would not make it physically valid. Dimensional consistency is a necessary but not a sufficient condition for physical validity.

(ii) The volume of a sphere of radius $R = (4\pi/3)\ R^3$. Dimensional consistency requires that the power of R is 3, but cannot help with pure numbers like π and 4, since they are dimensionless and therefore unconstrained by dimensional consistency.

(iii) There is deep circularity in some of these procedures, but the power of dimensional consistency across initially disparate topics, etc. attests to its validity as a principle.

Generality

Relationships are at their most general when expressed in the most dimensionless form. For example the pressure/height relation in an isothermal hydrostatically balanced atmosphere is given by

$$p = p_0\ e^{-z/He}$$

If this is rewritten by *normalizing* (i.e. non-dimensionalizing) pressure by expressing it in units of pressure p_0 at some datum level (i.e. $p^* = p/p_0$), and normalizing height by expressing it in units of the exponential scale height ($z^* = z/H_e$), then we have $p^* = e^{-z^*}$. A single curve will fit all such normalized pressure profiles, and indeed will fit normalized graphs for all other processes involving pure exponential decay. More sophisticated normalization helps greatly in the study of atmospheric turbulence, where observations cannot yet be fitted into a comprehensive theoretical framework (Section 10.9, Eqn 10.20).

7.10 Geostrophy

In the free atmosphere (i.e. above the ABL), observation and scale analysis of synoptic scale flow suggests that the friction terms in the equation of motion are relatively small. According to Eqn 7.17a and b and discussion, horizontal air flow is then described by the x and y component equations:

$$\frac{du}{dt} = fv - \frac{1}{\rho}\frac{\partial p}{\partial x}$$

$$\frac{dv}{dt} = -fu - \frac{1}{\rho}\frac{\partial p}{\partial y} \qquad 7.19$$

where $f(= 2\,\Omega \sin \phi)$ is called the *Coriolis parameter*, and represents the Coriolis effect arising from the component of the Earth's rotation about the local vertical. Scale analysis further suggests that the relative accelerations du/dt and dv/dt are often an order of magnitude smaller than the Coriolis and pressure-gradient terms.

The relative magnitudes of the relative accelerations and Coriolis terms are described by the dimensionless ratio (relative acceleration)/(Coriolis acceleration) which is called the *Rossby number* (Ro), after the Swedish meteorologist who focused modern understanding of the dynamic role of the Earth's rotation in the atmosphere. Ro quantifies the relative importance of the Earth's rotation in flows of air or water, in the same way as the Reynolds number quantifies the importance of fluid friction.

According to scale analysis (Section 7.9)

$$Ro \sim \frac{U^2/L}{\Omega U} = \frac{U}{\Omega L}$$

Using characteristic scales for synoptic scale behaviour (Section 7.9), Ro values are ~ 0.1, suggesting that Coriolis accelerations tend to dominate rather than overwhelm relative accelerations. Observations of real atmospheric and oceanic flows, and carefully contrived and scaled rotating laboratory models (Section 7.11), show that flow with $Ro \sim 0.1$ is dominated by broad, shallow vortices turning about the local vertical. When, however, $Ro >> 1$, as in the meteorological small scale (cumulus, etc.), Coriolis terms are negligible and there are few signs of organized rotation outside the extreme case of the tornado.

7.10.1 Geostrophic flow

If the Rossby number is truly zero, $\mathrm{d}u/\mathrm{d}t$ and $\mathrm{d}v/\mathrm{d}t$ disappear and Eqns 7.19 become

$$fv = \frac{1}{\rho}\frac{\partial p}{\partial x} \quad \text{and} \quad fu = -\frac{1}{\rho}\frac{\partial p}{\partial y} \qquad 7.20$$

defining what is called pure *geostrophic flow*—an ideal to which real synoptic scale air flow approximates but never completely conforms. Since the geostrophic approximation is not nearly so good as the hydrostatic approximation which dominates the vertical component of the equation of motion, it is better to say that observed air flow is *quasi-geostrophic* on the synoptic scale. It is nevertheless a very useful and informative approximation to outline and investigate in this and succeeding sections.

In the special case of a pure W wind (i.e. positive u and zero v) the *geostrophic equations* (Eqns 7.20) simplify further to

$$\frac{\partial p}{\partial x} = 0 \quad \text{and} \quad \frac{1}{\rho}\frac{\partial p}{\partial y} = -fu$$

which show that the pressure gradient is parallel to the y axis, with pressure decreasing in the positive y direction (i.e. N'ward—Fig. 7.13). Since isobars are perpendicular to the pressure gradient, they must lie E–W, parallel to the assumed W'ly air flow, and since lateral pressure gradient increases with geostrophic wind speed, spacing of chosen isobars *decreases* with increasing wind speed and vice versa, the constant of proportionality varying with air density and with latitude (through f).

The relationships between air flow and isobaric parallelism and spacing apply to any horizontal geostrophic flow, regardless of direction. It follows that

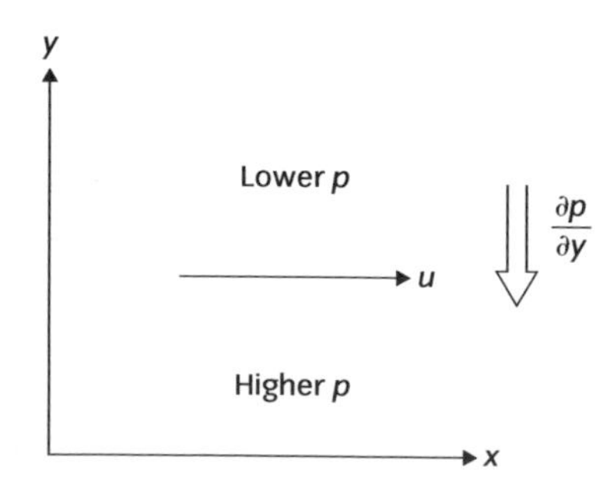

Figure 7.13 W'ly (E'ward) geostrophic flow in a NS pressure gradient (N hemisphere).

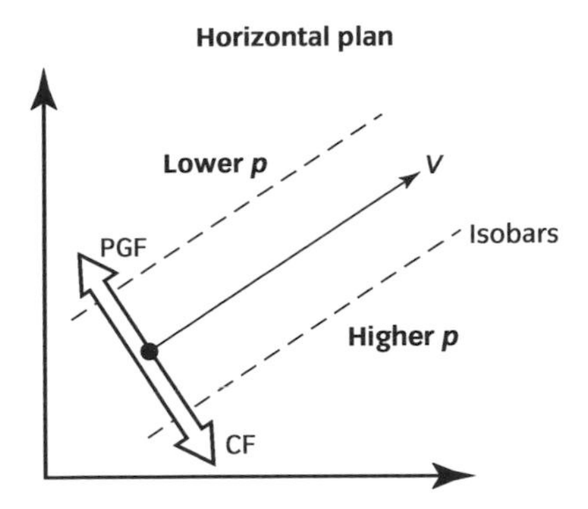

Figure 7.14 Generalized geostrophic flow in the N hemisphere. In the S hemisphere, the same pressure field would be associated with flow in the opposite direction, so that lower pressure would lie to the right looking down the flow.

horizontal wind speed V and pressure gradient $\partial p/\partial n$ in any horizontal direction n are related by

$$\frac{1}{\rho}\frac{\partial p}{\partial n} = f\,V \qquad 7.21$$

where $\partial p/\partial n = [(\partial p/\partial x)^2 + (\partial p/\partial y)^2]^{1/2}$ and $V = (u^2 + v^2)^{1/2}$ and directional relationships in the N hemisphere are shown in Fig. 7.14 and summarized in an empirical law named after the nineteenth-century Dutch pioneer meteorologist *Buys-Ballott*, which relates the relative directions of surface wind and pressure gradients in statements such as 'low pressure in the N hemisphere is on your left hand when standing with your back to the wind', on the understanding that left becomes right in the S hemisphere. It is customary to rearrange Eqn 7.21 in the form

$$V = \frac{1}{\rho f}\frac{\partial p}{\partial n}$$

and to define the *geostrophic wind* V_g as the wind which would satisfy this balance perfectly in magnitude and direction.

$$V_g = \frac{1}{\rho f}\frac{\partial p}{\partial n} \qquad 7.22$$

From this equation and Fig. 7.15 we see that geostrophic flow can be regarded as maintaining perfect balance between the pressure gradient force ***PGF*** and the Coriolis force (−***CA***). Since Coriolis acts towards the opposite side of the air flow in the other hemisphere, geostrophic balance reverses in direction.

Using isobaric contours

It is conventional (Section 4.6) in operational meteorology to describe horizontal pressure fields at all levels above sea level by plotting height contours, in metres above mean sea level, of convenient isobaric surfaces (1,000, 850 hPa, etc.). Since the distortion of isobaric surfaces out of the horizontal plane is always quite small in the atmosphere, horizontal gradients change so little in these small height differences that contours of an isobaric surface are almost exactly parallel to isobars on an adjacent horizontal surface, and Buys-Ballots' law holds when 'lower isobaric contours's are substituted for 'lower pressures'.'

In Box 7.7 it is shown that replacement of $\partial p/\partial n$ by $\partial Z_p/\partial n$ (the slope of the p isobaric surface) removes the variable ρ from Eqn 7.22, giving the isobaric equivalent

$$V_g = \frac{g}{f}\frac{\partial Z_p}{\partial n} \tag{7.23}$$

The disappearance of ρ and the near uniformity of g means that at any particular latitude (i.e. at any particular value of f) the same contour slope corresponds to the same geostrophic wind speed, regardless of altitude—i.e. regardless of whether we are dealing with observations made near sea level, or in the middle stratosphere where air density may be 20 times smaller.

We can use Eqns 7.23 and 7.22 to estimate isobaric slopes and horizontal pressure gradients geostrophically associated with typical actual winds in middle latitudes (where $f \approx 10^{-4}$ rad s^{-1}). For a wind speed of 10 m s^{-1} the isobaric slope is about 1 in 10^4 regardless of altitude, so that slopes are 8 in10^4 or so in the most powerful jet streams (80 m s^{-1}) in the high troposphere (NN 7.2). Associated pressure gradients are about 1 hPa per 100 km in the low troposphere (where air density is about 1 kg m^{-3}) but only about 3 hPa per 100 km in the 80 m s^{-1} jet core, because increased contour gradient is significantly offset by reduced air density. There is an obvious advantage in using isobaric contours rather than isobars on horizontal surfaces to indicate flow speed on maps.

7.10.2 How geostrophic is real flow?

The accuracy of the geostrophic approximation in any particular situation is shown by the extent to which observed winds resemble geostrophic winds in both

NUMERICAL NOTE 7.2 Isobaric slopes and pressure gradients

Since Earth's angular velocity Ω is 7.27×10^{-5} rad s^{-1} (NN 4.8), the Coriolis parameter at latitude 45 ° is 2 Ω sin 45° = $2 \times 7.27 \times 10^{-5} \times 0.707 = 1.03 \times 10^{-4} \approx 10^{-4}$ s^{-1}. Rewriting Eqn 7.22 in the form $\partial Z_p/\partial n = f\,V_g/g$, we find the isobaric slope corresponding to a geostrophic wind speed of 10 m s^{-1} to be $\doteq 10^{-4} \times 10/9.8 \approx 10^{-4}$, which is 10 m in 100 km. In a strong jet stream of 80 m s^{-1}, the isobaric slope is 80 m in 100 km.

According to Box 7.7 $\partial p/\partial n = \rho\, g\, \partial Z_p/\partial n$, so that the corresponding horizontal pressure gradient in the low troposphere (where $\rho \approx 1$ kg m^{-3}) $\approx 1 \times 9.8 \times 10^{-4} \approx 10^{-3}$ Pa m^{-1}, which is 10^{-5} hPa m^{-1} or 1 hPa per 100 km. At jet stream level (about 300 hPa in mid-latitudes), air density is about 0.4 kg m^{-3} (by the ideal gas equation with temperature 260 K), so that the horizontal pressure gradient corresponding to a V_g of 80 m s^{-1} is about 3.2 hPa per 100 km.

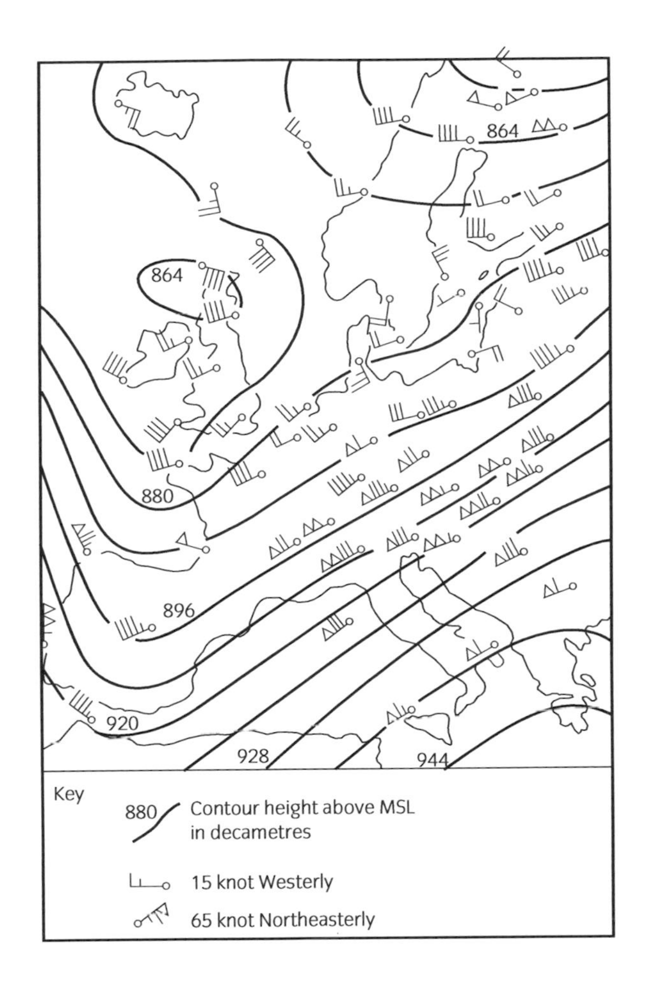

Figure 7.15 Observed quasi-geostrophic flow at the 300 hPa level over the British Isles and W Europe at 0000 Z, January 1, 1982.

strength (Eqn 7.23) and direction (Fig. 7.14). Typical correspondence between actual and geostrophic winds can be judged from Fig. 7.15 which shows isobaric contours and winds in the vicinity of a polar front jet stream over the British Isles. The data come from the synoptic radiosonde ascents on that occasion, and the contours were sketched by interpolating in the traditional manner by hand and eye between the point measurements of the heights of the 300 hPa surface. The wind vectors, measured by the speed of horizontal drift of the radiosondes as they rose through the 300 hPa surface, are parallel to the isobaric contours to within 20° everywhere, and 10° over most of the area. Actual wind speeds agree with geostrophic speeds calculated from the contour slopes (using Eqn 7.23) to within 20% everywhere and 10% in most places. This agreement is typical of the extent to which actual winds are geostrophic, provided we avoid the ABL, where friction interferes systematically with both wind speed and direction, and also avoid the immediate vicinity of the equator, where the Coriolis effect is too small (*Ro* is too large) to encourage even quasi-geostrophic balance.

BOX 7.7 Pressure gradient and contour slope

Figure 7.16 shows a vertical section through a region with horizontal pressure gradient directed to the right, as shown by the upward slope of the isobars. Two isobars are shown, with pressures p and $p + \Delta p$. The horizontal pressure difference Δp between them is obviously identical to the vertical difference, and the latter can be related to their vertical separation Δz through the hydrostatic relation $\Delta p = g\,\rho\,\Delta z$. It follows that

$$\frac{\Delta p}{\Delta n} = g\,\rho\,\frac{\Delta z}{\Delta n}$$

As Δn becomes vanishingly small, $\Delta p/\Delta n$ becomes the horizontal pressure gradient $\partial p/\partial n$ and $\Delta z/\Delta n$ becomes the slope of the p isobar $\partial Z_p/\partial n$. Dividing across by air density we have

$$\frac{1}{\rho}\frac{\partial p}{\partial n} = g\,\frac{\partial Z_p}{\partial n}$$

where the left-hand side is the horizontal pressure gradient force per unit mass (*PGF*) in the direction of the n coordinate (Box 7.1). The right-hand side is the *PGF* term expressed in terms of isobaric gradient, and shows its convenient independence of air density ρ.

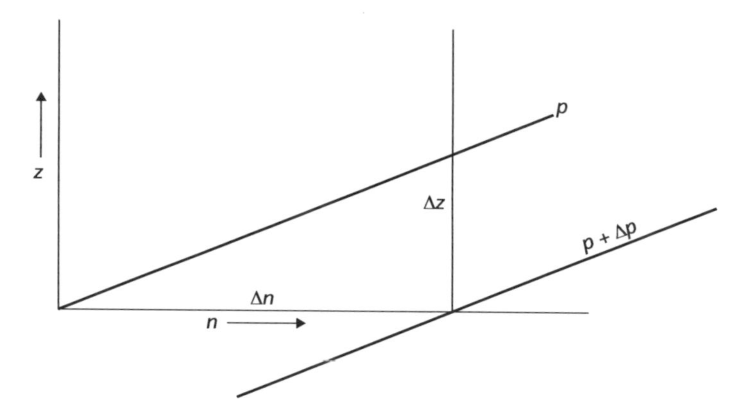

Figure 7.16 A vertical section through tilted isobaric surfaces p and $p + \Delta p$, with vertical axis z and horizontal axis n.

7.11 Why geostrophic?

The most surprising feature of geostrophic flow is that it is at right angles to the pressure gradient which must maintain it dynamically: air flows along the isobars rather than across them towards low pressure. In the vicinity of pressure minima like those in the centres of depressions, instead of flowing inwards to fill the low pressure, air flows round the low-pressure centre in broad vortices, anticlockwise looking down in the N hemisphere, and clockwise in the S hemisphere. As modern meteorology was beginning in the early nineteenth century, such wheel-like rotation attracted the generic name *cyclone* (Greek for wheel), but in time the term *cyclonic* narrowed to apply only to the direction of flow round a low-pressure centre in each hemisphere, with *anticyclonic* describing the opposite flow round a high.

Note that we have not explained why observed winds should be quasi-geostrophic on the synoptic scale. The scale analysis of the equations of motion, and estimated values of the Rossby number, are consistent in showing that geostrophic balance follows from the comparative insignificance of the relative accelerations and frictional terms, but they do not explain why these terms should be so small. Rewriting the Rossby number $Ro \sim U/(\Omega L)$ in terms of the time scale $t \sim L/U$ for a parcel moving through a flow pattern of scale L at speed U, we find $Ro \sim 1/(\Omega t)$. The criterion that Ro should be less than 0.1 (for at least quasi-geostrophic flow) then becomes

$$\frac{1}{\Omega t} < 0.1 \quad \text{or} \quad t > \frac{10}{\Omega}$$

which implies $t > 10^5$ s (≈ 1 day) for the terrestrial angular velocity $\Omega \sim 10^{-4}\ \mathrm{s^{-1}}$. This too is consistent with observation, but does not explain why synoptic scale systems should have time scales in this range. It must be associated with the relative quietness of the Earth's atmosphere on the synoptic scale: depressions take days rather than hours to form and die, and even the strongest jet streams flow at well below the speed of sound, so that air flow has time to settle into geostrophic equilibrium with pressure fields and vice versa. And it seems reasonable to link such quietness to the low strength of solar input; if this were to be 10 times larger than it is (say > 10 instead of 1 kW m^{-2}), then atmospheric motion would be much more vigorous, and the tendency towards geostrophic equilibrium might be continually frustrated by the eruption of very violent cumulonimbus, etc.

7.11.1 Starting from rest

Consider a very artificial approach to the establishment of geostrophic equilibrium—an air parcel starting from rest and moving freely in a steady synoptic-scale horizontal pressure field (Fig. 7.17). After initial movement towards low pressure, the parcel veers to the right, in the N hemisphere, in response to the growing Coriolis force, or equivalently in response to the hidden anticlockwise rotation of the weather map. If the parcel's tendency to overshoot and undershoot is critically damped, the parcel will approach and eventually reach geostrophic equilibrium as shown, with the parcel moving parallel to the isobars (or isobaric contours) at the speed required by Eqns 7.22 or 7.23. The parcel will take a finite time to reach this state—a time inversely related to the magnitude of the Coriolis parameter and hence to the Earth's rate of rotation.

If we consider a situation with a much shorter intrinsic time scale—for example air flowing at 10 m s^{-1} around a hill of horizontal dimension 10 km (so that an air parcel is within the distorted flow for a period $t \sim 10^3$ s), then there is clearly not enough time to develop geostrophic equilibrium between the flow

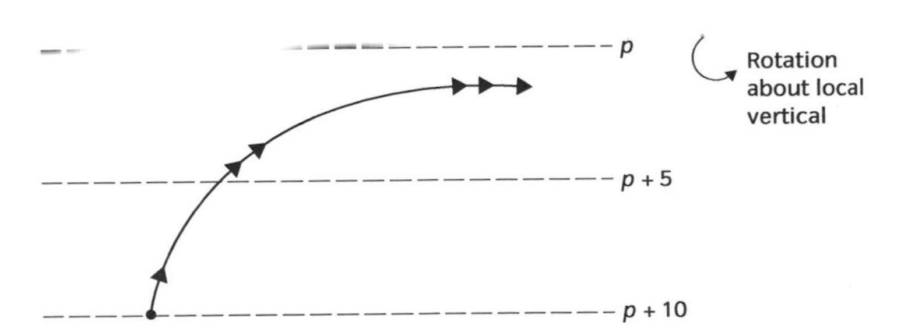

Figure 7.17 Idealized approach to geostrophic equilibrium from rest (N hemisphere).

and the pressure field distorted by the hill. Flow round the hill therefore cannot be expected to be even quasi-geostrophic, as is confirmed by observation, and by substituting the time or space scale from the flow in the expressions for the Rossby number quoted above and in Section 7.10. In terms of the categories of scale in Fig. 1.5, the geostrophic tendency appears only in air flow on the large meso-scale, synoptic, and larger scales.

This discussion is suggestive rather than persuasive, since its arguments are largely circular, and some are artificial. A sophisticated technical review of atmospheric geostrophy concluded after 20 pages that 'it seems unlikely that the near geostrophic balance in our atmosphere can be accounted for in any simple manner' *[29]*—an intriguing situation, given its widespread and important consequences.

7.12 Thermal winds

We have seen in Eqn 7.23 that geostrophic wind speed is proportional to isobaric contour slope. It follows that if contour slope varies with altitude, so must geostrophic wind speed. Figure 7.18 depicts a vertical section through a part of the troposphere where contour slope (and therefore geostrophic wind speed) increase consistently with height. For simplicity the pictured situation has no change of direction of contour slope (and therefore of wind direction) with height, so that the geostrophic flow is everywhere perpendicular to the cross-section; however, all relationships discussed in the following have simple vector equivalents when directions do vary with height.

Applying Eqn 7.23 to the isobaric slopes at pressures p_1 and p_2, we find the difference in geostrophic wind speed by subtracting the geostrophic equations for the two pressure surfaces:

$$V_{g2} - V_{g1} = \frac{g}{f}\left[\frac{\partial Z_{p2}}{\partial n} - \frac{\partial Z_{p1}}{\partial n}\right] = \frac{g}{f}\frac{\partial}{\partial n}\left[Z_{p2} - Z_{p1}\right] \qquad 7.24$$

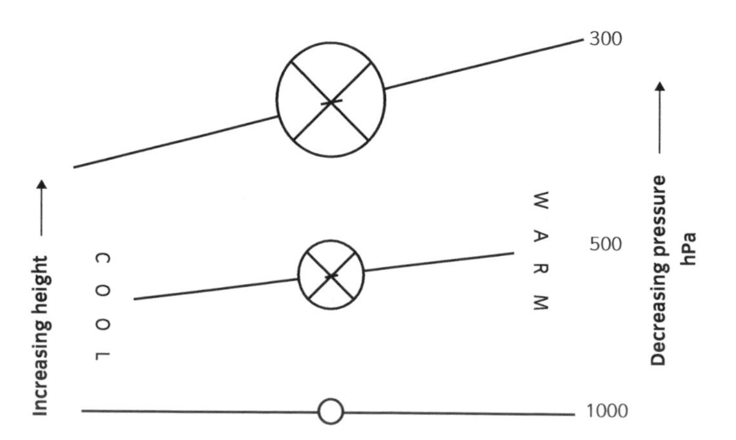

Figure 7.18 Vertical cross-section illustrating the thermal wind relation. Tails of geostrophic wind arrows are enclosed by circles whose diameters increase with wind speed (N hemisphere).

where $[Z_{p2} - Z_{p1}]$ is the vertical *thickness* of the layer bounded by the p_1 and p_2 surfaces, and $\partial[Z_{p2} - Z_{p1}]/\partial n$ is its horizontal gradient. In Fig. 7.18 the thickness of the layer between the 1,000 and 300 hPa surfaces increases sharply to the right, and the numerical value of its gradient can be found graphically by subtracting the slope of the lower surface from the slope of the upper surface. Multiplication by g/f will then evaluate the right-hand side of Eqn 7.24 and hence the increase in geostrophic wind with height through the layer.

Since the thickness of a layer bounded by any two isobaric surfaces is proportional to the layer mean temperature (Section 4.6) on the absolute scale, it follows that the thickness gradient in Eqn 7.24 and Fig. 7.18 corresponds to the horizontal gradient of the layer mean temperature. The relationship between horizontal gradient of layer mean temperature and the difference in geostrophic wind between the bottom and top of the layer is called the *thermal wind relation*. And since a horizontal temperature gradient in an isobarically bounded layer means that the layer is baroclinic, the steepness of the layer mean temperature gradient is a measure of the degree of *baroclinity* (Section 4.6), as is the vertical shear of geostrophic wind. The thermal wind relation is therefore highly relevant to the great weather systems of the systematically baroclinic mid-latitude troposphere.

In frontal zones and larger-scale systems, realistic isobaric slopes are usually between 10 and 100 times smaller than the slopes of isotherms (Sections 4.6 and 11.2), so that the temperature gradient of an isobarically bounded layer is very nearly the same as the temperature gradient of a horizontal layer. Equation 7.24 therefore strongly associates a vertical shear of horizontal geostrophic wind speed through a layer with a horizontal temperature gradient.

7.12.1 Thermal wind equation

Figure 7.18 shows that the temperature gradient lies at right angles to the wind shear and that an extension of Buys-Ballot's law (Section 7.9) connects the directions of vertical shear and horizontal temperature gradient: 'in the N hemisphere, low mean temperatures are on your left when standing with your back to the *thermal wind*' (the vertical hear of geostrophic wind).

Since the isobarically bounded layer is in hydrostatic equilibrium, it follows from Box 7.8 that the horizontal gradient of its thickness ($\partial[Z_{p2} - Z_{p1}]/\partial n$) is proportional to the horizontal gradient of its layer mean temperature ($\partial T_p/\partial n$) and that Eqn 7.24 becomes

$$\frac{V_{g2} - V_{g1}}{Z_{p2} - Z_{p1}} = \frac{g}{f\,T_p}\frac{\partial T_p}{\partial n} \qquad 7.25$$

This is the finite difference form of the *thermal wind equation*; the differential form follows when we allow the isobaric layer to become very thin, with a well-defined horizontal temperature gradient $\partial T_p/\partial n$ at any location.

$$\frac{\partial V_g}{\partial z} = \frac{g}{f\,T_p}\frac{\partial T_p}{\partial n} \qquad 7.26$$

Since the thermal wind relation follows from both geostrophic and hydrostatic balance, it will be realistic to the extent that these two apply to the real

BOX 7.8 Thermal wind relation

According to Section 7.10 and Eqn 7.24 the difference in geostrophic wind speed between two isobaric levels is directly related to the horizontal gradient of the thickness of the bounded layer:

$$V_{g2} - V_{g1} = \frac{g}{f}\frac{\partial}{\partial n}\left[Z_{p2} - Z_{p1}\right] \qquad \text{B7.8a}$$

Since the atmosphere is in nearly perfect hydrostatic equilibrium we can relate the thickness (the vertical depth) of a column to its mean density and therefore temperature, so that the right-hand side of Eqn B7.8a depends on the horizontal gradient of column mean temperature.

To examine this dependence, we begin with Eqn B4.3e (Box 4.3) which arises by integrating the hydrostatic relation through a vertical column of air to find the vertical thickness of a section bounded by pressures p_1 and p_2:

$$Z_{p2} - Z_{p1} = \frac{R\,T_p}{g}\ln\left(\frac{p_1}{p_2}\right) \qquad \text{B7.8b}$$

where the column mean temperature T_p is an average weighted by the logarithm of pressure

$$T_p = \int_{p_2}^{p_1} T\,\mathrm{d}\ln p / \ln\left(\frac{p_1}{p_2}\right)$$

Since everything on the right-hand side of Eqn B7.8b is constant except T_p (the isobars being fixed by initial choice), substitution in the right-hand side of B7.8a simplifies to

$$V_{g2} - V_{g1} = \frac{R}{f}\ln\left(\frac{p_1}{p_2}\right)\frac{\partial\,T_p}{\partial\,n} \qquad \text{B7.8c}$$

and we can use Eqn B7.8b again to replace the terms before $\partial T_p/\partial n$ on the right-hand side of Eqn B7.8c and rearrange to find

$$\frac{V_{g2} - V_{g1}}{Z_{p2} - Z_{p1}} = \frac{g}{f T_p}\frac{\partial T_p}{\partial n} \qquad \text{B7.8d}$$

If we consider a vanishingly thin layer, then the left-hand side of B7.9d tends to $\partial V_g/\partial z$, which is the vertical gradient (or shear) of the geostrophic wind, while on the right-hand side, $\partial T_p/\partial n$ is the horizontal temperature gradient in the very thin isobaric layer.

atmosphere. Hydrostatic equilibrium holds so nearly perfectly on the synoptic scale (Section 7.7) that we must expect the thermal wind relation to apply to the same extent that its flow is geostrophic, as summarized in Section 7.11. Two of the most important examples of the thermal wind relation in action are outlined in Fig. 7.19, and others appear in later chapters of the book.

7.12.2 Applications

(i) Figure 7.19a represents a meridional section of the W'ly circumpolar vortex which dominates tropospheric flow in middle and high latitudes (Section 4.7) of the N hemisphere. The thermal wind relation ensures that the meridional temperature gradient imposed by the Sun's unequal input is associated with the increase of W'ly winds with height. It also ensures that the strongest W'lies occur in the upper troposphere—the top of the layer with consistent poleward temperature lapse. The increase of W'ly wind speed with height is consistent with the thermal wind relation, just as the E'ly shear above the wind maxima (W'ly wind speed decreasing with increasing height) is associated with the equatorward temperature

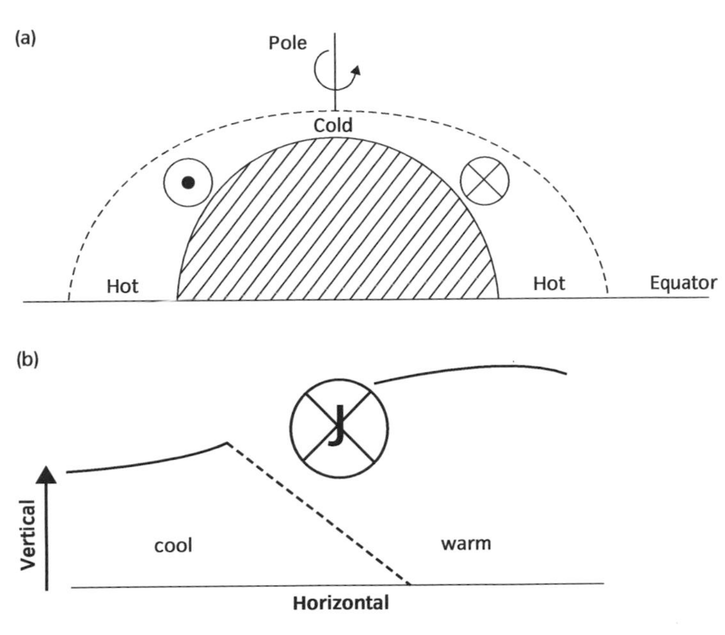

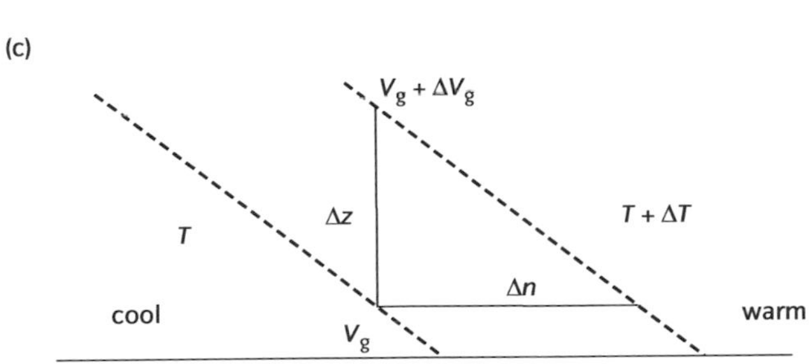

Figure 7.19 (a) The circumpolar vortex in section, with W'ly winds represented by arrow heads and tails, and temperature gradients between low and high latitudes implied by exaggerated thickness profiles (N hemisphere).
(b) A polar front in section, with jet stream arrow tail, zone of strongest horizontal temperature gradient (dashed line), and tropopauses (heavy lines), for the N hemisphere. (c) A close-up of the foot of the cold front in (b), showing the zone's finite horizontal breadth Δn and vertical thickness Δz.

lapse in the high troposphere (Fig. 4.10a). The reversal of the Coriolis directional relation across the equator ensures that the thermal wind relation and the great circumpolar W'ly vortex are reproduced in each hemisphere.

(ii) Figure 7.19b shows a vertical cross-section through an active front in middle latitudes (Section 12.3). The strong wind shear up to jet core level and the sharp lateral horizontal temperature gradient throughout this layer are basic features of any polar front, and their coexistence is obviously consistent with the thermal wind relation—colder air lying to the left of the jet stream in the N hemisphere (with back to the wind shear), as observed. In the S hemisphere, cold air lies to the right of the jet stream.

(iii) Equation 7.26b can be applied in finite difference form to a simple close-up of a sloping frontal zone (Fig. 7.19c) with horizontal width Δn, separating air masses with temperature difference ΔT along any horizontal. The *thermal wind* ΔV_g is the increase in geostrophic wind (flowing into the page in the N hemisphere) in the vertical depth Δz of the frontal zone.

$$\frac{\Delta V_g}{\Delta z} = \frac{g}{f\,T}\frac{\Delta T}{\Delta n} \qquad 7.27a$$

which can be rewritten

$$\frac{\Delta z}{\Delta n} = \frac{f\,T}{g}\frac{\Delta V_g}{\Delta T} \qquad 7.27b$$

According to Fig. 7.19c, $\Delta z/\Delta n$ is the slope of the frontal zone, which Eqn 7.27b shows is geostrophically consistent with a certain thermal wind and temperature contrast. Inserting realistic values for ΔV_g, ΔT, f and g, we find the associated frontal slope to be 1:60 in middle latitudes (NN 7.3), which is in the middle of the range of values found in special meso-scale observational analyses of fronts.

7.12.3 Baroclinic instability

These examples show that the thermal wind relation applies to a great deal of large-scale atmospheric behaviour. However abstract the equations may appear at first glance, they outline the related air flows and temperature fields of many important weather systems. Indeed advanced dynamical analysis *[35]* shows that the combination of vertical wind shear and lateral temperature gradient can be *baroclinically unstable*, in the sense that the slightest disturbance can give rise to poleward-and-upward and equatorward-and-downward exchanges of synoptic scale masses of air, just as convective instability gives rise to the essentially vertical exchange of much smaller masses of air. Such *slope convection* lies at the heart of the extratropical cyclone, just as vertical convection lies at the heart of convection on the meteorological small scale.

7.13 Geostrophic departures

7.13.1 Linear flow

On the synoptic scale, differences between actual winds and their geostrophic ideal are often systematic rather than random. For example, in the entrance region to the core of polar-front jet streams there is a tendency for the accelerating winds to be consistently *sub-geostrophic* (slower than geostrophic) and angled somewhat across the isobars towards lower pressure. And the sub-geostrophic tendency shows again as air decelerates out of a jet core, the flow now being angled towards higher pressure.

NUMERICAL NOTE 7.3 Slope of frontal zone

According to Eqn 7.27b, the thermal wind relation requires that the slope of the baroclinic frontal zone $\Delta z/\Delta n$ is given by $(f\,T/g)\,\Delta V_g/\Delta T$. Taking the wind shear ΔV_g vertically through the frontal zone to be 30 m s^{-1}, and the temperature difference ΔT horizontally across the frontal zone to be 5 °C, and using 265 K, 10^{-4} s^{-1} and 10 m s^{-2} for T, f and g respectively, we have $\Delta z/\Delta n = (10^{-4} \times 265/10)\,30/5 = 0.016$ (slightly less than 1:60)—an observably realistic slope for frontal zones.

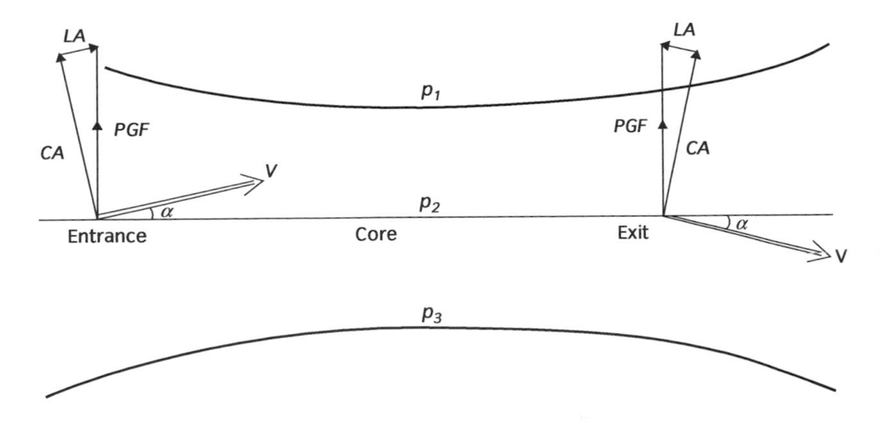

Figure 7.20 Plan view of jet entrance and exit regions with isobars p_1, p_2, and p_3 (highest), wind vectors *V*, pressure gradient forces *PGF*, Coriolis accelerations *CA*, and linear accelerations *LA* (in the N hemisphere).

Figure 17.20 contains plan views of the dynamics of both these situations. In the jet entrance, the linear acceleration *LA* along the line of flow adds vectorially to the hidden Coriolis acceleration *CA* perpendicular to the flow, to give the total acceleration. If there are no other forces or accelerations, the equation of motion requires that this total acceleration must correspond exactly to the pressure gradient force per unit mass (*PGF*). According to the geometry of Fig. 17.20, $CA/PGF = \cos\alpha$ where α is the angle of air flow across the isobars. Since *CA* is $f\,V$ and *PGF* is $f\,V_g$ (after Eqn 7.22), we have $V/V_g = \cos\alpha$, confirming what is graphically obvious: that the actual wind speed *V* is sub-geostrophic. Oblique flow across the isobars to slightly lower pressures provides for the linear acceleration into the jet core. In a realistic case where air accelerates steadily from 45 to 55 m s^{-1} along a 250 km path into the jet core, NN 7.4 shows that α is about 24°, and that the 50 m s^{-1} value for *V* is about 7% below the 55 m s^{-1} value for V_g, which corresponds geostrophically to a lateral pressure gradient of about 2 hPa per 100 km in the upper troposphere.

In the flow out of the jet exit, the diagram shows that the actual wind speed is again sub-geostrophic with $\cos\alpha = V/V_g$ but that the decelerating air (note the reversed *LA*) flows obliquely to slightly higher pressure as it exits from the core.

7.13.2 Gradient flow

Significant geostrophic departures also arise from *lateral accelerations* (accelerations across the line of flow) such as those associated with cyclonic and anticyclonic rotation of air. Figure 7.21a sketches the dynamics of cyclonic air flow round a horizontal circular path of radius *R* in the N hemisphere. Forces and accelerations are all along the radial axis, the air speed *V* must be such that V^2/R together with the Coriolis acceleration $f\,V$ corresponds exactly to *PGF*.

$$\frac{V^2}{R} = PGF - f\,V \qquad 7.28a$$

NUMERICAL NOTE 7.4 Ageostrophic flow in a jet entrance

Using the basic kinematic relationship ($v^2 = u^2 + 2as$), acceleration from 45 m s^{-2} to 55 m s^{-2} along a distance of 250 km implies a steady acceleration $[55^2 - 45^2]/(2 \times 250{,}000) = 2 \times 10^{-3}$ m s^{-2}, which is the linear acceleration LA in Fig. 17.20. For the following calculations we also assume $V = 50$ m s^{-1}, $f = 10^{-4}$ s^{-1} and $\rho = 0.4$ kg m^{-3}.

According to the geometry of Fig. 7.20, $CA/PFG = \cos\alpha$ and $LA/PGF = \sin\alpha$. Using $CA = f\,V$ and $PGF = f\,V_g$, we have $V = V_g \cos\alpha$, and $LA = f\,V \sin\alpha$. The last relation gives $\sin\alpha = LA/(f\,V) = 2 \times 10^{-3}/(10^{-4} \times 50) = 0.4$ which means that $\alpha \approx 24°$. From $V_g = V/\cos\alpha$ we have $V_g \approx 55$ m s^{-1}. Since $PGF = f\,V_g$ we have $PGF = 5.5 \times 10^{-3}$ m s^{-2}. From $PGF = (1/\rho)\,\partial p/\partial n$ we have $\partial p/\partial n = 0.4 \times 5.5 \times 10^{-3}$ Pa m^{-1} which is 5.5 hPa per 100 km.

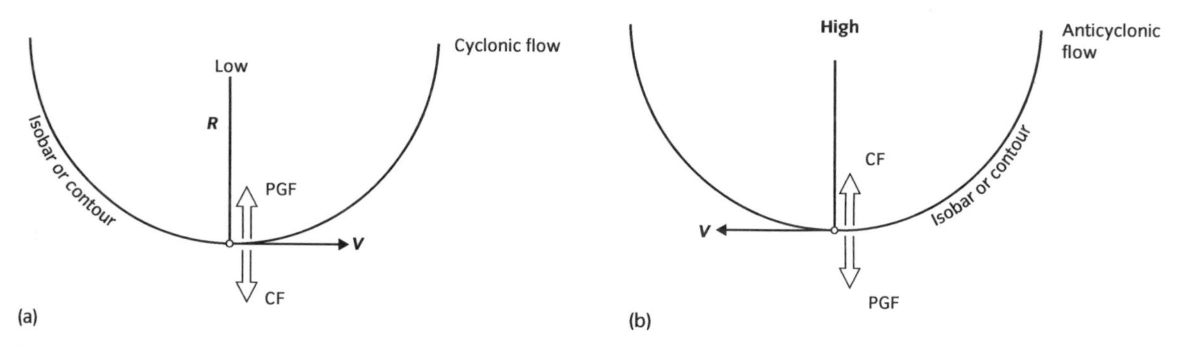

Figure 7.21 Plan views of circular cyclonic (a) and anticyclonic (b) flow, with wind velocities, and pressure gradient and Coriolis forces (N hemisphere).

Replacing PGF as usual with $f\,V_g$, we have

$$\frac{V^2}{R} = f\left(V_g - V\right) \qquad 7.28b$$

which confirms that the balancing wind speed V is sub-geostrophic, since ($V_g - V$) is positive. Winds satisfying this equilibrium are called *gradient winds*. When Eqn 7.28b is rearranged to express the *geostrophic departure* ($V_g - V$) as a fraction of the gradient wind speed V, we find

$$\frac{V_g - V}{V} = \frac{V}{Rf}$$

which is a Rossby number (Section 7.10). Actual winds are often observed to fall below geostrophic wind speeds by 20% or more in vigorous extratropical cyclones, and are to this extent apparently gradient winds, though in practice it is often impossible to measure the air flow curvature R accurately enough to assess the accuracy of the agreement (see below).

Figure 7.21b depicts anticyclonic flow in the N hemisphere, where the same reasoning shows that gradient wind speeds should be supergeostrophic ($V > V_g$), as is usually observed. In fact it is shown in Box 7.9 that if the equivalent of

BOX 7.9 Gradient balance in anticyclones

Figure 7.21(b) represents the gradient balance in anticyclonic circular flow in the N hemisphere. As in the case of cyclonic gradient flow (main text), the equation of motion is entirely radial, and the equivalent of Eqn 7.28b is

$$\frac{V^2}{R} = f\left(V - V_g\right)$$

confirming that the gradient wind speed is supergeostrophic rather than sub-geostrophic. Rearranging this as a quadratic equation we find

$$V^2 - R\,f\,V + R\,f\,V_g = 0$$

which has the standard solution

$$V = \frac{1}{2}\left[R\,f \pm (R^2 f^2 - 4R\,f\,V_g)^{1/2}\right] \qquad \text{B7.9a}$$

To be physically meaningful the term under the square root must be zero or positive (otherwise V contains an imaginary component)

$$4\,R\,f\,V_g < R^2 f^2$$

$$V_g < \frac{R\,f}{4}$$

showing that the geostrophic wind speed (and therefore PGF and pressure gradient) cannot exceed $R\,f/4$ – a value which increases with latitude (through f). If we substitute this maximum geostrophic wind ($V_g = R\,f/4$) in B7.9a the square root term vanishes, leaving

$$V = \frac{R\,f}{2} = 2\,V_{gmax} \qquad \text{B7.9b}$$

for the strongest possible anticyclone at that latitude (that f).

Strongest anticyclone

The linear increase of wind speed V with radius R in B7.9b means that air in the strongest anticyclone is rotating about the centre of the anticyclone like a solid wheel with uniform negative (since anticyclonic) angular velocity

$$\omega = -\frac{V}{R} = -\frac{f}{2}$$

Since $f/2$ is the component of the Earth's angular velocity about the local vertical, we see that the limiting anticyclone is one which is rotating at a rate which exactly cancels the rotation of the local surface, i.e. is not rotating when viewed from an external non-rotating reference

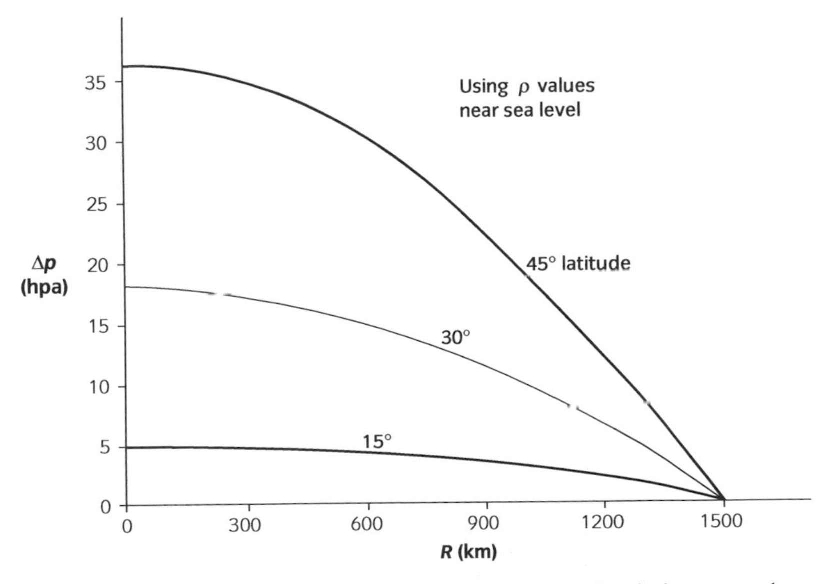

Figure 7.22 Radial profiles of pressure elevation on a horizontal in the low troposphere in the steepest anticyclones allowed by gradient balance at latitudes 15, 30, and 45°.

frame. As discussed in Section 7.16 this is expected to be the limiting result of pronounced horizontal divergence of flow with conservation of angular momentum.

Using the basic relation between V_g and radial pressure gradient $\{V_g = (\rho/f)\ \partial p/\partial R\}$ the radial pressure gradient in the strongest anticyclone is given by

$$\left(\frac{\partial p}{\partial R}\right)_{max} = -\rho\frac{R f^2}{4}$$

where the proportionality to $(-R)$ shows that pressure falls with increasing distance from the central pressure. Integrating this relation outwards from the centre to any radius R, we find an expression for the maximum pressure excess Δp at the centre over the value at R

$$\Delta p_{max} = \rho\frac{R^2 f^2}{8}$$

Radial pressure profiles at latitudes 15°, 30°, and 45° are sketched in Fig. 7.22, showing that substantial elevations of surface pressure are hardly possible at latitudes much lower than the subtropics so long as air flow is dominated by the gradient balance.

Eqn 7.28 is solved to find an expression for the gradient wind speed V corresponding to any particular geostrophic wind speed V_g (and therefore for any particular pressure gradient), then V_g can never be greater than $(R\,f)/4$, which implies that for any latitude and radius of trajectory curvature R there is a maximum pressure gradient which increases with R. Although in fact trajectory and isobaric curvatures are not identical (see below) they are obviously related, and the existence of such an upper limit to the pressure gradient therefore suggests that pressure gradients should die away towards the centre of an anticyclone, as is observed. No similar constraint applies in theory or observation to any kind of cyclonic flow.

7.13.3 Path curvature

Although this simple analysis is instructive and beloved of textbooks, it is very difficult to check by observation how close actual winds are to gradient equilibrium values, simply because it is very difficult to observe actual paths (trajectories) traced out by moving air parcels. The *streamlines* of flow which we can produce by careful analysis of observed winds on weather maps may be quite misleading in this respect for two important reasons. Firstly, the flow is usually fairly unsteady, so that streamlines are incessantly wriggling. And secondly, although trajectories can be established by connecting a succession of horizontal streamline segments from a series of maps at consecutive observation times, this resulting trajectory will be seriously misleading if air flow is tilted more than a very little out of the horizontal, as is known to be the case in large-scale flow. The same uncertainty prevents accurate assessment of the balance of linear accelerations and geostrophic departures in situations such as Fig. 7.20: the actual linear acceleration of a parcel rising slowly through the strong wind shear typical of a jet entrance may be very different from what appears on a standard isobaric chart. These are part of a very general problem in air motion analysis discussed in Section 12.4.

7.13.4 Antitriptic flow

Another important source of geostrophic departure arises from the enhanced turbulent friction of the atmospheric boundary layer. Figure 7.23 represents typical

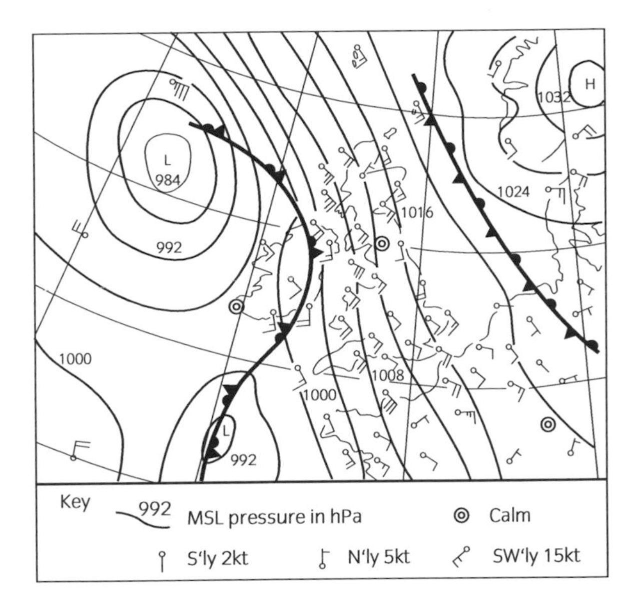

Figure 7.23 Synoptic surface winds and mean sea-level pressures over the British Isles at 0000 Z, 10 January 1969, showing winds tending to blow obliquely towards low pressure, most obliquely in lightest winds. The map shows SE'ly flow E of the occluded front of an old low (sickle shape left of centre), with a weak stationary front between Britain and Scandinavia—a typically non-standard situation!

air flow near the surface: winds measured at the standard observing height (10 m above the local surface) are consistently angled at between 10° and 30° across the isobars, varying with wind speed and local surface, and may have speeds which are less than 30% of geostrophic values. The situation is outlined in some detail in Chapter 10, but for the moment the role of frictional drag can be seen by considering the three-way equilibrium between Coriolis, pressure-gradient, and frictional forces (Fig. 7.24). The angular deviation and reduced strength of actual winds near the surface are consistent with the expected opposition of friction to motion over the surface, the rate of working by the pressure gradient on the parcel as it moves towards lower surface pressures being equal to the rate of working by the parcel against frictional drag.

The consistent motion of air across isobars has obvious implications for systems having closed isobars (Fig. 7.25), since friction then maintains a tendency for low-pressure centres to fill and high-pressure centres to empty—each of which is important for the decay of synoptic scale weather systems. The effect is quite strong near the surface, but is largely confined to the atmospheric boundary layer and is seldom significant more than a few hundred metres above the surface.

7.13.5 Low latitudes

Geostrophic equilibrium as defined by Eqns 7.19 is clearly impossible at the equator since the Coriolis parameter f is zero there. Very close to the equator, the Coriolis terms are observed to be so small at realistic wind speeds that large-scale flow is not even quasi-geostrophic. This has important practical consequences for weather forecasters in equatorial regions. There is little point in using the horizontal pressure fields (or isobaric contours) to consolidate wind fields, as is standard over the rest of the globe, since the two are not geostrophically related to a

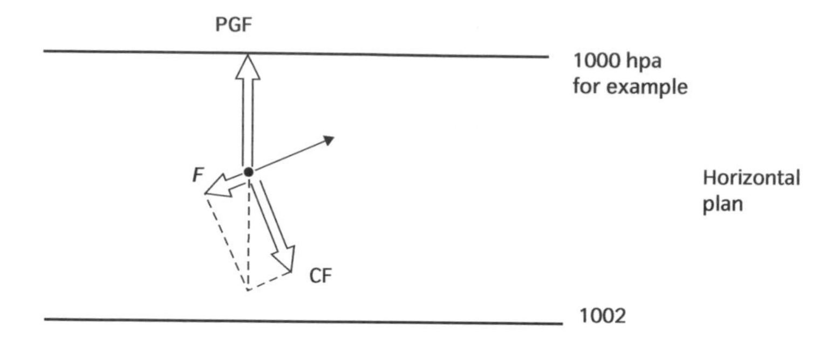

Figure 7.24 Horizontal balance of pressure gradient force, Coriolis force, and friction force in the atmospheric boundary layer, showing ageostrophic flow across the isobars towards lower pressure (N hemisphere). In the S hemisphere with the same pressure field, the vectors rotate 180° like rigid arms about the PGF axis.

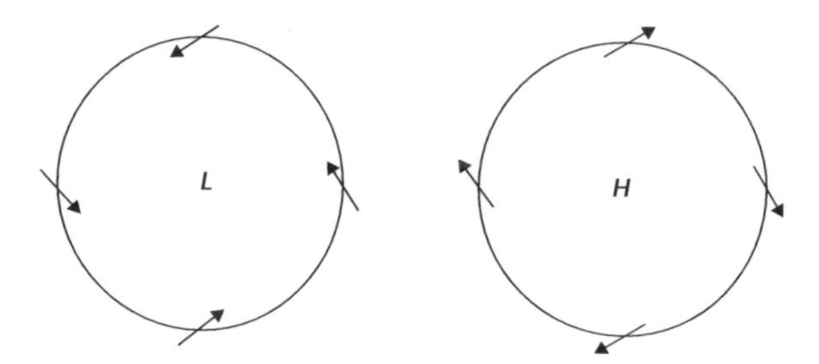

Figure 7.25 Frictionally induced convergence and divergence in the boundary layers of cyclones and anticyclones (N hemisphere).

useful degree. Synoptic-scale flow is therefore analysed by drawing streamlines through the observed winds. Away from the occasional very vigorous weather system the pressure gradients apparent on weather maps are much weaker than at higher latitudes, and are often dominated by a diurnal tide driven by the Sun's heating. As latitude increases, the local tendency to geostrophy increases; for example in the weather systems which thrive on the border between the inter-tropical convergence zone and the trade winds, there is a fairly obvious relation between pressure and wind fields (Section 13.3). At least this is so in the N hemisphere where this border is further from the equator than in the S hemisphere, because of the larger land areas in the north.

More fundamentally, the increasing irrelevance of geostrophic equilibrium as the equator is approached is apparent in the increasing Rossby number $U/(L\,f)$ as the Coriolis parameter f decreases with latitude. This should lead us to expect atmospheric flow in low latitudes to differ in character from that prevailing in middle and high latitudes. Although the detailed dynamical reasoning is subtle, it should seem at least plausible that the increase in this Rossby number with latitude is closely related to the striking difference between the Hadley circulation in low latitudes and the continually contorting circumpolar vortex of middle and high latitudes. This is nicely confirmed by experiments with a laboratory-scale analogue of the atmosphere quaintly called the *dish-pan [35]*. An annulus of water is trapped between a cold central pole and a warm equator (Fig. 7.26) in an open-topped pan rotating about an axis through the pole. This represents a flat Earth with only a single value for the Coriolis parameter (which remember

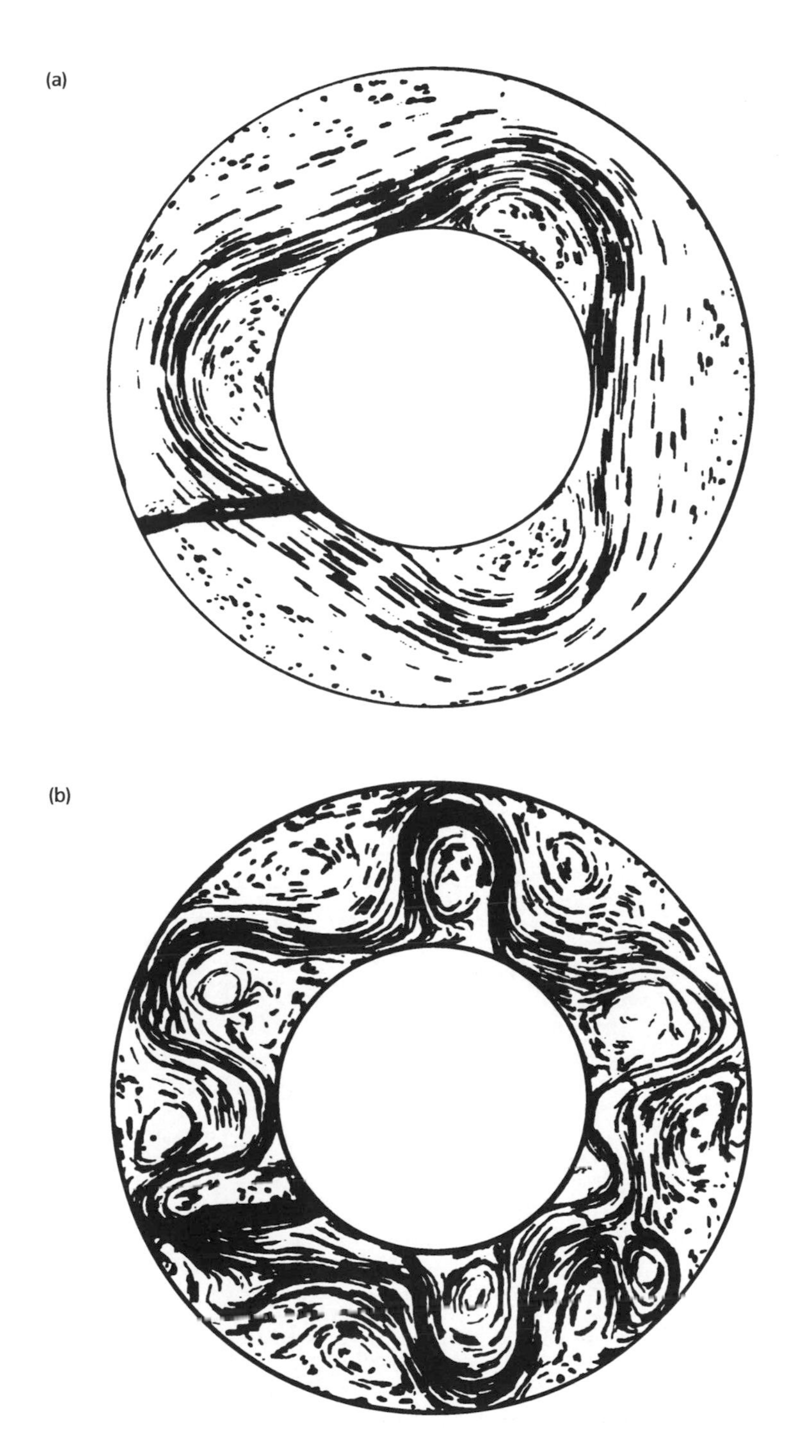

Figure 7.26 Circumpolar flows of water in a dish-pan operating in Rossby number regimes similar to those in the upper troposphere of (a) low and (b) middle latitudes in the N hemisphere. The dish-pan has a cold core and a warm rim, and surface water flow (showing jet streams) is observed by a long-exposure camera rotating with the pan.

is twice the component of rotation about the local vertical). Experiments with relatively slow rates of rotation therefore tend to simulate terrestrial behaviour at low latitudes and produce Hadley circulations, whereas wavy and slowly eddying circumpolar flow typical of higher latitudes is observed at faster rotation rates. The laboratory demonstration is clear, but the full theory of water flow in the dish-pan is a little simpler than it is for comparable atmospheric flow.

7.14 **Small-scale motion**

7.14.1 Scale analysis

Consider air flowing in the vicinity of a moderate-sized hill or shower cloud. The horizontal scale L is $\sim$ 10 km, which is effectively the same as the vertical scale set by the depth of the troposphere. This symmetry of scale is very different from the extremely flattened configuration of flow on larger scales. Horizontal wind speeds are mainly set by the synoptic-scale situation, so that 10 m s^{-1} is again typical for U. Vertical wind speeds are about an order of magnitude smaller, so that there is some asymmetry but very much less than on larger scales. Time scales for air being in the influence of such a system range from $\sim 10^3$ s for air passing horizontally through the flow over a hill, to as much as an order of magnitude larger for air rising through a shower cloud. Horizontal pressure variations are observed by special dense networks of sensitive barographs to be $\sim$ 1 hPa.

A list of basic and secondary scales is as follows, others being the same as in the list for synoptic scales (Section 7.9):

horizontal and vertical scale length	L	10 km	10^4 m
minimum time scale	t		10^3 s
horizontal pressure change	p	1 hPa	10^2 Pa
horizontal wind speed	U		10 m s^{-1}
vertical wind speed	w		5 m s^{-1}
horizontal acceleration	$\frac{U}{t}$		10^{-2} m s^{-2}
vertical acceleration	$\frac{w}{t}$		10^{-3} m s^{-2}
horizontal pressure gradient	$\frac{p}{L}$		10^{-2} Pa m^{-1}

Although turbulent friction is difficult to scale even roughly, we can set a limit to its magnitude by using KU/L^2 together with a K value from the atmospheric boundary layer (10 m^2 s^{-1}—Eqn 7.10 and discussion), increased 10-fold to offset the probable error in using L rather than the smaller subscale on which friction is especially effective (Section 10.9). This probably overestimates friction in small-scale flow in the free atmosphere, but may be about right for the interiors of turbulent convection. The resulting value for the frictional term (see below) shows that friction is relatively unimportant on the chosen scale, but becomes rapidly more important with decreasing scale.

The magnitudes of terms in the components of the equation of motion (Eqn 7.16) are now as follows:

$$\frac{du}{dt} = 2\,\Omega \sin\phi\, v - 2\,\Omega \cos\phi\, w - \frac{1}{\rho}\frac{\partial p}{\partial x} - FRF_x \qquad 7.29a$$

10^{-2} 10^{-3} 5×10^{-4} 10^{-2} 10^{-5}

$$\frac{dv}{dt} = -2\,\Omega \cos\phi\, u - \frac{1}{\rho}\frac{\partial p}{\partial y} - FRF_y \qquad 7.29b$$

10^{-2} 10^{-3} 10^{-2} 10^{-5}

$$\frac{dw}{dt} = 2\,\Omega \cos\phi\, u - g - \frac{1}{\rho}\frac{\partial p}{\partial z} - FEF_z \qquad 7.29c$$

10^{-3} 10^{-3} 10 10 10^{-5}

7.14.2 Horizontal motion

In the horizontal (Eqns 7.29a and b), the balance is mainly between acceleration and pressure gradient, with a marginally significant Coriolis effect. There is therefore little or nothing of the geostrophic tendency which dominates flow on larger scales, as suggested by the Rossby number

$$R_o = \frac{U}{\Omega L} \quad \text{or} \quad \frac{1}{\Omega t} \sim 10$$

which is about two orders of magnitude too large for quasi-geostrophic balance. Essentially the time period for air parcels in the influence of such small systems is so short that the Earth's rotation is largely irrelevant.

Local accelerations and pressure gradients are opposed, as shown by the opposing signs of the dominant terms in Eqns 7.29a and b. This opposition is often apparent on the meteorological small scale. For example, as air approaches the side of a hill and decelerates into a stagnation zone on its upwind side (Fig. 7.27), its pressure rises even if it moves perfectly horizontally, with the result that the pressure of the whole zone is raised slightly. Pressure is also raised

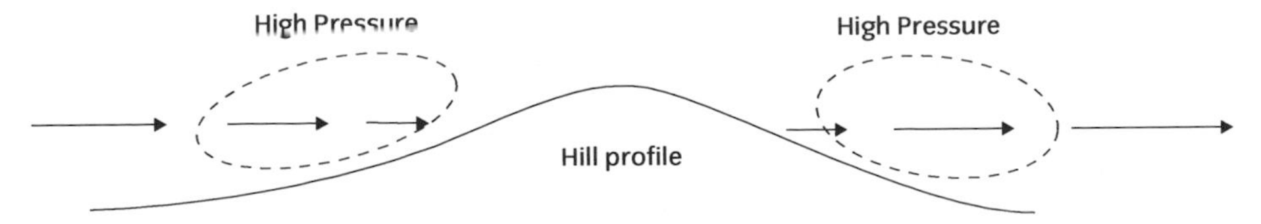

Figure 7.27 Air impinging on a hill and maintaining raised pressures in the stagnation zones immediately to windward and leeward. The pinching in at the surface indicates powerful convergence there.

a little in the lee of the hill (where the pressure gradient force acting away from the hill is associated with the acceleration of air out of the stagnation zone there), but usually by less than in the upwind stagnation zone because of the complex influence of friction, with the result that there is a net downwind drag of the wind on the mountain, and of course an equal and opposite drag of the mountain on the wind.

Cyclostrophic balance

Sometimes, as on larger scales, there are local accelerations which are perpendicular to the local wind, as air flows round tightly curved trajectories (round the sides of an isolated hill for example), which can maintain a near balance between lateral pressure gradient force and centripetal acceleration. Such flow is termed *cyclostrophic*, and it applies to a wide range of curved flows centred on our chosen 10 km scale: on a smaller scale it applies to air in the vicinity of tornadoes and dust devils (Section 11.6 and Fig. 7.28) , and on a larger scale it applies to air in the active annulus of high winds in a hurricane (Section 13.5.2). The equation for cyclostrophic balance in air moving at tangential speed V around a circular path of radius R is simply

$$\frac{V^2}{R} = \frac{1}{\rho}\frac{\partial p}{\partial R} \qquad 7.30$$

Air near the funnel of a tornado often blows at speeds well in excess of 50 m s^{-1} around paths with radii of curvature of about 100 m, which imply very large centripetal accelerations (> 25 m s^{-2} or 2.5 g). Corresponding pressure gradients are therefore > 30 hPa per 100 m (NN 7.5), and pressures in the funnel cores are depressed below ambient pressures by many tens of hPa. As described in

Figure 7.28 A dust devil—a dusty vortex induced by intense local horizontal convergence at the base of a powerful thermal. They are especially common in tropical dry seasons during the heat of the day. (see Plate 10).

NUMERICAL NOTE 7.5 Special small-scale accelerations

Cyclostrophic balance

Air moving at speed V round a circular path of radius R is suffering continuous centripetal acceleration V^2/R (Box 7.3), which has value 4.9 ms^{-2} (nearly 5 times gravitational g) in the case of a tornado funnel with circumferential wind speed 70 ms^{-1} and radius 100 m). If air density is 1.2 kg m^{-3}, then cyclostrophic balance (Eqn 7.30) implies $\partial p/\partial R = \rho\, V^2/R = 30$ Pa m^{-1}, i.e. 30 hPa per 100 m.

Foot of downdraft

Using the basic kinematic relation $v^2 = u^2 + 2\,a\,s$, a severe Cb downdraft of 20 m s^{-1} at height 1,000 m dying to zero at the ground implies a downward deceleration of $u^2/2\,s = 20 \times 20/2{,}000 = 0.2$ ms^{-2}. This deceleration requires an extra vertical PGF (on top of the hydrostatic part balanced by g) which we can write as $(1/\rho)\,\Delta p/s$. Rearrange this to find $\Delta p = 0.2\,\rho\,s = 0.2 \times 1.2 \times 1{,}000 = 240$ Pa (2.4 hPa), the implied elevation of surface pressure above the hydrostatically expected value.

Section 11.6, this heavily distorted pressure field has consequences which are visually dramatic and hazardous. As on larger scales, it is actually very difficult to measure air parcel trajectories in practice.

Note that the Coriolis term in cos ϕ in Eqn 7.29a, which was negligible on the synoptic scale, is now nearly comparable with the familiar sin ϕ term, because of the near equality of vertical and horizontal wind speeds. It is not clear whether or not the additional term plays a significant role in cumulonimbus, for example, because it is extremely difficult to make detailed observations of winds inside their turbulent, cloudy, and precipitating interiors, but the form of Eqn 7.29a suggests that there might be an E'ward tendency in updrafts and a W'ward one in downdrafts.

7.14.3 Vertical motion

Vertical motion on the small scale is usually dominated by hydrostatic balance between vertical pressure gradients and gravity, since these terms are usually four orders of magnitude larger than the next largest (vertical acceleration and Coriolis). However, much more rapid vertical accelerations are known to occur occasionally, at the bottoms and tops of vigorous cumulonimbus, for example, in which the length and time scales of accelerating and decelerating air are locally an order of magnitude smaller than the values listed above. For example, cold air pours downwards in shafts of heavy precipitation in cumulonimbus (Sections 11.4 and 11.6), and may impinge on the underlying surface so vigorously that the downdraft $(-w)$ reduces from 20 m s^{-1} to zero in the kilometre between cloud base and the surface. The corresponding value for dw/dt is $\sim$ 0.2 m s^{-2}, which is 2% of g and is associated with a measurable distortion of the hydrostatically expected value of surface pressure (a rise of over 2 hPa—NN 7.5). Transient pressure 'footprints' of around this magnitude are observed by arrays of microbarographs traversed by a big cumulonimbus, though analysis is complicated by the simultaneous presence of adjacent masses of warm and cool air.

This is a rather extreme example of deviation from hydrostatic equilibrium; in convection in particular it is much more common to have almost immeasurably small and very localized deviations which might seem to be trivial except that they are the driving force of all buoyant convection.

Buoyancy

The ambient air in which a convecting parcel is about to move can be considered to be in hydrostatic equilibrium, with only the convecting parcel out of balance. The ambient vertical pressure gradient is therefore related to the ambient air density by

$$\frac{\partial p}{\partial z} = -\, g\, \rho'$$

Substitution in the pressure gradient force term of Eqn 7.29c gives $g\,\rho'/\rho$ for the vertical pressure gradient force acting on the convecting parcel, where ρ is the density of the convecting parcel, which is slightly different from the density ρ' of the environment. The upward pressure gradient force and the downward gravitational force on the convecting parcel can be put together to produce the net upward force.

$$g\left(\frac{\rho'}{\rho}\right) - g = g\left(\frac{\rho' - \rho}{\rho}\right) = g\, B \qquad 7.31$$

This deviation from hydrostatic balance is more familiarly known as the *buoyant force* on the parcel, and in fact Eqn 7.31 follows from Archimedes' principle (Box 7.10). The term in brackets [ ] is known as the *buoyancy* B, since it contains the crucial proportional density difference between the parcel and its environment which gives the net buoyant force. When the parcel is less dense than the surrounding air, the buoyant force is positive and tends to produce upward acceleration. In the atmosphere, the density deficit of a buoyant parcel arises most often because it is warmer than its surroundings, though extra humidity can add significantly in some circumstances, and extra cloud burden can detract somewhat (Section 4.1 and Box 11.1).

The familiar dynamic effects of buoyancy are as described above, but in any particular situation the direction and magnitude of the associated acceleration depends on what other forces are acting in addition, and Eqn 7.29c shows that these can be the Coriolis and frictional forces. To see whether acceleration and buoyancy are closely matched, or whether there are other significant items in the dynamic balance, we compare vertical parcel acceleration with net buoyant force by forming the dimensionless ratio known as the *Froude number Fr*:

$$Fr = \frac{\text{acceleration}}{\text{buoyancy}} = \frac{dw}{dt}/g\, B \sim \frac{w^2}{g\, BL}$$

Strictly speaking this is the *internal* Froude number, to distinguish it from the number originally used by the pioneer fluid dynamicist William Froude in modelling ship wakes and waves. It is another important dimensionless number to rank with the Reynolds and Rossby numbers in meteorology, and describes the relative importance of gravity (reduced by *B*) in the dynamics of vertical motion.

When *Fr* is much less than unity, the buoyant force per unit mass is much larger than the observed vertical acceleration, showing that other additional factors must be significant in determining acceleration. Values for B in cumulus are very small ($\sim 1/300$) on account of the very small temperature excesses in rising air, and those for cumulonimbus are only a little larger (Section 11.4). It follows from such values and observed strengths and dimensions of updrafts that Fr values in atmospheric convection cover a considerable range centred on about 0.2, confirming that there are forces at work significantly offsetting the expected accelerating effects of buoyancy.

Factors offsetting buoyancy

Convection is visibly full of turbulent friction (Fig. 7.29). Cumulus of all sizes clearly must encounter considerable resistance as they push through their cloudless surroundings. If buoyant parcels were rigid, their drag could be modelled in wind tunnels or water flows provided we ensured *dynamical similarity* with the real atmosphere, i.e. similar values for the Reynolds and Froude numbers. Strain gauges could be used to measure the resulting drag. But buoyant masses of air are not rigid; they are barely distinguishable from their surroundings, and they deform and tumble as they rise, in ways which must considerably alter the drag from the equivalent rigid body value; and of course strain gauges cannot be used. Laboratory and theoretical models have been used to show that the drag is very substantial, but the work is incomplete and beyond the scope of present discussion *[36]*, so we have to make do with results from rigid bodies.

Simple observation of cumulus shows that another significant dynamic process is at work. Cloudy air expands very noticeably as it rises, showing that the buoyant, cloudy air is incorporating air from its immediate environment at a very considerable rate (Fig. 7.29). Such *entrainment* is a basic property of all convection, from small thermals in the atmospheric boundary layer to large cumulonimbus (Section 11.3), and its presence means that we must modify the equation of motion to allow for variation of air-parcel mass with time. This is done by returning to the original, general form of Newton's second law of motion (Eqn 7.2), whose vertical component can be written simply as

Figure 7.29 A cloudy turret rising out of a parent cumulus congestus, dissipating quickly as a result of rapid entrainment of unsaturated air.

$$\frac{d}{dt}(M\,w) = NUF \qquad 7.32$$

where NUF represents the net result of all vertical components of force on the parcel. The left-hand side of Eqn 7.32 can be expanded to make explicit allowance for the effects of variations in parcel mass M:

$$M\frac{dw}{dt} + w\frac{dM}{dt} = NUF \qquad 7.33$$

The first term is the familiar product of mass and (vertical) acceleration. The second term represents the rate of increase of momentum which is accounted for by the rate of increase of parcel mass rather than acceleration. If in any example we know the strength of the net force NUF and observe only the upward acceleration of the parcel (not its growth by entrainment) we will underestimate the true production of upward momentum by ignoring the momentum being used to set in motion large masses of initially static ambient air. As usual an effect ignored on the left-hand side of the equation of motion will seem like a opposing mysterious force on the right-hand side, in this case a force logically called *entrainment drag*. Observations and some theory suggest that the obvious frictional drag and the less obvious entrainment drag are of comparable importance in convective dynamics, and that together they offset the buoyant forces to such an extent that actual updraft speeds are only a tiny fraction of what they would be in a frictionless and non-entraining thermal. Only large cumulonimbus seem to have the capacity to operate considerably more efficiently than this (Section 11.6), by a dynamic economy of scale.

BOX 7.10 Archimedes, buoyancy, and centrifuge

Consider a parcel of air (or any body) of density ρ totally immersed in air (or any fluid) of density ρ'. According to an extended form of Archimedes' principle (c. 250 BCE) the immersed parcel experiences an upward thrust which is equal to the weight of the air displaced. Since the parcel experiences a downward force equal to its own weight, the net upward force F is given by the difference

$$F = \text{weight of displaced air} - \text{weight of parcel}$$

If the volume of the parcel is Vol, this expression can be rewritten

$$F = g\,(\rho' - \rho)\,Vol$$

Dividing through by the parcel mass $m = \rho\,Vol$ we find the net upward force per unit mass

$$\frac{F}{m} = g\,\frac{(\rho' - \rho)}{\rho} \qquad B7.10a$$

which is identical to Eqn 7.31 derived from the vertical component of the equation of motion, an agreement which confirms that Archimedes' principle is an elegant and extremely useful summary of an important consequence of hydrostatic balance.

For a completely immersed parcel (Archimedes' intuition actually dawned as he *floated* in a brimming bath), the upward force arises because the hydrostatic vertical gradient of pressure in the supporting fluid produces an upward force which will exactly support the weight of the buoyant parcel if it has the same density as its surroundings. If, however, the parcel has a lower density than its surroundings, its weight will be smaller than that of the same volume of ambient air, the upward pressure force on the parcel will exceed its weight, and the parcel will accelerate upwards. If the parcel is denser than its surroundings it will accelerate downwards.

The same thing tends to happen in the presence of any pressure gradient, vertical or horizontal, associated with a body force or equivalent acceleration. In the vortex of a tornado (Section 11.6.4) the horizontal pressure gradient may exceed the vertical by a factor of two or more. The net inward pressure force on a body immersed in the vortex will maintain a centripetal acceleration closely matching that of the surrounding air if the body has the same density as the ambient air. Dense debris picked from the surface and demolished buildings, etc. will be accelerated inwards by the ambient pressure gradient much more slowly than the ambient air, and therefore will be centrifuged outwards relative to the swirling air in a potentially lethal spray which scours the zone in and around the base of the vortex and its cloud funnel.

The density differences in Eqn B7.10a often arise from temperature differences. In the case of an ideal gas like air, the equations of state for the parcel $p = \rho R T$, and for the ambient air $p = \rho' R T'$, are used to replace the densities with the common pressure (assumed uniform across any horizontal) and R to give

$$\frac{F}{m} = g\frac{(T - T')}{T'} \qquad \text{B7.10b}$$

where the temperature excess of the parcel $(T - T')$ can be expressed in °C but the absolute temperature T' of the ambient air must be in kelvins.

7.15 Compressing and deforming air *[33]*

7.15.1 Continuity

Air, being fluid, deforms readily when it impinges on other bodies. Sometimes these are rigid, such as mountains, or nearly rigid, such as the sea surface, but most often they are other bodies of air. Although the emphasis so far in this chapter has been on the air parcel as an isolated, compressible packet (especially in vertical displacements), we must now combine that compressibility with its essential fluidity.

The expansion of rising air and the contraction of sinking air are large even in the relatively shallow depth of the troposphere, because air is easily compressed. Compressibility may suggest elasticity, but it is important to note that although sound waves and the continual small pressure adjustments which travel at the speed of sound (Section 4.3) depend crucially on the elasticity of air (its ability to spring outwards after compression), all other atmospheric dynamics related to weather operate as if the air had no elasticity, even though it is obviously compressible and deformable. Elasticity is unimportant in most atmospheric dynamics because solar power drives the atmosphere at speeds well below the speed of sound. For example, air sinking towards the surface in the downdraught of a large cumulonimbus (Section 11.4) compresses substantially as its ambient pressure rises, but nevertheless impinges on the surface so gently that there is no perceptible tendency to compress further and rebound upwards by re-expansion. Let us look more closely at the compression and deformation of air involved in synoptic and small-scale weather systems, as defined in Sections 7.9 and 7.14.

We can derive an expression which is very useful for this discussion by assuming that the initial mass of any body of air is conserved no matter how it may subsequently deform, expand, combine, or divide. The equation expressing the conservation of mass in flowing air is called the *continuity equation* because it is most easily derived by considering the continuity of mass flux as air blows through a volume fixed in space. Its derivation is outlined in Box 7.11:

$$\frac{\partial \rho}{\partial t} = -\left[\frac{\partial}{\partial x}(\rho\, u) + \frac{\partial}{\partial y}(\rho\, v) + \frac{\partial}{\partial z}(\rho\, w)\right] \quad 7.34$$

The left-hand side of this equation is the rate of change of density with time at a fixed position, known as its *local rate of change*. The terms in brackets on the right-hand side describe the net mass flux out of the fixed volume in question arising from longitudinal gradients of mass flux. There is a component of this *mass flux divergence* arising from each coordinate axis, as depicted in Fig. 7.30. A component (along the x axis for example) is positive if the mass flux in the x direction increases with x, because in that case the flux into the near side of a perpendicular cube face is less than the flux out of the opposite face. Such *divergent*

BOX 7.11 Continuity of mass

Figure 7.30 shows a very small frame fixed in space with air blowing through the open faces. Considering air flow in the x direction, the mass flux into the face perpendicular to the x axis at x is given by $(\rho\, u)_x\, dy\, dz$, where the subscript x means that the term in brackets is evaluated at x. A similar expression describes the mass flux out of the face at $x + dx$, and the net influx is given by their difference

$$\text{net mass influx} = (\rho\, u)_x\, dy\, dz - (\rho u)_{x+dx}\, dy\, dz$$

As in Box 7.1 we can replace the difference in the variable term by the product of its gradient and the small separation dx, to find

$$\text{net mass influx} = -\frac{\partial}{\partial x}(\rho\, u)\, dx\, dy\, dz$$

Equivalent expressions give the components of net mass influx in the y and z directions, and the resultant net flux of mass into the volume is the algebraic sum of the three components:

net mass influx

$$= -\left[\frac{\partial}{\partial x}(\rho\, u) + \frac{\partial}{\partial y}(\rho\, v) + \frac{\partial}{\partial z}(\rho\, w)\right] dx\, dy\, dz \quad \text{B7.11a}$$

Since mass is conserved, the net rate of mass influx must equal the rate of increase of mass in the volume bounded by the frame:

$$\frac{\partial}{\partial t}(\text{mass}) = \frac{\partial \rho}{\partial t}\, dx\, dy\, dz \quad \text{B7.11b}$$

since the volume bounded by the fixed frame is conserved. Combining B7.11a and 7.11b and simplifying, we find

$$\frac{\partial \rho}{\partial t} = -\left[\frac{\partial}{\partial x}(\rho u) + \frac{\partial}{\partial y}(\rho v) + \frac{\partial}{\partial z}(\rho w)\right]$$

which is one of the standard forms of the continuity equation.

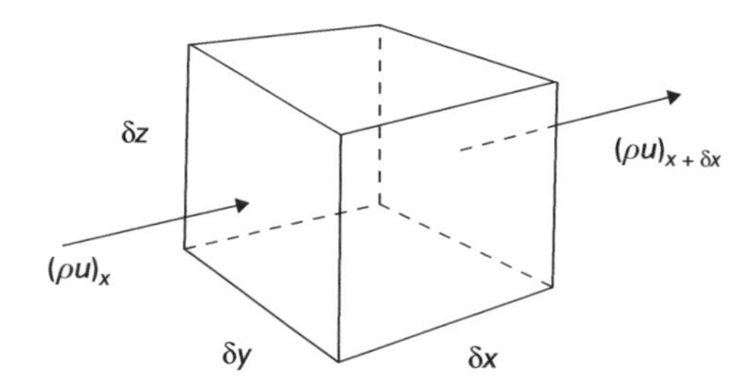

Figure 7.30 Flows of air through opposite faces of a fixed volume. When mass fluxes are unequal, they contribute to mass flux convergence or divergence.

mass flux contributes to a fall of air mass in the cube and hence to a decrease in the cube's air density, as ensured by the minus sign in Eqn 7.34.

7.15.2 Horizontal and vertical stretching

Detailed study of atmospheric mass fluxes on scales ranging from the synoptic to the meteorological small-scale shows that the horizontal mass-flux divergence (the first two terms on the right-hand side of Eqn 7.34) is always very nearly balanced by an opposing vertical mass-flux divergence $\partial(\rho\, w)/\partial z$, so that the local rate of change of air density (the left-hand side of Eqn 7.34) is a relatively small residual. In fact observed magnitudes of the opposing mass flux terms are such that if they did not largely cancel, the local change of air density would often approach 100% per day on synoptic scales, and 100% in a few minutes on small scales. Nothing remotely like this is observed: local percentage variations of air density are effectively equal in magnitude to local percentage variations in absolute temperature, and are never more than a very small fraction of the figures mentioned above. To this extent therefore

$$\frac{\partial}{\partial x}(\rho u) + \frac{\partial}{\partial y}(\rho v) + \frac{\partial}{\partial z}(\rho w) \simeq 0 \qquad 7.35$$

It seems from this that the atmosphere arranges its flow to minimize local concentrations and depletions of mass so that, for example, when mass is gathering horizontally in convergent mass flow (Fig. 7.31), as in the lower troposphere of an extratropical cyclone or a cumulonimbus, it is stretching (diverging) vertically. Since the vertical mass flux is zero at the Earth's surface, a vertical divergence of mass flux in overlying layers of air implies a vertical mass flux which increases with height above the surface, and hence a significant rate of uplift at the top of the horizontally converging layer (the middle troposphere in Fig. 7.31). We can investigate the magnitude of this uplift by integrating the terms of Eqn 7.35 over the height range from the surface to any particular level of interest, as sketched in Fig. 7.32. The integration of the first two terms gives the net horizontal influx of mass through the sides of the imaginary column from surface to the chosen level. The integral of the balancing third term

$$\int_b^t \left[\frac{\partial}{\partial z}(\rho w)\right] dz = [\rho w]_t - [\rho w]_b = [\rho w]_t$$

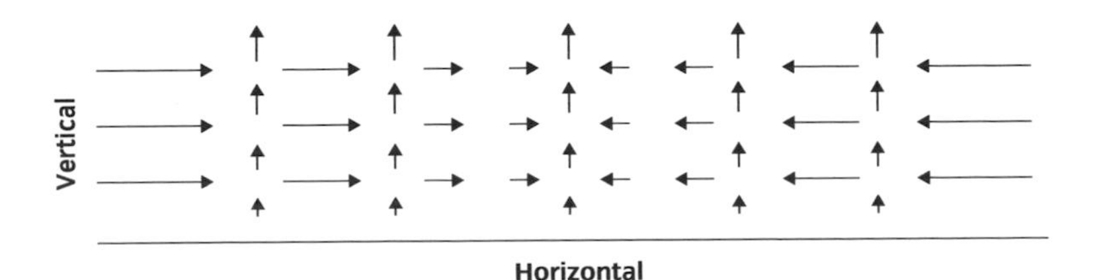

Figure 7.31 Horizontal (one dimensional) convergence and vertical divergence of mass flux on a vertical section through the lower troposphere. The pattern implies consistent uplift in the middle troposphere.

simplifies to the vertical mass flux out of the top of the column at the chosen level because there is no mass flux through the solid bottom surface. It follows from the near balance expressed by Eqn 7.35 that

$$w_t \simeq -\frac{1}{\rho}\int_b^t \left[\frac{\partial}{\partial x}(\rho\, u) + \frac{\partial}{\partial y}(\rho\, v)\right] dz \qquad 7.36$$

Values for the right-hand side of Eqn 7.36 have been determined by careful analysis of synoptic-scale observations, and they show that the horizontal convergence of mass in the lower troposphere is consistent with rates of uplift ~ 10 cm s^{-1} in the middle troposphere, which are known from other evidence (rates of steady rainfall for example—Section 12.4) to be quite typical. Mass convergence is difficult to measure accurately on the scale of a cumulonimbus, but measured updrafts ~ 5 m s^{-1} in middle levels imply that horizontal convergences are 50 times larger than synoptic scale values.

7.15.3 Vertical compensation

Atmospheric mass fluxes are reluctant to concentrate in vertical columns. We know this from the observed smallness of pressure tendencies—rates of rise or fall of pressure on fixed surface barographs. Consider air entering and leaving a vertical column extending from the surface to the top of atmosphere (Fig. 7.32). The net horizontal flux of mass into the column adds to the total mass of the column, and hence to the pressure at its base (the surface), by the excellent hydrostatic approximation. In fact the rate of rise of surface pressure (the *pressure tendency* measured by a barograph) is simply g times the net mass inflow to a vertical column resting on unit horizontal area.

$$\frac{\partial p_s}{\partial t} = g\int_b^t \frac{\partial M}{\partial t} = -g\int_b^t \left[\frac{\partial}{\partial x}(\rho\, u) + \frac{\partial}{\partial y}(\rho\, v)\right] dz \qquad 7.37$$

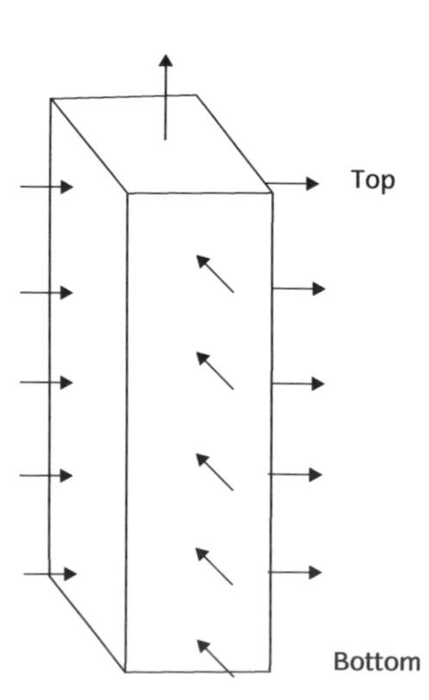

Figure 7.32 Net convergent (two dimensional) horizontal mass flux in an atmospheric column, with associated uplift at its top.

NUMERICAL NOTE 7.6 Uncompensated pressure tendencies

Putting together Eqns 7.36 and 7.37, the uplift w_t at the top of a layer of air suffering net convergence is related to the pressure tendency at the bottom of the layer by $\partial p_s/\partial t = g\rho w_t$. Taking the layer top to be in the mid-troposphere, we can assume $\rho \approx 0.5$ kg m^{-3}. Using $w_t \approx 10$ cm s^{-1} (0.1 m s^{-1}) for the uplift in a depression, the corresponding surface pressure tendency $\partial p_s/\partial t$ is given by $10 \times 0.5 \times 0.1 = 0.5$ Pa s^{-1}, which is 43,200 Pa (432 hPa) per day—nearly half an atmosphere! The same calculation with $w_t \approx 5$ m s^{-1} gives 450 hPa in 30 minutes at the base of a cumulonimbus.

But by Eqn 7.36 and in other ways we have been able estimate the net horizontal mass flows into the lower parts of depressions and cumulonimbus. If we substitute these into the right-hand side of Eqn 7.37 (NN 7.6) we find that in a depression, surface pressure tendencies would be about 500 hPa per day (assuming average air densities of about 0.5 kg m^{-3}), and in cumulonimbus the same enormous change should occur in about half an hour. Nothing like this is observed, even though pressure tendencies are often considerably increased by translation of horizontal pressure gradients embedded in moving weather systems. On the synoptic scale, pressure tendencies seldom exceed 20 hPa per day, and on the scale of cumulonimbus they seldom exceed a few hPa in the hour-long life of the cloud.

The simplest possible explanation for these puzzling results is that almost all of the net inflow of mass to the lower half of the tropospheric column is being offset by a net outflow from the upper half. As shown in Fig. 7.33 horizontal convergence of mass extends up the middle troposphere, above which there is horizontal divergence of mass up to about the tropopause. The balancing vertical mass flux (Eqn 7.36) increases with height up to the level at which horizontal convergence is replaced by divergence, and then diminishes steadily to about zero at the tropopause. The aggregated horizontal convergence and divergence *compensate* (nearly balance), and the relatively slight imbalance accounts for the small rates of change of pressure observed at the bottom of the column. The falling minimum surface pressure in the formative stages of an extratropical cyclone, for example, arises because divergence aloft slightly but persistently exceeds convergence beneath.

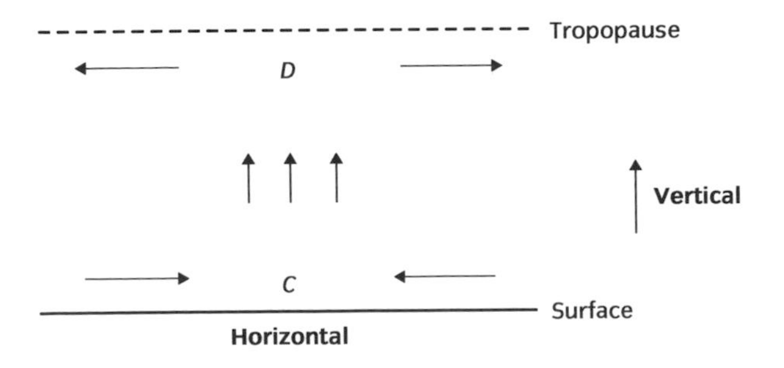

Figure 7.33 Schematic flow in a vertical section through a cloudy weather system. Assuming the same pattern in a perpendicular section, horizontal convergence below and divergence aloft are linked by complementary uplift in middle levels.

More elaborate schemes with several layers of alternating convergence and divergence are of course possible, and could equally well explain the small pressure changes observed at the surface, but observations suggest that the two-layer scheme accounts for most synoptic-scale and cumulonimbus-scale weather systems. On the synoptic scale, for example, an anticyclone is like an inverted cyclone, with convergence overlying divergence and maximum downward mass flux occurring at the mid-tropospheric interface. Such widespread and persistent subsidence is associated with many of the most striking properties of anticyclones (Section 13.2).

We have seen that the atmosphere arranges its flow on the synoptic and small scales so as to minimize the congestion and depletion of air mass in localized volumes and columns, and that each of these tendencies is associated with extremely basic and important aspects of weather behaviour. But although the analysis has been useful, it is important to note that we have not made use of any dynamics (forces and accelerations) in the process—the basic physical property assumed is simply the conservation of mass. We have not in any sense explained why the atmosphere arranges its flow in these ways, nor even discussed the mechanisms involved; but the examination has been valuable even so, and has pointed to quite basic types of behaviour which must be explained by full dynamic theory. Such theory exists now in respect of synoptic scale motion (the theory of baroclinic instability which is briefly mentioned in Section 12.6) but is far beyond the level of an introductory text; a theoretical framework barely exists in the case of cumulonimbus.

7.15.4 Synoptic scale flow

Though the basic form of the continuity equation has proved directly applicable to the real atmosphere, it is useful to develop it a little further for the particular case of synoptic-scale motion, to see how compression and decompression can be made explicit in the formal description of deformability begun by Eqn 7.35.

A serious snag emerges when we consider the horizontal mass convergence in geostrophic flow. We can do this by taking the geostrophic expressions for $\rho\, u$ and $\rho\, v$ from Eqn 7.20 and substituting them in the first two terms of Eqn 7.35. Ignoring the relatively small variation of f with y (i.e. with latitude), we find

$$\frac{1}{f}\left[-\frac{\partial^2 p}{\partial x \partial y}+\frac{\partial^2 p}{\partial y \partial x}\right]$$

which is zero because the terms in the brackets cancel exactly, showing that pure geostrophic flow has zero horizontal mass divergence. Physically this arises from the inverse relation between wind speed and pressure gradient which is basic to geostrophy.

Figure 7.34 depicts a situation in which the geostrophic flow is increasing to the E, which corresponds to positive $\partial(\rho\, u)/\partial x$. The associated narrowing of the N–S separation of the contours corresponds to a negative $\partial(\rho\, v)/\partial y$ which exactly balances $\partial(\rho\, u)/\partial x$, and gives zero net horizontal mass divergence. This result is disappointing, because it shows that the so far very useful geostrophic approximation completely fails to allow for the convergence and uplift of air which we know is essential in producing cloud, precipitation, and pressure

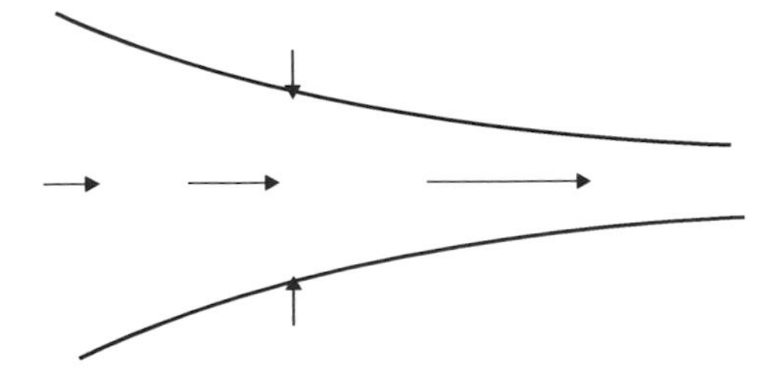

Figure 7.34 Plan view of balancing longitudinal velocity divergence and lateral convergence in confluent geostrophic flow. Strictly the airflows should angle as in Fig 7.20.

variations. It appears that geostrophic balance is too well balanced for the real atmosphere. It is still useful as described earlier, but it must not be used in ways which allow it to eliminate vertical motion.

We can make less drastic and more useful simplifications by expanding each of the first two terms of Eqn 7.35 into a term in the gradient of wind speed and another one in the gradient of density. Combining the latter from the two horizontal directions we have

$$u\frac{\partial \rho}{\partial x} + v\frac{\partial \rho}{\partial y}$$

which represents the contributions to density variation arising from motion in the presence of instantaneous horizontal density gradients. Since air density tends to move with the air itself, these terms can be regarded as representing the horizontal advection of existing density gradients by the flow. Detailed observations of synoptic-scale flow indicate that horizontal advection of density is about an order of magnitude smaller than the other terms in the expansion of the original mass flux divergence, which are

$$\rho\left[\frac{\partial u}{\partial x} + \frac{\partial v}{\partial y}\right] \text{ written } \rho\, D \qquad 7.38$$

and represent the contributions arising from longitudinal stretching of flow. The relative insignificance of the density advection terms is consistent with the tendency to minimize density concentrations (and therefore density gradients) already noted, and at least in part reflects the tendency of the thermal wind relationship to align horizontal winds and lines of constant density and hence minimize density (and temperature) advection.

7.15.5 Divergence

The bracketed expression in Eqn 7.38 is called the *horizontal divergence* of velocity, or simply the *divergence*, and is usually denoted by D. The approximate form of the continuity equation Eqn 7.35 can now be written in the form

$$D + \frac{1}{\rho}\frac{\partial}{\partial z}(\rho\, w) \simeq 0 \qquad 7.39$$

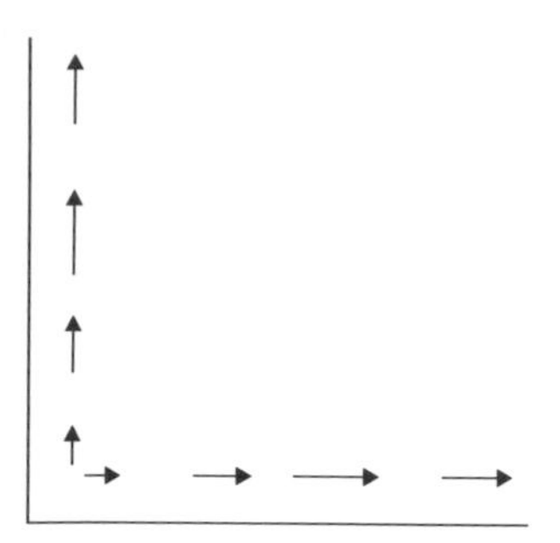

Figure 7.35 Plan view of flow diverging in both horizontal directions.

Velocity divergence is similar to the divergence of mass flux discussed above, but it is considerably simpler to interpret, depict and quantify. Figure 7.35 shows a flow with positive divergence D. If a patch of this air were marked with smoke at some initial time, the horizontal area A of the patch would increase with time for as long as D remained positive. The patch would also move bodily (NE'ward in the case shown), but this is irrelevant unless accompanied by a change in A. It is shown in more formal expositions *[35]* that D is equal to the fractional rate of change of horizontal area of the patch (i.e. following the patch).

$$D = \frac{1}{A}\frac{dA}{dt} \qquad 7.40$$

From a variety of evidence it is known that D values $\sim -10^{-5}\,\mathrm{s}^{-1}$ are typical of the zone of horizontal convergence which virtually fills the lower troposphere in an extratropical cyclone. If divided equally between the two terms of D, it follows that each of the stretching terms $\partial u/\partial x$, $\partial v/\partial y$ is similarly $\sim 10^{-5}\,\mathrm{s}^{-1}$. This amounts to a longitudinal gradient of wind speed of about $1\ \mathrm{m\,s}^{-1}$ in 100 km, which is barely measurable from standard synoptic observations, but yet corresponds to extremely significant convergence on the synoptic scale. It is a major problem in meteorology that such significant features are only marginally measurable by direct observation, and is compounded by the irrelevance of the geostrophic approximation in this context. Considerable effort has gone into finding alternative ways of estimating divergence from available observations; fortunately there is a useful direct connection with large-scale rotation, as indicated in the next section.

In a region such as the lower troposphere of an extratropical cyclone, according to Eqn 7.40 the convergence (negative D) in the lower troposphere is balanced at every level by

$$\frac{1}{\rho}\frac{\partial}{\partial z}(\rho\, w) \qquad 7.41$$

If the atmosphere were truly incompressible and uniform this would simplify to the vertical stretching term $\partial w/\partial z$, and the continuity equation would become simply

$$D + \frac{\partial w}{\partial z} = 0 \qquad 7.42$$

showing that at every level, vertical stretching balances horizontal convergence, or vertical squeezing balances horizontal divergence. If a D value of $-10^{-5}\,\mathrm{s}^{-1}$ prevails throughout the first 5 km of the troposphere, and is balanced at every level by positive $\partial w/\partial z$, then the cumulative uplift speed w at the 5 km level (found by integrating the vertical stretching term up to this level) becomes very simply

$$w = \int_0^5 \frac{\partial w}{\partial z}\,dz = -z\,D$$

$$= -5{,}000 \times \left(-10^{-5}\right) = 5 \times 10^{-2}\,\mathrm{m} \quad \text{i.e. } 5\ \mathrm{cm\,s}^{-1}$$

—a fairly typical value for rates of uplift associated with nimbostratus in extra-tropical cyclones (Section 12.4). The behaviour described by Eqn 7.42 is like that of a slack plastic bag containing a fixed volume of water: when squeezed in the horizontal it expands in the vertical and vice versa—simple incompressible deformability.

7.15.6 Compressibility

Because the atmosphere is in fact highly compressible, its behaviour is more complex than is allowed by Eqn 7.42. However, the observational studies mentioned

BOX 7.12 Compressibility and divergence

Although beyond the general level of this book, the following derivation is included (largely without commentary) because it does not appear explicitly in texts seen by the author.

The second term of Eqn 7.39 can be rewritten

$$\frac{1}{\rho}\frac{\partial(\rho\, w)}{\partial z}=\frac{\partial w}{\partial z}+\frac{w}{\rho}\frac{\partial \rho}{\partial z} \qquad \text{B7.12a}$$

Since $\rho=\dfrac{p}{RT}$ ideal gas

$$\frac{1}{\rho}\frac{\partial \rho}{\partial z}=\frac{1}{\rho\, RT}\frac{\partial p}{\partial z}-\frac{p}{\rho\, RT^2}\frac{\partial T}{\partial z}$$

$$=\frac{1}{p}\frac{\partial p}{\partial z}-\frac{1}{T}\frac{\partial T}{\partial z}$$

Now $\dfrac{\partial p}{\partial z}=-g\,\rho$ hydrostatic

and $\dfrac{\partial T}{\partial z}=-\dfrac{S\,g}{C_p}$

where S is the environmental lapse rate as a fraction of the dry adiabatic lapse rate g/C_p.

Hence $$\frac{1}{\rho}\frac{\partial \rho}{\partial z}=-\frac{g\rho}{p}+\frac{g\,S}{C_p\,T}$$

$$=-\frac{g}{C_pT}\left[\frac{C_p}{R}-S\right]$$

The value of C_p/R is almost exactly 3.5. Except in relatively shallow inversions or isothermal layers, S values range from about 0.4 (rather more stable than the warmest saturated adiabat) to the limiting value of 1 (a dry adiabat). If for simplicity we assume an S value of 0.5, the error in term [ ] is relatively small even in extreme cases, and less than 5% most of the time.

To the extent that S is 0.5, [ ] = 3 and

$$\frac{1}{\rho}\frac{\partial \rho}{\partial z}=-X \qquad \text{B7.12b}$$

where $X\simeq-\dfrac{3\,g}{C_pT}=1.2\times10^{-4}\pm0.1\times10^{-4}\ m^{-1}$

depending on the value of absolute T. Combining Eqn 7.39, B7.12a and B7.12b we have

$$D+\frac{\partial w}{\partial z}=X\,w \qquad \text{B7.12c}$$

The solution of B7.12c for the simple two-layer case depicted in Fig. 7.36 which includes uniform convergence C from zero z to $z=h$ and uniform divergence $-$C from there to $z=2\,h$. w is zero at zero z. If B7.12c is used in the form $\partial^2 w/\partial z^2=X\;\partial w/\partial z$, the following solution can be found by standard methods, or simply checked by substitution.

Up to $z=h$ $\quad w=\dfrac{C}{X}(e^{Xz}-1)$

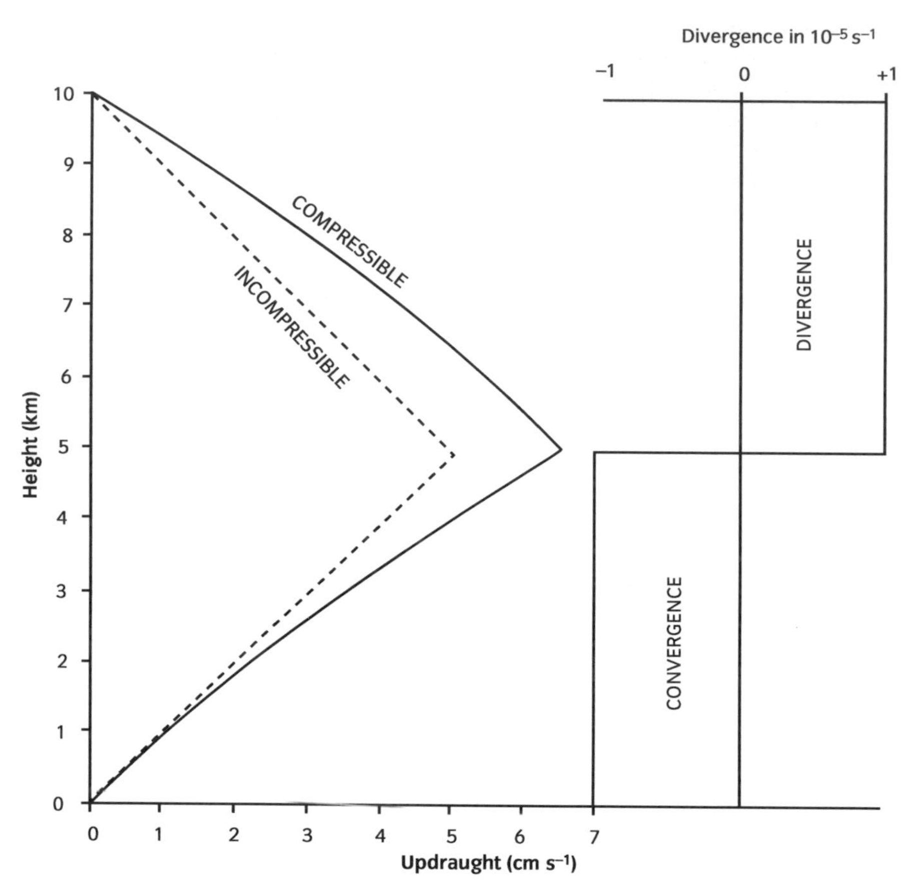

Figure 7.36 Vertical profiles of updraft associated with a simple two-layer model of typical synoptic-scale convergence (below) and divergence (aloft) in an incompressible atmosphere (dashed line) and a realistically compressible one (solid line).

at $\quad z = h \quad w = w_h = \dfrac{C}{X}(e^{Xh} - 1)$

up to $\quad z = 2h \quad w = w_h - \dfrac{C}{X}(e^{Xz'} - 1)$

where $\quad z' = z - h$

at $\quad z = 2h \quad w = 0$

A particular case of this solution is plotted in Fig. 7.36, where $C = 10^{-5}$ s^{-1} and $h = 5{,}000$ m, as found in extratropical cyclones, and X has been rounded to 10^{-4} m^{-1}.

above show that it is only the vertical compressibility and non-uniformity which have significant effects, at least on the synoptic scale. If we try to adapt Eqn 7.39 to the form of Eqn 7.42, we find (Box 7.12)

$$D + \frac{\partial w}{\partial z} = Xw \qquad 7.43$$

where X ($= -(1/\rho)\,\partial\rho/\partial z$) has a nearly constant value of about 10^{-4} m^{-1} in realistic conditions. The deviation this represents from incompressible behaviour is significant but not large, so long as no single layer of consistent convergence is as deep as the scaling thickness $1/X$, which is the case in the real atmosphere since $1/X$ is about 10 km (as above) and most layers of consistent divergence or convergence are no more than half this depth. For example, in the case of the 5 km layer with convergence (negative divergence) of about -10^{-5} s^{-1} quoted above, the balancing rate of uplift at the top of the converging layer is increased from 5 to 6.5 cm s^{-1} by the inclusion of the right-hand side in Eqn 7.43 (Box 7.12).

In advanced texts (*[35]*) it is shown that the complication of the extra term on the right-hand side of Eqn 7.43 can be avoided if the vertical coordinate z is replaced by pressure p. In fact all the complicating minor terms in Eqn 7.34 vanish and the continuity equation becomes

$$\frac{\partial u}{\partial x} + \frac{\partial v}{\partial y} + \frac{\partial \omega}{\partial z} = 0$$

where ω is dp/dt, the rate of change of pressure in the moving air parcel, and the horizontal stretching terms are actually evaluated on an isobaric surface. The x, y, p coordinate system is used very widely in theoretical and computational work because of such simplifications, but formal simplicity is achieved at the cost of conceptual subtlety.

7.15.7 Frictional convergence

We finish this section by applying the concept of virtually incompressible deformability to the case of frictional convergence in the atmospheric boundary layer of an extratropical cyclone. Because the wind vectors near the surface are

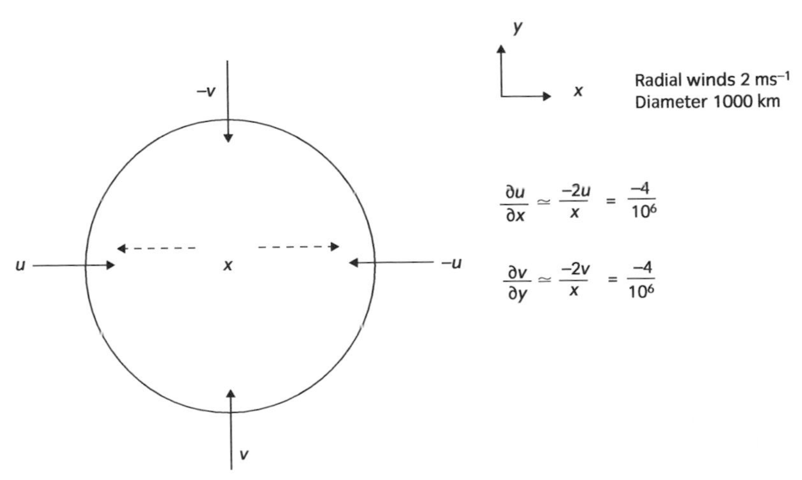

Figure 7.37 Frictional convergence in the base of an extratropical cyclone, with realistic light wind components perpendicular to circular isobars.

consistently angled a little inwards across the isobars by friction, there is an inward component of air flow which may typically average ~ 2 m s^{-1} in the first 500 m. According to Fig. 7.37 this produces a value of about -10^{-5} s^{-1} if it persists across a 1000 km-wide region such as the central zone of a mature extratropical cyclone. This must be balanced by the vertical stretching $\partial w/\partial z$, which is therefore 10^{-5} s^{-1}. Integration through the depth of the boundary layer shows that air at the top of the converging layer (i.e. at the 500 m level) is rising at 0.5 cm s^{-1}. This may seem very small, but becomes significant when you consider that it is lifting the overlying atmosphere over an area of 10^6 km^2. This lifting by frictional convergence influences a much greater depth of air than can be reached directly by surface friction, though the effects aloft have been transformed from the simple horizontal drag near the surface.

7.16 Rotational dynamics *[33]*

7.16.1 Angular momentum

The Earth's rotation affects atmospheric behaviour in several important ways, especially on the synoptic and larger scales, with the widespread quasi-geostrophic balance (Section 7.10), and the tendency to conserve angular momentum about the Earth's spin axis in low latitudes (Section 4.7). Like the related Coriolis effect, angular momentum needs careful consideration if the consequences of its conservation are not to seem like a conjuring trick. Fortunately we can learn much from the simplest possible case.

Unaccelerated motion

Consider a body P of fixed mass m moving with uniform velocity $\mathbf{V}$ while acted on by zero net force, and conserving its momentum $m\,\mathbf{V}$ according to Newton's second law of motion (Eqn 7.1). The crucial new step is to describe the body's motion relative to a point O by means of a radius vector $\mathbf{R}$ from O to P which swings about O and shortens and lengthens as P moves steadily along a straight line.

It is obvious from Fig. 7.38 that the product $(m\,r\,V)$ is conserved during this movement, since m, r, and V are the same everywhere along the path. This is defined to be the *angular momentum* of P about O, and we see that the constant length $r = R \sin\theta$ is the multiplier which converts its linear momentum $m\,V$ into its angular momentum $m\,r\,V$ about O. Several general aspects of angular momentum (AM) emerge from this special case.

(i) The AM of P about O has magnitude $(m\,R\,V \sin\theta)$, where V is the magnitude of the velocity vector $\mathbf{V}$ of P relative to O, and θ is the angle between $\mathbf{R}$ and $\mathbf{V}$.

(ii) From Fig. 7.38 we see that the AM of P at position 1 can be written as $m\,R_1\,V_t$, where $V_t\,(= V\sin\theta)$ is the instantaneous tangential speed of P as $\mathbf{R}$ swings around O. At position 2 (where $\mathbf{R}$ has its minimum length r) V_t is simply V, and the AM of P is $m\,r\,V$.

(iii) Comparing positions 1 and 2, we see that V_t increases from $V\sin\theta$ to V, while the radial distance from O decreases from R_1 to $r\,(= R_1 \sin\theta)$, so that the product $R\,V_t$ remains constant for P as it moves along its path. This inverse

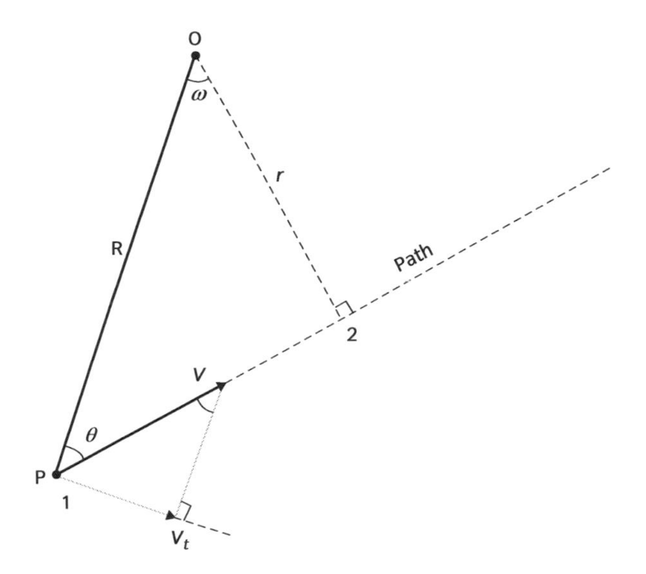

Figure 7.38 Angular momentum about O of body P (mass m) moving with steady speed V along the dashed straight path. The angular momentum (mass × $V \times r$) is obviously conserved along the path.

relation between V_t and R is a completely general signature of the conservation of angular momentum.

(iv) According to the definition of angular velocity (Box 4.4) we can regard V_t as the product of the instantaneous values of the radius R and angular velocity ω of P as seen from a non-rotating frame at O, so that $V_t = \omega R$ and $AM = m\,\omega\,R^2$.

Accelerated motion

In the special case of Fig. 7.38 angular momentum is conserved because linear momentum is conserved. Full treatments *[24]* show that when linear momentum is not conserved (i.e. when P is acted on by a net force $\boldsymbol{F}$), the vector equation of motion (Eqn 7.1) can be manipulated to show that the turning moment $\boldsymbol{TM}$ (*torque*) of $\boldsymbol{F}$ about O is equal to the rate of change with time of $\boldsymbol{AM}$ of P about O—Newton's second law of motion in angular form:

$$\boldsymbol{TM} = \frac{\mathrm{d}\,(\boldsymbol{AM})}{\mathrm{d}t} \qquad 7.44a$$

If $\boldsymbol{TM}$ is zero, then $\mathrm{d}(\boldsymbol{AM})/\mathrm{d}t = 0$ and $\boldsymbol{AM}$ is conserved, and this occurs not only when $\boldsymbol{F}$ is zero (as in Fig. 7.38), but also when non-zero $\boldsymbol{F}$ has zero torque about O, which is the case whenever the line of action of $\boldsymbol{F}$ passes through O (towards or away from O). Gravity is just such a *central* force when we choose to place O at the centre of the Sun in the case of planetary motion, or at the centre of the Earth in the case of air motion over its surface. However strong the force may be, if we choose an O anywhere along the line of its action on a body P, then the angular momentum of P about O is conserved, because the force has zero torque about O.

Circular and near-circular motion

The expressions for torque and angular momentum are simplest when a body is moving in a circular path radius R round O while acted by a tangential force F_t. The expression for AM is $(m\ R\ V)$ and the scalar equivalent of Eqn 7.44a is

$$R\,F_t = \frac{d}{dt}(m\,R\,V_t) \qquad 7.44b$$

When the left-hand side is zero (when the force is central—has no tangential component)

$$m\,R\,V = \text{constant} \qquad 7.45$$

so that when mass m is constant (as is usually the case)

$$R\,V = \omega\,R^2 = \text{constant} \qquad 7.46$$

highlighting the inverse relation between V and R which is the classic signature of AM conservation.

When motion is not perfectly circular, R will vary, and V_t (the tangential component of V) will vary inversely with R if angular momentum is conserved. In the case of air migrating poleward from low latitudes in the upper troposphere, relatively slow persistent poleward motion reduces its distance from Earth's axis of spin, and the large absolute V_t (relative to a non-rotating frame) becomes even larger according to Eqn 7.46.

It is tempting but very misleading to suppose that the acceleration (the increase of V_t) of migrating air with decreasing R is somehow 'caused by' the conservation of angular momentum; it is certainly constrained by it, but the associated acceleration of P (deceleration if R increases) is in fact related to forces doing work in the usual way (Box 9.1). As air creeps polewards in the high troposphere, work is done by the components of gravity and pressure gradient forces pulling and pushing the air in towards Earth's axis of spin, and it is quite easy to confirm that the work they do on the air supplies the energy needed to increase its V_t by exactly the amount required by conservation angular momentum (Box 7.13).

Many of these points are nicely illustrated by the popular example of spinning figure skaters. Firstly they generate angular momentum by carrying their initially linear momentum into a spiral trajectory which winds into a spin on a fixed vertical axis. Having done this with arms outstretched, the arms are dropped close to the body, obviously reducing the radial distance of the mass of their hands and arms from the axis of spin. If spinning with little friction on the skate points, angular momentum about the vertical axis through the body is nearly conserved during this process. In terms of Eqn 7.46 R is reduced and the resulting sharp increase of V_t, and even sharper increase of ω, produces the spectacular spinning blur. To slow the spin before attempting to set off over the ice again, the arms are stretched out to reduce V_t to manageable proportions by the reverse of the initial contraction. The energy needed to enhance the spin is supplied by the work done by the skater's muscles in drawing hands and arms inwards against the centrifugal tendency of their whirling mass. A very tired skater will not achieve as great an angular acceleration as a fresh skater, though each will conserve angular momentum during the manoeuvre.

BOX 7.13 Work done in conserving angular momentum

Figure 7.39a shows a body P being pushed centripetally by a central force from an initial circular orbit of radius R_1 about O to a final orbit of radius R_2. If its angular momentum is conserved (i.e. no other force is acting) its initial tangential velocity must increase from V_1 at radius R_1, through V at radius R to V_2 at radius R_2 in such a way that $V_1 R_1 = V R = V_2 R_2$. To keep P in a circular orbit of radius R there must be a central force F to maintain P's centripetal acceleration m V^2/R (Box 7.3) towards O. Whether F is gravity, tension in a radial string, pressure gradient forces, etc. it must have strength

$$F = m\frac{V^2}{R}$$

As P creeps in by a little distance dR (a negative change since R is shrinking) the force does work $-F\,\mathrm{d}R$ on P. F increases as R decreases and the total amount of work W done on P as it is pushed slowly (radially) from R_1 to R_2 is given by

$$W = -\int_{R_1}^{R_2} F\,\mathrm{d}R = -m\int_{R_1}^{R_2} V^2 R^{-1}\,\mathrm{d}R$$

$$= -m V_2^2 R_2^2 \int_{R_1}^{R_2} R^{-3}\,\mathrm{d}R$$

where the final form uses the angular momentum conservation rule $V R = V_2 R_2$. Integrating for all the little dRs covering the distance from R_1 to R_2 we find

$$W = -m V_2^2 R_2^2 \left[-\frac{R^{-2}}{2}\right]_{R_1}^{R_2}$$

$$= \frac{1}{2} m V_2^2 R_2^2 \left[R_2^{-2} - R_1^{-2}\right] = \frac{1}{2} m \left[V_2^2 - V_1^2\right]$$

where the final stage uses the angular momentum conservation rule again. Since the component of kinetic energy of the centripetal creep is assumed to be too small to matter, the final expression for W is exactly the kinetic energy needed to accelerate P from speed V_1 to V_2 (Eqn 9.2). The changes in kinetic energy associated with the conservation of angular momentum arise in the usual way from the work done by or against the central forces acting. The omission of such analysis (even by informal description) from simple treatments of the conservation of angular momentum can shroud its observed consequences with an unnecessary air of mystery.

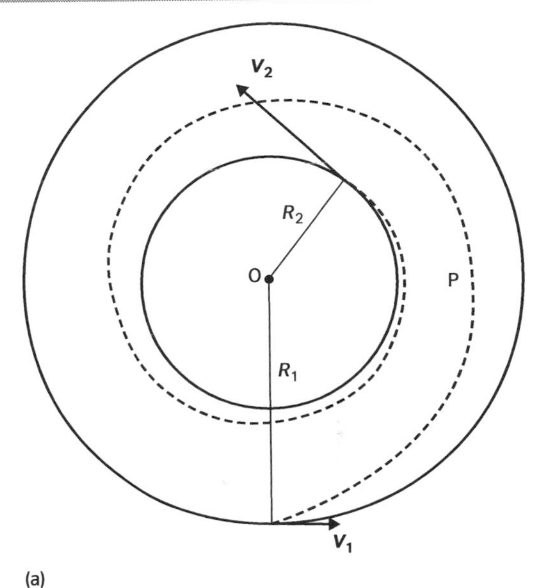

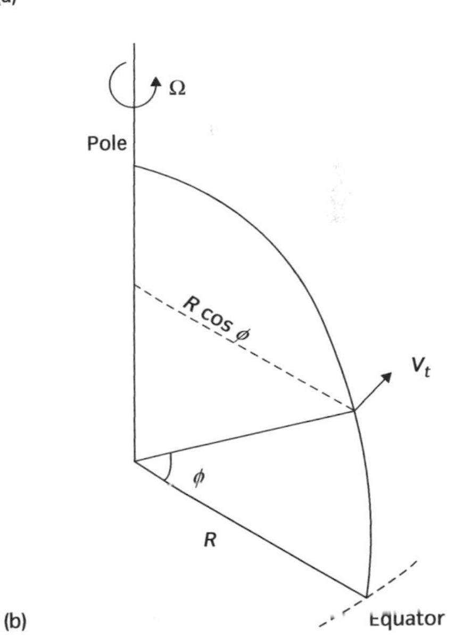

Figure 7.39 (a) Body P spiralling gently from position 1 to 2 under the influence of a central force (i.e. always acting towards O). As the distance from O shrinks, the central force does work on P, accelerating it from V_1 to V_2 but conserving angular momentum. Radial speed is assumed to be much lower than tangential speeds. (b) Radius of rotation of a point on the Earth's surface at latitude ϕ.

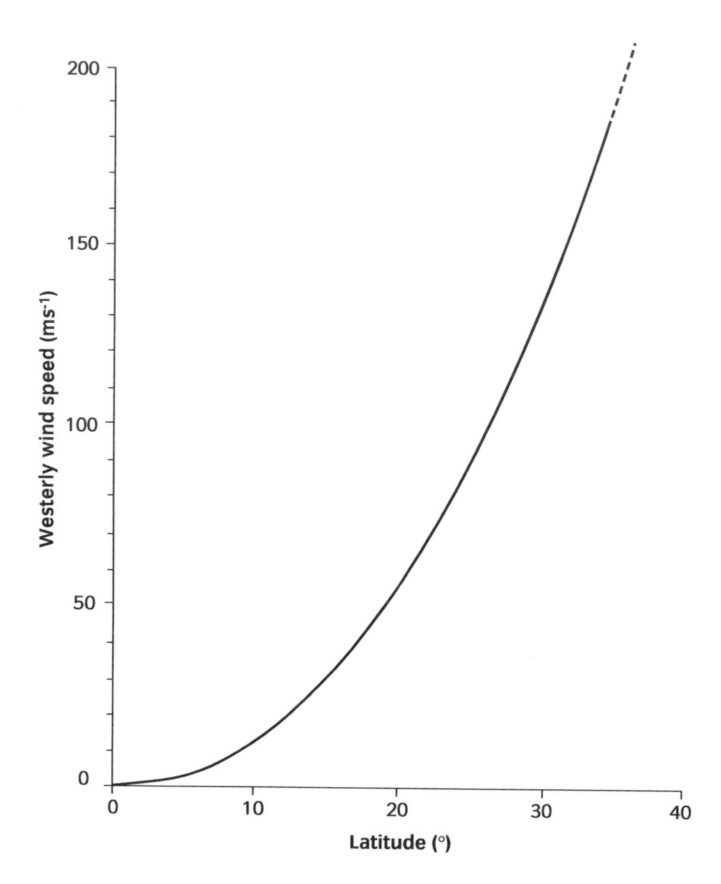

Figure 7.40 W'ly wind speed (relative to the Earth's surface) of air moving to various latitudes while conserving angular momentum from a state of zero wind speed at the equator.

7.16.2 Angular momentum budget

The tendency to conserve angular momentum in the Hadley circulation (Section 4.7), is considered further in the light of this introduction to angular momentum. Air rises to the high troposphere in the Inter-tropical Convergence Zone (ITCZ), usually fairly near the equator, and is then pushed gently polewards in the high-altitude branch of the circulation. At these heights there is little friction with the surface to produce a tangential torque on E'ly or W'ly flow. And since the migrating air circumnavigates the hemisphere several times in the course of its motion to the tropics, it can experience no overall working by E–W pressure gradient force—if it slides down a pressure hill in one part of its circumnavigation, it must slide up a balancing slope elsewhere. For these reasons there should be very little zonal torque acting on the ring of migrating air, so that the poleward pushing by pressure gradients will work as a central force and tend to accelerate the air (relative to a non-rotating frame) just enough to conserve the angular momentum it had when it first rose out of effective frictional contact with the surface in the ITCZ.

As justified in Box 4.4 and depicted in Fig. 7.39, if the air starts its poleward migration at the equator with zero zonal wind speed (relative to the Earth), then conservation of this initial angular momentum requires that its E'ward speed relative to Earth's rotating surface at some higher latitude ϕ is given by

$$U_{\phi} = \Omega R \left(\frac{1}{\cos \phi} - \cos \phi \right) \tag{7.47}$$

And as shown in Fig. 7.40, the implied increase of U_ϕ with latitude is marked, with values exceeding 55 m s^{-1} at latitude 20° and 130 m s^{-1} at latitude 30°. When averaged zonally, observed speeds seldom exceed 60% of these ideal values, but nevertheless the combination of poleward air motion in the high troposphere, and the consistent strong increase of W'ly wind speeds with latitude on the equatorial flank of the subtropical jet stream in each hemisphere, indicates that conservation of angular momentum is a major factor in maintaining these jet streams.

Exchanges of angular momentum

The lack of complete agreement between observed and predicted zonal speeds shows that angular momentum is not fully conserved. At least three mechanisms are at work.

(i) There is still some frictional connection with the surface: cumulonimbus locally connect the bottom and top of the troposphere, exchanging momentum and therefore communicating drag.

(ii) The argument about continuity of zonal pressure profile breaks down at levels which are pierced by mountains, the highest of which reach the upper troposphere. As sketched in Fig. 7.41, the tendency for lee pressures to be a little lower than windward pressures means that there is a net torque persistently opposing zonal air flow. In addition, air flowing over mountains can maintain patterns of gravity waves up to great heights (Section 11.8.2), which can extend some of this drag to heights far above the mountain tops.

(iii) On a hemispheric view (Fig. 7.42), the atmosphere is continually gaining W'ly angular momentum in low latitudes and losing it at high latitudes, in each case by momentum exchange with the underlying surface. Large masses of air take up the W'ly momentum of the local surface through boundary-layer friction and the mechanisms described above. They may then move away to another region as they become involved in a large weather system, where they meet and conform their flow to other masses from other source regions. In so doing they exchange momentum, even though they do not mix in the way that small parcels do. The general result of such exchanges is that relative W'ly momentum is transferred to high latitudes and relative E'ly momentum is transferred to low latitudes. The W'lies of mid-latitudes are maintained in this way, largely by the momentum exchanges associated with synoptic and larger-scale weather systems. The upper branch of the Hadley circulation is like a rim of equatorial angular momentum reaching polewards on nearly frictionless bearings near the equator, but being impeded more and more as it reaches towards the subtropics by momentum exchange with air reaching down from higher latitudes. In fact there is a transition from a smooth poleward flux of zonal angular momentum, in

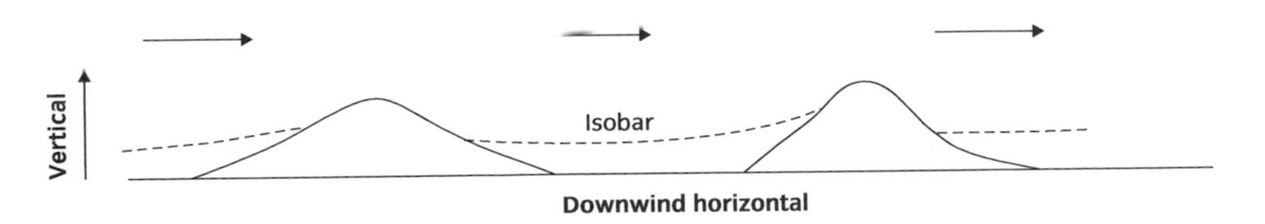

Figure 7.41 Large-scale drag arising from slight asymmetry of upwind–downwind pressure fields around hills.

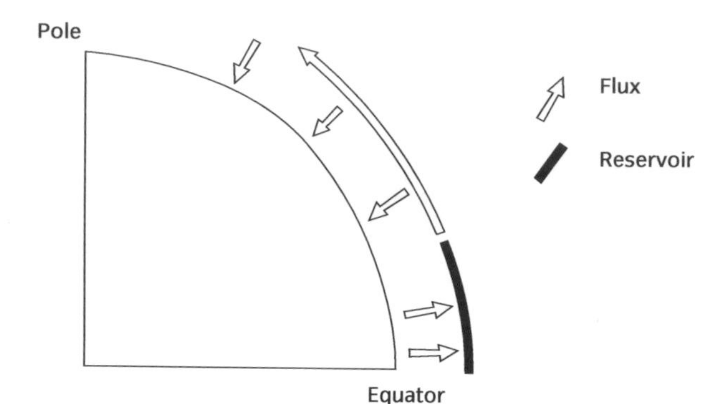

Figure 7.42 Vertical and meridional fluxes of W'ly zonal momentum.

which zonal equatorial angular momentum is nearly conserved, to an eddy flux in whose zonal averages fall far short of values corresponding to conservation of equatorial angular momentum. The core of the subtropical jet stream meanders in the transition zone.

Conservation of angular momentum lies at the root of the observed tendency of all poleward-moving air to increase its relative W'ly motion, or reduce its E'ward motion, and of all equatorward-moving air to do the opposite, and not just in the Hadley circulation. Examples are widespread in the atmosphere and are conspicuous in synoptic-scale weather systems: they give rise to the great cloud swirls of Fig. 1.1, in particular to the E'ward curvature of the swathes of frontal cloud moving polewards in middle latitudes from the polar flanks of the subtropical high-pressure systems (Section 12.4).

The tendency of air to veer to the right of its current direction of motion (in the N hemisphere) obviously resembles the Coriolis effect. In fact the two approaches are entirely equivalent, the conservation of zonal angular momentum in poleward-moving air being identical to the integrated effect of the E'ward Coriolis force associated with poleward motion. The Coriolis formulation is more useful for examining the shape of individual flows, while zonal angular momentum is more useful for looking at large-scale budgets and statistics.

7.16.3 Vorticity

The large, flattened air masses of synoptic-scale weather systems also tend to conserve angular momentum as they rotate about vertical axes located in their centres. The dynamics of this rotational motion is a very important part of the overall dynamics of large-scale motion systems, and is treated at length in advanced texts *[33]*. One of the most important results of such treatments emerges in the following simplified approach.

Consider a narrow horizontal circular ring of air rotating on a weather map as part of a synoptic-scale cyclonic or anticyclonic system (Fig. 7.43). If its angular velocity about a locally vertical axis through the centre of the ring is ω and its radius is r, then conservation of angular momentum about the axis requires (from Eqn 7.46) that

$$r^2\,\omega = \text{constant}$$

$$\text{or}\quad \frac{\mathrm{d}}{\mathrm{d}t}(r^2\omega) = 0$$

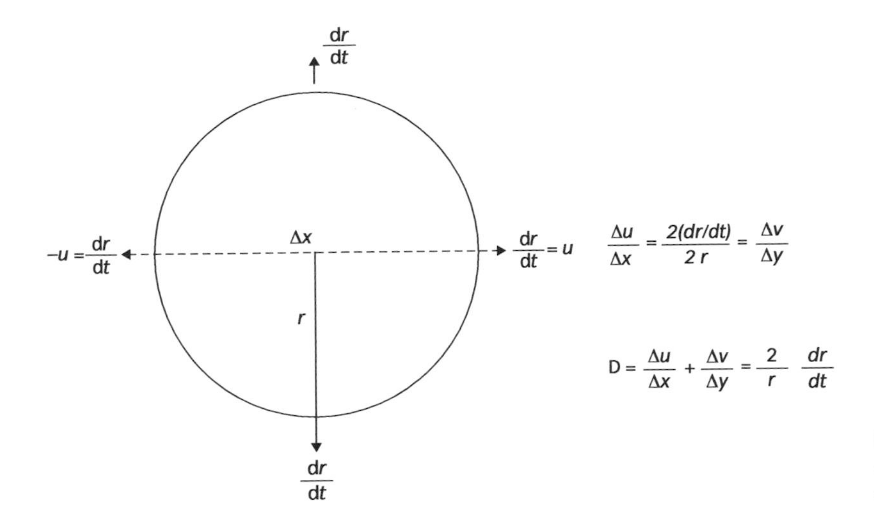

Figure 7.43 Expansion of a horizontal ring of air in symmetrically divergent flow.

We can expand the second form to find an expression for the rate of change of angular velocity of the ring ($d\omega/dt$) when it is expanding or contracting in response to synoptic scale divergence or convergence (Section 7.15).

$$r^2 \frac{d\omega}{dt} + \omega \frac{d}{dt}(r^2) = 0$$

Rearranging this to focus on the required rate of change of angular velocity, we find

$$\frac{d\omega}{dt} = -\frac{2\omega}{r}\frac{dr}{dt}$$

But it is apparent from Fig. 7.43 that the horizontal divergence D of the circular plate of air bounded by the thin ring is equal to $(2/r)\, dr/dt$ so that

$$\frac{d\omega}{dt} = -\omega D \qquad 7.48$$

So far we have considered the rotation of the air relative to the map, but since the map is actually rotating with angular velocity $\Omega \sin \phi$ about the local vertical, where Ω and ϕ are the Earth's angular velocity and latitude (Fig. 7.38), $\Omega \sin \phi$ should be added to ω in Eqn 7.48 to produce

$$\frac{d}{dt}(\omega + \Omega \sin\phi) = -(\omega + \Omega \sin \phi)\, D \qquad 7.49$$

Vorticity defined

It is conventional when dealing with rotation in fluids to describe localized rotation by means of *vorticity*, which is defined to be twice the instantaneous angular velocity of the fluid parcel. The usual sign convention means that cyclonic rotation is positive. The component of *relative vorticity* ($2\,\omega$) about the local vertical is always symbolized by ζ in meteorology. Since the doubled angular velocity of the map about the local vertical ($2\,\Omega \sin \phi$) is the familiar Coriolis parameter f, it follows that Eqn 7.49 can be rewritten

$$\frac{\mathrm{d}}{\mathrm{d}t}(\zeta + f) = -(\zeta + f)D \qquad 7.50$$

The Coriolis parameter is sometimes called the *planetary vorticity* on this account, and the total $(\zeta + f)$ is known as the *absolute vorticity* of the air.

A more thorough derivation of Eqn 7.50 must include torque terms. One arises from friction, which obviously exerts a torque tending to reduce vorticity relative to the underlying rough surface. Another very significantly arises from the baroclinity of the atmosphere, since an extended form of the buoyancy force always tends to align intersecting isobars and density isopleths and hence impose rotation. Other terms allow for the twisting of x and y components of vorticity into the local vertical, in particular the substantial component of vorticity about the local horizontal which is associated with any thermal wind there may be, since shear always implies vorticity about an axis perpendicular to the plane of the shear profile. However, the very simple Eqn 7.50 applies usefully to the development of synoptic-scale weather systems, and even more reliably to events on larger scales.

7.16.4 Applications

- Consider the development of cyclonic rotation in air initially at rest on the weather map. Viewed from the weather map the rotation seems to come from nowhere, but in fact the initially apparently static air has the planetary vorticity f of the map itself. In the presence of synoptic-scale horizontal convergence (negative D) in the lower troposphere, Eqn 7.50 shows that positive relative vorticity will begin to grow according to

$$\frac{\mathrm{d}\zeta}{\mathrm{d}t} = -f\,D \qquad 7.51$$

If D remains constant and none of the complicating terms (friction, etc.) becomes significant, the solution of Eqn 7.50 (Fig. 7.44) shows that the absolute vorticity $(\zeta + f)$ increases exponentially with a doubling time of $\ln 2/D$. Assuming a typical magnitude of $10^{-5}\ \mathrm{s}^{-1}$ for D, we find that absolute vorticity doubles (i.e. relative vorticity increases from zero to f) in nearly 20 hours, as is observed to be typical of developing extratropical cyclones.

- In the same way, Eqn 7.50 helps to explain how negative relative vorticity develops in the presence of synoptic-scale horizontal divergence. Starting from zero relative vorticity again (air turning with the underlying surface), divergence produces negative relative vorticity, which corresponds to anticyclonic rotation relative to the surface. If the D remains constant, and no other terms interfere, absolute vorticity decays exponentially towards zero, i.e. relative vorticity decays toward $-f$ as shown on Fig. 7.44. Such a limit to the magnitude of anticyclonic vorticity is amply confirmed by synoptic scale observation, and distinguishes anticyclones from cyclonic systems, which show no such restraint and often produce relative vorticities several times as strong as

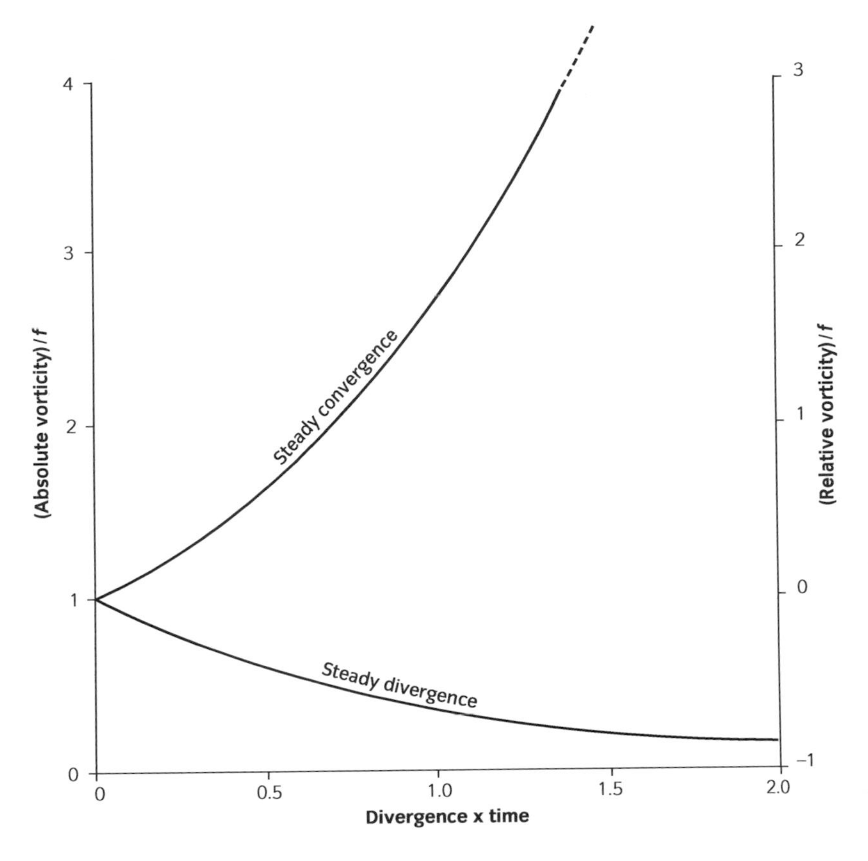

Figure 7.44 Exponential growth with time of absolute and relative vorticity in a steadily convergent part of a weather system, and exponential decay in a steadily divergent part. The initial relative vorticity is zero in each case, and axes have been normalized by f and $1/D$ for generality. In a realistic sysnoptic scale divergence of $10^{-5}\ s^{-1}$, the horizontal axis is 1.0 at 10^5 s (nearly 28 hours).

the planetary vorticity f. Indeed to the extent that all synoptic-scale rotation about the local vertical consists of concentrations or dilutions of the Earth's rotation, absolute rotation in the opposite direction is impossible, which is consistent with a detailed examination of gradient balance in anticyclonic flow (Box 7.9).

The association of divergence with anticyclonic vorticity and convergence with cyclonic vorticity is often quite obvious in a single weather system. In terms of the vertical profile of an extratropical cyclone sketched in Fig. 7.33, we should expect to find cyclonic vorticity in the lower troposphere as already discussed, and anticyclonic vorticity in the upper troposphere. The latter is usually much less obvious, but is evident nevertheless, as described in Section 12.2.

- In the previous section it was shown that synoptic-scale uplift and subsidence are so sensitive to convergence and divergence that routine observations of wind are barely able to detect convergent conditions capable of maintaining widespread, vigorous nimbostratus. We have now seen that synoptic-scale rotation about the local vertical is similarly sensitive. So although it is surprisingly difficult to measure the associated horizontal convergences and divergences of mass, even a casual glance at a satellite picture such as Fig. 1.1 shows striking evidence of both uplift and rotation. In fact the sensitivity of rotation to uplift and subsidence can be used to good effect in theory and computation.

- By contrast with its inability to cope with convergence or divergence, the geostrophic approximation is well able to represent synoptic-scale rotation. Advanced texts *[35]* show that the relative vorticity about the local vertical is given in terms of the usual horizontal wind components by

$$\zeta = \frac{\partial v}{\partial x} - \frac{\partial u}{\partial y}$$

Substituting the geostrophically associated pressure gradients for u and v according to Eqn 7.20, and ignoring gradients of ρ and ϕ in comparison with gradients of wind speed, we find

$$\zeta_g = \frac{1}{\rho f}\left[\frac{\partial^2 p}{\partial x^2} + \frac{\partial^2 p}{\partial y^2}\right]$$

The terms in the bracket [ ] do not cancel as they did in the case of geostrophic divergence. Each of the pair corresponds to the curvature of the horizontal pressure profile across a vertical section through the depression (the curvature of the equivalent isobaric surface), and a little thought will show that each is positive if the profile is concave upwards (Fig. 7.45). In other words a pressure minimum is associated geostrophically with positive (cyclonic) vorticity or rotation, and a pressure maximum with anticyclonic vorticity or rotation. In fact a synoptic-scale map analysed with the help of the geostrophic approximation provides a quite usefully accurate representation of large-scale vorticity, despite considerable ageostrophy.

If we describe the pressure field in terms of the slopes of isobaric surfaces (Eqn 7.23), we find an equivalent expression for ζ_g which does not require that we assume uniform air density.

$$\zeta_g = \frac{g}{f}\left[\frac{\partial^2 Z_p}{\partial x^2} + \frac{\partial^2 Z_p}{\partial y^2}\right]$$

- Finally note that the relationship between convergence and rotation also applies to very intense circulations on the small scale, the most extreme example being the tornado. Although detailed understanding of the formation of tornadoes is still very incomplete, it is clear that their intense rotation is

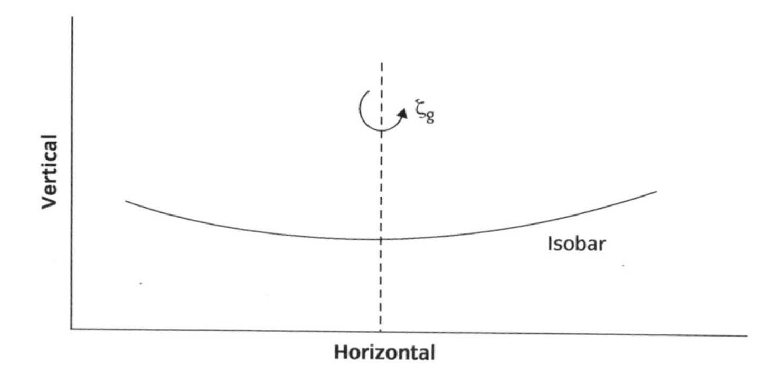

Figure 7.45 Vertical radial isobaric profile across a depression with cyclonic geostrophic vorticity (N hemisphere).

associated with intense convergence, and is described at least qualitatively by Eqns 7.49 and 7.51. As discussed in Section 11.6, tornadoes form beneath the bases of very vigorous cumulonimbus, where intense convergence feeds the powerful updrafts aloft. This convergence concentrates and intensifies background vorticity to produce the powerful localized vortex which is the tornado. However, in such tumultuous conditions it is not clear which of the many complicating terms missing from Eqn 7.49 may also be significant. Most tornadoes are observed to rotate cyclonically, but since conditions in a tornado (Section 7.14) imply vorticities $\sim 10^4 f$, it would be rash to conclude that they are merely extreme concentrations of planetary vorticity. As in the hoary legend of the wash-basin vortex changing direction in ships crossing the equator, there are many local factors which may be much more significant, though these should probably not include the tendency to produce cyclonic vorticity by road traffic driving on the right, as suggested once by some American meteorologists in playful mood.

Checklist of key ideas

You should now be familiar with the following ideas.

1. Equation of motion of an air parcel, with directions, magnitudes (and consistency of units and dimensions), and components in the meteorological reference frame.

2. Concepts of solving the equation of motion analytically and numerically, numerical forecasting in outline.

3. Formal expressions for pressure gradient and gravitational forces on an air parcel, with magnitudes.

4. Frictional forces on air parcels in laminar and turbulent sheared flow, molecular and eddy viscosity, all with magnitudes, frictional stress as momentum transfer.

5. Non-inertial reference and inertial reference frames, centripetal and Coriolis accelerations of a body observed from a rotating reference frame, Coriolis acceleration and Coriolis force as equivalent descriptions.

6. Fluid adjustment of the solid Earth to the centrifugal effect of spin, real and apparent g, mean sea level as the horizontal datum.

7. Coriolis effect in three dimensions by perpendicular turntables.

8. Scale analysis of full equation of motion for air parcels in synoptic scale flow, laminar and turbulent flow and Reynolds number, predominance of turbulent flow.

9. Dimensional analysis and its applications.

10. Geostrophic horizontal flow, geostrophic wind, Coriolis parameter, reality of geostrophic flow, Rossby number.

11. Thermal wind relation in concept and application.

12. Systematic geostrophic departures in linear flow, gradient flow in cyclonic and anticyclonic flow, antitriptic flow in the atmospheric boundary layer, flow in low latitudes.

13. Scale analysis of horizontal and vertical meteorological small-scale flow, cyclostrophic flow, buoyancy and Froude number, near balance between buoyancy and drag in convection, entrainment drag.

14. Continuity of mass flow, near incompressibility of large- and small-scale air flow, horizontal convergence and vertical stretching, vertical compensation in synoptic and small-scale cloudy systems.

15. The concept of conservation of angular momentum under action of central forces, application to air flow in subtropical jet stream, work needed to conserve angular momentum.

16. Vorticity related to local angular momentum, relative and absolute vorticity about local vertical, variation of vorticity with synoptic scale horizontal convergence and divergence, application to onset and growth of cyclonic and anticyclonic rotation, application to tornadoes.

Problems

Outline answers to these problems can be found on the Online Resource Centre. Answers to odd numbered problems can be found under Student Resources, answers to even numbered problems under Lecturer Resources.

Level 1

7.1 If pressure increases both N'wards and E'wards at 5 hPa per 100 km, show that the maximum rate of increase is about 7.1 hPa per 100 km towards the NE. What contribution does this make to the pressure gradient in the local vertical?

7.2 Sketch the sheared planetary boundary layer, including the directions of the shearing stresses on the top and bottom surfaces of an embedded horizontal layer, and add the direction of the associated momentum flux.

7.3 A person is standing on spring weighing scales lying on a rotating turntable. In what way is the reading of the scales affected by the rotation if (a) the scales are horizontal, and (b) they are tilted somewhat towards the axis of rotation?

7.4 What would be the effect on the accuracy of the geostrophic approximation near the Earth's surface if all turbulence there were to disappear?

7.5 Assuming zero geostrophic wind speed near the surface, use the thermal wind relation to find the wind direction in the high troposphere in the following situations in the N hemisphere: colder air to the NW; warmer air to the W; colder air to the N; warmer air to the SW.

7.6 Consider the usefulness of installing a convector heater in the interior of a spacecraft or space station.

7.7 What is the fundamental significance of the cyclonic direction of rotation?

7.8 What directions of rotation should be enhanced in the low troposphere of a cyclone, the low troposphere of an anticyclone and in the high troposphere of an anticyclone?

Level 2

7.9. A toy train is running along a straight track resting on a rotating turntable and passing through its axis of rotation (Box 7.4). Assuming that a lateral force of 1 N per unit train mass is sufficient to tip the train sideways off the track, find the maximum safe speed of the train along the track when the turntable is completing a revolution every 3 s. And what is this speed if the turntable turns twice as quickly?

7.10 In the low troposphere, where air density is 1 kg m^{-3}, what are the geostrophic wind speeds associated with a horizontal pressure gradient of 2 hPa per 100 km at latitudes 45° and 5°?

7.11 Repeat the calculations of Problem 7.10 in the high troposphere where air density is only 0.25 kg m^{-3}, and find the equivalent slopes of isobaric surfaces at 45° latitude for the low troposphere and high troposphere.

7.12 Apply the thermal wind relation to the case of a well-defined front with an isobaric (effectively horizontal) temperature gradient of 5 °C per 100 km throughout a 1 km deep layer with mean temperature 250 K. At latitudes where f is 10^{-4} s^{-1}, and taking g to be 10 m s^{-2}, find the increase in geostrophic wind speed along the front from bottom to top of the layer.

7.13 Consider the vertical motion of a parcel of constant temperature 300 K in air with temperature 299 K. Using Eqn B7.10b, find the buoyant force per unit mass of parcel. And using the standard relation ($w^2 = 2\ a\ z$) between speed w and distance z at uniform acceleration a, calculate the upward speed of the parcel after ascending 1 km from rest. Why is this answer unrealistically large?

7.14 A cumulus cloud is observed to be increasing its volume by 1% per second while rising at 1 m s^{-1}. Find its entrainment drag per unit buoyant mass and compare it with the typical buoyant force per unit mass estimated in Problem 7.13 (1/30 m s^{-2}).

7.15 A typical value for horizontal convergence (negative divergence, Section 7.15) in the lower troposphere of an extratropical cyclone is 10^{-5} s^{-1}. Show from Eqn 7.40 that the percentage decrease in the horizontal area A of a horizontal slab of embedded air is 3.6% per hour. If D remains constant from time t_1 to t_2, this equation can be integrated to show that A is shrinking exponentially according to $A_2 = A_1\ e^{-c(t2-t1)}$. Use this to calculate A_2/A_1 after 1, 2, and 3 days of consecutive decay.

7.16 In the case of Problem 7.15, find the associated rate of uplift at heights 1, 3, and 5 km above the surface, assuming that the divergence is uniform throughout this height range and that the atmosphere behaves like a simple incompressible fluid (i.e. use Eqn 7.42).

7.17 In the case of Problem 7.15, use the integrated form of Eqn 7.50 $A_2 = A_1 e^{c(t2-t1)}$ (where A_1 and A_2 now represent final and initial absolute vorticities—Section 7.16) to find A_2/A_1 after 1, 2, and 3 days of exponential growth. Assume zero relative vorticity initially (i.e. $A_1 = f$) to find relative vorticity after 1, 3, and 3 days.

Level 3

7.18 Since it makes no difference to the Coriolis force, why was it helpful in Problem 7.9 to specify that the track passed through the rotation axis? Ignoring the Coriolis force (or supposing that the train ran on super wide-gauge rails), there would be an effect on the train at fast enough rotation. Describe it and consider its sensitivity to train speed along the track. So this must happen in the atmosphere too. Why does it not matter?

7.19 Carry out a scale analysis (Section 7.9) of the horizontal equation of motion for air moving into a powerful mid-latitude jet stream, in which individual air parcels accelerate uniformly into the jet core from 40 to 80 m s^{-1} in 1,000 km. Look in particular at the implications for geostrophic balance, and discuss in relation to the beginning of Section 7.13.

7.20 Explain fully why the geostrophic relation fails in sufficiently low latitudes, using the scale analysis to suggest how low is sufficiently low. Consider whether or not this should have an effect on the occurrence of tornadoes in low latitudes.

7.21 If air has a tangential speed of 30 m s^{-1} at a distance of 300 m from the centre of a tornado, what tangential speed should it attain if it spirals in to a radius of 30 m while conserving its angular momentum? Assuming cyclostrophic balance, what is the implied horizontal pressure gradient in these two locations in the low troposphere, and express these as multiples of a typical strong synoptic scale value.

7.22 Sketch the directions of vertical fluxes of E'ly momentum in those parts of the trade winds where the E'ly flow is surmounted by W'lies in the upper troposphere, and outline the types of mechanism carrying these fluxes through the base and top of the trades layer.

7.23 There is a widespread popular belief that the direction of water flowing out of hand basins should tend to be anticlockwise in the N hemisphere and clockwise in the S hemisphere. Examine the likely validity of this belief by identifying the principle which is being assumed and assessing the relative importance of other factors involved.

Discussion topics

7A Discuss the roles of drag in all types of synoptic and small-scale meteorological dynamics, including cyclonic weather systems and the cumulus family.

7B Summarize and discuss all points at which atmospheric dynamics is influenced by the flatness of the atmosphere (the gross asymmetry between vertical and horizontal scales). Speculate on how things might differ if the atmosphere were much deeper.

7C Outline the influence of the Earth's rotation on atmospheric motion, discussing reasons for the difference between large- and small-scale motion systems in this respect. Speculate briefly on how things might differ if the Earth rotated (a) much faster than at present (as it did in the distant past before being slowed by marine tidal friction), and (b) much more slowly than at present (as it will do in the distant future for the same reason).

8 Radiation, convection, and advection

8.1 The Sun and solar radiation

8.1.1 Solar radiation

The Sun is a powerful emitter of electromagnetic radiation in wavelengths ranging from the ultraviolet (*UV*) to the near infrared (*IR*), as shown by the spectrum of its power output (Fig. 8.1). Apart from the relatively small *UV* component, human vision has evolved to use the shorter wavelength half of this range, leaving longer wavelengths beyond the red end of the visible spectrum, i.e. in the *IR*. Data for Fig. 8.1 come from radiometry from the Earth's surface, from special balloons tens of kilometres above sea level, and (since the 1960s) from artificial satellites just outside the atmosphere.

Almost all solar radiation (*SR*) comes from the *photosphere*, a relatively shallow shell of incandescent gases forming the visible surface of the Sun. The photosphere has an outer radius of nearly 700,000 km (over 100 Earth radii) and consists mainly of hydrogen and helium at temperatures of nearly 6,000 K, and at substantial pressures maintained by the weight of overlying, less visible gases. In typical laboratory conditions, spectroscopes show that incandescent gases emit *line spectra* (bright peaks of emission in very narrow wavelength ranges); but at the higher temperatures and very much higher pressures of the photosphere, radiating atoms collide with one another so frequently and violently that the individual emission lines broaden and merge into the relatively smooth hump shown in Fig. 8.1.

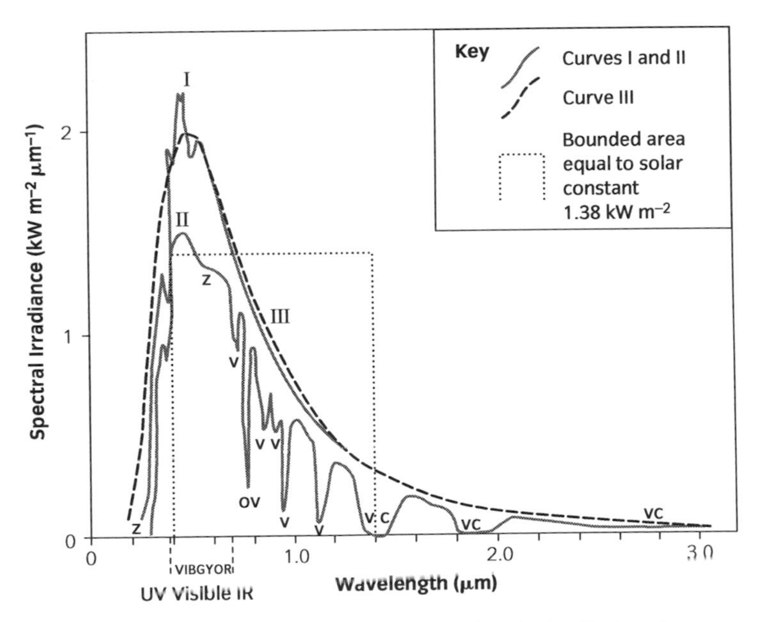

Figure 8.1 Power spectrum of solar irradiance of the Earth—the distribution of power (vertical axis) across the range of solar wavelengths (horizontal axis). Curve I is the irradiance just outside the Earth's atmosphere—the total area under the curve being the solar constant. Curve II is typical of the irradiance at sea level with the Sun in the zenith of a clear sky. Notches marked o, c, v, and z indicate atmospheric absorption by oxygen, carbon dioxide, water vapour, and ozone respectively. Curve III is the power specturm for a blackbody at 5,900 K reduced to allow for inverse-square attenuation betwen the Sun and Earth.

Observation and theory agree that the solar photosphere radiates *SR* with nearly the maximum efficiency possible for any material body at that temperature. Bodies which are such optimum emitters of electromagnetic radiation are also total absorbers of it, and are termed *blackbodies* since they reflect no incident electromagnetic radiation. Blackbodies and the *blackbody radiation* they emit are studied as an important part of thermodynamics, and some of their properties are outlined in Box 8.1, including the emission spectrum for a blackbody at temperature 5,800 K. After allowing for the geometrical divergence of light reaching Earth from the Sun (Eqn 8.1), agreement with the solar spectrum observed just outside Earth's atmosphere is good enough for us to regard the Sun as a blackbody emitting electromagnetic radiation at the temperature of the photosphere. Agreement is imperfect in detail, especially at the peak of the solar spectrum, and on its *UV* flank.

Visible *SR* is centred on wavelengths of about 0.5 μm, not far from the wavelength of maximum solar emission (per unit wavelength range)—the peak of the hump in Fig. 8.1. On their own, these dominant wavelengths would appear green to our eyes, but the presence of other visible *SR* wavelengths, the variation of eye response across the range of wavelength, and the extreme brightness of the whole, produces the familiar 'seething gold' appearance of the Sun when glimpsed well above the horizon. When viewed by spectroscope, dark *Fraunhofer lines* are seen 'bar-coding' the spectrum; these are produced by selective absorption of certain wavelengths by cooler solar gases overlying the photosphere. However, the dips in brightness corresponding to these lines are much too narrow to show on the wavelength scale of Fig. 8.1, whose much broader indentations are produced by selective absorption of *SR* by the Earth's atmosphere rather than by the Sun's.

The Earth's cloudless atmosphere modifies the solar spectrum quite significantly by selective absorption and scattering, and the effects of a standard cloudless atmosphere have been included in Fig. 8.1. When near the horizon, atmospheric absorption, and preferential scattering of blue light from the solar beam, dulls and reddens the solar disk. Clouds are much less wavelength-selective than air, but when thick enough can reduce the total power of *SR* reaching down to land and sea surfaces by as much as 90%, mostly by backscattering from cloud droplets rather than by absorption.

BOX 8.1 Blackbody radiation

Theory

The emission of electromagnetic radiation from the molecules of a solid or liquid body as they jostle in thermal agitation is controlled by a dynamic equilibrium between the emitted flux of *photons* (ballistic packets of radiant energy) and the radiating body. Thermodynamic analysis of this equilibrium *[22]* justifies

$$E_\lambda = a_\lambda f(\lambda, T) \qquad \text{B8.1a}$$

where

(i) E_λ is the *spectral radiant emittance* of the surface at wavelength λ (its emittance per unit wavelength range centred on λ).

(ii) a_λ is the *spectral absorptivity* of the surface (the fraction of incident radiation of wavelength λ absorbed by the surface). The fact that a_λ in B8.1a describes absorption as well as emission, is known as *Kirchhof's* law. When dealing with emission, a_λ is often replaced by the *spectral emissivity* ε_λ, since the two are identical according to Kirchhof's law, which holds at the Earth's surface and throughout the troposphere and stratosphere. At much greater altitudes, molecular collisions in the rarefied air no longer greatly outnumber emissions and absorptions of photons, so that absorptivity and emissivity are no longer identical, and Kirchhof's law fails.

(iii) $f(\lambda,T)$ is the *Planck function* which varies only with wavelength λ and absolute surface temperature T, being independent of the nature of the emitting material—as must be the case for blackbody emittance to have a single value at any surface temperature.

Kirchhoff's law states that the efficiency of absorption of certain wavelengths by a surface is exactly matched by its efficiency of emission of the same wavelengths, where maximum emission is defined by $f(\lambda,T)$. The most efficient emitter of electromagnetic radiation at a given temperature ($\varepsilon_\lambda = 1$) is a perfect absorber of all emitted wavelengths ($a_\lambda = 1$) and therefore a reflector of none—hence the term *blackbody*. By definition a blackbody has $a_\lambda = \varepsilon_\lambda = 1$ and $E_\lambda = f(\lambda,T)$. Some Planck functions $f(\lambda,T)$ are sketched in Fig. 8.2 for three absolute temperatures T_3, T_2, and T_1 in the ratio 11:9:7.

It is observed, and confirmed theoretically by integrating the Planck function across all λ, that the total emittance of a blackbody across all wavelengths varies with absolute temperature T according to *Stefan's law*

$$E = \int f(\lambda,T)\, d\lambda = \sigma T^4 \qquad \text{B8.1b}$$

where the universal constant σ is the *Stefan–Boltzmann constant*, value 5.67×10^{-8} W m^{-2} K^{-4}. Stefan's law gives the emittance from unit area of blackbody surface at temperature T, and corresponds to the total area under the Planck curve for that temperature (Fig. 8.2). Blackbody emittance increases very sharply with absolute temperature—multiplying by sixteen as absolute T doubles.

Observation, and further manipulation of the Planck function, agree in showing that the wavelength λ_m of maximum spectral emittance of a blackbody (the peak of the appropriate Planck curve) decreases with increasing absolute temperature according to *Wiens' displacement law*

$$T \lambda_m = Y \qquad \text{B8.1c}$$

where Y is a universal constant with value 2,897 μm K when λ is expressed in micrometres (microns – μm), as is usual in meteorology. The reddening radiant emission of a metal bar as it cools from white heat is a classic example of Wien's law in action.

Observation

Real materials approach the blackbody ideal if their molecules are sufficiently energetic and close together to interfere continually with each other's atomic and molecular energy levels, smoothing their individual line spectra into a Planck function curve. In terrestrial conditions, solids and liquids are nearly all effectively 'black' in the near and far *IR*, whereas gases are not. Water vapour and carbon dioxide have very complex lines and bands of lines in their far *IR* emission and absorption spectra, which arise from quantized molecular rotation and vibration. By contrast they are almost completely ineffective as emitters and absorbers in the visible and near *IR*. In visible wavelengths, a_λ is very small and therefore so is E_λ ($= a_\lambda f(\lambda,T)$). By contrast, in the thick cluster of lines in the far *IR*, there is enough vapour or carbon dioxide in most of the lower atmosphere to merge the lines, so that a_λ is nearly 1 and E_λ approaches the blackbody ideal $f(\lambda,T)$. Kirchhoff's law applies throughout.

By definition, a *greybody* is one whose spectral absorptivity and emissivity are uniform (and equal) across all wavelengths, and less than 1. The wavelength of maximum spectral emittance of a greybody is unaltered by comparison with a blackbody at the same temperature, but its emittance is reduced according to

$$E = a\, \sigma T^4 = \varepsilon\, \sigma T^4$$

Real solids and liquids are often nearly greybodies across substantial wavelength ranges. For example, water and ice are very pale grey (with small absorptivity a) in the visible range, but are very dark grey (absorptivity only slightly less than 1) in the far *IR*.

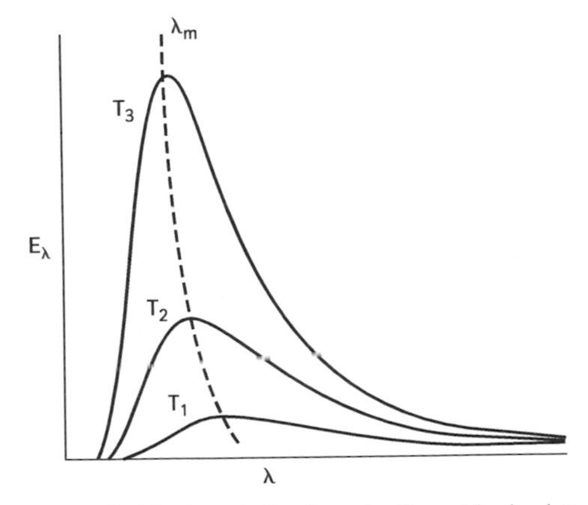

Figure 8.2 Blackbody emission from bodies with absolute temperatures in the ratios 11:9:7. The areas under the Planck curves follow Stefan's law, and the wavelengths of peak emission (dashed line) follow Wien's law.

NUMERICAL NOTE 8.1 Solar radiation

By Eqn B8.1a, photosphere emittance $E_s = \sigma T^4 = 5.67 \times 10^{-8} \times (5{,}800)^4 \approx$ 64.2 MW m^{-2}. Treating the photosphere as a sphere of radius $R_s = 695{,}000$ km, its total luminosity is $4\,\pi\,R^2\,E_s \approx 4\,\pi\,(695 \times 10^6)^2 \times 64.2 \times 10^6 = 3.90 \times 10^{26}$ W.

With Earth at 216 solar radii from the centre of the Sun, R/R_s in Eqn 8.1 is 216 and the solar irradiance there is $E_s/(216)^2 = 64.2 \times 10^6/(216)^2 \approx 1.38$ kW m^{-2}.

From Wien's displacement law (Eqn B8.1c) the wavelength of the peak output from the Sun is given by $\lambda_m = 2{,}897/5{,}800 \approx 0.5$ µm, with λ_m in micrometres because Y on the right-hand side of B8.1c is in µm K.

The mass flux corresponding to the solar luminosity E is found from $m = E/c^2 = 3.9 \times 10^{26}/(3 \times 10^8)^2 = 4.3 \times 10^9$ kg s^{-1}. During the Sun's life (4,500 Myr or 1.4×10^{17} s), the total mass radiated at the current rate is 6.1×10^{26} kg—about 0.03% of the Sun's current mass of about 2×10^{30} kg.

8.1.2 Irradiance and solar constant

We can use *Stefan's law* for blackbody emission (Box 8.1) to estimate the *emittance* E_s of the solar photosphere (the power radiated in all outward directions from each square metre of the photosphere) to be about 64 MW m^{-2} (NN 8.1). Multiplying by the surface area of the whole photosphere ($4\,\pi\,R_s^2$), we find the solar *luminosity* (its total radiant power output) to be about 3.9×10^{26} MW. Though vast by terrestrial standards, this output has radiated away less than 0.03% of the present solar mass throughout its 4,500 Myr lifetime (NN 8.1). The output is maintained by thermonuclear reactions in the solar core, and has probably changed little during much of the life of the Earth, though astronomical studies of the life-cycles of similar stars show that in its youth, the luminosity of the 'young dark Sun' was about 2/3 of its current mature value.

Solar radiation (*SR*) streams outwards through the solar system, its *irradiance* I_R (the rate of flow of radiant energy through unit area facing the solar beam) falling with increasing distance from the Sun. Since only trivial amounts of *SR* are absorbed by planets and interplanetary dust and gas, the total solar flux through a sphere of any radius R ($> R_s$) concentric with the Sun has the same magnitude as the Sun's luminosity L.

$$L = 4\,\pi\,R_s^2\,E_s = 4\,\pi\,R^2\,I_R \quad \text{so that} \quad I_R = E_s\left(\frac{R_s}{R}\right)^2 \qquad 8.1$$

This is the famous 'inverse square' relationship which describes the rapid decay of I_R with increasing R (I_R quartering as R doubles, etc.). Since Earth orbits the Sun at a distance of about 216 solar radii R_s, it follows that solar irradiance I_R at Earth's orbital distance is only $(1/216)^2$ of the photospheric emittance E_s, i.e. 1.38 kW m^{-2} (NN 8.1).

The Earth's orbit around the Sun is actually slightly elliptical (Box 14.2), their separation increasing by about 3.5% in the 6 months from *perihelion* (nearest the Sun—currently about January 4) to *aphelion* (furthest from the Sun). The inverse square dependence on R produces a corresponding 7% decrease in the irradiation of the Earth, with the maximum irradiance occurring shortly after N hemisphere midwinter in the present state of Earth's orbit and spin. Though the effects of this variation on seasonal climatic rhythms are swamped by the much larger effects of the 23.5° tilt of the Earth's equatorial plane relative to the plane of its orbit round the Sun (see Section 8.8), ellipticities have been much larger in various past times, with seasonal rhythms in irradiance apparently helping trigger glacials and interglacials during the present ice age (Section 14.7).

The annual average solar irradiance just outside Earth's atmosphere is called the *solar constant* (often represented by S), whose numerical value determines the state and behaviour of the atmosphere, by constraining the solar power supply to the atmospheric engine, and the surface temperatures which are crucial to the evolution and maintenance of life (Section 3.9). The solar constant has been measured repeatedly with increasing accuracy throughout the twentieth century, but only those measurements made from artificial satellites have been completely free from atmospheric interference. Current values for the solar constant are about 1.38 kW m^{-2}, with uncertainties of about 1%. Small variations seem to be linked with sunspot numbers (Section 14.6), though data are limited by the relative newness of satellite observations.

Note that the narrative sequence of this section reverses actual measurement procedure. Photospheric temperature is actually deduced from measurements of the solar constant, using Stefan's law for blackbody emission and assuming zero interplanetary absorption. However, such estimates agree very well with estimates and checks made by several other means, for example by using Wiens' law (Box 8.1) on measurements of the wavelength of maximum solar emission on Fig. 8.1, and by theoretical models of the interior and surface of the Sun. As usual in the geosciences in particular, consistency of results from different approaches is the strongest evidence of model validity.

8.2 Earth's energy balance

As described in Section 14.5, the Earth's atmosphere, land, and ocean systems undergo irregular vacillations between warm epochs and glacial events on time scales ranging from tens of thousands to hundreds of millions of years. Despite their dramatic effects on surface conditions, especially in middle and high latitudes, these changes are associated with quite small variations in the global average surface temperatures established by Earth's radiative equilibrium, and even smaller imbalances between the radiant energy fluxes to and from the planet Earth. On much shorter time scales, the big hemispheric seasonal variations are reproduced year after year with relatively small variations in global annual averages. Despite the observed variability in its weather and climate, Earth's overall energy budget is in balance to a very good approximation.

Solar radiation is the only substantial energy input to Earth's surface layers. The next largest is the *geothermal flux* from Earth's hot interior (heated by early gravitational collapse and sustained by subsequent radioactive decay), but its global average value of 0.05 W m^{-2} is nearly four orders of magnitude smaller than its *SR* equivalent. Starlight, and energy flows between Earth, Moon, and Sun through ocean tides and gravity, contribute even less. Since the energy budget of Earth's surface layers is closely balanced, there must be a continual energy output balancing the solar input, and like *SR* it must be in the form of electromagnetic radiation to be able to cross the near vacuum of interplanetary space. Calling this output *terrestrial radiation* (*TR*), Earth's energy balance is maintained between absorption of incoming *SR* and emission of *TR* to space. We will treat this balance in the simplest possible way initially to establish a context for later realistic elaboration.

8.2.1 Input and output

Since the Sun subtends only 0.5° when viewed from Earth (Fig. 8.3), the planet intercepts solar radiation at almost exactly the same rate as a Sun-facing disk with radius equal to Earth's radius R_E (a much larger Sun would illuminate some of Earth's far side). To this good approximation, the average rate of interception of *SR* (Earth's *insolation*) is

$$\pi R_E^2 S \tag{8.2}$$

where S is the solar constant. Dividing by Earth's surface area ($4 \pi R_E^2$) we find the average insolation per unit area to be $S/4$ (= 345 W m^{-2})—a useful benchmark, even though *SR* input varies greatly between equator and Poles.

Not all intercepted radiation is absorbed: significant amounts are scattered back to space by clouds, etc. and play no part in Earth's energy balance. The proportion lost in this way is called Earth's *planetary albedo* and conventionally denoted by the symbol a, so that only $(1 - a)$ of intercepted *SR* is absorbed. The rate of absorption of *SR* by Earth is therefore

$$(1 - a) \pi R_E^2 S \tag{8.3}$$

and the corresponding average rate of *SR* absorption per unit surface area is $(1 - a) S/4$. Using current values for S and albedo a (= 0.3), the average *SR* absorption rate is 242 W m^{-2} (NN 8.2)—another important benchmark. Though S has probably varied little during Earth's long maturity, its absorption is sensitive to planetary albedo, and therefore to varying amounts of atmospheric cloud and surface ice. For example, *SR* absorption must have been substantially lower during Snowball Earth (Section 14.7.3), because of reflection by extensive ice cover.

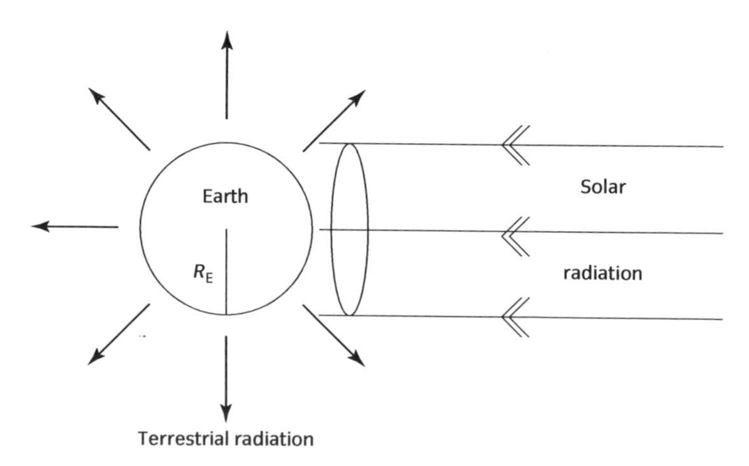

Figure 8.3 Nearly parallel solar rays impinging on the Earth, and terrestrial radiation rays diverging from it.

To model terrestrial radiative output we deliberately oversimplify and assume that Earth emits *TR* as a spherical blackbody of radius R_E and uniform absolute temperature T_E. Stefan's law (Box 8.1) then requires that total *TR* power output is

$$4 \pi R_E^2 \sigma T_E^4 \qquad 8.4$$

where σ is the Stefan–Boltzmann constant, value 5.67×10^{-8} W m^{-2} K^{-4}.

In writing expression 8.4 we are effectively defining an *equivalent blackbody temperature* T_E of the Earth—the temperature of an isothermal blackbody whose radiant power output is the same as the real Earth's. Though obviously too simple to be realistic, we will find shortly that this approach provides a surprisingly good framework for modelling *TR* and understanding its role in maintaining Earth's surface temperatures in their familiar range.

8.2.2 Radiative balance

Earth's radiant energy balance can now be stated by equating Expressions 8.3 and 8.4. After cancelling πR_E^2 we have

$$\sigma T_E^4 = (1-a)\frac{S}{4}s \qquad 8.5$$

whose right-hand side is Earth's average *SR* absorption per unit area. Using its current value (242 W m^{-2}), and the universal value for σ, we find T_E to be 256 K (−17 °C) (NN 8.2). This is so far below the observed average surface temperature of the Earth (288 K or +15 °C), that it might seem that our simple model for Earth's radiative equilibrium has failed its first test. In fact it will soon appear that the model is unrealistic in only one important respect, whose correction effectively removes the 32 °C discrepancy. However, to understand and make this correction, we need to know more about the nature of *TR*, and especially its interaction with Earth's atmosphere and surface (Section 8.3).

NUMERICAL NOTE 8.2 Radiative equilibrium

Input According to Expression 8.3 and the text, Earth's average insolation per unit surface area is $S/4$ (= 1,380/4 = 345 W m^{-2}). With current albedo $a = 0.3$, the average rate of *SR* absorption per unit surface area is $(1 - 0.3)\, S/4 = 0.7 \times 345 \approx 242$ W m^{-2}.
Output If Earth emits like a blackbody with uniform temperature T_E, then Earth's radiative equilibrium requires $\sigma T_E^4 = 242$ W m^{-2}, so that $T_E = [242/(5.67 \times 10^{-8})]^{1/4} = 256$ K or about −17 °C.
Sensitivity to albedo If albedo rises to 0.4, $(1 - a)\, S/4$ falls to $0.6 \times 1,380/4 = 207$ W m^{-2} and Earth's T_E falls to $[207/(5.67 \times 10^{-8})]^{1/4} \approx 246$ K.
Sensitivity to solar luminosity The luminosity of *the young dark Sun* (Section 14.6) was about 67% of its current value, so that Earth's solar constant S was then $0.67 \times 1,380 \approx 925$ W m^{-2}, and $(1 - a)\, S/4$ had a value of 162 W m^{-2} for an albedo of 0.3 (i.e. as now). Using this in Eqn 8.5 gives Earth's T_E as 231 K (−42 °C). When the Sun swells to become a red giant shortly before its death as a white dwarf, its luminosity will be ~ 1,000 times its present value, and so will Earth's solar constant. Use this in Eqn 8.5 to find Earth's T_E to be 1,440 K (1,167 °C).

(i) The balance expressed by Eqn 8.5 is stable towards variations in solar input to Earth because it has intrinsic negative feedback. If solar input rises (because the Sun's luminosity increases, or Earth's albedo decreases), *SR* input will initially exceed output, Earth will warm, and *TR* output will increase until global radiative equilibrium is re-established at some higher T_E. If solar input decreases, Earth will cool until reduced *TR* output re-establishes radiative equilibrium at a lower T_E.

(ii) Earth's radius R_E does not appear in Eqn 8.5, because it has cancelled out: the T_E value which balances Earth's *SR* input and *TR* output is the same for a 1 metre artificial satellite as it is for a large planet. In fact the only planetary factors determining T_E are the albedo a and the solar constant S.

Albedo

Equation 8.5 demonstrates the importance of planetary albedo in controlling terrestrial temperatures (T_E)—with T_E falling by about 10 °C for a 0.1 rise in albedo (NN 8.2). In Section 8.6 It is shown that Earth's albedo is determined mainly by reflection and scattering of *SR* by clouds, rather than by land and sea surfaces. Since clouds are maintained by updrafts which are ultimately driven by the solar input to the planet, we see that the atmosphere has an important means of internal self-regulation—more input, more cloud, more reflection offsetting the input increase. This is negative feedback at work within the atmosphere, reinforcing the negative feedback of the planet outlined in (i).

Solar constant

The young dark Sun (Section 8.1) is believed to have had a luminosity about one third less than current values, so that Earth's solar constant was then a little over 900 W m^{-2} compared with the current 1,380. The T_E value for radiative equilibrium with this reduced *SR* input is about 230 K (NN 8.2) for an Earth otherwise as at present, compared with the 256 K maintained by the current solar constant—a temperature fall of 26 °C (nearly10% on the absolute scale).

8.3 Terrestrial radiation and the atmosphere

So far the simple model of Earth's radiative equilibrium assumes that *TR* is emitted as if by a blackbody at temperature 256 K. Although this cannot be strictly true of real *TR*, it is useful to consider the power spectrum of radiation emitted by a blackbody at this temperature (Fig. 8.4). This shows significant emission across wavelengths ranging from about 4 to 100 μm, with maximum spectral emittance centred around 11 μm (by Wien's law—Box 8.1 and NN 8.3). Virtually all of such *TR* is in the *far infrared*—i.e. beyond the *near infrared* half of the solar spectrum—and the absence of significant overlap allows *SR* and *TR* to be considered as quite separate radiative fluxes. Compare the axes of Figs. 8.1 and 8.4 to see that the power spectrum of *TR* has a much flatter peak and a much greater spread of wavelength than the power spectrum of *SR*.

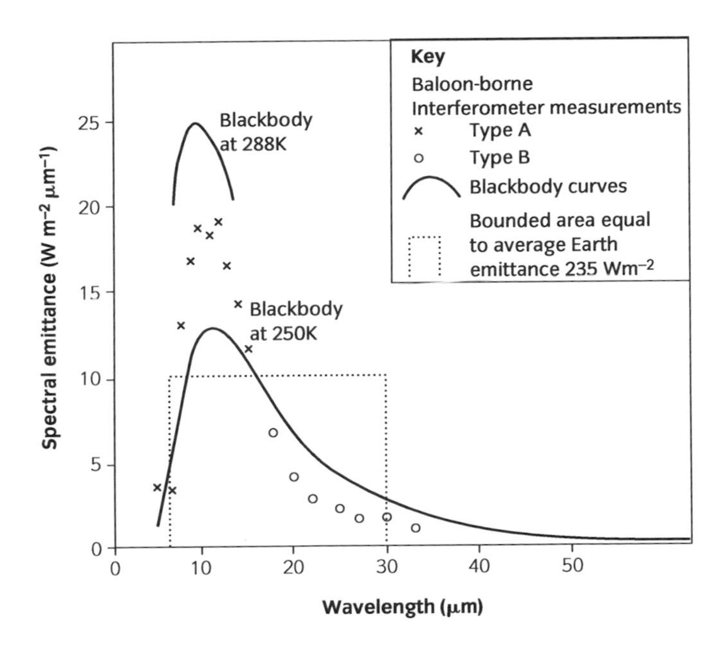

Figure 8.4 Power spectrum of terrestrial radiation. Measurements A and B were made by balloon-borne spectrometers about 16 km above sea level (above most of the atmosphere active in the far infrared). Observations on either side of the peak loosely fit the blackbody curve for 250 K (bounded area 221 W m^{-2}). Observations at the peak fall short of the level expected for radiation coming through the atmospheric window from a 266 K surface (see blackbody peak), consistent with reports of 65% cloud cover at altitude 7 km at the time. Allowing for reasonable variability, the area under the observed spectrum agrees with the 235 W m^{-2} needed for planetary equilibrium.

NUMERICAL NOTE 8.3 Terrestrial radiation

Surface

By Stefan's law the emittance of a blackbody at 288 K is given by

$\sigma T^4 = 5.67 \times 10^{-8} \times (288)^4 = 5.67 \times 10^{-8} \times 6.88 \times 10^9 = 390$ W m^{-2}. According to Wien's law the wavelength of peak spectral emittance is given by $\lambda_m = Y/T = 2{,}897/288 = 10.1$, the value being in µm, because the constant Y is in µm K.

Planet

If we replace surface temperature 288 K with Earth's equivalent blackbody temperature T_E (256 K), the terrestrial emittance falls nearly 40% to 244 W m^{-2} but the wavelength of peak spectral emittance shifts only slightly to 11.3 µm.

8.3.1 Detailed interactions

As detailed in Box 8.1, the radiative properties of many terrestrial materials differ quite markedly between solar and terrestrial wavelengths. Virtually all common solids and liquids, including water and all but the thinnest clouds of water or ice, act as almost perfect blackbodies in *TR* wavelengths, absorbing them completely (and therefore allowing no transmission or reflection), and emitting them like a blackbody at the body surface temperature. Furthermore, although the two main atmospheric gases (nitrogen and oxygen) are effectively transparent to *TR*, and therefore by Kirchhoff's law very poor emitters of *TR*, the minor component of water vapour, and the trace of CO_2, absorb *TR* so strongly that the atmosphere is almost opaque in parts of the *TR* spectrum.

The details of this complex behaviour are known from extensive laboratory and theoretical studies of the radiative behaviour of water vapour and CO_2. A greatly simplified picture (Fig. 8.5a) shows two main features:

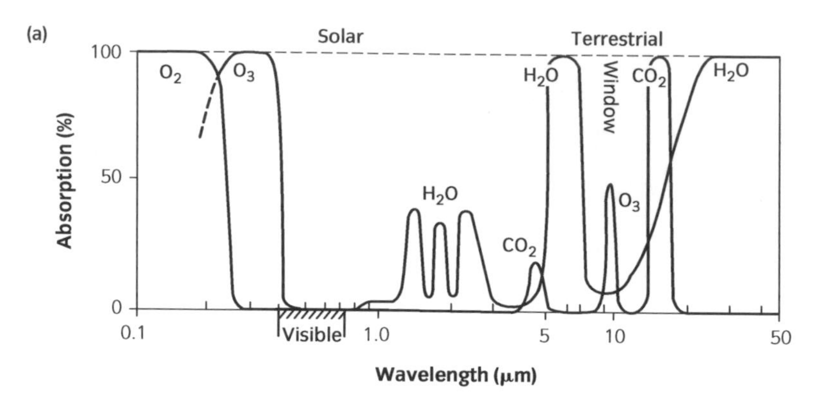

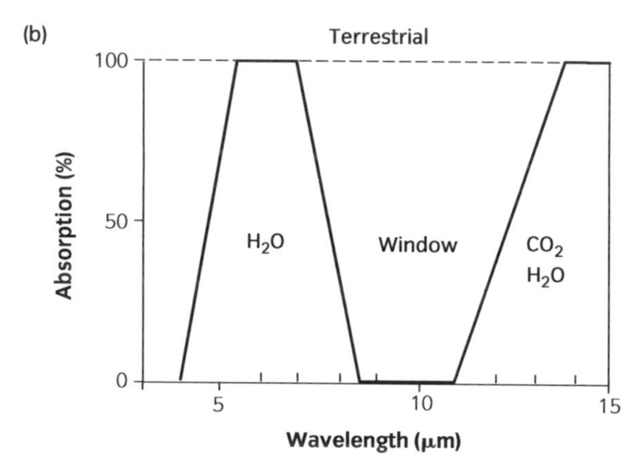

Figure 8.5 (a) Smoothed spectrum of atmospheric absorption of solar and terrestrial radiation. Percentages refer to typical absorptions by the full depth of a cloudless atmosphere. Gaseous absorbers are labelled where they are most effective. (b) Simplified profile of atmospheric absorption in and around the atmospheric window, used by Simpson in 1928 in the first detailed examination of Earth's radiative balance—an approach not bettered until ways of accounting for the microstructure of absorption bands and lines were developed in the 1950s. Absorption values refer to a layer containing 300 g of vapour per horizontal square metre.

(i) strong absorption by water vapour between wavelengths 5 and 8 μm, and again beyond 14 μm by water vapour and CO_2; and

(ii) near transparency in a wavelength range centred around 10 μm. In this *atmospheric window* there is some absorption by the much rarer gas ozone, which is unimportant for the troposphere, but is very important for the stratosphere on account of its concentration of ozone and relatively small heat capacity (Sections 3.4 and 8.6).

8.3.2 Opaqueness and transparency

The presence of the atmospheric window is important, especially for conditions at the Earth's surface, because its wavelengths include the peak of the power spectrum of *TR* (NN 8.4), allowing rapid surface cooling on cloudless nights (Section 10.3) by direct *TR* emission from the surface to space. However, it is the opaqueness of the atmosphere to wavelengths on either side of the atmospheric window which dominates the overall conditions of the atmosphere in the presence of *TR*.

The atmosphere is really very opaque indeed to wavelengths strongly absorbed by water vapour and CO_2. The simplified absorption spectrum in Fig. 8.5b shows that a layer of air deep enough to contain 300 g of water vapour in a column resting on a horizontal square metre completely absorbs wavelengths between 5 and 7 μm, and strongly absorbs wavelengths in another 1 μm wavelength range on

NUMERICAL NOTE 8.4 Minimum depth of an opaque air column

The mass M of air in a column of depth D resting on unit horizontal area is given by M = density × column volume = ρD. If the air has specific humidity q, the mass of water vapour $m = q M = q \rho D$, which is in kg m^{-2} if we use SI units and q as a pure ratio. If q is in g vapour per kg air, m will be in grams m^{-2}. Using $\rho = 1$ kg m^{-3}, $q = 10$ g kg^{-1} and 300 g m^{-2} for the minimum m value for total opaqueness, the corresponding minimum depth D for an opaque air column is $m/(q \rho) = 300/(1 \times 10) = 30$ metres.

either side. In conditions typical of the low troposphere (air density ~ 1 kg m^{-3} and specific humidity ~ 10 g kg^{-1}) such a layer is only about 30 m deep (NN 8.4). A similar estimation shows that the same thickness of air contains enough CO_2 and H_2O vapour to absorb all *TR* with wavelengths greater than 14 μm.

Consider a package of *TR* energy emitted from the Earth's surface in these heavily absorbed wavelengths. It will be completely absorbed by the lowest 30 m of air, warming the molecules of water vapour and CO_2, which then quickly share this heat with the surrounding air molecules. But by Kirchhoff's law, water vapour and CO_2 will emit these wavelengths with the same efficiency as they absorb them, so that a layer deep enough to be opaque to them will emit them as would a blackbody at the layer temperature. Note that this applies only to the absorbed wavelengths: by Kirchhoff's law the same layer of air will emit almost nothing in the wavelength range of the atmospheric window, because it absorbs almost nothing in those wavelengths, the window being transparent to them.

8.4 Cascade of terrestrial radiation

In the emerging picture, Earth's solid and liquid surfaces radiate *TR* as blackbodies, and all wavelengths outside the window are completely absorbed by layers of air no more than a few tens of metres deep in the low troposphere. By contrast a considerable proportion of all *SR* passes straight though the atmosphere to the surface, which is warmed on absorption, with a smaller amount being absorbed directly by the atmosphere. Is it possible that the Earth's surface and atmosphere are being maintained at steady temperatures solely by absorption of *SR*, and absorption and emission of *TR*? In other words can the surface and atmosphere maintain a purely radiative equilibrium? The question is considered throughout the rest of this section, and given an outline answer which is crucially important for the dynamic state and behaviour of the lower atmosphere.

For simplicity we assume that the atmosphere consists of a number of discrete layers, each of which is just thick enough to be opaque to *TR* in wavelengths outside the atmospheric window. Such layers are about 30 m deep in the low troposphere (NN 8.4), but considerably deeper at higher levels where the absorbing gases are (like the air) much less dense. To account for observed totals of CO_2 and water vapour, there should be over 100 of these opaque layers in the full depth of the atmosphere, but the situation can be outlined by considering just a few—the surface, three opaque layers, and an overlying transparent layer representing the stratosphere (Fig. 8.6).

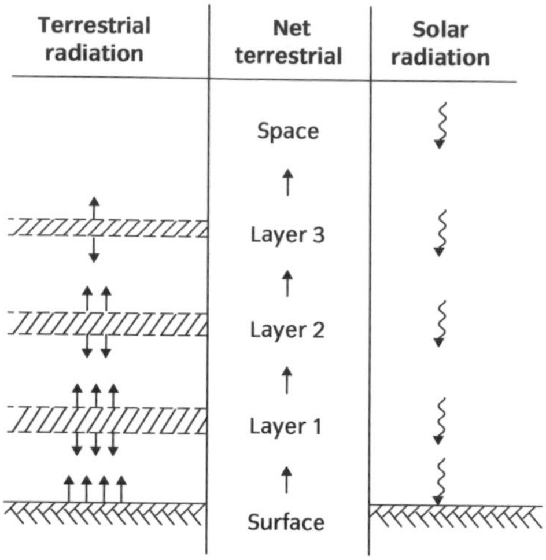

Terrestrial radiation | Net terrestrial | Solar radiation
Space
Layer 3
Layer 2
Layer 1
Surface

Figure 8.6 Schematic picture of vertical fluxes of terrestrial radiation in a three-layer model atmosphere in purely radiative equilibrium.

In an initial model we ignore all direct absorption of *SR* by the atmosphere, leaving the surface as the only absorber, and ignore too the transparent window in *TR* wavelengths, leaving the model atmosphere completely transparent to *SR* and completely opaque to *TR*.

8.4.1 Initial model

From Fig. 8.6 we see that the surface receives and absorbs *SR*, emits *TR* upwards, and absorbs *TR* coming downwards from the first overlying opaque layer of air. Each opaque layer of air emits *TR* upwards and downwards, and absorbs *TR* emitted by the immediately overlying and underlying layers. The only exception is the highest opaque layer, which absorbs *TR* coming up from beneath, emits *TR* directly to space, but receives no *TR* from above, because it is the highest layer containing significant quantities of CO_2 and water vapour, and because there are no *TR* wavelengths coming from the Sun—the photosphere being too hot for *SR* to overlap with *TR* in the infrared.

If the surface is in thermal equilibrium purely by radiative exchange, then it must lose as much energy by net output of *TR* (emission less absorption) as it gains by absorption of *SR*. And if each atmospheric layer is to be in radiative equilibrium separately, each must lose as much energy by emission of *TR* as it gains by absorption of *TR*. In addition, the net upward flux of *TR* must be uniform from just above the surface, to just above the top of the highest opaque layer—i.e. between surface and layer 1, between adjacent atmospheric layers, and above layer 3 in Fig. 8.6. Only in this way can the surface and atmospheric layers be separately and collectively in radiative equilibrium.

As shown by the arrows on Fig. 8.6, to sustain the net upward flux of *TR* through the opaque atmosphere, the emissions of *TR* by the atmospheric layers must decrease with increasing altitude to satisfy radiative equilibrium. And since decreasing emissions require decreasing layer temperatures (by Stefan's law), it follows that there must be a lapse of temperature from a maximum value at the surface to a minimum value in the highest opaque layer.

8.4.2 Improved model

We can now improve on the initial model by reinstating some direct atmospheric absorption of *SR*, and allowing for the presence of a transparent atmospheric window in *TR* wavelengths.

Significant absorption of *SR* in the lower layers of the atmosphere (Section 8.6) requires that the net upward flux of *TR* must decrease with increasing altitude, reaching the value found in the Initial model at levels above the highest layer absorbing *SR*. The effect is to reduce the temperature lapse required for pure radiative equilibrium there, holding it below what would be required if all *SR* absorption were concentrated at the surface. Above the highest layer absorbing *SR* the situation is as in the Initial model.

The inclusion of the transparent atmospheric window in TR wavelengths allows some of the net upward flux of *TR* from the surface to travel out to space independently of the cascade of atmospheric absorption and emission of *TR*. Again the result is to reduce the temperature lapse rates required in the Initial model, but this time throughout the layers interacting with *TR* (in wavelengths outside the window), since the window now carries a fraction of the net upward *TR* flux of the Initial model. Since detailed studies (Section 8.6) show that the window flux accounts for less than 10% of Earth's radiative output (the single net upward arrow in Fig. 8.6), the effect is not large.

8.4.3 Dynamic atmosphere

The Initial and Improved models show that the surface and atmosphere *could* maintain thermal equilibrium purely by absorption of *SR* and the cascade of *TR* absorption and emission, but only if there is a big enough temperature lapse between each atmospheric layer and between the surface and lowest atmospheric layer. To see if the lapse rate can be big enough we need a quantitative version of the qualitative model depicted in Fig. 8.6. This was first attempted by Emden in 1913 and has been repeated since with increasing refinement. All such studies indicate that the temperature profile required for purely radiative equilibrium of the surface and stratified atmosphere would be so steep as to be *convectively unstable* in the lower atmosphere—temperature lapse rates would exceed the dry or saturated adiabatic lapse rates, as appropriate (Section 5.9). Since the air is much too fluid to sustain convective instability without breaking into vigorous convection, if follows that it is impossible for the atmosphere with its observed distribution of gases (including water vapour and carbon dioxide) to maintain thermal equilibrium purely by absorption of *SR* and emission and absorption of *TR*. Thermal equilibrium requires convection as well as radiation, showing that Earth's atmosphere is necessarily rather than incidentally dynamic.

8.5 Greenhouse effect

We can now explain the anomaly highlighted towards the end of Section 8.2—the difference of over 30 °C between the observed average surface temperature of the Earth and the much lower effective blackbody temperature estimated for radiative equilibrium of the planet. Contrary to the original assumption, the Earth

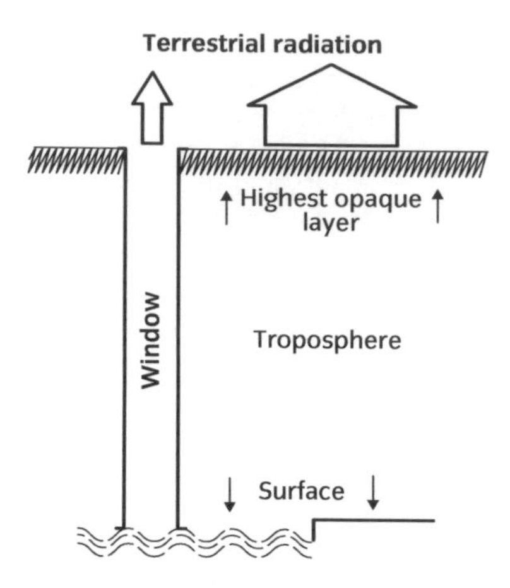

Figure 8.7 Schematic picture of terrestrial radiative fluxes leaving the Earth. About 20% (in power) comes through the atmospheric window from the warm surface or the tops of warm, cool, or cold clouds, while the remaining 80% comes from the cold top of the highest atmospheric layer opaque outside window wavelengths.

does *not* act as a blackbody with a single emitting surface situated at its solid and liquid surface: most *TR* emitted to space comes from the atmosphere (the upper troposphere, as described below), with only a little coming directly from the much warmer underlying land and sea surfaces (Fig. 8.7).

The details summarized in the next section show that about 90% of *TR* energy emitted to space comes from the atmosphere—10% by radiation coming through the window from cloud tops, and 80% by radiation outside the window wavelength range (from water vapour and CO_2 in the highest opaque layer of the atmosphere). This leaves only 10% coming through the window from the underlying land or sea. Detailed analysis also shows that the highest opaque layer (i.e. opaque outside the window) of cloudless air is in the upper troposphere, where temperatures are 40–60 °C below typical surface temperatures. Since cloud tops are often in much the same height and temperature range, we see that Earth's *TR* output is dominated by emissions from the upper troposphere, from water vapour, CO_2, and cloud there. When combined with the temperature lapse imposed by the cascade of *TR* absorption and emission (Section 8.4) and associated convection, this means that Earth's effective blackbody temperature T_E must be well below the temperatures familiar to us at the bottom of the troposphere in low and middle latitudes. Detailed analysis confirms that the resultant T_E is very close to the 256 K estimated in Section 8.2 from an isothermal model Earth in radiative equilibrium.

Looked at in another way, the effect of the atmosphere's near transparency to *SR* and near opaqueness to *TR* is to raise temperatures of land and sea surfaces well above those values which would prevail in the absence of the atmosphere (from 256 to 288 K). This elevation of surface temperatures has become known as the *greenhouse effect* because, like the atmosphere, the glass of a garden greenhouse is transparent to *SR* and opaque to *TR*. In fact, however, the name is quite misleading since a greenhouse stays warm mostly because its glass stops heat convecting from the Sun-warmed interior—*TR* blockage by glass having a much smaller effect. This was established a century ago by studies of the successful performance of a greenhouse paned with quartz, and is confirmed by the current popularity of polythene greenhouses, since both quartz and polythene are largely transparent to *TR* as well as SR, but stop convection from their warm interiors. Sunlit greenhouses, like car interiors, are effectively cooled by opening windows to let the hot air out—openings that are much too small to emit enough *TR* to make much difference. Despite all this, the term 'greenhouse effect' has become conventional, and will be used freely in this book, with the understanding that it applies to the behaviour of planetary atmospheres rather than greenhouses.

8.5.1 Global warming

The important role of the CO_2 trace in the atmosphere has come under intense scrutiny in the last 50 years, because of the rapid growth of this trace gas by domestic and industrial combustion in the course of global industrialization of human society. By the middle of the twenty-first century, atmospheric CO_2 levels will have doubled their pre-industrial values, increasing Earth's greenhouse effect by an amount which most specialists in the field now believe will have very significant effects on climate in this and future centuries (Section 14.9). For the moment it is important to realize that the *global warming* of popular debate is a small but potentially very significant human addition to the large natural terrestrial greenhouse effect which has maintained surface conditions favourable for life for most of Earth's history.

Venus

The planet Venus has an atmosphere about 100 times the mass of Earth's atmosphere, consisting almost entirely of CO_2 and a blanket of thick sulphuric acid cloud, both of which contribute to a huge greenhouse effect. The highly reflective cloud blanket maintains an albedo of about 0.7, making Venus a brilliant object in the night sky, and throwing away so much incoming solar power that Venus' effective blackbody temperature is only 245 K, compared with Earth's 256 K, even though Venus' solar constant is twice that of Earth's, being nearer the Sun (Sections 8.1 and 2). Very robust landing probes have measured surface temperatures of about 730 K, implying a greenhouse effect of nearly 500 K, over fifteen times Earth's 30 K. Estimates of total amounts of carbon in the crust, ocean, and atmosphere of Earth, and in the crust and atmosphere of Venus, show that the two planets have similar totals. However, most terrestrial carbon is chemically locked in carbonate rocks, and some is dissolved in the ocean, with only a trace in the atmosphere, whereas most Venusian carbon is in the form of atmospheric CO_2, having degassed from rocks cooked during a runaway greenhouse effect early in the planet's history. Recent observations have detected signs of a primitive Venusian ocean which evaporated in the warming process.

8.6 Global budget of radiant energy

Simple models of atmospheric interactions with *SR* and *TR* (like Fig. 8.6) are useful for outlining basic behaviour, but are inadequate for more detailed understanding. In the past century, people have developed ways of dealing with the lines and bands of lines, etc. making up the gross absorption spectrum sketched in Fig. 8.5a, so that interactions between the atmosphere and streams of *SR* and *TR* can be numerically evaluated *[17]* and used with observed distributions of radiatively active materials to calculate all significant items in the radiant energy budget for the Earth's atmosphere and surface. An immense amount of data is required, including appropriately averaged observed distributions of temperature, pressure, water vapour, carbon dioxide, ozone, cloud, and surface albedo.

A classic study of the global radiative energy budget was made by London in 1957 and has been refined since *[37]*. Data from a large number of geographical locations were averaged over many years, and further averaged over 10-degree latitude zones before being used in the extensive calculations needed to estimate radiant fluxes. The results reveal very important variations with latitude discussed later (Section 8.7). For the present global picture, data were averaged further to produce an annual global average of the radiant energy budget and its vertical distribution (Fig. 8.8). All values are quoted as percentages of the annual global average insolation per unit area at the top of the atmosphere -345 W m^{-2} (Expression 8.2 and NN 8.2)—so that each unit in Fig. 8.8 is 3.45 W m^{-2}.

The overall radiative equilibrium of the Earth is demonstrated in Fig. 8.8 by the absorption of 67 units of *SR* (100 incident less 33 scattered back to space, predominantly by clouds) and the emission of 67 units of *TR*, overwhelmingly from tropospheric cloud, water vapour, and carbon dioxide. Conservation of energy allows us to assume and confirm that the budget balances overall and in every part. For example, the equality of *SR* input and *TR* output is assumed because even a very small inequality would imply unrealistically fast changes of global temperature (NN 8.5). Likewise the large imbalance in the purely radiative budgets of Earth's surface and troposphere implies realistic complementary non-radiative (essentially convective) energy fluxes to avoid unrealistic surface warming and tropospheric cooling (below and Sections 8.9 and 8.10).

More recent satellite data (London's original work just predated the first meteorological satellites) confirms the *SR* and *TR* fluxes but suggests that Earth's albedo is about 30% rather than the 33% used from pre-satellite observations, which requires a reworking of internal fluxes. Since the original data are self-consistent and good enough for present purposes we will use them in the following discussion, but amend them later.

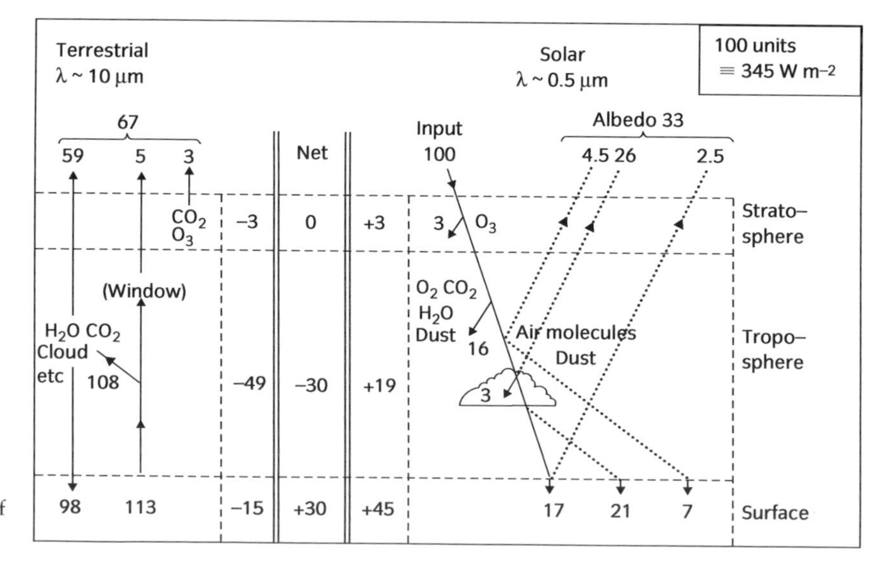

Figure 8.8 Global annual average flux densities of solar, terrestrial, and net radiation in the stratosphere, troposphere, and at the Earth's suface. On the solar radiation side (the right), dotted lines represent reflection and scattering, and solid lines transmission and absorption. On the terrestrial side (the left), solid lines represent emission, transmission, and absorption. The central column contains the net radiative input to each part of the system. Values are from *[37]*.

8.6.1 Surface radiation budget

Solar radiation (SR)

After traversing the atmosphere, *SR* reaches the surface as both *direct* and *diffuse* radiation. The major diffuser is cloud, though air molecules are also important. The scattering of light by air molecules was described theoretically by Rayleigh in the nineteenth century, who showed that, when the scattering bodies are much smaller than the scattered wavelengths (molecules ~ 10^{-4} μm compared to wavelengths ~ 0.5 μm), shorter wavelengths are scattered much more efficiently than longer wavelengths. This accounts for the blue appearance of a cloudless sky well away from the Sun's disk, since the light we receive is enriched in short (blue) wavelengths scattered laterally from passing sunbeams. It also accounts for the reddening of the solar disk when observed directly, since preferred scattering of blue light out of the line of sight leaves the direct sunlight reddened. Reddening is particularly obvious in the low Sun because the blue-biased molecular scattering accumulates along the long oblique path through the atmosphere.

By contrast, clouds, mists, and most hazes do not alter the colour of sunlight scattered by their droplets and particles because they are comparable in size with the wavelengths they scatter (called Mie scattering after the theoretician who described it). Clouds appear nearly white in normal sunlight, and are reddened at dawn and dusk only because they are being illuminated by direct sunlight already reddened by Rayleigh scattering by air molecules. The resultant scattering of all solar wavelengths is called *diffuse reflection*, and Fig. 8.8 shows that this accounts for 21 out of 47.5 units of *SR* reaching the surface, compared with only 7 units scattered by air molecules and dust. With a total of 28 out 47.5, nearly 60% of *SR* reaching the surface is diffuse rather than direct.

Although the distinction between the largely directionless diffuse component of *SR* and the highly directional direct component is obvious visually, and can be important for the climatology of hilly terrain (Section 10.2), it is largely irrelevant to the global energy budget of Earth's surface, where total absorbed insolation is all that matters. Globally, almost all *SR* incident on the surface is absorbed there, if not at first contact then soon after, by repeated scattering between surface elements (grass, etc.) and between the surface and low cloud (not depicted in Fig. 8.9). In the end only 2.5 units of the incident 47.5 are scattered back to space, giving an effective surface albedo of a little over 5%. In polar zones, surface albedos can be much larger, exceeding 70% for fresh snow. Albedos for various common surfaces are given in Table 10.1. Away from polar zones, the effective albedos of the Earth's surface are very much smaller than cloud albedos, which produces the darkness typical of cloud-free areas on satellite panoramas (Fig. 1.1).

Terrestrial radiation (TR)

Just above the surface, *TR* exists in large, opposing fluxes with a relatively small net upward flux (113 − 98 = 15 units in Fig. 8.8). The gross upward flux (113) is equivalent to the output from a blackbody with temperature 288 K (15 °C)—the average temperature of the Earth's surface (NN 8.3). Cloud, CO_2, and water vapour radiate strongly downwards from the overlying air, but do not quite match the upward radiation from the surface, partly because of the absence of any downward *TR* flux in atmospheric window wavelengths when there is no cloud, and partly because the lowest opaque atmospheric layers are usually cooler

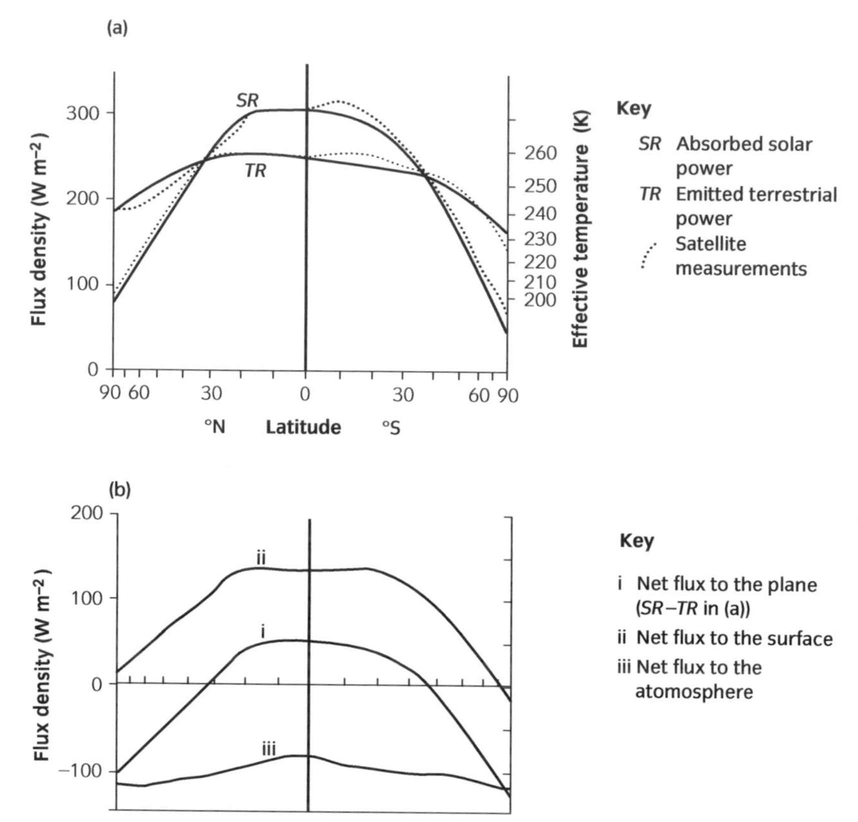

Figure 8.9 (a) Meridional profiles of zonally averaged fluxes of solar (*SR*) and terrestrial (*TR*) radiation absorbed and emitted by the Earth. The vertical axis shows the radiant flux density per unit area of Earth's surface. Figure 8.19 and the Appendix justify the sine function weighting of the horizontal axes. (b) Fluxes in (a) are analysed into (i) the net radiant flux to surface and atmosphere (the planet), (ii) the net radiant flux to the surface, and (iii) the net radiant flux to the atmosphere. From *[37]*.

than the surface. Although each of the opposing fluxes above the surface is larger than the solar input to the planet (a consequence of the greenhouse effect), the net upward flux is much smaller, and is significantly smaller even than the net solar input to the surface, with the result that the surface is in overall radiative energy surplus by 30 units (104 W m^{-2}). To maintain the surface energy balance which we know must exist in the long term, this radiative excess must be *convected* from the surface into the troposphere, since no other actual energy fluxes are remotely adequate.

8.6.2 Atmospheric radiation budget

Stratosphere

Three units of *SR* are absorbed, mainly by selective absorption of soft UV between 25 and 45 km above sea level (Section 3.4). Though this may seem a relatively small input, the air at those heights has so little mass (and therefore heat capacity), that it would warm very rapidly if the input were not closely balanced by the small net output of *TR* from stratospheric CO_2 and O_3. Water vapour is so scarce at these levels that it plays no role, despite its great importance at much lower levels. The stratosphere is therefore in radiative equilibrium overall. This does not mean that there are no convective heat fluxes within the stratosphere, or between stratosphere and troposphere, but it does mean that convection is not essential to the energy balance of the stratosphere in the way that it is for the surface and troposphere.

Troposphere

Of the 97 units of *SR* entering the top of the troposphere (Fig. 8.8), 16 are absorbed by dust particles and by molecules of oxygen, water vapour, and carbon dioxide, 30.5 are scattered back out to space, mainly by cloud, 3 are absorbed by cloud, leaving 47.5 units to pass through to the surface, either directly or after forward scattering. The strong scattering of *SR* by cloud droplets contrasts with their relatively weak absorption of *SR*, which contributes only 3 units to the tropospheric absorption total of 19 units. Including the 3 units absorbed in the stratosphere, the atmospheric absorption total of 22 units is substantial, being about one third of the total absorption of 67 units of *SR* by the planet. Our visual impression of the near transparency of the cloudless sky is misleading because much of the *SR* absorbed is in the infrared rather than the visible half of the solar spectrum (Fig. 8.5a).

Because of the vertical distributions and temperatures of water vapour, cloud, and CO_2, the output of *TR* from the troposphere to space is 59 units, much of it from the high troposphere. This large loss is only partly offset by the small net gain of 10 units maintained by the heavy *TR* exchanges with the surface. The *TR* flux in the atmospheric window plays no direct role in the tropospheric energy budget unless it is intercepted by cloud. All told, the troposphere suffers a net loss of energy of 49 units by *TR* exchange, and a gain of 19 units by absorption of *SR*, giving an overall net radiative loss of 30 units.

8.6.3 Non-radiative budget of surface and troposphere

All these radiative exchanges maintain a system which is in radiative equilibrium overall (the planet), and in the stratosphere, but which has a radiant energy imbalance between surface and troposphere, amounting to 30 units in Fig. 8.8, i.e. to 104 W m^{-2} on a global annual average. Despite this, surface and troposphere are kept in thermal equilibrium by complementary convection of heat from surface to troposphere, which is associated with the full range of weather systems from the smallest buoyant thermals to the great Hadley circulation. In addition, but obscured by the global averaging of Fig. 8.8, there is strong advection of heat from low to high latitudes. The radiative balance of the stratosphere suggests that these convective and advective fluxes are confined to the troposphere, as does the virtual confinement of cloud to the troposphere, and the widespread appearance of convectively very stable layers just above the tropopause. The operation of these non-radiative heat fluxes is outlined and quantified in the last three sections of this chapter.

8.7 Meridional distribution of radiative fluxes

The meridional averaging used to produce the global energy budget outlined in Section 8.6 buries the meridional distributions of the radiative fluxes to and from the Earth, and with them the meridional advection which their imbalances reveal. To uncover these, annual zonal averages of the radiative fluxes are presented in Fig. 8.9a and b. Values should be compared with 231 W m^{-2}, the 67 units of global average solar input per unit surface area on Fig. 8.8.

Nonlinear latitude scales have been chosen to make equal lengths along them represent zonal rings of equal surface area, so that equal areas under the curves represent equal flows of radiant energy to or from the Earth; linear scales would over-represent surface areas and fluxes at high latitudes, as a Mercator map projection does surface areas.

8.7.1 Planetary profiles of *SR* and *TR*

SR input

Figure 8.9a shows the meridional distributions of *SR* absorbed by the Earth's surface and atmosphere, and *TR* emitted to space. The solar input is concentrated in low latitudes because the Sun passes near the zenith there each day, whereas at higher latitudes it misses the zenith by an angle which increases with latitude (Fig. 8.10). The increasing obliqueness of illumination with latitude spreads the solar beam over greater surface areas, and increases path length through the atmosphere (Box 8.2). Obliqueness reduces *SR* absorption per unit horizontal area in high latitudes in the usual way, and contributes to the oblique forward scattering of *SR* to space from haze, cloud, and reflective surfaces like water (as seen in surface glitter beneath the rising or setting Sun). Increasing path length offsets these effects somewhat by increasing atmospheric absorption.

The resulting reduction of *SR* input to high latitudes would be even more marked than it is in Fig. 8.9a if it were not for the 23.5° angle between Earth's orbital and equatorial planes, which maintains the familiar seasonal march of the zenith noon Sun between the tropics, and the associated cycle of solar noon elevations at all latitudes (Fig. 8.10). The midsummer maximum in noon solar elevations is especially important in high latitudes, where it raises *SR* annual input well above the zero which persists throughout the polar winter above the Arctic and Antarctic Circles (Fig. 8.11). Though the *SR* maximum in low latitudes is in principle flattened and broadened by the same solar migration, the numerical effect is small in Fig. 8.9a.

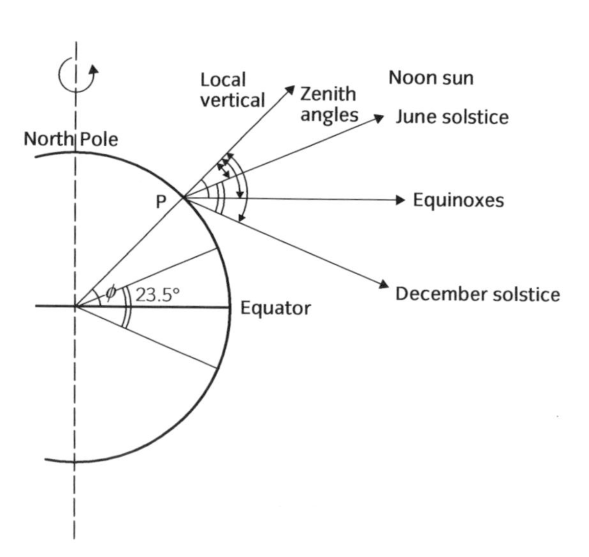

Figure 8.10 Zenith angles made by solar rays at noon at surface position P (latitude ϕ) at equinoxes and solstices, using Earth's equatorial plane as datum, so that the Sun seems to rise and fall seasonally by 47°.

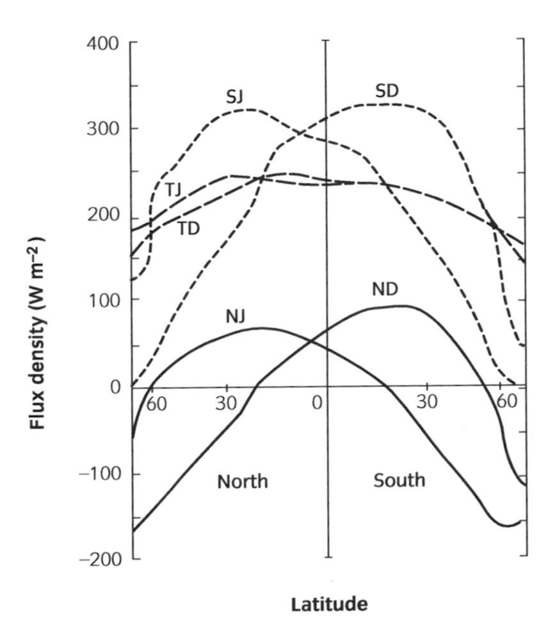

Figure 8.11 Meridional profiles of zonally averaged solar and terrestrial and net radiant input to Earth (S, T, and N respectively), averaged through the seasons beginning in June and December (J and D). For example, the curve NJ is net radiation in the seasons June, July, and August. Solar profiles differ greatly between summer and winter, whereas Terrestrial profiles differ only very slightly in the N hemisphere. Data from *[66]*.

The meridional distribution of *SR* is shaped a little by the systematically uneven distribution of cloud, which is concentrated in equatorial areas by the intertropical convergence zone, and depleted in the subtropics by the semipermanent anticyclones there. The steeper falling away of *SR* towards the Antarctic, as compared with the Arctic, is due to the presence of the great Antarctic ice cap, whose reflective surface scatters much more *SR* to space than does the much smaller Greenland ice cap, even with the reflective sea ice in the Arctic Ocean. In principle there must also be a slight bias of *SR* input towards the southern hemisphere because perihelion occurs in the southern summer, but the expected 7% effect (Section 8.1.2) is swamped by the effects already mentioned.

TR output

The meridional variation of absorbed *SR* is large and obviously related to the familiar zoning of climate. By contrast, the meridional variation of *TR* emission is very small because *TR* emission to space comes mainly from the upper troposphere (Fig. 8.7), especially from the water vapour there. Since convection keeps most of the upper troposphere close to saturation (saturation vapour density being determined by air temperature (Fig. 6.2)), and since the highest opaque layer must have a definite vapour density to be only just opaque, it follows that the temperature and *TR* emittance of the highest opaque layer are remarkably uniform across the globe. The smallish decline in *TR* with increasing latitude is largely due to the decline of emissions through the atmospheric window from the Earth's surface and the tops of low clouds, both of which are cooler in higher latitudes.

Net input and output

The very different meridional distributions of *SR* and *TR* in Fig. 8.9a reveal a gross imbalance between low latitudes, which gain much more radiant energy than they lose, and high latitudes which lose much more than they gain. There is no net gain (or loss) by the Earth as a whole, as we can confirm by careful measurement of the areas between the *SR* and *TR* curves (which are directly proportional to energy by choice of latitude scale). If this latitudinal imbalance of the radiation budget was not offset by heat carried by flows of air and water, high latitudes would cool and low latitudes would warm until their *TR* output reached the same steep meridional profile as the solar input, with very much colder Poles and warmer low latitudes than we experience now—in fact like the Moon.

Air and water flow polewards and equatorwards as well as zonally, transporting heat by horizontal *advection*. The familiar mid-latitude experience of relatively warm air moving polewards and cooler air moving equatorwards, implies just such net poleward advection of sensible heat, as does the great poleward flux of relatively warm sea water in the Gulf Stream in the western part of the North Atlantic, compared with the cool equatorward drift in the E North Atlantic. And net poleward fluxes of water vapour maintain very significant poleward fluxes of latent heat, as well as the heavy rainfalls of middle latitudes. The advective roles of the atmosphere and oceans are examined further in Sections 8.9 and 8.10.

8.7.2 Profiles of net radiative input to planet, surface, and atmosphere

The solar and terrestrial fluxes in Fig. 8.9a are regrouped in 8.9b to show the meridional profiles of net radiative (*NR*) inputs (i) to the planet, (ii) to the surface, and (iii) to the atmosphere.

The first of these (i) is *SR* – *TR* transcribed from Fig. 8.9a, and shows the meridional imbalance already discussed. The net input to the Earth's surface (ii) has a similar shape to (i), but is everywhere increased by a little over 100 W m^{-2}. Its value is a little under 150 W m^{-2} over the half of the Earth's surface lying between latitudes 30° N and S, and falls below 50 W m^{-2} only at latitudes greater than about 60°. By contrast, the net radiative input to the atmosphere (iii) is negative (i.e. it is a net radiative output) and remarkably uniform, remaining close to −100 W m^{-2} at all latitudes, in close agreement with the global tropospheric loss apparent in Fig. 8.8.

The combination of the steep meridional profile of the net radiative input to the surface (ii) in Fig. 8.9b and the near uniformity of the net radiative loss by the atmosphere (iii), means that in low latitudes the net radiative gain by the surface exceeds the net radiative loss by the troposphere, whereas the position is reversed in high latitudes. This strong meridional variation in the radiative imbalance between surface and atmosphere shows that there is consistent heat transport both upwards and polewards, taking heat from the radiatively warming surfaces of low latitudes to the radiatively cooling upper troposphere of all latitudes. In principle such transport could be accomplished by a combination of strictly vertical convection and strictly horizontal poleward advection, but in fact a significant portion in middle latitudes is effected by large-scale, slightly tilted air flow, called *slope convection* to distinguish it from the normal nearly vertical convection with which it cooperates, together with horizontal air and ocean currents.

8.8 Seasonal and diurnal variations of radiative fluxes

8.8.1 *SR*, *TR*, and *NR*

In middle latitudes, the seasonal rhythm of solar input is one of the most prominent features of weather and climate, even in relatively temperate western continental margins. In Fig. 8.11, the seasonal breakdown of the annual average profiles of Fig. 8.10 shows that the difference between winter and summer values of the *SR* input to the Earth's surface and atmosphere exceeds 100 W m^{-2} poleward of latitude 25°, and reaches 150 W m^{-2} over a considerable part of the middle latitudes. These large seasonal variations are driven by the seasonal march of the Sun between the hemispheres, and modified somewhat by uneven distributions of land, cloud, and ice caps as mentioned in Section 8.7. The seasonal variation in *TR* output is by contrast extremely small, being visible only the N hemisphere as a small reduction from July to December beyond about latitude 25°. Though numerically small, this reduction arises from a major seasonal feature—the relatively clear skies over the great continental interiors in N middle and high latitudes which reduce the local greenhouse effect, enhancing the seasonal swing in surface temperatures and driving a seasonal variation in the *TR* output through the largely open atmospheric window.

The combination of the relatively small seasonal variation of *TR* output and the very large seasonal variations in *SR* input drives large seasonal migrations in the net radiative balance, the zone of maximum net input shifting from about 15 °N in the N hemisphere summer to 25 °S six months later. Note that the southern hemisphere summer (i.e. December) maximum of net radiative input (*ND*) is significantly higher than the northern one (*NJ*), and peaks further from the equator.

8.8.2 Seasonality of *NR* profiles

The other major feature of Fig. 8.11 is the extensive meridional gradient of net radiative input, stretching from the maximum in the summer hemisphere to the minimum at or near the winter pole. The range between the positive maximum and the strongly negative minimum is nearly 500 W m^{-2}, and the steep and extensive gradient between them is a major factor in maintaining the intense weather activity of middle and high latitudes, especially in winter. We can trace the connection as follows: the gradient of *NR* extends and steepens the meridional temperature gradient of the troposphere in the winter hemisphere, especially in middle and high latitudes (Fig. 4.10), and with it the baroclinity which drives much of the large-scale weather activity there, including extratropical cyclones. In the summer hemisphere the difference between maximum and minimum *NR* is smaller and is largely confined to latitudes greater than 30°. In middle latitudes, it seems that this narrower and less strongly baroclinic zone is considerably less effective in maintaining synoptic and larger-scale weather activity, extratropical cyclones, etc. being less numerous and vigorous than in the winter hemisphere. Note that this tendency is confined to middle latitudes; on the equatorial flanks of the subtropical highs, large-scale tropical weather systems tend to be more vigorous in the summer, since local sea-surface heating is more important than baroclinity. And equatorial and tropical systems are dominated by the seasonal march of the Hadley circulation, with a range of seasonal effects (Sections 4.7 and 14.3).

BOX 8.2 Surface irradiance

Figure 8.12 shows a beam of direct sunlight falling on a horizontal surface from a solar *elevation angle* β above the horizon (a *zenith angle* of $90 - \beta$). According to the geometry of sunbeam and surface, the radiant flux density F_h on a horizontal surface (the *horizontal irradiance*) is related to the irradiance F_n of a surface normal to the beam by

$$F_h = F_n \sin \beta \qquad \text{B8.2a}$$

confirming that $F_h = F_n$ when the Sun is in the zenith ($\beta = 90$), and tends to zero as the Sun approaches the horizon ($\beta = 0$).

Other factors which vary with β are the path length of the solar beam through the atmosphere (Fig. 8.12), and the associated attenuation of the sunbeam by scattering and absorption. Though interactions between *SR* and the atmosphere are complex *[38]*, an important result emerges from a simple analysis.

The small reduction $\mathrm{d}F$ in irradiance arising from the presence of a small mass $\mathrm{d}m'$ (per unit beam cross-section area) of attenuating material in a length of sunbeam is proportional to both F and $\mathrm{d}m'$

$$\mathrm{d}F = -k\ \mathrm{d}m'\ F \qquad \text{B8.2b}$$

where the minus sign ensures that $\mathrm{d}F$ is a reduction, and k is an *attenuation coefficient* which can vary strongly with wavelength (Fig. 8.1) but which we suppose has been averaged to a single value across the spectrum. In many circumstances, attenuating material is horizontally stratified, like much atmospheric material, so that $\mathrm{d}m'$ is related to the element of mass $\mathrm{d}m$ in a vertical column (between the same horizontals as the oblique sunbeam) by

$$\mathrm{d}m' = \mathrm{d}m \operatorname{cosec} \beta$$

so that B8.2b becomes

$$\frac{\mathrm{d}F}{F} = -k\,\mathrm{d}m \operatorname{cosec} \beta$$

Summing the attenuation along the sloping sunbeam from the top of the atmosphere (where F_o is the solar constant) down to the surface (where F_n is the surface flux density normal to the angled beam), standard integration gives

$$\frac{F_n}{F_o} = e^{-km \operatorname{cosec} \beta} \qquad \text{B8.2c}$$

where m is the total mass (per unit horizontal area) of the active material in a vertical column through the atmosphere. This is a form of *Beer's law*, and the product ($k\ m$) is known as the *optical depth* of the atmosphere.

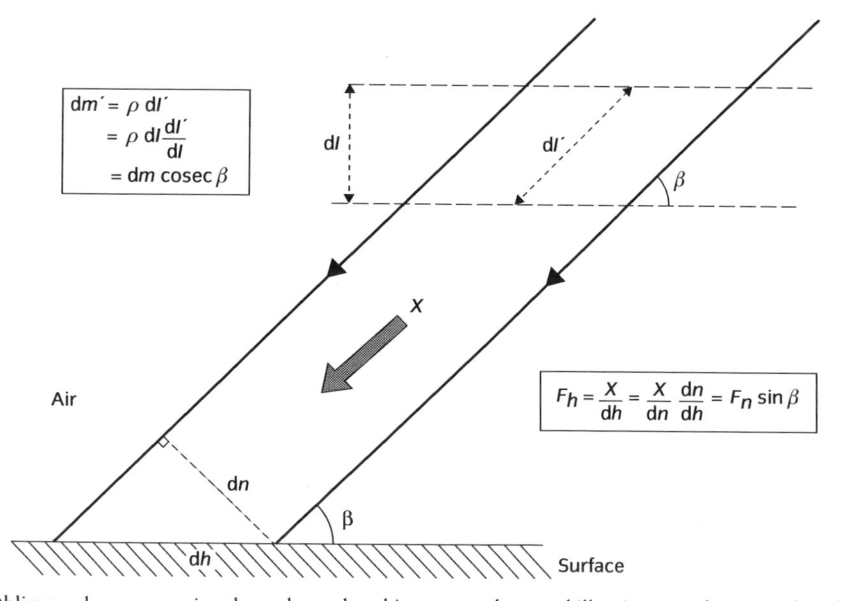

Figure 8.12 Oblique solar rays passing through an absorbing atmosphere and illuminating a horizontal surface.

The left-hand side of Eqn B8.2c is called the *transmissivity* T_β of the angled beam—the fraction of the radiation incident on the top of the atmosphere which reaches the surface in the direct beam. Since cosec 90° = 1/sin 90° = 1, the *zenith transmissivity* T_{90} is given by

$$T_{90} = e^{-km}$$

so that

$$T_\beta = (T_{90})^{\text{cosec}\,\beta} \qquad \text{B8.2d}$$

It is usual to describe the transparency of an atmosphere by its zenith transmissivity even at latitudes where the Sun can never reach the zenith. A clear sky in the British Isles can have a zenith transmissivity of as much as 0.8. When solar angles are 60° (near the maximum at those latitudes), transmissivity is reduced only a little below the zenith value, but it falls away more rapidly with decreasing solar elevation, so that at $\beta = 15°$ it is about half the zenith value.

The horizontal irradiance at the foot of an oblique sunbeam is found by combining Eqns B8.2a and B8.2d to produce

$$F_h = F_o\, T_\beta \sin \beta \qquad \text{B8.2e}$$

Figure 8.13 is a graph of horizontal surface irradiance F_h against solar elevation angle for a clear summer's day in Britain.

This treatment does not describe atmospheric transmission when attenuating material is *not* horizontally stratified (for example when the sky is dotted with cumulus), nor does it deal with diffuse radiation, where light scattered out of the direct beam may still reach the surface, adding to its irradiance.

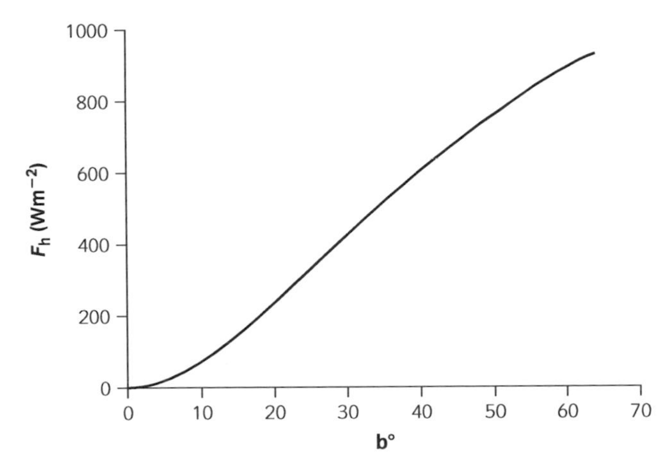

Figure 8.13 Irradiance of a horizontal surface near sea level in central England on a clear summer's day (zenith transmissivity 0.8). The horizontal axis is solar elevation, and maximum elevation corresponds to noon at the summer solstice.

8.8.3 Diurnal variations

(i) In *SR*: The average surface budget in Fig. 8.8 conceals large diurnal variations. The *SR* input, averaging 45 units overall, is obviously zero at night, and would average 90 units in daylight, assuming a 12-hour day and no variation with latitude. Assuming that dawn to dusk variation of *SR* is shaped like half a sine curve (Box 10.2), the noon maximum corresponding to a 90 unit average is 141 units (486 W m^{-2}), consistent with values estimated from solar elevation and atmosphere transmissivity (Box 8.2). Under a zenith Sun and clear sky, noon maxima can reach about 320 units (1,100 W m^{-2}). All values can be reduced to about one tenth under thick overcast. Maximum values fall with decreasing solar elevation (i.e. increasing latitude at a fixed time of the year) as outlined in Box 8.2.

(ii) In *TR*: Figure 8.8 shows that on average the Earth's surface loses 15 units by net output of *TR*. Since five of these are lost directly to space through the atmospheric window, which is closed by cloud for about 50% of the time, the *TR* loss through the window in cloudless conditions must be about 10 units, bringing the net *TR* loss by the surface to 20 units (nearly 70 W m^{-2}), with somewhat larger values in regions with less atmospheric water vapour (giving a 'cleaner' atmospheric window) and smaller values elsewhere—consistent with values derived in other ways in Section 10.3, and with radiometer measurements. Net *TR* loss by the surface must average about 10 units (nearly 35 W m^{-2}) in cloudy conditions. Diurnal variation of *TR* is blunted because the tendency to greater emission from hot sunlit surfaces is offset by the greater vapour content and cloud cover often associated with Sun-driven convection.

(iii) In *NR*: It follows from the above that there can be very large diurnal variations about the average value of 104 W m^{-2} (30 units) quoted in Fig. 8.8 for the net solar and terrestrial radiative input (*NR*) to the Earth's surface. On cloudless periods at low latitudes a maximum *NR* input of about 1,000 W m^{-2} at noon may give way to a net output of about 70 W m^{-2} during the ensuing night. Under a thick overcast, maximum net daytime input can be reduced to about 100 W m^{-2}, and nocturnal loss reduced to zero under a cloud base as warm as the surface (i.e. very low level). At higher latitudes the effects of the reduction in noon solar input is modulated by the increasing seasonal variability of day length (Section 14.1 and *[38]*), and net losses of about 70 W m^{-2} can continue throughout the long polar nights.

The considerable impact of these diurnal variations on surface climate is outlined in Sections 10.3 and 10.4, where we see that the small effective heat capacities of land surfaces encourage the large swings of temperature between early afternoon maxima and dawn minima which are such a familiar feature of cloudless conditions overland. Diurnal temperature variations at and over water surfaces are very much smaller, because of their very much greater heat capacities. In addition to their relevance to surface climate, these diurnal variations are obviously consistent with the tendency for overland convection in particular to peak in the afternoon and die away in the evening.

8.9 Non-radiative heat transport

We have seen in Sections 8.6–8.8 that interactions between solar and terrestrial radiation and the Earth maintain substantial radiative energy imbalances between the surface and troposphere, and between low and high latitudes, as well as imposing temporary imbalances between summer and winter and day and night. In the remainder of this chapter we quantify these imbalances, and begin to relate them to the conduction, convection, and advection of heat which underlies every aspect of our weather and climate.

Heat is *conducted* through apparently stationary solids, liquids, and gases, as more energetic molecules collide with their less energetic neighbours, sharing some of their extra energy, which spreads further as the process continuously repeats. Although molecules do migrate in the process (at least in liquids and gases—in solids they just vibrate) parcel migration is unnecessary for heat conduction. By contrast, it is essential for heat convection and advection (horizontal convection). In turbulence, though parcel motion is so random that there is little or no total mass flux on a scale much larger than the turbulent eddies, there is a strong directed heat flux whenever parcels moving in a certain direction are consistently warmer than parcels moving in the opposite direction, as happens in the vertical over a Sun-warmed surface. On larger scales, parcel motion is organized into updrafts, downdrafts, currents, and counter-currents, which will transport heat whenever these are consistently warmer in some particular direction—as in middle latitudes, where air flows from lower latitudes are consistently warmer than those from higher latitudes, and heat is pumped polewards in net consequence. On scales larger than the updrafts and downdrafts, etc. the net mass flux in any particular direction is usually small and transient, so that the distribution of atmospheric and ocean mass remains nearly steady.

8.9.1 Consequences of zero non-radiative heat transport

We can appreciate the climatic potential of global and meridional radiative imbalances in the surface and troposphere by estimating the rates of warming and cooling which would occur if there was no conduction, convection, or advection of heat to offset the gains and losses apparent in Figs. 8.8 and 8.9.

Global

Consider the rate of warming of the top 10 metres of the world ocean if it sustains a net radiative input of 100 W m^{-2} (rounding the value from Fig. 8.8) without balancing heat loss. Ignoring evaporation (which consumes over half of this imbalance to feed the latent heat flux from the surface—NN 8.7), the surface layer would warm at 0.2 °C per day, raising global surface temperatures by 6 °C in a month (NN 8.5). The estimate shows incidentally that even this shallow layer of ocean has enough heat capacity to cause significant diurnal and seasonal smoothing.

Land surfaces have very much smaller heat capacities (Section 10.4) than water surfaces, since only their top millimetres are reached by *SR*, as compared with the tens of metres of water directly warmed by *SR*. They have little smoothing capacity and respond quickly to heat inputs and outputs, as shown by the large diurnal and seasonal temperature variations over land.

A similar estimation finds the rate of cooling of the troposphere subjected to a net radiative loss of 100 W m^{-2}. With a heat capacity per unit horizontal area of about 7.5 MJ °C^{-1} m^{-2} (NN 8.5), the entire troposphere would cool by just over 1 °C per day! The estimate shows incidentally that the troposphere has enough heat capacity to smooth the daily variations of heat input from the surface, which have amplitude of order 100 W m^{-2}. Only the atmospheric boundary layer over land, especially its lowest tens of metres, warms and cools diurnally by much more than this, providing the familiar temperature cycle of our daily lives.

High and low latitudes

According to Fig. 8.9b, latitudes below 30° (i.e. half of Earth's total surface area) experience a net radiative gain of about 50 W m^{-2}, while higher latitudes experience a balancing net radiative loss to maintain global thermal equilibrium.

NUMERICAL NOTE 8.5 Warming and cooling by radiative imbalances

Since the density and specific heat capacity of water are 1,000 kg m^{-3} and 4,200 J kg^{-1} K^{-1}, the heat capacity of water per unit volume is 4.2 MJ K^{-1} m^{-3}, and the heat capacity C of a 10 m depth of water beneath unit surface area is 42 MJ K^{-1} m^{-2}. To find the rate of warming dT/dt of the 10 m deep water layer in response to a heat input of 100 W m^{-2}, use dT/dt = heat inflow/C = $100/(42 \times 10^6) \approx 2.4 \times 10^{-6}$ °C s^{-1} i.e. about 0.2 °C per day, or just over 6 °C in 30 days.

Taking the mass of troposphere per unit horizontal area to be 7,500 kg m^{-2} (corresponding to a layer from about 1,000 to 250 hPa), and using C_p = 1,000 J kg^{-1} K^{-1} for the specific heat capacity of air, tropospheric heat capacity is $7{,}500 \times 1{,}000 = 7.5$ MJ m^{-2}. Subject to a radiative rate of loss of 100 W m^{-2}, the troposphere would cool at $100/(7.5 \times 10^6) \approx 1.3 \times 10^{-5}$ °C s^{-1} or about 1.1 °C per day. Notice that the troposphere has a heat capacity equivalent to that of a global layer of water just under 1.8 m deep.

Combining these, the total heat capacity of the troposphere and top 10 m of water is nearly 50 MJ K^{-1} m^{-2}. If this is subject to a sustained radiative input or output of 50 W m^{-2} (as happens in the intertropical and higher latitudes respectively) the rate of warming/cooling is $50/(50{,}000{,}000) \approx 10^{-6}$ °C s^{-1}, or 31.5 °C per year.

Taking the low latitude input of 50 W m^{-2}, and combining the heat capacities of the troposphere and top 10 metres of water (as above), we find (NN 8.5) that the low latitude surface and troposphere would warm by over 30 °C per year, and higher latitudes would experience similar persistent cooling. Note that only poleward advection of heat by currents of water and air can offset this radiative imbalance, horizontal radiative transport being ineffective since horizontal fluxes of *TR* are blocked by the air's opaqueness to *TR*, and vertical convection is irrelevant.

8.9.2 Heat conduction

We have confirmed that the imbalances in the radiant energy budget shown in Figs. 8.8, 8.9, and 8.11 would cause catastrophically large and fast temperature changes if they were not smoothed by heat storage and balanced by heat conduction, convection, and advection in the atmosphere and oceans. Can heat conduction play a significant role?

Box 8.3 shows that sensible heat is conducted down a temperature gradient in still air at a rate which is proportional to the prevailing temperature gradient and the *thermal conductivity* of the air. In conditions typical of the low troposphere, a temperature gradient of 1 °C per metre will conduct heat at a rate of 25 milliwatts through a square metre perpendicular to the axis of the temperature gradient—over 1,400 times smaller than the 40 W m^{-2} needed to keep the surface and troposphere in thermal equilibrium as discussed above (40 rather than 100 because surface evaporative cooling accounts for just over 60 W m^{-2}—NN 8.7). To conduct 40 W m^{-2} upwards from the surface in still air would need a vertical temperature lapse rate of 1,600 °C m^{-1} (NN 8.6).

On a scale of metres or larger, nothing like this is observed, but a temperature gradient of 1.6 °C per millimetre will conduct heat at about the required rate, and something like this is found in the film of still air immediately overlying all solid and many liquid surfaces (Section 10.6). In thicker layers, atmospheric turbulence takes over, but conduction is still effective whenever and wherever such millimetric scale temperature gradients are generated between colliding eddies. Molecules of water vapour similarly diffuse down the sharp gradients of vapour density found in the first millimetres of air overlying water, moist Earth, or well-watered vegetation, initiating the flux of latent heat which is taken over by turbulence more than a few centimetres higher (Section 10.7).

In the top millimetres of water surfaces, heat is conducted upwards or downwards according to the product of the vertical temperature gradient and the thermal conductivity of water, which is about 25 times larger than the value for air (NN 8.6). Using the global average values from Fig. 8.8, 45 units (155 W m^{-2}) of *SR* warming the top tens of metres of ocean must be brought to the surface to feed the loss of 15 units (52 W m^{-2}) by net *TR* emission from the water surface, and the 30 units (103 W m^{-2}) to maintain heat loss and evaporation from the sea surface (Fig. 8.14). A lapse of about 1 °C in 4 mm will conduct heat to the surface at the required rate (NN 8.6). Observation shows that such conducting layers are very shallow, and that turbulent diffusion predominates at greater depths. In wind-disturbed sea surfaces there is no identifiable conducting layer, though conduction is still at work smoothing temporary, localized temperature gradients as in turbulent air.

We see that heat conduction in air and water is unable to cope with the heat imbalances imposed by solar and terrestrial heat fluxes, except in shallow, static surface films. In the rest of the atmosphere and ocean, these imbalances must be carried by moving air and water—by heat convection and advection.

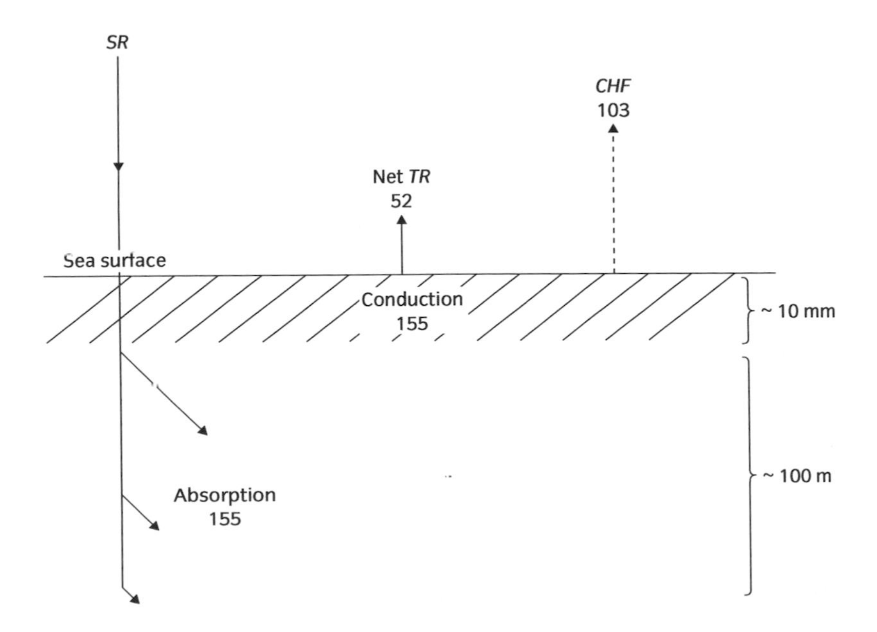

Figure 8.14 Schematic picture of fluxes of solar radiation (*SR*) to and under the ocean surface, net terrestrial radiation (*TR*) and convective heat (*CHF*) from its top surface, and heat conduction up through the laminar skin just under its surface. Numbers are in W m^{-2}, and come from Fig 8.8 assuming that the ocean behaves like the equilibrium Earth.

NUMERICAL NOTE 8.6 Heat fluxes by conduction

Air

According to Box 8.3 conducted heat flux H per unit area $= -k\,\partial T/\partial n$ where $k = 24.3 \times 10^{-3}$ W m^{-1} K^{-1} for dry air. When the temperature gradient $\partial T/\partial n$ is 1 °C per m, $H = 24.3 \times 10^{-3} \times 1 = 24$ mW m^{-2}. Conversely, to conduct $H = 25$ W m^{-2} requires $\partial T/\partial n = H/k = 25/(24.3 \times 10^{-3}) = 988$ °C m^{-1} or about 1 °C per millimetre.

Water

Thermal conductivity $k = 0.591$ W m^{-1} K^{-1}. The temperature gradient for $H = 155$ W m^{-2} is given by $155/0.591 = 262$ °C m^{-1} or about 1 °C in 4 mm depth of water.

8.9.3 Convective fluxes of sensible and latent heat

On a global annual average (Fig. 8.8) the convective flux must be about 30 units (about 100 W m^{-2}) close to the Earth's surface, decreasing upwards through the troposphere roughly as outlined in Fig. 8.16, to become zero at and above the tropopause.

Figure 8.15 Near-infrared satellite panorama from the equator to latitude 60° S centred on Australia, taken by a geostationary satellite and manipulated to expand higher latitudes. There are some very deep convective cloud masses (very cold, white tops in these wavelengths) in the equatorial zone, and extensive mostly lower (greyer) cloud over W Australia. The long plumes of extensive cloud coming from the NW to New Zealand, and to the sea area off SW Australia, are the counterparts of the warm, moist SW'ly flows of N hemisphere midlatitudes. A classic showery pattern in the cool SW'lies lies over and S of Tasmania.

If the Earth was completely arid, all this convected heat would have to be in the form of sensible heat, warmer parcels rising and cooler ones sinking to effect a net upward transport of heat (Box 8.3). But the rising branch of the hydrologic cycle (Sections 3.7, etc.) requires that rising air is consistently moister than sinking air, with the result that diffusion and convection of water vapour maintains a flux of latent heat between surface evaporation and cloudy condensation aloft. Indeed we can convert the vapour mass flux into its associated flux of latent heat simply by multiplying by the coefficient L of latent heat.

The Bowen ratio

The apportionment of total convective heat flux between sensible and latent heat has long been recognized as a very important measure of climatic type, the ratio of sensible to latent heat fluxes (*SHF/LHF*) being called the *Bowen ratio* β, after a pioneer in evaporation studies. In arid zones such as the subtropical deserts, β values are much larger than 1 (approaching 10), while in warm, humid zones they are much smaller than 1 (approaching 0.1). Figure 8.16 includes the division between fluxes of sensible and latent heat found in the study by London *[37]*, and shows that they are consistent with a global annual average β value of about 0.6. London's analysis is outlined below to illustrate the practical difficulties of establishing such basic data.

Because measuring turbulent transport is difficult (Section 10.8), it is impractical to make direct measurements of convective fluxes of heat and vapour on a routine basis. London's estimate is based on worldwide measurements of precipitation, which suggest that the global annual average precipitation (including the rainfall equivalent of snow and hail) is about 800 mm. Since the global hydrologic cycle

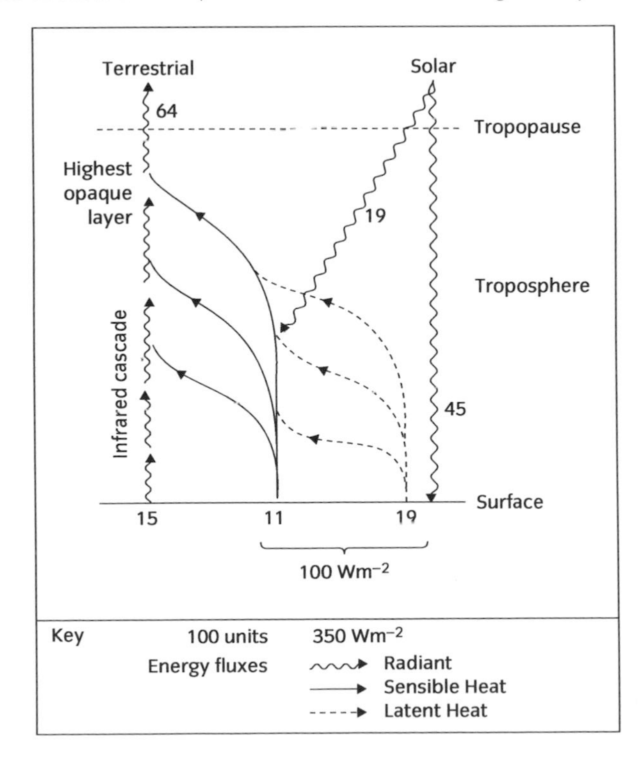

Figure 8.16 Global annual averages of convective and radiative energy fluxes in the atmosphere, showing progressive handing over with increasing height from latent to sensible heat fluxes and from these to the terrestrial radiative upward cascade. Values are from Fig. 8.8 and other parts of *[37]*.

NUMERICAL NOTE 8.7 Rainwater mass flux and latent heat and Bowen ratio

If annual precipitation is 800 mm, the annual rainwater volume standing on a square metre is 0.8 m^3, and the corresponding rainwater mass flux (water density being 1,000 kg m^{-3}) is 800 kg m^{-2} yr^{-1}. Multiply by the latent heat of evaporation of water (2.5 MJ kg^{-1}) to find the equivalent latent heat flux density $= 800 \times 2.5 \times 10^6 = 2.0 \times 10^9$ J m^{-2} yr^{-1}. Divide by the number of seconds in a year to get $LHF = 63.5$ W m^{-2}.

Total convective flux from surface = net radiative input to surface = 103.5 W m^{-2} (30 units of 3.45 W m^{-2} from Fig. 8.8). Sensible heat flux $SHF = 103.5 - LHF = 40$ W m^{-2}, and Bowen ratio $\beta = SHF/LHF = 40/63.5 \approx 0.63$.

is nearly steady over time periods of a couple of months (over five times the residence time for water in the atmosphere—NN 3.6), and since dew and frost are relatively unimportant sinks of atmospheric water vapour, this rate of precipitation to the surface must be very nearly equal to the evaporation rate from the surface. An evaporation rate of 800 mm per year corresponds to a latent heat flux of 63.5 W m^{-2} (NN 8.7). And since the total convective heat flux from the surface must balance the radiant energy flux to the surface (yearly average 103.5 W m^{-2} from Fig. 8.8), the sensible heat flux must be 40.0 W m^{-2}. This value is uncertain to at least $\pm$ 5%, because of the inherent unreliability of precipitation measurement, especially over the oceans (Section 2.6), but the numbers suggest a global Bowen ratio of about 0.65 ± 0.1.

Figure 8.16 depicts a progressive handing-over from latent to sensible heat fluxes, which begins in the low troposphere and is largely complete in the upper middle troposphere. This corresponds to the typical appearance of most cloud, with conversion from latent to sensible heat concentrated in the first few kilometres above the cloud-free layer. Of course there are many tall cumulonimbus, especially in the showery outbreaks of middle latitudes, and in the equatorial zone, but much of the cloud reaching the upper troposphere in these zones has already condensed and converted latent heat to sensible heat at or below the middle troposphere.

Figure 8.16 also depicts the progressive handing over from the convective heat flux to the cascade of *TR*, which is effectively complete in the high troposphere—the altitude of the highest layers which are opaque to *TR* (Section 8.4), above which the atmosphere's thermal equilibrium is dominated by radiation, with a relatively minor role for convection.

BOX 8.3 Atmospheric heat conduction, convection, and advection

Conduction

From extensive experiment and theory it is known that heat is conducted down a temperature gradient in a static material body at a rate given by

$$H = -k\frac{\partial T}{\partial n} \qquad \text{B8.3a}$$

where

- H is the heat flux density (the rate of heat flow through unit area perpendicular to the heat flow) in W m^{-2} in SI units,
- $\partial T/\partial n$ is the instantaneous temperature gradient normal to the H cross-section,

- the minus sign indicates that heat flows *down* the temperature gradient (from higher to lower temperatures), and
- k is the *thermal conductivity* of the conducting material.

Values for k are established by laboratory measurement for a wide range of materials and conditions. Gases in terrestrial conditions are much less efficient conductors than liquids and solids, largely because of their lower densities. In air in the low troposphere, $k \approx 24 \times 10^{-3}$ W m^{-1} °C^{-1}, whereas the value for water ≈ 0.59 W m^{-1} °C^{-1}.

Convection

Figure 8.17a depicts a horizontal plane of area A, a fraction N of which is filled with air of density ρ, temperature T, rising at speed w, while the remainder (fraction $1 - N$) is filled with air density ρ', temperature T', sinking at speed w'. Since w is numerically equal to the volume of air flowing through unit horizontal area per unit time, the upward mass flux through unit area is $\rho\, w$, and the upward and downward mass fluxes through A are given by

$$UMF = N\, A\, \rho\, w \quad \text{and} \quad DMF = (1 - N)\, A\, \rho'\, w'$$

If each is multiplied by the product of the air temperature and specific heat capacity, the net upward flux of sensible heat is given by

$$NUSHF = C_p\, T\, UMF - C_p\, T'\, DMF$$

When the net vertical mass flux is zero, as is often nearly the case in convection,

$$UMF = DMF$$

and

$$NUSHF = UMF\, C_p\, [T - T'] = N\, A\, \rho\, w\, C_p\, \Delta T$$

and the net upward heat flux density (i.e. $NUSHF$ per unit horizontal area within A) is

$$NUSHX = N\, \rho\, w\, C_p\, \Delta T \qquad \text{B8.3b}$$

where N is the fraction of A filled with updrafts of density ρ, speed w and temperature excess (over downdrafts) $[T - T'] = \Delta T$. Using typical observed values for convection $N = 1/10$, $\rho = 1$ kg m^{-3}, $\Delta T = 1$ °C, and the standard $C_p = 1{,}000$ J kg^{-1} °C^{-1}, we find that an updraft speed of 1 m s^{-1} will carry the required $NUSHX$ of 100 W m^{-2} (Section 8.10) in a convectively active area.

Similar analysis shows that the net upward flux of latent heat is given by

$$NULHF = UMF\, L\, [q - q'] = N\, A\, \rho\, w\, L\, \Delta q$$

$$NULHX = N\, \rho\, w\, L\, \Delta q \qquad \text{B8.3c}$$

where L is the latent heat of vaporization of water (≈ 2.5 MJ kg^{-1}) and $[q - q'] = \Delta q$ is the excess of specific humidity of the rising air (expressed as pure ratios, not g per kg). The ratio of sensible to latent heat transport

$$\frac{NUSHX}{NULHX} = \frac{\Delta T\, C_p}{L\, \Delta q}$$

which has numerical value $4 \times 10^{-4}\ \Delta T/\Delta q$. If the sensible and latent heat flux densities are equal, their ratio is unity and $\Delta q = 4 \times 10^{-4}\ \Delta T$, which becomes $0.4\ \Delta T$ when Δq is expressed in g per kg, rather than a pure ratio. This shows that in convection a Δq value of 0.4 g kg^{-1} is as effective as a 1 °C temperature difference in convecting heat.

Advection

Consider a straight East–West line of length D on the Earth's surface, a fraction N of which is crossed by warm air currents of temperature T moving polewards at speed V, and the remainder (fraction $1 - N$) being crossed by cool air currents of temperature T' moving equatorwards at speed V' (Fig. 8.17b). If the pressure 'thicknesses' of the warm and cool air currents are Δp and $\Delta p'$, the poleward and equatorward mass fluxes are given by

$$PMF = N\, D\, \frac{\Delta p}{g}\, V \quad \text{and} \quad EMF = (1 - N)\, D\, \frac{\Delta p'}{g}\, V'$$

If mass flow and counterflows balance to zero, as is usually nearly the case, the net poleward heat flux is given by

$$NPSHF = PMF\, C_p\, [T - T'] = PMF\, C_p\, \Delta T$$

and the net heat flux per unit length of the line of latitude (within D) is

$$NPSHX = N\, \frac{\Delta p}{g}\, V\, C_p\, \Delta T \qquad \text{B8.3d}$$

In slope convection and advection, observation suggests that N is about 1/2 (i.e. poleward and equatorward flows each occupy 1/2 of the active longitude range), Δp is about 75,000 Pa (corresponding to an active troposphere extending from 1,000 to 250 hPa), ΔT is about 10 °C, and g and C_p have their standard values. Then the value of V required to carry the globally estimated *NPSHX* of 300 MW m^{-1} (Section 8.10) is 8 m s^{-1}.

If we now use these values for N and ΔT in B8.3b together with $\rho = 1$ kg m^{-3} and the usual C_p, we find the vertical heat flux carried by slope convection. Taking the global estimate of 100 W m^{-2} for *NUSHX* (Section 8.10) the required uplift speed is 2 cm s^{-1}.

The ratio of uplift speed 2 cm s^{-1} and horizontal poleward speed 8 m s^{-1} would be the slope of the rising air (1:400) if all advection was by slope convection.

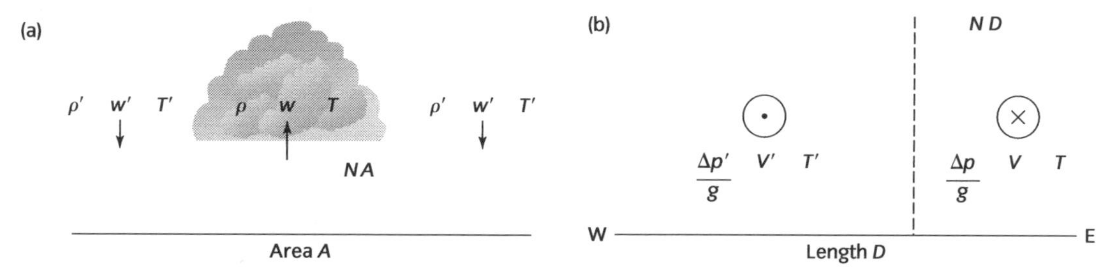

Figure 8.17 (a) Schematic picture of opposing vertical fluxes of mass and sensible heat in convection. Slightly warmer faster updrafts are embedded in larger areas of gentle subsidence. Mass fluxes balance, but there is a net upward heat flux. (b) Schematic picture of opposing poleward fluxes of mass and sensible heat in large scale advection in middle latitudes. Warmer poleward flows (cross tail of arrow) cooperate with cooler equatorward flows (point of arrow) to balance mass fluxes but maintain a net poleward heat flux.

8.9.4 Global advective heat fluxes

The annual average fluxes of solar and terrestrial radiation maintain a net heat input to the planet between latitudes about 32 °N and 38 °S, and a net output at higher latitudes (Fig. 8.9b, curve (i)). The observed nearly steady thermal state of the planet is maintained by a balancing advection of heat from low to high latitudes which must be the net result of meridional exchange of air and water, with warmer masses moving polewards and cooler ones moving equatorwards.

The observed steadiness of the atmosphere's thermal state means that annual average heat poleward advection across any latitude must nearly balance the net loss of radiant energy from all higher latitudes in that hemisphere—a quantity which can be calculated from curve (i) in Fig. 8.9b. As outlined in the Appendix for completeness, we can make successive calculations to produce the meridional profile of poleward advection of heat depicted in Fig. 8.18. Beginning at a pole

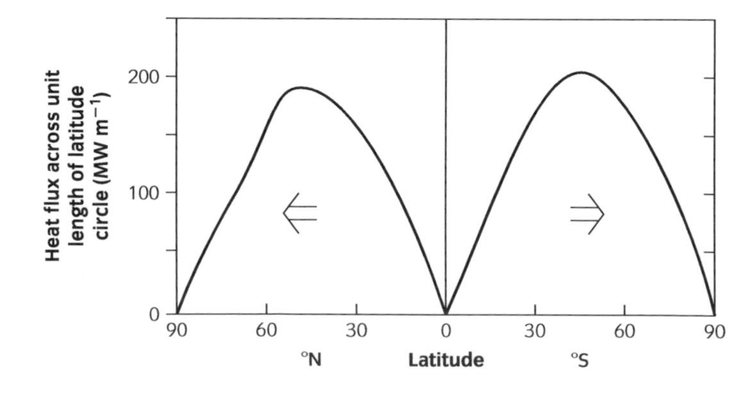

Figure 8.18 Meridional profiles of annually averaged poleward heat fluxes advected by the atmosphere and oceans, expressed per unit length of latitude circle.

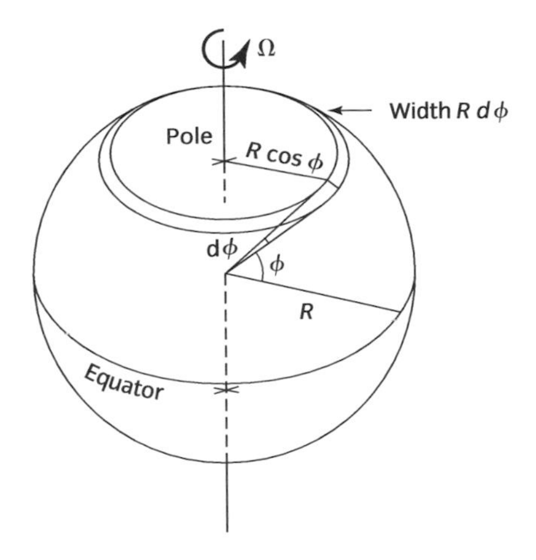

Figure 8.19 Oblique view of Earth, with narrow zonal ring at latitude ϕ. The surface area of the ring is ($2\ \pi\ R \cos \phi\ R\ d\phi$), which is A d(sin ϕ), where A ($= 2\ \pi\ R^2$) is the surface area of a hemisphere. The surface area between latitudes ϕ_1 and ϕ_2 is A ($\sin \phi_2 - \sin \phi_1$), which justifies the sin ϕ weighting in Figs 8.9 and 8.11. See also the Appendix.

(Fig. 8.19), where poleward advection must be zero, and choosing successively lower latitude zones, the calculated poleward heat advection increases progressively until we reach the latitude at which the net radiative flux changes from negative to positive, beyond which the advected flux begins to fall away again as the net input of radiant energy in each latitude zone lessens the need for heat advection from lower latitudes. In fact the poleward heat advection calculated in this way is found to reach about zero at the equator, showing that the annual radiant energy budget for each hemisphere is nearly balanced and does not depend on significant net advection between the hemispheres.

The advective fluxes in Fig. 8.18 have been divided by the lengths of the local latitude circles to produce average fluxes per unit length of latitude circle. Physically, such values represent the average advection of heat between imaginary goal-posts set one metre apart on a latitude circle, and reaching up to the tropopause and down to the sea bed. At any particular time and place the instantaneous heat advection between such goal-posts will be polewards or equatorwards depending on the local air and water flow, but after averaging around a latitude circle and over a substantial time period there is a systematic poleward heat flow with a reasonably definite value.

The maximum values apparent in Fig. 8.18 are very large indeed, approaching 200 MW m^{-1}, which means that the average poleward heat energy advection between goal posts 10 m apart is 2 GW—the generating capacity of a large power station. Note that displaying fluxes per unit length of latitude circle displaces the F maxima noticeably polewards from the latitudes of reversing radiant balance.

Although there is significant seasonal variation in the advected fluxes, it is much smaller than we might expect from the very large seasonal variations in net radiative balance (Fig. 8.11), because of the smoothing influence of the large heat capacities of the oceanic surface layers. For example, the middle latitudes of summer hemispheres would seem to require no poleward heat advection according to Fig. 8.11, but in fact in midsummer the ocean surfaces there are still warming towards their temperature maxima (to be reached in late summer or early autumn), significantly offsetting the radiative gain in the process. Despite such smoothing, winter advection in middle latitudes is noticeably stronger than summer advection, as suggested by the noticeably greater frequency and intensity of extratropical cyclones in winter.

8.10 Atmosphere and ocean systems convecting and advecting heat

The maintenance of the thermal state of Earth's surface layers demands continuing vertical and horizontal transports of heat, and on scales of more than centimetres these are carried out by vertical convection currents and horizontal advection currents in the atmosphere (especially the troposphere) and oceans. In the troposphere, convection is dominated by upright, turbulent *cumuliform* convection, associated with the whole cumulus family from thermals rising from a warm surface, through fair weather cumulus, to the largest cumulonimbus. Individually these are in the *small* meteorological scale (Fig. 1.5) but they are often patterned on the meso, synoptic, and even hemispheric scales by the larger weather systems of the troposphere. Advection in the troposphere occurs in horizontal air flows at all scales and heights, but especially in the synoptic and larger scales in the low and high troposphere. In addition, very significant tropospheric convection and advection occurs in the slightly but systematically tilted synoptic scale flows between the low and high troposphere associated with extratropical cyclones, which we will call *slope convection*, to distinguish it from the upright cumuliform convection.

In the oceans, there is turbulent convection in surface layers at all latitudes, and systematic downward convection of cold dense water into the ocean deeps in high latitudes. Advection is achieved by wind-driven currents in the top hundreds of metres (e.g. the Gulf Stream), and slow persistent density currents in the ocean deeps, fed by descent of cold dense waters in high latitudes.

The network of flows and currents is sketched in Fig. 8.20 and outlined in the following.

8.10.1 Troposphere

Convection

Most cumuliform convection begins as a turmoil of turbulent metre-sized up- and downdrafts over a relatively warm surface. At any instant the active area is divided equally between up- and downdrafts, but the rising air is systematically warmer and moister (except over deserts) so that there are significant net upward fluxes of sensible and latent heat, even though there is little or no net vertical mass flux.

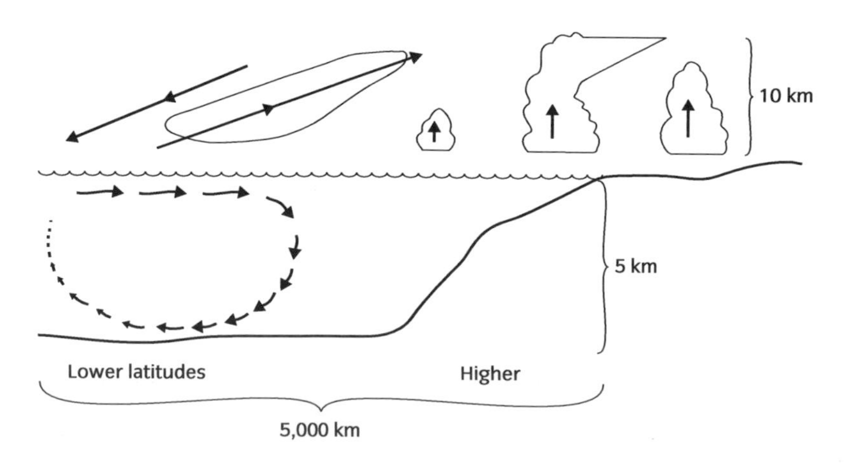

Figure 8.20 Convection and advection currents in the atmosphere and oceans, showing convection and slope convection in the atmosphere, wind-driven surface currents, and deep currents in the ocean with quick descent and slow ascent.

At greater heights in the convecting layer, the updrafts are narrower and stronger, so that at and just above cloud base updrafts of ~ 1 m s^{-1} typically occupy ~ 10% of the active area, the rest of which is filled with gently subsiding air. This asymmetry persists through all the large types of cumuliform convection, and maintains the typical sporadic distribution of cumulonimbus, and the showers they produce. Net upward mass fluxes of air remain small across the active zone, but updrafts from 1 to 5 m s^{-1} or more, with temperature excesses of 1–5 °C (rarely more), and specific humidity excesses of a few g kg^{-1}, are able to maintain net upward fluxes of sensible and latent heat ~ 100 W m^{-2} when averaged across the active area (Box 8.3).

In equatorial regions, deep showery convection is clustered in the Inter-tropical Convergence Zone (ITCZ) which girdles the sea-covered globe and is broken into seasonally mobile fragments over low-latitude land masses, giving local rainy seasons. The ITCZ maintains a strong vertical flux of latent and sensible heat from surface to the high troposphere, and a much weaker net upward mass flux which drives the Hadley circulation (Fig. 8.21) feeding air polewards and E'wards in the high troposphere to the Subtropical High-pressure zones, where it sinks gently but persistently back to the low troposphere, to feed the equatorward and W'ward flows of the Trade Winds. The Trades are populated by trade wind cumulus (Fig. 8.22) maintained by convection from warm sea surfaces.

The persistent concentration of cumuliform convection in the Trades and the ITCZ is a spectacular example of the patterning of apparently random updrafts by much larger scale flows. On the equatorial flanks of the Trades, the showery convection is patterned on the synoptic scale by the family of tropical cyclonically rotating disturbances (*tropical cyclones*) which develop sporadically and drift W'wards. In these systems the more or less deep cumuliform convection is patterned around mostly shallow low-pressure systems which move slowly W'wards in the Trade Wind flow. The most intense of these are the hurricanes, typhoons, severe cyclonic storms, etc. variously named in the different regions they afflict, whose very deep meso-scale low-pressure centre is surrounded by a ring of powerful cumulonimbus (Section 13.5).

Uplift in slope convection is confined to the *extratropical cyclones* of middle and occasional high latitudes, and is much more extensive and weaker than cumuliform updrafts. It maintains the large areas of stratiform cloud and

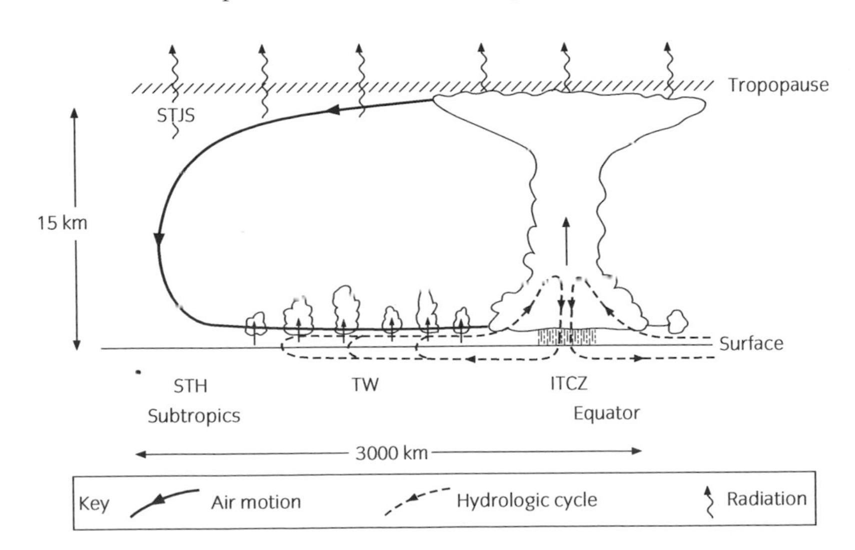

Figure 8.21 A vertical meridional section of the Hadley cell in one hemisphere, with Earth curvature ignored. In anticlockwise order from the equator, the main labelled features represent the intertropical convergence zone, the subtropical jet stream, the subtropical high-pressure systems, and the Trade Winds.

Figure 8.22 Trade- wind cumulus on the coast of Thailand. At first glance the tilt of the clouds suggests that they are leaning forwards, with a sheared wind from the sea. In fact the kite surfer is restraining his kite in a strong breeze blowing from the land, and the clouds are leaning back against the wind in a reversed shear, as is typical of the Trades. The flow is NE'ly near the surface in the N hemisphere, but it reverses direction in less than a kilometre and so is strongly sheared back against the wind, as the clouds show.

persistent precipitation associated with the fronts of extratropical cyclones (Section 12.3). Uplift speeds are too small to be measured directly, but from indirect estimates (such as steady rainfall rates) are known to be ~ 10 cm s^{-1} across areas which can exceed 10^5 km^2. Comparable areas and strengths of sinking air are interlaced with the rising air flows in complex three-dimensional flows (Section 12.4). Again uplift and subsidence produce little net vertical mass flux (though enough to help feed subsidence in nearby ridges of high pressure), but systematic temperature and vapour excesses in the rising air flows maintain net upward fluxes of sensible and latent heat which can be ~ 100 W m^{-2} when averaged over the large active areas of each extratropical cyclone.

As in low latitudes, cumuliform convection in middle latitudes is usually patterned by synoptic scale systems. The equatorward flow of cool air on the W side of each extratropical cyclone is characteristically dotted and patterned with shower clouds pumping sensible and latent heat into the cool air (Fig. 2.14). Within the cold fronts in particular, vigorous cumulonimbus are often embedded in the nimbostratus of the frontal cloud proper, adding cumuliform updrafts to the general uplift of the sloping air flow, and compounding the two types of convective heat transport.

Anticyclones in all latitudes contain large masses of very gently subsiding air, losing heat to space by net emission of terrestrial radiation, and allowing air which has risen in cloudy updrafts and uplift elsewhere to return to the low troposphere. In the low troposphere this subsiding air is in dynamic equilibrium with shallow cumuliform convection from the surface (Section 13.2), so that the surface-based convecting layer is warmed partly by convection from the surface and subsidence from above.

Advection

In middle and high latitudes the net poleward advection of heat is carried largely by the warm poleward air flow on the E side of each extratropical cyclone and the cool equatorward flow on its W side, whose contrasts are a familiar part of daily life in middle latitudes, especially on western margins. Less obviously, powerful localized slope convection parallel to the line of the cold front pumps sensible and latent heat polewards, upwards and E'wards, while slope convection in the lower parts of the warm front pumps heat polewards and upwards (Section 12.4). Blocking anticyclones (substantial high-pressure systems well poleward of the Subtropical Highs) have warm air flowing polewards on their W sides and cool equatorward flows on their E sides.

In lower latitudes there are still substantial meridional flows of heat, but the pattern lacks the obvious poleward drive of higher latitudes. The W ends and poleward flanks of STH cells feed warm moist air into higher latitudes, to become part of the poleward heat transport there, some of it in slope convection. Their E ends and equatorward flanks feed air into the equatorward and W'ward Trade Wind flows, which carry sensible heat equatorwards, together with a proportion of latent heat which increases equatorwards as the Trade Wind cumulus pump vapour into the initially dryish low-level flow (Fig. 8.21). This increasing flow of latent heat feeds taller and taller cumulonimbus, including those involved in tropical disturbances and cyclones, culminating in the giants of the ITCZ. As a result there is a significant net equatorward heat transport in the low troposphere which is reversed in the upper troposphere, where there is persistent poleward flow of air warmed by cloudy ascent in the ITCZ. This latter is largely sensible heat since virtually all the vapour has condensed in the cold tropical high troposphere. In this way the sensible and latent heat fluxes, so closely married in middle latitudes, are opposed in the Hadley circulation: equatorward flow of latent heat in the low troposphere feeds the convection in the ITCZ from which it flows polewards, as sensible heat in the high troposphere—the vapour having been precipitated by the torrential showers of the ITCZ.

The Hadley circulation is the only clear example of roughly steady flow and counterflows of contrasting air in the troposphere, but even it is broken by the necklace of semipermanent STH cells, and the seasonal disruption of the monsoonal migrations of the ITCZ fragments overland, especially the enormous outburst of the Asian monsoon. All other heat transporting air flows are characteristically unsteady, from the cumuliform convection which is rife in all latitudes, to the outbursts of slope convection of middle latitudes, where systematic heat transport is effected by unsteady flow and counterflow of systematically contrasting air (Box 8.3).

8.10.2 Oceans

The world ocean plays a huge role in the heat budget of Earth's surface layers by vigorous exchange with the low troposphere by means of conduction, convection, and terrestrial radiation to and from its surface. Its relatively low albedo helps it capture solar energy better than land, its heat capacity is over 1,000 times that of the atmosphere, and it is freely mobile. The last two imply a huge capacity to transfer heat in water currents, sketched in Fig. 8.23 and outlined below. Despite these enormous exchanges and transfers, the heat budget of the oceans is very nearly perfectly balanced when its surface and deep activity is accounted over a few integral years, because the only other heat fluxes are the very small geothermal heat flux up through the sea bed (about 50 mW m^{-2} on average), the frictional heating by tides in shallow waters (~ 10 mW m^{-2}),

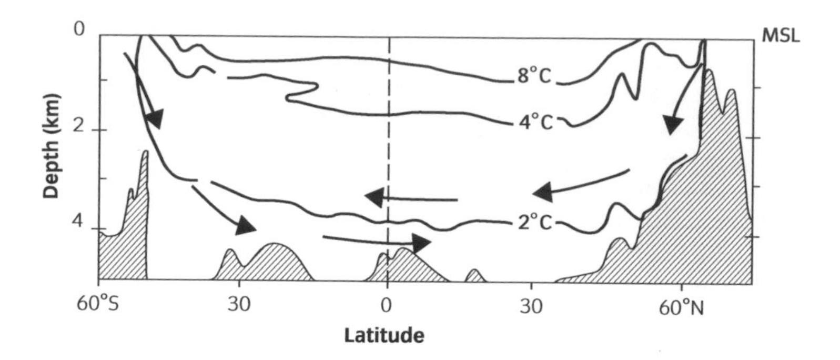

Figure 8.23 An idealized vertical meridional section through the Atlantic, showing cold water sinking into the deeps at high latitudes, and the permanent thermocline (centred on the 8 °C isotherm) separating the warm and cold water spheres. Notice how deep is the cold water sphere even in the equatorial zone—probably a characteristic of all ice ages.

and imbalances consistent with climate change (~5 mW m^{-2} at most). A small persistent heat flow from the atmosphere accompanies global warming and the opposite accompanies global cooling.

Convection

Across most of the oceans the main vertical heat flow is driven through the top tens of metres of water by wind-driven turbulence, upwards or downwards depending on whether the sea is warmer or cooler than the overlying air. Relative air–sea temperatures reverse seasonally in middle and high latitudes, and irregularly in all latitudes when weather systems produce temporary rises or falls of air temperature, for example in the cold air flows W of extratropical cyclones.

Most of these vertical transfers cancel out over a year or so, but a net diffuse downward heat flow in low latitudes balances the persistent downward flow of cold water (upward heat flow) localized in high latitudes (Fig. 8.23 and below). The downward heat flux is concentrated dynamically under the STHs, where the Ekman flow patterns (Section 10.10) in the turbulent top ~ 50 metres of the gyres of warm water under the STHs, maintain synoptic scale convergence of the warm surface waters, bulging them down many hundreds of metres *[39]*. In the range of meridians averaged to make Fig. 8.23, the effect shows most clearly in the downward bulge of the 8 °C isotherm at around 30 °N.

The downward flux of cold water in high latitudes (arrowed in Fig. 8.23) is maintained by strong surface cooling by net radiative loss and convection to the cold air. Seasonally aided by salt extruded from growing sea ice, this dense water falls slowly into the ocean deeps, where it meanders over the sea bed (typically 4 km below the sea surface) or the top of even colder, denser water. The coldest flows come from the edge of Antarctica. All this water is gradually forced upwards by the arrival of newer, colder water, until it begins to mix with surface waters in low latitudes and eventually (after decades or centuries) is returned to the high-latitude sink regions by the great poleward currents. This whole process has kept the ocean deeps full of very cold water since the long slide into the present ice age began tens of millions of years ago (Section 14.5)

Advection

Broad slow equatorward advection of cold water in the ocean deeps is more than balanced by concentrated narrow warm poleward advection of warm surface waters on the W sides of the ocean basins in the subtropics—the Gulf Stream (Fig. 8.24) in the NW Atlantic, the Kuroshio in the NW Pacific, and similar

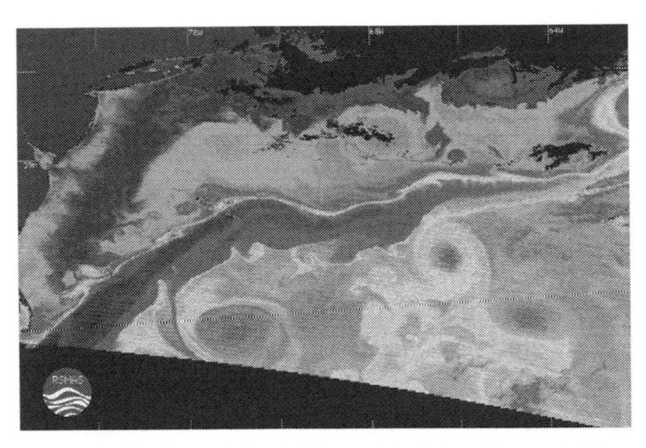

Figure 8.24 A satellite picture of the Gulf Stream. The infrared image has been coloured so that warm water looks red, cold water looks blue and purple, and impossibly cold surface (cloud tops) look black. Land and the area with no data has been systematically blackened. The Gulf Stream is warmer than the water to its E, and much warmer than the water to its W, some of which, showing near the black cloud masses, is very cold. The Gulf Stream is several hundreds of metres deep and flows at about 2 m s^{-1}, carrying vast amounts of heat polewards. (See Plate 11).

currents in corresponding areas in the S hemisphere. The full meridional mass flux balance also involves weak broad equatorward flows of cool surface waters on the E sides of the subtropical gyres. But though the mass fluxes of water balances overall, the counterfluxes of warm and cold water maintain very significant systematic poleward advection of heat in the usual way, accounting for about one quarter of the total required to satisfy the meridional radiative imbalance shown in Fig. 8.18, and leaving the remaining three quarters to tropospheric advection.

8.10.3 Wind speeds and heat fluxes

We have seen that heat is carried upwards and polewards in the atmosphere by cumuliform and slope convection, i.e. by weather systems associated with cumulus, cumulonimbus, and frontal stratiform clouds. These systems have typical properties which are well defined by observation and which can be checked for consistency with the heat fluxes which we know have to be maintained for meridional equilibrium.

Consider the situation at latitude 45° where the poleward heat flux per unit length of latitude circle is about 200 MW m^{-1}, and the vertical heat flux density is 100 W m^{-2} (as it is nearly everywhere over the globe). Assume that one quarter of the poleward heat flux is carried by ocean currents, leaving 150 MW m^{-1} to the atmosphere; and assume that one half of the vertical and poleward heat fluxes in the atmosphere are in the form of sensible heat. Then the flux densities of sensible heat which must be maintained by atmospheric weather systems are 75 MW m^{-1} polewards and 50 W m^{-2} upwards. Such equal partitioning between latent and sensible heat is apparently inconsistent with the surface data in Fig. 8.16, where latent heat predominates, but we are interested at present in transport in the lower middle troposphere, where some of the latent heat flux at the surface has already been transformed to sensible heat by cloud formation, rather than at the surface.

With some further simplification of the model of vertical and slope convection, we can estimate the vertical and horizontal wind speeds required to maintain the required fluxes of sensible heat.

Required wind speeds

Consider first the net poleward advection of sensible heat. We suppose that this is effected entirely by the slope convection and horizontal advection associated with extratropical cyclones and intervening ridges of high pressure, and suppose further that on average these effectively occupy one quarter of the 45° latitude circle. Taking these quadrants to be filled by equal and opposite meridional air currents which have wind speed V and temperature difference 10 °C, then according to the method of Box 8.3, the meridional wind speed V must have value 8 m s^{-1} to provide the flux density of 300 MW m^{-1} required in the active quadrant. This is an encouragingly realistic value, bearing in mind that horizontal wind speeds in the more active parts of the troposphere are $\sim$ 10 m s^{-1}. In fact our simplistic model is effectively combining weaker general horizontal flow with stronger sloped flow in a relatively small active zone, which consequently should be well over 10 m s^{-1} on average, as observed.

In considering the vertical transport of sensible heat, we assume that this is borne equally by cumuliform and slope convection, and that cumuliform convection occupies another quadrant of the latitude circle. This means that each contributes 25 W m^{-2} of sensible heat flux when zonally averaged, and 100 W m^{-2} in its active quadrant. However, they do this in very different ways, since the updrafts in cumuliform convection occupy only about 10% of the area of their active zone, and are only about 1 °C warmer than the subsiding air which fills the remaining 90%, whereas up- and downdrafts in slope convection cover roughly equal areas and differ in temperature by about 10 °C. Using the method of Box 8.3, cumuliform and slope convection will each produce vertical heat fluxes with density 100 W m^{-2} in their active quadrants if the updrafts have speeds about 1 m s^{-1} and 2 cm s^{-1} respectively. For a comparable upward flux of latent heat in cumuliform convection, the rising air should be about 0.4 g kg^{-1} moister than the sinking air. This is actually quite a small difference in vapour content, making the point that latent heat fluxes are potentially large in the moist low troposphere.

Again these values are reasonably realistic. Updrafts in cumuliform systems are known to range from about 1 m s^{-1} in small cumulus to over 10 m s^{-1} in large cumulonimbus. And updrafts in fronts are known from indirect evidence to range from a few centimetres per second in weak systems to several times this in vigorous ones. Notice that the ratio of vertical to horizontal wind speeds in slope convection inferred from required heat fluxes is consistent with air currents having slopes of about 1 in 400 compared with values ~ 1/100 derived from three-dimensional air trajectory analysis (Section 12.4). The discrepancy is reduced when we realize that horizontal speeds are perhaps twice as great in the zones of active slope convection.

Such very simple calculations are not meant to be conclusive; the realistic results for wind speeds and updrafts clearly depend on judicious selection of the various values assumed. Nevertheless these values are in themselves quite plausible, and the consistency of the whole analysis suggests that cumuliform and slope convection can account for the heat fluxes known to be advected and convected through the middle latitude troposphere.

Appendix Meridional heat flux profiles

According to the global geometry outlined in Fig. 8.19, the surface area of a narrow zone bounded by latitudes ϕ and $\phi + d\phi$ is $2\,\pi\,R^2 \cos\phi\, d\phi$. The surface area ${}_1A_2$ of a finite zone bounded by latitudes ϕ_1 and ϕ_2 is found by integrating the narrow zone expression between these limits.

$$ {}_2A_1 = 2\,\pi\,R^2\,(\sin\phi_2 - \sin\phi_1) \qquad \text{B8.4a} $$

Substituting values for ϕ_1 and ϕ_2 confirms that the surface area of a hemisphere is $2\,\pi\,R^2$, and that 50% of this area lies between the equator and latitude 30°, while only 13.4% lies poleward of latitude 60°, highlighting the visually misleading geometry of the Mercator map projection and justifying the cos ϕ weighting used in Figs. 8.9 and 8.11.

If we divide the hemisphere into latitude zones each 10° wide, and within each zone define a zonal average net influx F of radiant energy per unit horizontal area (from Fig. 8.9 or equivalent), then the total influx into a zone is the product of F and the zonal surface area (B8.4a). In a steady thermal state this heat influx must be exactly balanced by the net efflux of heat borne by meridional advection. If advection is described by meridional heat flux densities M_1 (average flux across unit length of latitude circle at latitude ϕ_1), then the advected flux across any latitude circle is $M_1\,2\,\pi\,R\cos\phi_1$, and the equation for the thermal equilibrium of a zonal strip bounded by latitudes ϕ_1 and ϕ_2 is

$$ M_2\cos\phi_2 - M_1\cos\phi_1 = F\,R\,(\sin\phi_2 - \sin\phi_1) $$

after cancelling the common $2\,R$. Since M at the poles is necessarily zero, we can work equatorwards zone by zone using observed values for F to calculate M values at each latitude.

Checklist of key ideas

You should now be familiar with the following ideas.

1. Solar radiation, the laws of blackbody radiation, photospheric emittance, the inverse square law, Earth's solar constant 1.38 kW m^{-2}.

2. Terrestrial radiation and radiative equilibrium, planetary albedo, atmospheric interactions with terrestrial radiation.

3. Vertical cascade of terrestrial radiation, necessarily dynamic troposphere, natural greenhouse effect, greenhouse roles of water vapour, cloud, and carbon dioxide.

4. Global annual budget of solar radiation, terrestrial radiation, and convected heat.

5. Meridional and seasonal distributions of solar and terrestrial radiation and convected and advected heat.

6. Surface irradiance, atmospheric optical depth and transmissivity, diurnal patterns of solar input and terrestrial output.

7. Convective and advective heat transport preserving planetary, meridional, and local thermal equilibria, near surface heat conduction.

8. Convective fluxes of sensible and latent heat, the Bowen ratio, expressions for conductive, convective, and advective heat fluxes.

9. Convective and advective heat fluxes related to regimes of weather and climate in low and middle latitudes, ocean heat fluxes and circulation.

10. Atmospheric heat fluxes related to typical tropospheric patterns of meridional winds, vertical air flow, and temperature contrasts.

Problems

Outline answers to these problems can be found on the **Online Resource Centre**. Answers to odd numbered problems can be found under Student Resources, answers to even numbered problems under Lecturer Resources.

Level 1

8.1 Some stars are much redder in appearance than others. What does this tell you about their surface temperatures? The orbital radius of Venus round the Sun is about two thirds of Earth's. Find the ratio of the solar constants of the two planets.

8.2 A black-and-white cat is lying in strong sunlight. Compare its visual appearance with its appearance when viewed by a radiometer sensitive to the infrared, whose display renders strong radiators black and weak radiators white. Why is this rendering used in meteorological satellites?

8.3 How would the shape of the terrestrial radiative spectrum observed in Fig. 8.4 have changed if, at the time of observation, there had been a complete overcast of clouds with tops at temperature 250 K?

8.4 Which of the following alterations would tend to enhance, and which reduce, the atmospheric greenhouse effect: widening the atmospheric window; making the flanks of the window more opaque to terrestrial radiation; having a hotter but smaller Sun with the same solar constant; increasing the direct absorption of solar radiation by the atmosphere? Give a brief reason for each choice.

8.5 Describe everyday observations which suggest that limitation of convection is crucial to the performance of actual greenhouses and similar constructions?

8.6 Assuming that the fluxes of terrestrial radiation in Fig. 8.8 are otherwise unchanged, what rate of heat loss do you expect at the surface at night (a) when the sky is clear, and (b) when the sky is overcast by low cloud?

8.7 Find the angle of solar elevation (above the local horizon) at noon at latitudes 25°, 45°, and 65° at the winter solstice, the equinoxes, and the summer solstice.

8.8 In Fig. 8.16 the net upward flux of terrestrial radiant energy increases with height through the troposphere, so that every parcel of finite depth loses more through its top than it gains through its base. How is this maintained without continual cooling of the air?

8.9 Consider the near-surface wind directions and air temperatures on the W and E flanks of a mid-latitude anticyclone straddling a W continental margin, and argue that the resulting net advection of sensible heat is likely to be equatorwards in summer and polewards in winter. Similarly consider the directions of the net fluxes of latent heat.

8.10 In Fig. 8.21, what happens to the latent heat released by condensation in the ITCZ, and what are the main differences in the middle troposphere between air ascending in the ITCZ and descending in the STH?

Level 2

8.11 Venus, Earth, and Mars orbit the Sun at average distances of 108, 150, and 228 million kilometres respectively. Given that the Sun is a spherical blackbody of radius 700,000 km and temperature 5,800 K, find the solar constant for each planet.

8.12 Using Eqn 8.5, find that the equivalent blackbody temperature of the Earth when its albedo is 0.3, and repeat this when the albedo is 0.35, as in the solstices of the last glacial maximum (Section 14.6). Repeat this in a more general way by rearranging Eqn 8.5 to show that the ratio of the equivalent blackbody temperatures T_{E1}/T_{E2} is given by

$$[(1-a_1)/(1-a_2)]^{1/4}$$

where a_1 and a_2 are the respective albedos, and inserting values for T_{E1}, a_1, and a_2.

8.13 Find the radiative emittance of the following surfaces, all except one of which are blackbodies: water at 25 °C, sand at 40 °C (emissivity 0.85), fog at 5 °C, cloud at −20 °C and at −40 °C. Consider a dark rockface (solar albedo 0.3) in the full glare of strong direct Sun (normal irradiance 1,100 W m^{-2}), and find its surface temperature at radiative equilibrium, using the sand emissivity value. Though this must be an overestimate (why?) people claim to have fried eggs in the Sahara.

8.14 In Fig. 8.8, the radiative inputs and outputs for the stratosphere look very small (3 units) compared with those of the troposphere (~ 30 units). But for a true comparison we need to compare their thermal capacities. Assume that the atmosphere has decadal scale height 16 km, that the troposphere fills 0–16 km, that the stratosphere fills 16–48 km, and show that the tropospheric heat capacity is ≈9.1 times larger.

8.15 In a certain clear sky the zenith transmissivity for solar radiation is 0.75. Assuming the normal solar irradiance above the atmosphere to be 1.38 kW m^{-2}, find the irradiance at sea level of normal (to the direct Sun) and horizontal surfaces when the Sun is 30, 60, and 90° and above the horizon (Box 8.2).

8.16 Consider the energy loss represented by net 30 W m^{-2} radiative loss over a three-month season, and express this in terms of the cooling of a 10 m layer of water, the cooling of the whole atmosphere, and the depth of ice formed by freezing of the ocean surface at 0 °C.

8.17 If 1% of the maximum poleward heat advection across the 45° line of latitude (peak of Fig. 8.18 multiplied by the length of the line of latitude there) were to be converted with 50% efficiency into usable electrical power, estimate the power available per head of world population.

Level 3

8.18. Use the approach outlined in the Appendix to show that if the net radiative input per unit surface area is F when averaged over all latitudes from the equator to 45°, and the average meridional heat flux to higher latitudes is M per unit length of latitude line at 45°, then $M = R\,F$ where R is the radius of the Earth. Use this expression to estimate a value for F from the maximum M in Fig. 8.18 and compare your result with Fig. 8.7b.

8.19 Consider the following very simple model of the atmosphere in purely radiative equilibrium. It consists of a single thin sheet of air at absolute temperature T_a which is transparent to solar radiation but acts as a greybody with absorptivity and emissivity a in the far infrared. The underlying surface acts as a blackbody with temperature T_g. Find the relation between T_g and the solar input, and between T_a and T_g, and calculate numerical values for T_a and T_g using a realistic value for solar input and $a = 0.8$.

8.20 The meteorological display mode for infrared imagery (darker for strong IR emitters, lighter for weak) makes many panoramas look much as they do in visible wavelength, but some differences do arise. Discuss the differences to be expected in the following cases: patchy sea fog; total cover of low stratocumulus; not very cold snow-covered low ground showing through high cloud; a hot pale desert; clouds in oblique sunlight; fresh snow or hail showers on previously snow-free ground at the same temperature; loose pack ice.

8.21 Check on the equatorial value for the solar input in Fig. 8.9a by considering the input there at an equinox. (*Hint:* in this situation solar radiation is collected continually on a diameter of the Earth's disk, but is spread around the Earth's circumference by its rotation.) The difference allows you to estimate the effective albedo there.

8.22 Modify the method for calculating heat advection in Box 8.3 to analyse the following simplified picture of poleward ocean surface currents such as the Gulf Stream at 30° N: zonal width 200 km; poleward speed 1 m s^{-1}; depth of current 500 m; temperature excess of warmer flow 5 °C over cooler return flow elsewhere. Divide the heat flux by Earth's circumference at 30° latitude to compare with Fig. 8.18.

8.23 The heat capacity of the Arctic Ocean obviously tends to limit the seasonal swings of surface temperature there (between polar day and polar night). How is this tendency affected by the presence of permanent ice there, and by the seasonal presence, freezing, and melting of sea ice?

9 The atmospheric engine

9.1 Introduction

The Earth's atmosphere and oceans act as a set of global, interacting, self-regulating *heat engines*, continually taking in solar heat energy and using a small part of it to move air and water (and the heat they carry) from warm *heat sources* to cool *heat sinks*, before exhausting all the input energy to space as terrestrial radiation (Section 1.5). Awareness of heat engines hardly existed before steam engines were

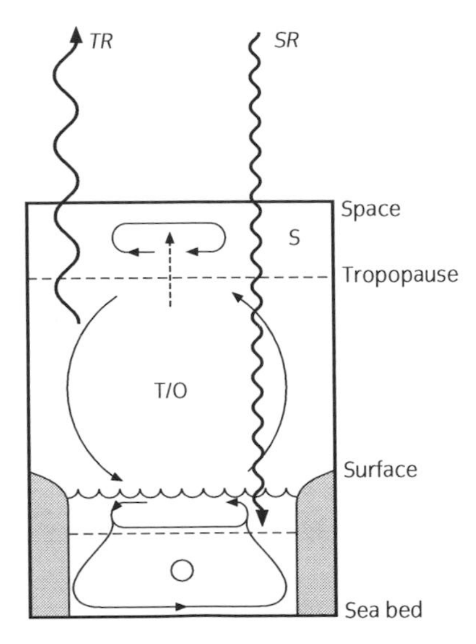

Figure 9.1 Schematic outline of the atmospheric and oceanic heat engines. Curved arrowed lines represent air or water flow in the troposphere/ocean heat engine T/O, the stratospheric heat pump S, and the deep ocean engine O. *SR* and *TR* are the solar radiant and terrestrial radiant energy inputs and outputs, and the vertical dotted arrow from T/O to S represents the injection of mechanical energy into the stratosphere.

invented in the eighteenth century, but subsequent development of these and other artificial heat engines has led us to recognize that heat engines are widespread in nature, and that the energy they process is the common currency of all physical, chemical, and biological activity. In this chapter we examine the atmospheric and ocean engines, identify their energy inputs, and trace their energy transformations, transports, and outputs.

Though natural and artificial heat engines can differ greatly in scale and detail, they have similar functional outlines and obey the same physical laws. Two of these laws (the *first* and *second laws of thermodynamics*) are amongst the most important known to science, and we apply them to simplified models of the atmospheric and oceanic heat engines as a framework for their mechanistic description (Fig. 9.1). The dominant *tropospheric* engine (which includes the wind-driven ocean surface currents) is driven by solar heat entering at warm heat sources concentrated at and near the Earth's land and ocean surfaces in low and middle latitudes. The *mechanical energy* generated keeps the troposphere and surface layers of the ocean in motion, distributing heat energy to cooler regions and eventually feeding it to the much cooler *heat sinks* located mainly in the high troposphere. A weaker *ocean engine* (Section 8.10.2) maintains a slow but persistent circulation through the ocean deeps, and interacts with the tropospheric engine, though it could probably operate without it. A very small proportion of the mechanical energy produced in the troposphere/ocean engine drives the weak *stratospheric heat pump*—a heat engine working in reverse to pump heat from cool stratospheric heat sources to slightly warmer sinks, where it becomes part of the total output of terrestrial radiation.

9.2 Heat and mechanical energy

By 1850, after a century of confusion and clarification, the many apparently different forms of energy (heat, kinetic, potential, electrical, radiant, etc.) were being recognized as different forms of a common physical property, exchanging in characteristic ways, and measurable through the performance and consumption of *mechanical work* (Section 5.2 and below). In the SI system the unit of work is the newton metre (N m), named the *joule* (J) after James Joule, who played a crucial role in quantifying the relationship between mechanical and heat energy in the 1840s. Energy *flow rates* or *powers* are measured in joules per second, named *watts* (W) after James Watt, the pioneer steam engineer.

Conversion from (solar) *radiant energy* to *heat energy* maintains the energy input to the terrestrial heat engines, and conversion from heat to (terrestrial) radiant energy maintains the balancing energy output. Within the atmospheric and oceanic engines two other types of energy are particularly important:

(i) *latent heat*—the heat energy which is apparently consumed when water evaporates and which is released again when water vapour condenses to water; a similar but smaller absorption of heat on melting and release on freezing plays a subsidiary role (Section 6.11);

(ii) *mechanical energy*—the energy of motion of an air or water parcel and of its position in Earth's gravitational field.

9.2.1 Mechanical energy

Kinetic energy

If a force F pushes a freely moving body of mass m a little distance ds in the direction of F, it does a little *work* F ds on the body. Using the equation of motion , we find (Box 9.1) that the work done produces a little increase in the body's energy of motion—its *kinetic energy* ($KE = 1/2\ m\ V^2$).

$$dKE = F\,ds = d(1/2\ m\ V^2) \qquad 9.1$$

The work done by F *on* the body *increases* its kinetic energy by transferring energy from the source of F to the body. If F and ds are opposed, F ds is negative, and the work done *by* the body moving against F *decreases* its kinetic energy by transferring energy from the body *to* the source of F. Doing work always transfers energy between bodies, in this case between the body and the origin of F.

If we accumulate all the little elements of work done along an extensive path of a freely moving body (Fig. 9.2), the total work done on the body as it traverses the path is equal to the increase in its kinetic energy between its beginning and end, regardless of intermediate values.

$$\text{total work done} = 1/2\ m\ (V_2^2 - V_1^2) = KE_2 - KE_1 = \Delta KE \qquad 9.2$$

This simple and very useful result applies to any body moving freely under the influence of any force or forces, including parcels of air or water moving in the terrestrial heat engines. If F and ds are not aligned, the element of work done in moving ds is F_s ds, where F_s is the component of F (Section 7.2) in the direction of ds.

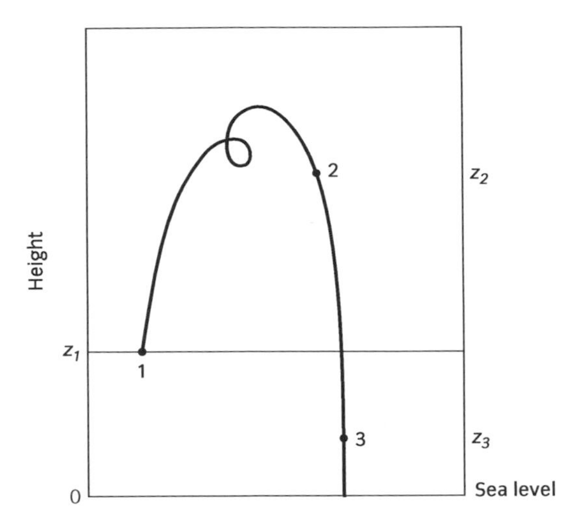

Figure 9.2 The vertical and horizontal trajectory of an air parcel of mass m acted on by several forces. Assuming uniform g, gravitational potential energies at points 1, 2, and 3 are $m\,g\,z_1$, $m\,g\,z_2$, and $m\,g\,z_3$ relative to mean sea level datum. Work done by the forces (including gravity) acting along the parcel path between any two points exactly equals the change in the parcel's kinetic energy ($1/2\ m\ V^2$) between the points.

Potential, kinetic, and mechanical energy

The atmospheric and oceanic engines work in the presence of Earth's gravitational field, which acts vertically downwards with nearly uniform strength in the small height range of meteorological interest. In this field a parcel of mass m experiences a downward gravitational pull (its weight) $m\,g$, where g is the gravitational acceleration (Section 7.4.2). If the action of some other force (e.g. buoyancy) raises the parcel from height z_1 to z_2 above sea level, the work done by that force increases the energy stored in the gravitational field. We say that the parcel's *gravitational potential energy* ($PE = m\,g\,z$) has been *increased* by an amount $\Delta PE = m\,g\,(z_2 - z_1)$ which is the work done in raising it *against* the pull of gravity. This change is unaffected by intervening vertical looping or horizontal motion, i.e. by any motion other than the height increase $(z_2 - z_1)$ between the end points of the parcel path (Fig. 9.2).

If the parcel's height falls from z_2 to z_3, we say that the parcel's gravitational potential energy has been *decreased* by the work done by gravity

$$\Delta PE = -\,m\,g\,(z_2 - z_3) = m\,g\,(z_3 - z_2)$$

regardless of any horizontal movement. Parcel PE acts like an energy reservoir which fills as the parcel rises, and empties as it falls.

In the simple case where gravity is the only force acting on a parcel, Box 9.1 shows that a change of PE is associated with an equal and opposite change of KE so that

$$\Delta PE + \Delta KE = 0 \quad \text{which means} \quad PE + KE = ME = \text{constant}$$

where ME is the *mechanical energy* of the parcel. As the parcel rises it slows as its KE converts to PE, and as it falls it accelerates as its PE converts to KE. We can regard the kinetic and potential energies of a body as complementary

BOX 9.1 Kinetic and potential energy

Work done and kinetic energy

Consider an air parcel of fixed mass m moving freely in a straight line at speed V under the action of a force F. According to the equation of motion (Eqn 7.3)

$$F = m\frac{dV}{dt}$$

In a little time interval the parcel moves distance ds and F does work $F\,ds$ on the parcel, where

$$F\,ds = m\,ds\frac{dV}{dt} = m\frac{ds}{dt}\,dV = m\,V\,dV \quad \text{B9.1a}$$

using the fact that ds/dt is the parcel speed V. By calculus or algebra (§)

$$V\,dV = d(1/2\ V^2)$$

so that

$$F\,ds = d(1/2\ m\ V^2) = dKE \quad \text{B9.1b}$$

showing that the work done by F on the parcel is equal to the increase of the parcel's kinetic energy KE ($= 1/2\ m\ V^2$).

In the general case where F and V are not aligned (F is always aligned with acceleration dV/dt, but acceleration and velocity are often out of alignment) F in B9.1b is replaced by F_s, the component of F in the direction of motion. If F is perpendicular to V, it does no work and speed V is unaltered, even though velocity may be changing direction, as in the case of circular motion under a centripetal force.

(§) If you are not familiar with $V\,dV = d(1/2\ V^2)$, rewrite ($V\,dV$) as the product of the mean speed $1/2(V_1 + V_2)$ and the speed increase $(V_2 - V_1)$, to find $1/2(V_2^2 - V_1^2)$, or $d(1/2\ V^2)$.

Any extensive parcel trajectory (however curved, looped, or knotted), and however variable the force acting along it, can be constructed from a very large number of very small elements ds. Totalling all the elements ($F\,ds$) of work done along the extensive parcel trajectory

$$\int F\,ds = \frac{1}{2}mV_2^2 - \frac{1}{2}mV_1^2 = KE_2 - KE_1 = \Delta KE \quad \text{B9.1c}$$

showing that the total work done on the parcel by F along the trajectory is equal to the increase in parcel KE between its end points, regardless of variations along the way.

Kinetic, potential, and mechanical energy

When a stone (mass m, weight $m\,g$) is thrown upwards at speed V in frictionless air, the work done by the stone against gravity as it rises a little height dz is given by $m\,g\,dz$, which by Eqn 8.1 and text corresponds to the little change in the stone's KE

$$dKE = -\,m\,g\,dz$$

where the minus sign ensures that dKE is negative (speed falls) when dz is positive (stone rises). Since ($m\,g\,dz$) is the little increase in parcel PE we have

$$dKE = -\,dPE \quad \text{or} \quad dKE + dPE = 0$$

The same relation applies to substantial changes

$$\Delta KE = -\,\Delta PE \quad \text{or} \quad \Delta KE + \Delta PE = 0$$

showing that

$$KE + PE = ME = \text{constant}$$

parts of its composite mechanical energy, and can continue to do this even when other forces are acting in addition to gravity, which is always the case with moving parcels of air or water (NN 9.1), as they push or are pushed by their neighbours.

Consider an air parcel of unit mass falling from rest completely freely (i.e. in the absence of any other air parcel) through a height interval of 10 km, the height of a fairly tall thunder cloud. Parcel PE reduces by 100,000 J (NN 9.1) which is

NUMERICAL NOTE 9.1 Mechanical energy in cumulo nimbus downdraft

A parcel of mass m at height H has $PE = m\ g\ H$, where g is gravitational acceleration. If m is 1 kg, g is 10 m s^{-2} and H is 10,000 m, then parcel PE is 1 × 10 × 10,000 = 100,000 J. In free fall this converts entirely to parcel $KE = 1/2\ m\ V^2$, so that parcel speed V after falling H is given by $1/2\ m\ V^2 = m\ g\ H$, $V = \sqrt{(2\ g\ H)}$, which is independent of parcel mass m, and has value 447.2 m s^{-1} for a fall of 10,000 m. In realistic downdraft descent in a vigorous cumulonimbus, parcel speed at ground level may be about 30 m s^{-1}, which corresponds to parcel $KE = 1/2\ m\ V^2 = 0.5 \times 1 \times 30 \times 30 = 450$ J for a parcel of unit mass. The parcel's ME has fallen from 100,000 J to 450 J, a loss of nearly 99.6%.

fully converted into KE, so that it hits the ground at about 447 m s^{-1}—about 30% faster than the speed of sound! Note that this speed is independent of the parcel mass, as Galileo proved by dropping different weights simultaneously from the leaning tower of Pisa, and that the parcel's total ME has remained constant at 100,000 J (relative to the ground) throughout the conversion.

In the real atmosphere, the falling air parcel continually collides with other parcels and typically has a maximum fall speed of only 30 m s^{-1} at the end of the fall, as shown by strong downdraft squalls at the bases of vigorous thunderstorms. The reduction of the parcel PE is still 100,000 J, but the increase in its KE is only 450 J (NN 9.1), so that over 99.5% of its ME has been lost by innumerable collisions on the way down, and by conversion to heat.

9.2.2 Heat energy

As clues about the atomic nature of matter were gathered during the nineteenth century, the conviction grew that the heat of a body was related to the degree of molecular agitation—vibration in a solid, slithering in a liquid, and ballistic flight between collisions in a gas. It was then only a step to surmise that heat was a form of energy relating to that agitation, and that there should be a relationship between such *internal energy* and the external forms we see in the gross movement of air or water parcels, etc. i.e. a relationship between molecular energy and parcel energy. In gases, kinetic energy of ballistic molecules is an obvious energy form to consider (see below), but all gases with more than one atom per molecule have rotational and vibrational energies as well. Though the interpretation was not completed until the early twentieth century (by Einstein and others), the crucial work had been done 60 years earlier by Joule, Clausius, Maxwell, and others, so that the role of heat as a form of energy was already fully recognized from the beginnings of modern meteorology, as was the nature of the troposphere and ocean system as a complex of heat engines (Section 1.5).

Joule's experiment

Joule investigated the relationship between internal (i.e. heat) energy and macroscale mechanical energy in a famous experiment sketched in Fig. 9.3. He injected a known amount of mechanical energy into a known mass of thermally insulated water by stirring, and carefully measured the very small rise in temperature as the kinetic energy of the stirred water was degraded to heat by friction. In doing this

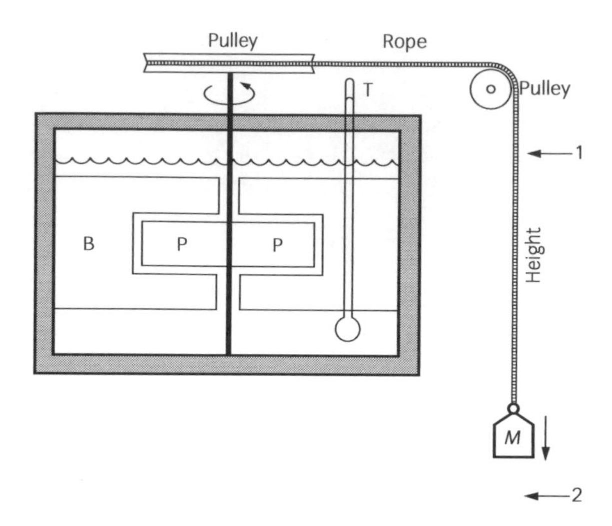

Figure 9.3 Joule's experiment. Water in an insulated box is stirred by paddles P turning between baffles B, as a weight mass M sinks from height z_1 to height z_2. When paddle and water motion have ceased, the rise in water temperature is measured by thermometer T and compared with the work $M\,g\,(z_2 - z_1)$ done by gravity on M.

he modelled two vital but unobtrusive aspects of the atmospheric and oceanic heat engines: the inevitable frictional destruction of the kinetic energy of moving material (water in his experiment, air and water in the atmosphere and oceans), and the production of heat in the process.

Joule arranged that mechanical energy was stirred into the water by paddles turned by a sinking weight. Close-fitting baffles inhibited gross water rotation and enhanced friction. In modern SI units, he found that it always took the input of the equivalent of 4,150 joules (the best modern value is 4,186) of *mechanical energy* to raise the temperature of 1 kg of water by 1 °C in domestic conditions (NN 9.2). This is the *specific heat capacity of water* in those conditions. Subsequent work by other methods, and thermal comparisons between all solids, liquids, and gases, has established that the equivalence between mechanical and heat energy is completely independent of the type of apparatus or material used, though of course each material has its individual specific heat capacity, and some experiments are more accurate than others for practical reasons. Similar work with air by other methods (since Joule's apparatus does not suit low-density fluids like air), shows that it takes 1,004 joules of mechanical energy to raise the temperature of 1 kg of air by 1 °C at constant pressure in typical laboratory temperatures.

Note the sequence of energy transformations occurring between the first and last stages of Fig. 9.3: gravity does work on the sinking weight, driving a conversion of *PE* into the same amount of *KE* of moving paddles and water (mostly water), virtually all of which is eventually degraded into the same amount of internal energy (heat energy) by viscous friction within the water and between the water and paddles and walls. It is this frictional transformation from parcel scale *KE* to heat which confirms the numerical and physical equivalence of mechanical and heat energy. Degradation of mechanical energy to heat by friction occurs in all heat engines, but is especially important in the atmospheric and oceanic engines, because friction riddles their whole mechanism in ways which are deliberately minimized and localized in artificial heat engines.

NUMERICAL NOTE 9.2 Joule's experiment

If Joule had used a sinking 10 kg weight, and 10 kg of water, and had carried out his experiment perfectly accurately (in fact he did very well, measuring temperature changes reliably to about 0.01 °C), he would have found that the weight had to sink 2,134 m to raise the water temperature by 5 °C. The work done by gravity on the sinking weight is $m\,g\,\Delta z = 10 \times 9.81 \times 2{,}134 =$ 209,300 J. If all this is converted into the kinetic energy of water and paddles, and then into the internal energy of the (eventually) still water, then the perfect experiment shows that it takes an energy input of 209,300 J to raise the temperature of 10 kg of water temperature by 5 °C, showing that the specific heat capacity of water is $209{,}300/(5 \times 10) = 4{,}186$ J kg^{-1} °C^{-1} in these conditions (atmospheric pressure 1,013 hPa and laboratory temperature 0 °C).

Heat as molecular kinetic energy

Figure 9.4 represents a moving air parcel and its swarming molecules. The *KE* of the parcel as measured by a fixed anemometer is ($1/2\,M\,V^2$), where *M* is parcel mass and *V* is its speed. The total *KE* of its constituent air molecules is very much larger, however, being the sum of the *KE*s of all the individual air molecules in the parcel, which are moving in all directions with speeds ~ 300 m s^{-1} (about the speed of sound in air). According to the kinetic theory of gases *[40]*, the average *KE* per molecule is directly proportional to the absolute temperature *T* of the air, as measured by a thermometer fixed to the anemometer as the parcel sweeps past. The temperature measured by a thermometer moving with the air parcel must be slightly less, since the air parcel motion has been subtracted (vectorially) from the motion of every air molecule, reducing the average and total molecular *KE* sensed by the thermometer.

Box 9.2 confirms that the total *KE* of the parcel molecules when viewed from the moving parcel frame is less than their total *KE* in the stationary frame by exactly the parcel *KE* in the stationary frame, so that the moving thermometer registers a slightly lower temperature (about 0.1 °C for a parcel speed of 10 m s^{-1}) than

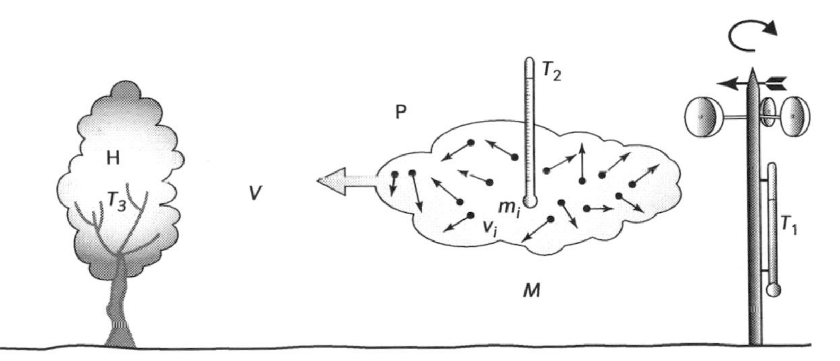

Figure 9.4 An air parcel P of mass *M* moves with speed *V* as measured by the rotation of a fixed cup anemometer. The parcel's temperature T_1 is measured by a fixed thermometer and a slightly lower temperature T_2 is measured by a thermometer moving with the parcel. If the parcel is then disrupted and stopped by a dense hedge H, the final air temperature T_3 will fall short of T_1 because some parcel-scale energy is used to increase the spin and vibration of air molecules instead of their ballistic kinetic energy, and to warm the hedge (Box 9.2 and Appendix).

the stationary thermometer. We will clarify this relationship later by parcel scale dynamics and thermodynamics (Eqn 9.9), but for the moment it is important to see that there is a direct relation between parcel scale and molecular scale energies, and especially between their changes. If the air parcel collides with a dense hedge downwind of the anemometer, it will be disrupted and lose all its parcel scale *KE* once residual eddies have died away. The parcel scale kinetic energy has not been lost of course: it has been transformed into molecular scale *KE*, sensed as a slight warming of the stopped air and hedge. Indeed we can calculate the degree of warming, provided we use the specific heat capacity of the air, because this allows for some parcel *KE* being used to increase the rotation and vibration of the dogbone-shaped diatomic air molecules (like N_2 and O_2), leaving less to increase the ballistic *KE* which is sensed as temperature (Numerical Appendix). Notice the complete equivalence of the air impacting the hedge and the swirling water impacting the baffles in Joule's experiment.

The molecular reality is much simpler for ballistic air molecules than for slithering water molecules, or the complex vibrations of leaf 'molecules', but the principle of energy conservation is general and perfect. The total molecular energy (ballistic, slithering, diatomic vibration and rotation, crystal vibration, etc.) in each case is called the *internal energy* whose changes are expressed in terms of temperature changes ΔT *[22]* in the form of changes of heat capacity $m\ C\ \Delta T$, where m and C are the mass and specific heat capacity of the material. Energy changes are easily accounted for in this way, though detailed understanding required the development of quantum mechanics in the early twentieth century.

BOX 9.2 Molecular kinetic energy

The following very simple treatment demonstrates the physical relationship between the molecular scale and parcel scale kinetic energies of a gas.

Figure 9.4 represents an air parcel of mass M moving past a stationary anemometer with velocity U. It has momentum $M\,U$ and kinetic energy $1/2(M\,U^2)$. Actually the parcel consists of a swarm of N molecules each with a different momentum ($m_i\,v_i$) and kinetic energy $1/2\ m_i\ v_i^2$, whose total kinetic energy is the sum of all the molecular kinetic energies $\Sigma 1/2\ m_i\ v_i^2$, and is much larger than $1/2\ M\ U^2$ because the molecules are moving much faster than the parcel. The total molecular momentum is much simpler: according to Newtonian dynamics the parcel momentum ($M\,U$) is the vector sum of all the little molecular momenta $\sum m_i\,v_i$:

$$M\,U = \Sigma m_i\,v_i \qquad \text{B9.2a}$$

If instead we observe and measure from a frame moving with the parcel (i.e. with velocity U), each molecule will now have velocity ($v_i - U$) relative to the moving parcel. Vector and scalar algebra together with B9.2a show that

$$\Sigma 1/2\ m_i\ (v_i - U)\bullet(v_i - U) = \Sigma 1/2\ m_i\ v_i^2 - 1/2\ M\ U^2 \qquad \text{B9.2b}$$

which means physically that total molecular *KE* as measured from the moving parcel (left-hand side of B9.2b) is less than the total as measured from fixed frame (first term on the right-hand side), the difference being the parcel scale *KE* in the fixed frame, as you might expect.

According to the kinetic theory of gases *[40]* the molecular kinetic energy of the air is directly proportional to its absolute temperature T—average *KE* per molecule being $3/2\ k\ T$, where k is Boltzmann's constant $= 1.38 \times 10^{-23}$ J K^{-1}. A thermometer strapped to the stationary anemometer will therefore record a slightly higher temperature than a thermometer moving with the air parcel—0.12 °C for a parcel of air moving at 10 m s^{-1} (Numerical Appendix).

9.2.3 Transformation from heat to mechanical energy

Joule's experiment focuses on the degradation of parcel kinetic energy to heat through the action of friction, which is inevitable in the operation of all heat engines. However, his experiment has no equivalent of the transformation *from* heat *to* mechanical energy which is the defining activity of all heat engines. In the atmosphere this usually begins when air is warmed by direct or indirect absorption of solar energy. The air is warmed, increasing its internal energy by ($C_v\,\mathrm{d}T$), and doing work ($p\,\mathrm{d}Vol$) by expanding against its surroundings (Section 5.2), and thereby transfering energy to them.

$$\mathrm{d}Q = C_v\,\mathrm{d}T + p\,\mathrm{d}Vol \qquad 9.3$$

As detailed later (Section 9.6), extensive layers of Sun-warmed air expand (inflate) vertically, increasing their own gravitational potential energy and that of the overlying atmosphere, and thereby transforming nearly 30% of the solar heat input into potential energy, a small proportion of which is sooner or later converted into the kinetic energy of air moving about in weather systems of all scales (Section 9.7).

In the molecular picture of the warming parcel, heat injection increases the internal energy of its air molecules, which bombard their surroundings more vigorously and push them back, exporting mechanical energy. You can sense, however, that such conversion from heat to mechanical energy can never be total, because long before the parcel molecules can exhaust their internal energy, they will be moving so feebly that they will be unable to do work on their surroundings. In this way conversion from heat energy to mechanical energy is restricted in a way that conversion from mechanical energy to heat is not, with the result that there is always residual heat which is *unavailable* for conversion into mechanical energy. In principle this leads through the second law of thermodynamics to an inescapable limitation on the production of mechanical energy by any conceivable heat engine—a limitation which is especially severe in the Earth's atmospheric and oceanic engines (Section 9.7).

9.3 Laws of thermodynamics *[25]*

9.3.1 First law

The overall balance of Eqn 9.3 is a particular example of the completely general *principle of conservation of energy* which states that energy is neither created nor destroyed, however often or variously or over what time period it is transformed, shared, or stored. An equivalent statement is that the energy budget of a system balances perfectly when all its energy inputs, outputs, conversions, and storages are properly accounted for. When heat energy is involved, the principle is known as the *first law of thermodynamics*.

In Figure 9.5a and Eqn 9.4, we express the first law in a general form applicable to the operation of a heat engine which is taking heat Q_1 from a warm heat source, converting some of this into mechanical energy ME, and moving the remaining heat energy Q_2 to a cooler heat sink, where it leaves the engine.

$$Q_1 = Q_2 + ME \qquad 9.4$$

The energy balance demanded by the first law applies to the overall balance of Eqn 9.4, and to every subdivision and/or conversion there may be within Q_2 and ME, and can be invoked to determine the size of a poorly defined item, provided all other items are accurately quantified (as was done in Section 8.9). It is a totally reliable and comprehensive accounting principle (Box 9.3).

Applied to natural engines

The format of Eqn 9.4 and Fig. 9.5a is more relevant to artificial engines than it is to natural ones, since artificial engines are usually designed to export mechanical energy for use outside the engine, so that its eventual degradation to heat by friction (in wheels, resistive heating in electrical power cables, etc.) does not occur within the engine proper. In the atmospheric and oceanic engines by contrast, ME is retained entirely within the engine and degraded there, as in Joule's experiment. A more appropriate form of the first law for the atmospheric and oceanic engines is shown in Fig. 9.5b and Eqn 9.5.

$$Q_1 = Q_2 + ME = Q_3 \qquad 9.5$$

Here Q_1 is the heat input by absorption of solar radiation, some of which is converted temporarily into ME before being converted back to heat within the engine, partly by PE loss at heat sinks and partly by frictional destruction of KE, so that the output Q_3 to space is equal to Q_1. Similar conversions take place within the connected ocean engines.

Energy fluxes

So far we have regarded the terms in Eqns 9.4 and 9.5 as packets of energy, measurable in joules. They can also be regarded as energy *fluxes* (flows), measurable in watts, which are obviously relevant to the continuous and fairly steady running of the atmospheric and oceanic engines. For example Eqn 9.5 could also apply to instantaneous rates of flow and rates of transformation of energy when

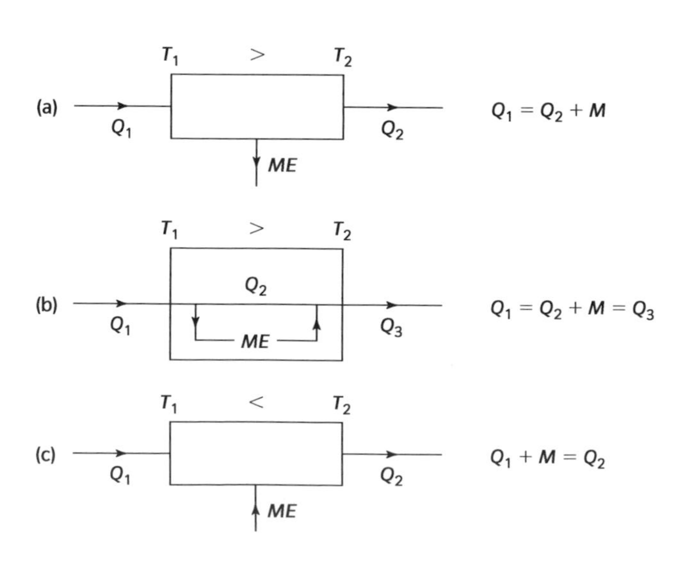

Figure 9.5 (a) A simple artificial heat engine with warm heat source and cool heat sink converting some of its heat input Q_1 into useful mechanical energy ME. (b) A typical simple natural heat engine with warm heat source and cool heat sink converting some Q_1 into ME which is then converted back into heat (mostly by friction) within the engine. (c) A simple heat pump which uses an input of mechanical energy to pump heat from a cool source to a warmer sink.

BOX 9.3 Conservation of mass and energy

The apparently separate principles of conservation of energy and conservation of mass were unexpectedly connected in the early twentieth century, when Einstein showed that mass is a measure of energy through $E = m\,c^2$. Though this implies that the mass (and weight) of a body or system increases and decreases with its energy, the changes are extremely small in the atmosphere and oceans, because speeds of molecules, water, and air flow are very much less than the speed of light c, and because nuclear energy transformations play quite marginal roles. For example, the increase of the mass of an air parcel when accelerated into a tornado funnel or heated by 10 °C is of order 10^{-11}%—far too small to be observable or dynamically significant. As a result, the conservation of mass assumed in earlier chapters is still an excellent approximation, even when there are significant local gains or losses of energy.

Note that the principle of conservation of energy remains absolute in relativistic theory, despite misleading glosses on $E = m\,c^2$ which can seem to imply that mass and energy are interconvertible and that it is only their sum that is conserved. The fundamentals are that energy is conserved (according to the principle) and that all forms of energy have mass. The first statement is crucial for understanding atmospheric activity (and nearly everything else in the universe), but the second is effectively irrelevant to terrestrial systems not involving radioactivity.

the Earth is in perfect (though dynamic) radiative equilibrium with the Sun and interplanetary space, as is very nearly the case. If the atmospheric engine represented by Eqn 9.5 is not in a perfectly steady state, as implied by past and present climatic change, constituent fluxes will balance as usual at any instant (the first law applies regardless), but some fluxes and reservoirs will be unsteady, and overall input and output may be temporarily imbalanced. For example, during the growth phases of ice ages, significant amounts of *PE* are stored in high-latitude continental ice caps (Box 14.1). And during persistent global warming, the big oceanic heat reservoirs are slowly filling.

9.3.2 Second law

As steam engines developed in the eighteenth and early nineteenth centuries, people began to suspect that their efficiency in converting heat into the desired mechanical energy was being limited in principle, as well as by engineering practicalities. That principle proved to be the *second law of thermodynamics*, now regarded as a cornerstone law of nature, fully comparable in importance and scope with the first law. Each of the two laws of thermodynamics rests on an impossibility: in the first law it is the impossibility of creating or destroying energy; in the second law it is the impossibility of extracting mechanical energy from heat by a cyclic (or indefinitely continuous) process unless the heat source is warmer than the heat sink—'cyclic', etc. because isolated infringements are not covered by the law. In terms of the temperatures T_1 and T_2 of the heat source and sink in Fig. 9.5a, the second law states that T_1 must be greater than T_2 if there is to be continued production of *ME*, as already assumed in Fig. 9.1, etc.

This particular statement of the second law (due to Kelvin) suits the heat engine approach of the present chapter, but is only one of several apparently different statements which between them cover a broad range of physical behaviour, from the molecular to the cosmic *[25]*. The contrast between order and disorder is basic to all, especially the difficulty of reversing the natural tendency to drift from order to disorder. Indeed one of the shortest statements of the second law, is that heat naturally flows from higher to lower temperatures. In simple heat conduction this may seem too obvious to need discussion, but Kelvin's formulation of the second law shows that nature demands the same direction of heat flow in the vastly more complex worlds of natural and artificial heat engines.

9.3.3 Mechanical efficiency

The most perfect heat engine imaginable would have no friction, and would run so sweetly and slowly that it could be stopped and run in reverse (i.e. as a heat pump) at any stage. If we use the first and second laws of thermodynamics to analyse the performance of such a *reversible engine,* using an ideal gas as its working material, we find *[20]* that the proportion of heat input converted to mechanical energy (the engine's *mechanical efficiency* $MEF = ME/Q_1$) is completely determined by the absolute temperatures of its heat source and sink (T_1 and T_2 respectively).

$$MEF = \frac{ME}{Q_1} = \frac{Q_1 - Q_2}{Q_1} = \frac{T_1 - T_2}{T_1} \qquad 9.6a$$

Here the third group of terms follows from the second group by using the first law

$$Q_1 = Q_2 + ME$$

and the fourth (the last) group is proved from the third by detailed thermodynamic analysis. An equivalent statement is that a proportion Q_2/Q_1 (i.e. $1 - MEF$) of the heat input Q1 is in principle *unavailable* for conversion into mechanical energy.

Equation 9.6a implies that

(i) when the heat source is warmer than the heat sink the reversible heat engine can operate as shown in Fig. 9.5a and Eqns 9.4 and 9.5;

(ii) when heat source and sink are at the same temperature (i.e. $T_1 = T_2$), none of the heat input Q_1 is available for conversion to ME, so that the engine's MEF is zero;

(iii) when the heat source is cooler than the heat sink (i.e. $T_1 < T_2$), the flows of heat and mechanical energy shown in Fig. 9.5a and Eqns 9.4 and 9.5 are forbidden by the second law. However, the negative ME required to balance Eqn 9.4 corresponds to Fig. 9.5c, which is a *heat pump* using an input of mechanical energy to pump heat from a cooler source to a warmer sink, just as a domestic refrigerator pumps heat from its cool interior into the warm kitchen but only while there is a continuing input of mechanical energy (usually maintained by conversion from electrical energy). This does not break the second law, and Eqn 9.6a applies to Fig. 9.4c in the form

$$\frac{ME}{Q_1} = \frac{T_2 - T_1}{T_1} \qquad 9.6b$$

The universal limit

The behaviour of highly idealized slow reversible engines and pumps might seem irrelevant to the real world of highly *irreversible* and vigorous engines like the troposphere and surface ocean. However, thermodynamic analysis shows that the *MEF* for a reversible engine is the maximum possible efficiency for any type of heat engine working between the specified temperatures, reversible or irreversible, regardless of working material (air, carbon dioxide, steam, water, etc.), or indeed of any detail of construction or operation. The physical reasoning is rigorous in detail *[20]*, but hinges on the fact that if any heat engine could do better than Eqn 9.6a (i.e. produce more *ME* from the same Q_1 with the same given temperatures T_1 and T_2), then it could be used to drive an idealized engine in reverse (i.e. as a heat pump, Fig. 9.5c), so that the composite real engine-ideal pump system would be able to pump heat sustainably from a cooler sink to a warmer source without requiring any mechanical energy input—an impossibility according to Kelvin's expression of the second law.

Further thermodynamic analysis shows that when a heat engine operates *irreversibly* (i.e. with friction and far from equilibrium, like all realistic engines), its mechanical efficiency is always less than the theoretical reversible optimum, sometimes by a long way. For realistic artificial or natural engines we therefore have

$$MEF = \frac{ME}{Q_1} < \frac{T_1 - T_2}{T_1} \qquad 9.7a$$

and for heat pumps

$$\frac{ME}{Q_1} < \frac{T_2 - T_1}{T_1} \qquad 9.7b$$

though the magnitude of the inequalities can be found only by detailed analysis of each particular real engine or pump.

Since these restrictions apply to the performance of all possible heat engines, we know that they must apply to the atmospheric and oceanic engines, and the stratospheric heat pump, even though many of their details are still imperfectly understood—a leap of faith typical of all the natural sciences, but well tested by experience.

9.4 Terrestrial heat engines in outline

Like all heat engines, each terrestrial engine has one or more *heat sources*, where heat energy enters the engine, has sites of *energy transformation* where energy changes from one form to another (crucially between heat and mechanical energy), and has one or more *heat sinks*, where heat leaves the engine. Figure 9.6 outlines these and other features of the major engines at work in the ocean and atmosphere, focusing especially on the dominant troposphere/ocean engine, details of which are discussed in the rest of this chapter.

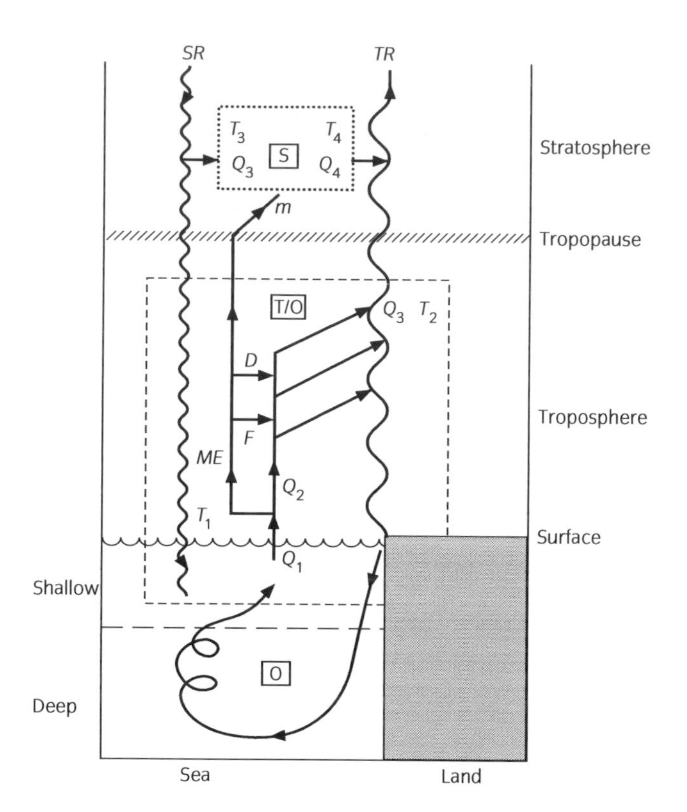

Figure 9.6 The troposphere/ocean engine T/O, ocean engine O, and stratosphere heat pump S in more detail. In O water sinks quickly into the depths and rises slowly again. In T/O most mechanical energy *ME* is degraded to heat by friction *F*, but a small amount *m* goes on to drive S, so that Q_4 is slightly greater than Q_3 although temperature T_4 is slightly higher than T_3. In T/O the series of oblique arrows represents the progressive handing over from heat to *TR* (Fig 8.16).

9.4.1 Troposphere/ocean engine

Heat inputs

For simplicity Fig. 9.6 shows solar heat Q_1 entering the tropospheric/oceanic engine (T/O) at a single temperature T_1—effectively the average temperature of the surface and low troposphere—at which solar radiant energy (*SR*) is converted directly or indirectly into the heat energy which warms the land, sea, and air. According to Figs. 8.8, 8.16, and the discussion, the global annual average heat input is divided almost equally between:

(i) liberation of latent heat at and above cloud base, maintained by *SR*-driven evaporation from land and sea surfaces—19 units in the figures;

(ii) direct *SR* absorption by the troposphere, mostly the dense low troposphere—also 19 units;

(iii) direct input of sensible heat (11 units) to the base of the troposphere from the warmed land and sea surfaces, and 10 units of terrestrial radiation (*TR*) from the warmed surfaces which is then absorbed by the very low troposphere. This ignores the 5 units of surface *TR* which pass through the atmospheric window to space without effect, and the 3 units of *SR* input to the stratosphere and balancing *TR* output. The total input to the surface and low troposphere is 59 units (each worth 3.45 W m^{-2} in Fig. 8.8) or nearly 204 W m^{-2}. Remembering that Fig. 8.8 assumes an output of 67 units instead of the satellite observed 70 (Section 8.6), this scales up to nearly 213 W m^{-2} in reality.

Assigning all tropospheric input to a single temperature is over-simple but not fundamentally misleading, and suggests that a reasonable value for T_1 is close to the global annual average surface temperature of 15 °C. The error in ignoring the fact that quite a lot of the heat input occurs significantly above the surface (some above cloud base) is offset by the error in ignoring the fact that most input occurs in lower latitudes where surface temperatures exceed 15 °C.

Heat outputs

In Fig. 9.6 the heat output ($Q_3 = Q_2 + ME$) from the troposphere/ocean engine is written as if it occurs at a single temperature T_2, though in reality (as shown by the adjacent oblique arrows on the diagram) it occurs across a range of temperatures, like the *SR* input in lower, warmer regions. T_2 must be close to the temperature of the highest layer of the atmosphere still largely opaque to *TR*—roughly 233 K (−40 °C). In fact we can do better by noting that the *TR* output from the engine is the same as the *SR* input, and so must be 213 W m^{-2}, which corresponds to the emittance from a blackbody at about 248 K (≈ −25 °C) by the method of NN 8.2. This layer is maintained in the upper troposphere by the vertical distributions of water vapour, cloud, and carbon dioxide (Section 8.4), which in turn are maintained by the stirring of the air by weather systems. Above this level, convection and advection are insignificant, and the upward energy flux is carried as *TR* without further air motion.

ME production, transformation, and destruction

Using a surprisingly small part of the solar heat input Q_1, the troposphere/ocean engine generates mechanical energy *ME*—some as kinetic energy *KE* of moving air, and some as gravitational potential energy *PE* of lifted air. There is repeated exchange between *KE* and *PE* in large-scale weather systems, such as the Hadley circulation, monsoons and tropical cyclones of low latitudes, and the extratropical cyclones and anticyclones of middle and higher latitudes, and in cumuliform convection in all latitudes—in fact in any weather system containing organized vertical motion. Very importantly, as we shall see later (Section 9.7), large amounts of atmospheric *PE* are stored in tall atmospheric columns, and converted in small amounts to *KE* in convective and slope convective weather systems.

Some tropospheric *ME* is transferred to the ocean surface, mainly in the form of *KE* transferred by friction between the turbulent air and the roughened water surface (Fig. 9.7). Wind waves spread some of the input *KE* across wide areas of sea surface, but some maintains ocean surface currents like the Gulf Stream against internal oceanic friction (Fig. 8.24). Vigorous storms transfer both *KE* and *PE* by raising synoptic scale storm surges (Section 12.3 and 13.5) in the oceans. Small amounts of tropospheric *ME* leak into the stratosphere to drive the heat pump there, as shown in Fig. 9.6 and mentioned below.

For clarity these details of *ME* are omitted from Fig. 9.6, which depicts only the ultimate fate of *ME*. Any particular element of *ME* exists for only a limited time. Column *PE* can be lost in heat sink zones like the upper troposphere, exported to space by *TR* as a column chills and vertically deflates (*D*). The *KE* of atmospheric and oceanic flow is sooner or later degraded to heat by viscous friction (*F*), especially in the turbulent atmospheric boundary layer (Sections 9.8 and 10.7), where the wind shear maintains turbulence and intense transient wind shears, like the baffles in Joule's apparatus. Although such

Figure 9.7 A moderately rough sea (Beaufort 6 to 7). The kinetic energy of the wind leaks into the sea surface layers by horizontal frictional drag and vertical turbulent buffeting to produce surface water currents and surface gravity waves. The currents carry huge fluxes of heat and lose a little of their kinetic energy by friction with sea beds and slower water, and the waves spread their mechanical energy (continually exchanging between kinetic and potential energy) quickly and with little loss (by "white horses") over wide areas which is degraded to heat by friction when they break on coasts.

frictional heating may generate further production of *ME* (for example in the atmospheric boundary layer), it will eventually be degraded to heat and leave the engine from a heat sink in the high troposphere or low polar troposphere, radiating through the open atmospheric window and contributing to Q_3 as depicted in Fig. 9.6.

Since ME is very much smaller than Q_2 (as we shall see), Q_3 is only marginally larger than Q_2; however, this relatively small *ME* plays an absolutely crucial role in maintaining the activity and state of the atmosphere and oceans, especially their ability to transport heat.

Heat fluxes

In the black-box pictures of heat engines we have used so far (Figs. 9.1–9.6), Q_2 is the very large proportion of the energy input which is not converted to mechanical energy within the engine. These huge amounts of heat are carried through the engine from source to sink, partly by horizontal and vertical currents of warm air, partly by ocean surface currents, and partly by the vertical cascade of terrestrial radiation (Section 8.4). About 25% of the total poleward heat flux is carried by surface ocean currents, some heat transferring back to the cooler troposphere while en route to high latitudes, and some staying in the ocean surface until it is radiated to space from near-polar regions. Such entanglement means that the troposphere and ocean surface layers form an integrated *troposphere/ocean engine* (T/O in Fig. 9.6), which we will simply call the *tropospheric engine* in later sections, since the troposphere is the driving agent.

Since atmospheric convection and advection are driven by tropospheric motion (and therefore related to *KE*), and surface ocean currents are driven by *KE* frictionally transferred from the troposphere, it is clear that Q_2 and *KE* are intimately connected, even though they are distinct energy forms, and that the energy transfer Q_2 achieved by the tropospheric engine is vastly greater than the relatively small *ME* production we examine in following sections.

Mechanical efficiency

Regardless of detail, the second law inescapably limits the overall mechanical efficiency *MEF* (Eqn 9.6a) of the troposphere/ocean engine through the temperatures of its heat source and sink. In our simplified outline, the temperature of the heat source is about 15 °C (288 K), whereas the effective temperature of the heat sink in the upper troposphere is about −25 °C (248 K). It follows that the maximum possible *MEF* of the engine is given by Eqn 9.7a i.e. (288 − 248)/288 or 14%.

Since friction is endemic throughout the engine, and atmospheric conditions are unsteady and far from equilibrium, the whole system is so deeply irreversible that its actual efficiency is far below this maximum. No simple analysis will say by how much, but detailed estimates of the troposphere's actual behaviour suggest that its *MEF* is an order of magnitude below the maximum allowed in a reversible engine operating between these temperatures. *ME* is therefore only a few per cent of Q_2, which is broadly consistent with our living conditions being dominated by heat and cold more than by wind force, despite occasional stormy exceptions, our keen interest in winds for industry and leisure, and our bodily sensitivity to wind through the sensation of wind-chill.

9.4.2 Ocean engine

Alongside the surface-based oceanic part of the troposphere/ocean engine, there is an intrinsic *ocean engine* (O in Fig. 9.6) which reaches into the ocean depths (Fig. 8.23, Section 8.10.2).

As in the troposphere/ocean engine, the heat source of the ocean engine is located in low latitudes, where the strong sunlight is absorbed in the top tens of metres of water. Some of this flows polewards in warm wind-driven currents like the Gulf Stream (itself part of the troposphere/ocean engine), being gradually chilled by net emission of *TR* and contact with the cooler air in higher latitudes. Eventually some especially dense water slides into the ocean depths in downward convection, breaking away from the tropospheric engine and driving the ocean engine proper from its surface heat sink.

The cold sunken water meanders equatorwards in slow persistent currents channelled by the topography of the ocean bed or sliding progressively upwards over layers of newer colder water until it eventually (often years later) comes near enough to the surface to be warmed by the Sun, indirectly at first (by downward stirring of Sun-warmed water), and finally in the top 100 m, directly by sunlight. Together with the surface currents (Gulf Stream, etc.) the complete current network has become known as the *oceanic conveyor belt* (Fig. 9.8 and *[39]*), but in the present approach the deep branch merits separate treatment. Note that in the deep ocean engine the mechanical impetus of the engine seems to be focused at the heat sink, rather than at the heat source as in the troposphere/ocean engine. All heat engines need both a heat source and sink, regardless of which seems be the locus of prime motion.

Mechanical efficiency

The mechanical efficiency of the deep oceanic engine is limited by the relatively small temperature difference between the low-latitude heat source, where water

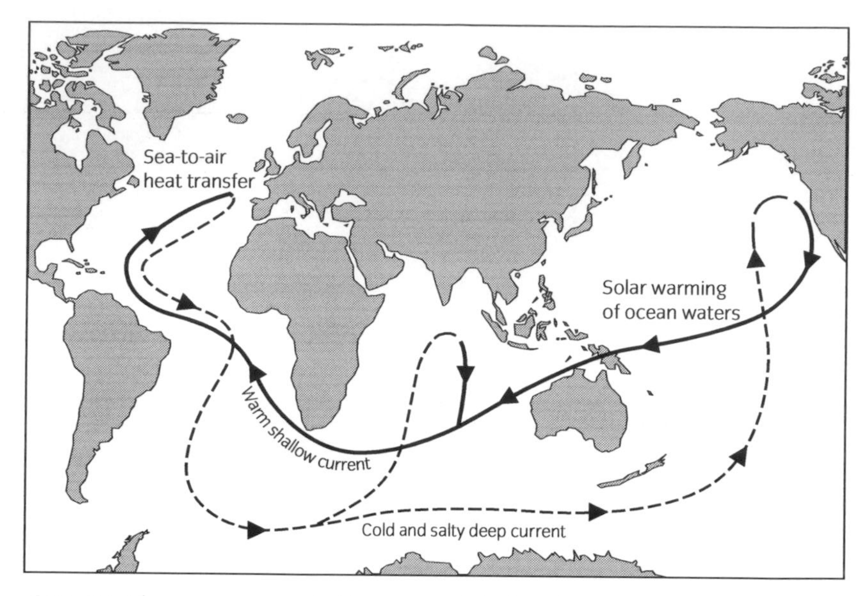

Figure 9.8 The *ocean conveyor belt* includes both the fast wind-driven poleward flows of warm surface water and the slow cold density currents threading the ocean depths, and connects the troposphere/ocean and ocean engines. The first connection occurs in polar oceans where some water, chilled and salinated at the surface, sinks into the deeps, while the rest remains on the surface to become part of the cool equatorward flows on the E sides of ocean basins (often omitted in modern diagrams like this one). The second connection occurs when sunken water rises very slowly into the surface layers in low latitudes where it is warmed by direct sunlight to become available for fast poleward flow again.

temperatures are about 25 °C (298 K), and the high-latitude heat sink where they are about 0 °C (273 K). According to Eqn 9.4, maximum *MEF* is (298 – 273)/298, which is about 8%, though as in the tropospheric engine, actual *MEF* is likely to be much less than this, because of irreversibility.

9.4.3 Stratosphere heat pump

Compared with the troposphere, the stratosphere is relatively inactive (Section 3.2). Its radiative budget (Fig. 8.8) shows that absorption of *SR* (largely *UV*) is nearly balanced by net emission of *TR*, and that its heat sources and sinks are at very similar temperatures. According to the second law of thermodynamics the stratospheric engine can therefore produce little or no mechanical energy (Eqns 9.6a and b).

This cannot be the complete picture, however, since much of the upper stratosphere keeps flowing quite quickly despite inevitable friction. Closer examination shows that the lower middle stratosphere acts as a relatively weak *heat pump*, driven by an input of m from the troposphere to pump heat from a heat source above the cold, high equatorial tropopause (temperature T_3 in Fig. 9.9) to a slightly *warmer* heat sink above the lower polar tropopause (temperature T_4). Some details are shown in Fig. 4.6. This poleward heat flux is believed to be carried by synoptic-scale eddies in the low stratosphere which are driven by upward leakage (m) of mechanical energy from the long waves of the troposphere (Section 12.6 and *[35]*). In principle, the heat pumped to the polar stratosphere must add a little to the outward flux of terrestrial radiation from the troposphere there.

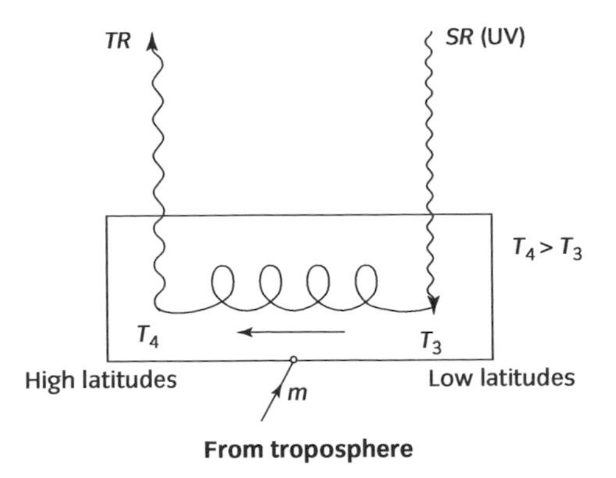

Figuer 9.9 The stratospheric heat pump takes in heat by absorbing solar UV in the cold equatorial mid-stratosphere and losing it to space by emission of terrestrial radiation from the somewhat warmer polar regions. The energy needed to sustain this flow against the temperature gradient is supplied by internal gravity waves propagating upwards from the dynamic troposphere.

Mechanical efficiency

The stratospheric heat source (temperature T_3) is probably no more than 10 °C cooler than its heat sink (temperature T_4), in a temperature range centred on 250 K. According to Eqn 9.6b, optimum *MEF* (= m/Q_3) is only about 4%, and realistic values are much smaller, as usual. Note that *MEF* is not the most appropriate measure of the performance of a heat pump. In assessing the efficiency of artificial heat pumps and refrigerators, 1/*MEF* is more relevant, since it expresses the heat flux emerging from the engine as a multiple of the *ME* flux (m) used to maintain it. If actual *MEF* is 0.4%, 1/*MEF* is 250, showing that a small input m achieves a relatively great heat flux.

9.5 Energetics of air parcels and columns

Before looking more closely at the workings of the tropospheric engine, we examine some detailed expressions of its *energetics*, found by combining the laws of dynamics and thermodynamics. These expressions show in detail how, and by how much, the component energies of a body of air is changed by adding or removing heat, and by the working of gravity, and the pressing and rubbing of other bodies of air, land, and sea. We begin with small parcels of air moving about in the active atmosphere, and then consider much larger volumes. As in Chapter 7, the equations and terms should be read for physical meaning rather than mathematical details, although these are outlined for completeness.

9.5.1 Air parcels

The *total energy* *TE* of an air parcel of mass m, and absolute temperature T, moving at speed V at a height z above sea level, is the sum of its internal energy *IE* (= $m\,C_v\,T$), its kinetic energy *KE* (= $1/2\,m\,V^2$) and its gravitational potential energy *PE* (= $m\,g\,z$).

$$TE = m\,C_v\,T + 1/2\,m\,V^2 + m\,g\,z$$

Because of interactions with radiation, gravity, and neighbouring air parcels, the total energy of a parcel is not conserved: if we follow a parcel moving through a weather system, its *TE* will increase if the parcel gains energy (from sunlight, surrounding air, etc.), decrease if it loses energy, and be conserved only if gains and losses balance exactly. The principle of energy conservation applies overall, and to every transfer and transformation, but the parcel is exchanging energy freely with the rest of the atmosphere, which includes Earth through g, and the solar system through *SR* and *TR*!

To quantify all these changes, we combine the equation of motion and first law of thermodynamics (Box 9.4) to derive an expression for the rate of change of parcel *TE* with time at any point on its journey through the weather system. This produces

$$\frac{\mathrm{d}TE}{\mathrm{d}t} = \frac{\mathrm{d}Q}{\mathrm{d}t} - TPW - FRW \qquad 9.8$$

where

(i) $\mathrm{d}Q/\mathrm{d}t$ is the net rate of heat input to the parcel by all relevant means: net radiative absorption, heat conduction from adjacent parcels and surfaces, release of latent heat by condensation of vapour within the parcel, and heating by frictional degradation of parcel *KE*. Overall $\mathrm{d}Q/\mathrm{d}t$ can be positive or negative or zero since each of its components can be so too, except for frictional heating, which is always positive.

(ii) *TPW* is the rate of *total pressure working* by the parcel on its surroundings as it moves up or down pressure gradients and contracts or expands. Positive *TPW* in Eqn 9.8 contributes to negative $\mathrm{d}TE/\mathrm{d}t$ by exporting some of the parcel's *TE* to its surroundings, by expanding and cooling, and/or by moving against the ambient pressure gradient and slowing. Negative *TPW* increases parcel *TE* by contracting and/or moving down the ambient pressure gradient. As shown in Box 9.4, for an ideal gas like air

$$TPW = m\,R\,\frac{\mathrm{d}T}{\mathrm{d}t} - \frac{m}{\rho}\frac{\partial p}{\partial t}$$

where R is the specific gas constant for air and $\mathrm{d}T/\mathrm{d}t$ is the rate of change of temperature of the moving parcel, ρ is air density, and $\partial p/\partial t$ is the rate of change of pressure at a stationary point being passed by the parcel (i.e. the pressure tendency there).

(iii) *FRW* is the rate of frictional working by the parcel on its surroundings. A major part of *FRW* can be positive or negative (exporting or importing the parcel's energy by rubbing against slower or faster moving parcels), but a part of it (*FDH* in Table 9.1) represents the frictional degradation of the parcel's *KE* to heat, which always contributes to a positive *FRW* and therefore to a loss of its *KE* (Box 9.8). *FDH* exactly balances the frictional heating component in $\mathrm{d}Q/\mathrm{d}t$ since it represents an energy transformation within the parcel.

Tropospheric example

A full statement of Eqn 9.8 for a moving parcel of air is

$$\frac{\mathrm{d}}{\mathrm{d}t}\left[m\,Cv\,T + \frac{1}{2}m\,V^2 + m\,g\,z\right] = \frac{\mathrm{d}Q}{\mathrm{d}t} - m\,R\,\frac{\mathrm{d}T}{\mathrm{d}t} + \frac{m}{\rho}\frac{\partial p}{\partial t} - FRW \qquad 9.9$$

Table 9.1 shows the terms of Eqn 9.9 evaluated (NN 9.3) for air parcels rising through the troposphere in deep weather systems. By typical observation, they may rise from sea level to an altitude of 10 km in the course of a day (in the frontal zone of an extratropical cyclone), or in an hour (in a cumulonimbus), in each case cooling by 60 °C and accelerating from 5 to 30 m s^{-1} (perhaps to 60 m s^{-1} in a frontal zone). Values are expressed in terms of energy changes ($\Delta TE = \Delta KE + \Delta PE + \Delta IE$) and exchanges ($\Delta Q$ and working terms like TPW) arising in an hour (3.6 ks) or a day (86.4 ks). Equivalent rates of change can be found by dividing the changes by the relevant time period (NN 9.3).

The terms in the left-hand side of Eqn 9.9 are shown in the upper half of Table 9.1, while terms on its right-hand side are shown in the lower half.

Discussion

The typical changes assumed for parcel temperature, speed, and height ensure that there is a large rise in parcel TE during parcel ascent, since the rise in PE is

Table 9.1 Energy changes and exchanges for a 1 kg air parcel rising and accelerating through the troposphere

Energy/work		Change in kJ		
		in 1 hr (3.6 ks)	in 1 day (86.4 ks)	
LHS Eqn (9.9)				
ΔKE	$\frac{1}{2}\Delta(V^2)$	0.4 kJ	3.6 kJ	
ΔPE	$g\,\Delta z$	98	98	
ΔIE	$C_v\Delta T$	−43	−43	
ΔTE	Σ	55	59	
RHS Eqn (9.9)				
TPW				
	$R\,\Delta T$	17	17	
	$-\frac{\Delta\rho}{\rho}$	~ 0.5 ±	~10 ±	see NN 9.3
ΔQ				
	latent heat	38	38	see NN 9.3
	net radiation	0.1 ±	2 ±	see NN 9.3
FRW		~ 0.1 ±	~ 3 ±	
	FDH	~ 0.02	~ 0.5	see NN 9.3
		55 ± 1	55 ± 15	

considerably larger than the fall in *IE*, and each is at least an order of magnitude larger than the change in *KE*. In these conditions, the net increase of *TE* ranges from 55 to 59 kJ (for a 1 kg parcel), depending on the parcel speed in the high troposphere.

The terms on the right-hand side of Eqn 9.9 are less clearly defined by observation, with the exception of the large term $R\ \Delta T$ which dominates *TPW*, and the latent heat item which dominates the ΔQ. $R\ \Delta T$ is determined by the assumed 60 °C temperature lapse. The large and positive latent heat term is determined by the amount of parcel water vapour condensed by chilling during ascent into the high troposphere. If 10 g of water vapour is condensed in the 1 kg parcel, the latent heat term $L\ \Delta q$ is 25 kJ. In fact the liberation of latent heat is already quite closely constrained by the assumed temperature lapse of 60 °C, which is 38 °C less than the dry adiabatic temperature lapse expected in a 10 km ascent, precisely because of the liberation of latent heat by cloudy condensation. According to NN 9.3 such 38 °C 'warming' corresponds to an evolution of 38 kJ, as tabulated, and the condensation of about 15 g of water vapour in the parcel.

Table 9.1 shows that Eqn 9.9 balances, at least to within the error of assessment, suggesting that the rise in parcel *TE* during ascent and cooling is at least reasonably accounted for by pressure working by the parcel, and by evolution of latent by cloudy condensation. However, the important ΔKE term is too small relative to other terms to be usefully constrained by the overall equation, especially as ΔKE must be heavily restrained by frictional working (*FRW*), which is effectively impossible to evaluate for individual air parcels. As a result the energy equation cannot be used to make a useful estimate of the change in parcel *KE* along an observed trajectory in a weather system, though we will return to it later (Section 9.to assess the production of *KE* by the global tropospheric engine (Section 9.8).

BOX 9.4 Parcel energy equations

Mechanical energy

As in Box 9.1, when a net force *F* acts in line with a moving parcel's short displacement d*s*, the small increase in parcel *KE* is given by the work done *F* d*s* by the force *F*.

$$dKE = F\,ds$$

Dividing across by the short time interval d*t* in which this happens, we get

$$\frac{dKE}{dt} = F\frac{ds}{dt} = FV$$

where *V* is the speed d*s*/d*t* of the parcel, and *F V* is the *rate of working* on the parcel by *F*. Replacing *F* by the components along the line of displacement d*s* of the various force terms acting on the parcel (Section 7.4), we get

$$\frac{dKE}{dt} = FV = PGW + GRW - FRW \quad \text{B9.4a}$$

where

(i) *PGW* is the rate of working by the surrounding pressure gradient on the parcel as it moves down the gradient.

(ii) *GRW* is the rate of working by gravity on the parcel as it loses height *z* above MSL, which is $(-\,w\,m\,g)$, where *w* is the parcel's rate of rise d*z*/d*t*. Since the parcel's gravitational potential energy *PE* is $m\,g\,z$, $GRW = -\,dPE/dt$.

(iii) FRW is the frictional rate of working by the parcel on surrounding air, land, or water, so that positive FRW reduces KE.

Note the absence of Coriolis terms (Box 7.4, etc.) in B9.4a; these make no contribution to dKE/dt because they always act at right angles to parcel motion and therefore do no mechanical work on the parcel as it moves.

Moving GRW ($= -dPE/dt$) to the left-hand side of B9.4a produces an expression for the rate of increase of parcel mechanical energy $ME = KE + PE$:

$$\frac{dME}{dt} = \frac{d}{dt}[KE + PE] = PEW - FRW \qquad \text{B9.4b}$$

This states that the parcel's rate of increase of mechanical energy is given by the rate of working of the ambient pressure gradient on the parcel less the rate of frictional working by the parcel on its neighbours.

Thermal energy

The first law of thermodynamics (Eqn 9.3) for a parcel can be rearranged to give the rate of change of parcel internal energy IE

$$\frac{dIE}{dt} = \frac{dQ}{dt} - PEW \qquad \text{B9.4c}$$

where

(i) dQ/dt is the net rate of heat input to the parcel by all effective means.

(ii) PEW is $p\,dVol/dt$, the rate of working by the expanding parcel on surrounding air.

This is one of two parts of the total rate of pressure working (TPW) by the parcel; the other part is the rate of working by the parcel against the ambient pressure gradient, which is the negative of PGW in Eqn B9.4b, the minus sign appearing because PGW is defined as the pressure gradient working on the parcel. We have

$$TPI = PEW + (-PGW) \quad \text{so that} \quad PEW = TPI + PGW$$

and B9.4c becomes

$$\frac{dIE}{dt} = \frac{dQ}{dt} - TPI - PGW \qquad \text{B9.4d}$$

Further analysis shows that for an ideal gas like air, TPW is

$$TPW = mR\frac{dT}{dt} - \frac{m}{\rho}\frac{\partial p}{\partial t} \qquad \text{B9.4e}$$

where m, R, T, and ρ are the mass, specific gas constant, absolute temperature, and density of the moving air parcel, dT/dt is the rate of change of parcel temperature, and $\partial p/\partial t$ is the rate of change of pressure at a fixed location (i.e. as would be measured by a stationary barograph as the parcel passes).

Total energy

Adding B9.4b and B9.4d gives the rate of increase of total parcel energy TE

$$\frac{dTE}{dt} = \frac{d}{dt}[IE + KE + PE] = \frac{dQ}{dt} - TPW - FRW \qquad \text{B9.4f}$$

where the pressure gradient working terms PGW have cancelled, since they represent exchanges between ME and IE and therefore do not alter TE. We see that the rate of change of the parcel's total energy is given by the rate of heat input to the parcel, less the total rate of pressure working by the parcel, and less the rate of frictional working by the parcel—a statement we could have made in general terms simply on the basis of the conservation of energy.

Using B9.4e to replace TPW, B9.4f becomes

$$\frac{dTE}{dt} = \frac{dQ}{dt} - mR\frac{dT}{dt} + \frac{m}{\rho}\frac{\partial p}{\partial t} - FRW \qquad \text{B9.4g}$$

Rearranging and writing out the energy terms in full, we get

$$\frac{d}{dt}\left[\frac{1}{2}mV^2 + mgz + mc_vT + mRT\right] = \frac{dQ}{dt} + \frac{m}{\rho}\frac{\partial p}{\partial t} - FRW$$

Using $C_p = C_v + R$ this becomes

$$\frac{d}{dt}\left[\frac{1}{2}mV^2 + mgz + mC_pT\right] = \frac{dQ}{dt} + \frac{m}{\rho}\frac{\partial p}{\partial t} - FRW \qquad \text{B9.4h}$$

NUMERICAL NOTE 9.3 Magnitudes of terms in Table 9.1

All energy values apply to a 1 kg parcel, and should be read as 'kJ per kg' before scaling up to larger masses. Assumed parcel behaviour: V increase 5 to 30 (or 60) m s^{-1}, T fall 60 °C, z rise 10 km, all in 1 hr (3.6 ks) or 1 day (86.4 ks).

Left-hand side of Eqn 9.9

$$\Delta KE = 1/2\,(30^2 - 5^2) = 875/2 \approx 438\ \text{J} \approx 0.4\ \text{kJ}$$
$$(3.6\ \text{kJ for 5 to 60 m s}^{-1})$$

$$\Delta PE = 9.8 \times 10{,}000 = 98\ \text{kJ}$$

$$\Delta IE = 717 \times (-60) \approx -43\ \text{kJ}$$

Right-hand side of Eqn 9.9

- When T falls by 60 °C, the term $R\,\Delta T$ in TPE (B9.4g) has value $287 \times (-60) \approx -17$ kJ, where R is the specific gas constant for air (= 287 J kg^{-1} K^{-1}).
- When surface pressure at a given location rises by Δp in a time interval in which surface air density is nearly constant, the contribution to parcel TPW for 1 kg air of density ρ is about $\Delta p/\rho$ at the Earth's surface (B9.4g). With increasing height above the surface, changes in Δp and ρ tend to be proportionate, so that the surface ratio probably still applies at least roughly. For a large pressure tendency of 5 hPa per hour, the term $\Delta p/\rho$ becomes $5 \times 100 = 0.5$ kJ (per kg parcel) in 1 hour and $24 \times 0.5 = 12$ kJ in 24 hours.
- If the parcel vapour content falls by 10 g kg^{-1} during cloudy condensation, evolution of latent heat is $10 \times 10^{-3} \times 2.5 \times 10^6 = 25$ kJ. Alternatively, note that parcel ΔT would be − 98 °C for a dry adiabatic height rise of 10 km, so that the assumed −60 °C implies a potential 'warming' of 38 °C by evolution of latent heat. This is equivalent to a warming of 38 °C at constant pressure, which implies a heat input of $\Delta H = C_p\,\Delta T = 1{,}000 \times 38 = 38$ kJ, which according to $\Delta H = L\,\Delta q$ implies a condensation of vapour and a fall in specific humidity of $\Delta q = H/L = 38{,}000/(2{,}500{,}000) = 0.0152 = 15.2$ g kg^{-1}.
- The global average net loss of terrestrial radiant energy from the troposphere is about 170 W m^{-2} (49 units in Fig. 8.8). Spread evenly among the typical 7,000 kg of tropospheric air resting on a square metre, this corresponds to about 25 mW per kg, i.e. nearly 100 J per hr (per kg), or 2 kJ per day, with values 10 times larger in extreme cases.
- Estimates of frictional degradation in the atmospheric boundary layer (ABL) suggest that parcel kinetic energy is typically being degraded at a rate of about 1 W in a vertical column of ABL resting on 1 square metre of surface. Taking the ABL to be 200 m deep (so that the column contains about 200 kg of air), the degradation rate per kg is ~ 5 mW, equivalent to nearly 20 J per hour and over 0.4 kJ per day (Box 9.8 and *[26]*).

9.5.2 Air columns

We have seen that analysis of the energy budgets of individual small air parcels is conceptually and numerically revealing, but involves important terms which we cannot feasibly evaluate, constrains kinetic energy only marginally, and, by focusing on a single parcel, ignores overall dynamic and static effects arising from interactions between the myriad other parcels of the fluid atmosphere. One of the most important of these comes from the nearly perfect hydrostatic equilibrium exhibited by vertical air columns in all but the most violent atmospheric activity. Although thorough treatment of columns and larger volumes of air is beyond the technical scope of this book *[41]*, a simplified examination of hydrostatically balanced air columns produces useful expressions for the total energies of air columns (CTE) and their components

$$CTE = CKE + CPE + CKE$$

which provide a framework for outlining the energetics of air columns and larger air masses.

Figure 9.10 shows a vertical column of air reaching from the Earth's surface to the edge of space. Though air may be blowing quickly through its open stationary framework, we know that it flows in nearly perfect hydrostatic equilibrium, so that at any instant, air pressure p falls exponentially with increasing height z in the column, and p at any z is equal to $M\,g$, where M is the mass of air in the column above unit horizontal area at z, and g is the gravitational acceleration, which we can assume is uniform at 9.8 m s^{-2}.

Column totals

Consider the contributions to the column total energy CTE made by the large number of thin horizontal slices making up a column, each slice of mass m_i (per unit area), lower face height z_i above sea level, containing air with wind speed V_i,

BOX 9.5 Centre of gravity of air columns

The atmospheric column shown in Fig. 9.10 consists of a vertical pile of many horizontal slices, each of small mass m_i per unit horizontal area, shallow depth Δz_i, density ρ_i, and height z_i above the column base. Since $m_i = \rho_i\,\Delta z_i$ for each slice (Δz_i and slice volume per unit area being numerically identical)

$$CPE = \sum m_i\, g\, z_i = g \sum \rho_i\, \Delta z_i\, z_i$$

Since the *centre of gravity* is defined to be the level at which the whole weight of the column appears to act, we also have

$$CPE = M\, g\, C$$

where M is the column's total mass ($\sum m_i = \sum \rho_i\, \Delta z_i$) and C is the height of the centre of gravity of the column above its base. Because we take g to be uniform, C is also the centre of mass of the column. Combining these expressions we have

$$CPE = g \sum \rho_i\, \Delta z_i\, z_i = M\, g\, C \qquad \text{B9.5a}$$

Rearranging the second and third parts of B9.5a, cancelling the common g and using $M = \sum m_i$ we find

$$C = \frac{CPE}{M} = \frac{\sum \rho_i\, \Delta z_i\, z_i.}{\sum \rho_i\, \Delta z_i}$$

If we let the slices become infinitely thin and numerous, C becomes

$$C = \frac{\int_0^\infty \rho\, z\, dz}{\int_0^\infty \rho\, dz} \qquad \text{B9.5b}$$

where integration is from column base to the top of the atmosphere, where both ρ and $(\rho\, z)$ are effectively zero in the near vacuum of space.

According to Eqn 4.4, an isothermal air column in hydrostatic equilibrium under uniform g, has a pure exponential profile of air pressure p described by $p = p_0\, e^{-z/H}$ where z is height above column base (at which $p = p_0$), $H\,(= R\,T/g)$ is the uniform exponential scale height, and T is the uniform column temperature. Since the ideal gas equation for air ($p = \rho\, R\, T$) applies everywhere, the uniform T cancels, $\rho = \rho_0\, e^{-z/H}$, and B9.5b becomes

$$C = \frac{\int_0^\infty e^{-Z/H}\, z\, dz}{\int_0^\infty e^{-Z/H}\, dz}$$

From standard integration $\int e^{-z/H}\, z\, dz = H^2$ and $\int e^{-z/H}\, dz = H$ [8] so that

$$C = H \qquad \text{B9.5c}$$

showing that the centre of gravity of an isothermal air column in hydrostatic equilibrium lies exactly one exponential scale height H above its base.

and absolute temperature T_i. The index i numbers the slices from 1 at the base of the column to a large number N at its top (Fig. 9.10), which is the top of the atmosphere. The energy contributions from slice i are

$$KE = \frac{1}{2} m_i V_i^2 \quad PE = m_i g z_i \quad IE = m_i C_v T_i$$

and the column totals are the sums of these for the N slices in the column.

$$CKE = \frac{1}{2} \sum m_i V_i^2 \quad CPE = g \sum m_i z_i \quad CIE = C_v \sum m_i T_i$$

(i) *CKE* This is simply the mass-weighted sum of the squared wind speeds from the observed wind profile through the column.

$$CKE = \frac{1}{2} \sum m_i V_i^2 \qquad 9.10$$

(ii) *CIE* If we assume that the column is isothermal (that all T_i are the same T), the column total of *IE* simplifies to

$$CIE = C_v T \sum m_i = M C_v T \qquad 9.11$$

where $M = \sum m_i$ is the total mass of the air column, whose typical value at sea level is a little over 10,000 kg m^{-2}.

(iii) *CPE* The column total of *PE* is

$$CPE = g \sum m_i z_i = M g C$$

where C is the height of the *centre of gravity* of the column above its base—i.e. the height at which all the weight (or mass, g being uniform) of the column appears to be concentrated. According to Box 9.5 the centre of gravity of an

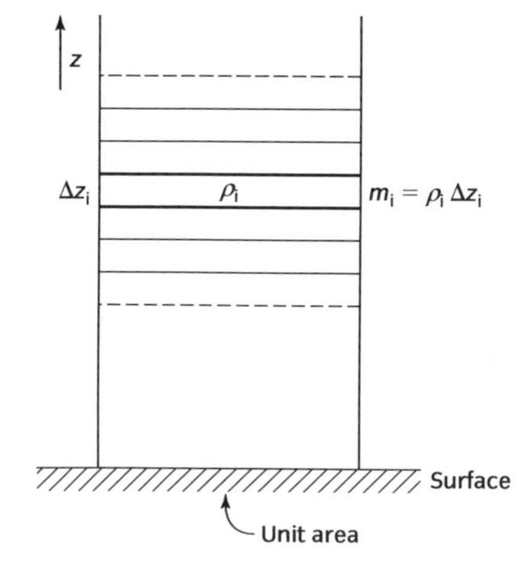

Figure 9.10 A vertical column of air with unit horizontal area treated as a pile of shallow discs. A disc of thickness Δz_i and air density ρ_i has volume Δz_i, mass $(\rho_i \Delta z_i)$ and weight $(g \rho_i \Delta z_i)$. The total weight of the column is the sum of the weights of all the constituent slices $\sum_i g \rho_i \Delta z_i$.

isothermal column in hydrostatic equilibrium is H (= $R\ T/g$)—the exponential scale height of the air in the column (Section 4.4). This leads to the very important result

$$CPE = M\,g\,H = M\,R\,T \qquad 9.12$$

The temperature dependence of *CPE* expresses an obvious physical reality: an air column with a certain mass per unit area will stand taller if it is warmer, and will therefore have a higher centre of gravity and a greater *CPE*. The particularly simple result for an isothermal column carries over to more realistic vertical temperature profiles when T is replaced by a pressure (or mass) weighted average temperature.

Collecting terms, the column total energy *CTE* is

$$CTE = \frac{1}{2}\sum m_i V_i^2 + M R T + M C_v T \qquad 9.13$$

or

$$CTE = \frac{1}{2}\sum m_i V_i^2 + M C_p T$$

using the familiar relationship $C_p = C_v + R$.

Note how the static terms in *CTE* have been simplified and related by aggregating parcel terms into column totals under the constraint of hydrostatic equilibrium.

Total potential energy

The formal similarity between the last two terms of Eqn 9.13 has led to them being paired under the title *total potential energy* (*TPE*) of the atmospheric column, or of any larger air mass made up of columns, where

$$TPE = CIE + CPE = M\ C_p\ T$$

The term 'potential' in *TPE* is used in the general sense of potentially capable of being converted into kinetic energy. To avoid confusion between *TPE* and *CPE*, remember that the full title of the latter is the column total of gravitational potential energy. The expression for *TPE* recognizes that both *CIE* and *CPE* are capable of being converted into *CKE* in the right circumstances. Hydrostatic equilibrium through the column ensures that they stay in strict proportion with each other, so that *CPE* is always 28.6% of *TPE* and 40% of *CIE*. In fact we will see below and later that only a very small proportion of *TPE* can be converted to *CKE* in realistic atmospheric disturbance.

The expression for column *TE* is now simply

$$CTE = CKE + TPE \qquad 9.14$$

Inserting realistic values in Eqn 9.14 we find the values in Table 9.2, whose precise values follow from the particular conditions assumed (NN 9.4) and are much less significant than their proportions. The ratio *CPE*/*CIE* is precisely 40% as it must be.

Table 9.2 Components of CTE for a typical air column with jet stream

CPE	746	MJ m^{-2}	see NN 9.4
CIE	1,864		
TPE	2,610		
CKE	≈7	corresponding to column average speed 37 m s^{-1}	
CTE	2,617		

NUMERICAL NOTE 9.4 Column energy totals

Consider a stationary vertical column in the flowing atmosphere of an extratropical cyclone, containing a jet stream with speeds of more than 50 m s^{-1} between 400 and 200 hPa, with a maximum of a little over 70 m s^{-1} centred at 300 hPa (about 9 km above sea level). If wind speeds in the rest of the column range from 10 to 30 m s^{-1}, *CKE* is estimated below in ten 1,000 kg m^{-2} layers (pressure thickness about 100 hPa each).

$$CKE = 1/2 \times 1{,}000 \times \Sigma V^2 = 500 \times [10^2 + 15^2 + 20^2 + 25^2 + 30^2 + 50^2 + 70^2 + 50^2 + 30^2 + 20^2] \approx 6.7\ \text{MJ m}^{-2}$$

$$V = \sqrt{(2\ CKE/M} = \sqrt{(2 \times 6.7 \times 10^6/10{,}000)} \approx 37\ \text{m s}^{-1}$$

If the column average temperature is 260 K, and $M = 10{,}000$ kg m^{-2} we have

$$TPE = CIE + CPE = M\,C_p\,T = 10{,}000 \times 1{,}004 \times 260 = 2{,}610\ \text{MJ m}^{-2}$$

If all this *TPE* could be converted into *CKE* for a column of (mass $M = 10{,}000$ kg per square metre), the resulting average wind speed *V* would be given by $1/2\ M\ V^2 = TPE$, so that $V = \sqrt{[2\ TPE/M]} \approx 722$ m s^{-1}, compared with the speed of sound ≈ 330 m s^{-1}.

As in the case of the air parcels of Table 9.1, *KE* values are relatively very small, *CKE* values being at least two orders of magnitude less than *CPE* and *CIE*. If all the *TPE* could be converted into *CKE*, the air in the column would be accelerated to about 722 m s^{-1} (NN 9.4), more than twice the speed of sound, and an order of magnitude faster than observed maximum winds. However, for this to happen we would have to collapse the air column to zero height (and volume), which would require cooling to absolute zero temperature. Since nothing like this is even conceivable, we see at once that only a part of *TPE* (in fact a very small part) is actually *available* for conversion to *CKE* in the real atmosphere.

9.6 Tropospheric heating, cooling, and response

We now look more closely at how a layer of air responds to the heat inputs and outputs outlined in Section 9.4. In the present section we examine the initial response which is to warm and inflate vertically (or cool and deflate), lifting (or lowering) the overlying unheated (or uncooled) air. In Section 9.7 we will see how this inflation and deflation can trigger tropospheric overturning and convert small but significant amounts of *TPE* into *KE*, maintaining the atmosphere's familiar mobility and heat flows.

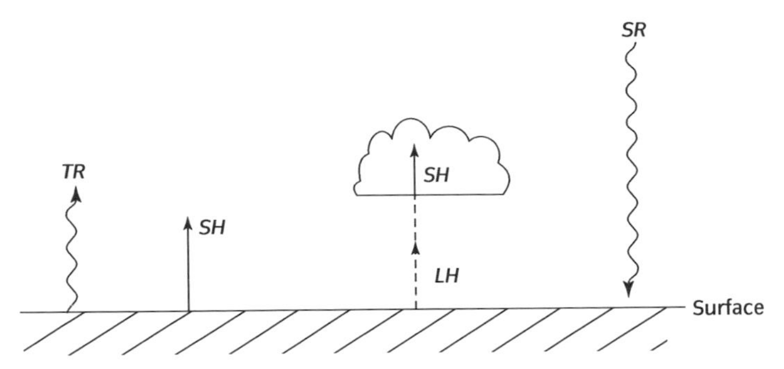

Figure 9.11 Heat inputs to a surface-based layer of the troposphere, by absorption of terrestrial radiation *TR* emitted from the sun-warmed surface, by absorption of sensible heat *SH* conducted and convected from the surface, and by liberation of latent heat *LH* when vapour evaporated from the surface condenses as cloud above the lifting condensation level for the local air. Some *SR* is absorbed directly in the layer of air, but most is absorbed at the surface as shown.

9.6.1 Heat inputs (Fig. 9.11)

Radiant energy from the Sun (*SR*) is absorbed in the top centimetres of land surfaces, and tens of metres of water surfaces, warming them and then warming the overlying air by upward conduction and convection of sensible heat (*SH*), and emission of terrestrial radiation (*TR*) which is absorbed by the lower atmosphere. Though layers of air warmed in this way may be shallow initially, in conditions encouraging convection they can reach up to include much of the lower half of the troposphere before completing the primary inflation stage and becoming ready for secondary overturning. This happens as small thermals rise out of the surface layer, carrying heat into the air above, where it is deposited as the thermal mixes out of existence, preparing the way for deeper convection. Throughout the process viscosity feeds on strong transient shears between adjacent eddies, smoothing them out of existence and degrading the turbulent *KE* into heat which adds very slightly to the warming of the layer (Table 9.1). On a larger view the sequence has warmed and inflated the layer in preparation for larger-scale convective rearrangement.

Extensive layers of air hundreds of metres deep overlying a heated land surface can be warmed by many °C in a few hours on a sunny morning (Section 5.3), without generating any motion on scales much larger than the individual turbulent eddies, except for the gentle inflation considered below. When convection becomes deep enough to generate cloud (more than 300 metres deep in typical near-surface humidities), much more solar heat is released into the cloudy air as latent heat (*LH*) is evolved—solar heat that has been latent since it was used to evaporate water from the Sun-warmed surface. This has the effect of deepening and invigorating the input of heat from the surface to the convecting layer of air. Direct absorption of incoming *SR* by the low troposphere adds to the heating of the lower half of the troposphere (Section 9.4).

Regardless of microscale details, the effect of all this heat input is to warm an extensive layer of air (horizontally meso-scale or even synoptic scale) and inflate it vertically, as if it were contained in a relatively shallow loose inflatable mattress.

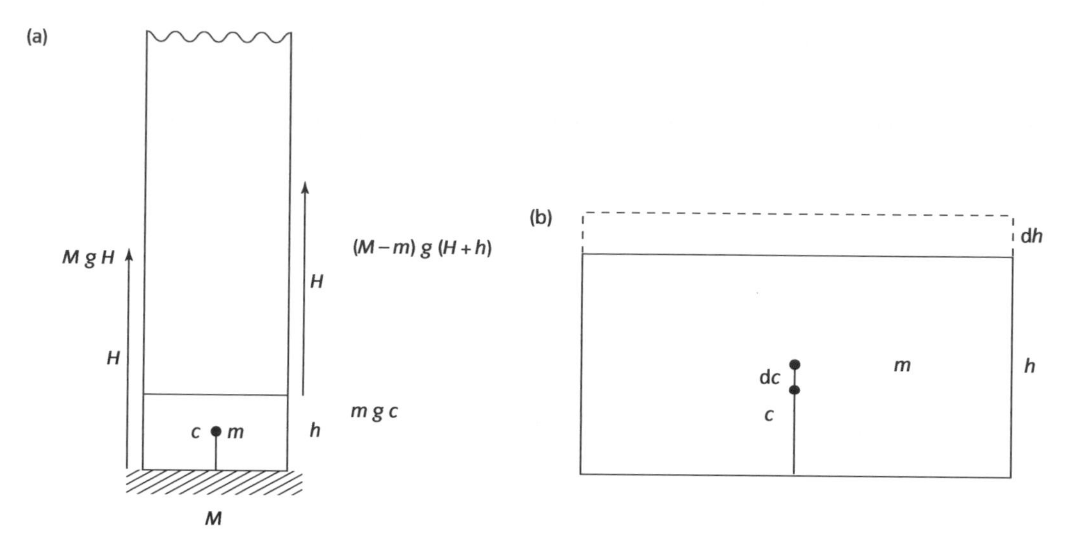

Figure 9.12 (a) The centre of gravity of a horizontal slab in an isothermal atmospheric column in hydrostatic equilibrium. The centres of gravity of the whole column, the warming slab, and the overlying layer are respectively H, C, and $H + h$ above the base. (b) Close up of the base slab of (a), showing the small vertical inflation of the heated slab and the associated small rise in its centre of gravity.

9.6.2 Inflation

We can examine the vertical inflation layer of air (calling it a *slab* to distinguish it from unheated layers) as it is warmed, by extending the first law of thermodynamics (Section 5.3) to include gravitational potential energy (*PE*), recognizing explicitly that *PE* is a part of the mechanical energy whose production by the tropospheric engine we wish to trace.

The situation is depicted in Fig. 9.12 and detailed in Box 9.6. We can assume that the heat input to the slab, with its warming and inflation, takes place at constant pressure in a hydrostatically balanced air column; that the warmed slab expands (*inflates*) vertically but not horizontally, because it is very much broader than deep; and that the rising temperature of the slab is kept horizontally and vertically uniform by internal stirring. The last assumption ignores the vertical temperature lapses typical of deeper slabs, which can be modelled as multiple sandwiches of thin isothermal layers with slightly differing temperatures.

The response of the warmed slab is described by Eqn 9.15, in which M is the mass per unit horizontal area of the whole atmospheric column, and m, h, and c are the mass per unit area, thickness, and height of the centre of gravity of the slab which is warming and inflating in response to a little heat input dQ per unit horizontal area.

$$\underset{}{dQ} = \underset{\text{(i)}}{m\, C_v\, dT} + \underset{\text{(ii)}}{m\, g\, dc} + \underset{\text{(iii)}}{(M - m)\, g\, dh} \qquad 9.15$$

The labelled terms represent:

(i) the little increase in the slab's internal energy as its temperature rises by dT;

(ii) the little increase in the slab's *PE* as inflation raises its centre of gravity by dc; and

(iii) the little increase in the *PE* of the overlying atmosphere as it is raised a distance dh on the top of the inflating slab.

As shown in Box 9.6, terms (ii) and (iii) collapse into $m\,g\,\mathrm{d}H$, where H (= $R\,T/g$) is the exponential scale height of the air in the slab, so that $m\,g\,\mathrm{d}H = m\,R\,\mathrm{d}T$. This is the slab version of (Eqn 9.12) for the whole air column, with slab mass m replacing total column mass M. Using $C_p = C_v + R$, Eqn 9.15 now becomes

$$\mathrm{d}Q = m\,C_v\,\mathrm{d}T + m\,R\,\mathrm{d}T = m\,C_p\,\mathrm{d}T \qquad 9.16$$

which is formally identical to the expression for the first law of thermodynamics written in Eqn 5.3 without reference to gravity. Equation 9.16 shows that the heat energy input $\mathrm{d}Q$ to the inflating slab is used in proportions completely determined by the thermal properties of air—the proportion C_v/C_p (= 71.4%) being used to increase the slab's internal energy (warming) and proportion R/C_p (= 28.6%) being used to increase the *PE* of the whole column, by lifting the overlying air and by raising the centre of gravity of the slab itself. The right-hand side of Eqn 9.16 is in fact the slab inflation's contribution to the increase in the *TPE* of the whole atmospheric column, some of which is at least potentially available for subsequent conversion into kinetic energy by subsequent convection and advection (Section 9.7).

BOX 9.6 Heat input and slab inflation

We can adapt the treatment of Box 9.5 to find the centre of gravity of a horizontal slab of air at the base of a full air column (Fig. 9.12) in hydrostatic equilibrium at uniform temperature T. The centre of gravity of the overlying air (mass $(M - m)$ per unit area) is at height H (= $R\,T/g$) above the top of the slab which is to be heated (i.e. $H + h$ above its base), whereas the centre of gravity of the whole column (mass M) is at height H above the slab's base. If c is the height of the slab's centre of gravity above its base, then applying B9.5a to the total column, slab, and overlying air gives

$$M\,g\,H = m\,g\,c + (M - m)\,g\,(H + h)$$

so that

$$c = H - \frac{M-m}{m}\,h \qquad \mathrm{B9.6a}$$

Though B9.6a was found by assuming that the overlying air had the same temperature as the slab, it must be valid regardless of the state of the overlying air, since that has no effect on the distribution of mass within the slab. If the slab in Fig. 9.12 warms by $\mathrm{d}T$, the exponential scale height H of its pressure profile increases by $\mathrm{d}H = R\,\mathrm{d}T/g$, and slab thickness h inflates by $\mathrm{d}h$. Rewriting B9.6a in terms of little changes in c, H, and h (m and M are fixed by assumption) we have

$$\mathrm{d}c = \mathrm{d}H - \frac{M-m}{m}\,\mathrm{d}h$$

Substituting this in Eqn 9.15, the result describes the behaviour of the slab as it warms and inflates in response to a small heat input $\mathrm{d}Q$.

$$\mathrm{d}Q = m C_v\,\mathrm{d}T + m\,g\left[\mathrm{d}H - \frac{M-m}{m}\,\mathrm{d}h\right] + (M - m)\,g\,\mathrm{d}h$$

$$= m\,C_v\,\mathrm{d}T + m\,g\,\mathrm{d}H \qquad \mathrm{B9.6b}$$

since the terms in $(M - m)$ cancel. Using $\mathrm{d}H = R\,\mathrm{d}T/g$ we find

$$\mathrm{d}Q = m\,C_v\,\mathrm{d}T + m\,R\,\mathrm{d}T \qquad \mathrm{B9.6c}$$

whose last term is the little increase in column PE ($dCPE$) resulting from the inflation of the slab *and* its raising of the overlying air.

Slab inflation and column lifting

- The pure exponential profile of pressure within the isothermal slab relates its top and bottom pressures and overlying masses according to $p/p_0 = (M - m)/M = e^{-h/H}$ so that $h/H = \ln[M/(M - m)]$, and $dh/dH = \ln[M/(M - m)]$, since the masses M and m are unchanged by warming. We see that the increase dh in slab thickness h is directly proportional to dH ($= R\, dT/g$) and hence to the increase in slab absolute temperature, as expected.
- The precursor line to B9.6b shows that the work done by the inflating slab in lifting the overlying column is $(M - m)\, g\, dh$, which when expressed as a proportion of $dCPE$ ($= m\, g\, dH$) is

$$\frac{(M-mg)dh}{mg\,dH} \quad \frac{M-m}{m}\frac{dh}{dH} \quad \left(\frac{1}{X}-1\right)\ln\left[\frac{1}{1-X}\right]=Y(X)$$

where $X = m/M$ is the ratio of slab mass to total column mass, and Y is the function of X which specifies the proportion of $dCPE$ which is used in lifting the air overlying the slab. The proportion of $dCPE$ used to raise the centre of gravity of the inflating slab is $1 - Y$, since the two must add up to 1. Table 9.3 presents values of Y and $1 - Y$ calculated for a range of $X = m/M$, using $\ln x = \text{loge}\, x = 2.303 \log_{10} x$

Table 9.3 Overlying column $dPE/dCPE$ vs. m/M

$X=\frac{m}{M}$	0.1	0.3	0.5	0.7	0.9	0.99
Y(X)	0.95	0.83	0.69	0.52	0.26	0.05
1 – Y(X)	0.05	0.17	0.31	0.48	0.74	0.95

9.6.3 Inflation and lifting

Although Eqns 5.3 and 9.16 are apparently identical, there is an important difference between the situations they describe. In the simple gravity-free case, all the energy not used to warm the air is used to do work on the surroundings of the heated air, which means moving the overlying air in the situation of Fig. 9.12. In the real gravity-bound atmosphere, that same proportion of work done is divided between raising the centre of gravity of the inflating slab, and lifting the overlying unheated air. The two components of potential energy increase (inflating the slab and lifting the overlying air) can play quite different roles in subsequent atmospheric activity, as shown by the simple models of convection and slope convection considered in Section 9.7.

Table 9.3 in Box 9.6 shows that the proportion of the increase in CPE which arises within the warming slab depends only on m/M, the ratio of the mass m of the warmed slab to the total column mass M. When this ratio is very small ($m << M$), the proportion of $dCPE$ used to raise the centre of gravity of the warmed slab is $m/(2\,M)$, and the rest is used to lift the overlying atmosphere. When m is only very slightly less than M, almost all of $dCPE$ is used within the warmed slab, since there is very little overlying air to lift. When m is about $0.7\,M$ (a typical proportion for the troposphere in the whole atmosphere), the apportionment of $dCPE$ between inflating and lifting is about 0.5. This means that even when the whole troposphere is warmed (for example by a population of tall cumulonimbus), about 50% of the resulting increase in CPE is stored in the lifted stratosphere, which corresponds to about 14% of the increase in TPE of the whole atmosphere. Patchy heating (and cooling) of the troposphere can therefore produce significant inputs and removals of stratospheric PE, in addition to the injection of tropospheric ME into the stratosphere by the upward propagation of gravity waves from the troposphere (Section 11.8).

Shallow heating (NN 9.5)

Early morning heating overland is often confined to surface-based slabs of order 100 m deep (i.e. with a pressure thickness of order 10 hPa), capped by residual low-level nocturnal inversions (Section 5.3). With a corresponding $m/M \approx 0.01$ (slab mass $\approx$ 1% of the total atmospheric column), Table 9.3 in Box 9.6 shows that only $\approx$ 0.5% of the increase of column *PE* is confined to the warming slab, 99.5% being stored in the lifting of the unheated 99% of the whole column. If heat is injected into the warming slab at 200 W m^{-2} for an hour, the input of 720 kJ m^{-2} warms the slab by over 7 °C , which becomes about 2.5% less dense as a result, and inflates by about 2.1 metres (NN 9.5). The rest of the atmosphere is lifted by this amount, accounting for virtually all of the increase in *CPE*, and therefore nearly 29% of the input heat energy and increase in *TPE*. Although very significant, such a small inflation is quite unobservable in the convective turmoil of the warming slab. In the same two hours the convecting layer could easily deepen by tens or hundreds of metres as convection burrows upwards into the overlying air (e.g. Fig. 5.2).

Deep heating (NN 9.5)

If conditions encourage much deeper convection, heat can be injected through a large part or all of the troposphere. Taking for example a layer 700 hPa deep, $m/M \approx 0.7$ and according to Table 9.3 in Box 9.6 the injection of *PE* is divided equally between inflating the troposphere and lifting the overlying stratosphere. If heat is injected at 300 W m^{-2} for 12 hours, as might happen when cool air sweeps over a warmer ocean, the whole troposphere will warm by nearly 2 °C, raising its own centre of gravity by over 27 m and lifting the overlying stratosphere by 63 m (NN 9.5).

9.6.4 Cooling and deflation

Heat is continually being lost from the high troposphere by net *TR* emission to space, and also from the proximity of the low troposphere in high latitudes to land and ice surfaces which are losing heat to space by net *TR* through the

NUMERICAL NOTE 9.5 Slab inflation

Shallow heating

A slab of air bounded by pressures 1,000 and 990 hPa (mass $m = \Delta p/g \approx 10 \times 100/10 = 100$ kg m^{-2}) receives heat from the underlying surface at 200 W m^{-2} for an hour ($\Delta Q = 200 \times 60 \times 60 = 720$ kJ m^{-2}). The slab warms by $\Delta T = \Delta Q/(m\, C_p) = 720{,}000/(100 \times 1{,}000) = 7.2$ °C using C_v/C_p (71.4%) of the energy input, and leaving 28.6% ($\approx$ 206 kJ) to increase *CPE*. Since $m/M = 0.01$, $Y = 0.995$, so that 99.5% of the 206 kJ is used to lift the overlying atmosphere by $\Delta h = \Delta Q/[(M - m)\, g] = 205{,}000/[9{,}900 \times 10] = 2.1$ m. The same answer emerges more roughly from $\Delta h/h = \Delta T/T \approx 7.2/288 = 2.5\%$ of about 100 m.

Deep heating

A deep slab of air bounded by pressures 1,000 and 300 hPa (mass = 7,000 kg m^{-2}) receives heat at 300 W m^{-2} for 12 hours ($\Delta Q \approx 13$ MJ m^{-2}). The slab warms by about 1.8 °C, increasing its *CIE* by about 9.3 MJ m^{-2} (= 0.717 ΔQ) and *CPE* by about 3.7 MJ m^{-2} (= 0.286 ΔQ), of which 50% (about 1.9 MJ m^{-2}) is stored in the lifting of the overlying unheated stratosphere by 63 m ($\approx 1.9 \times 10^6/(3{,}000 \times 10)$). The 50% follows from $Y = 0.5$, since $m/M = 0.7$.

NUMERICAL NOTE 9.6 Cooling and deflation

High troposphere slab between 300 and 200 hPa

About 100 W m^{-2} for a day ($\approx 9 \times 10^6$ J m^{-2}) is lost by a 100 hPa slab ($\approx$ 1,000 kg m^{-2}) bounded by 300 and 200 hPa in 24 hours, producing a cooling of $\Delta Q/(m\, C_p) \approx 9 \times 10^6/(10^3 \times 10^3) = 9$ °C. With a total column mass of about 3,000 kg m^{-2} above the base of the cooling slab and 2,000 kg m^{-2} of stratosphere above its top, the slab mass is 1,000 kg m^{-2}, $m/M = 0.333$ and $Y \approx 0.80$ in Table 9.3, showing that 80% of the *CPE* loss ($9 \times 0.286 \times 0.8 \approx 2.0$ MJ m^{-2}) comes from slab deflation $\Delta h = \Delta PE/$(stratosphere mass $\times$ g) $= 2.0 \times 10^6/(2{,}000 \times 10)$, i.e. a subsidence of 100 m.

Low troposphere slab between 1,000 and 900 hPa

If the same 9 MJ m^{-2} is lost by a 100 hPa slab in the low troposphere, the cooling will be identical (9 °C), and so will be the overall loss of *CPE* ($= 0.286 \times 9 \approx 2.6$ MJ m^{-2}), but the corresponding deflation will be very different because the mass of overlying air is 9,000 kg m^{-3} and $m/M = 1/10$. In this case nearly all the *CPE* loss is effected by the subsidence of the overlying air ($Y = 0.95$ in Table 9.3), so that $\Delta h = 2.6 \times 10^6/(9{,}000 \times 10) \approx 29$ m.

atmospheric window. The resulting cooling and *deflation* of these tropospheric heat sinks is described by the same thermodynamics as the warming and inflation described above. Equations 9.1 and 9.16 apply with negative dQ, dT, dc, and dh, and energy flowing from right to left across the equals sign.

Heat losses from the high troposphere are outlined in Fig. 8.8. The highest layer of the cloudless troposphere which is opaque to *TR* outside the transparent window, radiates to space at a global average rate of 59 units on Fig. 8.8, which corresponds to 213 W m^{-2} as discussed in Section 9.4. In the longer term this loss is made good from beneath by convected and advected heat and by the net upward *TR* flux, so that the heat sink zones do not progressively cool below their typical temperatures, but at any particular time and place the imbalance between output to space and upward flux from beneath may correspond to a loss rate of ~ 100 W m^{-2} (nearly 9 MJ m^{-2} per day).

Upper and lower tropospheric cooling and deflation (NN 9.6)

If upper tropospheric cooling and deflation occurs in a 100 hPa slab (for example a layer about 2 km thick centred on 350 hPa, about 8 km above MSL) then a net rate of heat loss of 9 MJ m^{-2} per day (about 100 W m^{-2}) produces a cooling of about 9 °C and a deflation of about 100 m, lowering the overlying atmosphere (almost all of it stratosphere) by this amount.

If the same rate of heat loss occurs in a 100 hPa slab overlying a winter surface in high latitudes, the much shallower layer (now only 800 m thick, being in the low troposphere) cools by the same amount as before in the upper troposphere, but the much smaller m/M value (0.1 compared to 0.33) means that almost all *CPE* is lost by the subsidence of the much larger mass of overlying atmosphere (the stratosphere and most of the troposphere) by about 30 m.

9.7 Atmospheric motion

We have seen how inputs of heat to an atmospheric column at constant pressure increase its total potential energy (*TPE*)—the sum of its column internal energy

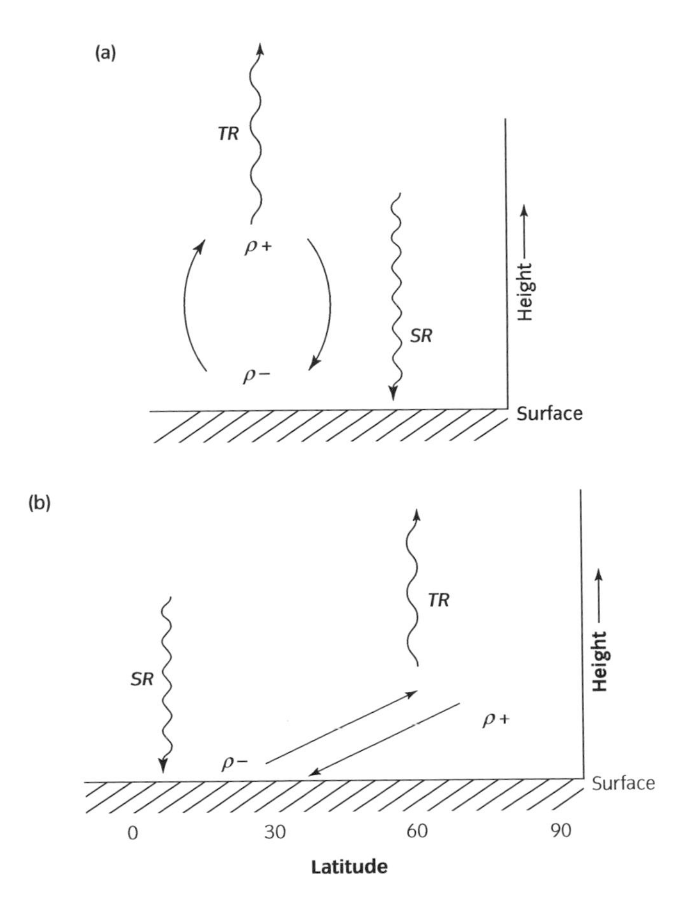

Figure 9.13 (a) Schematic outline of atmospheric convection. Solar heating maintains low (potential) density air in the low troposphere, and net *TR* emission maintains high (potential) density air in the high troposphere. Convective overturn lowers the gravitational potential energy of the column. (b) Schematic outline of slope convection in the middle latitudes of a flat Earth. Solar heating in lower latitudes maintains a zone of low (potential) density in the low troposphere, and net *TR* emission maintains a zone of high (potential) density in the high troposphere in high latitudes. Slope convective exchange takes place across an extensive range of latitude and longitude, though the latter is not shown in this meridional diagram.

(*CIE*) and column (gravitational) potential energy (*CPE*)—by warming and inflating at least the low troposphere, the average rate of solar energy input being about 200 W m^{-2} (Section 9.4). Though this is obviously an important first step in the conversion of heat to mechanical energy by the tropospheric engine, the process is far too slow to account directly for the kinetic energies corresponding to air motion observed in weather systems. For example the associated rate of injection of *PE* at about 57 W m^{-2} into the lowest 1,000 kg m^{-2} of the troposphere (the first 100 hPa) raises the top of the inflating slab at about 0.6 mm s^{-1} (NN 9.7), which is unobservably small in comparison with observed convective updrafts $\sim$ 1 m s^{-1}, and three to five orders of magnitude smaller than wind speeds observed in the upper troposphere (which often exceed 50 m s^{-1}). How is such enormously faster motion generated from such slow beginnings?

The answer is demonstrated in a wide range of atmospheric activity: the atmosphere is always so close to the tipping point of convective and larger-scale instabilities (Section 5.11), that persistent inflation, even though it is very slow, is continually setting up conditions in which small but significant amounts of the huge atmospheric reservoir of *TPE* become *available* for rapid, sporadic conversion to kinetic energy. The situations are sketched in Fig. 9.13: heat sources vertically beneath heat sinks can trigger convection, and adjacent warm and cool air masses are prone to the slope convection which underpins synoptic scale weather

NUMERICAL NOTE 9.7 Inflation speeds

If heat enters an inflating slab at rate at 200 W m^{-2}, then column *PE* is increasing at 57 W m^{-2} (0.286 × 200). According to Box 9.6, if the slab has mass 1,000 kg m^{-2}, almost all of the *PE* is used to lift the overlying 9,000 kg m^{-2}. Taking a one second interval and using $\Delta PE = (M - m)\, g\, \Delta h$, with $\Delta PE = 57$ J m^{-2}, $(M - m) = 9{,}000$ kg m^{-2}, the little rise Δh is given by $\Delta PE/[(M - m)\, g] = 57/[9{,}000 \times 10] \approx 0.6 \times 10^{-4}$ m, i.e. 0.6 mm, showing that the top of the inflating layer is rising at 0.6 mm per second.

activity in middle latitudes. The crucial factor in each case is the uneven distribution of heat sources and sinks.

Calculations of the quantities of *TPE* made available for conversion to *CKE* in these ways have been made by many meteorologists since they were first attempted by the Austrian pioneer Margules in 1903, when he established the important concept of *available potential energy* (*APE*) and showed that only a very small fraction of atmospheric *TPE* becomes available for conversion to *KE* in realistic conditions. In fact the ratios *APE*/*TPE* which he and others have estimated then and since are even smaller than the small values we will find in the following simple incompressible models, and correspond well with the small air speeds (and kinetic energies) actually observed (Table 9.1).

9.7.1 Kinetic energy production

Convection

If an air column is in a convectively neutral state (Section 5.10), a very small further warming of its lower parts, and/or cooling of its upper parts, can trigger rapid convective overturning in which the potentially less dense air below changes places with potentially more dense air above, in the process converting a small proportion of *TPE* into the *KE* of convective motion, as sketched in Fig. 9.13a. We say that prior warming and/or cooling has produced some *TPE* which is *available* for conversion to *KE* in convection—i.e. it has produced some *APE*. We observe that such convection can occur on a wide range of height and time scales, from metre scale thermals to 10-kilometre-tall cumulonimbus, lasting from seconds to hours.

Slope convection

Figure 9.13b sketches the mechanism of a very simple model of *slope-convective* exchange between adjacent masses of air, which may have much the same surface pressure (and therefore the same air masses per unit horizontal area), but significantly different vertical distributions of mass because they have experienced different prior inflation histories. Such differences are continually produced by the gradient of solar heat input between lower and higher latitudes.

When the contrasting masses are adjacent, at any chosen height above the common base level, pressure in the more highly inflated (warmer) column is greater than in its neighbour, so that air will tend to flow across to the cooler column. But as pressures equalize aloft in this way, pressures at lower levels in the cooler column will exceed those on the same horizontal in the warmer column, and

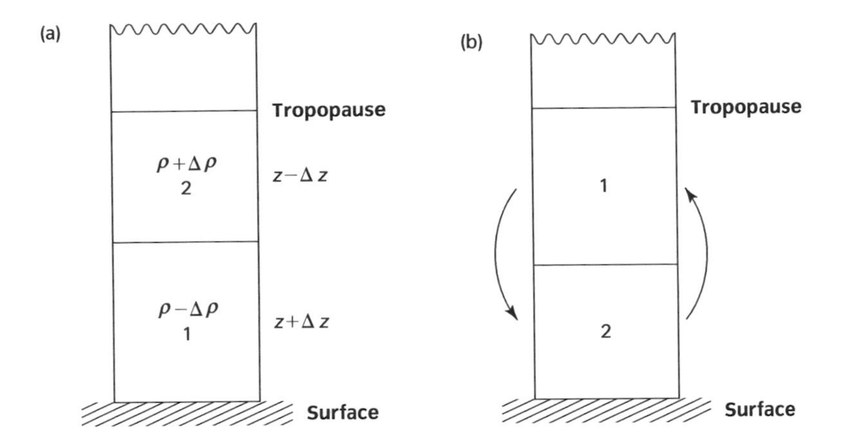

Figure 9.14 Incompressible model of unmixed convective overturn. (a) Before overturn, block 2 overlies the less dense block 1, together filling the troposphere under the stratosphere *S*. The arrangement is convectively unstable. (b) After convective overturn, the blocks have exchanged places, lowering the centre of gravity of the troposphere and ready to be destabilized again by warming below and cooling above. The stratosphere is unmoved.

cooler air will flow across, undercutting the warmer air. The rearrangement will be complete when the air from the warmer column has filled the upper half of the resulting double column, and air from the cooler column has filled the lower half of the double column, so that pressures are once again uniform across every horizontal from base to top. The rearrangement combines the vertical exchange typical of upright convection, with extensive lateral exchange, so that the result is a slightly but systematically sloped exchange.

Slope-convective exchanges generate *KE* by converting *TPE* which became *available* when the two columns were placed side by side by large-scale atmospheric motion—the conversion initiated and maintained by temporary pressure gradients. Such conversion of *APE* to kinetic energy takes place mainly in meso and synoptic scale weather systems, and produces large areas of gentle but persistent uplift and sinkage which combine with much stronger horizontal flows to produce slopes $\sim 1/100$.

9.7.2 Convective cycle

In this and the following subsections we use very simple models to examine the production of *APE* by localized atmospheric inflation and deflation, and its conversion to *CKE* by convective and slope-convective overturning. The models use incompressible blocks of air, with discontinuities of air density rather than density gradients, with no friction or mixing, and with just enough thermodynamic behaviour to give realistic response to heat inputs and outputs.

The assumption of incompressibility means that the models produce *KE* by converting *CPE* rather than *TPE*, and that model *APE* is actually available gravitational potential energy. Model *CPE* is therefore a proxy for real *TPE*, and model available gravitational potential energy (which we will continue to call *APE*) is a proxy for true *APE*. The assumed absence of mixing (and friction) obviously makes the models much more efficient than reality.

Figures 9.14a and b outline the 'before and after' stages of one cycle of convective conversion of *APE* to kinetic energy. The third stage (not shown) is the recovery of *APE* by differential heating and cooling which sets up the next cycle, allowing the process to repeat indefinitely or until interrupted by a change in circumstances.

Stage (i) About to convect (Fig. 9.14a)

A two-part column of model air, each part with mass m per unit horizontal area, is in a convectively unstable state, with more dense air overlying less dense air, because the lower half has been inflated by prior heating to occupy a depth $z + \Delta z$ (numerically identical to volume per unit horizontal area), whereas the upper half has been deflated by prior net cooling to occupy a smaller depth (volume) $z - \Delta z$. As shown in Box 9.7 the proportional differences in magnitudes of air density, depth, and absolute temperature (temporarily treating the model air as an ideal gas at constant pressure) are given by

$$\frac{\Delta \rho}{\rho} = \frac{\Delta z}{z} = \frac{\Delta T}{T}$$

Since temperature differences ($2\,\Delta T$) between rising and sinking air are about 2 °C (in about 260 K) in real cumulus and cumulonimbus convection (Section 11.2), $\Delta T/T$, etc. are about 0.4%. The differences in column depth in Fig. 9.14 have obviously been greatly exaggerated for clarity. Note that the model makes no attempt to explain why ΔT is about 1 °C; it simply appeals to observation. In reality such temperature differences must arise from 'stickiness' in the system (inertia and friction) which allows ΔT to build up to this level, rather than leak away sooner in feebler convection.

Stage (ii) Convective overturn (Fig. 9.14b)

In the model convective exchange, the half columns move through each other without mixing or friction, so that the denser air falls a distance $z + \mathrm{d}z$ to become the lower half of the column, while the less dense air rises a distance $z - \mathrm{d}z$ to become the upper half. Box 9.7 shows that the resultant change in CPE is given by

$$\Delta CPE = -2\, m\, g\, \Delta z$$

where the minus sign means that CPE has reduced, feeding conversion to KE. As a proportion of average CPE ($= 2\, m\, g\, z$) this APE ($= 2\, m\, g\, \Delta z$) is

$$\frac{APE}{CPE} = \frac{\Delta CPE}{CPE} = \frac{\Delta z}{z} = \frac{\Delta T}{T} \qquad 9.17$$

Inserting the observed 0.4% for $\Delta T/T$ shows that this convective model converts less than 0.5% of its CPE to KE at each convective overturn.

Stage (iii) Recovery

To return the column to stage (i) for another convective overturn, APE has to be rebuilt by inflating the lower half of the column by net heating, and deflating its upper half by net cooling. As confirmed in Box 9.7, when the lower half inflates from thickness $z - \Delta z$ to $z + \Delta z$ and the upper half deflates from thickness $z + \Delta z$ to $z - \Delta z$, the total increase in CPE is $2\, m\, g\, \Delta z$ – the amount of CPE converted to KE in the last convective overturn. Inflation below and deflation aloft have now restored the column's APE to the level reached just before the last convective overturn, and we can envisage cycles of APE loss and restoration continuing for as long as the heating and cooling continues. Since heating from the underlying sunlit surface and cooling by net emission of terrestrial radiation to space is maintained over most of the globe by the continual throughput of solar energy, convective cycles are a global feature of the troposphere, and a continuing source of air motion and associated atmospheric *KE*.

Note that if the upper half had not deflated, inflating the lower half and lifting the upper half would have produced $\Delta CPE = +3\ m\ g\ \Delta z$. Upper deflation has therefore used up one energy unit ($m\ g\ \Delta z$) which otherwise would have been available for conversion to *KE*, showing that *APE* is not necessarily converted to *KE*—it can be consumed by deflation and exported from Earth as *TR*.

BOX 9.7 Incompressible convection and slope convection

In Fig. 9.14a the cooler, denser, shallower upper part of the column has temperature $T - \Delta T$, density $\rho + \Delta\rho$ and thickness $z - \Delta z$, compared with $T + \Delta T$, $\rho - \Delta\rho$ and $z + \Delta z$ for the warmer, less dense, thicker lower part. Since each half column has mass m (per unit area) and volume = thickness (for unit area), from mass = density × volume we have

$$m = (\rho + \Delta\rho)(z - \Delta z) = (\rho - \Delta\rho)(z + \Delta z)$$

For small proportionate differences (realistic $\Delta z/z$ values are ~ 1% or less) we have

$$\frac{\Delta\rho}{\rho} = \frac{\Delta z}{z}$$

If the incompressible model air behaves like real air at constant uniform pressure, $p = \rho\ R\ T$ is uniform throughout the column, so that $\Delta\rho/\rho = \Delta T/T$, and we have

$$\frac{\Delta\rho}{\rho} = \frac{\Delta z}{z} = \frac{\Delta T}{T}$$

Convection

In convective overturn, the upper, denser slab falls distance $z + \Delta z$ to replace the lower less dense slab which rises $z - \Delta z$ to rest on the fallen slab (Fig. 9.14b). Treating the air slabs as low-density homogeneous 'bricks', the work done by gravity on the slabs during their exchange is

$$m\ g\ [(z + \Delta z) - (z - \Delta z)] = 2\ m\ g\ \Delta z$$

and this is the *APE* of the set up (available *CPE*), now being converted to *KE*.

$$\Delta CPE = -2\ m\ g\ \Delta z$$

Dividing by full *CPE* ($\approx 2\ m\ g\ z$, since the total column mass $2\ m$ has its centre of gravity very close to height z), we find the proportionate fall in *CPE* during convective overturn

$$\frac{\Delta CPE}{CPE} \approx \frac{\Delta z}{z} = \frac{\Delta T}{T} \qquad \text{B9.7a}$$

Restoration

To restore *APE* for another convective overturn, the lower slab must inflate by $2\ \Delta z$ from $(z - \Delta z)$ to $(z + \Delta z)$, raising its centre of gravity by Δz, and increasing its contribution to overall *CPE* by $m\ g\ \Delta z$. The upper slab deflates by $2\ \Delta z$, so that its centre of gravity falls by Δz relative to its base, which corresponds to a *rise* of Δz relative to sea level since the upper base is raised $2\ \Delta z$ by the lower slab. Each half of the column therefore contributes $m\ g\ \Delta z$, so that the total ΔCPE is $+2\ m\ g\ \Delta z$, making good the loss in convection and setting up the next convective overturn.

Slope-convection

Figure 9.15a has adjacent full columns, with heights $2\ z - 2\ \Delta z$ and $2\ z + 2\ \Delta z$. The shorter acts like two half columns each of mass m, thickness $(z - \Delta z)$, density $(\rho + \Delta\rho)$, and temperature $(T - \Delta T)$, while the taller acts like two half columns each of mass m, thickness $(z + \Delta z)$, density $(\rho - \Delta\rho)$, and temperature $(T + \Delta T)$. The proportional differences $\Delta\rho/\rho = \Delta z/z = \Delta T/T$ are related as in the convective model.

Slope-convective rearrangement in Fig. 9.15b is a lateral overturn, with the half columns moving counter-clockwise by one block: slab 1 moving horizontally right, slab 2 rising $z - \Delta z$ to rest on top of slab 1, slab 3 moving left and sinking by $2\ \Delta z$ to come to rest on slab 4 which has fallen $z - \Delta z$ to base level. The horizontal movement of slab 1 does not change its PE, the movements of slab 2 and 4 exactly cancel in their effects on PE (masses m rising and falling $z - \Delta z$), and slab 3 reduces total CPE by $m\,g\,2\,\Delta z$ as it moves left and sinks $2\,\Delta z$. Total *CPE* for the two-column system therefore changes by

$$\Delta CPE = -2\,m\,g\,\Delta z$$

Since the combined *CPE* of the two columns $\approx 4\,m\,g\,z$ (mass 4 m at centre of gravity height $\approx$ z), the magnitude of the proportional fall in *CPE* is

$$\frac{\Delta CPE}{CPE} \approx \frac{\Delta z}{2z} = \frac{\Delta T}{2T} \qquad \text{B 9.7b}$$

which is half the corresponding ratio for convection (B9.6a), because it takes two columns rather than one to make $2\,m\,g\,\Delta z$ available by slope-convective rearrangement.

Restoration

To restore *APE* for another exchange, we divide and separate the two-layer double column remaining after slope-convective rearrangement, producing two identical columns, each with inflated air overlying deflated air. One column goes to lower latitudes where its lower half is warmed and inflated from $z - \Delta z$ to $z + \Delta z$, raising its centre of gravity by Δz, and lifting its upper half by $2\,\Delta z$. The total energy needed to do this is $3\,m\,g\,\Delta z$. The other column goes to higher latitudes where its upper half is deflated from $z + \Delta z$ to $z - \Delta z$, allowing gravity to do work $m\,g\,\Delta z$ as its centre of gravity falls by Δz (half the thickness deflation). The net energy input needed to restore *APE* for further slope-convective rearrangement is therefore $3\,m\,g\,\Delta z - m\,g\,\Delta z = 2\,m\,g\,\Delta z$ as expected.

Model convection and reality

- In the model, convective overturn is assumed to be instantaneous. In the real atmosphere we observe that the time take for convective overturn varies from a few seconds to an hour or so, depending on vertical scale.
- In the model, the speed of *APE* recovery is determined by the time taken to warm and reflate the lower air at the prevailing rates of heat input from the surface. Rates of cooling in the upper troposphere will be comparable. According to the thermodynamics of inflating air at constant pressure (NN 9.8), a heat input of about 8 MJ m^{-2} is needed to warm the lower half of the troposphere by 2 °C. If this is supplied by surface-based warming at 100 W m^{-2} (half the $\approx$ 200 W m^{-2} average from Section 9.4 to allow for the inefficiency of shallow convective warming), recovery will take about a day. The convective cycle is therefore quite asymmetrical in time: brief convective overturn, with simultaneous *APE* collapse, is followed by much slower *APE* recovery.
- During convective overturn, a strong net upward heat flux is carried by rising warm and sinking cool air. Typical up and down drafts of a few metres per second, with temperature differences $\sim$ 2 °C, are consistent with upward fluxes of sensible heat of a few kW m^{-2} (Box 8.3, etc.). This is over an order of magnitude larger than the global annual average of the convective flux through the troposphere (about 100 W m^{-2}), confirming that convection acts in scattered, short, and powerful bursts.
- The model is unrealistically efficient because it ignores the mixing which is known to be very important and often visually obvious in real convection (Section 7.14). If descending and ascending model air parcels were to collide and merge in the mid-troposphere, their positive and negative buoyancies would cancel, having converted only half the potential energy into kinetic energy they would have accomplished in a full overturn. If all parcels did this, the *APE* wastage would be 50%, with further losses by viscous friction and degradation of *KE* to heat.

NUMERICAL NOTE 9.8 APE recovery times (Figs. 9.13, 9.14 and 9.15)

Convective

If 4,000 kg m^{-2} of lower troposphere (roughly the layer between 1,000 and 600 hPa) has been left 2 °C cooler by convective overturn, then the heat input ΔQ needed to return it to the point of convective overturn is given by $\Delta Q = m\,C_p\,\Delta T = 4{,}000 \times 1{,}000 \times 2 = 8$ MJ m^{-2}, where the C_p value for real air has been used for the model air. At 100 W m^{-2}, the time taken to input 8 MJ is $8 \times 10^6/10^2 = 8 \times 10^4$ s or c 22 hours.

Slope-convective

After slope-convective overturn, an air column whose lower troposphere is 10 °C cooler than its upper troposphere moves over a surface heat source of strength 100 W m^{-2}. If the cool troposphere has mass 4,000 kg m^{-2}, the heat input needed to return it to its condition before slope-convective overturn is five times the convective figure, i.e. 40 MJ m^{-2}, which at 100 W m^{-2} takes time $t = 40 \times 10^6/100 = 40 \times 10^4$ s or 4.6 days. In reality it will take longer because of continuing cooling aloft.

- In Fig. 9.13a the dead weight of the stratosphere on the troposphere plays no energy role because the tropopause neither rises nor falls during the perfectly balanced model of convective overturn and *APE* recovery. More realistic asymmetry of motion or heating will lift or lower the stratosphere, exchanging *PE* with the troposphere.

9.7.3 Slope-convective cycle

The meridional separation of heat sources and sinks keeps the bulk of the troposphere significantly warmer between the tropics than it is in middle and high latitudes, so that equally massive air columns stand considerably taller in low latitudes than they do elsewhere (Fig. 4.9). When brought together by large-scale air flow, this contrast in *CPE* (the model proxy for *TPE* in the real atmosphere) drives lateral and vertical rearrangement, in which the upper parts of the taller columns spread over the undercutting lower parts of the shorter columns, lowering overall *CPE* and generating *KE* from the difference. Figure 9.15 shows the slope-convective cycle in the simple incompressible model.

Because of the great horizontal extent of these movements and masses, the cycle of preparation and rearrangement takes much longer than in the case of convection. In the long term a dynamic and thermodynamic balance is maintained, but in the shorter term the elements of the cycle are localized and sporadic on large scales of space and time (thousands of kilometres and days), as we see in the middle latitudes where their activity is concentrated.

In practice the slope-convective rearrangement occurs in the two stages. In each hemisphere, a zonally homogeneous pattern of meridional temperature gradient tends to break down as vast fingers of warm air move polewards separated East–West by fingers of cool moving equatorwards. This reduces the horizontal separation of contrasting air masses, encouraging shorter range (but still extensive) E–W advective rearrangements between adjacent contrasting fingers, as observed in the cooperation of *long waves* with smaller *cyclone waves* which occur in the families of depressions and ridges of high pressure in middle latitudes (Section 12.6). The following model cycle applies most directly to cyclone waves.

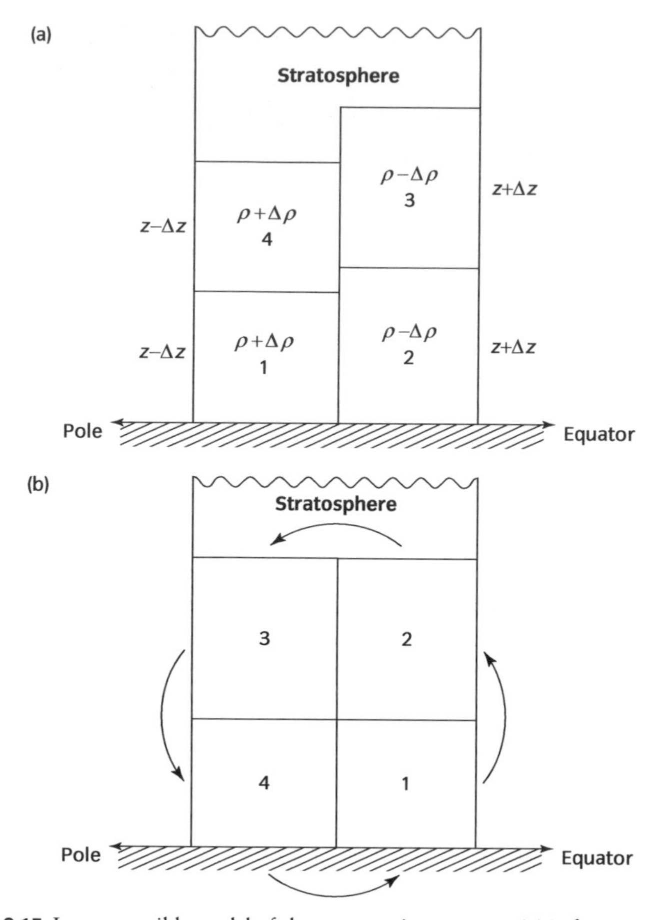

Figure 9.15 Incompressible model of slope-convective overturn. (a) Before overturn, two taller, less dense blocks (2 and 3) of troposphere stand on the equatorward side of two shorter, denser blocks (1 and 4). The arrangement is slope-convectively unstable. (b) After partial overturn (anticlockwise—looking E in the N hemisphere), we have two identical, convectively stable, tropospheric columns, ready to be separated by large-scale air flow and destabilized in lower and higher latitudes for later recombination. The stratosphere has been levelled without changing its overall potential energy.

Cyclone waves

Figure 9.15 outlines two of the three stages of the simple model of the slope-convective cycle.

(i) *Confluence* (Fig. 9.15a) Two columns of incompressible air stand by side by side with identical masses, each of mass 2 m per unit area. They have been brought together by confluent air currents—the taller ($2z + 2\Delta z$), warmer ($T + \Delta T$), and less dense ($\rho - \Delta\rho$) column coming from lower latitudes, and the shorter ($2z - 2\Delta z$), cooler ($T - \Delta T$), and denser ($\rho + \Delta\rho$) one coming from higher latitudes. Each column consists of two identical half columns, which correspond to the half columns in the convective model, so that Δz, ΔT, and $\Delta\rho$ terms relate as in the convective model.

$$\frac{\Delta\rho}{\rho} = \frac{\Delta z}{z} = \frac{\Delta T}{T}$$

Temperature differences are considerably larger than in convection, with the column temperature differences ($2\ \Delta T$) being at least 10 °C (in about 260 K), so that $\Delta T/T$, etc. $\approx 2\%$, at least four times that in convection.

(ii) *Slope-convective rearrangement* (Fig. 9.15b) Pressure differences at all levels above base level drive air both laterally and vertically, the less dense column rising and spreading over the more dense column, which sinks and spreads underneath, each column halving its depth and doubling its area to form the two-layer composite column shown. The related algebra (Box 9.7) shows that, as in the case of convection, work is done by gravity overall, the centre of gravity falling so that total *CPE* changes by

$$\Delta CPE = -2\ m\ g\ \Delta z$$

which is available for conversion to *KE*. This is the same as the expression for convective overturn, but has been generated from two columns (i.e. twice the amount of *CPE*), so that the proportional fall by advective rearrangement is

$$\frac{APE}{CPE} = \frac{\Delta CPE}{CPE} = \frac{\Delta z}{2z} \tag{9.18}$$

which is half the expression for convection (Eqn 9.17). However, substituting 2% for $\Delta z/z$, $\Delta CPE/CPE$ is about 1%, which is just twice the convective value. We see that proportionately more *CPE* is made available in the model slope-convective exchange, though the difference is not huge.

(iii) *Recovery* Now the composite double column is split by large-scale air flow and separated into two initially identical convectively stable columns . One is taken polewards where its upper half cools and deflates from $z + \Delta z$ to $z - \Delta z$, producing a cool column with total height $2\ z - 2\ \Delta z$. During deflation, the centre of gravity of the upper layer falls by Δz and gravity does work on it ($= m\ g\ \Delta z$). The other portion of the divided composite column is taken equatorwards where its lower half inflates from $z - \Delta z$ to $z + \Delta z$, producing a warm column with total height $2\ z + 2\ \Delta z$. In the process the centre of gravity of the lower half rises by Δz, and the upper half is lifted by $2\ \Delta z$, so that the total work done against gravity is $m\ g\ (\Delta z + 2\ \Delta z) = 3\ m\ g\ \Delta z$.

The final result is to recover a short, cool column in high latitudes and a tall, warm column in low latitudes, ready to be brought together in another slope convective cycle in middle latitudes. Low-level inflation produces 3 units (each of $m\ g\ \Delta z$) of *CPE* from the low-latitude surface heat source, and high-level deflation loses 1 energy unit of *CPE* at the high-latitude upper level heat sink, so that net production is 2 units. Viewed over two consecutive slope-advective cycles, the 3 units recovered at low latitudes in the first cycle are exported polewards in the next slope-convective overturn, where 1 unit is lost by upper level deflation during recovery. As in convection, not all *APE* is converted to *KE*: some is deflated in the high-troposphere heat sink and lost to space by net *TR* radiation.

Model slope convection and reality

- The slope convective cycle operates on a longer time scale than the convective cycle, and shows little of its asymmetry in time—the slope convective rearrangement taking much the same time as the *APE* recovery which prepares the next cycle. Although horizontal motion must add some delay to both phases, the time taken for *APE* to recover enough to drive further rearrangement is probably determined mainly by the time taken to recover the temperature changes (2 ΔT in Fig. 9.15) from available heat inputs and outputs. If these are about 100 W m^{-2}, as in convection, the recovery should take about 4 days (NN 9.8), which is consistent with the observed time intervals between consecutive members of a family of cyclone waves. The time scale of the rearrangement which constitutes a particular cyclone wave (especially the sloping exchange in fronts—Section 12.4) is dynamically constrained, as it is in convection, though over much larger horizontal scales and larger masses, and with continual lateral Coriolis deflection.

• The horizontal heat fluxes maintained during slope-convective rearrangement are much larger than the vertical heat fluxes found in convection. Taking 10 °C for the temperature difference between opposing flows in fronts, each of which might have a speed of 20 m s^{-1}, the net horizontal heat flux in the direction of movement of the warm air will be about 100 kW m^{-2} which is consistent with the heat flows needed to balance Earth's meridional radiative imbalance, and the vertical imbalance too, assuming realistic slopes of 1:50 (Section 8.9).

• Although the simple model of the slope-convective cycle is unrealistic in ignoring turbulent mixing and friction, the omission is probably not so important as it was in the case of convection. The synoptic scale flows which dominate cyclone waves are visibly smoother than the meteorologically small and microscale motions which dominate convection, which suggests that they are systematically less prone to smaller scale mixing and associated drag, and therefore more efficient in their conversion of *APE* to *KE*, as the presence of jet streams suggests.

9.8 Global atmospheric engine

We complete this outline of Earth's surface heat engines, by looking at the global atmosphere, averaged over a few integral years to reduce it to a nearly steady, though very dynamic, state. The feeble stratospheric heat pump is buried in its diffuse upper regions, and the ocean surface beneath injects, removes, and transports a great deal of heat, while staying close to separate energy equilibrium. In fact on these time scales the relatively large heat capacity of even the ocean's surface layers stabilizes the atmosphere by heat exchanges which hold its potentially fast response to the gentler pace of climate change (see term (ii) below).

To apply the principle of conservation of energy to the atmosphere, we extend the energy equation for a moving air parcel (Eqn 9.8 copied below in Eqn 9.19) to cover the whole atmosphere. Though this might seem to compound the difficulties encountered with a single air parcel, in fact dynamic and thermal interactions between parcels greatly simplify the result. As usual, magnitudes are expressed per unit area of Earth's surface – W m^{-2} for energy flows and rates of change, MJ m^{-2} for energies.

$$\frac{d}{dt}[KE + PE + IE] = \frac{dQ}{dt} - TPW - FRW \qquad 9.19$$

(i) (ii) (iii) (iv)

Numbered terms in Eqn 9.19

(i) $\frac{d}{dt}[KE + PE + IE]/dt$

the rate of change of the atmosphere's total energy TE

This is the rate of change of the total of the three energy terms summed over all the air parcels in each vertical column (Eqn 9.13), summed again over all the columns making up the global atmosphere, and averaged over a few integral years. The hydrostatic relation allows *PE* to be written as $M\ R\ T$

$$\frac{dTE}{dt} = \frac{d}{dt}\left[\frac{1}{2}M\ V^2 + M\ R\ T + M\ C_v\ T\right] \approx 0$$

and to be combined with *IE* ($= M\ C_v\ T$) to form *TPE* ($= M\ C_p\ T$)

$$\frac{dTE}{dt} = \frac{d}{dt}\left[\frac{1}{2}M\ V^2 + M\ C_p\ T\right] \approx 0$$

The rate of change of *TE* is very nearly zero because the atmosphere's nearly steady state ensures that temporal averages of global *V* and *T* are very stable over

NUMERICAL NOTE 9.9 Growth in TPE, and radiative imbalance

- If the global average atmospheric temperature T rises by ΔT, global average *TPE* per unit area ($M\ C_p\ T$) increases by ($M\ C_p\ \Delta T$). Taking global average atmospheric T to be about 260 K, a rise of 1 °C corresponds to an increase of 1/260 (about 0.4%) in *TPE*. If global average T rises by 1 °C in 3 years, $\Delta T/\Delta t = 1/(3 \times 365 \times 24 \times 60 \times 60) \approx 1 \times 10^{-8}$ °C per second and the corresponding value of $dTPE/dt$ ($= M\ C_p\ dT/dt$) is about $10{,}000 \times 1{,}000 \times 1 \times 10^{-8} = 0.1$ W m^{-2}. If this is driven solely by the radiative imbalance $SR_i - TR_o$, this will have to average 0.1 W m^{-2} over the 3 years.
- To repeat the exercise for observed sustained fast climate change of 1 °C per century ($\approx 3.2 \times 10^{-10}$ °C s^{-1}), we need to recognize that on these longer time scales, heat exchanges freely with the top 500 m of ocean (equivalent to 350 m for a global ocean, since oceans cover 70% of the globe), increasing the effective heat capacity by $350 \times 1{,}000 \times 4{,}200 \approx 1{,}500$ MJ m^{-2} from the purely atmospheric value of 10 MJ m^{-2} ($= M\ C_p$ as above). To maintain warming at 1 °C per century, the radiative imbalance would have to be $(1{,}500 + 10) \times 10^6 \times 3.2 \times 10^{-10} \approx 0.5$ Wm^{-2}.

a few integral years. Smoothing out seasonal and slightly longer temperature changes, residual changes of global temperature are usually much less than 1 °C in 3 years, which corresponds to a change of about 1 in 260 in $M\,C_p\,T$ or 0.4% (assuming global average atmospheric temperature to be 260 K), or a rate of change of 0.1 W m^{-2} (NN 9.9). We can say that global d*TE*/d*t* is observed to be zero to within this limit most of the time.

(ii) d*Q*/d*t* *the net rate of heat input to the atmosphere*

The full expression for an individual air parcel contains heat exchanges between other air parcels by conduction, mixing, and *TR*, as well as inputs directly and indirectly from the Sun (including sensible and latent heat and net *TR* from the surface) and net *TR* losses to space. Exchanges with other parcels cancel out when summed over all the parcels of the atmosphere, because a conserved energy total (definite though not reliably calculable) is simply being redistributed. For the whole system we are left with

$$\frac{\mathrm{d}Q}{\mathrm{d}t} = SR_i - TR_o + \frac{\mathrm{d}q}{\mathrm{d}t}$$

where SR_i is the inflow of solar energy from the Sun (warming the air directly and via the warmed surface), TR_o is the outflow of *TR* from the atmosphere (emitted directly to space and via surface layers cooled by emission to space), and d*q*/d*t* is the rate of frictional heating aggregated through the atmosphere.

The observed temperature variations discussed under term (i) imply temporary radiative imbalances of about 0.1 W m^{-2} for the atmosphere (NN 9.9). In periods of fast climate change like the present (Section 14.9), observed rates of sustained global warming can reach or exceed 1 °C per century, which imply sustained radiative imbalances of about 0.5 W m^{-2}, after allowing for the very large heat capacity of surface ocean layers exchanging heat with the atmosphere on this longer time scale (NN 9.9). However, most of this imbalance is taken up by the large oceanic heat reservoir so that effective radiative imbalances for the atmospheric engine (i.e. true radiative imbalances less compensating heat flows into the oceans during global warming) are kept to about 0.1 W m^{-2} in 200 W m^{-2}.

As will be seen later, these effective imbalances are more than an order of magnitude smaller than d*q*/dt. To the extent that the effective imbalance $SR_i - TR_o$ for the atmosphere is zero, term (ii) becomes

$$\frac{\mathrm{d}Q}{\mathrm{d}t} = \frac{\mathrm{d}q}{\mathrm{d}t}$$

(iii) $-TPW$ *the total rate of pressure working by the atmosphere*

When *TPW* is aggregated over all the air parcels of the atmosphere, the pressure workings between air parcels cancel out by an extension of Newton's third law of motion: the working of parcel *A* on parcel *B* is equal and opposite to the working of *B* on *A*, which is the mechanism for the conservation of mechanical energy within the atmosphere. Cancellation leaves only the rate of pressure working on the bottom and top surfaces of the system, which is zero on largely inflexible land surfaces and sea beds, and identically zero at the top of the atmosphere because there is no air pressure there to do work. We ignore a small amount of working done by wind pressing on the ocean surface which is eventually degraded to heat by water viscosity and returned to the atmosphere. With this small discrepancy (which we could account for in Eqn 9.19 by a leakage to dq/dt in term (ii), the pressure working term TPW vanishes on global aggregation.

$$-TPW = 0$$

(iv) $-FRW$ *the rate of frictional working on the atmosphere*

When $-FRW$ is aggregated for all the air parcels of the atmosphere, most of the frictional workings between air parcels cancel out for the same reason as internal TPW, leaving only the viscous degradation (FDH) of parcel KE to heat which takes place in the myriad intense transient wind shears between turbulent eddies. This FDH takes place throughout the atmosphere, but is especially concentrated in the chronically turbulent atmospheric boundary layer.

$$-FRW = -FDH$$

As in term (iii) we are ignoring a small exchange with the oceans: frictional working on ocean currents like the Gulf Stream is degraded to heat within the oceans by water viscosity and eventually returned to the atmosphere. And as in the case of term (iii) this could be accounted for by a small leakage to dq/dt.

9.8.1 Global atmospheric energetics

The energy equation for the global atmosphere can now be expressed very concisely in three composite terms, each of which is very nearly zero:

$$\underset{=0}{\frac{dTE}{dt} = \frac{d}{dt}[KE + TPE]} = \underset{=0}{[SR_i - TR_o]} + \underset{=0}{\left[\frac{dq}{dt} - FDH\right]} \qquad 9.20$$

The observed nearly steady state of the atmosphere requires that the rate of change of its total energy with time is nearly zero. The nearly perfect radiative equilibrium of the atmosphere requires that SR_i and TR_o cancel almost exactly. And since the frictional degradation of atmospheric (parcel) KE to heat occurs entirely within the atmospheric engine (unlike typical artificial engines) this conversion is described by a precisely balanced pair of terms.

$$\frac{dq}{dt} = FDH \qquad 9.21$$

Discussion

- Equation 9.20 describes an apparently very static balance between three zero-valued clusters of terms. However, the continual frictional conversion of

KE to heat described by Eqn 9.21 implies that the d*KE*/d*t* term on the left-hand side of Eqn 9.20 remains zero only because there is a balance between continual production of *KE* by weather systems and its continual destruction by *FDH*. In fact similar exchanges between most of the terms of Eqn 9.20 form a network of energy flows, as outlined below.

- Despite its appearance as the only uncancelled heating term in Eqns 9.20 and 9.21, frictional heating d*q*/d*t* does not drive the atmospheric engine: that role is obviously played by the balanced pair of solar input SR_i and terrestrial output TR_o which maintain flow of energy through the system. In fact d*q*/d*t* is an almost passive product of the movements of air and water, like the heat produced in Joule's experiment (Section 9.2), which was the end result of the energy input by the working of the sinking weight. In principle, heat produced by friction in the atmosphere and ocean can play a role in driving further activity, as happens in surface-based convection (Section 9.6), but that is a small secondary effect.

9.8.2 Atmospheric energy networks

Each term of Eqn 9.20 is steady because the energy flows to and from it are almost perfectly balanced. Figure 9.16a presents the equation again, with an arrowed network of energy flows which we can trace from term to term.

SR_i and TR_o

Radiant energy from the Sun is absorbed by the atmosphere at a global average rate of 242 W m^{-2} in the ways and proportions described in Section 9.4, with values scaled up from Fig. 8.8 to allow for London's underestimation of Earth's albedo. Of these 242 units, about 29 units 'short' across to TR_o in the stratosphere and by surface *TR* emission through the atmospheric window, leaving an input SR_i of 213 units to flow through the other terms of the equation, eventually reaching the *TR* term, where it adds to the 'shorted' 29 to reconstitute the balancing 242 W m^{-2} of *TR* emission from the planet to space.

d*TPE*/d*t*

This is held at zero by five energy flows:

(i) input from SR_i (by atmospheric warming and inflation);

(ii) input from d*KE*/d*t* as air slows by fluid impact with the rest of the atmosphere and returns its *KE* to the global *TPE* reservoir;

(iii) output to d*KE*/d*t* as weather systems convert some of the available part of their *TPE*—the small, crucial conversions which keeps the atmosphere dynamic rather than static;

(iv) output to TR_o, as cooling by net *TR* emission deflates *TPE* at the atmosphere's heat sinks; and

(v) input from dq/dt as a small part of frictional warming reinflates the atmosphere.

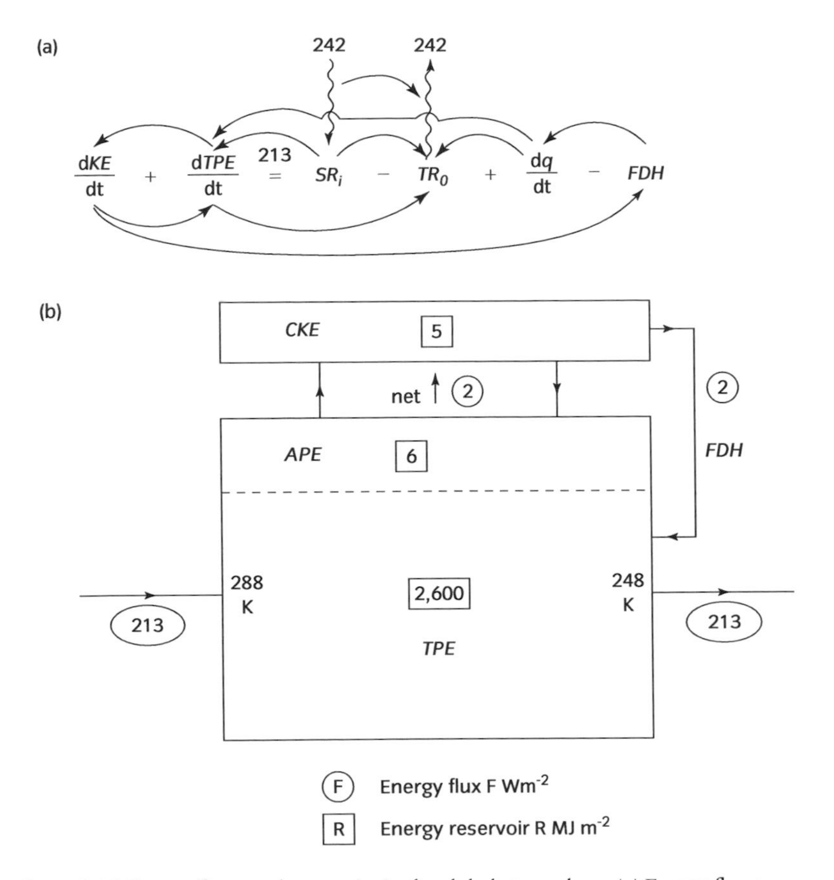

Figure 9.16 Energy flows and reservoirs in the global atmosphere. (a) Energy flows between the terms of Eqn 9.20. Average planetary inputs and outputs are 242 W m^{-2}, but of this only 213 W m^{-2} properly enters the troposphere engine by supporting the reservoir of *TPE* and then cycling through the mechanical and thermal energy components of the system. The remaining 29 W m^{-2} has 'shorted' across to the outgoing *TR*. (b) Energy flows (circled in W m^{-2}) and energy reservoirs (boxed in MJ m^{-2}) in the troposphere engine. *TPE* and *CKE* values come from basic observations, but *APE* and *FDH* values come from advanced studies.

dKE/dt

This is held at zero by a balance between conversion from *TPE*, return flow to *TPE*, and the small but relentless degradation of *KE* by friction (*FDH*).

—FDH represents the myriad localized intense shears receiving a small amount of parcel *KE* and degrading it to heat (dq/dt) by viscous friction.

dq/dt is the frictional heating (by *FDH*), some of which goes straight to TR_o, while the rest goes in to *TPE*, acting as a small secondary heat source distributed through the engine, though concentrated in the convecting boundary layer (Section 9.6).

9.8.3 Energy reservoirs and flows

Equation 9.20 gives no information about the magnitudes of the reservoirs of atmospheric energy (*TPE* and *KE*), since it describes only their zero rates of change. We can use typical observations from Table 9.2 to evaluate the reservoirs in Fig. 9.16b. This is now a detailed version of the outline heat engine first presented in Fig. 9.5 and elaborated in Fig. 9.6, and it provides a framework for a final summary of the state and activity of the atmospheric (essentially the tropospheric) heat engine.

TPE

Solar energy enters the tropospheric engine as SR_i (213 W m^{-2} as above), is absorbed in the low troposphere at temperatures of about 288 K, and eventually leaves it from the high troposphere at temperatures of about 248 K (Section 9.4). This throughput of energy maintains a dynamic atmosphere with a mass of just over 10,000 kg m^{-2} (99.7% dry air by mass), with a mean temperature of about 260 K, a reservoir of global *TPE* of about 2,600 MJ m^{-2} ($M\ C_p\ T$), and a decadal scale height of about 17 km ($2.3\ R\ T/g$). Water vapour and cloud may seem trivial by mass (0.3%), but they are major players through their domination of the atmosphere's greenhouse effect (as both vapour and cloud), albedo (as cloud), and heat fluxes (by evaporation and condensation).

CIE

Nearly perfect hydrostatic equilibrium of the vertical profile of atmospheric mass ensures that the global *TPE* reservoir consists of global *CPE* and global *CIE* in the proportions R/C_v (= 40%)—proportions established in the initial warming and inflation of air by SR_i (Section 9.6), and maintained as water vapour chills and condenses, converting its latent heat into *CPE* and *CIE* in the same proportions. In these ways over 71% of all incoming heat sooner or later becomes *CIE* in the familiar proportions, and the global mean production of *CIE* per unit area is just over 150 W m^{-2} (71% of 213) which corresponds to a global production of about 75 PW (75 million GW)—equivalent to the output of 75 million large power stations (NN 9.10).

CPE

If we regard the continual conversion of the proportion R/C_p of incoming heat to *CPE* (= $M\ R\ T$) as a production of mechanical energy, this percentage (29%) is double the maximum conversion of 14% allowed by the second law of thermodynamics, according to the temperatures of the tropospheric engines heat sources and sinks (Section 9.4). However, almost all of this *CPE* is molecular scale *PE* rather than parcel scale—a ballistic spray pressed upwards by their collisions and pulled downwards by gravity, inseparable from *CIE* (= $M\ C_v\ T$) under the umbrella of *TPE*. Only a small proportion of this *CPE* is made available by the convective and slope-convective activity of the atmospheric engine, with the result that the production of *APE* is well within the limits set by the second law—very well within, given the wastage associated with the atmosphere's pronounced irreversibility.

APE

Most of this vast store of *TPE* is held in a passive reservoir, swelling and shrinking slightly with seasonal rhythms and climate change, but a small proportion is always available for conversion to *KE* by convective or slope-convective overturn. Our simple analysis in Section 9.7 suggests that the *APE* proportion might be of order 1%, which would correspond in Fig. 9.16b to an *APE* reservoir of about 26 MJ m^{-2}. However the network of energy steady flows between *TPE*, *APE*, and *CKE* has five unknown flows and three equations, so that their values are not defined by this approach.

Detailed mechanistic studies in recent decades *[35]* suggest that the *APE* reservoir is actually much smaller, being between 5 and 6 MJ m^{-2}. Such a discrepancy is not surprising given the simplicity of the incompressible model (Section 9.7) and the difficulty of defining *APE* exactly. The same study suggests that *APE* is produced at about 3 W m^{-2}, of which about 25% is wasted by radiative deflation, leaving just over 2 W m^{-2} as the rate of production of atmospheric *KE* from *APE*, as noted on Fig. 9.16b.

CKE

Though by far the smallest reservoir of atmospheric energy (reducing the jet stream value from Table 9.2 to a more normal 5 MJ m^{-2}), atmospheric *CKE* is enormously important for the atmosphere, and for living conditions throughout the surface layers of the planet. Its biggest role is in effecting the heat transports needed to bridge the radiative imbalances imposed by the geometries and motions of the Sun and Earth, since these transports depend absolutely on the bulk movement of air (and water) parcels rather than molecular diffusion (Section 8.9), as organized in the unsteady air flows and cloud masses of the weather-filled troposphere.

As mentioned under *APE*, net production of *CKE* from *APE* probably proceeds at an average rate of just over 2 W m^{-2}, most of which is degraded to heat by viscous friction, largely in the atmospheric boundary layer (ABL)—the first few hundred metres of chronically turbulent air above land and water surfaces (Fig. 9.17).

Frictional destruction of *KE* becomes more and more noticeable as we descend below the top of the ABL (gradient level). Increasing drag holds winds further and further below the quasi-geostrophic values of the free atmosphere, and twists them increasingly towards lower pressure, so that working by the local pressure gradient balances working against friction and maintains a steady average flow at each level. Special studies (Section 10.10) show that at any given location and time turbulent frictional stress T increases from nearly zero at the gradient level (~300 m above the surface), to T_s at the top of the surface boundary layer (SBL), at most a few tens of metres above the surface. Within the SBL it stays at T_s right down to the surface as wind speed falls steeply to zero. As shown in Box 9.8 we can use these observed profiles of frictional stress and mean wind speed to make a rough but useful estimate of the rate of frictional destruction of the *KE* in the ABL, finding values of order 2 W m^{-2} for typical wind speeds, which is roughly comparable with values found in detailed studies.

Comparing this rate of dissipation with the average reservoir of atmospheric *CKE*, we see that it corresponds to a rate of destruction of about 0.15% per hour. If this rate were to continue without the usual continual replenishment by conversion from *APE*, the *CKE* reservoir would be half emptied in about two weeks, and the motion of the atmosphere, which has been incessant throughout the life of the Earth would be well on its way to extinction.

BOX 9.8 Frictional dissipation

Figure 9.17a shows a shallow horizontal slice of air of unit horizontal area being dragged forwards (i.e. in the direction of the ambient mean wind) by frictional stress exerted by faster-moving air just above the slice, and dragged backwards by friction with the slower-moving air just beneath. According to Sections 10.9 and 10.10, throughout the ABL the stress T on the lower surface of the slice is slightly larger than the stress on its upper surface, and increases downwards from nearly zero at the top of the ABL (the gradient level) to a maximum value T_s at the top of the surface boundary layer (SBL tens of metres deep), through which it remains effectively uniform at the value of the surface drag T_s, which can be measured directly by horizontal drag plates let into the surface.

The net frictional force dT on the shallow slab in Fig. 9.17a opposes the slice motion. If the speed of the slice is V, the net rate of working by friction against the slab motion is $V\,dT$, and the aggregated rate of frictional dissipation of kinetic energy in a column extending through the ABL is given by the sum of the rates of frictional working in all the constituent slices.

$$FDH_{ABL} = \int_0^{T_s} V dT \qquad \text{B9.8a}$$

Because of the strong concentration of turbulent and viscous friction in the ABL, this must account for much of the *FDH* term in Eqn 9.20 and Fig. 9.16.

Figure 9.17b shows a plot of V against T which is broadly consistent with this situation. The geostrophic wind speed V_g is reached at the gradient level, where T is zero, and wind speeds fall with decreasing height throughout the underlying ABL as T builds up towards the surface stress T_s which is reached at the top of the SBL, within which V falls steeply to zero at the surface. The simplest $T{:}V$ profile is a straight line from the top of the SBL to the top of the ABL, the area under which is

$$\int_0^{T_s} V dT = T_s V_s + \frac{1}{2} T_s (V_g - V_s) = \frac{1}{2} T_s (V_s + V_g) \qquad \text{B9.8b}$$

by simple geometry. We can use Eqn 10.25 to express the frictional stress T_s in the SBL in terms of the square of the 10 m wind speed V_{10} and the corresponding drag coefficient C_{10} (Section 10.9). Using also the useful empirical approximation $V_{10}/V_g \approx 0.4$, and the rather sweeping assumption that $V_s \approx V_{10}$, we obtain

$$FDH_{abl} \approx 0.11\, \rho\, C_{10}\, V_g^3 \qquad \text{B9.8c}$$

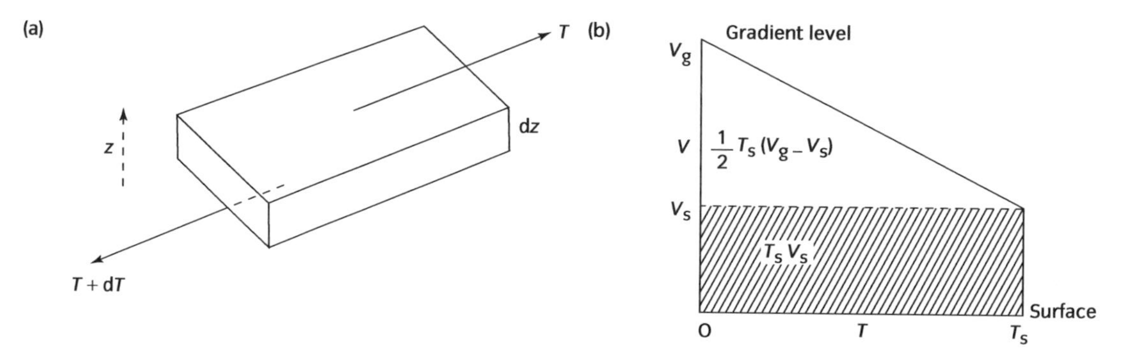

Figure 9.17 Friction in the atmospheric boundary layer (ABL) (a) Frictional stresses on the top and bottom faces of a slab of air embedded in the atmospheric boundary layer. The turbulent shearing stress is smaller on its upper surface than it is on its lower surface, except in the surface boundary layer, where they are equal. (b) A simplified vertical profile of average wind speed V against average eddy stress T through the ABL at a certain location and occasion. The rate of frictional dissipation of kinetic energy is given by the area under the profile. The surface boundary layer is hatched. Terms are described in Box 9.8.

Substituting a typical 1.2 kg m^{-3} for the air density ρ, and 3×10^{-3} for C_{10} (which corresponds to a typical surface roughness length of 1 cm—Box 10.6), we find

$$FDH_{abl} \approx 4 \times 10^{-4} V_g^3 \qquad \text{B9.8d}$$

Though only very approximate in absolute terms, this relation shows great sensitivity of frictional dissipation rate to the ABL to wind speed. The implied rate is about 0.05 W m^{-2} for a geostrophic wind speed of 5 m s^{-1}, over 3 W m^{-2} for a geostrophic wind speed of 20 m s^{-1}, and about 50 W m^{-2} for a force 8 gale (in which $V_{10} \approx 19$ m s^{-1} corresponds to V_g nearly 50 m s^{-1}). The middle value is consistent with detailed estimates to within an order of magnitude.

Mechanical efficiency

We established earlier (Section 9.4) that the maximum possible mechanical efficiency of the atmospheric engine is about 14%, given the temperatures of its heat sources and sinks, but that realistic values are likely to be much less, given that the atmospheric engine runs in conditions far removed from reversible perfection. The values quoted from detailed studies of the tropospheric engine (essentially the atmospheric engine) confirm how very inefficient the atmospheric engine is in this thermodynamic sense. Taking the estimated rate of production of *APE* (3 W m^{-2}) to be the actual rate of production of mechanical energy in the engine (i.e. ignoring wastage by radiative deflation), and comparing this with the effective energy input of about 210 W m^{-2} to the engine, its apparent mechanical efficiency of about 1.4% is only one tenth of the thermodynamical optimum allowed by the temperatures in the low and high troposphere (Section 9.4). This large discrepancy is a measure of how far the fast, unsteady, unbalanced behaviour of the real atmospheric engine departs from the slow, balanced reversible state of a perfect heat engine operating between the same source and sink temperatures.

9.8.4 Epilogue

We have found that the incessant, impressive, and occasionally destructive activity of the atmosphere is maintained by a conversion to kinetic energy of only a very small fraction of the total flux of energy driven through the atmosphere by the Sun, which in turn is only a minute part of the Sun's prodigal output. The enormous power of the hurricane, the Trade Winds which dominate over one third of the Earth's surface, the gales which trouble the seas and western margins of middle latitude continents—all are manifestations of an average rate of production of kinetic energy in an atmospheric column resting on a domestic table which could barely light a very small light bulb if converted fully to electrical power.

This is not as unlikely as it sounds: the atmosphere is like a large flywheel which is maintained in nearly steady rotation by a balance between power input from a weak engine and power loss by slight friction. The balance achieved is such that there is a substantial reservoir of kinetic energy in the flywheel at any time—enough to keep it 'turning' for a few weeks. Moreover all the more spectacular motion systems of the atmosphere represent considerable concentrations of kinetic energy: depressions, jet streams, cumulonimbus, tornadoes, and so on, all occupy only a very small fraction of the available atmospheric volume, and the most intense (tornadoes in this list) generally occupy the smallest fraction. In these small volumes, kinetic energies rise far above the global average. The flywheel is kept spinning mainly by continual nudging from little, transient, energetic wheels which represent individual weather systems. Even the Hadley circulation is kept turning by the continual eruption of hundreds of cumulonimbus in the intertropical convergence zone.

Because the atmosphere's dynamic state is nearly steady, the rate of destruction of atmospheric kinetic energy by friction must balance its rate of production by solar heating. If that production were suddenly to cease, by some unimaginable solar catastrophe, the atmospheric flywheel which has been spinning for over four thousand million years would grind to a halt in a period of weeks. Precisely how many depends on how we suppose the drag of the largely surface-based friction would reach the middle and upper troposphere, and on how the heat stored in the oceans would leak into the cooling atmosphere, but whatever the details, it is bound to be almost infinitely short compared with the vast time scale on which the Sun has been keeping our atmosphere active. Since prehistory, people have sensed that the Sun is very important, and have even worshipped its power and influence, but modern understanding shows that even the most extravagant response probably does less than justice to its absolutely crucial role in maintaining the Earth's surface and atmosphere in their familiar, homely, and biologically essential conditions.

NUMERICAL APPENDIX **Temperatures of moving and stopped air**

The following explores further the relationship between molecular and parcel *KE* depicted in Fig. 9.4 and discussed in Section 9.2 (Heat as molecular kinetic energy) and Box 9.2.

Suppose that an air parcel of unit mass is moving at 10 m s^{-1} past the fixed anemometer. Parcel *KE* is $1/2 \times 1 \times 100 = 50$ J relative to the anemometer, which is the excess of molecular *KE* relative to the fixed thermometer over molecular *KE* relative to the thermometer moving with the parcel. According to the kinetic theory of gases the temperature excess ΔT registered by the fixed thermometer is found from $(3/2)\ k\ \Delta T = 50/N$ where N is the number of air molecules in the air parcel (Box 9.2), and 50/N is the excess *KE* in joules per molecule. Noting that air behaves like a swarm of molecules, each of molecular weight 29 (Box 4.1), 1 kg of air contains $1000/29 \approx 34.5$ moles of air. Since 1 mole always contains 6.026×10^{23} molecules (Avogadro's number), the air parcel contains $N \approx 208 \times 10^{23}$ molecules, and $\Delta T = 100/(3\ k\ N) = 100/(3 \times 1.3 \times 208) = 0.12$ °C. The moving thermometer reads this amount below the fixed one.

When the air parcel is stopped by the hedge, some of the parcel's 50 J is used to make the air molecules rotate and vibrate as well as fly about faster, so that if the stopping occurs at constant volume, the air temperature will rise by a smaller $\Delta T = \Delta Q/C_v = 50/717 \approx 0.07$ °C, while at constant pressure it will rise by an even smaller $\Delta T = 50/C_p = 50/1004 \approx 0.05$ °C, because some thermal energy is used as the warmed air expands slightly and does work against its surroundings. In reality the warming will be even less, since some will be shared with the hedge.

Checklist of key ideas

You should now be familiar with the following ideas.

1. The concepts and numerical realities of mechanical working and changes in kinetic, potential, and mechanical energies of a rising and falling air parcel.

2. Heat as a form of energy, Joule's experiment and degrading kinetic energy to heat, heat input divided between internal energy and external mechanical work.

3. The first and second laws of thermodynamics and their application to a simple heat engine, absolute temperatures of heat sink and source and the universal limit to its mechanical efficiency.

4. The troposphere/ocean engine, the ocean engine and the stratosphere heat pump outlined, with their heat sources and sinks, their activity and their mechanical efficiencies.

5. The total energy of an air parcel and the meteorological and other processes which change it, with realistic values for a parcel rising through the troposphere.

6. Air columns and their centres of gravity, internal, potential, and total potential energies, with realistic values for the troposphere; the relative smallness of observed kinetic energies.

7. Realistic heat inputs into the low troposphere, with warming, and potential energy increase divided between internal inflation and external lifting, contrasts between behaviour of shallow and deep layers; cooling and deflation at heat sinks.

8. Atmospheric motion as a secondary response; convection and slope convection cycles outlined and simply modelled; the concept and reality of available potential energy, deflation and wastage of APE, low mechanical efficiencies of convection and slope convection.

9. The global atmospheric engine outlined and its energetics simplified to a nearly steady state with energy flowing through a network of energy reservoirs; the magnitude and significance of frictional degradation of kinetic energy to heat; the very low mechanical efficiency of the atmospheric engine.

Problems

Outline answers to these problems can be found on the Online Resource Centre. Answers to odd numbered problems can be found under Student Resources, answers to even numbered problems under Lecturer Resources.

Level 1

9.1 In a heat engine, what happens to the heat which is not converted into mechanical energy, and what happens to the heat which is converted?

9.2 Arrange the following energy types and transformation in sequence to represent a possible sequence in the working atmospheric engine: kinetic energy of moving air, solar radiant energy, terrestrial radiant energy, gravitational potential energy of lifted air, internal energy of warmed air, work done by expanding air, work done against friction, work done by gravity, absorption of radiant energy, emission of radiant energy.

9.3 James Joule made careful measurements of water temperatures at the bottoms and tops of alpine waterfalls. What do you suppose he was hoping to measure? It seems that he was not satisfied with the work. From your meteorological perspective, what might have complicated the issue?

9.4 Draw and label simple diagrams of a heat engine and a heat pump. Add to the latter to represent a domestic refrigerator to show why the net effect of its operation is to warm the room it is in.

9.5 Joule's experiment models the atmospheric engine perfectly in one way, but lacks two crucial ingredients, only one of which is mentioned in the text. Briefly mention the three items.

9.6 What happens to the mechanical energy of an air parcel as it comes down in a thunderstorm downdraft, and why?

9.7 Briefly distinguish between internal energy, kinetic energy, potential energy, mechanical energy, total energy, total potential energy, available potential energy.

9.8 Suppose a slab of air of mass 10 kg m^{-2} receives heat energy from the underlying surface at 100 W m^{-2} for 10 minutes. Find the amount of energy input, use Eqn 9.16 to find the rise in air temperature, and use it again to partition the energy input between increase in internal energy and increase in (gravitational) potential energy. What meteorological variable have we assumed to remain constant in this process?

Level 2

9.9 In Problem 9.8, virtually all the increase in gravitational potential energy (17,160 J m^{-2}) occurs as the slab lifts the overlying 10,000 kg m^{-2} of air. Find the very small increase in slab thickness, and discuss the implication for subsequent behaviour of this large warming (6 °C in Problem 9.8) and small expansion.

9.10 Consider the first 100 m of air over sunlit ground as a 'boundary layer heat engine' taking in heat energy at temperature 15 °C K and giving it out at the top of the layer at temperature 13.5 °C (allowing for a slightly superadiabatic temperature lapse rate through the layer). Assuming that this engine operates at the thermodynamic optimum, find the proportion of the heat input which remains in the form of heat. What does this tell you about the effective result of such surface heating?

9.11 Find the minimum amount of work needed to boil 1 litre of water from 20 °C (at sea level) in Joule's experiment. At least how far will a 10 kg wait have to sink to do this amount of work? Why 'minimum', 'sea level', and 'at least'?

9.12 Consider an isothermal atmosphere of total mass 10,000 kg m^{-2} at temperature 250 K. Find its column internal energy, (gravitational) potential energy, and total potential energy. If 1% of this were to be converted into the motion of the whole atmosphere at uniform speed, what would that speed be?

9.13 Considering the convective part of the tropospheric engine to operate between local heat source at the surface, and heat sink in the upper troposphere, compare the maximum mechanical efficiency in the low-latitude troposphere, where source and sink temperatures are 25 and −60 °C, and in the mid-latitude troposphere, where temperatures are 15 and −40 °C.

9.14 If the whole atmosphere is moving at 20 m s^{-1}, how much will it warm if all its KE is degraded to heat by internal friction, and by how much will its centre of gravity then rise (if it remains an isothermal atmosphere).

9.15 The deep ocean engine currently operates between 25 and 0 °C. Compare its current maximum mechanical efficiency, with the value 100 million years ago, when the operating temperatures were more like 28 and 10 °C. What does this say about the global importance of heat transfer in the older ocean?

9.16 Find the components of the total energy (relative to sea level) of a 1 kg air parcel moving at 80 m s^{-1} in the subtropical jet stream, 12 km above sea level, with a temperature of −60 °C.

Level 3

9.17 Why analyse the terrestrial engine system in terms of troposphere/ocean and intrinsic (deep) ocean engines, rather than troposphere and ocean engines? By way of an answer, outline the connection between the troposphere and the ocean surface layers, and the connection between the surface and deep ocean currents.

9.18 Take the case where the lower half (by mass) of a 10,000 kg m^{-2} atmosphere is heated at 100 W m^{-2} for 6 hours, and find its temperature rise, and the increase in (gravitational) potential energy of the atmosphere. Unlike Problems 9.8 and 9.9, the heated layer is now a significant proportion of the whole atmosphere, so that the potential energy increase is partitioned between the thickening of the slab, and the lifting of the upper half of the atmosphere. Use Table 9.3 in Box 9.6 to find this proportion, and hence find the slab thickening.

9.19 In Problem 9.13 the loss of convective mechanical efficiency with increasing latitude suggests that something else offering greater efficiency might become significant. What is that, and show in principle how it realizes the advantage.

9.20 Friction is an important sink of mechanical energy, but it is not the only one. Show how an opposing pressure gradient can convert kinetic to potential energy, and deflation at a heat sink can remove PE.

9.21 Consider a simple incompressible model of deep organized convection, such as occurs in a severe tropical cyclone. The deep troposphere is divided into two blocks, each of mass 4,000 kg m^{-2}, the lower being 5°C warmer (i.e. about 5/260 less dense). Find the proportional conversion of CPE to CKE on each overturn according to Box 9.7, and then use this same ratio on an estimate of TPE for the storm's atmosphere to find an upper limit to the production of CKE on each overturn. What realistic factors must reduce reality below this maximum?

9.22 Use the simple expression for rate of destruction of kinetic energy in the atmospheric boundary layer to estimate the rate in a severe tropical cyclone in which V_{10} is 50 m s^{-1}, and discuss the consequences of concentrating KE destruction heavily in KE maxima. It would be more realistic to use a larger drag coefficient in these conditions. Why?

10 Surface and boundary layer

10.1 Introduction *[42]*

10.1.1 Surface

The lower boundaries of the atmosphere are the solid and liquid surfaces of the Earth. The sea surfaces are relatively simple in light airs, but become complex and mobile in the presence of surface waves and currents. Land surfaces are highly irregular on space scales ranging from sand grains to mountains, and are studded with elements that can be mobile on time scales ranging from leaf flutter to sand dune migration. Dense vegetation is particularly complex and can blur the boundary between surface and atmosphere by elevating the sites of heat and momentum exchange well above the local ground surface. Water surfaces have very large effective heat capacities (Section 8.10), and the much smaller heat capacities of land can differ widely according to surface type and condition. Changes in ground cover by vegetation or snow can produce large seasonal changes in thermal and other behaviour of the surface and overlying air, and even a few centimetres thickness of surface ice can make a lake or ocean behave like snow-covered land.

10.1.2 Boundary layers

In most fluid flows over a solid or liquid surface, it is observed that there are one or more zones, close to the solid or liquid surface, in which flow is so strongly affected by the proximity of the surface that they are called *boundary layers*, to distinguish them from the relatively free fluid beyond. Several boundary layers are distinguishable at the base of the atmosphere, together making up the *atmospheric boundary layer* (ABL). Dimensions shown in (Fig. 10.1) are only notional since actual values vary greatly with circumstance.

Laminar boundary layer

Within a few millimetres of any surface, no matter how irregular, air motion is so strongly restrained by friction with the surface, and by the intermolecular friction (*viscosity*) of the air itself, that gross movement is insignificant, and

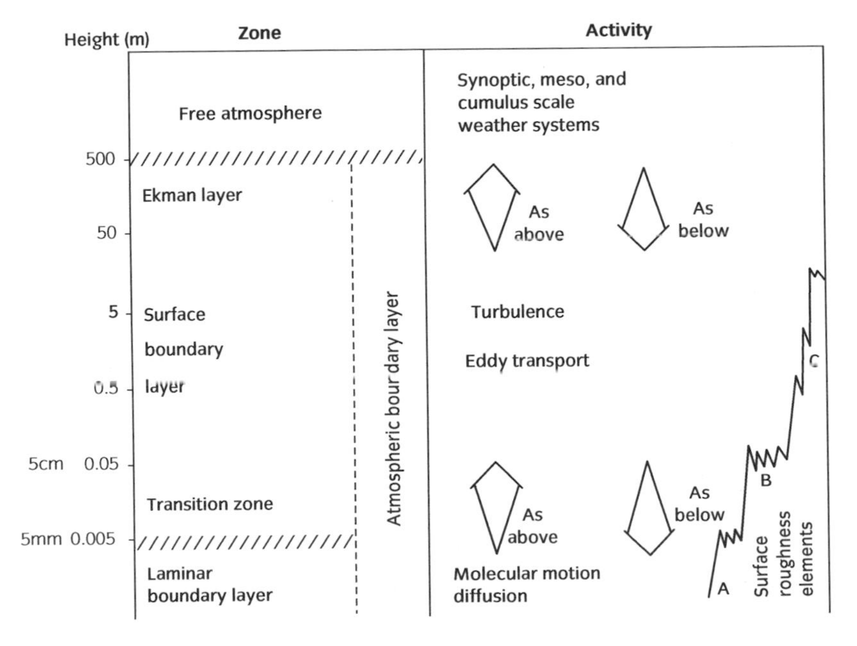

Figure 10.1 The structure and activity of the atmospheric boundary layer displayed against a natural height scale. Arrow widths increase with relative importance. A, B, and C represent small, medium, and large surface roughness elements respectively.

momentum, heat and matter, such as water vapour, are transported mainly by molecular diffusion (Section 3.1). This is the *laminar boundary layer* (LBL), so called because the only significant flow resembles the differential sliding of a sheaf of thin oiled plates, each of which is able to move only very slowly between its neighbours (Fig. 7.4). Although occupying only a minute fraction of the total volume of the ABL and the rest of the troposphere, the LBL controls exchanges between surface and atmosphere by acting as a universal membrane which is only sluggishly permeable to fluxes of heat and matter.

Surface boundary layer

As shown in Fig. 10.1, another boundary layer known as the *turbulent* or *surface boundary layer* (SBL) extends upwards from the LBL for a very variable and poorly defined distance. Just as the LBL is the region strongly influenced by the surface through molecular diffusion, so the SBL is the region which the local surface influences strongly by turbulent or eddy diffusion (Section 7.4.3). In fact it is the most conspicuously turbulent part of the whole atmosphere, and is the only part in which turbulence often tends towards a relatively well-ordered dynamic equilibrium. But despite this degree of order, despite its significance for large-scale exchanges between surface and troposphere, despite being the atmospheric environment of humankind, and despite over a century of close scrutiny, the turbulent transport of momentum, heat, and matter through the SBL is still poorly understood.

Ekman layer

Above the SBL there is a relatively deep transition zone in which the direct influence of the surface weakens with increasing height, to the point where it becomes indistinguishable from the rather indirect, though still quite strong, influences which the surface exerts on the rest of the troposphere. At this point (the *gradient level*) geostrophic departures induced by turbulent friction (Section 7.13.4) are unobservably small, having decayed upwards in a characteristic way first explained by Ekman. The inverted Ekman layer just under the ocean surface was historically the first to be analysed.

Atmospheric boundary layer

The total depth of the nest of boundary layers (laminar, surface, and Ekman) comprising the atmospheric boundary layer (ABL), is $\sim$ 300 m and highly variable in time and space. Above the ABL the atmosphere is considered to be so free of direct surface influence that it is called the *free atmosphere*, though there is evidence throughout this book that the whole troposphere is yet another boundary layer. In this chapter we examine the ABL as defined and depicted in Fig. 10.1.

10.2 Surface shape and radiation

The Earth's surface can be regarded as a horizontal plane with projections and indentations ranging in size from millimetric to mountainous. These affect the overlying boundary layers in various ways, among the most important of which

Figure 10.2 As the surface air chilled on a clear winter's night, it became slightly supersaturated, and vapour condensed fastest down the steep vapour pressure gradients over to any solid with tight curvature, such as leaf points and edges, spider's webs, building spikes (dendrites) of hoar frost.

are the redistribution of heat sources and sinks arising from fluxes of solar and terrestrial radiation (*SR* and *TR*).

10.2.1 Direct solar input

Consider first the effects of surface shape on direct solar input. A simple model of a solitary projection (Fig. 10.3 and Box 10.1) establishes some important points.

(i) *SR* is concentrated on the Sun-facing side at the expense of the other, and the concentration of direct *SR* is total when the other side is in shadow. When the Sun is low, sunlight falls nearly perpendicularly on steep Sun-facing slopes and very obliquely on gentle slopes. Since surface irradiance is proportional to the sine of the angle between the plane of the illuminated surface and the line of the incident sunbeam (Box 8.2), all perpendicularly illuminated slopes are receiving the maximum irradiance from the available *SR*. It is shown in Box 10.1 that the ratio of sloping surface irradiance to horizontal irradiance is given by

$$RI = \sin(\alpha + E)/\sin E$$

where α is the angle of slope and E is the angle of solar elevation above the horizon.

BOX 10.1 Direct sunlight on projections

Figure 10.3 shows a beam of direct sunlight with solar elevation angle E falling on a symmetrical projection with base angle α. The construction shows that the beam makes an angle $\alpha + E$ with the plane of the forward slope. Since the beam makes an angle E with a horizontal, the ratio RI of the irradiances of the slope and horizontal surfaces is given by the sine rule for oblique irradiance (Box 8.2):

$$RI = \frac{\sin(\alpha + E)}{\sin E}$$

For any chosen E, this expression shows that the maximum slope irradiance occurs when $\alpha + E = 90°$, i.e. when the beam is normal to the slope. Values of RI are plotted in Fig. 10.4 for three representative slopes and a range of solar elevations. To find an expression for the absolute irradiance AI of the slope, we can use Eqn B8.2d of Box 8.3 for the horizontal irradiance HI at the base of an atmosphere with zenith transmissivity T_{90}, together with the identity

$$\begin{aligned} AI &= HI \times RI \\ &= S\,[\sin(\alpha + E)]\,(T_{90})^{\operatorname{cosec} E} \end{aligned}$$

where S is the solar constant. For simplicity, values plotted on Fig. 10.4 are for AI/S. All are calculated for the case where zenith transmissivity is 0.8, which corresponds to a very clear sky.

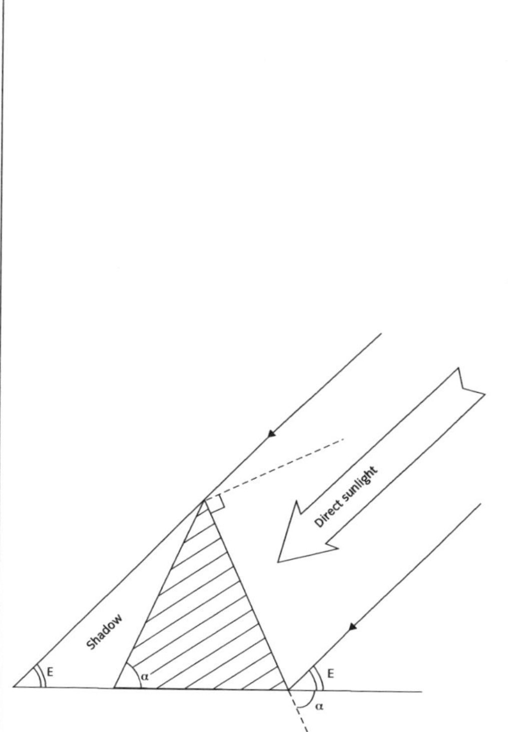

Figure 10.3 A beam of direct sunlight falling on a ridge with triangular cross-section.

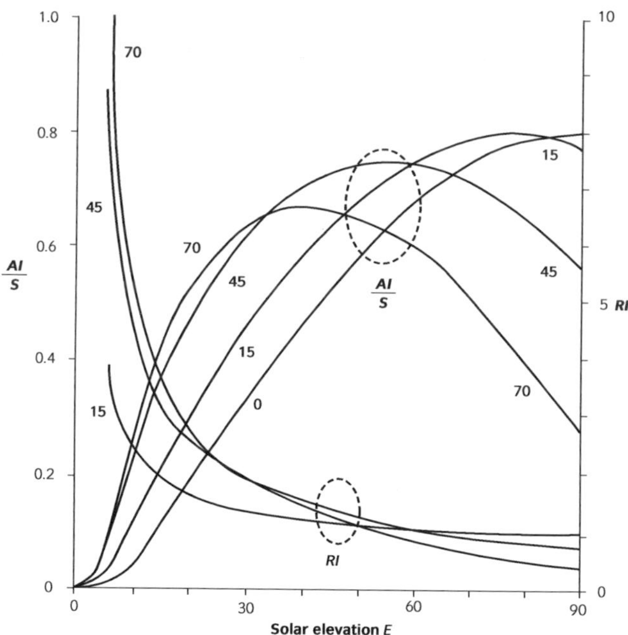

Figure 10.4 Direct solar irradiance on surfaces making angles of 0, 15, 45, and 70° with the horizontal. Curve *RI* represents the ratio of slope irradiance to horizontal irradiance. Curves *AI/S* represent absolute slope irradiance under a clear sky with zenith transmissivity 0.8 (Box 10.1), normalized by the solar constant S.

Values of RI are plotted in Fig. 10.4 for solar elevations ranging from 0° to 70° and slight, steep, and nearly vertical slopes. These confirm that steep slopes are at a great advantage when illuminated by the Sun at low angles of elevation. Although absolute irradiances may be quite low because of strong attenuation of the oblique sunbeams

Figure 10.5 A street of cumulus drifting downwind from a sunlit nursery slope.

in their long paths through the atmosphere (Box 8.2), the relative advantage can be important. Upright humans benefit greatly near dawn and dusk, our vertical bodies being well warmed while the solar warming of horizontal ground is virtually zero. Cumulus convection can be triggered over hilly country, as slightly warmer Sun-facing slopes give the overlying air a convective advantage which can produce lines (*streets*) of cumulus drifting downwind from the nursery slopes (Fig. 10.5).

In the examples plotted in Fig. 10.4, the absolute values of insolation are greatest for all but the slightest slopes when the Sun is moderately high in the sky—i.e. near the midday maximum elevation in mid-latitude summer. The very considerable preferential warming of Sun-facing slopes is then quite obvious to anyone walking in hilly terrain. On small projections, such as small buildings or vegetation, the concentrations of input are too localized to have any significant effect on larger scales, but they can be very important for conditions on the scale of the projections themselves, and plants and insects, etc. living there.

(ii) A projection which is steep enough in relation to the solar elevation angle, will leave a considerable area in shadow. In the case of the simple projection in Fig. 10.3, shadowing extends beyond the base when the solar elevation is less than the angle of slope. It is obvious that a steep mountain illuminated by a fairly low Sun can intercept *SR* which would otherwise have been spread over a horizontal area twice as large as the base of the mountain, and a very steep solitary projection such as a tree can intercept even more.

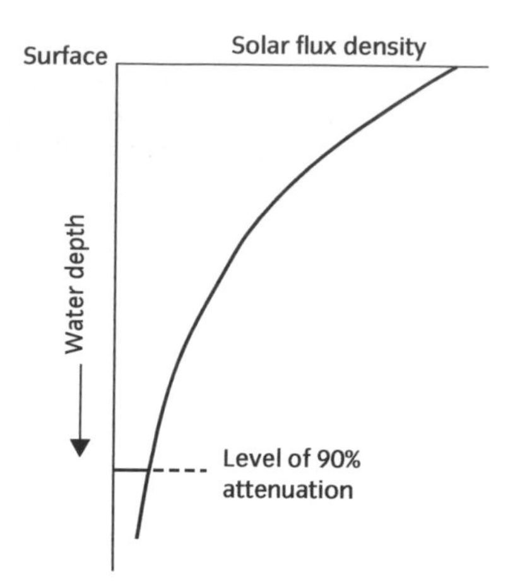

Figure 10.6 Exponential decay of solar irradiance with increasing depth in water.

(iii) When there are several projections close together, each one may shade its neighbour to some extent, so that solar input is concentrated on their upper parts, and little direct sunlight reaches basic ground level. This can occur in hilly terrain when the Sun is low, and is typical of dense stands of vegetation such as a field of cereal or a forest. In these, *SR* dies away with increasing depth below the top of the stand, much as it does below a water surface (Fig. 10.6)—the leaves, stems, branches, and trunks playing the same role as the absorbing and reflecting material in the water. We see from this how a complicated shape can blur the interface between atmosphere and surface, the radiatively active layer being spread through the upper parts of the projections.

(iv) In the simplest picture, the presence of projections redistributes the insolation without altering the total input. However, if projections are so steep and close together that *SR* suffers repeated scattering from adjacent surfaces, repeated absorption enhances the total capture by the surface. This happens markedly when *SR* is steeply incident on a deeply indented surface such as a forest, in which light passing below the treetops is repeatedly scattered between adjacent trees and leaves, with absorption increasing at each scattering. As aerial and satellite photographs show, coniferous forests have very low bulk albedos as a result, and are amongst the darkest of all common land surfaces (Table 10.1). The same effect is observed with water, for the same reason, with the result that albedos are very low, except when oblique illumination produces very high *specular* (mirror-like) reflectivity.

All of the above has assumed direct sunlight with geometrically edged zones of illumination and shadow. In most parts of the world, more than half of all surface irradiance by *SR* is diffuse, having been scattered from the incident direct beam by cloud, haze, and air molecules. Each of the above effects become less marked in diffuse sunlight, because diffuse sunlight is very much less unidirectional than direct sunlight, although not necessarily fully isotropic.

10.2.2 Terrestrial radiative output

Output of *TR* is complicated by surface shape in ways similar to those applying to the input of *SR*, although there is no equivalent of the direct solar beam and its sharp-edged distributions, since very extensive emitting areas ensure that *TR* is always diffuse. In essence the cooling by net emission and absorption of *TR* is governed by how much sky can be seen by a fish-eye lens at the surface in question. Any point on a plane unbounded horizontal surface 'sees' the maximum amount of sky (a hemisphere), whereas at the bottom of a narrow cleft between two projections, only a small amount of sky is 'seen' between the projecting walls. Since the effective blackbody temperatures of the walls are usually much higher than those of a clear sky, the net loss is greatly reduced in the cleft. Hoar frost distributions show these effects quite graphically in mid-latitude winters: an exposed car roof may be heavily frosted, whereas a neighbouring car beside a high vertical wall remains frost free. Gardeners fearing frost damage are sensitive to these effects, without necessarily understanding their cause. Overnight radiative chilling in a dense stand of vegetation is concentrated in the upper parts of the stand, though not necessarily in precisely the same height zone as that of the solar input.

10.3 Surface heat input and output

Consider quantitatively the radiative and convective processes injecting and removing heat from representative parts of the Earth's surface.

10.3.1 Solar radiation (*SR*)

Albedos of terrestrial surfaces range very widely from a few per cent, for damp dark soils, to over 90% for fresh snow (Table 10.1). High values are localized in snowy high altitudes and latitudes, and the relatively pale land deserts of the subtropics. Over the rest of the globe, values are held low by the presence of water and vegetation. Overland albedos are mostly between 5 and 20%; and over the sea they are mostly even lower, though rising to beyond 90% at glancing incidence. Rough surfaces generally have lower albedos than equivalent smooth surfaces because of the multiple absorption effect mentioned above.

Table 10.1 Radiative properties of surface materials

Surface material	Solar albedo	Terrestrial emissivity
Soils	0.05–0.40	0.90–0.98
Grasses	0.16–0.26	0.90–0.95
Forests	0.05–0.20	0.97–0.99
Desert	0.20–0.45	0.84–0.91
Snow	0.40–0.95	0.82–0.99
Water		
low solar elevation	0.10–1.0	0.92–0.97
high solar elevation	0.03–0.10	0.92–0.97

Albedos quoted in Table 10.1 are averages over the whole solar spectrum, and may be applied directly to total solar fluxes to estimate energy inputs to irradiated surfaces. The strong colorations of many terrestrial surfaces show that reflectivity often varies considerably with wavelength, but though this may be important sometimes (as in plant photosynthesis), it largely averages out in dealing with the heat economies of surfaces on human and larger scales.

To establish typical values for *SR* input, consider some typical values of *daily horizontal insolation* (*SR* incident on a horizontal surface totalled over the hours of daylight) tabled for a range of latitudes in Box 10.2. Under clear skies in midsummer these can lie between about 35 MJ m^{-2} in low and mid-latitudes, tailing off slowly into high latitudes as daylight lengthens to 24 hours but illumination

BOX 10.2 Daily insolation

For many climatological purposes it is useful to have the daily total input of solar energy incident on unit horizontal surface area (i.e. not yet absorbed, transmitted or reflected)—the so-called *daily insolation* (Fig. 10.7). If the zenith transmissivity is steady during the day, or if several more variably sunny days are averaged, the profile of solar horizontal irradiance from dawn to dusk nearly fits a half sine curve, and the daily insolation (the area under the curve) can be found by standard integration.

Figure 10.7 shows a half sine curve fitted between dawn and dusk to represent the daily profile of horizontal irradiance *I*. Ignoring small inaccuracies arising because the slight bell shape of the real curve is not perfectly fitted, the shape of *I* is approximated by

$$I = I_n \sin\left(\frac{\pi t}{N}\right) \qquad \text{B10.2a}$$

where I_n is the maximum (noon) irradiance, t is the time measured from dawn, and N is the time interval

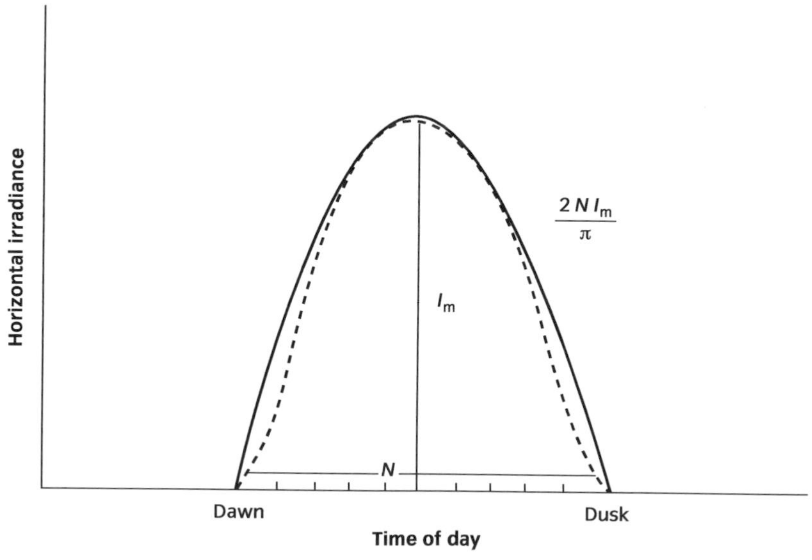

Figure 10.7 Idealized symmetrical diurnal time-profile (dashed) of horizontal surface insolation. The solid line is a fitted half of a sine curve.

between dawn and dusk. The daily insolation DI is found by integrating the expression for I between dawn and dusk.

$$DI = \int I\,dt = I_n \int \sin(\pi t/N)\,dt = 2\,N\,I_n/\pi \qquad \text{B10.2b}$$

Note that average daylight irradiance DI/N is $2/\pi$ times (i.e. nearly 64%) the insolation which would result if the Sun remained at its noon elevation throughout daylight. This allows easy estimation of daily insolation from $2\,I_n/\pi$ and day length N.

Estimating daily insolation

At latitude ϕ, the angle of solar elevation at noon is $E_n = (90° - \phi)$ at the equinoxes, and E_n plus or minus 23.5° at midsummer and midwinter respectively (Fig. 8.10). In very clear skies the zenith transmissivity T_{90} is about 0.8, and the corresponding noon transmissivity $T_n = T_{90}{}^{\text{cosec } E}$ (Box 8.2). Using the solar constant $S = 1380$ W m^{-2} (i.e. ignoring the 6% variation in solar intensity arising from Earth's elliptic orbit), the noon horizontal irradiance I_n is given by

$$I_n = (S \sin E)(0.8)^{\text{cosec } E}$$

If the day length in seconds is N, the daily horizontal insolation $DI = 2\,I_n\,N/\pi$ as above.

Cloud

Suppose cloud introduces an extra transmissivity factor on top of the clear sky value T_{cloud} for half of the daylight hours. Then the cloud-reduced daily insolation DI_r is related to the clear sky value DI by

$$DI_r = \frac{1}{2}\,T_{cloud} + \frac{1}{2}\,T_{cloud}\,DI = \frac{1}{2}\,(1 + T_{cloud})\,DI \qquad \text{B10.2c}$$

Using a value of 0.5 for T_{cloud} (implying that cloud halves the clear sky transmissivity)

$$DI_r = 0.75\,DI$$

so that 35 MJ m^{-2} is reduced to about 26 MJ m^{-2} and pro rata.

Table 10.2 Daily insolation

		E_n/°	T_n	I_n/Wm^{-2}	N/s	DI/MJ m^{-2}
Latitude 23.5°	midsummer	90	0.80	1,104	4.9 × 10^4	34.4
	midwinter	43	0.72	678	3.7 ...	16.0
Latitude 55°	midsummer	58.5	0.77	906	6.2 ...	35.8
	midwinter	11.5	0.33	91	2.6 ...	1.5
Latitude 80	midsummer	23.5*	0.57*	314*	8.6 ...	27.0*
	midwinter	< 0				0.0

Note: * Note that in midsummer inside the Arctic/Antarctic Circles the Sun does not set; its angle of elevation ranges daily about 23.5°, at latitude ϕ rising $90° - \phi$ above this at noon, and sinking $90° - \phi$ below it at midnight. For simplicity I have ignored the asymmetry introduced by atmospheric transmissivity, and calculated the insolation from the product $I_m\,N$, where I_m is the mean horizontal irradiance

NUMERICAL NOTE 10.1 Surface irradiances

As a spot check on Table 10.2 consider the daily solar insolation at latitude 55° in midsummer. Solar elevation is 90° − 55° + 23.5° = 58.5° at noon (Fig. 8.10). At this angle atmospheric transmissivity $T_{58.5} = (T_{90})^{\text{cosec } 58.5} = 0.8 \times 1.173 = 0.77$, and the noon solar irradiance of the surface is 1380 × sin 58.5 × 0.77 = 906 W m^{-2}, where 1380 represents the solar constant, and sin 58.5 represents the spreading of the foot of the angled sunbeam (Box 8.2). With a daylight length estimated roughly from standard graphs to be 17.3 hr (6.2 × 10^4 s), the simple rule of Box 10.2 gives daylight insolation = 906 × 6.2 × 10^4 × 2/π = 35.8 MJ m^{-2}, as in Table 10.1.

Cloud cover reduces this to 26.9 W m^{-2} (Box 10.2). Of this, 15% is scattered/reflected by a standard albedo surface, leaving 85% to be absorbed, giving a daily absorption of about 23 MJ m^{-2}.

becomes more oblique. In midwinter they can lie between about half this value in low latitudes, tailing off to a few MJ m^{-2} in mid-latitudes, and to zero poleward of the Arctic and Antarctic Circles. Under the cloud which covers most parts of the Earth for half the time, these values are reduced substantially, falling to one tenth of clear sky values under thick overcast. As a rough summary we can say that a typical daily insolation might be 25 MJ m^{-2} in summer, with half this in low-latitude winters and one tenth in mid-latitude winters. Assuming a surface albedo of 15%, 85% of the insolation is absorbed, implying typical maximum daily inputs (by absorption) of 20 MJ m^{-2} with proportionately lower values in the other circumstances.

10.3.2 Terrestrial radiation (*TR*)

As described in Section 8.3, terrestrial surfaces behave very nearly as blackbodies in *TR* wavelengths. Their closeness to blackbody behaviour is apparent in the nearness of their emissivities (Table 10.1) to the value 1.0. In fact most surfaces have emissivity values of between 0.9 and 0.98. Apart from old snow, only deserts have emissivities less than 0.9, because of the presence of quartz sand which is partly transparent in terrestrial wavelengths. Note that water is almost completely opaque to *TR* (which means that its emissivity is close to 1.0) although it is nearly transparent to *SR*—a striking example of how radiative properties can differ in different wavelength ranges. *TR* from the sea comes from the topmost millimetres of water which form an opaque layer radiatively separating the atmosphere from deeper waters.

For present purposes we will simply assume that terrestrial surfaces emit and absorb *TR* as if they were blackbodies. We can therefore use Stefan's law (Box 8.1) together with typical values of surface temperature to show that radiant emittances from most terrestrial surface lie within 90 W m^{-2} of 390 W m^{-2}—the value corresponding to the global annual mean surface temperature according to Fig. 8.8. Much of this output is offset by absorption of terrestrial radiation coming downwards from the sky, which has an average value of nearly 340 W m^{-2} (Fig. 8.8). This too varies considerably, exceeding 390 Wm^{-2} under warm, low clouds, and falling below 200 W m^{-2} under cold, clear skies. As a net result of the emission and absorption of *TR*, we can say that a typical piece of Earth's surface loses energy by net *TR* emission at a rate of 50 W m^{-2} or about 4.3 MJ m^{-2} in 24 hours.

10.3.3 Net radiative balance

Combining *SR* gains and *TR* losses we find that on a typical summer day in middle latitudes a net gain of about 20 MJ m^{-2} in daylight is followed by a net loss of about 2 MJ m^{-2} during the night. In winter there may be a continuous net loss as the weak *SR* input is more than offset by the continuing net *TR* output, and the daily total may be as much as 10 MJ m^{-2}.

Particularly in summer, convection tends to offset the daytime input very significantly by carrying sensible and latent heat into the atmosphere, and on all but the calmest nights, mechanical stirring of the atmospheric boundary layer feeds some sensible heat from the air down to the chilling ground. Dew and frost formation mark the return of some latent heat, but amounts are usually relatively small. The net result of convective and radiative exchanges is that many surfaces suffer a daily alternation between heat gains and losses, each of which can give a daily total of ~ 5 MJ m^{-2}. In middle and high latitudes there is also a large annual alternation between persistent daily gains in summer, when daytime gains outweigh night-time losses, and persistent daily losses in winter, when night-time losses exceed daytime gains. Allowing for convective and advective offsets, the summer gains and winter losses may each be ~ 500 MJ m^{-2}.

10.4 Surface thermal response

We have seen that there are daily and annual rhythms of heat gain and loss by any part of the Earth's surface, and have estimated their orders of magnitude. We know that surfaces respond by heating and cooling, and know also that their responses vary widely. In this section we look at the mechanisms involved in such responses.

10.4.1 Water

Since water is fairly transparent to solar wavelengths, especially in the visible range, sunlight is attenuated progressively as it travels downwards below the surface—only a small fraction of the incident energy usually reaching down more than a few tens of metres. Most of the attenuation ultimately leads to absorption and heating of the water, though some backscatter to the surface contributes a very little to the oceanic albedo. The contrast between the transparency of water to sunlight, and the almost total opaqueness of more than a few millimetres of land surfaces, accounts for much of the enormous difference in the thermal responses of land and sea—the heat capacity of a typical layer of illuminated ocean being enormously larger than that of a land surface, and its temperature response correspondingly smaller.

Despite the fact that sea water absorbs different solar wavelengths with different efficiencies, extinguishing the near infrared more rapidly than the visible, the depth profile of overall solar flux density often follows the ideal exponential shape expected in a homogeneous absorber (Fig. 10.6 and Box 10.3), with depths for 90% attenuation varying from only a few metres in *turbid* water, usually found near coasts, to as much as 50 m in the relatively clear waters often found in the open oceans (though plankton may contribute to turbidity there). Absorption in turbid waters is principally by suspended sediment and organic matter, and some associated dissolved substances, whereas in clear water it is principally by the water itself.

As mentioned in Section 8.9 turbulent mixing in ocean surface layers increases the depth of water warmed by *SR* and cooled by *TR*, enhancing their effective heat capacity. Observations suggest that depths subject to daily and annual cycles of warming and cooling are ~ 10 and ~ 50 m respectively, with corresponding heat capacities ~ 40 and ~ 200 MJ °C^{-1} m^{-2}. If the warming and cooling were distributed uniformly throughout these depths, then daily gains and losses of 5 MJ m^{-2} would produce temperature rises and falls of 0.12 °C, and seasonal

BOX 10.3 Sunlight in water

Adapting the method of Box 8.2 to the penetration of solar irradiance I to depth z under water, we have an expression for the change $\mathrm{d}I$ produced as sunlight passes a little further into the depths:

$$\mathrm{d}I = -\, k\, \mathrm{d}m\, I$$

where k is the absorption coefficient (in a less simple model it varies with wavelength), $\mathrm{d}m$ is the mass of water per unit horizontal area in the little range of depth $\mathrm{d}z$, and the minus sign ensures that I reduces with increasing depth. Replacing $\mathrm{d}m$ by $\rho\, \mathrm{d}z$, we find a relationship between $\mathrm{d}I$ and $\mathrm{d}z$ which can be integrated, in the simple case where k and ρ are uniform, to give an exponential decay of irradiance with increasing depth,

$$I_z = I_0\, \mathrm{e}^{-k \rho z}$$

where I_0 is the irradiance at the water surface, and I_z is the irradiance at depth z.
According to Box 4.3, the exponential and decadal scale depths for the irradiance are respectively $1/(k\, \rho)$ and $2.3026/(k\, \rho)$.

NUMERICAL NOTE 10.2 Warming of water columns

A vertical water column of depth d, unit horizontal area, density ρ, and specific heat capacity C has heat capacity $H = \rho\, d\, C$. Using 1,000 kg m^{-3} for ρ, and 4.2 kJ °C^{-1} kg^{-1} for C, $H = 4.2\, d$ MJ °C^{-1} m^{-2}, where d is in metres. This shows that the 10 m column subject to daily heat inputs and outputs of 5 MJ m^{-2} has a heat capacity of 42 MJ m^{-2} and will warm or cool by 5/42 = 0.12 °C. The 50 m column subject to seasonal gains and losses of 500 MJ m^{-2} has heat capacity 210 MJ °C^{-1} m^{-2} and will warm or cool by 500/210 = 2.4 °C.

gains and losses of 500 MJ m^{-2} would produce temperature changes of 2.4 °C (NN 10.2). Because temperature changes are largest at the surface, despite the effects of mixing, changes in surface temperature tend to be larger than these estimates, but not by very much. We see that marine temperature cycles are very much smaller than those familiar to us on land.

Some mixing in ocean surface layers is effected by mechanical turbulence reaching down from the troubled surface, but some is convective; in fact convective stability exerts an overall control on vertical exchange just as it does in the atmosphere. The difference is that in the oceans surface warming enhances convective stability, whereas in the atmosphere it diminishes it. For example, the depth of water warmed in summer is limited by the warming itself, since warmed surface water floats on the cooler, denser water beneath, so that the temperature gradient of the intervening *thermocline* maintains a density gradient which inhibits further vertical mixing. Cooling in the oceans is localized at the surface in the first instance, since terrestrial radiation, evaporation, and wind chill are concentrated there. By increasing the density of water at the surface, cooling encourages convective instability and hence the downward extension of the cooling by vertical mixing. In very sheltered waters such as lakes, the destruction of the summer thermocline in the autumn can be quite sudden, playing an important role in the lake's annual biological cycle by redistribution of nutrients, etc., but in the oceans it is masked by constant commotion. However, after the removal of the summer thermocline, further cooling in the oceans is distributed through a much deeper layer of water than before, which may even extend to the sea bed kilometres below.

10.4.2 Land

The response of a complex, vegetated land surface to surface heating and cooling is very difficult to analyse (Fig. 10.8), but some important features can be seen by considering the relatively simple case of a homogeneous layer of rock or soil responding to a regular cycle of surface warming and cooling. This is a standard problem in the theory of heat conduction, with a well-known and tested solution which we now examine and compare with observation (Fig. 10.9).

At any depth below the surface, we should expect that the rate of rise of temperature after the onset of surface heating will increase with the thermal conductivity k of the intervening layer and decrease with its heat capacity. Full theory *[20]* shows that the net effect is determined by the *thermal diffusivity*

$$\kappa = \frac{k}{\rho c} \qquad 10.1$$

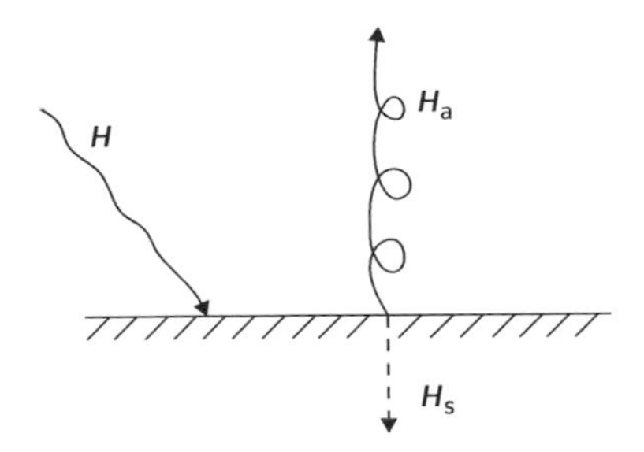

Figure 10.8 Net radiative heat input H to a land surface, partitioned between convected heat H_a to the atmosphere and conducted heat H_s to the subsurface.

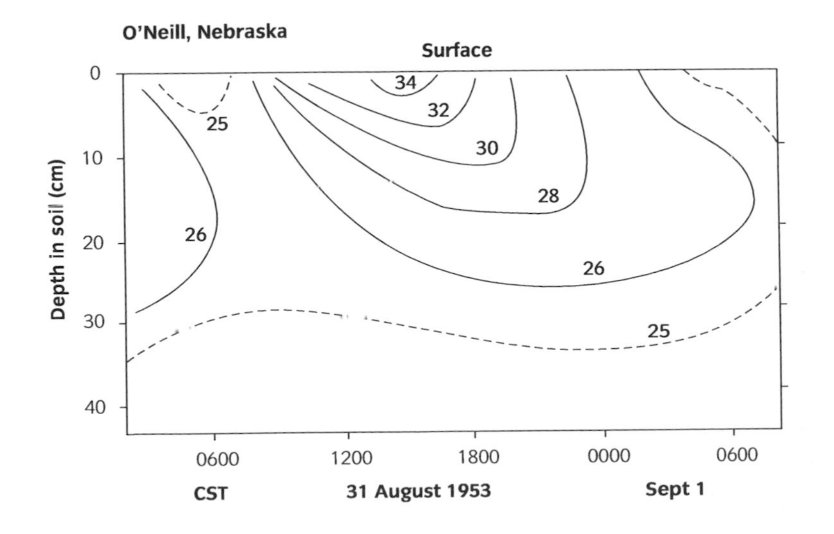

Figure 10.9 Depth–time section with isotherms labelled in °C, showing downward progress of the diurnal temperature wave under a prairie.

Table 10.3 Thermal properties of surface materials

Material	Dry peat	Dry sand	Wet sand	Water	Air	Units
Density	300	1600	2000	1000	1.2	kg m^{-3}
Specific heat capacity	1920	800	1480	4180	1000	J kg^{-1} K^{-1}
Thermal conductivity	0.06	0.3	2.2	0.57 (4000	0.025 6000)	W m^{-1} K^{-1}
$10^6 \times$ thermal diffusivity	0.10	0.23	0.74	(~ 10^3	~ 5×10^6)	m^2 s^{-1}
Damping depth						
diurnal	5.2 cm	7.9 cm	14 cm	(5 m	300 m)	cm and m
annual	1.0 cm	1.5 cm	2.7 cm	(100 m	7×10^3 m)	cm and m
Heat capacity						
diurnal	0.04	0.14	0.60	(31	0.63)	MJ K^{-1} m^{-2}
annual	0.81	2.75	11.4	(590	~ 12)	
				(eddy	values)	

where ρ and C are respectively the density and specific heat capacity of the ground material. The larger the value of κ, the more freely is a change in surface temperature conducted downwards, and the deeper is the layer affected in a given time. Values of κ for typical surface materials are listed in Table 10.3, and range from 0.1 for dry peat to 0.75 for wet sand (in units of 10^{-6} m^2 s^{-1}). Trapped air in dry soils is such a poor heat conductor that it ensures low κ values despite smaller values of ρ and C.

The depth of material affected also depends on the time period of the surface warming and cooling cycle. If this is short, then heat which has not penetrated far enough during the surface warming cycle is conducted back to the surface and lost during the succeeding cooling cycle. If the period is long, then a greater depth of soil is significantly warmed and cooled.

The detailed solution shows that if the surface heating and cooling cycle has period P, then there is an important depth parameter D known as the *damping depth* and given by

$$D = \left(\frac{\kappa P}{\pi}\right)^{1/2} \qquad 10.2$$

The damping depth defines the depth of the soil which is significantly affected by the surface heating and cooling in three different but related ways.

(1) If the amplitude of the surface temperature wave is ΔT_0 (Fig. 10.9), then the amplitude ΔT_z at depth z below the surface is given by

$$\Delta T_z = \Delta T_0 \, e^{-z/D} \qquad 10.3$$

which shows that the temperature wave dies away exponentially with increasing depth, its amplitude falling to 37% of the surface value at depth D, and to 10% at 2.3 D (Box 4.3).

(2) As the temperature wave is conducted down into the layer, it lags further and further behind the surface wave and is exactly in antiphase (out of step—minima coinciding with maxima and vice versa) at depth πD (Fig. 10.9).

(3) The amplitude ΔT_0 of the temperature wave at the surface is related to the heat H entering unit area of surface during the heating phase by

$$\Delta T_0 = \frac{H}{\sqrt{2}\,\rho C D} \qquad 10.4$$

from which it appears that the effective heat capacity of the affected layer (per unit surface area) is $\sqrt{2}\,\rho\, C\, D$, and that the depth of the layer which heat input H would warm uniformly by ΔT_0 is $\sqrt{2}\, D$.

It is clear from the values of D in Table 10.3 that even the relatively penetrating *annual* temperature wave affects a soil layer to a depth of no more than a few metres. Even in a responsive wet soil, a worm living at a depth of about 6 m would experience only 10% of the annual temperature range experienced by its cousin living at the surface, and their temperature calendars would differ by 6 months. The daily temperature wave penetrates only tens of centimetres. The near constancy of temperature at quite shallow depths has long been exploited in the construction and use of cellars.

Values for the effective heat capacity of typical ground layers are included in Table 10.3 and show that even relatively responsive wet soils have heat capacities of no more than a few per cent of the values quoted earlier for the sea surface, all of which reinforces the point that heat conduction through terrestrial surfaces is so poor that only very shallow layers are significantly affected by daily and yearly surface heating and cooling. Land surfaces are dominated by this effect, since there is no other mode of heat transport land; but in water surface layers, transparency to *SR* input and turbulent diffusion of sensible heat are important, as discussed already.

10.5 Partitioning between surface and air

Equation 10.4 represents an incomplete description of the situation at the surface because it ignores the finite effective heat capacity of the air. Using subscripts *s* and *a* for surface and air respectively, the total heat input H during the warming phase of the surface temperature cycle must be partitioned according to

$$\frac{H_a}{H_s} = \frac{(\rho\, C\, D)_a}{(\rho\, C\, D)_s} \qquad 10.5$$

where the symbols are as defined in the previous section for heat conduction. In fact heat transfer in the atmospheric boundary layer is by convection (and some terrestrial radiation) rather than by conduction, and in the surface layers of the seas it is by solar radiation and turbulent mixing, but for present purposes all such transport is assumed to be effectively equivalent to conduction—very rapid conduction in the cases of turbulent and radiative transport. Note that the relevant heat input

H is the excess of net radiant warming over evaporative cooling, since this is the sensible heat to be redistributed by conduction, turbulence, and radiation.

By an obvious extension of Eqn 10.4 we can now relate the amplitude of the surface temperature wave to the heat input H by

$$\Delta T_0 = \frac{H}{\sqrt{2}\,[(\rho C D)_s + (\rho C D)_a]} \qquad 10.6$$

Consider now the extremely different cases of heat input to desert and ocean surfaces, to assess the different partitioning of input between surface and atmosphere.

10.5.1 Desert surfaces

Diurnal

As the Sun's rays begin to strike the dry soil of the desert in the early morning, the surface warms rapidly because the surface heating is conducted downwards very inefficiently through the air-filled soil. According to Table 10.3, the damping depth for the daily heating cycle is only a few centimetres, and the effective heat capacity is ~ 0.1 MJ °C^{-1} m^{-2}. The gradual elimination of the nocturnal inversion in the overlying atmospheric boundary layer (Section 5.11) means that the heat transfer there is complex and changeable, but observations show that turbulent transfer often operates as if the eddy thermal diffusivity (the turbulent equivalent of κ) was within an order of magnitude of 5 m^2 s^{-1}. This is equivalent to an effective heat capacity of about 0.6 MJ °C^{-1} m^{-2} for the warming layer of air, showing that the great bulk of the heat input to the surface passes into the air rather than into the soil beneath, and that the total effective heat capacity of ground and the air together is ~ 1 MJ °C^{-1} m^{-2}. In a subtropical desert, despite the relatively high albedo and net loss by terrestrial radiation (clouds being scarce), the heat input H to the surface during the warming phase can easily exceed 20 MJ m^{-2}, giving a temperature cycle at the surface with amplitude exceeding 20°, which means a range from maximum to minimum of double this value (Fig. 10.10). In fact these estimates are quite conservative, and real surface conditions can be even more extreme: violent swings of temperature between blistering daytime heat and freezing towards dawn can shatter rocks by repeated violent expansions and contractions, and impose severe conditions on any organisms living there. However, the excellent thermal insulation of the dry soil (a major feature contributing to the extreme surface conditions), offers a refuge which is taken up by many burrowing animals and dormant plants.

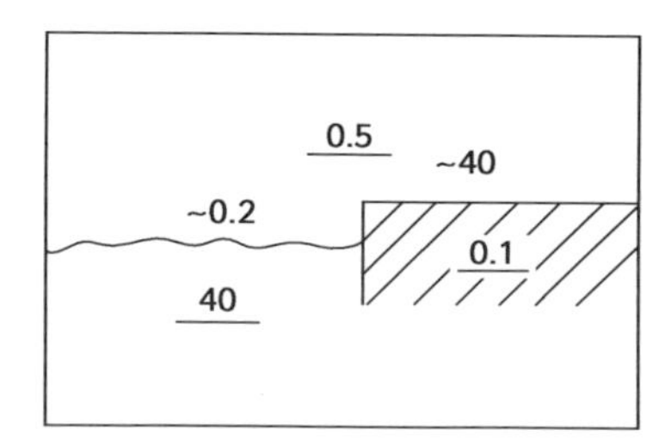

Figure 10.10 Typical heat capacities (underlined, in MJ °C^{-1} m^{-2}) and associated diurnal temperature ranges (~ in °C) for subtropical desert and ocean.

Annual

Estimates for the annual heating cycle in deserts show similar partitioning between air and ground, but at 15 MJ °C^{-1} m^{-2} the total effective heat capacity is much larger because of the greater penetration of heating and cooling into both air and ground. However, the annual cycle of heat input and output in the subtropics is quite muted because the Sun is still quite high in the sky even in mid-winter, with the result that the seasonal temperature cycle is smaller than the diurnal cycle, though still quite large. In middle and high-latitude deserts however, many of which lie in the rain shadows of mountain ranges, the much larger annual range of H values, together with the small effective heat capacities typical of all deserts, give rise to very large seasonal swings of temperature from searing summer heat to sustained sub-zero temperatures in winter. These too make great demands of landscape and life, for example severely testing the pioneers opening up the North American Midwest in the nineteenth century.

10.5.2 Ocean surfaces

At the other extreme we have the case of the ocean–air interface. The estimates of ocean heat capacity made in Section 10.4.1 will do well enough for the effective surface heat capacity in Eqns 10.5 and 10.6, i.e. 40 MJ °C^{-1} m^{-2} for the daily heating cycle. The effective heat capacity of the atmospheric boundary layer is assumed to be the same as in the case of the desert (0.6 MJ °C^{-1} m^{-2}), although reduction in convection may not be completely offset by increased stirring by the freer ocean winds. Regardless of detail, the heat input to the surface is obviously overwhelmingly taken into the sea rather than into the air, and the total effective heat capacity of the sea and air together is several tens of times larger than in the case of the desert. Since in addition the sensible heat input to the surface is significantly smaller than in the case of the desert (the large evaporative cooling more than offsets the effect of the much lower albedo), the result confirms the explanation for the observed very small oceanic surface temperature range given in Section 10.4.1.

10.6 Laminar boundary layer

10.6.1 Thickness

Air molecules in contact with a liquid or solid surface are so effectively stuck to it by molecular attraction that there is no relative motion. At a very little distance from the surface, molecules can move, but only very smoothly and sluggishly on account of viscous friction, and this laminar flow is always parallel to the local surface. This is the laminar boundary layer (LBL). With increasing distance from the surface, the flow speed increases, as do the accelerations and decelerations of air in response to disturbing factors such as surface shape, turbulent jostling by air beyond the LBL, etc. At some small critical distance δ from the surface, such unsteadiness overcomes the smoothing and damping effects of viscosity, and the air flow ceases to be laminar. The value of the thickness δ of the LBL in any particular situation is very important, since resistance to the diffusion of momentum, heat, and water vapour through the laminar boundary layer increases with its thickness, and becomes very high indeed for thicknesses of more than a few centimetres.

The outer limit of the laminar boundary layer is associated with a critical value of the appropriate Reynolds number ($Re = U\,L/\nu$), where U and L are characteristic speeds and dimensions of the air flow in the LBL, and ν is the kinematic viscosity of the air. The dimension L is clearly δ, but the choice of U is less obvious since the air flow in the laminar boundary layer is heavily sheared. As will become clearer in the next two sections, there is a very important velocity parameter in the turbulent air beyond the LBL which is called the shear velocity or *friction velocity* u_* and is defined by

$$u_* = \left(\frac{T}{\rho}\right)^{1/2} \qquad 10.7$$

where T is the flux of tangential momentum towards the surface (and the tangential stress on the surface) borne by turbulence outside the LBL and by viscosity within it, though the latter statement has to be qualified when the surface is rough (see below).

In the case of the LBL, observations in many different situations, from water channels to the atmosphere, show that its depth δ corresponds to a critical value for the Reynolds number

$$\frac{u_* \delta}{\nu} \sim 10 \qquad 10.8$$

which means that if we can assess atmospheric values of u_* we can assess the thickness of the LBL from $10\,\nu/u_*$.

Values of T can be measured directly, by attaching delicate strain gauges to horizontal plates set in the ground, and indirectly from atmospheric turbulence (Section 10.8) When combined in Eqn 10.7 with realistic values of surface air density ρ they form a wide range of values of u_* centred on ~ 0.3 m s^{-1}. It follows from laboratory values for ν ($\approx 1.5 \times 10^{-5}$ m^2 s^{-1}) that LBL thicknesses δ are ~ 0.5 mm, being thicker when the overlying air is less turbulent and vice versa. The layer in which transport is dominated by molecular rather than turbulent diffusion is therefore extremely thin, and is really no more than a viscous membrane of air adhering to all solid and liquid surfaces.

10.6.2 Fluxes through the LBL

Although very thin, the LBL covers all exposed surfaces, and is only significantly bypassed by momentum (Section 10.6). All other fluxes passing to or from the Earth's surface (heat, water vapour, carbon dioxide, etc.) are forced to pass through it, and its considerable resistance often maintains sharp gradients of temperature, vapour density, etc. For example, sensible heat diffusing from the surface obeys the equation for heat conduction (Box 8.3), which can be written in simple difference form

$$H = \frac{k\,\Delta T}{\delta} \qquad 10.9$$

NUMERICAL NOTE 10.3 Gradients through laminar boundary layer

Temperature

Rearrange Eqn 10.9 to find $\delta = k\ \Delta T/H$, and use $k = 0.025$ W m^{-1} K^{-1} (Table 10.2), $\Delta T = 2$°C and $H = 100$ W m^{-2} to find $\delta = 0.025 \times 2/100 = 5 \times 10^{-4}$ m or 0.5 mm. If the LBL is horizontal, the lapse of 2°C in 0.5 mm corresponds to a vertical lapse rate of 4°C per mm, or 4×10^5 °C per 100 m, which 4×10^5 times the DALR (≈ 1 °C per 100 m).

Flow speed

Using $T = \mu\,\Delta V/\delta$ in the form $\Delta V = \delta\,T/\mu$, and substituting 0.5 mm for δ, 10^{-1} N m^{-2} for T, and 18×10^{-6} Pa s for μ (Eqn 7.8 and text) we find $\Delta V = 0.5 \times 10^{-3} \times 10^{-1}/(18 \times 10^{-6}) \approx 3$ m s^{-1}. The value of T is typical of direct observation, and with values of 0.3 m s^{-1} for u_*, and 1.2 kg m^{-3} for ρ in Eqn 10.7 ($T = \rho\, u_*^2 = 1.2 \times 0.09 \approx 0.1$ N m^{-2}).

where H is the sensible heat flux per unit area of LBL, k is the thermal conductivity of still air, and ΔT is the temperature drop across the LBL, depth δ. A value of 100 W m^{-2} for H is typical when fairly dry land is being heated strongly by the Sun, and it follows from Eqn 10.9 that a fall of over 2°C is needed to drive sensible heat at this rate through a half millimetre thick LBL (NN 10.3). Expressed as a vertical temperature gradient this is over five orders of magnitude greater than the dry adiabatic lapse rate. However, the DALR limit applies only when air parcels are able to exchange freely in the vertical—a condition not met in the very viscous LBL, and not met either, as we shall see, in the lowest parts of the overlying turbulent layer. Though such sharp shallow temperature gradients are not obvious to us in daily life, they explain some of the extra warmth we feel on lying down to sunbathe in a sheltered position. They are very important to plants and small insects, and contribute to the substantial overestimation of air temperature by standard-sized thermometers exposed to direct sunlight.

The LBL diffuses tangential (most importantly horizontal) momentum from the moving atmosphere to the surface, and Eqn 7.8 for the tangential surface drag can be expressed in the form

$$T = \frac{\mu\,\Delta V}{\delta} \qquad 10.10$$

where ΔV is the shear of tangential flow speed across the LBL from zero at the surface to ΔV at the edge of the LBL. Assuming $T \sim 10^{-1}$ N m^{-2} (consistent with the u_* value assumed earlier) it follows that $\Delta V \sim 3$ m s^{-1} (NN 10.3)—a very substantial shear across such a shallow zone. But although shears of this intensity may occur occasionally they cannot be common, remembering that almost all natural surfaces, including all but the calmest water, have roughness elements which are much deeper than the depth of the laminar boundary layer. For example, each blade of grass in a grassy surface is enclosed in its LBL, but the air flowing over and through the grass is impeded by the physical obstruction of the grass stems (slightly enlarged by their laminar envelopes), with the result that the vertical profile of horizontal wind speed amongst the grass is determined principally by the distribution, size, and shape of the grass stems, rather than the film of LBL. Such surfaces are said to be *aerodynamically rough*, and their roughness is influential to levels well above the tallest roughness element, as we see below. Virtually all natural surfaces are aerodynamically rough, the only exception being water in zero wind and with no waves coming from elsewhere.

10.7 Turbulence

Above the very shallow LBL, the atmospheric boundary layer is chronically turbulent. The effects of turbulence are many and varied: the gustiness of all but the lightest winds is familiar (Fig. 10.11), as is the unsteady dispersion of smoke plumes (Fig. 10.12), but many other significant properties are much less obvious. And though turbulence in the atmosphere and other fluids has been studied intensively for over a century, it has proved surprisingly difficult to define and describe. Sophisticated experimental and theoretical work has confirmed the subtlety of the topic, so that even an elementary treatment is beyond the scope of this book *[42]*. The following is therefore a selective and mainly descriptive outline of some of its important properties.

- Turbulent flow is highly irregular in both space and time. For example, the wind speed in Fig. 10.11 varies continually and widely but shows no trace of regular wave-like oscillations. Though recent studies of the nocturnal atmospheric boundary layer in particular show that gravity waves are often present at some height above the surface, these are quite distinct from turbulence, including the turbulence they sometimes produce on breaking. Virtually every meteorological measurable varies in this way in the presence of turbulence, though wind shows by far the largest percentage range. Variations in temperature and humidity content are relatively small but are very important because of their contribution to heat and vapour fluxes, as we shall see.
- The unsteadiness and irregularity typical of turbulent flow represents a large-amplitude, highly chaotic response to inherent instability, which forbids useful prediction of the instantaneous flow pattern at any future time. Quantitative description and prediction therefore involve statistics ranging from the simple arithmetic mean and root mean square deviation to sophisticated power spectra,

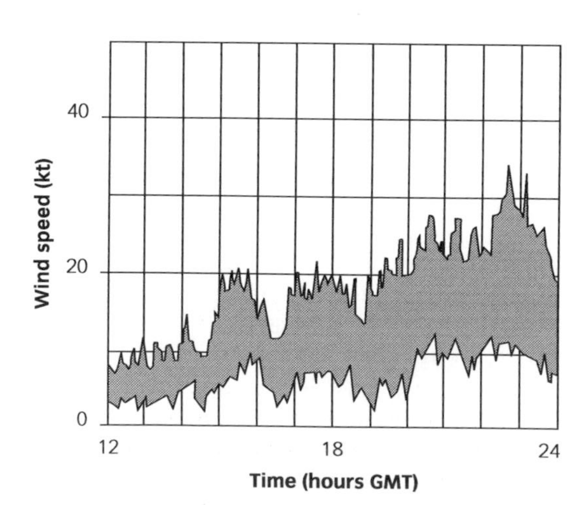

Figure 10.11 Anemograph for part of a typically breezy day in NW England.

Figure 10.12 A smoke plume undulating and spreading in the turbulent daytime boundary layer of air flowing along a steep-sided valley.

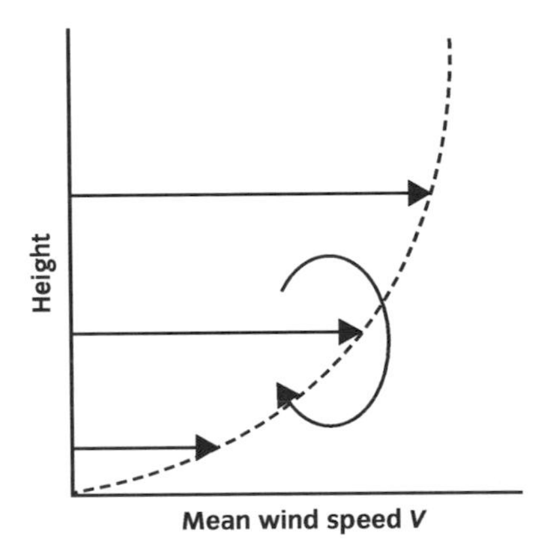

Figure 10.13 Air rolling and tumbling in strong wind shear near the surface.

which show the contribution of different periodicities to the total variance, and beyond. In qualitative description we can still make use of the concept of temporarily distinguishable eddies, though with the tacit assumption that they are too numerous and transient to be usefully observable in practice, except as correlations between wind components, etc.

- Turbulence is intrinsically three-dimensional and cannot be adequately described in fewer dimensions. This handicaps pictorial description and mental conception (which is strongly two-dimensional), though of course flat pictures like Fig. 10.13 are useful for limited description. If there is a dominant describable type of motion it is a continual stretching and folding which seems to be common to the topology of chaos *[43]*. And although a considerable proportion of atmospheric turbulence is *isotropic* (without preferred direction or axis), structures seen in longitudinal section (along the flow) often differ systematically from structures seen in lateral section.

- Turbulence is hierarchical. Inherent instability produces large eddies which in turn produce smaller eddies, and so on down a cascade of diminishing scales until the eddies are small enough to be literally rubbed out by viscosity. According to a famous jingle by a pioneer in the study of atmospheric turbulence, L.F. Richardson:

  'Big whirls have little whirls which feed on their velocity,
  And little whirls have smaller whirls, and so on to viscosity.'

The largest eddies have scales comparable with the depth of the ABL (~ 300 m), and the scale of the smallest is ~ 1 mm, comparable with the depth scale of the LBL. At the large end of the scale it is sometimes difficult to distinguish between large eddies and small structured systems, for example the 'smoke ring' vortex believed to be common in buoyant convection *[36]*.

The fact that the largest recognizable turbulent eddies in the ABL have scale ~ 100 m carries the implication that most turbulent variations (a cycle of overshoot and undershoot about a notional mean) last less than ~ 100 s at a fixed position on the surface, assuming typical horizontal wind speeds. If averages are taken over time periods several times this length, their values are often reasonably stable over somewhat longer time periods. The meteorological synoptic convention is to take the ten-minute averages recorded in the standard hourly observations, variability being one-sidedly assessed by recording the highest wind speed (the strongest gust) in the observation period. There are of course variations on scales longer than a few minutes, in fact there is a continuous range or spectrum up to time periods corresponding to the passage of major weather systems such as depressions, but the ten-minute average is especially useful because there is significant dip in the spectrum separating turbulence proper from longer period variations. In the following sections a ten-minute averaging period is assumed for all averages unless otherwise stated.

10.8 Origins and role of turbulence

Turbulence persists because air flow is continually disturbed on scales which are too large to be smoothed away by viscosity. If disturbances of scale 100 m are being introduced into an air flow of average speed 1 m s^{-1} (a very modest wind speed in the atmospheric boundary layer), the very large value of the associated Reynolds number ($\sim 10^6$) indicates that viscosity is quite incapable of damping this towards laminar flow. The disturbed air flow therefore continues jostling in an irregular fashion, so that turbulence continues even if the disturbing factor operates only intermittently.

10.8.1 Mechanical

The single most important source of disturbance is the wind shear which is endemic near the surface unless there is total calm (Section 7.4.3). Air parcels embedded in the wind shear tend to roll forwards about a lateral horizontal axis,

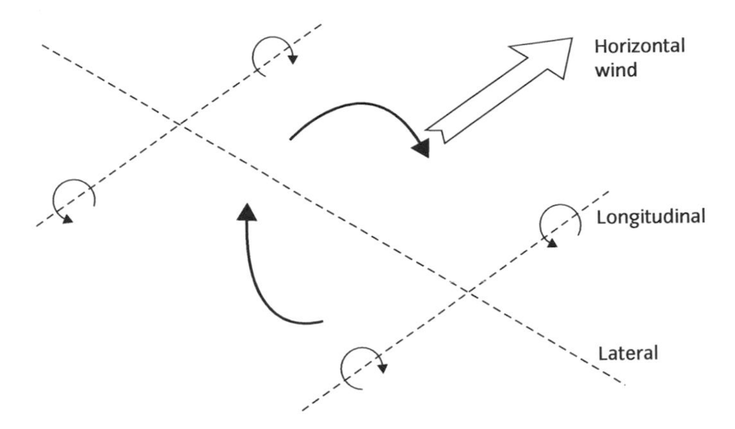

Figure 10.14 An air parcel rolling about a lateral axis with secondary vortices rotating in both senses about longitudinal axes.

temporarily forming a rotating eddy. Rotation is not confined to the longitudinal section portrayed in Fig. 10.13, because as a parcel begins to roll about a lateral axis, it tends to generate by friction opposing rolls about longitudinal axes on either side (Fig. 10.14).

All rolling eddies have air descending on one side and ascending on the other, but if the air is convectively stable, as is often the case near the ground at night, for example, vertical motion is inhibited by convective stability (Section 5.10). It seems that maintaining turbulence by shear in these circumstances must depend on a balance between encouragement by shear ($\partial V/\partial z$), where V is the average horizontal wind speed) and discouragement by stability $(g/\theta)\ \partial\theta/\partial z$ (from Eqn 7.31), where θ is the average potential temperature, and $\partial\theta/\partial z$ is positive when the air is convectively stable. Richardson showed theoretically that turbulence should tend to break out when the magnitude of the dimensionless number (later called the *Richardson number* and denoted by Ri)

$$Ri = \frac{\dfrac{g}{\theta}\dfrac{\partial\theta}{\partial z}}{\left(\dfrac{\partial V}{\partial z}\right)^2} \qquad 10.11$$

falls below 0.25. Much atmospheric observation confirms that critical values of Ri are indeed of this order. Substitution of realistic values for g and θ (in kelvins) shows that very considerable stability is needed to prevent generation of turbulence by typical observed wind shears ($\partial V/\partial z$) near the surface. This is consistent with the observed persistence of turbulence even on nights with quite pronounced surface cooling, provided the wind speed is kept up by the large-scale weather situation. If, however, surface cooling is sufficiently strong to raise the value of $\partial\theta/\partial z$ to the point where Ri exceeds the critical value, turbulence is damped out by the convective stability. Then turbulence no longer brings down V momentum from air at higher levels, and friction with the ground slows the air near the ground to a virtual standstill. Since surface cooling (assuming this to be a fairly clear night) is no longer even partly offset by the flux of sensible heat brought down by turbulence from the warmer air aloft, the surface cooling and overlying stability increase, and the air remains effectively still until the Sun comes up or a windy weather system comes in.

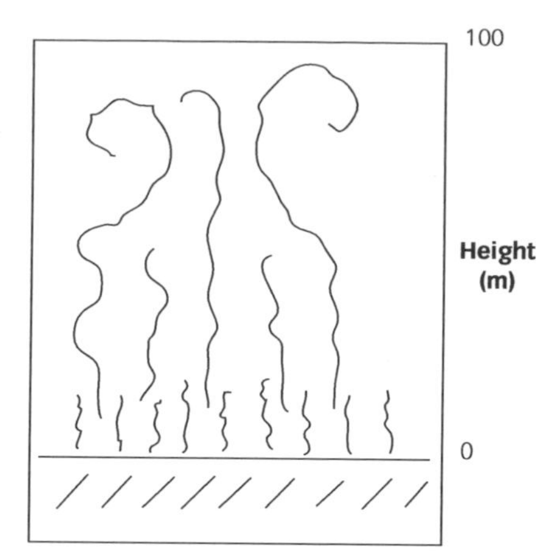

Figure 10.15 Outline of a plausible transition from submetric scale 'shimmer' just above a strongly heated land surface, to 100 m scale thermals aloft.

10.8.2 Thermal

Thermally driven convection also generates turbulence, provided the motion generated is too vigorous for viscosity to maintain laminar flow. In a layer of air of depth Δz and potential temperature lapse $\Delta\theta$, this balance is described by the dimensionless *Rayleigh number* (*Ra*)

$$Ra = \frac{g\,\Delta\theta\,(\Delta z)^3}{\kappa\,\nu\,\theta} \qquad 10.12$$

In all atmospheric conditions in which a relatively warm surface maintains a lapse $\Delta\theta$, *Ra* far exceeds the known critical value for the onset of turbulent convection ($\sim$ 50 000). Although very common, the resulting convection is not well understood. Very near the surface it is quite small-scale, and the turmoil of centimetric and metric-scale eddies rising and falling produces the shimmering of distant objects seen by line of sight just above a strongly heated land surface. At greater heights it seems that these small incoherent movements cooperate to produce larger-scale thermals which rise up through the atmospheric boundary layer, leaving a turbulent wake as they go. The latter stage corresponds well to Richardson's jingle, but the former suggests a reverse sequence of building from smaller to larger scales (Fig. 10.15). Regardless of detail, the effect is to maintain turbulence on a wide range of scales in the convecting layer.

10.8.3 Turbulent transport

As mentioned already, turbulence transports momentum, heat, and material such as water vapour and aerosol, and this occurs whenever air parcels (eddies) moving randomly carry systematically different quantities of momentum, heat, etc. depending on their direction of motion. Like the molecular diffusion which it overwhelms, turbulence always carries momentum, etc. from regions of excess to regions of deficit, i.e. down-gradient. In many cases the gradients are effectively vertical and so therefore are the transports, momentum always being carried down to the surface, and water vapour almost always being carried up.

Over land, sensible heat is often transported up by day and down by night, though the nocturnal fluxes are usually smaller.

Momentum

Consider, for example, the downward turbulent transport of V momentum (i.e. horizontal downwind momentum) through a fixed horizontal unit area some height above a horizontal surface, visually outlined by a thin wire frame. Since most of the effective eddies are larger than the frame itself, even in the SI system where the unit frame is 1 m square, an average downward flux arises because, in the sequence of up and downdrafts passing through the frame, the downdrafts systematically carry more V momentum, since they are coming down from the faster-moving air above (Fig. 10.13). The instantaneous downward flux of V momentum is given by $-\rho\, w\, V$, where the minus sign ensures that downdrafts (negative w) contribute positively to the downward momentum flux. If ρ, w, and V are sampled simultaneously, every second for example, the triple product can be evaluated every time, and averaged over 10 minutes, for example, to find the average downward flux of horizontal momentum:

$$-\overline{\rho\, w\, V} \qquad 10.13$$

which, as described in Fig. 7.5a and text, is identical with the average horizontal shearing stress T within the layer and at the underlying surface.

As shown in Box 10.4, because air density variations are proportionately small, and the average updraft is zero over a horizontal, impermeable surface, Eqn 10.13 simplifies to

$$T = -\rho\overline{wV'} \qquad 10.14$$

BOX 10.4 Turbulent momentum transport

According to Eqn 10.13, the average vertical downward flux of horizontal momentum through horizontal unit area is given by

$$T = -\overline{\rho\, w\, V} \qquad \text{B10.4a}$$

where the overbar represents averaging over a chosen time period (often 10 minutes) and the group of terms under it represents the instantaneous momentum flux density. Since variations in ρ are proportionately very much smaller than those in w and V, we can remove it from under the overbar and treat it as a constant. The instantaneous horizontal downwind speed can be written as the sum of the average speed and the instantaneous deviation from it:

$$V = V' + \overline{V}$$

The vertical speed can be analysed in the same way, but $\overline{w}$ is virtually zero in most realistic conditions in the surface boundary layer so that

$$w = w'$$

Substituting these in B10.4a gives

$$T = -\rho\,\overline{w'\,(V' + \overline{V})} = -\rho\,(\overline{w'V'} + \overline{w'}\ \overline{V})$$

Since $\overline{V}$ is constant in the averaging period by definition, the second pair of terms in the brackets includes $\overline{w'}$, the average of the deviations of w from its average, which is zero as assumed above. Hence

$$T = -\rho\,\overline{w\,V'} \qquad \text{B10.4b}$$

dropping the unnecessary prime from the updraft speed.

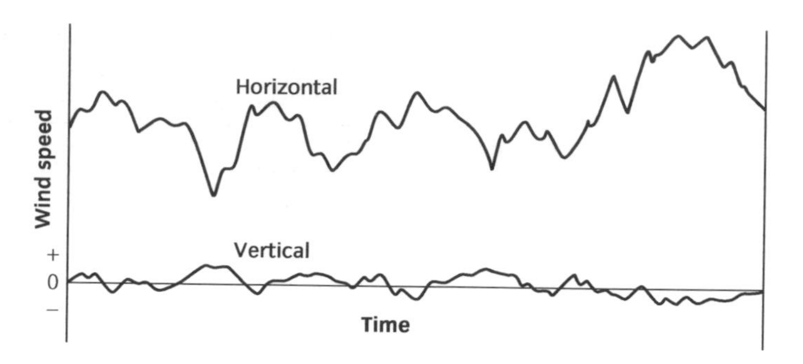

Figure 10.16 Idealized anemograph of vertical and horizontal wind speeds over a period of a few minutes in relatively steady conditions, showing the negative correlation which is normal near the surface.

where V' is the instantaneous deviation of V from its average V, w is the instantaneous updraft speed, and ρ is the average air density. The term is the *covariance* of vertical and horizontal winds, and is non-zero and positive when their fluctuations are negatively correlated, as shown in Fig. 10.16, and as they must be with faster flow above.

Combining Eqns 10.14 and 10.7 we see that

$$u_*^{\,2} = -\overline{w\,V'} \qquad 10.15$$

confirming the interpretation of u_* as a velocity related to the gustiness of winds.

This result shows that if we describe air flow in terms of average rather than instantaneous values, we must allow for correlations between variations occurring within the averaging period otherwise we will miss potentially significant momentum fluxes, which means that we may ignore important forces. The equations of motion (Eqn 7.16) are always written in terms of average values, since it would be useless to apply them to instantaneous observations of quantities as variable as winds, and hence they need correction. A thorough treatment of the equations of motion in terms of averages and instantaneous deviations *[35]* shows that terms such as $\partial(\overline{w\,V'})/\partial z$ must be included along with viscosity terms such as $\nu\, \partial^2 u/\partial z^2$ to represent friction terms such as FRF_x. The extra terms are the gradients of terms such as $\rho\overline{wV'}$ (known as *Reynolds stresses*), and their inclusion completes the treatment of frictional forces begun in Section 7.4.

The particular Reynolds stress represented by Eqn 10.14 is often the largest of the many possible components because $\partial V/\partial z$ is usually the largest shear, being the vertical gradient of horizontal wind. It can be measured by recording simultaneously the outputs from a cup anemometer (measuring V) and a propeller on a vertical shaft (measuring w) and calculating their covariance—easily done now using a dedicated microchip. Each anemometer must be able to respond accurately to the smallest eddies contributing significantly to the stress, and this requires a cup anemometer with a much faster response than the robust and therefore sluggish instrument (Fig. 2.4) used for the synoptic observation.

Heat and water vapour

A very similar treatment shows that the vertical turbulent flux of sensible heat H is given by

$$H = \rho\, C_p\, \overline{w\,T'} \qquad 10.16$$

where $\overline{wT'}$ is the covariance of updraft and temperature, and the minus sign is absent this time because it is conventional to consider an upward heat flux to be positive. At night, turbulence persisting above a cooling surface (in a strong enough wind shear) maintains a negative covariance of updraft and temperature, and hence a downward heat flux, whereas by day vigorous turbulence over strongly heated ground maintains a strong positive covariance and hence an upward heat flux.

The corresponding expression for the turbulent flux of water vapour is

$$F_v = \rho \overline{w\, q'} \tag{10.17}$$

where q is the specific humidity. Multiplied by the coefficient of latent heat L, this vapour flux becomes the turbulent flux of latent heat.

10.9 Surface boundary layer (SBL)

Turbulence as outlined in previous sections dominates the distributions and vertical transports of momentum, heat, and matter, such as water vapour and carbon dioxide, throughout the atmospheric boundary layer except for the laminar boundary layer. In the lower parts of this turbulent layer turbulence is so dominant that some striking regularities emerge.

Consider the vertical profile of average downward momentum flux (drag) T in a period in which there is no substantial change of meteorological conditions. The horizontal components of the equations of motion (Eqn 7.16a and b) give rise to a simple balance along the flow between the gradients of the Reynolds stresses and the longitudinal component of the pressure gradient force

$$\frac{1}{\rho}\frac{\partial p}{\partial n} = \frac{\partial}{\partial z}(\overline{w\, V'}) = \frac{\partial}{\partial z}\left(\frac{T}{\rho}\right) \tag{10.18}$$

where n is a horizontal axis pointing upwind. In a layer shallow enough to ignore vertical variations of air density, this relation becomes

$$\frac{\partial T}{\partial z} = \frac{\partial p}{\partial n} \tag{10.19}$$

showing that the vertical gradient of T has the same value as the horizontal downwind gradient in pressure. As shown in Box 10.5, this is usually small enough to ensure that the percentage change in average T in the vertical is less than 10% in a depth of 10 m—as confirmed by observation, and one of the most striking regularities of the *surface boundary layer* (SBL), otherwise termed the *constant flux layer* because of the vertical near uniformity of the vertical fluxes of heat and matter as well as of momentum.

Complications arise because the upper limit of the SBL varies considerably with time and position, especially over land. Using the 10% criterion again, the layer is only ~ 1 m deep in nearly still nocturnal conditions when turbulence and

BOX 10.5 The constant momentum flux layer

Rewriting Eqn 10.19 to apply to a layer of depth Δz in which the average turbulent flux T of horizontal momentum towards the surface varies by ΔT along a vertical through the layer, we find

$$\frac{\Delta T}{T} = \frac{\partial p}{\partial n} \frac{\Delta z}{T} \qquad \text{B10.5a}$$

where the pressure gradient $\partial p/\partial n$ along the flow is effectively determined by the synoptic scale pressure field. In pure geostrophic flow $\partial p/\partial n$ would be zero (air flowing along the isobars), but turbulent friction near the surface twists the surface flow by about 30°, so that the pressure gradient along the flow is about half (sin 30°) the radial gradient, and might be 1 hPa per 100 km, 10^{-3} Pa m^{-1}.

Rewriting B10.5a to get an expression for Δz

$$\Delta z = 10^3\, T \left(\frac{\Delta T}{T}\right) \qquad \text{B10.5b}$$

we see that the layer depth should be less than 100 T if we stipulate that $\Delta T/T$ should be less than 10% (fraction 0.1) in the vertical through the layer. Using a typical T value of 0.1 N m^{-2} (see later in this section), suggesting that a layer can be 10 m deep and still have a nearly constant turbulent vertical flux of horizontal momentum from top to base.

T have died away to low levels, but it may be ~ 100 m deep when strong wind shear or convection encourage very vigorous turbulence. But though the change from one extreme to the other can be quite rapid, as from dawn to mid-morning, for example, it is usually sufficiently slow for T to be well-defined in ten-minute averages, and it is in these that the vertical uniformity of profiles is observed.

10.9.1 Log wind profile

The relative simplicity of the SBL, and the fact that it is the part of the atmosphere in which we spend most of our outdoor lives, has attracted great attention since the study of the atmospheric boundary layer began in earnest early last century. In the absence of an approach from the first principles of the theory of turbulence, people looked for simple relationships involving readily measureable properties, in particular ten-minute average values. In the case of wind, the vertical profile of average wind consistently shows a characteristic shape with very strong wind shear ($\partial V/\partial z$) near the surface and diminishing values at increasing heights (Fig. 10.13). The simplest dimensionally consistent expression for the shear ($\partial V/\partial z$) is of the form

$$\frac{\partial V}{\partial z} = \frac{A\,U}{z}$$

where A is a pure (i.e. dimensionless) number and U is a velocity characteristic of the flow in the layer. A great deal of experimental evidence in the atmosphere, wind tunnels and water flows, confirms that this simple relationship actually holds in the form

$$\frac{\partial V}{\partial z} = \frac{u_*}{k\,z} \qquad 10.20$$

where u_* is the friction velocity introduced in Section 10.6 and k is an apparently universal constant with value close to 0.4 (experimental difficulties prevent more accurate measurement) named after *von Kármán*. It is obviously reasonable that the velocity parameter on the right-hand side of Eqn 10.20 should be u_* since this describes the turbulent activity which dominates the layer.

Equation 10.20 matches the shape of the mean wind profile, but implies infinite shear at zero height. However, the turbulent SBL gives way to the LBL (or the layer penetrated by roughness elements such as grass stems) before this unrealistic extreme is approached. Nevertheless the presence of very strong shear at very small heights, together with the role of wind shear in maintaining turbulence (Eqn 10.13 and discussion) is consistent with observation of vigorous small-scale turbulence near the surface for much of the time. Over land this is quenched by strong temperature inversions which develop on still, clear nights, but these are inhibited by the great heat capacity of the surface layers over lakes and seas, so that strong shear and turbulence are present nearly all of the time, encouraging and maintaining wind waves *[39]*.

If we assume that T and ρ are uniform along a vertical through the SBL at any particular location and time (nearly but never precisely true) then it follows that u_* too is uniform, and we can integrate Eqn 10.20 along a vertical to obtain an expression for the vertical distribution of average wind speed V:

$$V = \frac{u_*}{k} \ln z + B \qquad 10.21$$

where ln is the natural logarithm (i.e. to the base e) and B is a constant. Such a simple linear relationship between average wind speed and the logarithm of height is so widely observed, at least to a reasonable approximation (Fig. 10.17), that it is known simply as the *log wind profile*. It implies (Box 10.6) that wind speed increases in equal steps for equal multiples of height increase, the speed increment for a doubling of height being $(u_* \ln 2)/k$, i.e. about 1.73 u_*. Because of the straight-line relationship in Fig. 10.17, extrapolation suggests that there is a non-zero height z_0 at which V is zero. This (z_0) is called the *aerodynamic roughness length*, and it has a fairly well-defined value for any particular situation in which the surface does not deform much with increasing wind speed. It appears as the lower height limit in the particular form of Eqn 10.21:

$$V = \frac{u_*}{k} \ln\left(\frac{z_2}{z_0}\right) \qquad 10.22$$

According to Eqn 10.22 the average wind speed at any height in the SBL is completely determined by the aerodynamic roughness length z_0 and the friction velocity u_*. A slight rearrangement produces a general relationship between *normalized wind speed* V/u_* and normalized height z/z_0:

$$\frac{V}{u_*} = \frac{1}{k} \ln\left(\frac{z}{z_0}\right) \qquad 10.23$$

In recent studies of all parts of the atmospheric boundary layer, sophisticated use is made of such general, normalized relationships, most of them much more complex than Eqn 10.23. It is possible to conduct quite general and subtle discussions

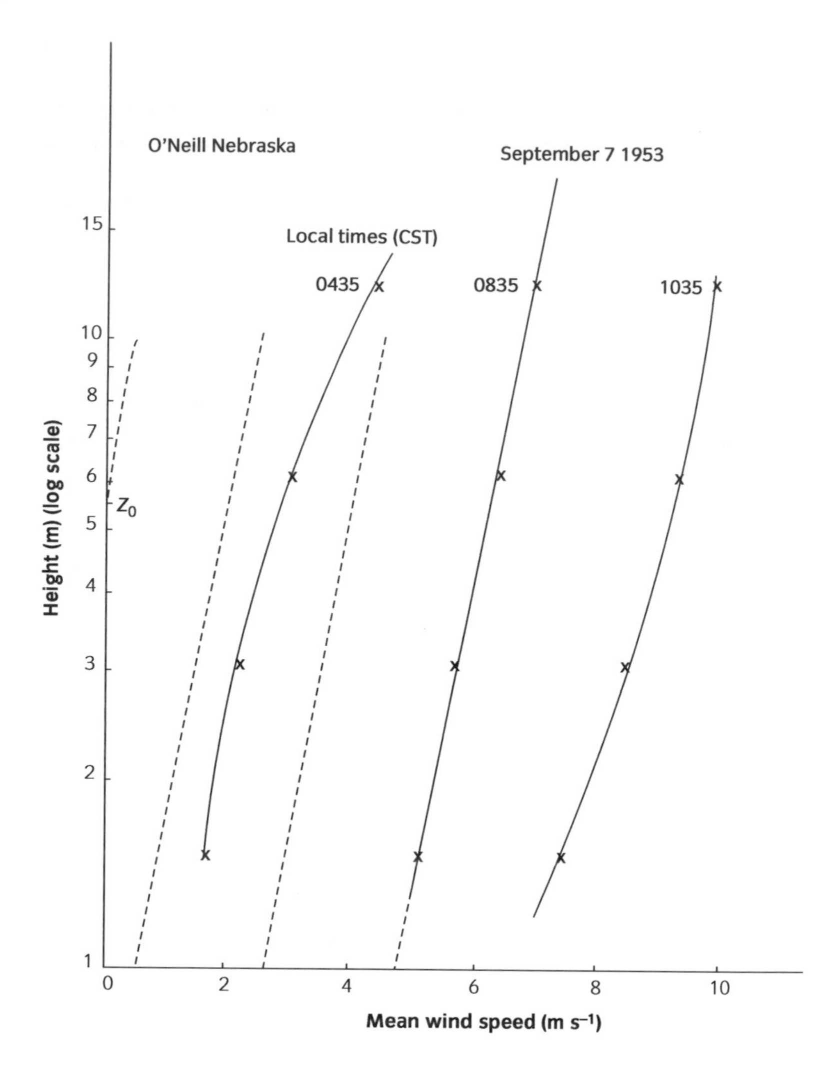

Figure 10.17 A nearly pure log wind profile (time 0835 CST) of mean horizontal wind over a prairie, including aerodynamic roughness length (5.4 mm) found by extrapolation to zero wind speed (dashed line through several cycles of the vertical scale). Curved lines show slight deviations observed in earlier and later convectively stable (time 0435, before dawn) and unstable (time 1035) situations. Data from *[44]*.

in such terms, and hence derive relationships to be checked against observation. For example, the logarithmic function on the right-hand side of Eqn 10.23 is a particular case of the more general function $F(z/z_0)$ which applies to the surface boundary layer in a much wider range of atmospheric conditions.

10.9.2 Roughness and drag

There is no simple relation between the value of z_0 and the average height l of the surface roughness elements (e.g. lengths of grass stems in a meadow), because shape and distribution density must also be important, but as a very rough guide $z_0/l \sim 1/10$ in many cases. Because z_0 is typically much smaller than l it follows that the actual shape of the wind profile must deviate from the idealized log profile below the roughness tops (well above a height z_0 above the ground

BOX 10.6 Log-linear wind relation

According to Eqn 10.21 in the neutrally buoyant SBL we have

$$V = \frac{u_*}{k} \ln z + B$$

If we choose two heights with ratio X

$$z_2 = X z_1$$

then it follows from the log wind profile that the difference in wind speed is

$$V_2 - V_1 = \frac{u_*}{k}[\ln z_2 - \ln z_1] = \frac{u_*}{k}\ln\left(\frac{z_2}{z_1}\right) = \frac{u_*}{k}\ln X \qquad \text{B10.6a}$$

Since k is a universal constant (0.4), and u_* is constant for a particular profile, we see that the difference in mean wind speed on any particular profile depends only on the choice of X. If we choose a series of heights in geometrical progression then it follows that the wind speeds will be in arithmetical progression. As an example, consider the difference between wind speeds at 2 m (head height) and 10 m (standard anemometer height). According to B10.6a

$$V_2 - V_1 = 1.61\frac{u_*}{k}$$

showing that in a gale (17 m s^{-1} at 10 m height) with friction velocity $u_* = 2$ m s^{-1} (a reasonable value in windy conditions), the wind speed at head height (2 m) is only 9 m s^{-1}. Notice that in the same conditions the wind speed at 50 m (at road level on a high bridge, or in the high rigging and spars of a tall sailing ship) is 25 m s^{-1} (49 knots).

z/m	2	10	50	geometrical
V/m s^{-1}	9	17	25	arithmetical

surface), since we can hardly expect flow among the roughness elements to be similar to flow clear of their tops. It follows that the only way to measure the z_0 of a particular surface is to observe the profile of average wind in the SBL above the grass, etc., using an array of anemometers on a mast, and to find z_0 by extrapolation as on Fig. 10.17. Since air flowing from one surface to another with a very different z_0 value adjusts quite slowly to the change in underlying surface roughness (as a rule of thumb the depth of the fully adjusted layer is one hundredth of the *fetch*—the horizontal downwind distance from the boundary between the surfaces (Fig. 10.18), it is important that the mast be positioned with adequate fetch. In small-scale terrain like much of the British Isles, homogeneous fetches are so short that the surface boundary layer is seldom fully adjusted, and the log-wind profile holds only approximately.

When roughness elements are tall and densely packed, as they are in mature cereal crops and dense forests, heights must be measured relative to a datum significantly raised above the ground to find a satisfactory fit with the log wind type of profile. If the required *zero plane displacement* is d, then

$$V = \frac{u_*}{k}\ln\left[\frac{(z-d)}{z_0}\right] \qquad 10.24$$

fits observed data quite well down to levels just above the crop or canopy. Wind flow within the crop or forest follows no general profile, being controlled by the vertical distribution within the particular vegetation in question (Fig. 10.19).

Forests with dense crowns and nearly bare trunks below can even have a secondary wind speed maximum well below the crowns.

Rearranging Eqn 10.23 to find an expression for the tangential stress T, we see that it is proportional to the square of the average wind speed V at any particular level.

$$T = \rho\, C_D\, V^2 \qquad 10.25$$

The constant of proportionality C_D is the *drag coefficient*, which depends only on the normalized height z/z_0 at which V is measured.

$$C_D = \left[\frac{k}{\ln\left(\dfrac{z}{z_0}\right)}\right]^2$$

The sensitivity of stress T to V which is apparent in Eqn 10.25 shows the expected proportionality to the kinetic energy (Section 9.2) of the wind blowing over the

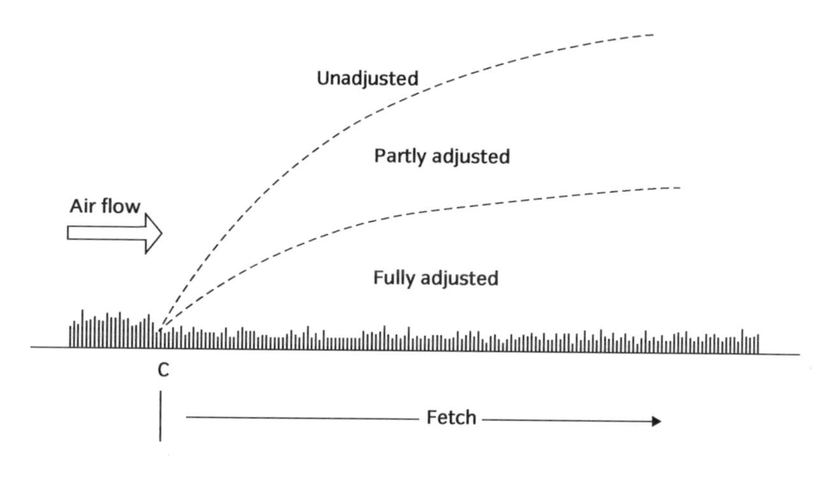

Figure 10.18 Sketch of downwind adjustment of a surface boundary layer to a change in surface roughness at C. The vertical is selectively enlarged, since the partly adjusted layer should be ~ 10 times deeper than the fully adjusted layer.

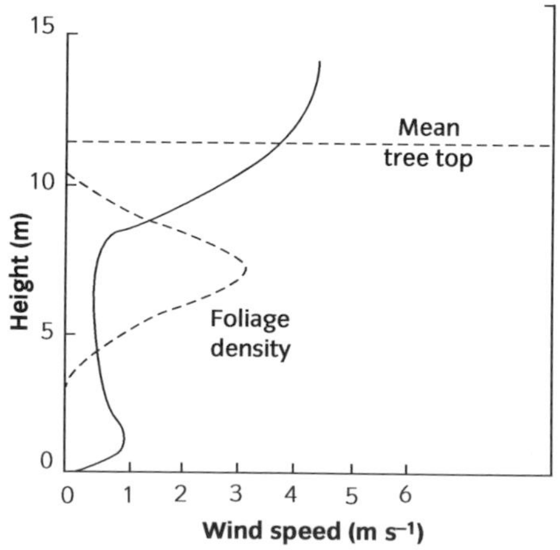

Figure 10.19 Profile of hourly mean wind speed in a forest of Sitka spruce in the middle of a sunny day. The top part of the profile represents measurements over open ground above the level of maximum foliage density.

surface, and highlights the importance of rare windy extremes in imposing high stresses on the surface, and causing damage. Forests can bear scars for decades which were caused by only an hour or so of unusually high average winds and their associated even briefer gusts.

10.9.3 Diurnal cycle

Strictly speaking the log wind profile is to be expected only when the air is neutrally stable for convection, otherwise a buoyancy term should appear in Eqn 10.20 and its derivatives. In convectively stable situations there are slight but systematic deviations from the log wind profile, because the damped turbulence allows a greater wind shear aloft. In convective instability the opposite occurs, and the wind shear aloft is reduced by the enhanced turbulent friction. However, these deviations from the pure log wind profile are quite small and are detected only by careful observation.

A much more obvious result of the variation in turbulent intensity occurs on most nights and days over land—the diurnal cycle of average wind speed at chosen low heights, from wind minima at night to maxima in daytime. This mechanism is apparent in Eqn 10.22, where V is directly proportional to friction velocity u_* for fixed z and z_0, and u_* increases as the square root of the turbulent momentum flux T, which is the horizontal shearing stress. Clearly when turbulence is vigorous, as it is when enhanced by thermally driven convection, momentum from faster air aloft is transferred down to the SBL, speeding up the average wind there. At night this downward transfer may die away almost completely, if surface cooling is strong enough to form a strongly stable layer somewhere above the shrunken SBL, so that the SBL and other layers below the stable one are brought nearly to a halt by friction with the underlying surface. The resulting

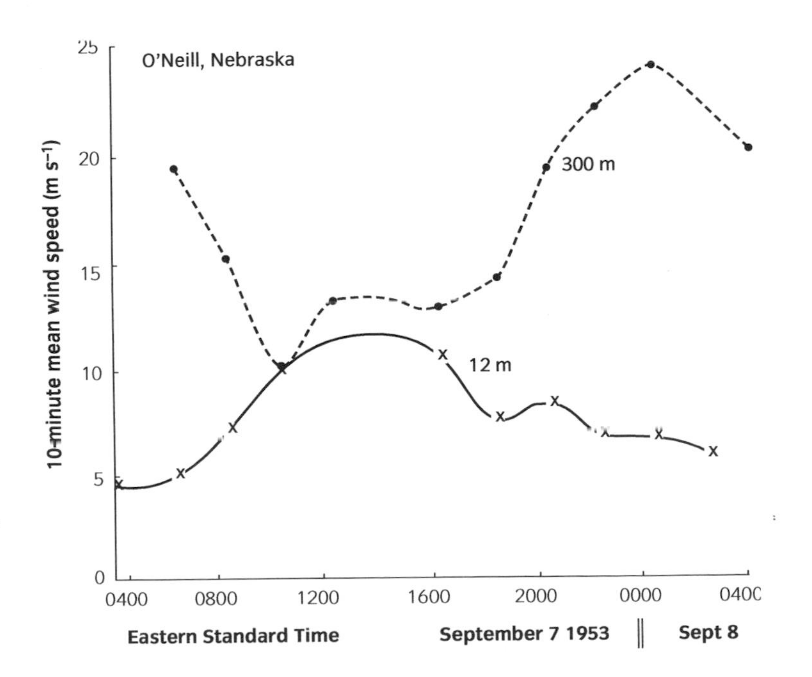

Figure 10.20 Typical diurnal variation in mean wind speed at heights of 12 and 300 m above a prairie site. Notice the related daytime speeding of surface air and slowing of the upper boundary layer (measured from a tethered balloon). Data from *[44]*.

diurnal variation in wind speed near the surface is one of the most familiar features of diurnal climate over land (Fig. 10.20). Over the sea, the diurnal rhythm in thermally driven convection is heavily muted by the large thermal capacity of the sea, so that the effect is weaker and much less closely tied to day and night.

The extra momentum delivered to the SBL and surface daily over land is of course brought down from higher levels, where the average wind speed is correspondingly reduced. Figure 10.20 shows this happening quite clearly at the 300 m level, well above the SBL. At such levels the result of enhanced turbulence in daylight is to extend upwards the effects of the frictional drag at the surface.

10.9.4 Eddy transports

By multiplying each side of Eqn 10.20 by u_* and rearranging, we find another expression for the horizontal shearing stress:

$$T = \rho(k u_* z)\frac{\partial V}{\partial z} = \rho K \frac{\partial V}{\partial z} \qquad 10.26$$

Comparing this with the equivalent expression for wind stress and shear in laminar flow (Eqn 7.8), we see that K ($= k\, u_*\, z$) plays the same role as the kinematic coefficient of viscosity ν. However, unlike its molecular counterpart this *coefficient of eddy viscosity* (K) is not simply determined by the thermodynamic state of the air; it depends on both the turbulent intensity (through u_*) and the height z, and it is the increase of K with z which maintains the inverse relation between wind shear $\partial V/\partial z$ and height z which is so characteristic of the surface boundary layer.

Similar analysis leads to an equivalent expression for the turbulent vertical flux of sensible heat

$$H = -\rho C_v (k u_* z)\frac{\partial \theta}{\partial z} \qquad 10.27$$

where H is the sensible heat flux density in W m^{-2}, and the minus sign is included because it is conventional to regard an upward heat flux as positive. Positive H occurs when there is a lapse of potential temperature θ with increasing height, i.e. a superadiabatic lapse rate (Section 5.9). In this simple analysis the eddy coefficient for heat transport (the turbulent equivalent of the thermal diffusivity) is assumed to be the same form as the eddy coefficient for momentum transport ($k\, u_*\, z$). More detailed analysis and observation suggest that this is not strictly true, and that heat is transported more efficiently than momentum when the layer is convecting, while the opposite is true when the layer is convectively stable.

10.9.5 Heat and vapour fluxes

Just as the average vertical flux of horizontal momentum T is nearly uniform throughout the depth of the SBL, so is the average vertical flux of sensible heat H, provided no stratifying mechanism is at work, such as a shallow fog or haze layer, interfering with net radiant and convective heat fluxed near the surface. If H is nearly uniform along a vertical axis, then according to Eqn 10.27 the vertical gradient of potential temperature must be inversely proportional to height z, which is consistent with the common observation of strongly superadiabatic lapse rates close to strongly heated land surfaces (Section 5.11), and sub-adiabatic lapse rates close to cool surfaces.

We can estimate the lapse $\Delta\theta$ of potential temperature between heights z_1 and z_2 by integrating Eqn 10.27 to obtain

$$\Delta\theta = \frac{H}{\rho k C_v u_*} \ln\left(\frac{z_2}{z_1}\right) \qquad 10.28$$

Substituting 100 W m^{-2} for H (corresponding to fairly strong surface heating) and 0.3 m s^{-1} for u_*, it follows that the potential temperature lapse $\Delta\theta$ between the base of the SBL and screen level (say heights 1 cm and 1.5 m) is about 4 °C. Although Eqn 10.28 is probably not very accurate in these conditions, because of the presence of significant buoyancy, the implication that temperature lapses of this order can occur is well supported by observation.

The temperature lapse switches over to a temperature inversion in the presence of significant surface cooling. This behaviour, together with similar behaviour in the laminar boundary layer overlying the surface, highlights the marked contrast in diurnal temperature range between the modest values measured at screen level and the much larger values observed very close to the underlying ground (Fig. 10.21)—an increase which has a marked influence of the flora and fauna living there.

Similar analysis leads to the equivalent expression for the average vertical transport of water vapour by turbulence:

$$F = -(k u_* z)\frac{\partial \rho_v}{\partial z} \qquad 10.29$$

where F is the vapour flux density and ρ_v is the vapour density. There are reservations about applicability and accuracy similar to those associated with Eqns 10.26 and 10.27. Multiplied by the coefficient of latent heat L, this vapour flux becomes the flux of latent heat

$$H_L = -L(k u_* z)\frac{\partial \rho_v}{\partial z} \qquad 10.30$$

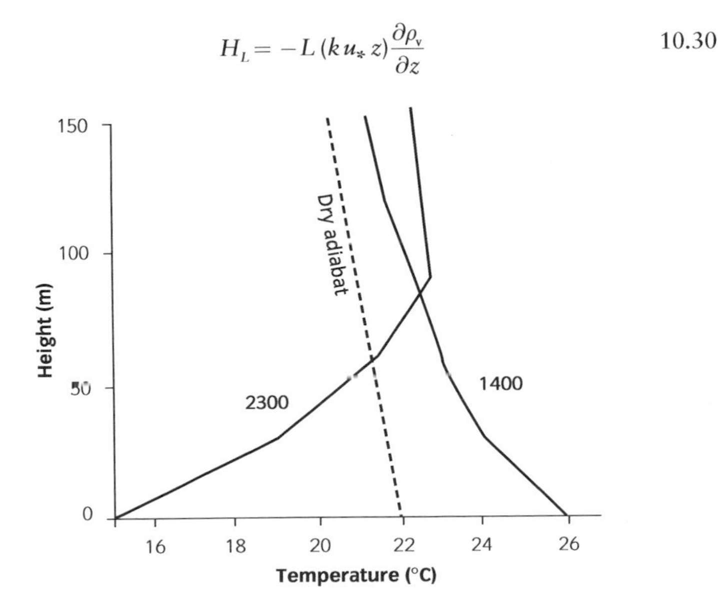

Figure 10.21 Daytime and night-time temperature profiles near the surface. The large diurnal range at the surface is typical of continental mid-latitudes in summer (Paisley, Ontario, 19 July 1962). On this occasion it is largely confined to the bottom 50 m.

Bowen ratio

Note that to the extent that the eddy diffusion coefficients are identical in Eqns 10.26 and 10.30) (though in principle buoyancy must enforce some differences ignored in this simple treatment), they cancel in the expression for the Bowen ratio

$$\beta = \frac{H}{H_L} = \frac{\rho C_p}{L}\frac{\partial \theta}{\partial z}\Big/\frac{\partial \rho_v}{\partial z} \qquad 10.31$$

with the result that this very important climatological parameter can be estimated simply by comparing vertical profiles of mean temperature and vapour density. Although this must be severely complicated by the prevalence of large diurnal variations in buoyancy effects, especially over land, it is a measure of the difficulty of finding better methods that this approach has been used extensively to partition sensible and latent heat fluxes in major studies of the atmospheric energy budget, such as those discussed in Chapter 8.

10.10 Above the surface boundary layer

10.10.1 Gradient level

Rising above the SBL, turbulence weakens in the diminishing wind shear, and eddies become larger and in some cases almost too structured to be considered turbulence proper (for example, the vortex-ring type of thermal). The average horizontal stress T arising from the Reynolds stresses reduces gradually from its value in the local SBL, and eventually becomes insignificant in comparison with horizontal pressure gradient forces. At such levels therefore the remaining Reynolds stresses have negligible effect on the balance between pressure gradient and Coriolis forces, and the quasi-geostrophic balance prevails (Section 7.10). The lowest level at which this happens is known as the *gradient level*, and though seldom sharply defined by observation, the gradient level is considered to be the upper limit of the atmospheric boundary layer (ABL) and the lower limit of the free atmosphere.

Rather subtle arguments suggest that the height of the gradient level above the local surface should be a modest fraction of the space scale u_*/f, where u_* is the friction velocity in the local SBL and f is the Coriolis parameter. Observations in the ABL bear this out reasonably well, and show that the gradient level is ~ 300 m above the surface on many occasions, being much lower when SBL turbulence is relatively weak (as on clear, still nights over land), and considerably higher when it is strong.

10.10.2 Ekman layer

Consider the forces acting on a layer of air somewhere between the SBL and the gradient level. They combine to give a net force opposing the motion of the layer because the turbulent drag forwards by the faster air just above the layer in question is less than the turbulent drag backwards by the slower motion just below the layer (since the shearing stress T diminishes upwards, as in Fig. 10.22). As discussed in Section 7.13.4 (antitriptic flow), this net drag pulls the air flow

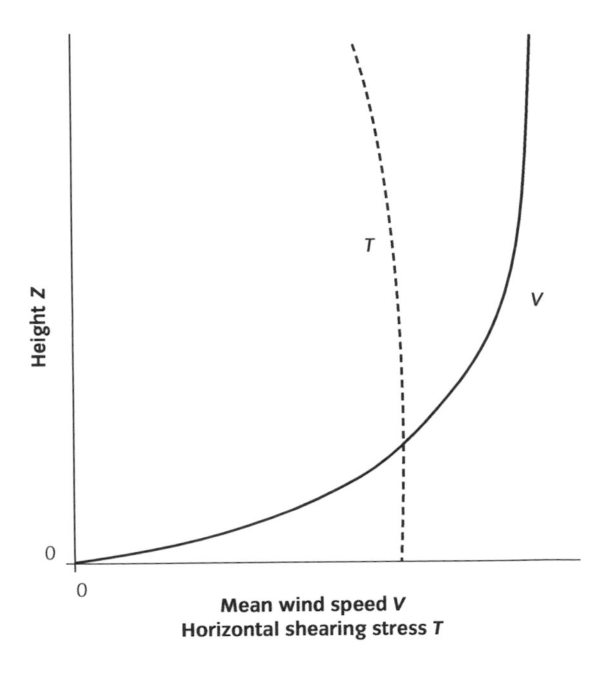

Figure 10.22 Idealized vertical profiles of mean wind speed and shearing stress in and just above the surface boundary layer. The SBL has a very strong wind shear and nearly uniform frictional stress T, which then begins to reduce.

away from its geostrophic alignment with the isobars so that it has a component towards low pressure and is sub-geostrophic in speed. This behaviour was first analysed by Ekman in 1905, and his theory outlines the elegant exponential spiral of wind vectors depicted in Fig. 10.23. The cross-isobar angle α increases with decreasing height above the surface, giving rise to winds which *back* with decreasing height in the northern hemisphere (wind direction rotating anticlockwise in plan view) and *veer* with decreasing height (rotating clockwise) in the southern hemisphere. The Ekman spiral is confirmed in outline by observations, but the maximum value of α hardly ever reaches the 45° predicted by his over-simple treatment of turbulent diffusion, and seldom exceeds 30°.

The maximum value of the cross-isobar angle α is reached at the top of the SBL and is maintained from there down to the surface because the relatively small differences between the large values of T at different heights within the SBL (previous section) are able to balance the pressure gradient and Coriolis forces without requiring significant turning of the wind vector with height. Of course the wind speed in the SBL falls away more and more rapidly with decreasing height (Fig. 10.23), in accordance with the log wind profile (previous section), but wind direction is uniform. Observation and theory indicate that at the 10 m level wind speeds are typically about 40% of the geostrophic value, the percentage increasing with increasing cross-isobar angle and decreasing surface roughness.

10.10.3 Convective or stratified ABL

Although turbulence becomes less frantic above the SBL, the larger eddies aloft seem to be well able to maintain the upward fluxes of sensible and latent heat which power the great cloudy weather systems of the free atmosphere. When the

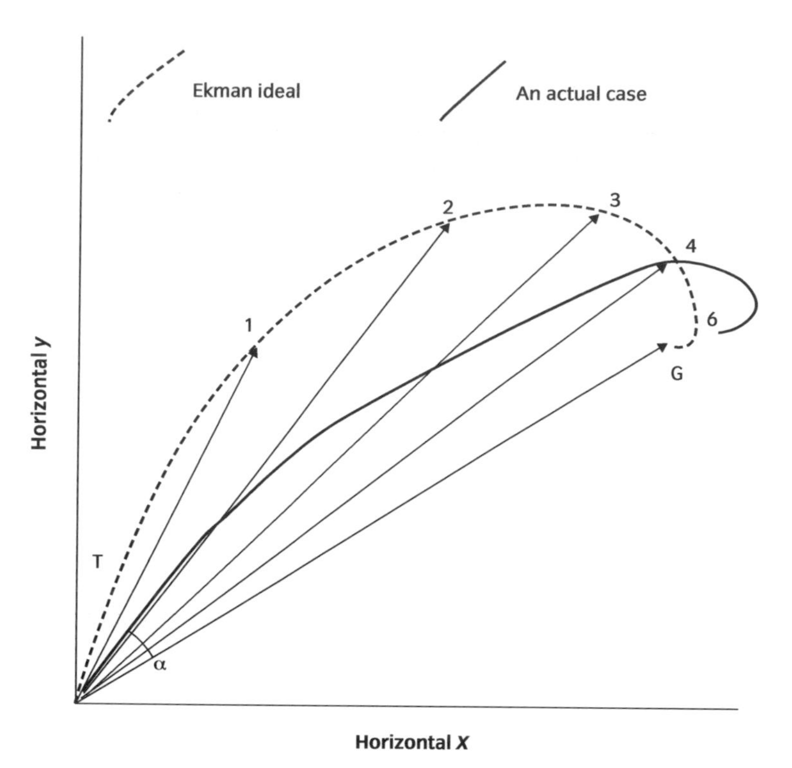

Figure 10.23 Ideal Ekman spiral (dashed) of mean horizontal wind vectors (numbered arrows) between the top T of the surface boundary layer, and the gradient level G. Numbers are heights in sixths of the level at which the wind is first parallel to the geostrophic wind (the gradient wind by definition). An actual profile (solid line) is plotted for comparison.

whole ABL is being stirred by buoyant convection, its upper parts are the site of rising, structured eddies often loosely entitled *thermals*. The precise nature of such thermals has not yet been clearly observed, partly because they are too large, mobile, and transient (~ 100 m, ~ 1 m s^{-1}, ~ 100 s) to be investigated by instruments on masts, and too small to be safely traversed by more than one instrumented aircraft at a time. Indirect evidence from birds and glider pilots who use their gentle updrafts to stay aloft with minimum effort can be interpreted either in terms of plumes fountaining from the strongly superadiabatic surface layer, or of discrete ring vortices (like giant invisible smoke rings) floating up after disconnecting from their surface origins.

However, the rest of the ABL above the SBL is often layered to an extent which prevents both the maintenance of the Ekman spiral and vertically uniform convective fluxes, especially over land where diurnal changes can be so large. For example, during the destruction of the nocturnal inversion discussed in Section 5.11, the deepening convecting layer is bounded above by the convectively very stable remains of the nocturnal inversion, which may reach down to within a few metres of the surface in the early morning. The resulting stratification of turbulence is bound to lead to sharp vertical gradients of vertical fluxes of heat and momentum which significantly shape profiles of temperature and wind speed. Layering is also produced by the differential flow associated with sea breezes and slope winds, both of which are important examples of processes which can strongly influence medium- and small-scale conditions in the SBL and the rest of the ABL over land.

10.11 Air flow over uneven surfaces

A number of different processes give rise to patterns of atmospheric behaviour near the surface on horizontal scales ranging from ~ 10 m to as much as 100 km. Some are mobile patterns associated with shower clouds and systems of shower clouds (Section 11.7), and some are locked onto terrain features for their lifetimes which may range from hours to days. Nearly all locked patterns occur over or near land rather than water surfaces, although land-locked water often has its own distinctive effects. Consider a few examples of land-based effects.

10.11.1 Slope winds

When a Sun-facing slope is warmed, a proportion of the buoyant forces generated can cooperate on the scale of the sloping surface to produce an overall up-slope (*anabatic*) wind. This may arise in conditions which are otherwise calm, or it may act to modify a larger-scale flow. Anabatic winds are often quite gentle (~ 2 m s^{-1}) because the overall buoyant force is largely offset by the turbulent friction generated by the winds, and because convection from the sloping surface penetrates the anabatic layer. However, in extreme terrain much stronger winds can be generated.

By the same mechanism in reverse, nocturnal cooling on a sloping surface can generate down-slope (*katabatic*) winds (Fig. 10.24) which can be rather stronger and shallower than anabatic winds because of weaker turbulent friction with the overlying air and the underlying surface. Violent katabatic winds can sweep down snow-covered coastal slopes, sometimes triggered by a shift in the larger-scale air flow, and threaten small boats coasting round Greenland and Iceland, for example; and the climatology of the coastal fringe of Antarctica is dominated by prolonged katabatic gales which sweep off the central plateau in all but the summer season. The diaries of pioneer explorers of Antarctica are full of descriptions of the resulting blizzards, one of which fatally trapped Scott and his South Polar party only 10 miles from the relative safety of a large supply dump *[45]*.

The gentler katabatic flows of small-scale hill country can produce ponding of cool air in hollows producing overnight quite sharp temperature contrasts between lower slopes collecting katabatic subsidence, and upper slopes in the warm air between the hill crests. Relative humidities are highest in the coldest air, even if there is no stream or lake to contribute more by evaporation, and fog will therefore form first in the hollows and valleys. If there is a lake in the hollow, the ponding cold air may well become considerably colder than the water surface (because of the latter's much larger heat capacity), giving rise to *steam fog* as the warm surface air convects into the cooler air ponding on its surface and is chilled to patchy saturation by the resulting mixing. Katabatic flows onto the sea from glaciers and ice sheets in high latitudes have a similar effect, but the greater temperature contrast makes the resulting *arctic sea smoke* (Fig. 10.25) deeper and thicker than most steam fog.

In more extreme terrain and in the presence of very strong nocturnal or even seasonal cooling, the ponding of cold air can lead to *frost hollows* whose minimum temperatures may be tens of degrees below those of the surrounding local area. On a larger scale, some of the lowest screen temperatures on Earth

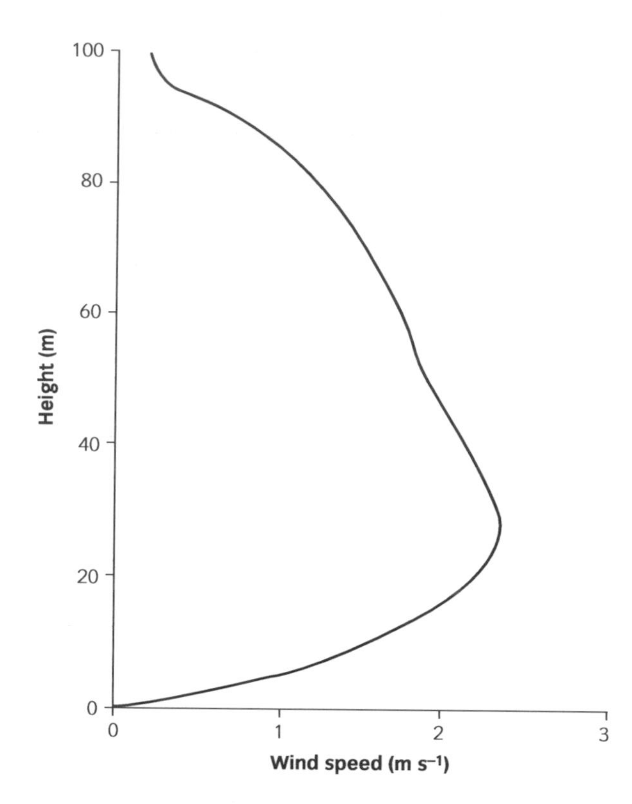

Figure 10.24 Vertical profile of wind speed in gentle katabatic flow down a 40° slope on the Nordkette, near Innsbruck. Notice as usual the strong wind shear near the surface.

Figure 10.25 Arctic sea smoke over a lead of open water in a frozen lake. The water beyond the foreground ice is probably just above freezing but is very much warmer than the overlying air. The vigorous small scale convection maintained by this instability is made visible by the cloud forming as the rising moist air is chilled to saturation by mixing with the cold air.

have been recorded in winter in the Siberian town of Verkhoyansk (−51.9 °C monthly mean in December 1904), which lies in a valley just inside the Arctic Circle surrounded by a horseshoe of mountains rising extensively above the 1000 m level.

10.11.2 Sea breezes

When coastal land is warmed by the morning Sun, it can quickly become considerably warmer than the adjacent sea, especially if there is no strong overall wind to minimize the contrast by rapid advection. In addition to the normal convection over land, there is a tendency in such conditions for the air there to rise relative to the adjacent cooler and denser air over the sea, in a weak but persistent large-scale buoyancy effect. The uplift may occur by an enhancement of the convective updraughts compared with downdraughts over the coastal margins, encouraging development of cumulus and even cumulonimbus there. The rising air is replaced by the cool, moist air which flows inland over the coast as a *sea breeze*. The flow is seldom more than a few hundred metres deep, above which there is often a more gentle and diffuse seaward flow. The boundary between the sea and land air as the former reaches inland can be very sharp, with marked contrasts in temperature, humidity, and haziness being ascribed to the presence of a *sea breeze front* (Fig. 10.26). Sometimes there is a definite line of enhanced convection along the front which is exploited by glider pilots, and lifts insects and the birds which feed on them.

In middle latitudes the most favourable conditions for well-marked sea breezes occur in summer when there is a light synoptic-scale air flow along the coast with the sea on the left hand (N hemisphere). The sea-breeze front may move tens of kilometres inland in such conditions before the waning of the Sun in the late afternoon begins to remove the driving agency of unequal heating. Even so, a

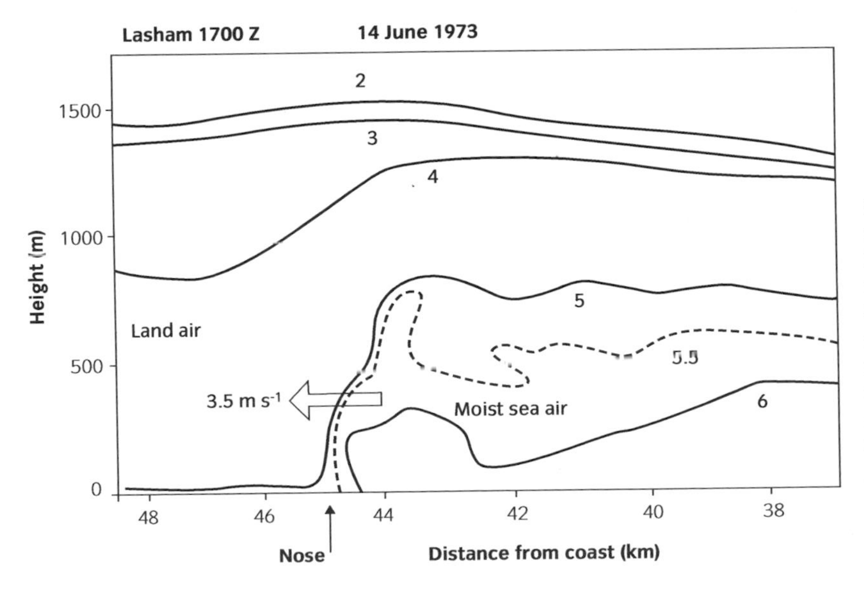

Figure 10.26 The nose of a sharp-edged, fast-moving sea breeze front 45 km inland from the south coast of England, with isopleths of specific humidity in g kg^{-1}. Birds and glider pilots use the uplift generated above the nose of dense moist sea air. The dry air aloft is subsiding gently in an anticyclone. Redrawn from *[73]*—a study by the late John Simpson who was a keen glider pilot in this area in his younger days.

well-developed sea-breeze front may continue rolling inland under the influence of its established density contrast until late evening. When a sea breeze persists for more than a couple of hours there is usually a rotation of the breeze vector in the direction to be expected from the operation of the Coriolis effect, which produces a veering in the N hemisphere. Even though well-marked sea-breeze fronts may occur on only a few days per year in any particular coastal region, the much more frequent sea-breeze tendency may have a very considerable effect on local coastal climates, lowering temperatures and reducing cloud and showery rainfall along the coastal margins compared with more than a few kilometres inland. Indeed the resulting enhanced freshness and brightness of coastal climates has contributed greatly to the rise of the seaside resort in Britain and similar countries. At night the sea-breeze effect tends to reverse, with the production of cooler, denser air over land leading to a seaward flow (a *land breeze*) which is known and used by coastal sailing craft.

10.11.3 Hill and mountain waves

When an air flow impinges on a hill or mountain, air is diverted both horizontally and vertically by the obstruction. In the typical slight convective stability of the lower troposphere, the forcible elevation of air passing over a hill, and of higher layers, can set up vertical oscillations of the air about its undisturbed level, which continue downwind of the obstruction, often for substantial distances. The consequences for the free atmosphere are mentioned again in Section 11.8, but at and near the surface the most important result is that wind speeds can be greatly enhanced or reduced at several zones on and downwind of a hill. A typical situation is depicted in Fig. 10.27, in which relatively high winds are induced by congestion of air flow near the crest of the hill (H on Fig. 10.27), and again several kilometres downwind (H), with lower wind speeds where air flow is reduced by *stagnation* (L_1) or *separation* (L_2). In very windy weather the enhanced winds can become damaging: in a famous W'ly gale over N England in 1962, enormous damage was done to forests and to the city of Sheffield, in which over two thirds of all buildings were officially registered as having been damaged. All this damage occurred in regions in the lee of the Pennines where subsequent detailed analysis of the air flow in the gale indicated that wind speeds were enhanced by being at the base of a lee wave *[46]*, like the H downwind of L_2 in Fig. 10.27.

In some cases the influence of topography on air flow is just as might be expected: a valley which is nearly parallel to the prevailing wind has a channelling effect which raises wind speeds above values found in otherwise identical valleys lying across the prevailing wind, though particular parts of a crosswind valley may be especially windy because of the presence of lee waves.

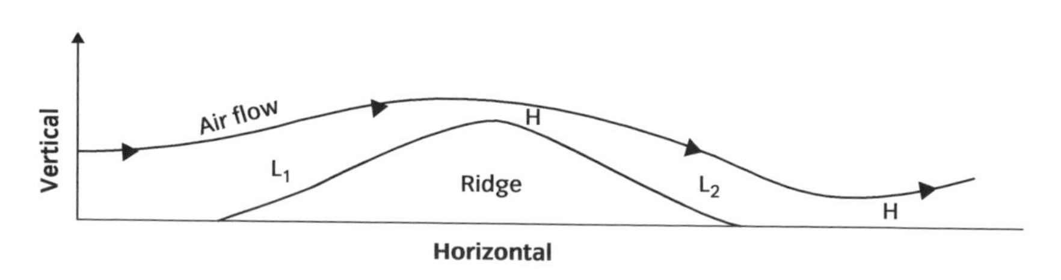

Figure 10.27 Strong airflow over a ridge which is extensive across the flow. Highest and lowest wind speeds occur in positions H and L respectively. Precise positions vary with ridge shape and profiles of wind speed and temperature to considerable heights.

10.11.4 Cloud and precipitation

Consider the flow of moist air over a hill or mountain. The enforced rise will be rapid (in the simplest view updraft speed is the product of surface slope and horizontal wind speed) and therefore nearly adiabatic. Unsaturated air therefore tends to rise, preserving its lifting condensation level and producing cloud at that level if the total rise is large enough.

In the particular case of a layer which is well stirred by mechanical or thermal convection from the surface up to cloud base and beyond, the level of cloud base over the surrounding low ground should be maintained over the hill, though there may well be more and thicker cloud cover over the hill in response to larger and stronger areas of uplift there (Figs. 10.28a and 6.3). This is often observed in hilly country, where the nearly uniform level of nearby and more distant cloud bases is particularly obvious when the observer has climbed close to cloud-base level, and is a nice confirmation that the combination of turbulent mixing and adiabatic eddy motion produces the same effect as wholesale adiabatic elevation.

In the presence of prolonged rainfall, however, it is normal to see low cloud blanketing the upwind slopes of hills at levels far below the base of the nimbostratus producing the general rain (Fig. 10.28b). It seems that air in the lower parts of what is the sub-cloud layer over the low ground is significantly moistened by evaporation from the wet land and the falling precipitation, and has in consequence a lower lifting condensation level which, however, does not become apparent until the air is forced to rise over the hill, though it may contribute to scraps of *scud* over the lower ground—Fig. 11.5.

At higher levels, moist air may be raised to condensation giving humps of cloud outlining the distorted air flow, and in the middle and upper troposphere

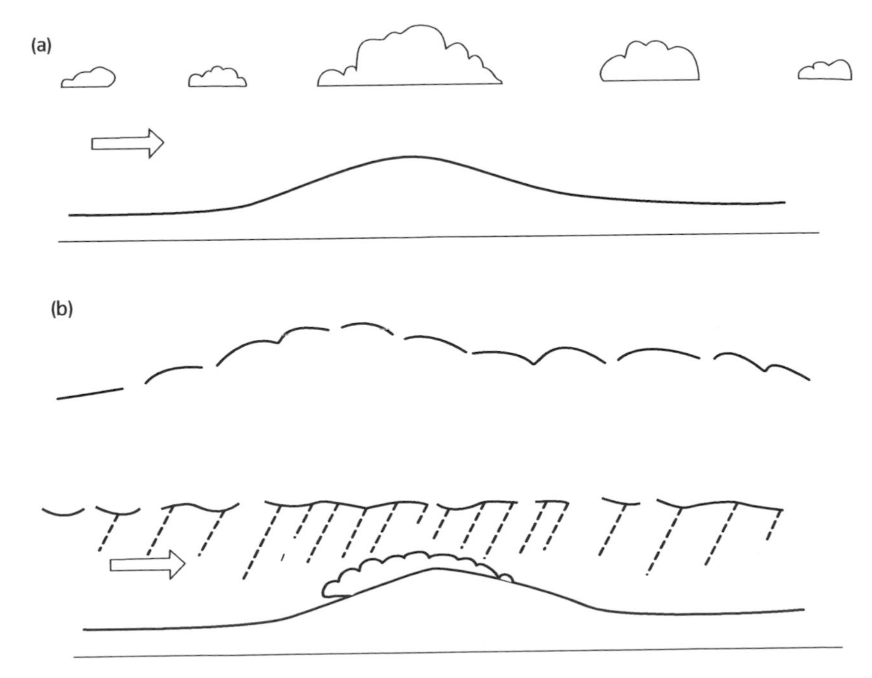

Figure 10.28 Cloud enhancement over hills. (a) Cumulus enhanced by buoyant or forced updraught, but cloud base level is uniform throughout the well-mixed subcloud layer, as in Fig 6.3. (b) Thicker nimbostratus over the hill, with a collar of low cloud (hill fog) where dampened surface air is forced to rise. Rain falling through the hill fog can grow rapidly by collision and coalescence.

these may be particularly smooth and lens-like (*lenticular*, Section 11.8). If the moist layer is already cloudy, the cloud layer will be deepened by the enforced rise (the low cap in Fig. 10.28b); and if there is a deep layer of nimbostratus, then it is deepened and its rate of precipitation enhanced over and slightly upwind of the hill in ways examined in Section 11.8. There is often a corresponding reduction in cloud on the downwind side of the hill, and a reduction in precipitation which may produce a climatological *rain shadow* if the orography is extensive enough and the paths of prevailing weather systems well enough aligned. A pronounced zone of sinking flow on the steep lee of a hill can suppress low cloud which otherwise blankets the sky.

Cloud patterns

Note that the three types of patterns mentioned above are patterns of flow. Clearly the patterns of radiant input discussed in Section 10.2 can have effects on small-scale climate which are at least as important. Flow and radiation can interact when the undulations of air in the low troposphere produce patterns of cloud which persist for hours or longer, shrouding some areas under cloud in the crests of waves, over or downwind of hills, while leaving other areas in bright sun beaming through holes in the cloud cover marking the unsaturated troughs of the undulating air flow. The presence of cloud holes is most marked in the case of otherwise complete layers of low cloud such as stratocumulus or stratus. In idle boyhood moments the author noticed that his part of the city of Belfast was often bathed in sun amidst surrounding gloom, as low cloud fell slowly down the steep lee slopes of hills to the west, dissolving and not reforming until the air permanently regained its lifting condensation level again several kilometres downwind. Such local effects are commonplace in non-uniform terrain, and their influence on local deviations from overall weather patterns is often considerable, as necessarily observant locals such as farmers and coastal fishermen are usually aware.

10.11.5 Föhn and chinook

The best known of all such local effects are the *föhn* winds of the European Alps, and the corresponding *chinook* winds on the eastern flanks of the Rocky Mountains of North America. They appear suddenly as unusually warm, dry, and strong winds blowing down the lee slopes of the mountain massifs and continuing for as much as 100 km to leeward. Once established they can continue for days, until a change in the large-scale pattern of weather and flow cuts them off. In winter the sudden warming can melt snow so rapidly that there is local flooding. In summer the desiccating winds can dry the surface and vegetation so much that there is serious risk of bush or forest fires. In the Italian Lake District in summer, the troublesome haze of the Po valley is swept away in a trice, replaced by unusually clear air, which persists throughout the föhn. Temperature rises range from a few °C to over 20 ° in some famous chinooks. Relative humidities usually fall below 50% and have gone below 10% on occasion. Winds often reach gale force, and can be magnified by local topography to damaging strengths.

Föhn and chinook events are conventionally and very commonly explained in terms of forced ascent and descent of air flowing over a mountain range which is reasonably high, and extensive both along and across the impinging air flow—an explanation first advanced by Hann in the infancy of meteorology (1866).

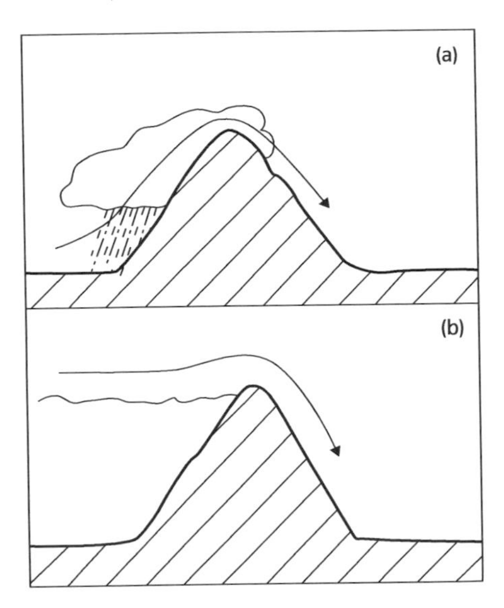

Figure 10.29 Contrasting air flows associated with föhn conditions in the lee of mountains. In (a) there is saturated ascent upwind, whereas in (b) air slides over air dammed upwind of the mountain. In each case the descending air tends to warm dry-adiabatically and become considerably warmer than at the corresponding level upwind of the mountain.

The dryness and warmth of the descending flow is explained by invoking cloud and precipitation on the upwind side, and dry descent on the downwind side. In the simplest picture (Fig. 10.29a) this would impose saturated adiabatic ascent and dry adiabatic descent on the air stream crossing the mountains, which could produce a resultant temperature rise of about 4 °C for every kilometre of intervening mountain height (6 °C fall in saturated ascent followed by 10 °C rise in dry descent). Detailed case studies confirm that this certainly happens on some occasions. Indeed the famous clarity of the N'ly föhn in the Italian Lake District shows that much industrial and natural haze has been washed out of the air during cloudy ascent of the N slopes of the Alps. However, there are other occasions when a föhn or chinook occurs when there is no precipitation on the windward side of the mountains. It seems that on such occasions the impinging air flow is largely dammed below the top of the mountain barrier, leaving the air above to slide over the top and descend on the lee side (Fig. 10.28b). The temperature rise on the lee side is then a measure of the stability (the increase in potential temperature with height) of the air to windward. Neither of these models accounts specifically for the considerable wind speeds which are common in föhn and chinook, and for these we must look to the dynamics of lee flow as mentioned above and again in Section 11.8. Active research is continuing on these interesting and important examples of the interaction of topography and meteorology *[21]*.

10.12 Surface microclimate

10.12.1 Vegetation

Conditions near the ground surface are strongly influenced by the presence of vegetation.

• Short grass establishes the value of the aerodynamic roughness length z_0 and therefore scales the log wind profile in the overlying surface boundary layer. The presence of reasonably well-watered grass also ensures that the latent heat flux from the surface is relatively large (i.e. that the Bowen ratio is relatively small) and that the daytime temperature rise is restrained. The binding of the soil by grass roots prevents the production of dust storms, which are such a feature of arid lands and can temporarily but drastically transform conditions by decimating solar input. It seems likely that before the colonization of Earth's land areas by plants (a relatively recent event on the timescale of Earth history—Section 3.9), dust storms were a major climatic feature, possibly even to the extent that they are still on the planet Mars, where hemisphere dust storms shroud the surface for weeks on end.

• Long grass and dense shrubbery interfere with fluxes of solar and terrestrial radiation to such an extent that the radiatively effective surface is raised above the ground surface and spread over a significant height range, the spread being greater for more open vegetation. This means, for example, that it is the upper parts of the vegetation which are warmed during the day and cooled at night, and that these gains and losses of heat are only subsequently communicated downwards by turbulence, radiation, and even by the falling of rainwater from wetted leaves. The upward communication to the overlying atmosphere is much the same as for a surface covered in short vegetation. The effect on daytime and night-time temperature profiles in a stand of vegetation is outlined in Fig. 10.30 for the case of a cereal crop, but effects in other types of vegetation are similar in principle. The consequence of the elevation of the radiatively active layer above the ground surface is to make the microclimate of the layer of the region between the active layer and the ground much less extreme, elevating nocturnal temperature minima and depressing daytime maxima to produce a much smaller diurnal variation there. This moderate climate then encourages colonization by plants and animals which could not tolerate the more extreme conditions prevailing before vegetation.

• The most dramatic effects of this type occur in forests, where the radiatively active layer may lie 10 m above the ground in temperate forests and several times this height in the great tropical rain forests which lie mainly in equatorial regions and are watered by precipitation from the intertropical convergence zone. The microclimate of the surface region beneath the dense, lofty canopies of tropical rain forests is familiar from the many films of wildlife there: the humid gloom, with its near absence of wind and diurnal temperature variation, contrasts sharply with conditions at and above the canopy, and is inhabited

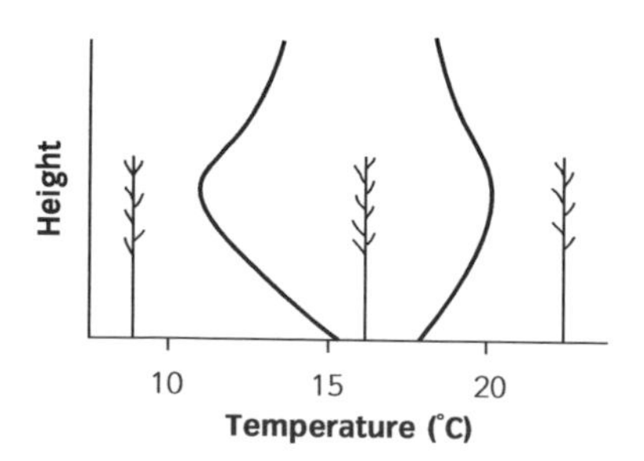

Figure 10.30 Idealized daytime and night-time temperature profiles in a barley crop. The crop is very much denser than shown, so that sunlight is strongly shaded by day, and nocturnal cooling is concentrated near the barley tops.

by a local biosphere which is irreversibly eliminated when the trees are felled wholesale. Even the ground surface is then unable to cope with the change of conditions, being alternately baked by the Sun and scoured by the deluges of rain from great equatorial cumulonimbus. Bearing in mind the considerable role played by this lush vegetation in the oxygen and carbon cycles of the Earth's atmosphere and surface (Sections 3.3 and 3.5), the ongoing wholesale destruction of large areas of tropical rain forest has become an icon of negative human impact on the planet.

10.12.2 Fog and smog

Fog forms when air close to a surface becomes slightly supersaturated and produces a layer of cloud in contact with the surface. The steam fog already mentioned is the only type of fog which is formed over a relatively warm surface; all other true fogs are formed over relatively cool surfaces. (*Hill fog* is not a true fog, being simply cloud which is low enough to envelope the upper parts of hills or mountains.)

Radiation fog forms in shallow layers, usually only a few metres deep, when radiative cooling of a ground surface takes the overlying air below its dew-point. It often appears shortly after dusk on a calm evening, under clear or nearly clear skies, over land which is fairly moist. In these conditions the air in the first few metres above the surface is almost still, and the fog when it forms outlines what is essentially a huge laminar boundary layer (Fig. 10.31). Local inhomogeneities may stratify the humidity and give rise to horizontal leaves or fingers of denser fog which enhance the laminar appearance. Over uneven surfaces radiation fog tends to form in hollows, as they fill with dense, moist air flowing from neighbouring slopes. Once formed, the fog layers tend to persist and thicken by direct cooling of their top surfaces by net long-wave radiation, but unless they grow

Figure 10.31 Radiation fog about 2 m deep forming in a meadow about dusk. It is smooth because turbulence is quenched by strong convective stability in the fog layer.

to a substantial depth (10 m or more) they are easily removed by an increase of wind, or by solar heating next day. Fog formation shields the underlying surface somewhat from further radiative cooling, though very gentle convection spreads the top surface cooling downwards through thicker fog layers.

Advection fog occurs when relatively warm air is chilled to saturation by over running a sufficiently cool surface. If the air is close to saturation initially, then only a little further cooling is necessary to form a fog layer which may be tens of metres thick. Pulses of relatively warm, moist air often move over land in autumn, winters and spring in mid-latitudes, in weak, slow-moving synoptic-scale weather systems. In late autumn and winter in middle and high latitudes, the land is usually much cooler than the sea, so that the air flowing from sea to land is cooled quite sharply. Even if this is not sufficient, radiative cooling under clear night skies often completes the production of saturation and fog, so that advection fog over land is often produced jointly by advection and radiation.

Advection fog also forms over the sea. The fogs of the Grand Banks fishing grounds off Newfoundland (graphically described by Rudyard Kipling in the novel *Captains Courageous*) are formed by the advection of warm, moist air from the Gulf Stream over the cold, fish-laden waters of the Labrador Current. Fogs form in summer in the coastal waters round the British Isles when warm, moist air from nearby land is cooled as it flows over the cool sea. This sea fog or *haar* plagues the eastern coasts of Britain, particularly in fine weather, when light E'lies associated with an anticyclone maintain large sheets of fog over the North Sea, some of which drift onto the British coasts. The fog is dispersed by flowing over a few kilometres of sunlit land, but the coastal strip may remain shrouded and chilled for days while the rest of the country is basking in the sun.

If the overall wind speed in and above a deep fog layer begins to increase, the fog-top inversion may survive and rise while the fog layer is deepened by stirring. In these circumstances, the lowest part of the layer may become slightly unsaturated, producing a layer of low stratus cloud, which in the context is simply lifted fog, overlying a shallow sub-cloud layer. Something like this can occur in the poleward extremity of the warm sector of an extratropical cyclone (Section 12.3).

Fog layers produced wholly or in part by advection are usually substantially thicker than those produced purely by radiation, with the result that the top surface of a substantial fog layer replaces the shrouded underlying surface as the new radiatively effective surface. This does not much affect the output of terrestrial radiation, since thick fog behaves like a blackbody in much the same way as any terrestrial surface. However, it transforms the input of any available solar radiation, since fog behaves like any cloud in backscattering strongly and absorbing very weakly. On a clear night therefore the top of the fog layer loses heat rapidly by net emission of terrestrial radiation, and this cooling is spread throughout the fog layer by the gentle convection which it encourages, so that the sharply concentrated surface cooling typical of an exposed land surface is spread out vertically, and reduced by the addition of the heat capacity of the foggy air (a 20 m layer of air having much the same heat capacity as a centimetre layer of soil). In late autumn and winter, the weak Sun when it rises may be incapable of removing the fog layer, the direct warming being minimal and the diffuse warming of the underlying surface being very small if the fog layer is deep enough. In fact all the daylight may do is reveal the gently convecting fog layer to an overflying observer (Fig. 10.32). Once a substantial fog layer is established in these conditions it tends to maintain itself, until the synoptic situation allows

Figure 10.32 A weakly turbulent fog blanket enveloping the Golden Gate bridge in San Francisco. The central span is about 70 m above water. The fog is formed as warm moist W'lies are chilled to saturation as they cross the cold Californian Current flowing equatorwards just off the coast, bringing spectacular fogs to the city and the local coastal strip.

increasing wind to disperse the fog by shear-driven turbulent mixing with the overlying air.

Smog In industrial or urban regions with significant numbers of short chimneys, the smoke and gases emptied into the fog layer (which is usually bounded above by a weak inversion maintained by radiative cooling of the fog top) can rapidly produce unpleasant and even dangerous concentrations of pollutants. London has been prone to such smoky fogs (*smogs*) since the burning of coal in domestic open fires became widespread there in the 16th century. The old local name for them is *pea soupers*, which nicely describes the greeny-brown obscurity in which they enveloped buildings and people. The dirtying and corroding effects on buildings have long been obvious (St Paul's Cathedral was already blackened before its completion in the early eighteenth century), but the effects on people were largely ignored until it became clear from hospital records that a spectacular but not untypical smog which lasted for five days in December 1952 had killed ten thousand people made vulnerable by age or chest infection. The ensuing outcry led to the introduction of controls on the production of smoke by inefficient combustion of fuels which have transformed the appearance of British cities in subsequent decades. The numbers of foggy days in cities have also dropped, showing that fog formation was being positively encouraged by the availability of condensation nuclei in smoky air.

When large urban areas began to appear in lower latitudes, the combination of air pollution and strong sunlight began to produce photochemical smogs in Los Angeles (the archetype), Athens, Beijing, and many other locations. These are more difficult to deal with than the London type, because they do not largely disappear when smokes are eliminated by more efficient combustion, and because many are in areas affected by subtropical anticylcones (Section 13.2).

Checklist of key ideas

You should now be familiar with the following ideas.

1. Atmospheric boundary layers: laminar boundary layer (LBL), surface boundary layer (SBL), atmospheric boundary layer (ABL), Ekman layer, free atmosphere.

2. Effects of surface geometry and structure on radiative inputs and outputs, albedos of common materials.

3. Direct solar surface irradiance and its variation with solar elevation and air clarity, daily insolation and its variation with latitude and season, surface materials as black or greybodies in terrestrial wavelengths.

4. Solar and terrestrial radiative penetration of land and sea surfaces and their very different thermal responses, partitioning of net input between surface and air, desert and ocean surface climates as opposite extremes.

5. LBL and factors determining its thickness, action as a surface film limiting flows of sensible heat, vapour (latent heat) to and from the atmosphere.

6. Momentum flow as surface drag, from turbulent air direct to surface via surface roughness bypassing the LBL.

7. Turbulence as chaotic, three-dimensional, hierarchical activity of a spectrum of eddies, encouraged by wind shear and convective instability, discouraged by convective stability and viscosity, efficiently transporting momentum, heat, and vapour down their average gradients.

8. SBL as shallow highly turbulent layer with locally uniform downward momentum flux, log wind profile, surface roughness length, surface drag, and drag coefficient.

9. Diurnal cycle of turbulent activity, eddy viscosity and other eddy transport, coefficients, Bowen ratio, Ekman layer as transition between SBL and the free atmosphere.

10. Anabatic and katabatic winds, sea breezes, hill waves and patterns of cloud and precipitation, föhn and chinook winds.

11. Microclimate below, in, and above vegetation, fog layer formation and maintenance, London type smog and its elimination.

Problems

Outline answers to these problems can be found on the **Online Resource Centre**. Answers to odd numbered problems can be found under Student Resources, answers to even numbered problems under Lecturer Resources.

Level 1

10.1. Why is the laminar boundary very significant for an elephant despite the gross discrepancy in their volumes?

10.2. If a typical incident solar ray makes three successive impacts on the interior of a stand of vegetation before escaping back to the atmosphere, and each element of vegetation surface has an albedo of 0.5, what is the gross albedo of the stand?

10.3. Why should there be less hoar frost round the bole of a leafless tree than in the middle of a field on the same occasion?

10.4. Which of the following usually contributes most to the very large heat capacities of lakes and seas: the large specific heat capacity of water; its turbulent mixing; its transparency to sunlight?

10.5. While illuminated by the sun, ground passes less sensible heat into the atmosphere when it (the ground) is moist than when it is dry. Suggest three reasons for this.

10.6. Suppose a complex computer model of the nocturnal boundary level at an open site in winter could forecast values for the Richardson number near the surface on the basis of conditions at the previous dusk. What values would you look for to give a warning of ground frost?

10.7. In a pure log wind profile, how do you expect (a) the wind speed and (b) the wind shear, to vary at equal multiples of height?

10.8. Because of the Ekman spiral, the wind direction at 10 m needs to be altered in a consistent sense to represent the direction of the gradient wind. What is this sense in the N hemisphere, and why is it fairly obvious when you think of the pressure field?

10.9. People living near hills often remark that in overcast conditions the sky is especially threatening in the direction of the hills. Why should this be so?

10.10. Why should fog often gather in hollows and valleys, leaving even very low hills protruding into the clear air?

Level 2

10.11. A beam of direct sunlight with flux density 1,000 W m^{-2} falls on several surfaces. Given that the angle of solar elevation is 35°, find the solar irradiance on (a) a surface tilted towards the Sun at an angle of 30° to the horizontal, (b) a surface tilted away from the Sun at the same angle, and (c) a horizontal surface.

10.12. Calculate the direct-beam irradiances for the surfaces and solar elevation of Problem 10.11 on an occasion when the sky was cloudless but rather hazy, zenith solar transmissivity 0.65, with 1,380 W m^{-2} and normal irradiance above the atmosphere.

10.13. Given that Problems 10.11 and 10.12 represent conditions at noon on a day in which the clarity of the atmosphere remained uniform throughout the full 9 hours of daylight, find the total insolation on a horizontal surface for the day.

10.14. Using a simple finite-difference form of the Richardson number, estimate the difference in wind speed across the layer between 2 and 3 m above the surface, which would be just sufficient to initiate turbulence despite a temperature inversion of 2 °C across the same layer.

10.15. In a very simple model of wind variations in the surface boundary layer, horizontal wind speeds V are 3 m s^{-1} when there is a downdraft (of 1 m s^{-1}), and they fall to 1 m s^{-1} when there is an updraft w (again of 1 m s^{-1}). The up- and downdrafts alternate regularly and account for all the time in the observation period. Sketch V and w on a common graph against time, and calculate their covariance and hence the friction velocity and average downward eddy flux of momentum (assuming air density to be 1.2 kg m^{-3}).

10.16. The average wind speeds at 8, 4, and 2 m height on a certain occasion are 7, 6, and 5 m s^{-1}. Find the implied roughness length of the underlying surface, the wind speed at 10 m, and the friction velocity.

10.17. In a certain föhn event, the air descending the lee slopes is observed to be up to 5 °C warmer than air at the same level on the upwind slopes. Find the loss of vapour content (specific humidity) implied by an adiabatic model of the ascent and descent. (Use Eqn 5.22 with reasonable values.)

Level 3

10.18. A clear calm night has established a marked temperature inversion near the surface. Now the wind begins to increase in response to an approaching synoptic scale weather system. Describe what you would observe with simultaneous measurements by a fast response thermometer and a vertical propeller anemometer before and after turbulence breaks out. What will happen to the average air temperature at the same time?

10.19. Estimate the solar, terrestrial, and net radiative balances for midsummer's day on the Arctic Circle, when the surface solar albedo is 0.2, the zenith transmissivity for solar radiation is 0.7, the surface temperature remains constant at 10 °C, and the surface emissivity is 0.95 in the far infrared. Use a simple approximation for the solar input curve.

10.20. Compare and contrast the climatic environment of a small crawling insect, an Irish wolfhound, and a man, each living in a mid-latitude continental prairie. Compare the experiences of the man and dog, given that they have the same body weight.

10.21. Find the correction factor to be applied to wind speeds measured at 2 m to bring them to the equivalent value at 10 m, assuming a log wind profile and an aerodynamic roughness of 1 cm over the surrounding area. Note that this is used to estimate synoptic standard wind speeds when there is no anemometer at 10 m. Consider the likely sources of error in such estimates.

10.22. The relationship between potential temperature lapse and sensible heat flux density H in Eqn (10.27) rather unhelpfully contains the friction velocity u_*. However, u_* can be replaced using the log wind profile. Do this and derive the useful relationship

$$H = \rho\, C_p\, C_D\, \Delta\theta\, \Delta V$$

where $C_D = [k/\ln(z_2/z_1)]^2$ is the drag coefficient C_D in Eqn 10.25.

Derive a similar expression for the latent heat flux, and use the two to show that the expression for the Bowen ratio is

$$\beta = \rho\,(C_p/L)\,(\Delta\theta//\Delta\rho_v)$$

which is the most widely used expression for the Bowen ratio.

10.23. Give an account of ways in which local meteorological conditions in hilly terrain can be influenced by the terrain.

11 Smaller-scale weather systems

11.1 Introduction

In the present chapter we concentrate on weather systems which are not individually large enough to show clearly on a synoptic weather map. The shower cloud (cumulonimbus) is a typical example of such *sub-synoptic scale* systems. Individual showers are revealed by the synoptic observation network when a shower occurs at a station at the time of observation or in the previous hour, but their structure is not seen, and showers occurring between observations or in the spaces or between stations are missed completely, though they may appear on satellite images.

The sub-synoptic range of scales is so wide that it is often subdivided into the *micro-scale*, *small scale*, and *meso-scale* (Fig. 1.5). Micro-scale refers typically to systems no larger than large individual eddies in the atmospheric boundary layer ($\sim$100 m) which affect surface conditions mainly through the turbulent variability described in Section 10, and will not be discussed further. Small scale usually includes systems no larger than a single large cumulonimbus, say 10 km in horizontal and vertical dimensions, though strong winds aloft can stretch a mature anvil several tens of kilometres from the main cloud. Meso-scale usually refers to systems, or patterns of systems, intermediate in scale between the small and synoptic scales, i.e. between ~10 and ~100 km in the horizontal. Synoptic meso and small scales often cooperate: for example synoptic scale showery air masses show meso-scale patterning of their small-scale shower clouds (Fig. 2.14).

The following sections cover the most important types of small and meso-scale tropospheric weather systems, the most widespread examples of which belong to the cumulus family—small cumulus (*cumulus humilis*), growing cumulus (*cumulus congestus*), and precipitating cumulus (*cumulonimbus*) (Box 2.3).

11.2 Small cumulus

The most numerous member of the cumulus family is the small cumulus, which over land typically begin to appear in mid-morning in previously clear skies, usually shortly after a convectively stable layer left by the previous night's cooling has been eliminated by solar surface heating (Section 5.3). Being typical of fine summer weather, they are often called *fair-weather cumulus,* but the title can be misleading on those occasions when they are the forerunners of cumulonimbus and showery afternoons.

11.2.1 Beginnings and growth

An individual cumulus is often ~100 m across when it first becomes visible, initially irregular but soon developing its characteristic shape (Figs. 11.1 and 11.2). The nearly flat base corresponds reasonably to the local lifting condensation level (LCL—Section 6.4) for buoyant *thermals* rising from the surface layer, according to the simplest model of the well-mixed sub-cloud layer, though it is usually a little higher than the LCL deduced from screen level measurements (Section 6.5). When seen from about their own altitude, cloud base levels are usually quite consistent across the field of view (Fig. 11.1), but closer inspection (Fig. 11.2) reveals a raggedness of cloud base on a scale of metres, and sometimes a slight but consistent upward tilt in cloud base level in the downwind direction. The raggedness reflects the known turbulent inhomogeneity of the air in and below the cloud, but the tilt suggests the action of an unknown mechanism on the scale of the cloud itself.

The sides and tops of individual cumulus are characteristically domed and knobbly. The dome shape attracted the generic name '*cumulus*' (Latin for hill) when cloud types were being named in the early nineteenth century, and is clearly consistent with an extensive, rounded parcel of buoyant moist air rising through and just above its LCL (Fig. 11.1). The knobbly appearance, which can look remarkably like the heart of a ripe cauliflower, shows the presence of turbulence on a range of scales smaller than the dome.

If an individual cumulus is observed for a minute or more, it is seen to be in a state of slow commotion—the dome swelling visibly, and knobs emerging from its

Figure 11.1 A field of fair weather cumulus over a coastal plain in NW England, with some cirrus in the high troposphere.

Figure 11.2 A small cumulus seen at close range, showing its domed upper parts and its ragged base tilted slightly upwards in the direction of drift.

sides and top, only to dissolve or be overtaken by another from within. Although less obvious, the rate of rise of the cloud top is somewhat less than the maximum updraft speed inside the cloud, because the rising mass of air is continually turning itself inside out like a rising smoke ring (a *ring vortex*—Fig. 11.3) with a smoke-filled core. Updrafts seldom exceed 2 m s^{-1}, because at these values their weak buoyancy is effectively balanced by drag. Drag arises partly as rising air forces its way through its surroundings (like a very large fluffy balloon), and partly from the

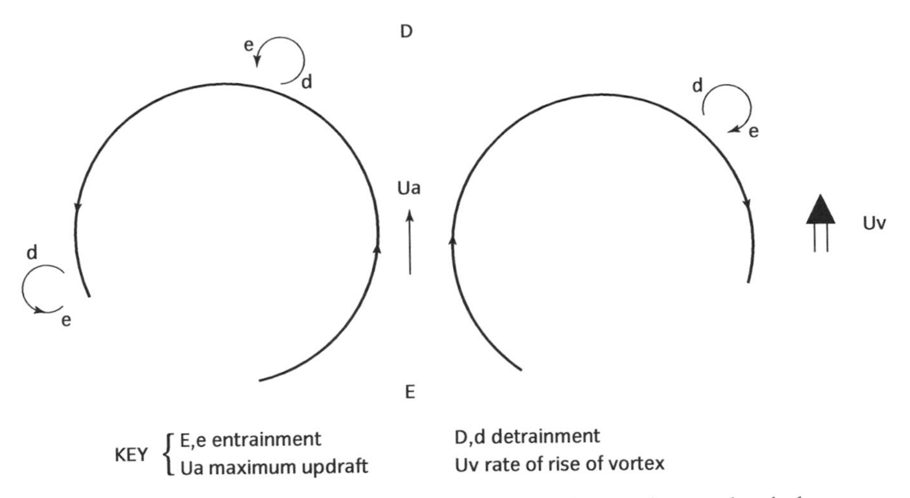

Figure 11.3 A cross-section through a buoyant ring vortex, showing the central updraft *Ua*, the smaller rate of rise of the vortex *Uv*, regions of vortex-scale entrainment *E* and detrainment *D*, and microscale entrainment *e* and detrainment *d*.

dynamic effect of *entrainment* of previously static ambient air into the growing dome (Section 7.14). Entrainment and its reverse (*detrainment*—Fig. 11.3), are basic features of the whole cumulus family, from cumulus humilis to cumulonimbus, with entrainment dominating during growth, and detrainment dominating during decay.

The visible sharpness of the edge of the maturing cumulus shows that cloud droplets are forming and disappearing so quickly that any individual eddy near the cloud side or top is either densely clouded or cloudless, depending on whether or not it is saturated. Such rapid droplet formation or evaporation is consistent with the known small size (~10 μm) and behaviour of cloud droplets (Section 6.9).

11.2.2 Maturity and decay

An individual cumulus humilis usually continues growing horizontally and vertically for five minutes or so, drifting in the ambient wind. By then it may have broadened to several hundred metres, reaching above the unchanging cloud base by rather less than this, and drifting several kilometres downwind from its point of origin. After another few minutes it will often lose vitality and sharpness of outline, and begin a process of dissolution which is completed in a few more minutes, leaving no visible remains. During dissolution, the cloud usually becomes less white and less sharp, as decreasing desaturation selectively evaporates first the large numbers of small droplets, leaving the cloud dominated by smaller numbers of larger droplets. As in a thinning fog, the scattering of sunlight is reduced and diffused, as the cloud heads for extinction by *terminal detrainment*.

Two factors can affect overall cloud shape at maturity.

(i) If the effective LCL lies only a few tens of metres below the base of a convectively stable layer (such as a temperature inversion), the cumulus will be vertically stunted by loss of buoyancy as cloudy air rises into the stable layer, and will preserve this flattened 'pancake' shape for the rest of its brief life. This is typical of cumulus forming beneath the base of the subsidence inversion of an anticyclone (Section 11.5.1). The persistent formation of flattened cumulus in anticyclones often maintains large areas of *stratocumulus*, in which individual clouds are so closely packed that they form a sheet patterned on a scale of ~100 m by narrow gaps between adjacent pancakes, effectively maintaining total cloud cover (Fig. 11.4). The pattern is probably maintained by a cellular organization of convection (Section 11.7). In industrial regions the trapping of smoke and other pollution under such a stable layer can maintain unpleasantly gloomy and noxious conditions at the surface.

(ii) Cumulus can be shaped by a vertical shear of horizontal wind in the cloud layer. Though never as marked as in the surface boundary layer, this shear can be strong enough to make the cumulus lean quite noticeably downwind. Though the direction of wind shear at cloud level need not parallel the wind direction there (which need not parallel either surface wind), on many occasions these shears are nearly parallel, so that cumulus lean in the same direction as the surface wind and shear. In very windy conditions the shear may be strong enough to disrupt the orderly development of cumulus, producing a ragged, tumbling cloud, called *fractocumulus* or *scud* when accompanied by raining cloud at higher levels, as is often the case in vigorous fronts (Fig. 11.5 and Section 10.11)

In the Trade Wind zones there are often large populations of cumulus, many of them taller than the small fair weather cumulus of middle latitudes, because of the more encouraging vertical profiles of temperature and humidity. These tend to lean backwards (i.e. against the surface winds—Fig. 8.22) in a way which looks 'wrong' to mid-latitude observers. This happens because Trade Winds extend only a kilometre or two above the ocean surface: in the N hemisphere,

Figure 11.4 A substantial deck of stratocumulus, with strong sunbeams coming down through a hole, suggesting that there is little cloud above the stratocumulus itself. The beams are visible because they are being scattered sideways in the dusty air.

Figure 11.5 Fractocumulus produced by tumbling low-level air moistened by rain falling from the cold front overhead as the picture was taken. The cloud is silhouetted against the bright clearance (at the back edge of the front) which is approaching from the W.

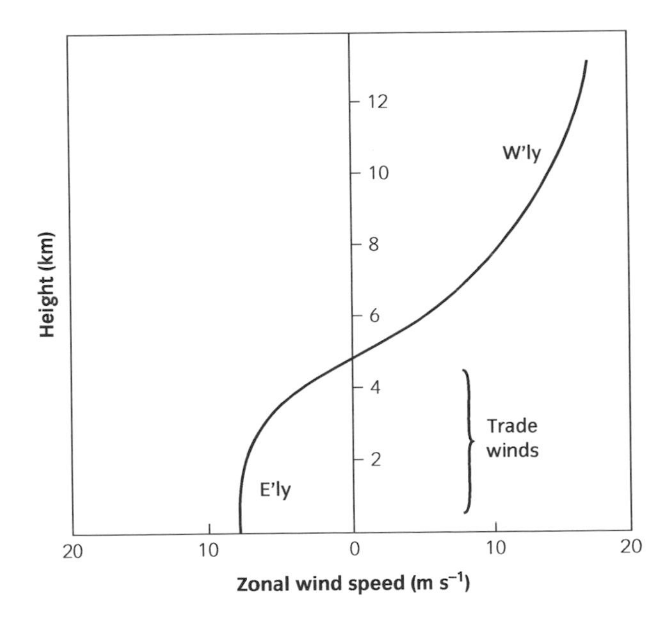

Figure 11.6 A vertical wind profile of zonal wind in and above winter Trade Winds. In summer the upper W'lies may be replaced by E'lies.

for example, the characteristic NE'ly flow fades rapidly with height, becoming westerly in and above the middle of the low troposphere (Fig. 11.6). Trade-Wind cumulus are therefore rising through a layer in which wind shear and surface winds are in nearly opposite directions, so that their upper parts are being sheared backwards relative to their bases, unlike their counterparts in higher latitudes which are sheared forwards.

11.3 Cumulus development

Unless there is a convectively stable layer aloft, some small cumulus avoid the dissolution described in the previous section, and carry on growing until they resemble mountains of cloud (Fig. 11.7). In this section we consider the development of such *cumulus congestus* (swelling cumulus) from small cumulus, and the factors which can encourage it.

Updrafts in small cumulus, and in the cloud-free thermals which feed them from beneath cloud base, are weak because buoyancy is weak and easily balanced by drag (Sections 7.12 and 11.2). Buoyancy is weak below cloud base because the temperature excess in thermals is usually less than 1 °C, because of their readiness to erupt from the heated surface layer rather than build up a 'head' of buoyancy before rising. Above cloud base, though the extra buoyancy produced by release of latent heat by condensation is partly offset by the collapse of low-density vapour into very dense cloud droplets (Box 11.1), it remains a potent driving force.

In the same way, the net loss of buoyancy associated with evaporation of cloud is very important, and is the primary way in which all clouds lose buoyancy. Evaporative cooling and loss of buoyancy occurs all over the sides and upper surfaces of a cumulus cloud throughout its life, and is especially destructive of small cumulus because of their large surface to volume ratio (which is inversely proportional to cloud dimension for any chosen cloud shape). Clouds which can reduce or delay the onset of such loss of buoyancy therefore have a considerable advantage over their fellows.

11.3.1 Formation and growth of cumulus congestus (Cg)

There are several conditions which can encourage the formation of small cumulus and their subsequent growth into larger clouds.

Figure 11.7 Cumulus congestus showing typical mountainous shape, with a broad base encouraging new thermals to rise through the protective cloud body to emerge at the summit as dense cauliflowers rising into the middle and upper troposphere.

BOX 11.1 Buoyancy

Atmospheric convection is driven by small density differences $\Delta\rho$ between convecting air parcels and their surroundings. According to Eqn 7.31 and Box 7.10 the net buoyant force F per unit parcel mass is related to the density deficit $\Delta\rho$ of a parcel compared with its surroundings

$$F = g \frac{\Delta\rho}{\rho}$$

Temperature excesses ΔT are the obvious source of such small density deficits, and the behaviour of air as a perfect gas ensures that $\Delta\rho/\rho \approx \Delta T/T$, where T is the absolute temperature of the surrounding air (Box 7.10). However, differences in vapour and cloud content are also significant sources of density difference, since vapour is less dense than dry air, and cloud water is a thousand times more dense.

Condensation

Consider a 1 kg parcel of moist air, and suppose that some of its vapour condenses, warming the parcel and condensing into cloud droplets which stay within the air parcel, since they sink so slowly through the air. Warming, and condensing from vapour, to cloud affects the air parcel density in three ways.

- If the process is adiabatic, the latent heat released warms the mixture of air, vapour, and cloud. The temperature rise ΔT is given by

  $$\Delta T = L\,m/(\text{parcel heat capacity})$$

  where m is the mass of vapour condensed, L is the specific latent heat of evaporation of water, and $L\,m$ is the latent heat released to the parcel. The parcel's heat capacity is actually the sum of the heat capacities of the dry air, vapour, and cloud water, but the proportions of vapour and cloud are so small that we can ignore them in comparison with the heat capacity of dry air, which is C_p per unit parcel mass. The parcel therefore warms by

  $$\Delta T = \frac{L\,m}{C_p} \qquad \text{B11.1a}$$

  Inserting values for L and C_p we find (NN 11.1) that parcel temperature rises by about 2.5 °C for each gram of cloud water condensed in a kilogram parcel—a very significant warming given that temperature excesses in convection below cloud base are only ~1 °C.
- The rise in parcel buoyancy by warming (by cloudy condensation) is offset by the decrease in the relatively low-density vapour component (molecular weight only 18/29 that of dry air—Box 4.1). As shown in the Appendix below and NN 11.1, the increase in density caused by the loss of vapour is equivalent to a cooling of 0.17 °C for each gram of water vapour condensed in a kilogram of air, and the retention of the gram as cloud water is equivalent to a further cooling of 0.27 °C, so that the total effect is equivalent to a cooling of 0.44 °C at 273 K, and slightly less at lower temperatures.
- Combining these three results, we see that the total effect of condensing 1 gram of vapour into cloud in a 1 kilogram air parcel is equivalent to a warming of about 2.1 °C—a very significant source of buoyancy for atmospheric convection.

Evaporation

If cloud is evaporated these effects are simply reversed, and the same analysis shows that evaporation of 1 g per kg of cloud water vapour produces a total buoyancy loss equivalent to a cooling of 2.1 °C. Since cloud evaporates when cloudy air mixes with cloudless air, the cooling effect is reduced in proportion to the mixing, but the loss of buoyancy can still be very significant.

Freezing or melting

The warming or chilling which follows melting of ice cloud or freezing of water cloud can be found from Eqn B11.1a by treating L as the latent heat of fusion (melting) of ice. Because the latent heat of fusion is so much smaller than the latent heat of vaporization, the temperature changes associated with the freezing or melting of 1 g of cloud are about 0.3 °C (NN 11.1). There is no offsetting correction to apply because the vapour content is unchanged and the slight difference in density between water and ice has quite negligible effects on overall parcel density.

Deposition or sublimation

Direct sublimation from ice to vapour involves an L which is the sum of the Ls for fusion and evaporation at the temperature involved. Chilling by sublimation could therefore be especially destructive of buoyancy if the very small differences in vapour density between cloud and environment at the low temperatures involved (below −25 °C) did not make the process so weak and slow. Corresponding warming by direct condensation from vapour to ice, as envisaged in the Bergeron–Findeisen process, is less inhibited and can be an effective source of buoyancy, as discussed in Section 6.11.3.

Appendix on buoyancy effects of vapour condensation and cloud retention

To relate air density to pressure p and absolute temperature T, rewrite Eqn B4.1 of Box 4.1 in terms of specific humidity q

$$p = \rho\,[q\,R_v + (1 - q)\,R_d]\,T = \rho\,[q\,(R_v - R_d) + R_d]\,T \qquad \text{B11.1b}$$

$$p = \rho\,R_d\,[q\,X + 1]\,T$$

where $X = (R_v - R_d)/R_d$ and T_v (= $[q\,X + 1]\,T$ is the *virtual temperature* of the moist air—the temperature at which dry air would have the same density as the given moist air at the same atmospheric pressure. Since $X = (461 - 287)/287 = 0.61$

$$T_v = [1 + 0.61\,q]\,T \qquad \text{B11.1c}$$

and we see that moist air has the same density as slightly warmer dry air (confirming that it is less dense than dry air) and that the temperature difference is

$$T_v - T = 0.61\,q\,T$$

where T and q are the absolute temperature and specific humidity of the moist air. At a typical q of 10 g kg^{-1} at 273 K, moist air has the same density as 1.7 °C warmer dry air.

NUMERICAL NOTE 11.1 Cloud condensation and buoyancy

From Eqn B11.1a, the increase in temperature when mass m of water vapour condenses in unit mass of cloudy air is $\Delta T = L\,m/C_p$. If m is 10^{-3} (1 g per kg), then using $L = 2.5 \times 106$ J kg^{-1}, and $C_p = 1{,}000$ J kg^{-1} K^{-1}, $\Delta T = 2.5$ °C. Condensing 1 g of vapour in 1 kg of air warms the air by 2.5 °C.

When 1 g of water vapour condenses to cloud in a 1 kg air parcel, according to B11.1c its virtual temperature falls by $\Delta T_v = 0.61\,\Delta q\,T = 0.61 \times 1 \times 10^{-3}$ T, which is 0.17 °C when $T = 273$ K. The density of moist air rises by an amount equivalent to a cooling of 0.17 °C at 0 °C. The 10^{-3} factor converts 1 g per kg to a pure ratio.

When the 1 g of cloud water remains within the 1 kg air parcel, the increase in overall parcel density is 10^{-3}, which according to $p = \rho\,R\,T$ is equivalent to a cooling of 10^{-3} T, which is ≈ 0.27 °C at $T = 273$K.

Cloud melting or freezing: If 1 g of water cloud freezes in 1 kg of cloudy air, the injection of latent heat of fusion warms the air by $\Delta T = L/C_p = 0.33 \times 10^6/(1000) = 0.33$ °C.

Vapour deposition to ice: If 1 g of water vapour condenses directly to ice cloud, the latent heat is about 2.5 + 0.33 ≈ 2.9 MJ kg^{-1}. At the low temperatures where this occurs, vapour contents are < 1 g kg^{-1} even at saturation.

(i) There must be a good supply of buoyant moist air rising from the surface layers—buoyancy to drive convection, and moisture (vapour) to feed cloud production and minimize evaporative cooling between clouds and their cloudless (but moistened) environment.

(ii) There must be convective instability, induced by solar surface warming, by advection over a warmed surface, and/or by direct or indirect cooling of the upper part of the convecting layer. These can be encouraged by several processes which work in extratropical cyclones: vertical overflowing of airstreams from different source regions (Section 12.4); gentle but persistent wholesale uplift of an extensive thick layer whose lower parts are significantly more humid than their upper parts (Section 11.6); and vertical stretching (associated with synoptic-scale horizontal convergence—Section 7.15) whose effects can be seen by vertically stretching a layer with a sub-adiabatic lapse rate on a tephigram.

(iii) When such conditions are encouraging the growth of many small cumulus, some will be reinvigorated by new thermals rising into their bases before their initial impetus is lost by evaporation and detrainment. A new thermal will rise through a pre-existing cumulus with little or no evaporative loss of buoyancy until it emerges from the top or side of the protective cloud body, and it will reach this stage with greater buoyancy than the pioneering thermals which first produced the cloud. Such reinvigorated cumulus are advantaged, and will tend to grow larger and live longer as a result. Indeed they may receive repeated additions of buoyancy from beneath cloud base by the same mechanism, each new influx being protected from destructive evaporation by rising through successively greater depths of accumulated cloud before reaching the edge of what is now a cumulus congestus. In this way a favoured minority of small cumulus grow much larger than their fellows, often reaching into the middle troposphere (~5 km above the local surface). In favourable circumstances they may dominate the total cloud population by volume, though heavily outnumbered by their smaller fellows.

Cloud shape

Careful time-lapse photography of individual cloudy thermals in cumulus congestus shows that as they rise they tend to expand as if bounded by an imaginary inverted cone, whose vertex has a semi-angle of about 15° (Fig. 11.8). From this, from laboratory work with similar convection in water, and from theory, it is known that such conical expansion results from net entrainment. The familiar

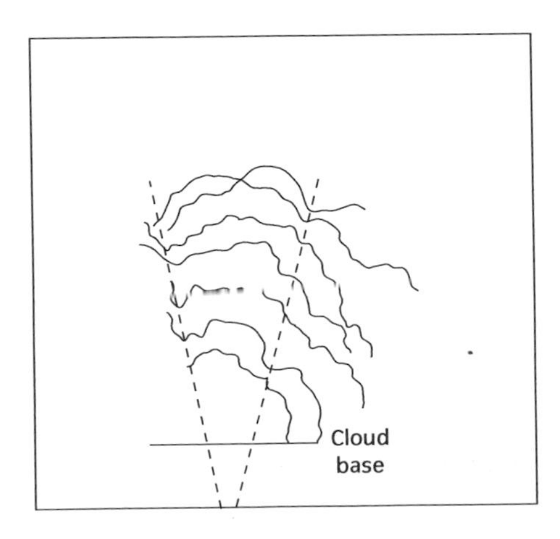

Figure 11.8 Outlines from a sequence of photographs of a cumulus congestus taken at one-minute intervals. Over the whole sequence the chosen cloudy thermal grows from being a small lump on the edge of the congestus to become its main turret.

outline of the cumulus congestus, on the contrary, shows a marked narrowing with height (Fig. 11.7). The reasons for this inversion of the individual behaviour of cloudy thermals, lie again in the role of the established cloud body as protector of new cloudy thermals: those on the flanks of the cloud suffer much more evaporative dissolution than those in its core, so that the congestus as a whole converges upwards despite the diverging tendency of its constituent thermals. The sacrifice of the flanks allows the central core to rise relatively efficiently into the middle and upper troposphere, a prerequisite for the production of cumulonimbus.

11.4 Cumulonimbus (Cb)

Though occasionally a light rain shower may reach the surface from a small cumulus congestus, and low stratocumulus may produced slight drizzle, it is a familiar fact that most showers fall from clouds which have already grown to a very substantial height, and hence warrant the title cumulonimbus. Indeed it is their dark bases, shadowed by the great depth of overhanging cloud, which make their visually obvious threat of rain.

The reasons for the association of cloud thickness with capacity for precipitation follow directly from the discussion of Section 6.11. It takes a considerable length of time, probably at least 20 minutes, for significant numbers of cloud droplets (or crystals) to grow selectively to the millimetric size at which their fall speeds are big enough to exceed updraft speeds, and their size enables them to survive evaporation as they fall through the sub-cloud layer. Only fairly substantial clouds last as long as this, and even if a shallower cloud managed to avoid dissolution, fall speed and updraft would have to be nicely balanced to give time for precipitation to develop.

Most showers begin to fall from Cb soon after they have developed from Cg. Where cloud base is reasonably high and the shower is seen against a pale background sky, the *virga* or dark trails of precipitation reaching down below cloud base can be visually obvious (Fig. 6.17), particularly if curved by wind shear, as is often the case in the blustery winds which accompany showers in cool equatorward air flows in middle latitudes. The visible thinning of the precipitation towards the surface occurs as smaller precipitation droplets are evaporated to extinction faster than larger ones, because of the great sensitivity of extinction distance to droplet size (Box 6.5). If the air is so cold that even the low troposphere is below 0 °C, the precipitation will fall as snow, which has a characteristically pale and diffuse appearance (Fig. 11.9).

As a Cb drifts with the lower middle troposphere wind over land, it leaves a trail of precipitation (rain or snow, often accompanied by hail) which is as wide as the active core of the cloud (usually no more than a few kilometres) and as long as the distance the cloud travels during its active life. Since this life is often about an hour, the length of the affected strip may vary from a few kilometres in very light winds to more than 100 km in a gale. Comparing clouds with similar precipitation rates and lifetimes but different drift speeds, we see that the rainfall collected by a gauge anywhere in the wetted strip is inversely proportional to the length of the strip and is greatest when the shower is stationary. Showers occurring in weak air flows are therefore particularly sporadic, and heavy in the

Figure 11.9 Snow falling from a small winter cumulonimbus in NW England. A second cumulnimbus is visble behind and to the right, and there are much smaller cumulus over the low hills.

affected areas. Even in strong air flows, showers may be widely separated, raising difficulties in observation and forecasting (Section 2.8). Typical rainfalls (or the equivalent rainfalls of snow and hail showers) range from 1 to 10 mm in showers of normal intensity. (Unusually intense storms are discussed in Section 11.6). Though such totals are small, being comparable with rainfalls from weak to moderate fronts, they usually fall in a small fraction of an hour, producing short bursts of moderate to heavy rain (rates of rainfall ~30 mm hr^{-1}), contrasting with the sustained periods of mainly light rain which are typical of frontal rainfall. In weak air flow the persistence of a heavy rain shower over a limited area can easily overload the local drainage, causing localized flooding, especially in urban areas.

11.4.1 Tower and anvil

Observation often shows that a Cg, having taken 20 minutes or so to reach the point of becoming a Cb by building slowly in height and volume, begins to grow much faster just as precipitation begins. As mentioned in Section 6.11, in middle latitudes this transformation seems to be associated with widespread *glaciation* of the upper parts of the Cb. The rapid release of latent heat of fusion as large amounts of supercooled water cloud freeze must tend to enhance buoyancy and further growth (Box 11.1), as must the loss of some of the dead weight of water through precipitation. The increase in vertical growth rate often produces a cloud tower in the upper troposphere which, at least initially, is considerably narrower than the pyramidal Cg beneath. Rates of rise of 5–10 m s^{-1} are observed to be quite normal in the tower, which are consistent with typical updraft speeds measured in the cloudy interior by a variety of direct and indirect methods. Soon veiled by glaciation, the tower is often quite quickly

Figure 11.10 A distant summer cumulonimbus in the same area as Fig 11.9, with its anvil spreading from left to right in the faster moving air in the upper troposphere.

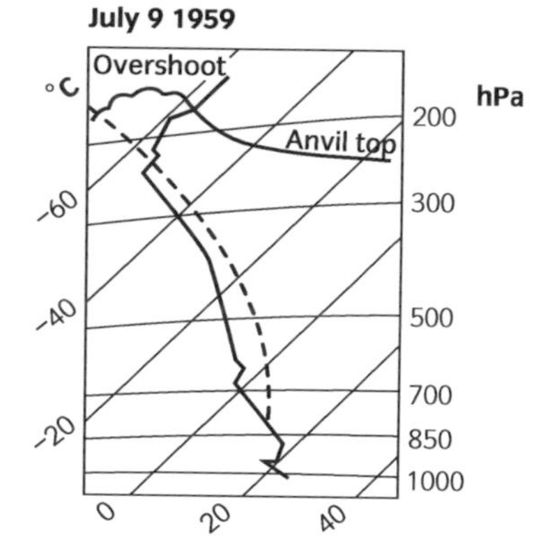

Figure 11.11 A tephigram of conditions in the troposphere on the occasion of an intense cumulonimbus (a severe local storm) in southern England. Radar and aircraft observations showed tower penetration into the stratosphere (marked overshoot) and anvil tops at the levels sketched. The dashed line is a saturated adiabat.

stretched horizontally and asymmetrically by strong wind shear in the high troposphere, producing the anvil-shaped top which is so characteristic of the mature Cb (Fig. 11.10).

The top of the anvil is usually fairly flat, suggesting that the updrafts are being halted at a fairly definite altitude which is often observed to coincide with the tropopause (Fig. 11.11). A simple treatment of the loss of buoyancy (Box 11.2) confirms aircraft observations that updrafts can overshoot their ultimate equilibrium level by hundreds of metres on account of their vertical momentum, before sinking back and spreading out. Sometimes a stable layer well below the tropopause can act as the upper limit of Cb development. In any case, the collision between towering updrafts and a stable layer feeds large volumes of ice cloud laterally into an anvil for the duration of the cloud. In the course of an hour's activity, huge amounts of ice cloud may be disgorged into the high troposphere. Since anvil cloud is relatively slow to evaporate, because differences in vapour density

BOX 11.2 Convective overshoot

Figure 11.12 represents a simplified vertical temperature profile through the tropopause, in which an isothermal stratosphere surmounts a dry adiabatic high troposphere.

Consider an air parcel in a cumulonimbus updraft which is neutrally buoyant as it rises through the nearly dry adiabatic temperature profile in the very cold high troposphere. As it rises above the tropopause, the parcel continues to cool dry adiabatically, becoming quickly colder than the surrounding stratosphere. At height z above the tropopause the temperature deficit of the parcel will be $z\,\Gamma_d$. According to Box 7.10 the resulting net downward buoyant force per unit parcel mass will therefore be $g\,(z\,\Gamma_d)/T$, which increases with increasing penetration of the stratosphere, as expected. This force opposes the continued rise of the overshooting parcel and eventually brings it to a halt before sending it back down again. We use the vertical component of the equation of motion to relate the overshoot to the updraft speed at the tropopause.

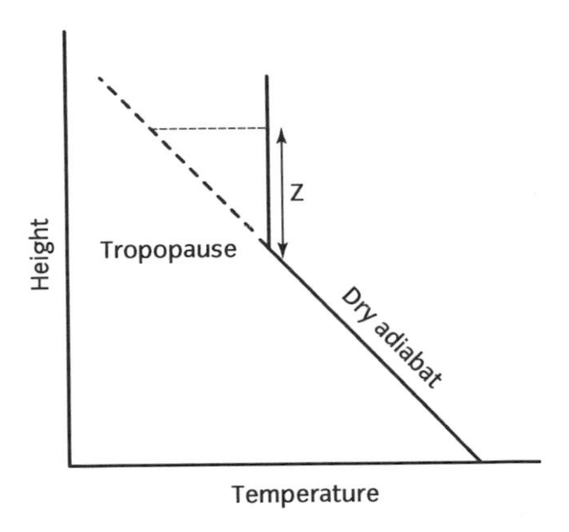

Figure 11.12 Idealized temperature–height profile through the tropopause. Solid lines depict ambient temperatures and the dashed line traces an overshooting air parcel. Continued overshooting can maintain a temperature inversion in the base of the isothermal stratosphere, enhancing the sharpness of the tropopause, and gradually raising its altitude.

Analysis

Ignoring frictional and entrainment drag, the vertical component of the equation of motion is

$$\frac{dw}{dt} = -\frac{g\,z\,\Gamma_d}{T} \qquad \text{B11.2a}$$

To solve this to find the variation of w with z, note that

$$\frac{dw}{dt} = \frac{dw}{dz}\frac{dz}{dt} = w\frac{dw}{dz}$$

Rearranging B11.2a to separate the w and z terms, and integrating through the ranges of w and z

$$\int w\,dw = -\frac{g\,\Gamma_d}{T}\int z\,dz = -\frac{g\,\Gamma_d}{T}\frac{z^2}{2} + C$$

We can ignore the proportionately small variations in absolute temperature T, and evaluate the integrals from the tropopause ($z = 0$, updraft w) to the height z at which the speed of updraft w has been reduced to zero. The result gives the depth of penetration z into the stratosphere by an updraft with speed w at the tropopause:

$$z = w\left(\frac{T}{g\,\Gamma_d}\right)^{1/2} = w\left(\frac{T\,C_p}{g^2}\right)^{1/2} \qquad \text{B11.2b}$$

where the second form comes by substituting (g/C_p) for the dry adiabatic lapse rate Γ_d. Substituting familiar values for C_p and g, and assuming a realistic value of 220 K for T, we find (in SI units)

$$z \approx 48\,w$$

which suggests that an updraft of 10 m s^{-1} should penetrate nearly 500 m into the stratosphere. This is an overestimate, since drag tends to reduce the penetration by an amount that cannot be reliably estimated. Observations of the tops of severe local storms indicate occasional penetrations of more than 1 km, which imply updraft speeds at the tropopause of 21 m s^{-1} in the simple model, and considerably more in reality.

between the cloud and ambient clear air are very small in the low temperatures of the high troposphere (even when the ambient air is very dry, as can be seen by extrapolating Fig. 6.2 to these temperatures), the resulting anvil may stretch many tens of kilometres down the direction of the wind shear. The lower surface

Figure 11.13 Mammatus on the underside of the overhang (probably but not definitely an anvil) of a cumulonimbus which has recently past over the camera location (see wet foreground).

of the anvil rises progressively with increasing distance from the parent tower as smaller and smaller ice crystals sink more and more slowly into the underlying clear air. Sometimes the undersurface of the anvil near the tower may hang down in smooth breast-like shapes known as *mammatus* (Fig. 11.13), which indicates the presence of pockets of negative buoyancy probably produced by anvil material evaporating into the dryer air of the middle troposphere.

11.4.2 Life cycle and strength

Like the much smaller cumulus humilis, cumulonimbus have a fairly clearly defined life cycle. Initial growth as Cg is followed by the transition to Cb, frequently accompanied by rapid extension into the high troposphere. Precipitation falls in intense shafts through the interior of the cloud body and into the clear air beneath cloud base. As these develop, air in their vicinity begins to fall, partly through frictional drag, and partly by chilling of unsaturated air by evaporating precipitation (Box 11.1). When they reach the underlying surface these downdrafts are felt as the chilly gusts which accompany all substantial showers. Eventually the updrafts are overwhelmed by the downdrafts and the cloud begins to die, essentially as a delayed result of its earlier success in producing cloud and precipitation. The main trunk of cloud dissolves into a temporary veil of precipitation, which looks much less solid than cloud because it consists of a relatively very small population of very large droplets and particles. After another 15 minutes nothing may be left of the recent glistening mountain of cloud apart from a tuft of anvil drifting and slowly dissolving in the high troposphere and the long track of surface which has received precipitation—residues of the vertical exchange which underlies all convection.

Many Cb are associated with hail and thunder. As mentioned in Section 6.11, hail is produced by the relatively rapid accretion of ice and supercooled water (mainly the latter) onto a falling frozen embryo. The growing hail stone soon has a relatively high fall speed through the ambient cloudy air, and in fact is able to grow to a substantial size only if the updraft of cloudy air is strong enough to hold it inside the body of the cloud for some minutes. However, Cb are typified by their strong updrafts, which can exceed 10 m s^{-1} in large clouds. Hail is therefore typical of larger Cb, though in the British Isles comparatively weak winter Cb often produce little conical pellets of spongy hail (*graupel*). The association of maximum particle size with updraft strength explains why the largest hail fall from summer Cb in middle latitudes, the potential for the necessary buoyancy being greatest then. Hailstones more than a few millimetres across remain unmelted even after falling through a kilometre or more of warm air and often appear as a narrow shaft of hail in or beside the core of heaviest rain (at least some of which is probably melted smaller hail). In lower latitudes the low troposphere is so warm that even quite large hail is melted before it reaches the surface.

The same strong updrafts which encourage the growth of hail usually separate electrical charge efficiently enough (Section 6.12) to cause lightning within the Cb and between its lower parts and the underlying surface. Most substantial Cb produce thunder in this way, and even those small ones which do not thunder audibly produce crackles of radio noise which indicate that weak sparks are jumping between regions of separated charge. The frequency of lightning strokes, like the maximum size of hail, is a measure of updraft strength, which is the core measure of the vigour of the parent cloud.

11.4.3 Practical experience

Civil aircraft are normally routed to avoid Cb because of the discomfort and even danger of flying quickly through consecutive up- and downdrafts. Sudden downward accelerations in particular (widely but quite misleadingly attributed to the presence of air pockets) can throw coffee cups, and occasionally even unstrapped passengers, upwards relative to the aircraft's interior—a dramatic example of how an unexpected acceleration seems like a force in the opposite direction. In a very few cases such loads have caused catastrophic structural failure of the aircraft. Although a few accidents have been attributed to lightning strikes on aircraft, this risk is relatively small, since the aircraft tends to keep close to the electrical potential of the ambient air. Aircraft icing is potentially a more important hazard (Section 6.13). Larger aircraft carry a small radar which can detect the shafts of heavy precipitation which are associated with the active cores.

Consider a typical sequence of events experienced at a position on the ground lying in the path of a well-developed summer Cb. The air is warm and humid, both factors contributing to the convective potential of the situation. The sky becomes gradually overcast as the anvil of the advancing cloud reaches overhead and beyond, though the details are often visually very unclear if the low troposphere is hazy, as it often is in such conditions. The wind, which may have been blowing gently in much the same direction as the storm movement, dies as a persistent draw of air towards the base of the updraft sets in and locally offsets the overall air flow near the surface. Sometimes there is no obvious relation between surface flow and storm motion, which is essentially with the air flow several kilometres above the surface, and occasionally the two are opposed. As yet there is no obvious evidence of the storm: there is no precipitation and usually no evidence of thunder or lightning unless it is already after dusk, but there is often awareness of the 'lull before the storm', which is confirmed as the main cloud mass begins to loom overhead. The depth of cloud is such that it may become dark enough to trigger automatic street lights.

Quite suddenly there is a sound of wind, and nearby trees begin to move a few seconds before the squally downdraft reaches the observer. If there is time to notice it before the onset of precipitation, the squall is obviously cooler than the sultry air it has replaced. The rain begins very suddenly, almost as if switched on, and may be heavy and contain bursts of small hail. At the same time the sky overhead begins to lighten considerably, giving the appearance of a pale grey vault. This is the tall column of downdraft where the cloud has been largely replaced by the optically much less dense veil of precipitation. This too is the region where the electrical activity is most obvious, with lightning strikes to ground, and less distinct flickering and rumbling from lightning in the middle troposphere. The sky lightens even further as the rain begins to ease off, and the squall winds subside. As the storm passes away downwind it may become clearly visible for the first time as a glistening mountain of cloud hiding the anvil beyond. The deposited carpet of cool, fresh downdraft remains, to be warmed again by the Sun if enough daylight is left.

11.5 Clouds and their environment

So far we have distinguished between cloudy and cloud-free air as if they were as separate as they appear to be visually. But in Sections 5.10 and 5.11 we considered how convection, like any mixing, tends to produce an environment which is unaltered by further convection. To the extent that this tendency is realized, the ambient air around thermals below cloud base, and around the cloudy thermals above, cannot be independent of the buoyant masses which are currently active—it has probably already been processed by many such thermals through successive entrainment and detrainment. If the individual thermals are scarce and weakly interactive with their surroundings, the ambient air may wait so long between successive interactions that it is seriously out of equilibrium with the thermals, and the distinction between thermal and environment may be quite marked. But in fact most types of cumulus convection are so vigorous that thermals and environment are only slightly out of equilibrium. This is why the temperature excesses inside clouds are usually so small, less than 1 °C in cumulus humilis and usually less than 5 °C in even quite big cumulonimbus. It is also why the various types of cumulus convection are associated with characteristic vertical profiles of temperature and humidity which we now briefly summarize.

Figure 11.14 depicts typical vertical profiles of potential temperature in the presence of various types of cumulus convection. Potential rather than actual temperature is used to remove the persistent temperature lapse of ordinary temperature, and the vertical scale has been expanded at lower levels to enlarge the details of structure in the lower troposphere which would otherwise be inconveniently small. The solid profiles represent conditions outside the cloud masses which, remember, occupy most of the total horizontal area at any level above cloud base, and therefore represent what is almost always recorded by radiosondes in these conditions, since only a small minority of sondes actually ascend through any substantial depth of convective cloud. The dashed lines represent

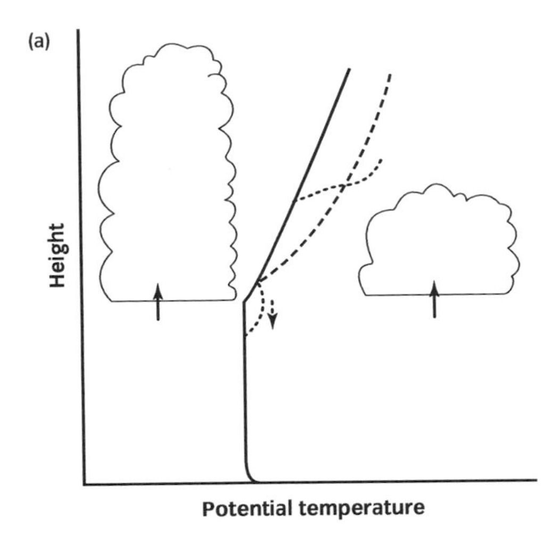

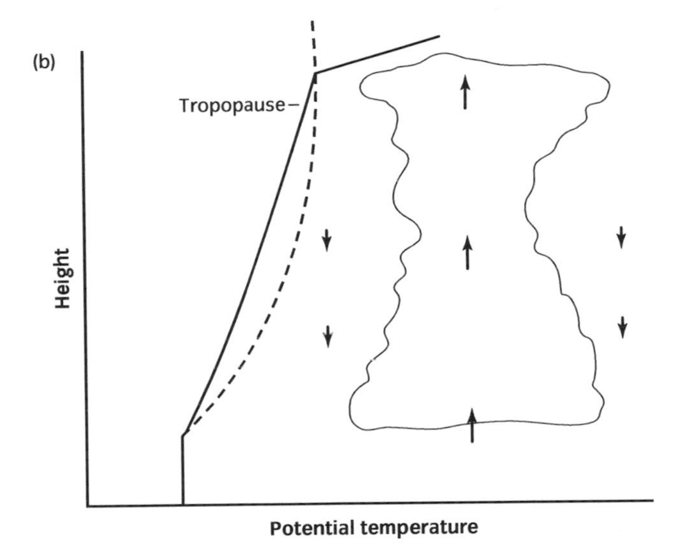

Figure 11.14 (a) Vertical profile of ambient potential temperature (note the vertical profile of the subcloud layer) in the presence of cumulus and flattened cumulus, with a saturated adiabat shown dashed. Dotted portions correspond to flattened cumulus. (b) As a but with vertical scale adjusted to accommodate the full height of a cumulonimbus.

saturated adiabats from cloud base. Conditions inside a cloud of the cumulus family (dashed) are always much closer to saturation, and are therefore moister than the air outside the cloud, and are crucially a little warmer, as discussed above.

11.5.1 Small cumulus

Below and around cloud base

Figure 11.14a represents conditions accompanying small cumulus, either in the early stages of a day overland which will later give rise to Cg, or (fine dotted) confined beneath a subsidence inversion not far above cloud base. The shallow sharp lapse of θ is typical of the first few metres over land (Section 10.6), and the nearly dry adiabatic layer (nearly uniform θ) reaching up to cloud base is the signature of sub-cloud convection. Careful measurements suggest that this layer is not exactly dry adiabatic, in that θ lapses by ~1 °C in the first few hundred metres above the surface boundary layer, and recovers by the same amount up to cloud base. If this is a general result, it must arise from details of the way in which thermals vary with height which we do not yet understand.

Near cloud base the profiles differ somewhat, depending on whether or not we are directly beneath a cloud. Rising into cloud base the temperature profile converts from a dry adiabat to something close to the appropriate saturated adiabat, as expected (Section 5.7). However, rising towards the level of cloud base in a gap between clouds, θ often begins to rise before cloud base level is reached (small dotted fillet), for reasons which depend on an important aspect of cloudy convection we now consider.

The cloudy updrafts are only part of an unsteady convective cycle which must include subsidence as well. In small cumulus, almost all of the subsidence occurs in the clear air between clouds. These and the cloudy updrafts must maintain a close balance of mass flux, which is only very slightly biased by the presence of any large-scale convergence or divergence—Section 7.15). But the rising air tends to follow a saturated adiabat, whereas most of the sinking air must tend to follow a dry adiabat. Air sinking dry adiabatically after cloudy ascent will become warmer than the rising cloudy air at the same level, by the release of latent heat in cloud production. Overall this tendency is held in check by the frequency and vigour of mixing between clouds and environment, but it seems that around cloud-base level such mixing is sufficiently weak to allow slightly warmer air to subside a little way down into the sub-cloud layer between clouds.

Above cloud base

Here the potential temperature profile between clouds deviates slightly but significantly from the expected saturated adiabat, being cooler (to the left in Fig. 11.14a), but still increasing with height. This too is a consequence of the different adiabatic processes being followed by the rising and sinking air. If both branches followed the dry adiabatic, then according to the general law of mixing and Section 5.10, the equilibrium profile approached by continued mixing of this type would be a dry adiabat, as is very nearly the case in most of the sub-cloud layer. If both branches followed the wet adiabatic process, then that would be the equilibrium profile. But since the branches differ, the equilibrium must lie somewhere between the dry and wet adiabatic profiles, at a position determined by details of the mixing effected by any particular population of clouds, including the vertical profile of rates of entrainment and detrainment. This is borne out by observation, though incompleteness of current theory and observation prevents detailed matching. Observed environmental lapse rates in layers populated by small cumulus often lie about one third of the way towards the dry adiabatic from the relevant saturated adiabatic, the fraction decreasing with increasing cloud cover. Temperatures inside the clouds differ very little from those outside, and some small cumulus seem to be capable of operating on zero temperature excess, relying on their excess water vapour for buoyancy.

Above the layer of really small cumulus the potential temperature profile typical of cloudy convection may extend through a deep layer populated by small Cg, or it may rise sharply, indicating the presence of a convectively stable layer (dotted on Fig. 11.14a). In the latter case the cloudy updrafts lose buoyancy very quickly as they rise into the warmer ambient air (like a small version of a Cb updraft penetrating the stratosphere—Box 11.2). Terminal detrainment quickly follows the cessation of upward motion, but an important exchange of air between the convecting and stable layers occurs in the process.

When the stable layer in Fig. 11.14a is maintained by anticyclonic subsidence, it may persist for days in a dynamic near-equilibrium with the underlying convection. When convection has the advantage, as when air moves over warm land or warm water, the convecting layer eats upwards into the stable layer, but at a rate which diminishes as erosion proceeds and sharpens the 'elbow' in the temperature profile and increases the stability of the immediately overlying air. When subsidence has the advantage, either through weakness of convection or enhanced anticyclonic vigour, then the convecting layer is squeezed downwards into a progressively shallower layer, producing shallow stratocumulus (Section 11.2), or even extinguishing all cloud as the base of the stable layer sinks below the lifting condensation level. The dynamic interaction at the base of the stable layer exchanges sensible heat and water vapour between the subsiding and convecting layers, the subsiding layer being cooled and moistened and the convecting layer being dried and heated as a result. Estimates have shown that an appreciable fraction of the total heat input into the surface-based layers of an anticyclone over land is brought down from the subsiding air in this covert way, the remainder coming through direct absorption of sunlight by the ground.

11.5.2 Congestus and cumulonimbus

Figure 11.14b represents conditions accompanying Cg and Cb convection. The details up to levels reached by small cumulus are essentially the same as in Fig. 11.14a, which is consistent with the observation that large cumulus are normally accompanied by small cumulus, though not the stunted type associated with anticyclones. Above these lower levels the potential temperature profile remains intermediate between the dry and saturated profiles, for reasons already discussed, but tends in the upper troposphere to return towards the saturated adiabat through conditions at cloud base. In fact the equilibrium buoyancy level indicated by drawing a saturated adiabat upward cloud base often gives a useful estimate of the level of cumulonimbus tops (Box 11.3).

The tendency for ambient temperature profiles to recover towards the saturated adiabat from the lowest limit of the convection no doubt reflects the concentration of detrainment and dissolution by Cb in the upper troposphere. It is as if the middle troposphere were partly bypassed by the deep convection currents linking low and high troposphere, so that the environment in these intermediate levels is influenced only at second hand by the tumult of small cumulus rapidly processing the low troposphere, with the scattering of cloudy towers reaching towards the top of the troposphere to disgorge their fans of icy anvil. The intertropical convergence zone is densely populated by many of the largest Cb on Earth, and it is significant that these are often technically termed *hot towers*, on account of their role in piping latent and sensible heat from the warm surface (most of it ocean) to the vicinity of the equatorial tropopause about 15 km above. In fact such hot towers collectively represent the ascending branch of the Hadley circulation (Sections 4.7 and 8.10).

The existence of a layer of minimum interaction between updrafts and their middle-troposphere environment is associated with a basic property of entrainment and detrainment: that they are least effective as a proportion of the mass of rising air when updrafts are largest. Basically this is yet another example of the importance of the decline of surface-to-volume ratio with increasing scale which applies to phenomena as diverse as the heat economies of mice as compared with elephants, and the slowing rate of growth of cloud droplets growing by condensation (Box 6.3): the mass of a thermal is proportional to its volume whereas its rates of entrainment and detrainment are proportional to its surface area. The insulating effects of large scale are confirmed by more technical arguments which follow from an extension of the treatment of entrainment begun in Section 7.14. These indicate that small cumulus should mix themselves out of existence in the short time taken for their tops to rise a few hundred metres, as indeed is usually observed, whereas updrafts more than a kilometre or so in breadth should be able to rise through the full depth of the troposphere without losing more than a fraction of their mass by mixing with the environment. The advantage of scale is apparent too in the mechanism outlined for the development of cumulus congestus (Section 11.3).

The other necessary aspect of such minimal interaction is that it encourages maximum differences between clouds and environment: temperature excesses in cloudy interiors are largest in the mid-troposphere and often amount to several °C in cumulonimbus. Buoyancy is thereby enhanced and this together with the reduction in entrainment drag per unit mass (Section 7.14) encourages strong updrafts. Compared with the fretful inefficiency of small cumulus, Cb are relatively very efficient in all that they do, and are capable of producing quite vigorous weather locally, as described in the previous section. When Cb become strong enough to cause local surface damage, as a small but significant fraction of them do, they are called *severe local storms*.

BOX 11.3 Cumulonimbus tops

Figure 11.15 represents a sounding on a showery autumn day in NW England. The showers were forming over a wide area of the NE Atlantic, as a cool NW'ly airstream over-ran the relatively warm sea (which is warmer there than even the western margins of the land throughout the winter). The sounding therefore portrays an air mass which is well adjusted to the showers it contains.

The cloud base at Aughton is very low, and the situation at low levels is complicated by a weak convectively stable layer which represents the influence of local land-based cooling (the Sun had set over an hour earlier). Over the slightly warmer sea, the lifting condensation level was probably as depicted. On the simplest adiabatic model of the showery convection, the rising air follows the saturated adiabat from the LCL, and is therefore warmer than the ambient air mass up to the level marked CT, where the saturated adiabat and environmental temperature profiles re-cross. On this model the air accelerates to produce a maximum updraft speed at CT, and then overshoots and returns to detrain at CT, as discussed in Box 11.2.

In fact the inevitable entrainment must dilute the rising air with increasing height, so that it drifts to the cold side of the saturated adiabat, before

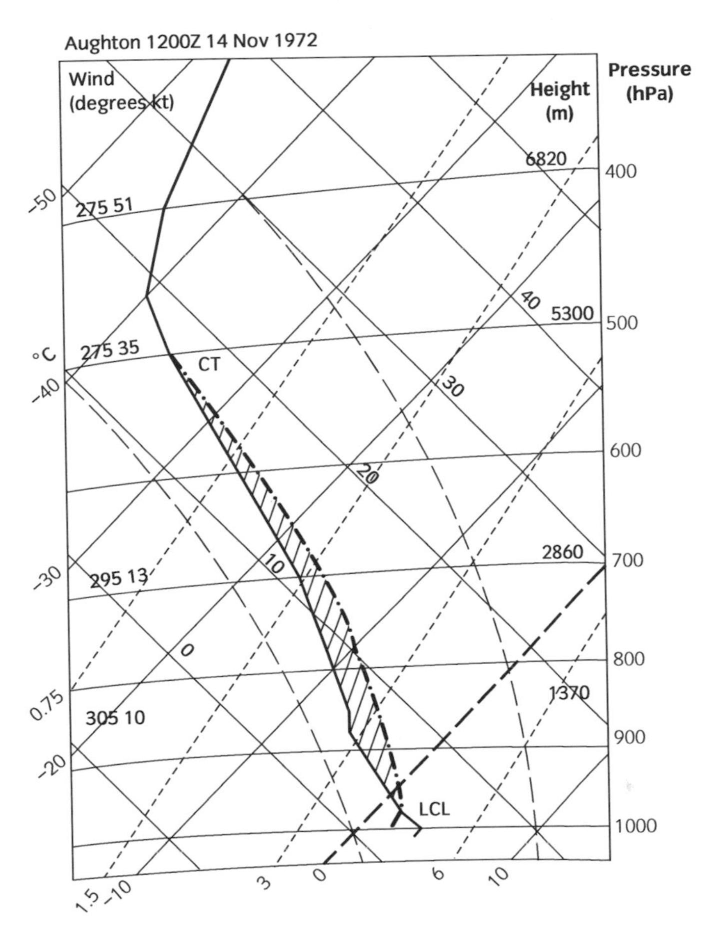

Figure 11.15 Tephigram for a showery day. The solid line traces the radiosonde temperature profile, and the dash-dotted line shows the construction used to estimate the altitude of cumulonimbus tops (Box 11.3).

widespread glaciation and consequent release of latent heat (Box 11.1—the evening sky was full of fresh and decaying anvils a little earlier) tends to reverse the trend in the upper part of the convecting layer. Because mixing and glaciation effectively cancel in this respect, the level predicted by the simple model agrees reasonably with observed heights of the tops of the highest cumulonimbus. The presence of both entrainment and frictional drag keeps updraft speeds well below the frictionless values (20 m s^{-1} on this occasion according to the hatched area on Fig. 11.15), and ensures that they are strongest in the upper middle of the convecting layer, because of the entrainment minimum there, rather than at the top.

11.6 Severe local storms

11.6.1 Origins

Certain land areas are seasonally subject to unusually intense cumulonimbus, as judged by rates of precipitation, maximum size of hail, squally wind speeds, and electrical activity. Such storms also quite often produce one or more tornadoes (Fig. 11.16): intense localized vortices extending from cloud base to surface, where they produce damaging winds (but not always the utter devastation focused on by the media—see later).

Severe local storms have been studied extensively in recent years because of the very considerable damage they can do. Much of this study has been carried out in North America, partly because of the frequency and severity of storms there, but also because that particularly affluent and sophisticated population demands detailed forecasting of such events, and supports a greater *per capita* number of forecasting and research meteorologists than anywhere else in the world. Despite this, it is important to realize that severe local storms occur in many regions scattered throughout the world, including the British Isles, Europe and European Russia, North India, North Indochina and China.

Despite gross differences of geography and climate between different affected regions, it seems that two features are usually present when such storms occur.

(i) There is a mechanism for accumulating and then quickly releasing substantial convective instability. Without such a temporary bottling up of energy, the atmosphere releases its instability as soon as it appears, in weak, unspectacular Cb, which do little more than contribute some local precipitation.

(ii) There is significant wind shear distributed through a good depth of the troposphere. Though less obviously relevant than (i), wind shear appears to play

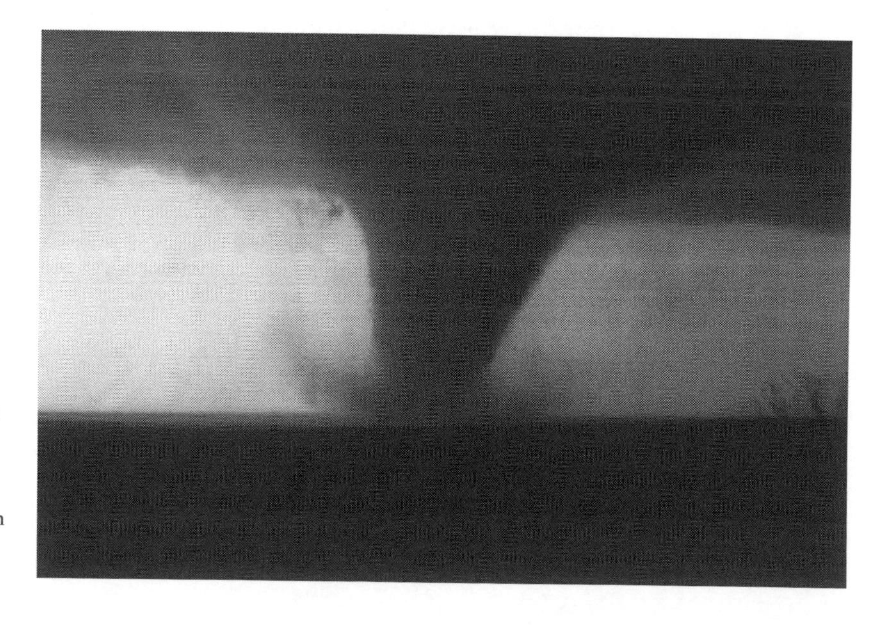

Figure 11.16 A powerful tornado in the United States (10 April 1979) showing the very dark parent cloud base tapering downwards into the vortex core. The great breadth of the cloudy core shows that the pressure in its centre is very low. Near the ground, violent winds maintain a spray of any debris significantly denser than air. From the US National Severe Storms Laboratory.

a crucial role in making the storms especially vigorous and long-lasting (and therefore severe), and it does so in two quite distinct ways which correspond to two different types of storm—the *multicell* and the *supercell* storms. We now consider these two features, and the ways in which they are believed to contribute to the appearance of severe local storms.

11.6.2 Accumulation and release

The accumulation and sudden release of convective instability occurs in different ways in different geographical regions, and usually depends on quite specific local distribution of sea, land, and topography, though the release of conditional instability (Section 5.10.2) is often a common feature. Consider, for example, the severe local storms which occur in the central United States in the spring and summer. These are amongst the world's most spectacular storms, and often occur ahead of weak cold fronts which have crossed the Rocky Mountains and are moving slowly across the Great Plains. A typical situation is depicted in Fig. 11.17, which shows a weak cold front orientated roughly SW to NE. The associated jet stream is similarly orientated (according to the thermal-wind relationship—Section 7.12) and provides the necessary wind shear throughout the troposphere in a fairly broad zone on either side of the jet core. As will be outlined in Section 12.4, the flow of air in and around fronts is much more complex than is apparent on horizontal or isobaric surfaces. There is always a three-dimensional interlacing of air flows from source areas which can be hundreds or even thousands of kilometres apart, and which can look quite unconnected with the system on purely horizontal or isobaric charts.

In the case depicted in Fig. 11.17 the air in the low troposphere well ahead of the advancing cold front is flowing N'wards from the Gulf of Mexico, and is consequently very warm and humid. Such air is potentially very good convective fuel, on account of its high wet-bulb potential temperature, but the potential would be frittered away in ordinary Cb if its release were not temporarily inhibited by the presence of an overlying 'lid' of warmer air which confines any convection in the warm, Gulf air below a convectively stable interface between the two layers (Fig. 11.18). The warmer air is dry and seems to originate over the hot, dry highlands of New Mexico and Mexico to the SW, where the strong Sun

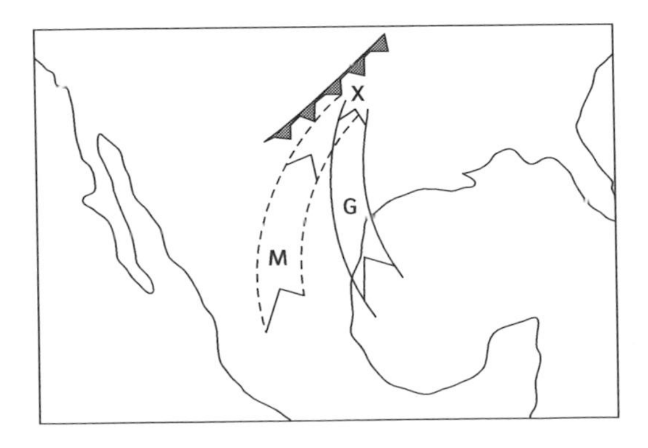

Figure 11.17 Airflows associated with outbreaks of severe local storms in the southern central United States. Warm moist air from the Gulf of Mexico is labelled G, and hot dry air from the Mexican Highlands is labelled M.

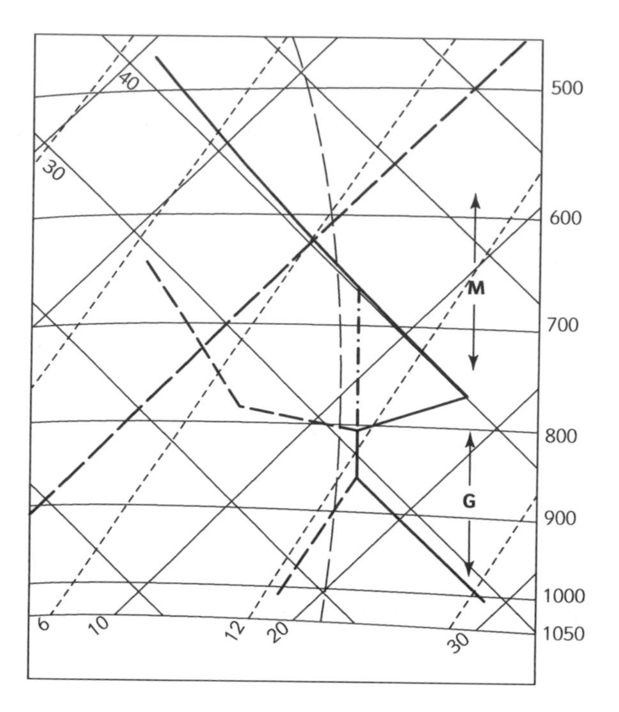

Figure 11.18 Tephigram of Gulf (G) and Mexican (M) air streams at position X in Fig 11.17. The chained line shows how the saturated Gulf air would rise in general uplift.

shining on the dry, elevated surfaces produces air with relatively high potential temperature, but low wet-bulb potential temperature. This Mexican air is drawn into the confluent sandwich of air streams which is typical of the vicinity of fronts, and overlies the Gulf air.

The release of convective potential occurs when the underlying moist Gulf air becomes warm enough to convect through the last vestiges of the weakening stable layer of the lid. This warming seems to come about partly as a result of solar heating of the Gulf air as it over-runs the warmed Plains in the middle or latter part of the day, and partly as a result of the general uplift of the whole sandwich of air flows in the lower middle troposphere which is associated with persistent large-scale convergence in the vicinity of the frontal system (Sections 7.15 and 12.2). The general uplift begins to warm the moist Gulf air faster than the dry Mexican air as soon as cloud begins to form in the Gulf air, since the latter then tends to rise wet-adiabatically while the unsaturated Mexican air continues along a dry adiabat (Fig. 11.18). In fact as the lid fades away, the steep temperature lapse rate of the Mexican air directly overlies the cloudy Gulf air beneath, producing a composite layer which is convectively extremely unstable. The cloudy condition for conditional instability(Section 5.10.2) is met and clouds begin to tower up into the Mexican air, accumulating buoyancy at least to the level of the upper middle troposphere, as they rise and become progressively warmer than their immediate environment, until the most favoured reach the tropopause. These towers offer efficient paths for the rapid ascent of more warm, moist Gulf air into the high troposphere, with consequent copious production of cloud, rain, hail, and thunder. It is now that the presence of wind shear through a deep layer begins to have its effect, achieving rather similar results by interestingly different means in the cases of the multicell as compared with the supercell storms.

11.6.3 Wind shear

Multicell storms

In the storms outlined above, the wind shear between lower and middle levels (above the disintegrating lid) is outlined in Fig. 11.19. The individual cumulonimbus tend to move with the flow at middle levels, so that we can find the air flow at low levels (and any other level of interest) relative to the travelling cumulonimbus by vectorially adding an overall movement which exactly opposes the middle-level flow, and brings it to a halt. When this is done we find that, from the viewpoint of an observer riding on an individual Cb, the low-level flow of warm, moist air comes in from a direction which is distinctly to the right of the direction of the motion of the storm. Here the humid air rises into the powerful updraft, producing the great column of cloud, precipitation, and electrical activity which constitutes the currently active convective cell. However, as the inevitable downdraft (Section 11.4) begins to choke the updraft, the inflowing surface air begins to develop a new updraft on the right flank of the previous one, climbing over the spreading flood of dense downdraft. This new cell flourishes at the expense of its parent until it too is replaced by another on its right flank. The sequence depicted on Fig. 11.19c can continue for as long as the supplies of potentially warm and cool air continue. Although close inspection reveals the sequential development of new cells as described, the overall impression is that there is a single, powerful storm which is moving distinctly to the right of the flow at middle levels. The composite storm may run on until evening cooling of the surface air reduces its convective potential, or it may break down earlier

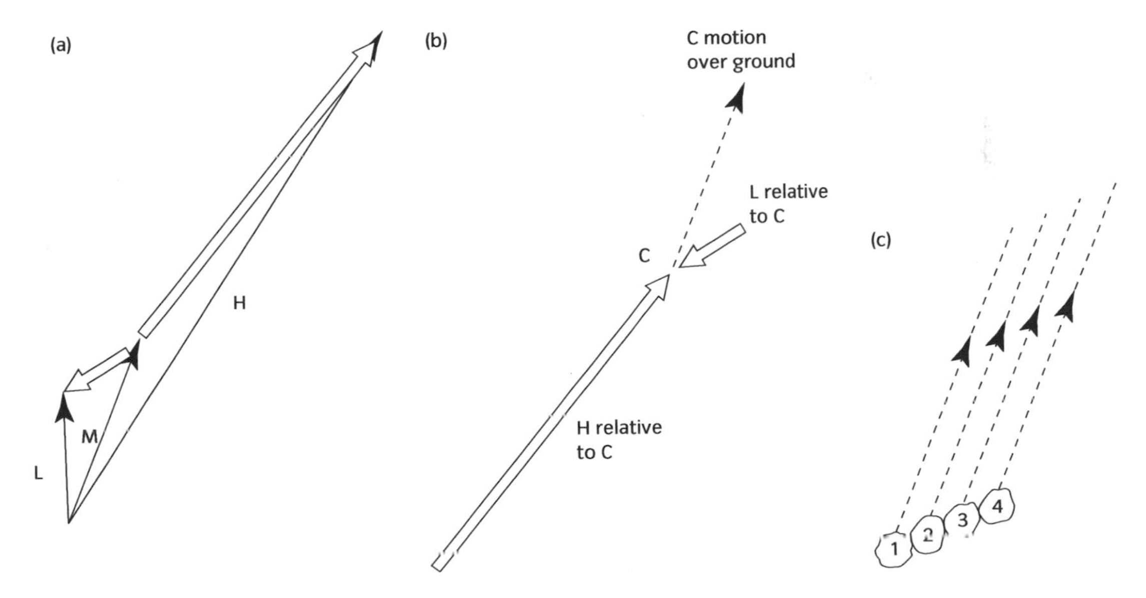

Figure 11.19 (a) Horizontal wind vectors in the low, middle, and high troposphere (L, M, and H as observed from the surface) on an occasion such as shown in Fig 11.17. The vector difference between M and H agrees in direction with frontal alignment through the thermal wind relation. (b) The same winds relative to a storm cell C moving with the middle troposphere flow. (c) Consistent development of new cells on the right flank pulls apparent storm motion to the right of individual cell motion.

because some local topographical feature inhibits the development of a new cell and breaks the sequence. Many quite ordinary Cb appear to have the multicell structure too, but when the convective potential is high enough to produce severe storms, the multicell structure ensures that the severity is persistent and distributed over substantial tracts of surface.

Figure 11.20 shows deep instability and light wind shear on a summer afternoon which produced very intense rainfall with severe local flooding on the E slopes of the Pennines in N England. The shaded area shows that a parcel ascending along a saturated adiabat from cloud base was warmer than the ambient air (measured by radiosonde about 100 km W of the storm) by an average of about 2 °C up to the cloud top at 300 hPa. The energy put into a buoyant air parcel rising from base to top is about 650 J per kg (NN 11.2) which is unusually large for Britain and would produce a maximum updraft of about 35 m s^{-1} in the absence of friction. The flooding was concentrated by the slow movement of the storm (only 9 knots at 700 hPa on Fig. 11.20) and the hilly terrain, which can

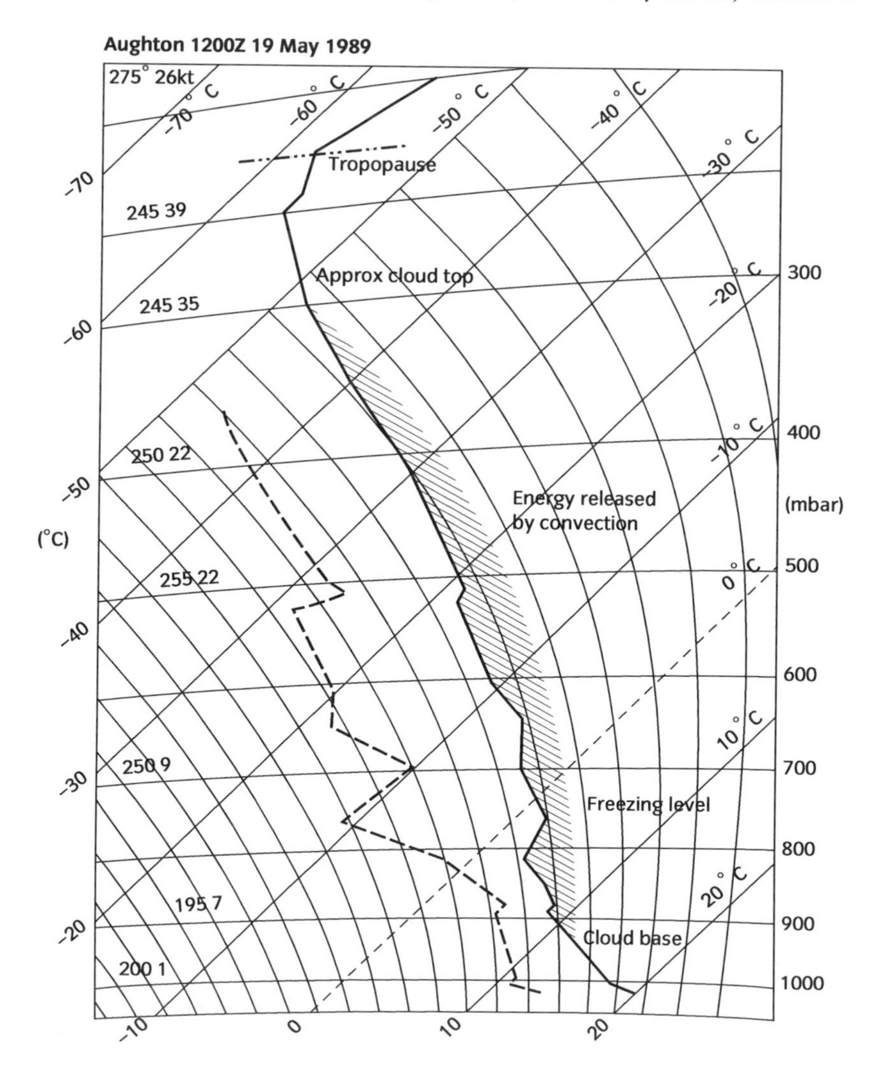

Figure 11.20 The troposphere on the occasion of unusually heavy rainstorms near Halifax (north-central Britain). The area of the shaded zone on the tephigram is a measure of the potential energy available for conversion in air rising wet-adiabatically from cloud base, which remains strongly buoyant up to the 300 hPa level. A conglomeration of cells of heavy precipitation was observed by radar moving slowly for several hours.

NUMERICAL NOTE 11.2 Buoyant work in a cumulonimbus updraft

The little work dE done by buoyancy as a parcel with temperature excess ΔT rises a little distance dz through its environment is given by $g\ (\Delta T/T)\ dz$ (Box 7.10, Eqn B7.10b). Using the hydrostatic relation to replace dz by change of pressure dp

$$dE = -R\,\Delta T\, dp/p$$

where R is the specific gas constant for dry air and the minus sign ensures that positive work is done by buoyancy as pressure falls. When this is integrated through a pressure slab from pressure p_1 to lower pressure p_2, the buoyant work

$$\Delta E = R\,\Delta T \ln (p_1/p_2).$$

Close inspection of the original of Fig. 11.19 shows that the temperature excess ΔT was distributed through the air column as follows

Pressure slab/hPa	ΔT/°C	$R\,\Delta T \ln (p_1/p_2)$
900 – 700	2	144
700 – 500	3	290
500 – 400	2	128
400 – 300	1	83
		645 J kg^{-1}

In a frictionless atmosphere this would be entirely converted to kinetic energy $V^2/2$ (the parcel mass being 1 kg) producing an updraft V of nearly 36 m s^{-1} at 300 hPa.

lock slow-moving storms as well as collect and funnel the precipitation. Radar images showed many precipitation cells in its two-hour life.

Supercell storms

In a *supercell* storm the deep shear is strong enough to allow a single powerful convective cell to become organized into a nearly steady, travelling storm. This usually happens because the winds aloft are much stronger than those beneath, but it can happen that they are opposed, giving a sheared slow-moving storm.

Though supercell storms originate in much the same ways as multicell storms, by accumulation and release of convective potential, the process of setting up a sheared, steady, moving structure is still not well understood. One probably important factor is a steady version of a process which is intermittent in multicell storms. The downdrafts associated with falling and evaporating precipitation originate in the upper middle troposphere, where the air is moving quite quickly relative to the air near the surface, because of the pronounced shear. The downdraft tends to maintain its initial momentum as it sinks, sloping forwards (in the direction of storm motion) and flooding out over the ground in the same direction. This dense wedge then helps to 'shovel' the incoming warm, moist air into the base of the updraft, which adopts a corresponding backward slope relative to the storm's direction of motion (Fig. 11.21).

However it is established, the slope of updraft and downdraft in at least the lower half of the troposphere ensures that the rising and falling air flows are adjacent and parallel, rather than entangled as they tend to be in the more vertical, ordinary Cb, with the updraft climbing over the downdraft (Fig. 11.21). The burden of precipitation no longer falls back through the updraft; instead it falls out of the updraft, unburdening it, and falls into the downdraft, enhancing its strength by drag and evaporative cooling. In the absence of the self-choking tendency of more vertical cumulonimbus, the single supercell could in principle continue to operate for as long as the convective potential of the overall

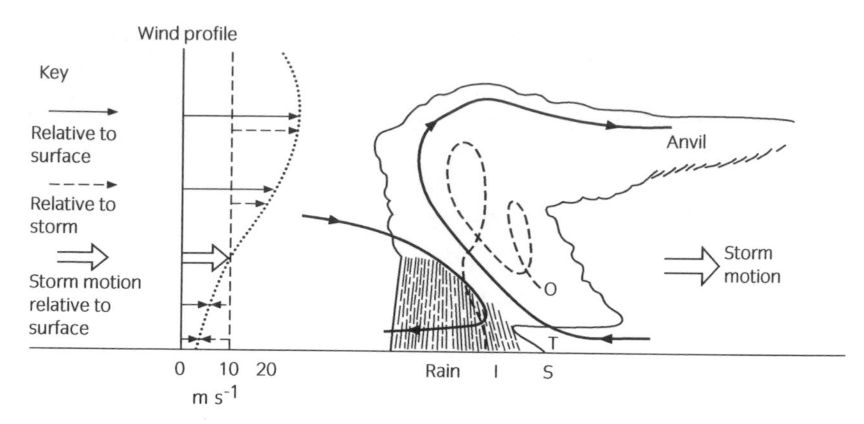

Figure 11.21 Vertical section through a supercell storm showing up- and downdrafts and inflows and outflows, as observed from a frame moving with the storm (i.e. with the flow in the lower middle troposphere). The dashed line in the cartoon indicates the path of a large layered hailstone from effective origin O to impact I on the surface. The squall or gust front S is the leading edge of the cool downdraft, and tornadoes are most likely in the region T.

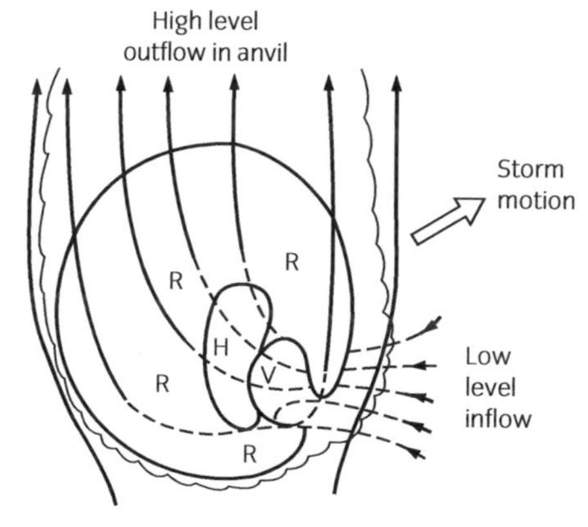

Figure 11.22 Plan view of a right-moving supercell storm, showing low-level inflow, climbing flow (dashed), and high-level flow and outflow (solid). Concentric envelopes contain areas of cloud, rain (R), and hail (H). The vault V is largely free of radar echo because the updraft there has few precipitation particles though it is full of dense cloud. A hook-shaped radar echo is often observed beside V just before tornadoes appear in the same area.

situation was maintained. In practice individual storms usually last for at least two hours.

We might expect the supercell storm to travel with the air in the middle troposphere (as implied in Fig. 11.21), as is the case with ordinary Cb and the individual cells of the multicell storms, but this is not always the case. The movement is sometimes distinctly to the right of the mid-tropospheric flow, but leftward motion has also been observed, and neither of these tendencies is well explained. The three-dimensional pattern of relative flow in the case of a particular right-moving storm is shown in Fig. 11.22, where it is apparent that the single supercell is maintaining a configuration of flow and storm motion which is similar to that of the composite multicell shown in Fig. 11.19. The pattern of air flow is intrinsically three-dimensional in ways which are the norm in extratropical cyclones on a vastly larger horizontal scale (Section 12.4).

11.6.4 Weather in severe local storms

Severe local storms usually travel at least tens of kilometres in their lifetimes of two hours or more, leaving carpets of precipitation, chilled air, and more or less serious damage. As always, faster storms (like some supercells) have longer tracks but spend a shorter fraction of their lifetimes over any location. The smooth continuity of supercells makes them more efficient and potentially more destructive than the intrinsically intermittent multicell storms, which have to build new towers every 20 minutes or so, but multicell storms can be very persistent and powerful even so. However, the efficiency of the supercell mechanism means that virtually all supercell storms are very severe.

Updrafts and precipitation

Prolonged updrafts have the opportunity to grow to great breadths (up to 10 km in supercells), minimizing drag relative to buoyancy by their small surface to volume ratio. Updrafts of as much as 50 m s^{-1} (well over 100 miles per hour) are believed to occur occasionally, on the basis of some risky traverses by military aircraft (civil aircraft are warned off), and indirect evidence such as updraft overshoot into the stratosphere (Box 11.2). Strong updrafts maintain high rates of production of cloud and precipitation, and permit the growth of giant hail (up to baseball or cricket ball size according to country), whose fall speeds (Fig. 6.10) indicate the presence of nearly comparable updrafts. The internal onion structure (Fig. 6.16) of some giant hail suggests a vertical oscillation of the growing stone, which can be explained by the tilted updraft of the supercell as depicted in Fig. 11.21. The young small stone is lifted in the updraft, spilled forwards in the anvil base, dropped back into the updraft core, lifted again, growing all the while, and so on through several cycles until it is too large to be supported by the updraft core, and falls to the surface. The large variations in temperature and cloud conditions between the bottoms and tops of the looping path are consistent with the alternating zones of clear and opaque ice (Section 6.10).

Although individually less spectacular than giant hail, much greater damage is done by the deluges of moderate-sized hail (pea to bean size) which often form in the core of each precipitation shaft (Fig. 11.23). Their impact can totally destroy fields of mature cereal crops, almost as if they had been visited by locust swarms, and the economic danger in vulnerable parts of the United States requires farmers to insure against hail damage. Insurance claims for hail damage have even been used to plot hail swathes from severe storms. In Russia and N Italy there have been persistent but largely inconclusive attempts to reduce hail damage by firing explosive rockets or shells into threatening clouds to check the growth of hailstones to damaging size. Initially done more in anger than expectation, the addition of freezing nuclei (Section 6.13) to the projectiles may have helped by increasing the numbers of hail and so reducing their individual size and impact.

Figure 11.23 (a) A dense fall of moderate-sized hail, probably no more than 1 cm across. However, their large numbers per unit area can make them very destructive of crops. (b) Tobacco plants ruined by a deluge of hail (melted away) as in picture (a). (See Plate 12).

Downdrafts

Downdraft too can attain great strength in severe local storms, by the drag effects of the very heavy precipitation, and by the separation of up- and downdrafts in supercells. It is not unusual to record gusts of hurricane strength ($> 33\ \mathrm{m\ s^{-1}}$) where falling air bursts onto the surface. Though usually less intense than tornadoes (Table 11.1), these winds can cause damage to crops, which can become irreversibly *lodged* on the ground by being battered down when wet.

Strong downdrafts can be very dangerous to aircraft when landing and taking off, when the loss of lift by sudden downward push, or when forward air motion (in the direction of flight), dangerously reduces air speed relative to the wings.

Table 11.1 Enhanced Fujita scale of tornado strength

Category	Wind speeds m s^{-1}	Relative frequency %	Damage
EF 0	29 – 38	53.5	light
EF 1	38 – 50	31.6	moderate
EF 2	50 – 61	10.7	considerable
EF 3	61 – 74	3.4	severe
EF 4	74 – 89	0.7	devastating
EF 5	89 –	0.1	total destruction

Example of damage description: EF 3 Severe damage.
Entire stories of well-constructed houses destroyed; severe damage to large buildings such as shopping malls; trains overturned; trees debarked; heavy cars lifted off the ground and thrown; structures with weak foundations blown away some distance.

These and milder *downbursts* from Cb not even in the severe category have become a major concern in civil aviation following a number of fatal accidents.

Some fairly dry parts of central N India are subject to Cb vigorous enough in at least some respects to be classed as severe local storms. The downdrafts are enhanced by strong evaporative chilling of rain falling through the very deep sub-cloud layer, which can even entirely evaporate the rain, like torrential virga tapering to nothing (Fig. 6.17), leaving the surface dry, but blasted by cold downdrafts which raise vast dust clouds from the arid land.

Tornadoes

The very rapid convergence of air at the base of the powerful updrafts of severe local storm can generate tornadoes by concentrating residual synoptic or meso-scale vorticity by the spinning-skater effect (Section 7.16). In a quasi-cyclostrophic balance between the inward acceleration of the whirling air (Eqn 7.30) and the inward pressure gradient force, pressures in the vortex core can be up to 100 hPa lower those on the same *horizontal* plane outside the tornado, as little as a few hundred metres away. Since this is comparable with the *vertical* lapse of pressure in the first kilometre of the atmosphere, the base of the cumulonimbus cloud is drawn right down to the surface in the centre of the vortex, as air spirals into its axis, decompresses, chills, and saturates before ascending (Fig. 11.16). Air whirls around the tornado vortex with tangential speeds very occasionally reaching 100 m s^{-1} (Table 11.1), though direct measurement is extremely difficult in such extreme and localized conditions.

The diameter of the violent core is usually somewhat larger than the cloud trunk, as seen in spectacular video records of flying debris, and may range from a few tens of metres to over a kilometre in the huge and hugely damaging category 5 tornadoes (Table 11.1). The Fujita scale of damage by tornadoes was intro duced in 1971 as an extension of the Beaufort scale of wind strength (Table 2.1) to these particular conditions. It has recently been amended with wind speeds reduced (by reappraisal of wind data) somewhat, but the match between category number and damage has been maintained to preserve statistics derived from the original scale. Most buildings suffer severe damage in the core of tornados in category 1 and higher, with roofs taken off and walls without iron frames demolished. Those stout enough to remain intact at the first blast may rupture outwards if they do not leak internal pressure fast enough to match the sudden fall of external pressure as the vortex core passes over. The detailed damage descriptions (the heart of the scale) become awesome in the higher categories (e.g. Table 11.1). Debris is whirled away by the fierce winds of the core, forming a lethal spray of fast-moving missiles centrifuging outwards on account of their high density compared with air (Box 7.10).

The Fujita scale shows that in the United States over 50% of recorded tornadoes are in the relatively mild category 0, but 35% are in category 1, with significant damage to buildings and trees. The blockbuster category 5s are very rare but utterly devastating, and it seems to be only a matter of time before a modern city centre will fall victim to one of these. In Britain, hundreds of tornadoes are reported each year by an enthusiastic group of amateur observers *[47]*. Their data seem to show that the vast majority are in category 0, and that category 1 is a rare limit so far.

Practical steps

The combination of hurricane-strength winds, torrential rain, giant hail, intense thunder and lightning, and the possibility of tornadoes, makes severe local storms some of the most damaging types of weather system, and this together with their intrinsic interest attracts the attention of researchers, forecasters, and recently dedicated bands of amateur storm chasers. The spectacular North American experience has encouraged the development of local forecasting and warning systems, geared especially to situations likely to produce tornadoes. Synoptic, radar and satellite data and personal reports are combed for signs of imminent or actual tornado formation (e.g. the radar *hook echo* in Fig. 11.22). Local broadcasting is then interrupted to give very specific warnings to areas at risk, which are repeated and amplified as tornadoes are tracked. Little can be done to protect against damage by category 1 and higher, but taking cover in storm cellars or sheltered rooms in solidly constructed houses very considerably reduces the risk of injury by falling or flying debris.

11.7 Convective systems

We have considered two distinct ways in which individual cumuliform elements can combine and cooperate to produce more extensive and persistent systems. Individual cloudy thermals tend to combine to form the aggregate which we call cumulus congestus when there is sufficient convective instability through a substantial depth of the troposphere (Section 11.3). And in the presence of wind shear through a similar layer, individual cumulonimbus can cooperate to produce the multicell sequence (Section 11.6). We now briefly and selectively review other examples of convection elements cooperating to produce distinctive systems on scales ranging from ~100 m (minimum small scale) to more than 1000 km (synoptic scale).

11.7.1 Cloud streets

Photographic reconnaissance by aircraft over warm seas shows that small cumulus often appear in remarkably straight, parallel lines called *streets* (Fig. 11.24),

Figure 11.24 Streets of cumulus aligned with low-level airflow.

which can continue downwind for as much as 100 km, with lateral separation of adjacent ~1 km. Over land, such lines are more irregular and almost always run downwind from a Sun-facing slope or some other favoured source of thermals, but the ocean surface is normally too homogeneous to allow a corresponding origin for maritime streets. Although the mechanism is still not fully understood, it is clear that maritime cumulus streets are a symptom of a larger-scale dynamic structure in the convecting layer which has the effect of encouraging convection along long parallel lines, and discouraging it along intervening lanes. Given the marginal buoyancy of small cumulus, only slight encouragement and discouragement is needed to produce a very marked degree of alignment in what would otherwise presumably be a random cloud distribution.

There is some evidence that opposing wind shears above and below a wind maximum in the low troposphere plays a significant role, with repulsion between the opposing horizontal vorticities of rising and sinking air currents encouraging their lateral separation *[48]*. Observation of soaring sea birds suggests that similar streets of thermals are quite widespread in the maritime boundary layer even when there are no cumulus to mark them out, and this has been supported by recent observation by special radar which is able to detect cloudless thermals by backscatter from their patterns of micro-scale turbulence.

11.7.2 Convective cells

Satellite pictures of showery air streams over middle latitude oceans almost always show a characteristic distribution of ring-like patches or cells (Fig. 11.25) which was quite unsuspected before the advent of weather satellites, since the cells are too small to be resolved by the very incomplete synoptic network there, and too large to be coherently viewed by normal aircraft. Each cell seems to consist of an extensive cloudless area surrounded by a roughly hexagonal ring

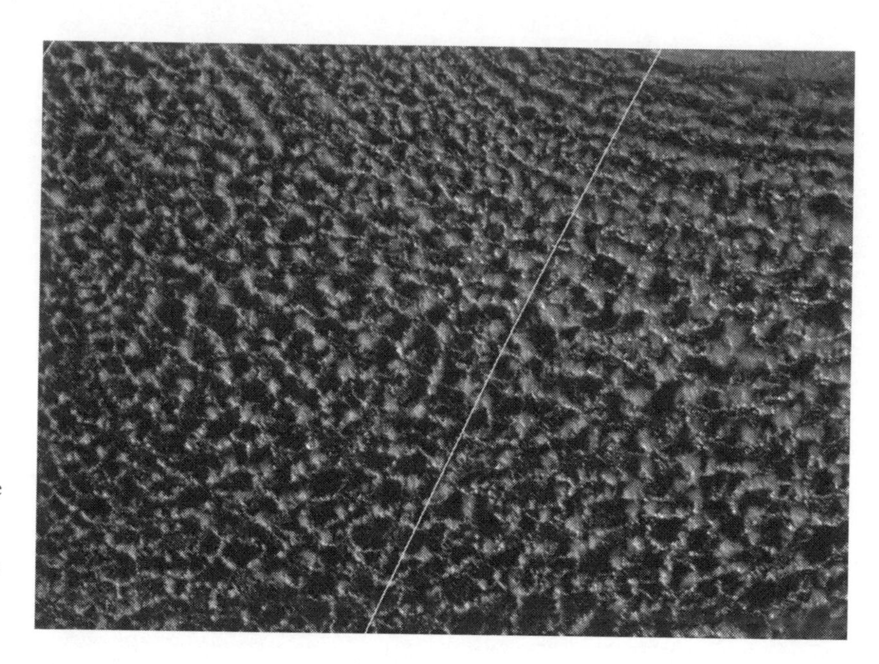

Figure 11.25 Close-up of open cells of showery convection in the centre of Fig. 2.14 . Cells are 20 to 30 km across in the centre and left of the picture, but are smaller and more linear in the upper right. Many cells include anvils.

of well-developed cumulus, including Cg and Cb, and the whole showery zone is made up of a close-packed mesh of such cells. Cells range in diameter from about 20 to 200 km, depending on larger-scale meteorological conditions, especially the depth and instability of the convecting layer. Similar cellular structure is observed on the western margins of the great maritime subtropical anticyclones, where gradually destabilizing air flows polewards and E'wards. All such cells are termed *open* on account of their open centres, and to distinguish them from the related but contrasting *closed* variety.

Oceanic areas covered by extensive stratocumulus are observed to be patterned by *closed* cells which are somewhat smaller horizontally and much smaller vertically than open cells. These consist of extensive 'pancakes' of thicker stratocumulus bounded by roughly hexagonal cloud-free margins. Again the whole zone is covered by a fairly close-packed mosaic of these meso-scale elements, each of which consists of an assembly of small-scale stratocumulus elements. Closed cells are most often observed on the eastern flanks of maritime anticyclones, where the subsidence inversion is relatively low and pronounced, and where the ocean surface is often kept cool by upwelling of cold water from the ocean depths (Fig. 11.26).

Open and closed cellular convection is also observed on a scale of centimetres in laboratory experiments pioneered by Bénard at the end of the nineteenth century. Small tanks of water heated gently from below are observed to form closed cells, while air in the same situation forms open cells. In pioneering theoretical studies of convection, Rayleigh was able to show that the crucial difference between the two types is the vertical profile of viscosity: in the air tank the viscosity is largest near the warm base, while in the water tank the viscosity is smallest there, because of the opposite dependence of viscosity on temperature in gases and liquids. On the very much larger scales of atmospheric cellular convection, the effects of molecular viscosity are overwhelmed by eddy viscosity (Eqn 7.10 and text), but the fact that the *Rayleigh numbers*

Figure 11.26 Closed cells in low-level cloud over the S Atlantic. The cloud is probably very extensive stratocumulus, which often has closed cells with horizontal scale~ 100 m, but these are much larger, at ~10 km or more to judge from individual cumulus and high cloud fragments (white).

(Eqn 10.12) are similar, if molecular viscosity is replaced by realistic estimates of eddy viscosity in the atmospheric case, strongly suggests that the two systems are dynamically similar despite their different scales. The Rayleigh numbers lie in the range known to be associated with *cellular* convection (laminar and regularly patterned), rather than the range of larger values known to be associated with *penetrative* convection (turbulent and chaotic). Though atmospheric cellular convection is far from laminar on the meteorological small scale (their convective elements are the intrinsically turbulent cumulus), on the meso-scale of the cells themselves the convective turbulence is unobservably small. Moreover the vertical profiles of eddy viscosity resemble their molecular counterparts: in open cells there is an upward lapse of eddy viscosity from a maximum close to the relatively warm sea surface (the normal state of affairs in and above the surface boundary layer); while in the case of closed cells, radiative cooling of the upper parts of the stratocumulus layer produces negative buoyancy which enhances turbulence there. It seems likely that the small laboratory models with their associated simplified (but still far from simple) dynamics mimic the meso-scale dynamics which gently but persistently constrain the cumulus elements to form the much larger patterns of cells in the atmosphere *[49]*.

11.7.3 Banded precipitation

Detailed studies of precipitation distribution in fronts show that there are quite pronounced patterns on the meso-scale (Fig. 11.27). Bands of much more intense precipitation are embedded in the synoptic-scale areas of frontal precipitation, often quite closely aligned with the nearest front. Although these bands are too small scale to be resolved by the synoptic surface network, their presence is consistent with the large variations in rainfall rate which are such a familiar

(a)

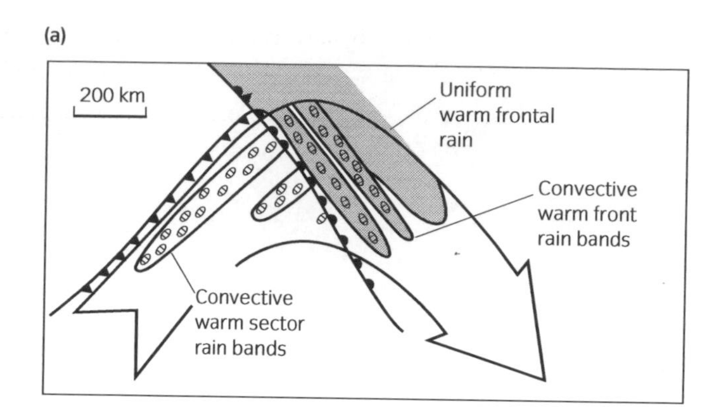

(b)

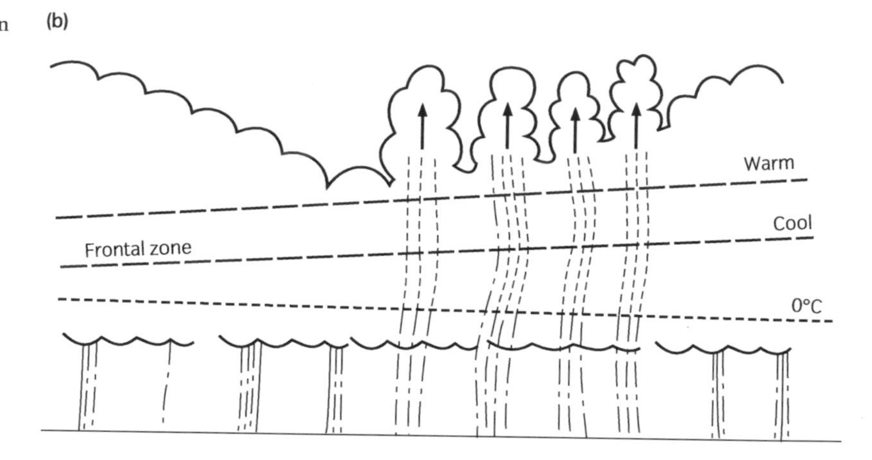

Figure 11.27 (a) Meso-scale patterns of enhanced precipitation in and near the fronts of a mid-latitude depression. The broad arrow is the warm conveyor belt (Ch 12.4) of ascending air in which bands of moderate rain are embedded (light stipple), which in turn contain patches of heavy convective rain (dark). (b) A vertical section across a warm front, showing embedded weak vertical convection whose precipitating ice and snow crystals seed the underlying cloud sheet.

feature of most fronts (Fig. 2.7). Weather radars show that these bands coincide with bands of Cb (whose shafts of precipitation show up as vertical stripes of radar echo on a range–height display) embedded in the frontal nimbostratus. The convective instability producing these showers arises from the differential advection typical of flow in the vicinity of fronts (Fig. 12.16), in which different air streams are brought together in a complex vertical sandwich, as mentioned in the previous section. Direct instability (vertical lapse of potential temperature) does not occur, but if a dry stream should flow over a saturated stream with much the same potential temperature but higher wet-bulb potential temperature, then the composite layer is unstable for cloudy convection, and clouds will mushroom up from a layer which becomes saturated by widespread ascent. Unlike the cases of severe storms, the instability here may be quite gentle, adding a cumuliform pattern to an existing stratiform system (Fig. 11.27). Though the Cb are often quite feeble, they are still capable of enhancing precipitation rates several-fold at the surface, either directly, or by their precipitating ice crystals seeding underlying layers of supercooled water cloud associated with the general nimbostratus. In cold fronts the embedded Cb tend to be much more vigorous, producing bursts of torrential rain and hail, vigorous gusts and lightning. The alignment of meso-scale rainbands parallel to the ambient or nearby fronts (some bands form outside the frontal cloud masses) presumably arises because the air flow at middle levels, in which the convection is embedded, is similarly aligned by the thermal wind effect, but the precise mechanism is not well understood.

11.7.4 Squall lines

One type of linear alignment of Cb has been known about for a considerable time, because it is big enough to show clearly on synoptic weather maps, and because the Cb are often extremely vigorous, sometimes being severe local storms. This is the *squall line* of the central and E United States. Typical conditions for the formation of a squall line are identical with those already described in Section 11.6 and Fig. 11.17 in relation to the formation of severe local storms in the same region. In such a situation Cb begin to erupt along or just ahead of the cold front which is moving from the W or NW. By some sort of dynamic cooperation they become clearly and closely aligned, to the extent that the very blustery leading edges of their squalls merge into a nearly continuous line, giving the system its title. The line may be several hundreds of kilometres long, acting like a very sharp forward extension of the cold front. However, the ability of the advancing wedge of cool, dense air to 'shovel' the warm air ahead into the updraft seems to encourage the squall line to advance faster than the parent cold front, so that it can become a detached line of powerful Cb. Over the day or two in which a particular *line squall* (as they are sometimes called when especially vigorous) is identifiable it may advance several hundreds of kilometres into the warm sector as a narrow band of intense rain and hail (Fig. 11.28). Forecasters are particularly interested in the development and progress of line squalls because of their tendency to be or become sites of very severe weather, including tornadoes. It seems that updrafts and downdrafts can become separated in much the same way as in the supercell storm (Section 11.6). Lines of Cb appear in other parts of the world (W Africa and N India for example), and although their mode of formation varies from place to place, and few are as energetic as the American line squalls, they all appear to derive dynamic advantage from their linear cooperation.

11.7.5 Tropical cyclones

There is considerable observational evidence to show that the most severe types of *tropical cyclones (hurricanes* in the Caribbean*)* can be regarded as spiral assemblages of very vigorous Cb clustered round the relatively quiet *eye*. The spiralling arms and the massive cloud wall surrounding the eye are composed of very large and vigorous Cb. The violent winds spiralling inwards towards the eye wall feed the powerful updrafts which sustain the very high rainfall rates and the copious production of ice cloud which fans out aloft to produce the dense white shield observed by satellite (Fig. 13.11). Other types of tropical weather system seem to be cooperating assemblages of Cb too, although the modes of cooperation are largely unknown. Tropical weather systems in general and hurricanes in particular are outlined in Sections 13.4 and 13.5. Even the vast intertropical convergence zone is in essence a planetary-scale girdle of very tall cumulonimbus linearly constrained (and constraining, since as usual there is no clear chicken or egg) by the Hadley circulation (Fig. 8.21).

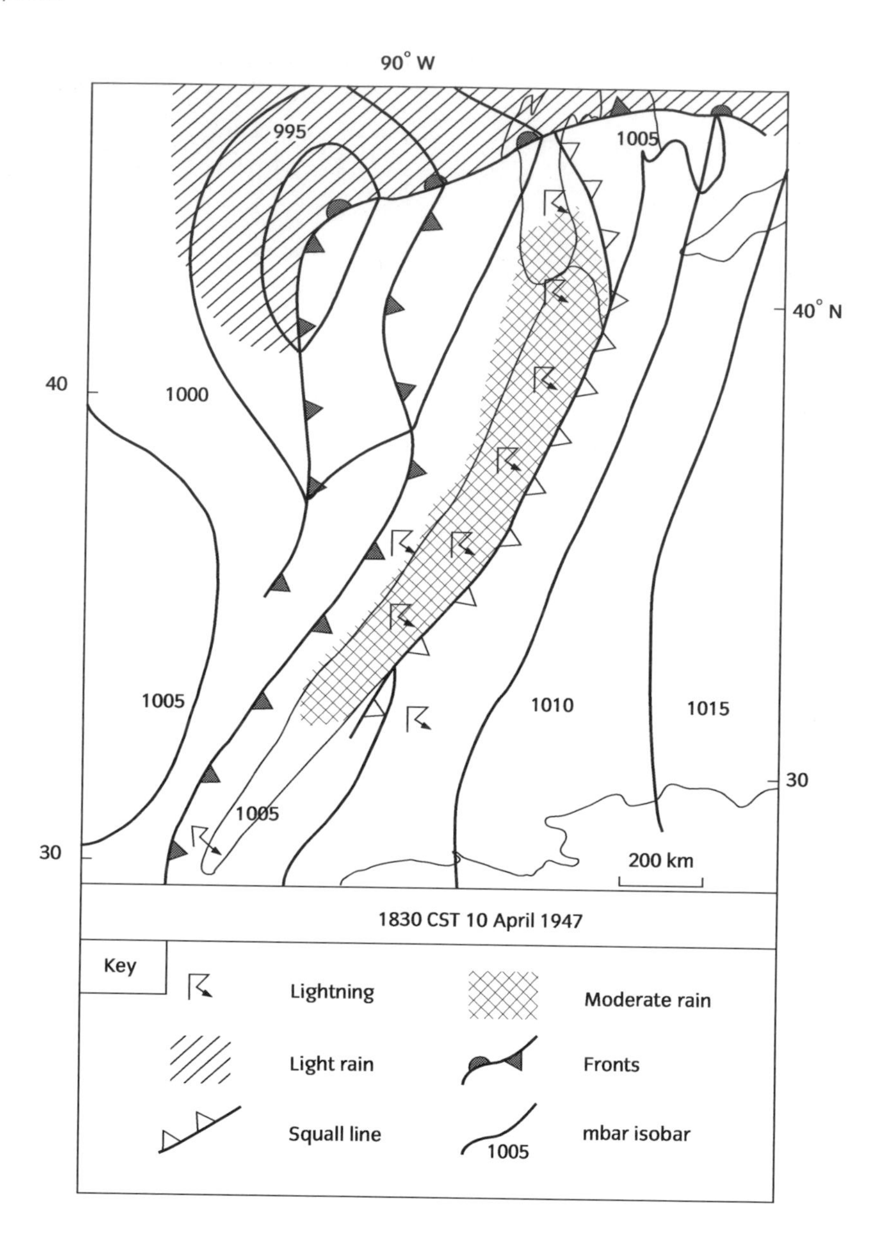

Figure 11.28 A classic case of a North American squall line, extending over 1500 km from Texas to Michigan and lying between 200 and 500 km ahead of a doublet of cold fronts.

11.8 Atmospheric waves

11.8.1 Atmospheric boundary layer

There are many examples of small and meso-scale atmospheric waves which are at least superficially similar to waves on the ocean surface. Acoustic echo sounders have been used in recent years to probe the overland atmospheric boundary layer by detecting backscatter from variations in air density on the scale of half the acoustic wavelength used (~10 cm). These are produced by micro-scale turbulence, but are patterned on larger scales which show the pictorially 'grassy' roots of convection during the day (Fig. 11.29a) and persistent waving shapes in convectively stable layers, usually at night. The latter were largely unsuspected before acoustic sounders enabled effectively continuous monitoring of the atmospheric boundary layer, but have now been observed to be typical of overland convectively stable boundary layers from the tropics to the poles.

Gravity waves

Many of the waving patterns are obviously *gravity waves*, in which vertical displacement from equilibrium is restored by gravity, with subsequent undershooting and overshooting of the equilibrium level giving rise to a series or *train* of travelling waves. Though they resemble surface water waves, they are best

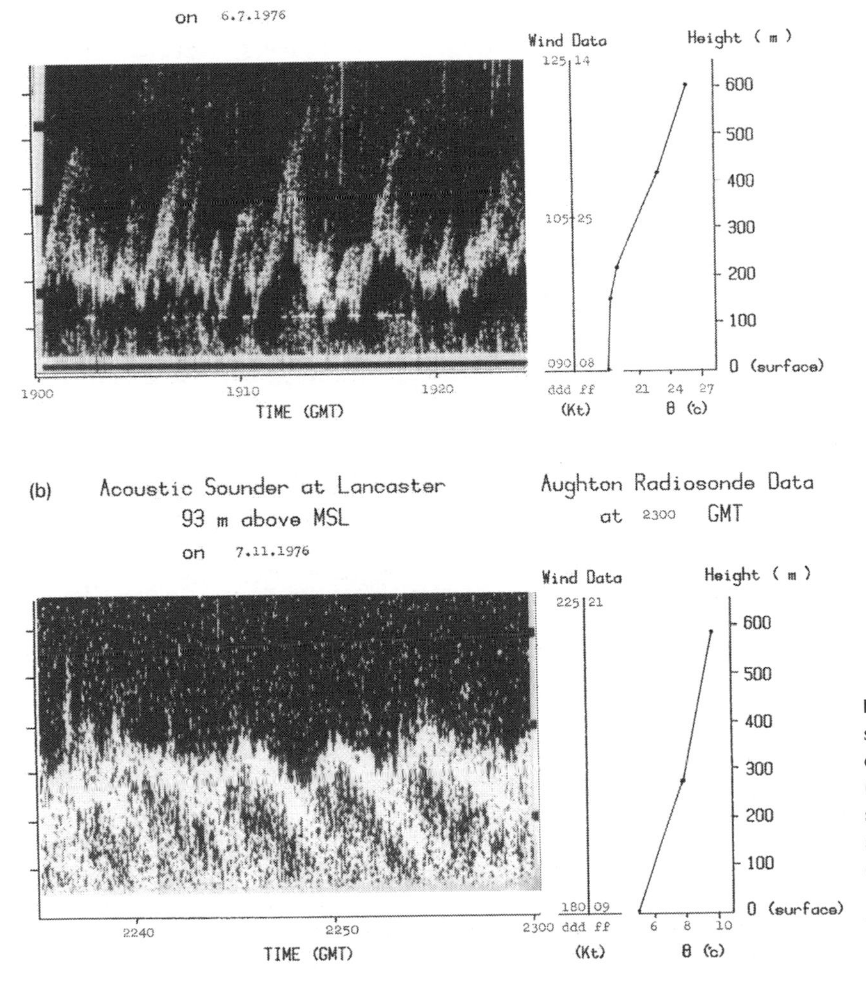

Figure 11.29 Height–time sections from an acoustic echo sounder in NW England. (a) Gravity waves show in a stable layer around 200 m, above the grassy signature of weak evening convection. (b) Braided shear waves in weak flow in the nocturnal boundary layer.

compared with the much slower *internal* waves observed under water on horizontal boundaries between water of slightly different densities. In the atmosphere such boundaries are always diffused by mixing, but the dynamics are essentially the same. A layer in which density lapses with increasing height is convectively stable, as discussed in Sections 5.10 and 5.11. It is shown in Box 11.4 that if an air parcel is displaced adiabatically and vertically in such a layer and then released, it will oscillate about the level of neutral buoyancy with what is called the *Brunt–Väisälä frequency* $N/(2\pi)$, where

$$N = \left[\frac{g}{\theta}\frac{\partial\theta}{\partial z}\right]^{1/2} \qquad 11.1$$

and $\partial\theta/\partial z$ is the vertical gradient of ambient potential temperature, which expresses the degree of convective stability of the layer (Box 5.5). This expression confirms what we might expect: that the frequency of oscillation increases with stability, just as the rate of vertical oscillation of a weight hanging on a spring increases with the stiffness of the spring.

On the summer evening shown in Fig. 11.29a, the grassy signature of weak convection from the surface reached up through a shallow dry adiabatic layer (uniform θ) and was surmounted by a rather spiky waving layer of turbulent echo embedded in a convectively stable layer ($\partial\theta/\partial z \approx 0.4$ °C per 100 m according to the nearest radiosonde). It seems likely that bursts of shallow convection at the end of a summer day were setting the overlying stable layer in gentle vertical motion. The corresponding Brunt–Väisälä period ($2\pi/N$) was about 525 s or 8.8 minutes (NN 11.3). On many other occasions at this location, smoother gravity waves were observed through the night, no doubt triggered as the stable nocturnal boundary layer flowed over the local lumpy topography.

BOX 11.4 Oscillations in a stable atmosphere

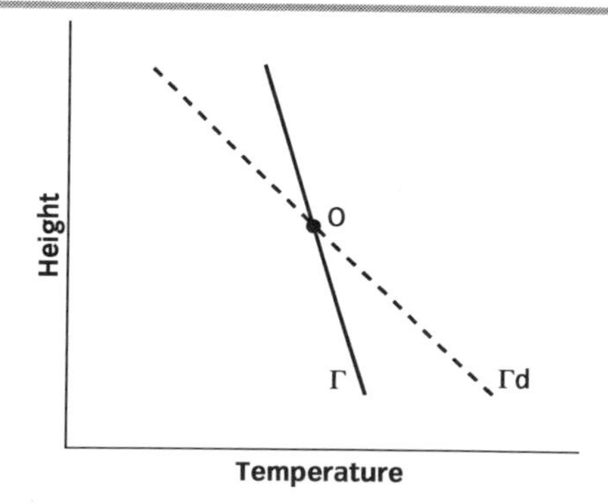

Figure 11.30 Temperature–height diagram of vertical oscillation of an air parcel (dashed) in a convectively stable environment (solid).

Figure 11.30 depicts a convectively stable layer of air in which a parcel is displaced vertically and dry adiabatically from its initial state of neutral buoyancy at O. Since the parcel cools more rapidly with height than its convectively stable surroundings, it becomes cooler than its surroundings (negatively buoyant) when raised above O, and warmer and buoyant when depressed below O. If the parcel is released after initial displacement it will accelerate towards O, overshoot, slow to a halt, accelerate back towards O again, and continue oscillating about O until the motion is damped out by drag, leaving the parcel at rest at O. We can examine the dynamics of such oscillation by adapting the method used in Box 11.2.

Analysis

If the sub-adiabatic lapse rate of the stable layer is Γ, and the dry adiabatic lapse rate of the moving parcel is Γ_d, then the temperature excess ΔT of the parcel at any height z above or below O is given by

$$\Delta T = z\,(\Gamma - \Gamma_d). \qquad \text{B11.4a}$$

Since $\Gamma_d > \Gamma$, Eqn B11.4a confirms that upward displacement (positive z) produces a parcel temperature deficit (colder than its environment), while downward displacement (negative z) produces a temperature excess, as expected and shown in Fig. 11.30.

Algebraic manipulation of the relation between small potential temperature changes $d\theta$ and dT in Eqn B5.5f re-expresses B11.4a in terms of the vertical gradient of potential temperature $\partial\theta/\partial z$ (as measured by a radiosonde for example) so that

$$\frac{\Delta T}{T} = -\frac{z}{\theta}\frac{\partial\theta}{\partial z} \qquad \text{B11.4b}$$

Multiplied by gravitational g, the left-hand side becomes the buoyancy term which we can use in the vertical component of the frictionless equation of motion (as in Box 11.2).

$$\frac{dw}{dt} = -\left(\frac{g}{\theta}\frac{\partial\theta}{\partial z}\right)z$$

Since upward speed w is dz/dt, $dw/dt = d^2z/dt^2$, and the equation can be written as

$$\frac{d^2z}{dt^2} = -N^2 z \qquad \text{B11.4c}$$

where $N^2 = \dfrac{g}{\theta}\dfrac{\partial\theta}{\partial z}$ is a constant.

Solution

Equation B11.4c describes an oscillating air parcel that is continually decelerating away from O, or accelerating back to it, at a rate which is directly proportional to its displacement from O. This is *simple harmonic motion*—the oscillatory motion which fits small-amplitude swinging pendulums, and serves as the model for many natural and man-made oscillating systems. Standard procedure *[8]*, or simple substitution, shows that its solution is of the form

$$z = A \sin N t \qquad \text{B11.4d}$$

where A is the *amplitude* of the oscillation (the maximum distance the parcel moves above or below O), which is the initial vertical displacement of the parcel in this case, and which remains constant because we have ignored the friction which would reduce it gradually with time. The behaviour of the sine function in Eqn B11.4d is such that one full oscillation is completed in a time period $2\,\pi/N$ known as the *Brunt–Väisälä* period P—the period of adiabatic oscillation of air parcels under gravity in the convectively stable layer.

$$P = 2\pi\left[\frac{g}{\theta}\frac{\partial\theta}{\partial z}\right]^{-1/2} \qquad \text{B11.4e}$$

The period is very long (the frequency $1/P$ is very low) if the layer is only slightly stable (small $\partial\theta/\partial z$), and becomes shorter (frequency higher) as the layer becomes more stable.

Note that we are ignoring the fact that B11.4d places the parcel at O at the start time, rather than at its maximum displacement. This is unimportant for the result we are examining and can be corrected by adding the necessary phase shift.

Shear waves

A very different type of wave is apparent in Fig. 11.29b, from acoustic sounding of a nocturnal boundary layer overland. A wide range of observations in the atmosphere (including Fig. 11.31 in the upper troposphere) and laboratory suggests that these interlocking wave shapes result from the *Kelvin–Helmholtz* instability of a sheared layer. KH instability is responsible for the flapping of a flag, feeding on the difference in wind speeds on either side of the flag, and restrained by the stiffness of the cloth. In the atmospheric case the sheared layers are horizontal and the restraining stiffness is the convective stability of the

NUMERICAL NOTE 11.3 Hill waves in a stable layer (see Fig. 11.30, 11.31, and Box 11.4)

- Consider a parcel of air which is neutrally buoyant in the middle of a stable layer with the waving layer in Fig. 11.29a. Potential temperature θ increased by 1.25 °C between 100 and 300 m above the surface, giving $\partial\theta/\partial z \approx 4.2 \times 10^{-3}$ °C m^{-1} or about 0.4 °C per 100 m. Using Eqn B11.4e with $g = 10$ m s^{-2}, $\theta = 20$ °C (293 K) we find

  $P = 2\pi\,[(10/293) \times 4.2 \times 10^{-3}]^{-1/2} = 525$ s or about 8.8 minutes.

- If $\partial\theta/\partial z$ is only 0.2 °C per 100 m (corresponding to a T lapse rate of 80% of the dry adiabatic—Box 5.5), the contents of [ ] are halved, and the Brunt–Väisälä period increases by a factor of 1.4 to over 12 minutes.

In Fig. 11.29b, potential temperature θ increased by 5.1 °C in 600 m, giving $\partial\theta/\partial z \approx 8.5 \times 10^{-3}$ °C m^{-1} and Brunt–Väisälä $N^2 = (10/280) \times 8.5 \times 10^{-3} \approx 3 \times 10^{-4}$ s^{-2}. The wind shear was about 6.2 m s^{-1} (12 knots) in 650 m, so that $\partial V/\partial z$ was 9.5×10^{-3} s^{-1}. According to Eqn 11.2 the Richardson number $Ri = N^2/(\partial V/\partial z)^2 \approx 3 \times 10^{-4}/(10^{-4}) = 3$.

sheared layer. Dynamic analysis of the balance between flapping by shear and smoothing by convective stability is beyond the scope of this book, but yields as a critical parameter the Richardson number Ri quoted in the discussion of the origins of turbulence in Section 10.8:

$$Ri = \frac{g}{\theta}\frac{\partial\theta}{\partial z}\Big/\left(\frac{\partial V}{\partial z}\right)^2 = \left(N\Big/\frac{\partial V}{\partial z}\right)^2 \qquad 11.2$$

When Ri falls below 0.25 (according to detailed theory) the convective stability is unable to restrain the tendency to flap, and shear waves grow in amplitude to the point where orderly flow breaks down in turbulence. The whole process converts some of the kinetic energy of the original orderly flow into turbulent kinetic energy (and ultimately heat), and considerably deepens the originally shallow sheared layer.

As shown in NN 11.3, on the occasion of Fig. 11.29b the wind shear ($\partial V/\partial z$) averaged 10^{-2} s^{-1} between the surface and 650 m, and the vertical gradient of potential temperature $\partial\theta/\partial z$ averaged about 8×10^{-3} °C m^{-1}. According to Eqn 11.2, Ri for the layer was about 3. Though an order of magnitude larger than the critical value of 0.25, the true value was probably lower on this occasion, since the shear was probably concentrated in the lower half of the layer by the greater stability there (and a doubling of shear gives a quartering of Ri). And it may be that this braided structure is an intermediate stage between stillness and full KH 'flapping'. The full range of behaviour of the atmospheric boundary layer is still very incompletely described and modelled.

11.8.2 The free atmosphere

Gravity and shear waves are also important at much larger scales and heights in the atmosphere.

Hill waves and lee waves

As mentioned in Section 10.11, air flow over hills and mountains often maintains standing waves in their lee. The theory of such waves is quite well developed but is subtle in principle and sensitive to the great range of atmospheric situations and terrain. A number of points arise from Fig. 11.31 which shows fixed (*standing*) hill waves and lee waves with air flowing through them.

(i) The dominant wavelength λ induced by an obstacle is found approximately by assuming that air flows through the steady standing waves at the undisturbed wind speed V, swinging up and down with the Brunt–Väisälä period $2\,\pi/N$. Since a parcel will complete one cycle while travelling a distance $2\,\pi\,V/N$

$$\lambda = \frac{2\pi V}{N}$$

If we substitute values corresponding to a lapse rate of about 80% of the dry adiabatic value (NN 11.3), then the Burnt–Väisälä period is about 12 minutes and we have the useful rule of thumb that the dominant wavelength in kilometres is related to the average wind speed in metres per second by

$$\lambda \approx 0.72\ V$$

the numerical factor (and wavelength) decreasing with increasing stability. Typical wavelengths of 3–20 km follow from typical wind speeds.

(ii) The wavelength of the lee waves is independent of terrain shape, but this is not the case for their amplitude—the magnitude of the vertical oscillations in the waves. The largest-amplitude waves are produced by terrain whose extent along the flow is considerably smaller than the dominant lee wavelength according to (i), but whose extent across the flow is very long. A hill or range which is very long in the direction of flow is ineffective because the air merely climbs slowly up one side and sinks down the other, without further vertical motion. And a hill which is short in the cross-flow direction is partly evaded by lateral air flow. Amplitude obviously tends to increase with terrain height, but quite low hills can be surprisingly effective if other conditions are optimum.

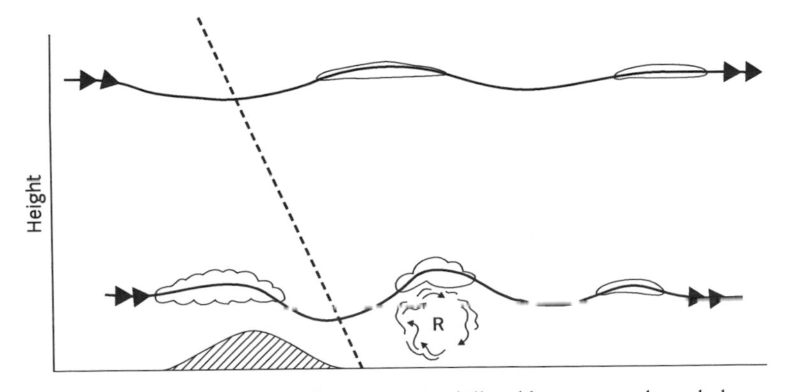

Figure 11.31 Vertical section of airflow containing hill and lee waves and a turbulent rotor R turning in a zone of nearly separated flow downwind of a hill. Note the typical increase of wavelength with height, upwind shift of leading peaks and troughs, and less typical decrease of vertical displacement (see text). Flow is usually visually more turbulent in the lowest clouds, and is often vilent in the rotor.

(iii) For a given terrain, the largest-amplitude waves arise when there is a significant increase with height in wind speed towards the barrier, and a significant decrease in stability. Largest amplitudes often appear in the vicinity of a stable layer and may be much larger than the height of the mountain or hill at heights far above its summit.

(iv) Motion is everywhere smooth and wavelike except in the immediate lee of a sharply scarped hill, and under the first wave crest downwind, where there may be a violently turbulent *rotor* in which trapped air is continually rotating and tumbling.

(v) For subtle dynamic reasons the axes of the wave crests and troughs shift upwind with increasing height throughout the region in which the wave train is exporting energy upwards. On the very large scale this is believed to be significant in communicating surface drag to the middle and upper troposphere, just as a wake of waves increases the drag on a boat by spreading its disturbance through a large water mass.

Large-scale patterns

Hill and lee waves become conspicuous when a layer of air is lifted above its lifting condensation level in wave crests, producing lenticular clouds whose smoothness and stability amid the flowing air highlight the standing wavy nature of the flow (Fig. 11.32). The dependence on atmospheric stability often enhances the number and size of lenticular clouds at dusk, producing clouds kilometres above hills which are only a few hundred metres high. Updrafts and downdrafts in large-amplitude wave clouds are sufficient to make flying conditions uncomfortable or even dangerous. The danger can be compounded by severe turbulence in the vicinity of a steep hill or rotor, and aircraft accidents have been attributed to such encounters, including the fatal crash of an airliner which broke up while allowing its passengers a view of Mt Fujiyama (height 3.8 km) in the lee of that spectacularly steep, conical mountain.

Figure 11.32 Evening Lenticular clouds in NW England, including cross-flow gravity waves. The hills of the English Lake District are about 25 km away in the line of sight.

Lenticular clouds have been noted and wondered at for centuries, but the meteorological satellite shows that vast trains of them are commonplace in the vigorous cloudy airstreams associated with extratropical cyclones (Fig. 11.33). Their extent would be unrecognized by a ground observer, and when buried in carpets of nimbostratus (invisible from the surface) they may give rise to the strong patterns of rain and snowfall observed in hilly terrain. Detailed mechanisms are complex and difficult to confirm, but it is clear that cloud production is enhanced on the upwind side of a wave crest (Fig. 11.31). If wavelengths are long enough, like the ones apparent in Fig. 11.33, then there is time enough for precipitation to be enhanced before the cloud is thinned in the lee of the wave crest. A related possibility is that precipitation falling through the denser cloud on the upwind side is able to grow much more rapidly by collision and coalescence, or that ice crystals falling through air suddenly forced towards water saturation by uplift grow rapidly by the Bergeron–Findeisen mechanism (Section 6.11). Mechanisms such as these are believed to be involved in the enhancement of rainfall on the

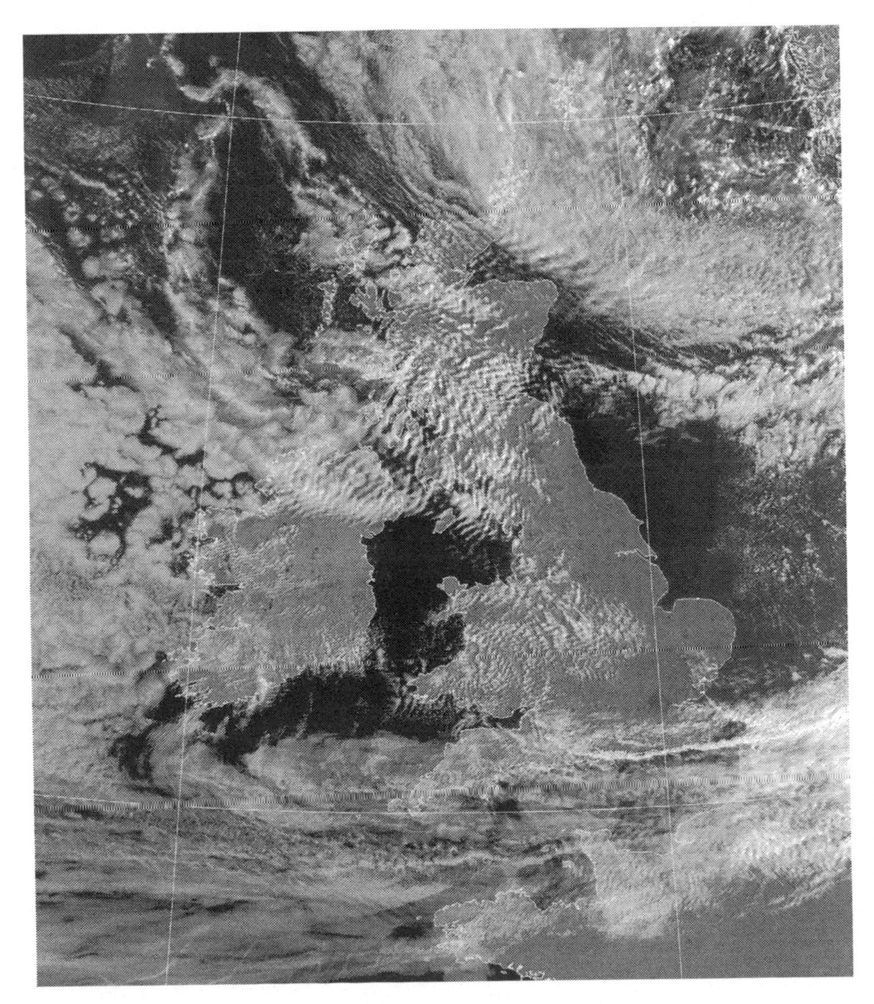

Figure 11.33 Satellite view of NW'ly flow over the N British Isles, showing extensive wave trains associated with hills and mountains there. There are some large closed cells NW of Ireland. Infrared image from sun-synchronous NOAA-6, 0836 Z, 21 August 1980.

upwind flanks of hills and mountains, and its depletion on the near downwind side, but pictures like Fig. 11.33 suggest that the effects could be widespread, though hidden amid the inherent structure and variability of weather systems. It seems clear from some detailed studies of rainfall patterns that they may trigger the formation of some of the meso-scale rain bands mentioned in the previous section.

Shear waves too can be important at much greater heights than the boundary layer in which they were described above. In the strongly sheared zones below the core of a jet stream in particular, the Richardson number may be held at about 1 throughout substantial volumes of the high troposphere by the large-scale structure. Local enhancements of shear or reductions in stability may then produce sheets of vigorous turbulence. When first encountered by high-flying aircraft, these were judged most surprising in cloudless air, where there was no obvious convective source of turbulence, and became known as *clear-air turbulence* (CAT), but they are also found in the cloudy parts of similarly sheared zones. Turbulence of small scale and amplitude is widespread in the high troposphere, as apparent on commercial flights in extensive periods of slight but persistent vibration, which feels as if the aircraft were rolling over soft cobblestones. CAT can be violent on occasion, presumably as aircraft encounter large shear waves or their turbulent aftermath, and the problem has encouraged serious study of ways of avoiding severe CAT by forecasting risky zones, or detecting current CAT by special aircraft radar. The problem remains, however.

Shear waves occasionally produce clouds which pick out the crests of the turbulent wavelets arranged across the shear (Fig. 11.34). These are sometimes called *billow clouds* because of their resemblance to breaking sea surface waves, though the dynamics are very different. Shear waves and CAT are important on a global scale because they maintain frictional destruction of the energy of large-scale air flows through substantial volumes of the troposphere, rather than have it concentrated entirely in the turbulent boundary layer, or communicated from there by gravity waves.

Figure 11.34 Billow clouds in strong wind shear in the high troposphere downwind of Mt Teide, Canary Islands. Continued observation shows that each wave has only a limited life before it breaks up in turbulence. Reproduced from *[29]* as acknowledged.

Checklist of key ideas

You should now be familiar with the following ideas.

1. The observed life cycle for small cumulus, buoyancy, with entrainment and detrainment; fractocumulus and stratocumulus as special cases.

2. Cumulus congestus as cooperating conglomerates of small cumulus, with mechanisms encouraging development, including the buoyancy effects of condensation and evaporation.

3. Development and life cycle of cumulonimbus, including tower, anvil, and convective overshoot, and associated weather at the surface and aloft.

4. Interactions between convective clouds and their ambient atmosphere from below cloud base to the troposphere.

5. Description and mechanism of severe local storms including accumulation and release of buoyancy, the role of wind shear in multicell and supercell storms, and the choking effects of precipitation; downdrafts and tornadoes.

6. Patterns in convective systems: cloud streets, open and closed convective cells, banded precipitation, squall lines, and hurricane eye-wall.

7. Gravity and shear waves in the atmospheric boundary layer and free atmosphere, including Brunt–Väisälä oscillation and the Richardson balance between shear and convective stability, hill waves, lee waves, lenticular clouds and billows.

Problems

Outline answers to these problems can be found on the **Online Resource Centre**. Answers to odd numbered problems can be found under Student Resources, answers to even numbered problems under Lecturer Resources.

Level 1

11.1 Soaring birds and gliders are often seen to be turning continually. Why do they need to do this?

11.2 Given that the top of a small cumulus rises at half the speed of the fastest updraft within it, estimate the typical speeds of such updrafts from the observation that many small cumulus grow to a depth of 100 m in 5 min.

11.3 Fractocumulus often appears at lower levels than any cloud observed before the arrival of the overlying nimbostratus. Suggest a reason for such behaviour.

11.4 In a certain NW'ly flow in the N hemisphere, showers are observed to arrive in small groups separated by clear periods which range from 5 min to an hour. Interpret these observations in terms of the motion of a regular pattern of open cells, and deduce their breadth and shower separation given that the pattern is moving at 15 m s^{-1}.

11.5 As a cumulonimbus slides overhead it becomes very dark. Where has the light gone?

11.6 An aircraft flies through a train of billows with a fairly uniform wavelength of 1 km at an air speed of 250 m s^{-1}. Given that the billows are moving with the ambient air, find the time period between successive jolts of the aircraft.

11.7 What is the minimum downward acceleration of an aircraft which can begin to throw passengers up out of their seats?

Level 2

11.8 A certain cumulonimbus is moving at 20 km hr^{-1}, and is raining evenly over a zone which is 2 km long in the direction of motion, with zero rainfall outside that. Given that 5 mm of rain is measured by all gauges receiving the full length of the shower, find the rainfall rate in mm per hour during the life of the cumulonimbus.

11.9 Find the rise in air temperature which results from the following situations: (a) 1 g kg^{-1} water cloud condenses from vapour; (b) 1 g kg^{-1} supercooled cloud water is frozen; (c) 1 g kg^{-1} ice cloud condenses from vapour.

11.10 Using the tephigram in Fig. 5.8 and referring to Fig. 11.17, trace the thermodynamic path of air moving adiabatically from a surface source in the Mexican Highlands (temperature 30 °C, pressure 900 hPa) to the 800 hPa level over position X in the United States. Compare this with air which rises in dry and saturated convection from a surface source at X with surface pressure 1.000 hPa and temperature and dew-point 26 and 23 °C respectively. What do you conclude about convection extending to greater heights?

11.11 Find the time period of vertical oscillation (the Brunt–Väisälä period) of an air parcel embedded in air with a temperature inversion of 1 °C per 50 m.

11.12 The core updrafts of certain cumulonimbus are observed to overshoot the tropopause by 500 m, penetrating a low stratosphere which has a temperature inversion of 5 °C km^{-1}. Find the implied updraft speed at the tropopause in the absence of drag.

11.13 A large area of wave clouds is seen in the high troposphere by a satellite on a certain occasion, with a coherent wavelength of about 50 km apparent for many wavelengths downwind from a hilly source region. Using the rule of thumb for lee wavelengths, estimate the wind speed in the high troposphere. How would this compare with a more stable atmosphere?

Level 3

11.14 Consider critically the likelihood of a report of 8 octals of cumulus humilis being accurate.

11.15 Discuss in detail the role of condensation and evaporation in creating and destroying buoyancy in small cumulus.

11.16 Before the outbreak of a certain summer thunderstorm, the oppressive air has temperature 30 °C and dew-point 24 °C. In the cool squall accompanying the main precipitation, these fall to 22 °C and 19 °C respectively. By following each up to the 700 hPa level on a tephigram (Fig. 5.7), at first dry adiabatically and then wet adiabatically, find the implied temperature difference there. This is clearly a measure of the buoyancy driving the cumulonimbus. What surface observations give us this temperature difference most directly?

11.17 A certain cloudy air parcel is 1 °C warmer than ambient cloud-free air. If the cloudy and cloudless parcels mix in equal masses, and 1 g kg^{-1} of cloud water is evaporated in the process, find the resultant temperature of the mixed air compared with the ambient air. Find the effective temperature of the mixed air as far as buoyancy is concerned.

11.18 Discuss the microphysical processes at work in the upper parts of a cumulonimbus as it glaciates and produces an anvil, mentioning their relevance to buoyancy and precipitation.

11.19 Use the method of Problem 6.20 to consider the growth of a giant hailstone of radius 5 cm in a cloud with 10 g kg^{-1} of water substance in collectable form, explaining how such stones can grow in a cloud which is seldom more than 12 km tall.

11.20 Describe and discuss mechanistically the tendency of cumulus of all types to appear in non-random patterns.

11.21 A passenger aircraft flies from vertically static air into a downdraft of 20 m s^{-1}. Assuming that the aircraft adopts the new ambient motion in a horizontal distance of 500 m, and that the plane is flying at 300 m s^{-1}, find its implied downward acceleration and describe the sensations of the passengers. If the pilot finally zeros the aircraft's sink rate (by climbing through the sinking air) just as the aircraft flies out of the downdraft into a 20 m s^{-1} updraft, repeat your calculations and descriptions.

11.22 Surface air contaminated by the Chernobyl reactor fire reached Northern England on 3 May 1986 just as thunder showers were breaking out near the centre of a slow-moving weak depression. Assuming convective updrafts of 3 m s^{-1} 1,000 m above the surface, use the method of Problems 7.15 and 16 to find the associated convergence in this sub-cloud layer, and estimate the reduction in horizontal area in 30 min. Consider the likely fate of contaminated aerosol particles in the converging layer.

Large-scale weather systems in mid-latitudes

12

12.1 Historical introduction

Although people have known for centuries that the great storms of middle latitudes must be much larger than can be assessed from any single point on the Earth's surface, that knowledge remained fragmentary until comparatively recently. While the European climate deteriorated from its thirteenth-century optimum, several storms caused very widespread damage and loss of life by coastal flooding. But given the slow communication between affected locations, awareness that these local disasters were caused by a smaller number of very large storms developed only in retrospect.

The conspicuous fall of atmospheric pressure before and during the arrival of a large storm was recognized as soon as barometers were developed in the seventeenth century, but it took determined journalism by Daniel Defoe to compile the first comprehensive account of the effects of a very severe gale (the great storm which struck S Britain on 26 November 1703) and publish it a year after the event *[50]*. On 21 October 1743 Benjamin Franklin was prevented from observing an expected lunar eclipse in Philadelphia by the sudden arrival of a NE'ly gale and associated cloud. He was surprised to learn later that the eclipse had been visible at Boston, 300 km to the NE, and that the gale did not arrive there until the next day. His curiosity aroused, he wrote to people living along the connecting path and was able to show from their reports that the storm had moved from Philadelphia to Boston, *against* the associated surface winds *[51]*, publishing the first recognition that storms did not simply blow along with their own surface winds.

During the next hundred years patient reconstruction of scattered observations of a few storms revealed the vast wheel-like (*cyclonic*) distribution of winds around a low-pressure centre which is now the iconic feature of weather maps, and which Franklin had perceptively recognized. Gradually the term cyclonic began to be used as a technical term to mean wheel-like rotation anticlockwise in the N hemisphere and clockwise in the S—the observed sense of rotation around low-pressure centres. Rotation in the opposite sense around high-pressure centres, became known as *anticyclonic*. Piddington and others incorporated such behaviour into an empirical 'law of storms' which worked well enough more than about 5° of latitude from the equator to allow mariners to plot a safe course around a potentially dangerous storm.

Until the development of the electric telegraph in the middle of the nineteenth century, there was no point in trying to organize networks of people to observe and possibly forecast the movement of such storms, because they often moved and changed at least as quickly as a galloping horse—the fastest available means of communication, apart from military heliographs. However, the arrival of the electric telegraph encouraged the formation of early synoptic observation networks (Section 2.8), which quickly established and extended the picture of cyclonic storms, especially those found in middle latitudes. In Britain the young Meteorological Office flourished briefly under Admiral Fitz Roy, who years earlier had captained the survey ship Beagle as it carried the young Charles Darwin around the world. Fitz Roy was a good organizer and perceptive observer (despite his fanatical opposition to Darwin's later theories), and Fig. 12.1 shows that he recognized that the contrast between warm and cold *air masses* was an essential feature of extratropical cyclones. The recognition that these contrasts were concentrated in narrow bands of especially active weather (*fronts*) had to wait 50 more years, for a small group of gifted observers, analysts, and theoreticians who worked in Norway between 1910 and 1930. Cut off from many European observations during the First World War, this group looked closely and perceptively at their observations and derived what has become known since as the *Norwegian cyclone model* (Fig. 12.2).

Although the Norwegian (sometimes called Bergen) group made good use of visual observations of middle and high clouds to observe behaviour throughout the depth of the troposphere, their models depended almost entirely on the surface synoptic network (Section 2.9). As aircraft flights become more common and longer, the need grew for direct observations of the upper air. The Second World War triggered the establishment of radiosonde stations (Section 2.10) on a global scale, which quickly confirmed the presence of a jet stream in the upper troposphere in the vicinity of every well-marked surface front, and placed both in the context of vast circumpolar vortices of W'ly flow in the middle and upper tropospheres of the middle and high latitudes of each hemisphere. The related patterns of air flow and cloud in the low, middle, and high troposphere

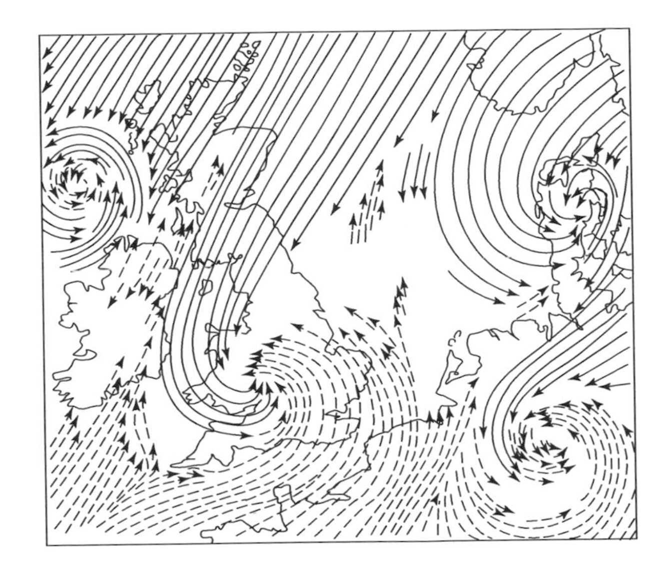

Figure 12.1 Fitz Roy's picture of contrasting warm (dashed) and cold (solid) flows in cyclonic disturbances around the British Isles.

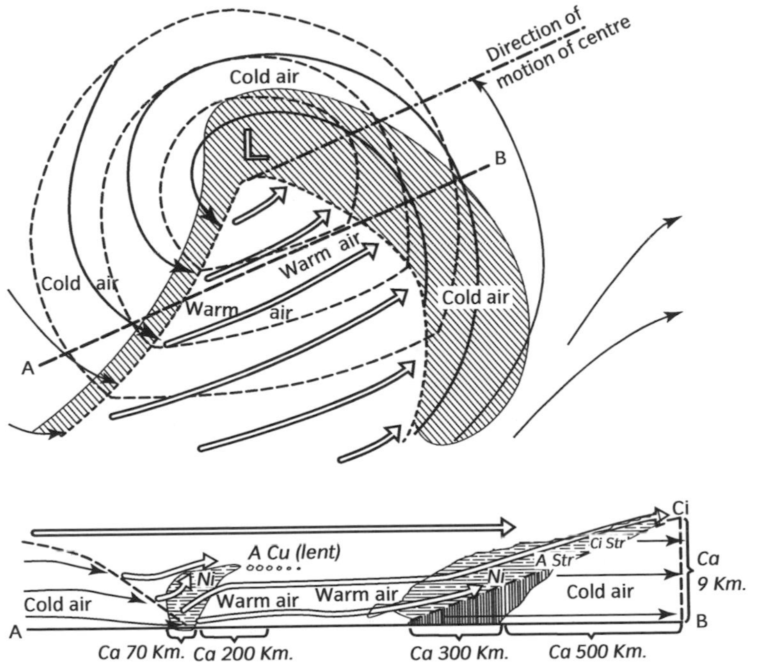

Figure 12.2 The classical Norwegian cyclone model after Bjerknes and Solberg, in *[80]*. The upper part is a plan view of surface conditions, with airflow, isobars, fronts, and areas of cloud and rain. The lower part is a vertical section from A to B, showing flow in the plane of the section, cloud, and precipitation. Subsequent amendments have added the familiar triangles and humps on the cold and warm fronts respectively, raised the tops of the deepest cloud masses, added jet streams aloft (Fig. 12.10) and altered cloud type abbreviations (Section 2.7).

which emerged from the surface and radiosonde observations were magnificently confirmed when meteorological satellites began to operate in the early 1960s, and their combination of high resolution and wide fields of view revealed a wealth of meso-scale structure which had been missed by the much coarser synoptic-scale observational network. Despite extensive subsequent work by meteorologists, the meso-scale structures revealed by satellites and weather radars are still poorly integrated with synoptic-scale structures of the Norwegian cyclone model and its developments.

12.2 Extratropical cyclones

12.2.1 Origins

The weather map in Fig. 12.3 shows a typical synoptic scale weather pattern in the N Atlantic: a front is strung out from W to E across several thousand miles of ocean, with several synoptic-scale weather systems in various stages of development along it, each marked by a wave-like distortion of the front. This is a *family* of extratropical cyclones, each of which has a similar characteristic structure and life cycle. Since each member moves E'ward along the front from a common *cyclogenetic* (cyclone forming) zone near its W end, the series from W to E on Fig. 12.3 matches the stages of the cycle for any one member, as detailed in the idealized series of Fig. 12.5.

The front which links these depressions (a widespread common name) is called the *polar front* in the Norwegian terminology which has become universal, and is a region of relatively sharp temperature contrast separating warmer air from lower latitudes from colder air from higher latitudes. Several factors can localize and intensify the overall meridional temperature gradient imposed by the unequal distribution of solar input (Section 8.7), continental coastlines being one of the most effective in the N hemisphere. For example, the N Atlantic in winter maintains a particularly sharp contrast between the cold *polar continental* air in the N American interior, and the warm *tropical*

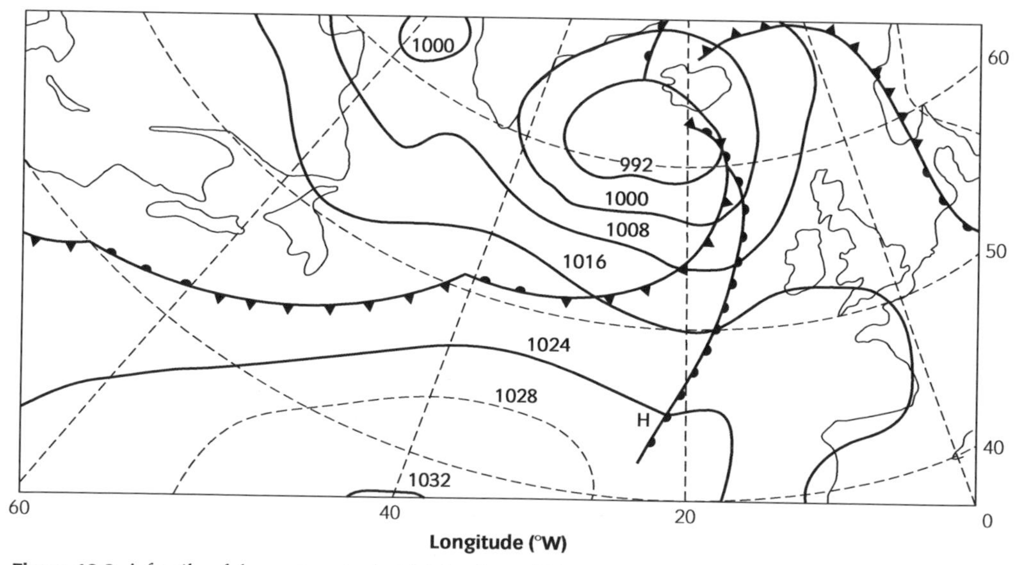

Figure 12.3 A family of depressions in the N Atlantic at 1200 Z on 6 July 1972. Like all actual cases it departs from the ideal, there being no really mature member between the old occluding system in the E and the young waves to the W. Fronts in the far NE are debris of the earliest members of the family. Notice the ridge of high pressure over the British Isles ahead of the warm front.

maritime air over the W Atlantic warmed by the powerful poleward flow of warm water in the Gulf Stream. Although such contrasting *air masses* are well defined only in the lower troposphere (below the free-flowing W'lies of the upper troposphere) they are a useful descriptive concept, partly because they define surface atmospheric conditions, and partly because the strong thermal contrasts between them are observed to be favoured sites for *cyclogenesis*. For example, the NW Atlantic breeds depressions which drive across the high N Atlantic to arrive in their maturity in the Western Approaches of Europe. And the NW Pacific, and the N Mediterranean (in winter), similarly breed depressions in the mid-latitude W'lies (Fig. 12.4). The Norwegian school believed that the polar front generated depressions through a wave-like instability, like a huge version of the KH instability of waves on a rippling flag (Section 11.8.1). Though current observation and theory does not support this view (Section 12.6), it is still widely used as a descriptive device. Whatever its detailed dynamics, it is clear that cyclogenesis depends on the juxtaposition of contrasting air masses.

12.2.2 Life cycle and structure

The stages in the life cycle of an extratropical cyclone are shown in Fig. 12.5—another classic diagram from the Norwegian school. In stage (a) the pre-existing front begins to develop a large elongated mass of deep stratiform cloud in response to persistent uplift throughout the troposphere. As low-level air converges to replace the rising air, the planetary vorticity (f) is concentrated (Section 7.16) and the air begins to rotate cyclonically, twisting the warm air polewards to the E of the centre, and the cold air equatorwards to its W, and sharpening the horizontal temperature gradients, especially in the converging low troposphere.

These developments continue in stage (b), by which time large areas of continuous precipitation have appeared close to the deforming front. Satellite pictures (Fig. 12.6) show a large area of middle and high-level cloud, often with a spume of cirrostratus fanning NE'ward over the crest of the frontal wave. The poleward curvature of this high cloud shows that the flow there is twisting anticyclonically in strong divergence above the main uplift (Fig. 7.33). By this stage sea-level pressures are falling sharply near the wave crest as divergence in the upper troposphere removes air from the centre of the depression faster than it is replaced by convergence in the low troposphere.

As is conventional, the fronts in Fig. 12.5 are marked in their surface positions. In fact the regions of strongest horizontal temperature gradient (the *frontal zones*) slope upwards at such very small angles (Figs. 7.19 and 12.2) that in the upper troposphere they lie hundreds of kilometres on the cold side of their surface positions. As the depression develops, the frontal zones become more and more sharply defined, especially in the low troposphere to the W of the crest of the frontal wave. Temperature gradients of at least 5 °C per 100 km are observed where the cold polar air undercuts the warm tropical air at the *cold front*, forming a very sharp thermal boundary between the *warm sector* and the cold air to its NW. The *warm front* acts as a weaker E boundary of the warm sector.

The depression develops with further growth in the meridional amplitude of the frontal wave and reduction of the pressure minimum at its crest. Figures 12.2 and 12.5c are regarded as beginning the mature stage of the life cycle.

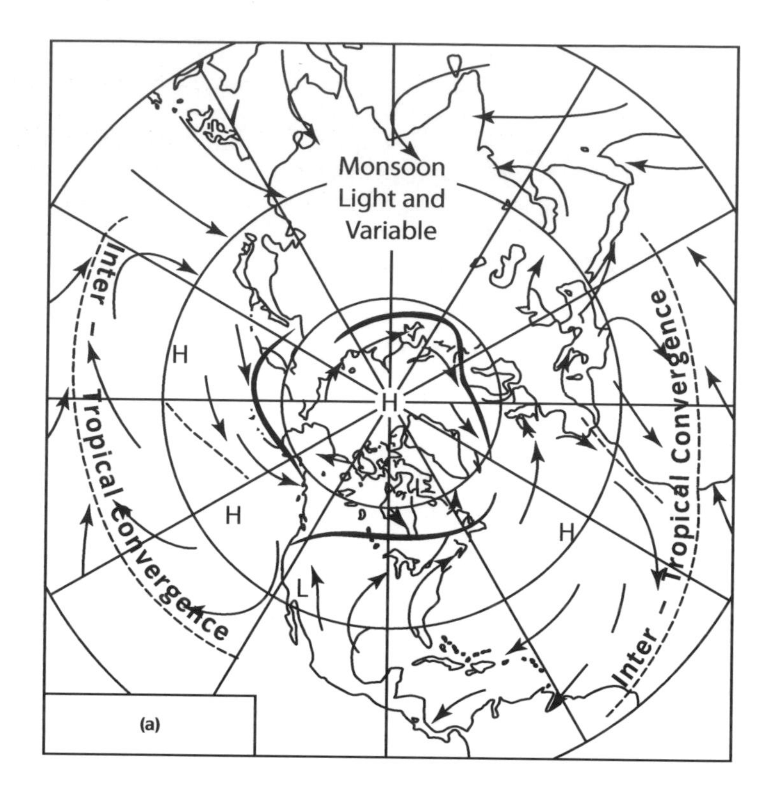

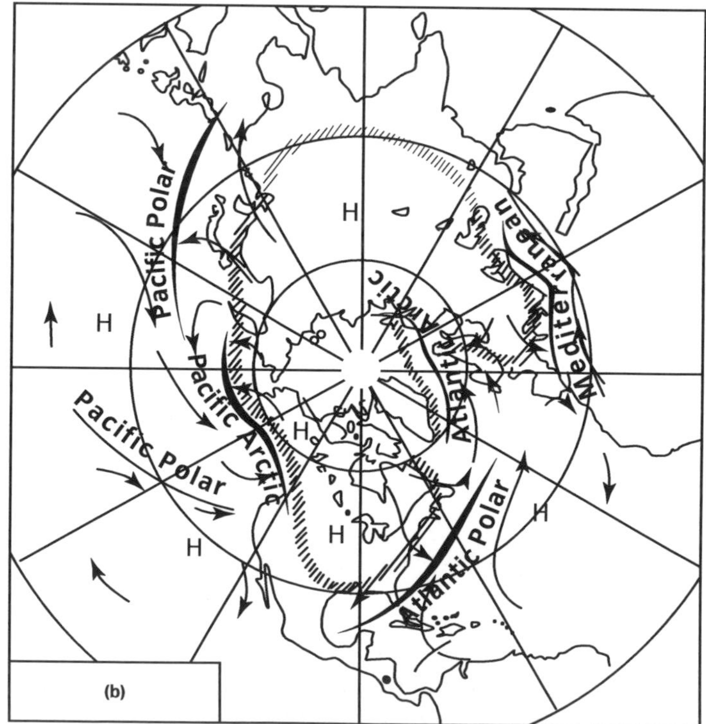

Figure 12.4 Zones of cyclogenesis in the N hemisphere in (a) summer and (b) winter. Note the N'ward migration in summer and the associated collapse of the Mediterranean cyclogenetic zone, on whose absence the local tourist industry depends.

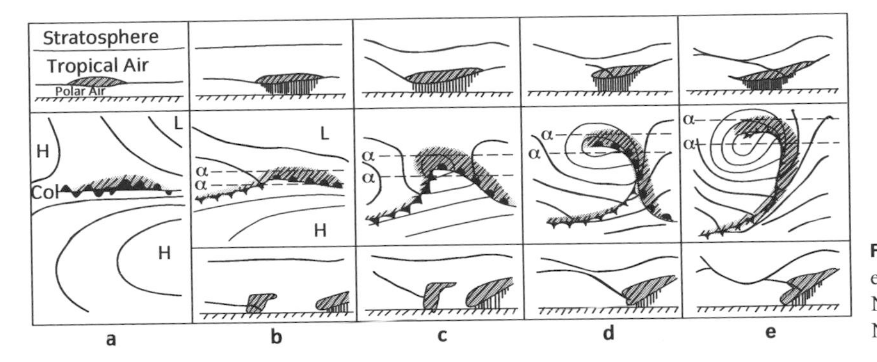

Figure 12.5 The life cycle of an extratropical cyclone in the N hemisphere as depicted by the Norwegian school.

Figure 12.6 Infrared satellite picture of young and old depressions in the NE Atlantic. A young system still in the open-wave stage lies over Spain and France, moving NE along the front trailing from the old occluded system whose comma-shaped cloud pattern is centred ever Scotland. Notice the large meso-scale swirl of cloud to the W of Ireland and the mottled cloud patterns in the surrounding showery airstreams (1917 Z, December 14, 1980). (See Plate 13).

The minimum surface pressure may now be tens of hPa below its undisturbed value, and surface winds may be at gale force over much of the area of cyclonic flow. Large masses of nimbostratus blanket the warm and cold frontal zones, giving substantial falls of rain or snow to the surface, depending on season and latitude. An extensive flange of altostratus and cirrostratus reaches polewards and E'wards ahead of the warm front, enlarging the 'spume' first noted in Fig. 12.6. Associated with the well-marked warm and cold frontal zones, there are powerful cores of the W'ly *polar front jet stream* in the high troposphere. The whole situation is described more fully in Section 12.4.

In the earlier stages the meridional extent of the frontal wave was smaller than its zonal extent. Now, however, the two dimensions are comparable and the warm sector is narrowing quickly, like an ocean swell steepening as it approaches the beach (though the cyclone 'wave' is *not* a vertical gravity wave like ocean waves). This narrowing leads to the *occlusion* of the warm sector, as the cold front overtakes the warm front and forms a composite or *occluded* front (Figs. 12.5d and 12.7). During occlusion the still-deepening low-pressure centre moves polewards and W'wards from its previously central position in the depression, drawing out the occluded front which connects it to the rump of the warm sector. The jet stream core is still aligned with the trailing cold front to the W, but not with the occluded front, which it crosses somewhat polewards of the shrinking warm sector (Fig. 12.19). Through the thermal wind relation, this lack of alignment shows that the thermal contrast across the occluded front is much less than it was in the precursor warm and cold fronts. There is still some contrast, since the fresh polar airstream W of the occluded front is a little colder than the cool air ahead of it.

After occlusion the depression progressively loses the marked temperature contrasts which were such a feature during its growth to maturity. In fact the circulation centre is now predominantly cold (and hence termed *cold core*), and spinning well polewards of the nearest warm air mass. The occluded front may continue to maintain widespread cloud and precipitation for several days, during which time its poleward end may spiral round the low-pressure centre, producing a striking 'comma' shape on satellite pictures (Fig. 12.7). This is about the only time that the low-pressure centre, which is such an obvious feature of surface synoptic weather maps, appears clearly in satellite pictures. The cyclonic rotation in the low troposphere may persist, maintaining strong winds for several more days, but it is like a vast eddy thrown off after vigorous growth, inexorably slowing by friction and filling as convergence below exceeds divergence aloft (Fig. 7.33).

12.2.3 Families of depressions

The E'ward and poleward motion of the depression from birth to occlusion is not well represented in Fig. 12.5. The open wave in particular (stages a and b) may move and develop rapidly in the prevailing W'ly flow, testing the skill of forecasters, especially in data-scarce areas like the central Atlantic. E'ward motion continues through maturity but dies away after occlusion, leaving the frontless swirl of Fig. 12.5e rotating about a nearly fixed centre. However, the residual cold front, which was such a prominent feature of earlier stages, still lies across the Atlantic like a string unwound from the equatorward rim of the depression as it rolled E'wards. This residue contains considerable contrast between polar air, flowing equatorwards down the W side of the depression, and tropical air at lower latitudes, often on the poleward flank of a subtropical anticyclone. This thermal contrast may encourage further cyclogenesis, beginning to bend the front again as in Fig. 12.5a and passing through the same life cycle as it works E'ward along the track of the old trailing front to merge with its decaying parent after occlusion. The second wave in turn leaves a trailing front which can encourage the growth of a third, and so on through a family of six or more depressions, all decaying into and re-invigorating the same terminal swirl begun by the founding parent. The sequence comes to an end when no further cyclogenesis occurs on the trailing front, which often happens as a ridge of high pressure

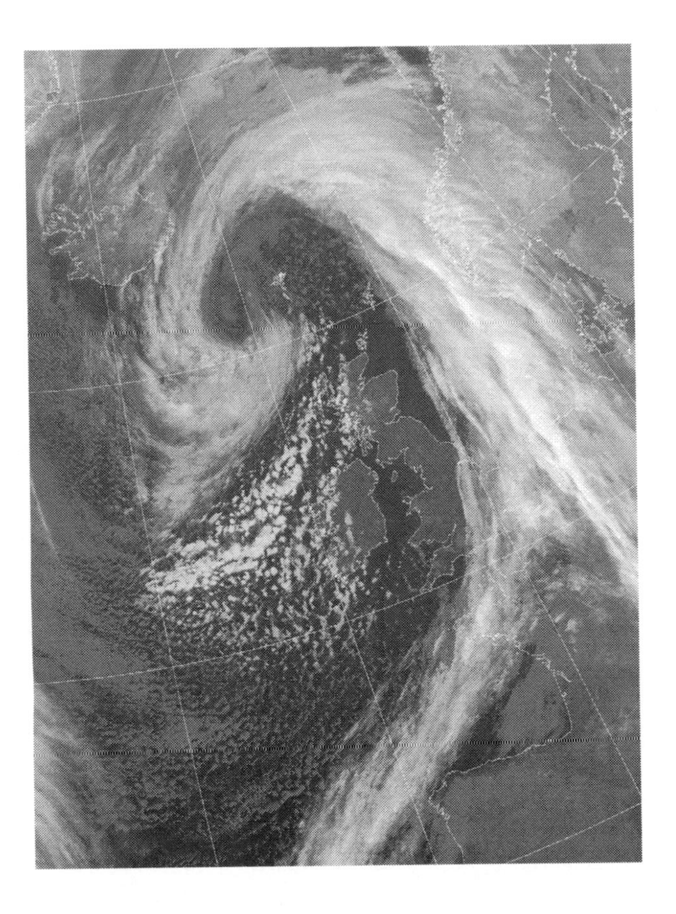

Figure 12.7 Infrared satellite picture of an occluded depression. The warm sector remnant lies over France, with an extensive occluded frontal cloud mass lying meridionally over the North Sea, S Norway, and the Norwegian Sea. The great W'ward hook of cloud surrounds the low-pressure centre at sea level, and lies to the N of the mottled pattern of shower clouds in the cool westerlies. (0910 Z 4 January 1980).

begins to build there, reaching polewards from the local subtropical anticyclone. The old terminal vortex, cut off from further invigoration, finally dwindles and dies, leaving little trace of the enormous weather systems which have troubled a considerable fraction of the hemisphere's middle latitudes for two weeks.

12.3 The mature depression

We can now detail the structure of an extratropical cyclone in its mature stage. As shown in Fig. 12.5c, this has well-developed cold and warm fronts separating the warm sector from cold air to the E and even colder air flowing equatorwards to the W. The structure is summarized in the classical plan and vertical cross-sections of the Norwegian Cyclone Model first published in the 1920s (Figs. 12.2a and b).

12.3.1 Pressure and wind

The low-pressure centre lies close to the apex of the surface fronts, and the isobars form a large pattern with many closed isobars when drawn at intervals of one or two hPa, as is normal practice. The isobars curve fairly smoothly in the cold air but run almost straight across the warm sector. The pattern of fronts and their associated weather usually moves parallel to the warm sector isobars—a useful rule of thumb in forecasting. The isobaric curvature absent from the warm sector appears to be concentrated at the two surface fronts, especially at the cold front where a deep kink in the isobars is often observed.

These sharp kinks are hydrostatically consistent with a boundary between cold dense air and warm less dense air that slopes upwards from the position of the surface front (Fig. 12.8). Looking horizontally towards the cold air, the increasing depth of denser air produces a relatively sharp gradient of surface pressure. In the presence of the overall fall of pressure towards the low-pressure centre, this sharp pressure gradient appears as a sharp clockwise turn of the isobars as we go from the warm sector across the surface cold front and into the cold surface air (N hemisphere). Of course there are no such discontinuities of air density in the real atmosphere, but the boundaries can be quite sharp on the synoptic scale, particularly at the cold front.

The quasi-geostrophic balance of large-scale air flow ensures that these kinks in the isobars are associated with winds turning sharply clockwise as we go across the surface cold front from its warm to its cold side. Clockwise rotation of wind direction is known as *veering*, whereas anticlockwise rotation is termed *backing*. In Fig. 12.2a the geostrophic wind veers from azimuth 260° to about 310° in going across the surface cold front, which is roughly equivalent to a veer from a SSW'ly to a NW'ly in compass directions. Of course surface winds are not accurately geostrophic because of friction (Section 7.11), but although reduced and backed by the effects of friction, they too veer sharply across the cold front.

In fact the surface position of a vigorous cold front is often the site of vicious squalls from cumulonimbus embedded on the warm edge of the nimbostratus, so that the sharp veer can be concentrated into a few hundred metres on the edge of the squalls. This behaviour was well known to sailors in square-rigged sailing vessels, who often had to go aloft to reset the angles of the spars, and to reef in the larger sails, just as the squally wind veer arrived. Struggling with heavy wet canvas cracking like gunshots under the strain of winds gusting well above gale force while lashed by bursts of rain and hail (and lightning—Section 6.13) would be highly uncomfortable in any situation, but 30 m above a deck heaving in the confused sea produced by the rapid veer of wind, it was an unforgettable

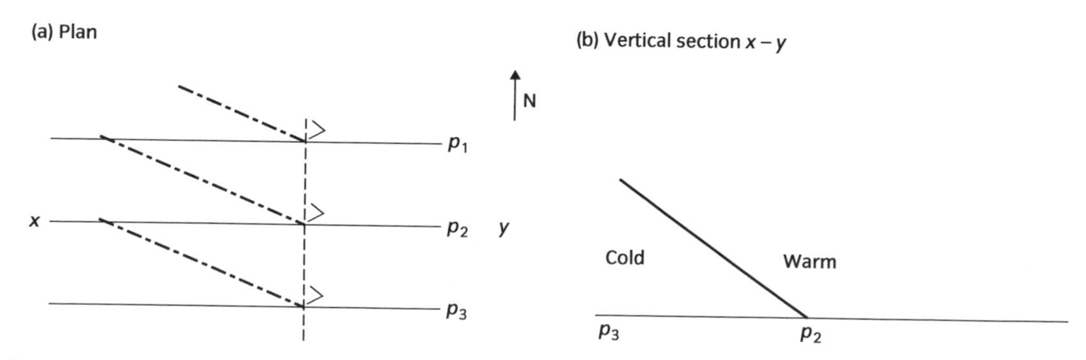

Figure 12.8 Horizontal pressure patterns near a front. The solid lines in (a) show the isobar pattern associated with uniform W'ly geostrophic flow in the N hemisphere. These change to the dash-dot pattern when we add a cold front as shown (b), to accommodate the associated wedge of denser air.

Figure 12.9 The famous 22° halo is sometimes more easily seen round the moon than round the sun, where it is easily lost against the bright sky. In either case it indicates the presence of a thin veil of cirrostratus which often comes second (after hooked cirrus, Fig. 2.12) in the sequence warning of an approaching warm front. (See Plate 14).

and highly dangerous experience which exacted a heavy annual toll of death and serious injury for the centuries in which such ships were widely used. By contrast the wind veer associated with warm fronts (as we cross from the cold to the warm side of the surface front) is usually smaller and more gentle. Such veers of wind and changes in pressure tendency are familiar to fixed observers as fronts drive across from the W, and the sharp rise of pressure after the passage of the surface cold front helps it to be positioned accurately on a synoptic chart.

Notice that all these directional relationships between pressure and wind are expressed for the N hemisphere and reverse in the S.

12.3.2 Cloud and precipitation

Cloud

Figure 12.2b represents a vertical section through a warm sector and its bounding fronts, along the line A–B in the plan view (Fig. 12.2a). Working W'wards from its E boundary, we have the sequence of observations and weather which would be experienced at a fixed location as the whole structure moves across from the W. Firstly (well to the NE of B) there is fair weather in the ridge of high pressure which usually precedes the warm front, with stunted cumulus if there is any convection at all. Then the outlying edge of the frontal cloud sheet is heralded by the appearance of hooked cirrus (mare's tails)—patchy tufts of ice cloud in the high troposphere (Fig. 2.7). Their progress across the sky indicates the speed of approach of the depression while their individual movement along the overall pattern reveals the relative flow of the air in the high troposphere. According to the thermal wind relation (Section 7.12) this relative

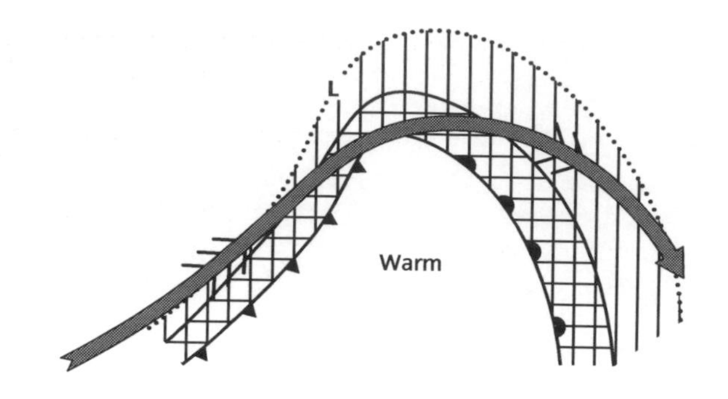

Figure 12.10 An idealized picture of a mature extratropical cyclone (Fig. 12.2), showing plan positions of the centre of low pressure (L), the warm and cold fronts, areas of stratiform cloud (hatched) and precipitation (cross-hatched), and the axis of the associated polar front jet stream centred on the 300 hPa level, with barbs in regions of highest speed.

motion should have cooler air on its left looking down the relative flow in the N hemisphere, i.e. out of the paper on the right-hand side of Fig. 12.2b, which is from right to left when viewed by an observer ahead of the approaching front, watching the hooked cirrus as they invade the sky. This is readily confirmed by observation, since the mare's tails are usually unobscured by lower cloud. The curved hooks are formed as streamers of larger ice crystals in the cloud fall into the slightly slower air below.

The hooked cirrus is usually followed by a film of cirrostratus which fills the high troposphere and diffuses the sunlight (Fig. 12.9) like thin ground glass. The shallow layer of geometrically regular, hexagonal ice crystals can refract sunlight to produce a faint but dramatic halo around the Sun or Moon, whose angular radius is 22° *[52]*. The cirrostratus thickens steadily as its bottom edge lowers, and by the time this reached middle levels (i.e. below the 7 km level) the Sun is barely visible. The cloud layer now consists of cirrostratus and altostratus, though quite often only the altostratus is visible from the surface, the cirrostratus being obscured. Aircraft observations suggest that such thick layers are usually built up of several thin layers rather than a single thick one. When seen over hilly land, there are often darker patches where layers of moist air in middle levels form lenticular clouds in standing waves over the hills (Section 11.8). Shortly after this the cloud thickens and darkens to nimbostratus and the precipitation begins. Nearer the surface position of the front, most of the lower troposphere too is filled with cloud, so that there is a deep multi-layered sandwich of cloud producing gloom and moderate to heavy precipitation at the surface.

Precipitation

Frontal precipitation consists of relatively large ice crystals sifting down through a deep layer of gently rising frontal air. If the air in the very low troposphere is cold enough, no more than a degree or so above freezing, precipitation will reach the surface as snow. This tends to happen in winter if the depression is at fairly high latitudes (or middle latitudes in winter continental interiors), or if the surface is high enough above sea level. In a maritime climate such as Britain's, even in winter the precipitating snow normally melts at least several hundred metres above sea level and falls as rain on low ground. The large wet flakes in the melting zone produce a strong horizontal echo pattern on weather radars (Fig. 6.29) known as the *melting band*. There are usually large quantities of supercooled water cloud in the lower troposphere, some of which is seeded by ice crystals precipitating from above to form meso-scale patches of heavier rain at the surface (Section 11.7). Apart from these, precipitation in warm fronts tends to be more uniform and steady than in any other type of weather system.

As precipitation falls into the sub-cloud layer it tends to moisten the layer by evaporation. Although there is no sunlight near the surface in these conditions strong enough to drive convection, the strong winds and wind shears can maintain mechanical convection (Section 10.8). Tumbling masses of air moistened in this irregular way reach their various lifting condensation levels to produce the low ragged cloud fragments called *fractocumulus* (Fig. 11.5) typical of such conditions. In quieter conditions, rain falling into cold air just above the surface near the surface front may saturate the cold air to produce *frontal fog*.

The traditional vertical cross-section (Fig. 12.2b) is probably unrealistic in indicating that the *top* surface of the warm frontal cloud mass descends towards the warm sector, though such is the respect for the Norwegian model that the feature is reproduced faithfully in many texts. Aircraft and satellite observations show that the top surface is fairly horizontal across the front, although meso-scale sculpting can cast dramatic shadows at dawn or dusk. The frontal cloud mass is therefore wedge-shaped in section with maximum thickness close to the surface front.

12.3.3 Jet stream

The Norwegians had little upper-air data originally, and therefore had to rely on visual observations from the surface. Though they knew that there were high winds in the upper troposphere near fronts, they had very few direct measurements, and consequently did not emphasize this feature. However, data from the radiosonde network changed all that, and the polar front jet stream is now an integral part of the model, and is of great practical importance to the large amount of air traffic routed at these levels. The core of the jet stream is shown in Fig. 12.10, and outlined in the patterns of Fig. 12.2. Notice that the jet cores lie in the warm air (although at these levels there is nothing like the thermal contrast typical of the low troposphere) roughly vertically above the locations of the frontal zones in the middle troposphere, consistent with the thermal wind relation. Core speeds of as much as 75 m s^{-1} are quite typical of vigorous cold fronts. The hooked cirrus ahead of the warm front is moving on the E flank of the jet core there: hence the visible equatorward motion noted above.

From the appearance of Fig. 12.2b it might seem that the great wedge of warm frontal cloud is produced by air creeping up the frontal slope in the plane of the frontal cross-section, and this is shown in diagrams in oversimplified texts. Though there is some motion in that direction, it is very much weaker than the component of motion along the front—perpendicular to the plane of the cross-section. In fact it is the flow up the slight but extensive slopes of air flows *along* the fronts which begin and maintains the production of cloud and precipitation. In the case of the cold front, air is climbing as it travels polewards (into the page in Fig. 12.2b), whereas in the warm front the air in the upper troposphere is climbing as it flows equatorwards (out of the page). The three-dimensional motion of the air is difficult to deduce accurately from analysis of synoptic data, and is very complex insofar as it is known, as will appear in the next section.

12.3.4 Warm sector and cold front

Continuing W'ward along Fig. 12.2, we have now reached the warm sector. This is often largely free of substantial cloud sheets, at least outside the poleward extremity where stratus and even fog may form as the humid air moves polewards over colder and colder surfaces (a classical example of advection fog). Being a tropical air mass, the relative warmth of the warm sector can be very noticeable at and near the surface. In winter this is an important source of heat for middle latitudes, and is an obvious example of the warm air advection considered more generally in Sections 8.9 and 8.10. On some occasions there may be much more than the typical scattering of small cumulus, with substantial patches of nimbostratus and cumulonimbus. Convective instability often increases on the W side of the warm sector (the N American squall line being an extreme example (Section 11.7). When winds are strong, as they often are in winter, warm sectors can feel chilly because of wind chill, even though air temperatures may be well above the seasonal average, as they usually are.

The temperature contrasts between the air before, in, and after a warm sector, can obviously be masked or accentuated by local cooling or heating, especially overland in summer. The dew-point temperature is then a much safer indicator of air-mass difference, being unaltered by simple isobaric heating or cooling (Section 5.4). The dew-point is highest in the warm sector, indicating its origin in the humid warmth of lower latitudes.

The cold front may arrive vigorously, with heavy bursts of precipitation, squally winds and thunder, and an abrupt wind veer and pressure kick (Fig. 12.8). On other occasions it may arrive more quietly, though the sky generally fills with cloud much more rapidly than it does in a warm front. Precipitation rates are typically higher and more variable than in warm fronts, because of the cumulonimbus often embedded in the nimbostratus, usually in groups as mentioned in Section 11.7, and often with hail and thunder.

The cold frontal cloud mass is usually narrower than the warm frontal cloud mass, because of the steeper frontal zone associated with the stronger temperature contrast in the lower troposphere (Eqn 7.27a). Its unpleasant weather therefore does not last too long, though note its very different appearance in young depressions mentioned below. The clearance at the back edge of the frontal cloud mass can also be much faster than the advance of the leading edge of the warm front, and occasionally offers a panoramic view of a sharp nimbostratus edge stretching from horizon to horizon as it passes over the observer (Fig. 12.11)—one of the few occasions when a surface-based observer gets a sense of the vast horizontal scale of large weather systems. Facing the retreating edge, small-scale structures such as aircraft condensation trails, etc. embedded in the upper part of the edge are seen to move quickly from right to left with the jet stream flow there. The observer is now in the cold polar air behind the front, which is kept in a convectively unstable state by equatorward flow over progressively warmer surfaces. These NW'lies (as they usually are) are therefore showery, with cumulonimbus hurrying along singly or in groups in the brisk flow of cool clear air. Sometimes there are larger areas of merged cumulonimbus or nimbostratus-like segments of cut-off occluded front which give short bouts of unpleasant weather, and of course it may become very cold in winter as the winds veer gradually to become N'ly.

Figure 12.11 The clearance behind a cold front coming in from the W off the Irish Sea at winter sunset. The cloud is still deep and dark over the camera, but its base slopes upwards towards the approaching clearance, whose brightness suggests that there is little cloud beyond, except presumably for some cumulonimbus in the cool NW'lies behind the front, flowing over the sea, which is warmer than the land at this time of year.

12.3.5 Diversity of weather

If this depression is not the last member of a family, the showers will gradually die away as they are suppressed by the ridge of high pressure preceding the warm front of the next depression advancing from the west. The observer is now about to repeat the sequence of observation and experience recounted above as another system moves across, but of course no two systems are identical, and the next one may differ very considerably in intensity, path, speed, and maturity. The orientation of the fronts relative to their direction of motion can strongly affect conditions experienced at any particular geographical location. In the mature state, fronts usually make quite large angles with their direction of propagation, with the result that they pass breadth-wise across any location, and do not last very long if they are moving quickly, as they often are. However, in an immature state, when the frontal wave is still very open (Fig. 12.5b), they tend to move nearly lengthwise, so that most of the local surface is unaffected by cloud, precipitation, etc. while an unlucky central stripe passes under almost the entire length of both the warm and cold fronts, experiencing one or more days of nearly incessant rain or snow which may fade and re-intensify as the slightly buckled fronts meander meridionally overhead.

In the vicinity of the terminal vortex of a depression family, which may persist with little motion (other than rotation) for a week or two, areas may suffer conditions which have very little to do with the sequence associated with the mature depression. If the location is to the E of the centre of the old low, winds are persistently S'ly, and there are periods of rain and cloud as the debris of old fronts pass over, with little thermal contrast or organized cloud structure. If the location is to the W of the low centre, a persistent showery N'ly flow is experienced, with longer periods of rain or snow as frontal debris swirls round from the N and E.

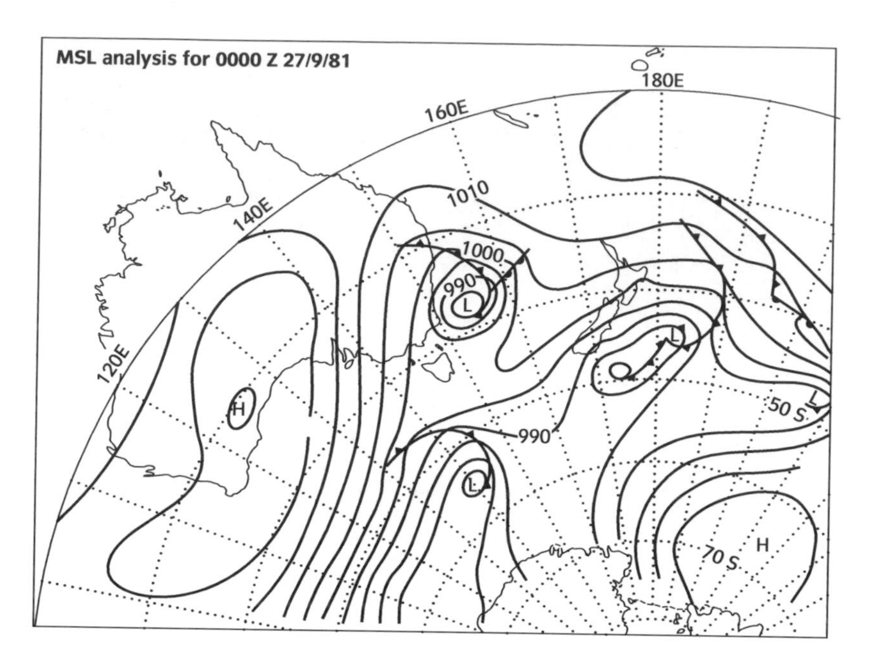

Figure 12.12 Sea level pressure at 0000 Z, September 27, 1981, showing a quadrant of the S hemisphere with zones of high pressure in the subtropics and high latitudes and depressions in mid-latitudes. Airflow is anticlockwise round the anticyclones, and clockwise round the depressions.

Synoptic weather maps show that a huge variety of conditions and sequences can arise depending on the orientation of the depression track, which has been assumed in Fig. 12.5 to be nearly W to E, but in fact it may have almost any orientation, though E'ly motion is rare and usually sluggish. The classic patterns and motions depicted in Fig. 12.2 must not be allowed to mask the full range of possible behaviour, as was recognized by the Norwegians themselves, and by professional forecasters since, but is sometimes lost in the enormous success of the Norwegian model and the few classic diagrams used to summarize it.

Note that the descriptions in this and the previous section have been written for the N hemisphere. It is a useful exercise to re-examine them to see how they should be changed to apply to the S hemisphere, bearing in mind that the direction of map rotation about the local vertical, and all related phenomena, reverses there. Figure 12.12 outlines a depression and its fronts in southern hemisphere middle and high latitudes.

12.3.6 Wind waves and storm surges

Waves

The large areas of strong winds in a mature depression have obvious and familiar effects on underlying surfaces. Trees, buildings, and crops are liable to extensive damage in the more vigorous storms, damage which increases rapidly as average winds rise above gale force (about 19 m s^{-1} ten metres above the surface) and the strongest gusts rise much higher. The sea surface becomes increasingly disturbed, with waves increasing in height, wavelength, and speed of propagation, and more and more crests are whipped into foaming turbulence. This response, and the difficulties posed for mariners, provided early motivation for the development of meteorology, but though modern ships are able to withstand an angry sea much better than the small and vulnerable ships of earlier times, they are by no means immune. At the end of January 1953 a vigorous depression passed E'ward to the

Figure 12.13 Heavy seas at Morecambe (NW England) during the storm surge of February 26, 1990, which caused the damage shown in Fig. 14.16a and some flooding. The huge burst of water is caused by collision between incident and reflected waves, from which high winds whipped water inland over the sea wall and road, breaking windows in some cases.

N of the British Isles. Very strong winds raised mountainous short seas in the narrow North Channel between Scotland and Northern Ireland just as the car ferry *Princess Victoria* began to make her way W'wards. As she turned back to shelter, an unusually large wave struck the vessel's stern and buckled the car loading door. The engine room began to flood and the ship drifted helplessly to the SW, calling for assistance. Sea conditions and visibility were so bad that despite the best efforts of lifeboats and several searching ships, the ferry was still alone several hours later when she foundered within a few miles of the Irish coast with heavy loss of life. The coxswain of one searching lifeboat described how his boat was thrown bodily out of the water by the steep high waves raised by the storm in a confined sea area.

Surges

Coastlines are significantly shaped by short periods of unusually violent waves, and the scouring currents they indirectly drive, and this role of the great storms has often exposed unwisely sited coastal settlements to discomfort, damage, and even disaster. Some of these difficulties also arise from an interaction between wind and water which is much less obvious, but potentially more damaging than the wind-driven waves of a storm—the *storm surge*. As a vigorous depression moves across a body of water, the sea over a considerable area may respond in ways which considerably increase (or decrease) the local water levels expected on the basis of the astronomical tide.

Storm surges contain two quite different types of response to the disturbed air: a static response in which the sea surface behaves like an inverted water barometer (rising under low pressure and sinking under high pressure), and a dynamic response in which shallow water in confined zones is set in large-scale lateral motion by the moving and changing horizontal wind stress, and then piled against a coast.

(i) The static effect is simple in principle. If an area of sea surface is subject to lower atmospheric pressure than surrounding sea surfaces, it will tend to rise, and surrounding surfaces sink, until there is uniform total pressure on the highest horizontal which is everywhere below the water surface (Fig. 12.14). If this equilibrium is reached (which takes many hours, water adjustment on this scale being slow), the sea acts as a hydrostatic manometer relating the deficit Δp of atmospheric pressure to the sea surface elevation Δh through the barometric equation for sea water, which becomes $\Delta h \approx \Delta p$ when Δh is in centimetres and Δp is in hPa (NN 6.1). Since minimum surface pressures in vigorous systems can be more than 50 hPa below surrounding values, water levels can be raised by more than half a metre—a very significant factor for those in charge of coastal defences if the storm surge coincides with a high astronomical tide. These effects will cause increasing concern as average sea levels rise because of global warming (Section 14.9).

(ii) The dynamic effects of a storm can be even greater. If, for example, a N'ly gale blows into the constricted North Sea between N Europe and the British Isles, such as happens when a vigorous depression moves E'wards over Scandinavia, then its large area of S'ward horizontal stress sets up a gross S'ward water movement which may eventually affect all the coasts of the North Sea. The water response is actually a ponderous wave of extremely long wavelength, and like the slopping movement of water in a shaken wash-basin, can be amplified if there is resonance between the imposed shake and the natural period of the water basin. Such time periods are many hours long in substantial water bodies like the North Sea, during which time the Coriolis effect imposes rightward water motion in the N hemisphere, just as it does on large-scale atmospheric flow (Section 7.10). As a consequence, in a S'ward surge in the North Sea, the highest water levels move S along the E coast of Britain before swinging anticlockwise around the Belgian, Dutch, and Danish coasts—like water swinging anticlockwise round a shaken water basin and piling against its walls. The height of the surge varies greatly from place to place along the coasts because of subtle interactions with coastal shape and bottom topography.

Just such a storm surge occurred on 1st February 1953, as the vigorous depression whose wind-waves had sunk the *Princess Victoria* on the previous day moved E'wards towards Norway. As it moved down the E coast of Britain the surge coincided with the peak of a high spring tide. The extra 2–3 m of water topped by an angry sea breached sea defences in East Anglia and the

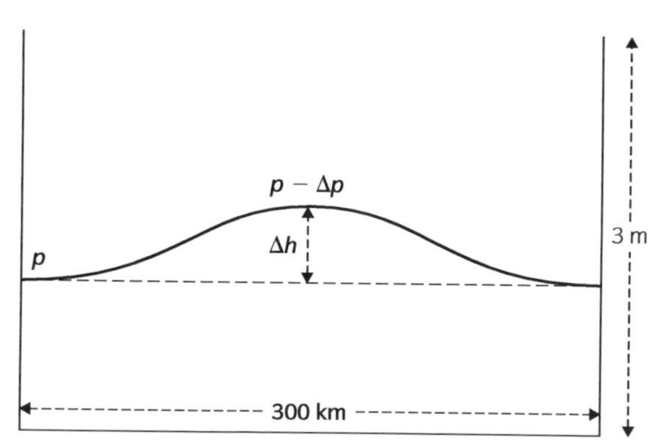

Figure 12.14 An idealized cross-section through a humped sea surface fully adjusted to an atmospheric low-pressure system. Total pressure (atmospheric and top of sea) is uniform on the dashed horizontal. On the height scale of the diagram, the hump of sea water under the central atmospheric low is just over 0.5 m, which corresponds to an atmospheric depression of just over 50 hPa.

Thames estuary, before going on to inundate tens of per cent of the land area of Holland, drowning several thousand people in all. The devastation and loss, and the realization that it could easily happen again, spurred improvement to local sea defences, including, thirty years later, an adjustable barrage in the Thames to protect the heart of London, which had escaped inundation in 1953 by less than half a metre of stone embankment. The storm surge of February 1953 has become the most intensively studied event of its type, though much less destructive of life than the cyclone-driven surges of the Bay of Bengal (Section 13.5).

12.4 Three-dimensional air flow

Consider the paths (*trajectories*) traced out by air parcels as they move in large-scale weather systems. Horizontal winds are so much stronger than vertical ones that you might suppose the motion to be effectively horizontal, so that we could trace air parcel trajectories simply by examining conventional horizontal or isobaric weather maps and using isobars or isobaric contours as streamlines of flow. For example, air moving in the low troposphere around the equatorial side of a mature depression would be expected to describe a cyclonically curved path determined by the curvature of the instantaneous pattern of isobars (as in Fig. 12.2) and the rate of E'ward translation of that pattern. Such an approach ignores the gross asymmetry between vertical and horizontal scale with which this book began (Section 1.1). In fact the atmosphere is so grossly flattened that unobservably weak vertical movements are just about as important in relation to the shallow depth of the troposphere as are its strong horizontal winds in relation to its enormous breadth. For example, a typical updraft in a front is estimated in Box 12.1 to be about 10 cm s^{-1}, at which rate it would take an air parcel about 28 hr to rise 10 km (i.e. through most of the depth of the mid-latitude troposphere), which compares well with the time taken for an air parcel moving at 20 m s^{-1} to traverse a system with horizontal dimension 2000 km. Unobservably slight vertical movements are just as important as the much more obvious horizontal ones if we wish to examine the three-dimensional motion of air in large-scale weather systems.

12.4.1 Estimating vertical motion

The large-scale vertical motions in large weather systems (ignoring embedded small-scale movements outlined in Section 11.7) are very slow indeed, and we must consider briefly how they can be measured. It is quite impractical to measure a rate of rise of a few centimetres per second by a vertical anemometer suspended kilometres above the surface in the disturbed conditions typical of an active front. Such weak movements have to be estimated by indirect methods, the most physically obvious of which uses observed background rates of rainfall. Warm fronts in particular produce fairly steady precipitation rates for long periods over extensive areas, in which much more vapour is condensed and precipitated than is instantaneously present in the full depth of the troposphere. It follows that there must be a close balance between the upward flow of water vapour in the rising air and the

BOX 12.1 Uplift and rainfall

Consider a steady-state weather system in which the upward mass flux of water vapour is exactly balanced by the downward mass flux of precipitation. We examine this equality in a vertical column of unit horizontal cross-section. The fact that air may be rising in a gentle slope rather than straight up the column is irrelevant provided there is no significant net horizontal divergence or convergence of water substance. We need only consider the vapour flux into the base of the cloud, since the details of updraft and vapour content at higher levels make no difference to the overall balance at cloud base.

Upward

Suppose that vapour rises into the cloud base in air of density ρ, specific humidity q (g kg^{-1}) and vertical speed w. The upward mass flux MF_v of water vapour in units of kg m^{-2} s^{-1} is given by

$$MF_v = 10^{-3}\,\rho\, q\, w$$

where the factor 10^{-3} converts specific humidity from g kg^{-1} to kg kg^{-1}.

Downward

If rain is falling at rate r (mm hr^{-1}), since 1 mm of rainfall corresponds to 1 kg of rainwater on 1 horizontal square metre (rainwater density being 1,000 kg m^{-3}), the corresponding downward mass flux MF_p of precipitation in kg m^{-2} s^{-1} is given by

$$MF_p = 2.78 \times 10^{-4}\, r$$

where the numerical factor converts mm per hour to kg per second.

Balance

Equating these fluxes, and rearranging to isolate updraft w, we find

$$w = 0.278\frac{r}{\rho q}$$

Air enters most cloudy updrafts in the low troposphere where air density is close to 1 kg m^{-3}. To this approximation

$$w = 0.3\frac{r}{q}$$

where w is in m s^{-1} when r is in mm hr^{-1} and q is in g kg^{-1}.

In middle latitudes the specific humidity of air rising into the base of a cloudy mass is typically about 10 g kg^{-1}. It follows that a typical frontal rainfall rate of 3 mm hr^{-1} implies a frontal scale updraft of about 9 cm s^{-1}. On this large scale the assumed balance between MF_v and MF_p is probably quite accurate.

A typical rainfall rate of a heavy shower is 30 mm hr^{-1}, which implies a w value of about 0.9 m s^{-1}. However, evaporative loss from the sides of cumulonimbus, and significant efflux of unprecipitated cloud into the anvil, means that MF_v is likely to be significantly greater than MF_p, and that the method will significantly underestimate updrafts from rainfall rates in small-scale systems.

downward flow of condensed water in the precipitation. As shown in Box 12.1, a typical background rate of rainfall (3 mm per hour) is consistent with an uplift speed of just under 10 cm s^{-1}, assuming a vapour content typical of the low troposphere. This is a useful order-of-magnitude estimation, but it assumes too much to be used accurately in any particular case.

Dynamic reasoning can be applied to observations of large-scale convergence and divergence, etc. to calculate the associated large-scale rates of rise or sink of air (implicit in Section 7.15 and Fig. 7.33), with similar results. However, this requires subtle estimation of horizontal convergence if it is not to demand impossibly precise flow measurements. Such methods underlie the forecasting of rates and amounts of precipitation, though these are still some of the most difficult forecasts to make despite the sophistication of modern numerical methods. However, a thermodynamic method has been used occasionally in recent decades to glimpse large-scale vertical air motion, and with it a picture of the three-dimensional flow in large systems. Though used for research rather than routine forecasting, it is relatively straightforward, at least in principle, and depends on the assumption introduced in Chapter 5, that vertical motion in large-scale weather systems is sufficiently fast for the succession of air parcel states to be nearly adiabatic (often termed *isentropic* since entropy is conserved in an adiabatic process). When the air is unsaturated the process is nearly dry adiabatic, and when the air is cloud-filled the process approximates to a saturated adiabatic one. These assumptions have already been used in the analysis of cumulus convection (Chapter 11), where air trajectories are effectively vertical, but now we are applying them to air motion which is intrinsically slightly but extensively sloped.

12.4.2 Isentropic analysis

In *isentropic analysis* an array of radiosonde data from the weather system under examination is analysed to pick out winds and other observations on each of several isentropic surfaces (i.e. each of uniform potential temperature). The most generally conservative label to use is the wet-bulb potential temperature (Section 5.8), though in practice the ordinary potential temperature is used in cloudless parts. Figure 12.15 outlines a simple vertical cross-section through isentropic surfaces. At any geographical location higher values of potential temperature correspond to greater heights, otherwise the air would have a lapse of potential temperature and be convectively unstable (Section 5.10). Note that in the presence of a horizontal temperature gradient, isentropic surfaces will slope upwards towards the colder air, since potential and actual temperatures on an intersecting isobaric surface must fall towards the cold air.

Horizontal air flow (the winds) on an isentropic surface will accurately represent the flow in the tilted plane of the surface because their angles of tilt are observed to be very small (<1%) in large-scale weather systems, even in the vicinity of fronts. However, it would usually be misleading to use the observed winds directly to draw streamlines of flow, since in most weather systems isentropic surfaces move with much the same speeds as the systems themselves (~10 m s^{-1} in the case of many depressions). Just as in the analysis of air flow in a super-cell storm (Section 11.6), the system velocity must be subtracted vectorially from each observed wind velocity to obtain the wind velocities relative to the moving weather system. Provided the system is not changing too rapidly, these can be used to draw streamlines of air flow through the travelling system relative to the system. Isentropic analysis is mostly applied to mature systems because they are relatively steady.

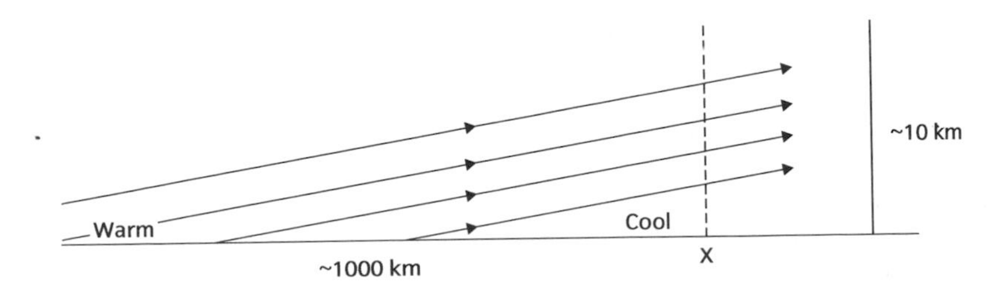

Figure 12.15 Idealized vertical section showing how sloping synoptic scale flow can produce a layered vertical profile at X. In reality, such flows are always three dimensional, as shown in Fig. 12.16.

Figure 12.16 depicts the typical air flow through a mature depression, as revealed by isentropic analysis of many systems in N America and the N Atlantic *[53]*. The main feature is the extensive flow of air upwards and polewards in and ahead of the advancing cold front, rising from the low troposphere in the warm air in the equatorward base of the warm sector, and turning gradually E'wards in the high troposphere in the forward and poleward flange of the warm front. This has become known as the *warm conveyor belt* because the air seems to be rising along an invisible sloping ramp fixed in the moving system, and because this is the warmest of a number of such flows. Since the air entering the conveyor belt is quite moist, it does not travel far up the slope before it reaches saturation and fills with stratiform cloud. The air continues rising, but rather more steeply now that it is moving on a saturated adiabat, and soon becomes part of the nimbostratus deck of the cold front. The crowding of the isobars on the ramp shows that it is steeply sloped and that air on it must be rising quickly through the

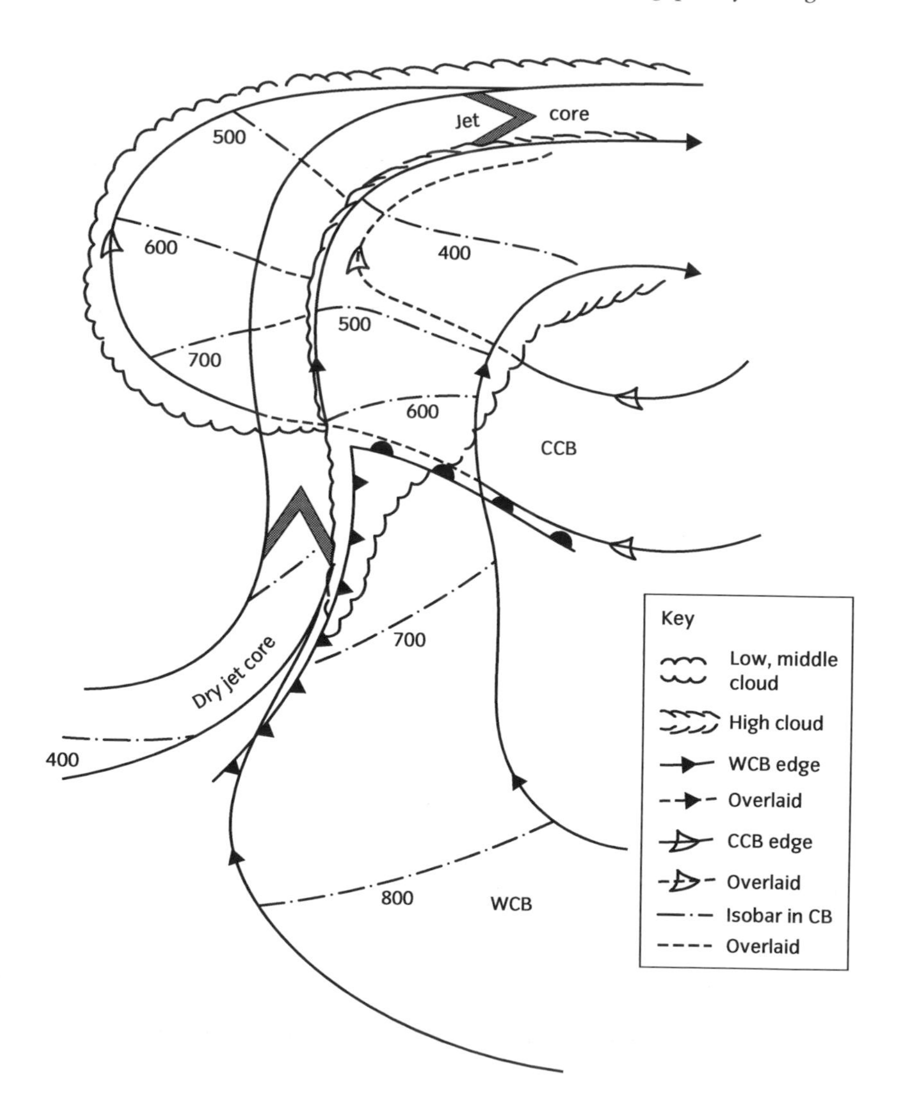

Figure 12.16 Major sloping airflows and cloud decks associated with a mature extratropical cyclone in the N hemisphere. WCB and CCB are the warm and cold conveyor belts. Intersections of climbing flows with isobaric surfaces are lines labelled in hPa. The coldest air in the system is descending equatorwards off the W edge of the diagram.

lower middle troposphere (around 750 hPa), producing cloud and maintaining the background precipitation of the cold front. Ascent continues into the upper parts of the warm front, contributing to some of the precipitation from its upper levels and maintaining the anticyclonically curving plume of cirrostratus noted in Fig. 12.6. Notice that the updraft speed at any point on the trajectory is given by the product of the relative wind speed and the trajectory slope. Typical values in the middle troposphere are 20 m s^{-1} and 1/100, giving an updraft of 20 cm s^{-1}, which is consistent with values found by other methods (above), bearing in mind that the present method probably concentrates on the fastest rising air.

In the low troposphere ahead of the warm front, air ascends in the *cold conveyor belt*, initially parallel to the warm front, but then swinging round in a grand anticyclonic arc which sometimes produces a pronounced flange of middle-level cloud reaching polewards and W'wards of the apex of the warm sector. The cold conveyor belt contributes mainly to the production of cloud and precipitation in the lower-middle troposphere of the warm front. Other features of air flow in Fig. 12.16 are the largely horizontal flow of air in the middle troposphere to the SW, catching up rapidly with the system and sinking slightly before rising to join the NW flank of the SW'ly jet stream associated with the cold front. Out of sight to the W in Fig. 12.16, air in the middle troposphere is sinking and turning cyclonically as it flows to lower latitudes, forming part of the equatorward sink of cool air which, together with the poleward rise of warm air, accomplishes the meridional and vertical exchange of air required for hemispheric energy balance (Sections 8.9 and 8.10).

This examination shows that the pattern of air flow in a depression is quite different from what might appear on a series of isobaric or horizontal charts. It is more complex and asymmetric, with the cold front looking simpler and more fundamental to the basic scheme (the meridional and vertical exchange of air) than the warm front does. And the pattern is inescapably three-dimensional, which greatly complicates any attempt to portray it on a two-dimensional page. The presence of important gently sloping flows is obviously consistent with the term *slope convection* used throughout this book to describe this vast and structured combination of convection and advection, and generally but not yet universally in the meteorological community.

12.5 Anticyclones

Even casual barometric observation in mid latitudes shows that fine weather is associated with high (especially rising) surface pressure, just as wet and windy weather is associated with low (especially falling) surface pressure. By the early nineteenth century it had become clear that high surface pressure occurs in the middle of anticyclonically rotating air masses (Box 7.9), and the name *anticyclone* began to be used. Although much less intense than cyclonic weather systems, anticyclones are quite distinctive, with characteristic structure and behaviour, and several variants of the basic type. The basic features of most types of anticyclone, including the slowly subsiding air which fills its broad core, and the subsidence inversion which separates this from the underlying convecting boundary layer, are seen most clearly in the subtropical anticyclone, which is fully described in Section 13.2.

12.5.1 Ridges

In middle latitudes the *ridges* of high pressure which separate members of a family of depressions are relatively narrow poleward extensions of a subtropical anticyclone and slide E'wards along its poleward flanks (Fig. 13.1). As experienced at a fixed location being approached and over-run by a ridge, the poleward limb of subsidence stunts and then suppresses the showers in the cool air behind the previous cold front, often giving a few hours of cloudless sky before the advance of hooked cirrus from the west heralds the approach of the next warm front (Fig. 12.3). In the summer in particular, the associated brief sunny spell, combined with the fresh, clean low-level air, can produce some of the most pleasant weather experienced in the industrial areas of Britain and Western Europe, now that more persistent anticyclonic weather is often marred by the accumulation of thick industrial haze in the surface layer. But the effect is always short-lived, as the ridge moves E'wards in step with the preceding and following members of the family of cyclonic systems.

12.5.2 Blocks

Sometimes an anticyclone may form in mid-latitudes well poleward of the normal position of subtropical highs. The development is associated with a temporary halt to the normal E'ward procession of depressions in middle latitudes, with the result that it has become known as a *blocking high* or *block*.

Once established, a block may persist virtually motionless for days or even weeks, causing large departures from the seasonally typical climate over a wide area. In winter in particular, the presence of a pronounced, low-level subsidence inversion can trap smoke and other pollution from large industrial areas and increase surface concentrations of noxious materials to the point of irritation or even danger. Combined with a deck of stratocumulus thick enough to shroud the weak winter Sun, the result is sometimes nicknamed *anticyclonic gloom*. The situation is worsened if fog forms, as discussed in Section 10.12. In summer the stronger convection from the surface tends to maintain deeper convecting layers, but encourages the photo-chemical reactions underlying the Los Angeles type of smog (Section 13.2). On a smaller scale, after a night's cooling under clear skies, smoke is often seen trapped in the stable morning air (Fig. 12.17).

Climatic anomalies can be very marked in the vicinity of a prolonged block. For example, the whole of the British Isles lay under and slightly E of a block throughout January, February, and part of March 1963, and the combination of loss of warm air advection, absence of thick cloud cover (which normally shields the surface from cooling through the long winter nights), and the advection of very cold air from the interior of Europe, reduced monthly mean temperatures over much of S Britain by about 4 °C. Populations of small birds were decimated, transport and industry were disrupted in January, and the E sides of evergreen shrubs bore brown freezing scorch marks for a decade afterwards. In summer on the other hand, a block is likely to lead to unusually high temperatures, since solar warming through the long, bright days far exceeds the net long-wave cooling. In July and August 1976 the British Isles lay under a block which came at the end of a period of more than a year in which

Figure 12.17 Industrial haze top in stagnant air probably in quiet anticyclonic weather, though the clouds over the mountains show no sign of the presence of a subsidence inversion (as in Fig. 13.4).

rainfall had been consistently well below normal because of an unusually high incidence of blocking. With reservoirs not fully restocked by the preceding rather dry winter, and the prolonged excess of evaporation over precipitation in the hot, dry summer, Britain's normally very heavy domestic and industrial consumption of water could not be sustained, and an unusually severe agricultural and hydrological drought ensued. Variations in the frequency of blocking are believed to underlie some of the climatic variations observed in recent centuries and decades (Section 14.5).

12.5.3 Continental highs

Anticyclones develop regularly in winter over continental interiors, the most pronounced being the Siberian High which extends over Siberia and N Asiatic Russia, and influences a considerable portion of the Asian continent in winter. These are known as *continental highs*, but though they exhibit the usual anticyclonic circulation in the low troposphere (geostrophically consistent with the pressure pattern), they differ in several ways from the other types of anticyclone. In particular they are relatively shallow systems, rarely extending their anticyclonic circulation above the lower middle troposphere. Radiosonde profiles through their centres show that air temperatures are substantially lower than those of the surrounding air masses throughout the low troposphere, rising only slowly with increasing height from the very low temperatures of the snow-covered land surface. As a result, unlike the common warm-cored types of anticyclone, the pressure excess in the centre of the cold-cored continental type fades quickly with increasing height.

In early winter, the cooling of a land surface is greatly accelerated by the first snow falls, which practically eliminate further solar warming but maintain the net loss by terrestrial radiation. The surface and overlying air cools progressively day by day, as in a polar night. The whole mass of overlying air sinks on top of the shrinking layer of cooling air near the surface, and the air aloft converges faster than the air near the surface air diverges, so that the surface pressure rises. After setting up, the high surface pressure is maintained by a dynamic equilibrium between convergence and divergence which is common to all types of anticyclone, though not easily explained. Despite persistent subsidence, the absence of an underlying convecting boundary layer prevents the appearance of anything like the elevated subsidence inversions of other anticyclones; instead a relatively smooth inversion can extend several kilometres from the surface (Fig. 12.18). Persistent subsidence keeps the low troposphere free of cloud, enhancing the surface cooling and making the anticyclone self-maintaining. Cloud and even precipitation can occur in the unaffected middle and upper troposphere, but the cyclonic weather systems, which would be the primary source of cloud in the absence of an anticyclone, are discouraged by the inert mass of cold air overlying the surface, so that continental interiors tend to be largely cut off from the W'lies for much of the winter. The persistence of continental highs throughout much of the middle and later parts of the Siberian winter gives rise to extremely low mean surface temperatures, which can be lower even than Antarctic values when accentuated by local topography (Section 10.11.1).

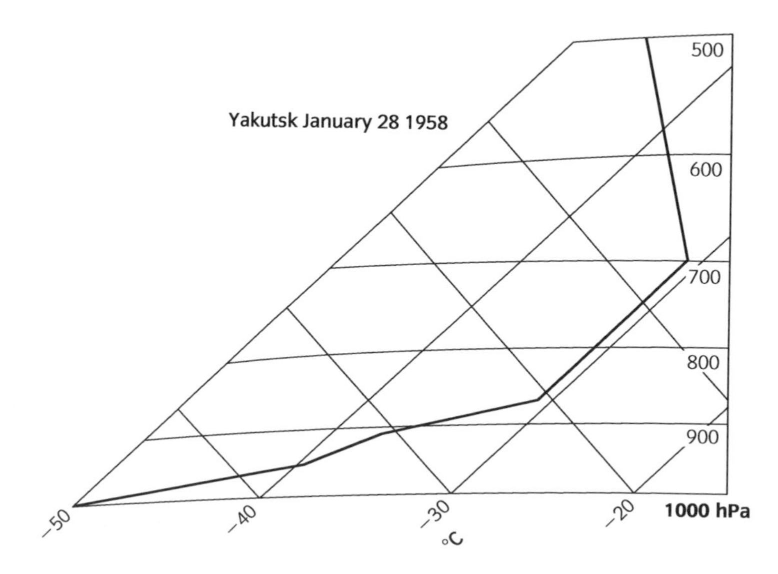

Figure 12.18 A vertical temperature profile in the Siberian winter anticyclone (Yakutsk, 62° N, 129° E), showing a very strong and deep surface-based temperature inversion. The standard British tephigram has had to be extended to include the very low temperatures observed in the low troposphere.

12.6 Waves in the westerlies

The great mid-latitude weather systems look like flat waves on weather maps, waves which almost always travel E'wards as they develop. The Norwegian school looked for their origins in an instability of the Polar Front, similar to the Kelvin–Helmholtz instability which makes a flag flap in response to a difference in wind speed on either side of its fabric (Section 11.8). Though that particular dynamical theory was fruitless, the general idea of treating such weather systems as wave-like disturbances of the much broader baroclinic zone of middle latitudes has flourished since the late 1930s *[35]*.

12.6.1 Cyclone waves

Analyses of mid-latitude depressions have provided many examples of its wave-like structure, especially in its formative stages (Fig. 12.5). This is especially obvious when isobaric contours are plotted in the middle and upper troposphere; indeed a sequence of such pictures of a developing depression vividly suggests an amplifying travelling wave (Fig. 12.19). Theoretical treatment of the dynamics of the baroclinic W'lies by Charney and Eady *[80]* in the late 1940s showed that they are unstable in a way which leads to the formation of waves which, at least in their early stages, strongly resemble those seen in weather maps of the middle and upper troposphere.

The theoretical methods used are subtle and mathematically complex, but in essence they explore the stability of the initial purely zonal baroclinic flow by imposing small-amplitude meridional perturbations and looking for subsequent development. The wavy perturbations grow exponentially at an initially realistic rate, in a range of wavelengths (W-to-E separations of adjacent troughs or crests) centred on values (a few thousand kilometres) confirmed by weather maps. One of the most important destabilizing factors is the vertical shear of zonal wind in the initially undisturbed flow. When this exceeds about 1 m s^{-1} per kilometre, a range of unstable wavelengths opens and broadens with increasing shear. The process is known as *baroclinic instability*, since according to the thermal wind relation (Section 7.12) such shear is associated geostrophically with a deep temperature gradient across the flow, i.e. with its N–S baroclinity. As shown in NN 12.1 a shear of this magnitude is associated with a meridional temperature gradient of about 1 °C per 400 km in middle latitudes, or nearly 6 °C across 20° of latitude. Since this is considerably smaller than the temperature gradient maintained by the meridional gradient of net radiative warming (Section 4.6), it appears that the middle latitude troposphere in its normal state is *baroclinically unstable*.

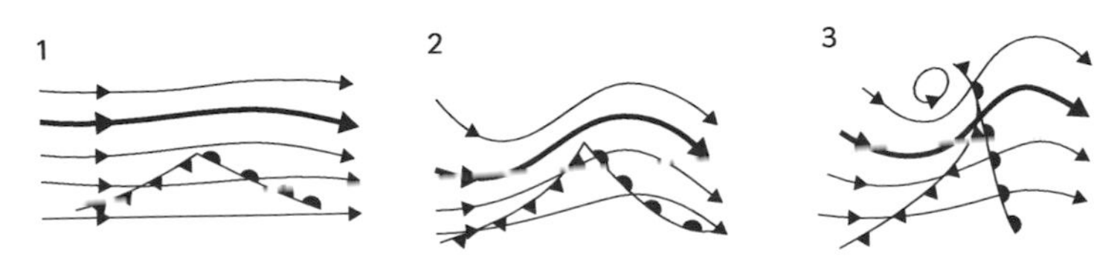

Figure 12.19 A sequence of simplified pictures of flow at the level of the polar front jet stream during the development of an extratropical cyclone. Sequences at all levels down to the lower troposphere similarly show an amplifying wavy disturbance of mainly W'ly airflow.

NUMERICAL NOTE 12.1 Horizontal temperature gradient in a frontal zone

According to the thermal wind relation (Eqn 7.27a), the lateral temperature gradient $\Delta T/\Delta n$ across a slab of air of average depth Δz, in which longitudinal geostrophic wind increases by ΔV_g from slab base to top, is given by $\Delta T/\Delta n = (f\,T/g)\,\Delta V_g/\Delta z$.

Taking the geostrophic parameter f to have a typical middle latitude value of $10^{-4}\,\mathrm{s}^{-1}$, gravitational g to be $10\,\mathrm{m\,s}^{-2}$, and the slab mean temperature T to be 260 K, the lateral temperature gradient corresponding to a longitudinal shear of geostrophic wind of $1\,\mathrm{m\,s}^{-1}$ in a vertical rise of 1 km is given by $\Delta T/\Delta n = (10^{-4} \times 260/10)\ 1/10^3 = 2.6 \times 10^{-6}\,°\mathrm{C\,m}^{-1}$, which is only 0.26 °C per 100 km, or nearly 6 °C in 20 degrees of latitude (20 × 111 km).

As already mentioned, the theoretical model baroclinic waves have observably realistic features, one of the most important of which is the offset or phase shift between the wavy patterns of temperature and flow. Figure 12.20 is taken from Eady's treatment, and shows that the crest of the thermal wave (i.e. the N–S axis of the warmest air) lags by one quarter of a wavelength behind (W of) the crest of the wave in mid-tropospheric isobaric contours, which represents the flow wave. This amounts to warmer air being carried polewards and cooler air being carried equatorwards, so that the system contributes to the poleward advection of heat required to balance the meridional heat budget. The phase shift of Fig. 12.20 is observed in the real atmosphere too. It is directly observable in satellite pictures such as Fig. 12.6, where the great plume of cloud indicating the presence of the warm conveyor belt pours NE'wards on the W side of the wave in the middle troposphere. And it is indirectly observable in the W'ward shift of wave axis with increasing height apparent in isobaric contours on upper air charts. For example, in Fig. 12.21, the deep layer of cold air to the W of the low-pressure centre at sea level ensures that the axis of low pressure in the upper troposphere is shifted W'wards by hundreds of kilometres. This has been a familiar feature of upper air charts since they were first drawn routinely, and was noticed but not recognized by observers in the nineteenth century who reported that the transit of pressure troughs on barographs in mountain-top stations lagged systematically behind those in nearby stations near sea level. The W'ward tilt of the axes of both theoretical and observed systems is very pronounced, axes making angles of only a degree or so to the horizontal, further justifying the name quasi-horizontal for such systems.

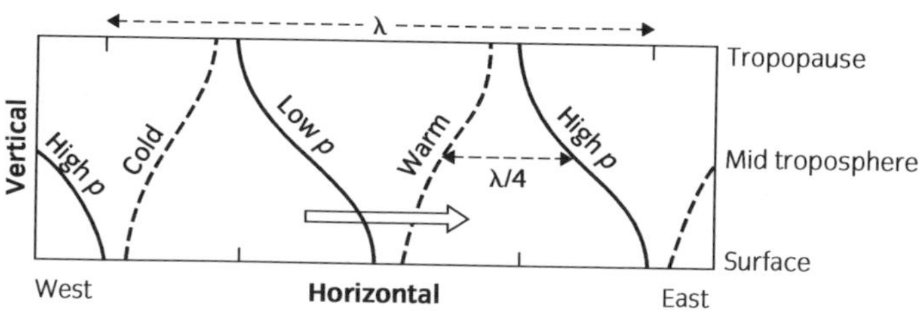

Figure 12.20 A vertical zonal section through the zonally displaced axes of waves of temperature and pressure (and therefore flow) in Eady's model of the troposphere of a developing extratropical cyclone. The arrow shows the direction of system propagation.

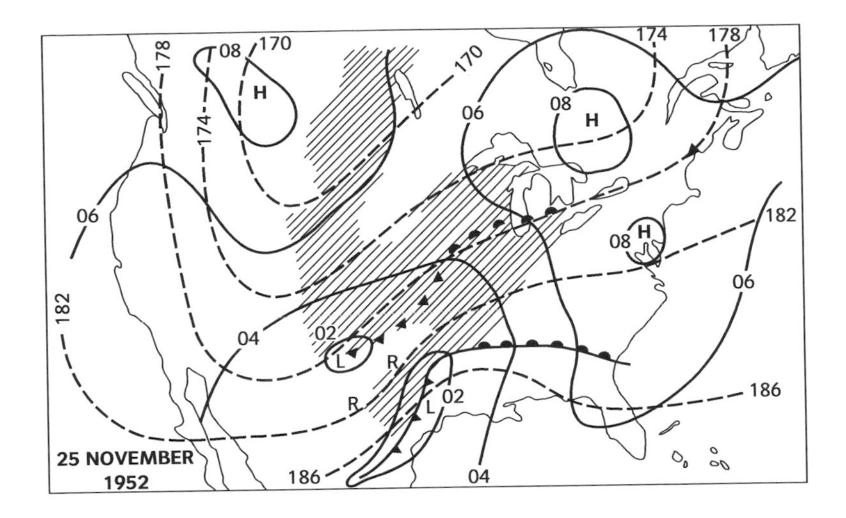

Figure 12.21 W'ward shift of low pressure axis with increasing height. The surface low pressure is centred at the apex of the warm centre while the axis of the 500 hPa trough is located about 1200 km further W.

12.6.2 Long waves *[33]*

A factor which complicates the identification of weather systems with individual cyclone waves is the simultaneous presence of several identifiable wavelengths in any actual situation. There are indications that these include smaller wavelengths which may relate to meso-scale structures observed in frontal cloud and precipitation, and may even represent a frontal instability of the type initially suspected by the Norwegians. However, there has been no doubt for several decades that there are waves which are considerably longer in zonal wavelength than those associated with individual depressions and ridges. These are known generally as *long waves* or planetary waves, and are sometimes named after Rossby who first discussed their significance and origin, though strictly speaking the term *Rossby wave* is now reserved for a particular kind of long wave.

Long waves do not depend on the presence of baroclinity, although they may be modified by it. But they do depend crucially on the tendency of absolute vorticity to be conserved in very large-scale flow in the middle troposphere (where there is little horizontal divergence for reasons discussed in Section 7.13 and apparent in Fig. 7.33. On such vast scales there is little relative vorticity about the local vertical, so that the main component is the planetary vorticity, which is simply the Coriolis parameter f (Section 7.16). If we imagine the very large-scale flow to be perturbed so that it is angled slightly polewards, the air will move to higher latitudes, where f is larger. To conserve absolute vorticity (planetary plus relative vorticity) the air develops balancing negative relative vorticity, which at its simplest means anticyclonic curvature (Fig. 12.22). The flow therefore curves back towards its original latitude, overshoots it and then curves back again cyclonically as it develops positive relative vorticity in lower latitudes. The oscillation about the original latitude continues indefinitely in the absence of any damping, and we can identify the zonal separation of adjacent crossings of the initial latitude as the wavelength L. It seems clear that the meridional variation of the Coriolis parameter must play an important part in determining the wavelength and other features of these long waves. The rate of increase of f with poleward meridional distance is represented by β, and is shown in Box 12.2 to have

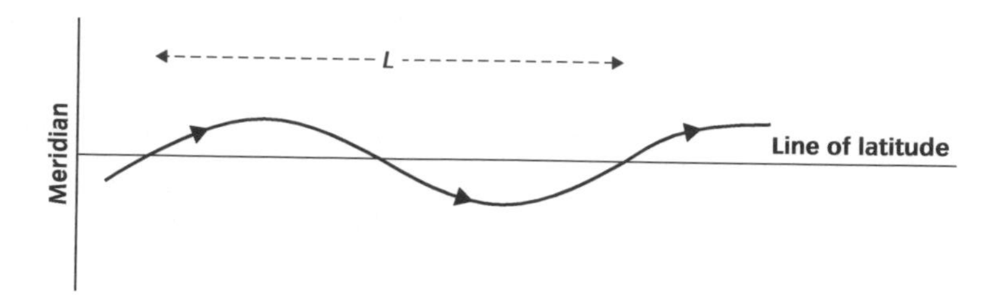

Figure 12.22 Idealized plan of a Rossby wave with wavelength L in W'ly flow.

values of about 1.6×10^{-11} m^{-1} s^{-1} in middle latitudes. In the especially simple case analysed by Rossby, there are no other important factors, and dynamical analysis *[33]* shows that such Rossby waves will propagate W'wards relative to the prevailing air flow with *wave speed* (i.e. speed of propagation of wave shape through the air) $\beta L^2/4\pi^2$. This means that the E'ward wave speed C relative to the Earth's surface is given by

$$C = U - \frac{\beta L^2}{4\pi^2}$$

where U is the speed of the initially undisturbed W'ly flow of air. This relationship shows that Rossby waves propagate E'wards over the Earth's surface at speeds which decrease with increasing wavelength L, and that there is a certain L at which the waves are stationary ($C = 0$), which is given by

$$L = 2\pi\left(\frac{U}{\beta}\right)^{1/2}$$

If U is assumed to be 20 m s^{-1}, then the wavelength for stationary waves in middle latitudes is about 7,000 km, four of which would girdle the Earth there (NN 12.2). Wavelengths longer than this propagate W'wards over the Earth's surface.

Observation is complicated by the presence of the shorter baroclinic waves which abound in middle latitudes, but when the flow in the middle or upper troposphere is averaged over a few days, the fast-moving baroclinic waves are largely smoothed away and the slow-moving long-wave pattern is exposed (Fig. 12.23). These shift majestically with time, usually moving E'wards at a few degrees of longitude per day, but transforming from time to time by extensive decay and re-growth in different meridians.

12.6.3 Topographical locking

There is clear evidence that there are preferred positions for long waves in relation to land masses and topography, especially in the N hemisphere. One mechanism for such linkage can be seen by considering large-scale zonal air flow impinging on a long meridional barrier like the Rocky Mountains (Fig. 12.25). As the air reaches the W (upwind) side of the mountains, it is progressively squeezed into a shallower layer by the rising land surface beneath. The squeezing effect is reduced by bodily ascent of the whole atmosphere, and by ponding on the

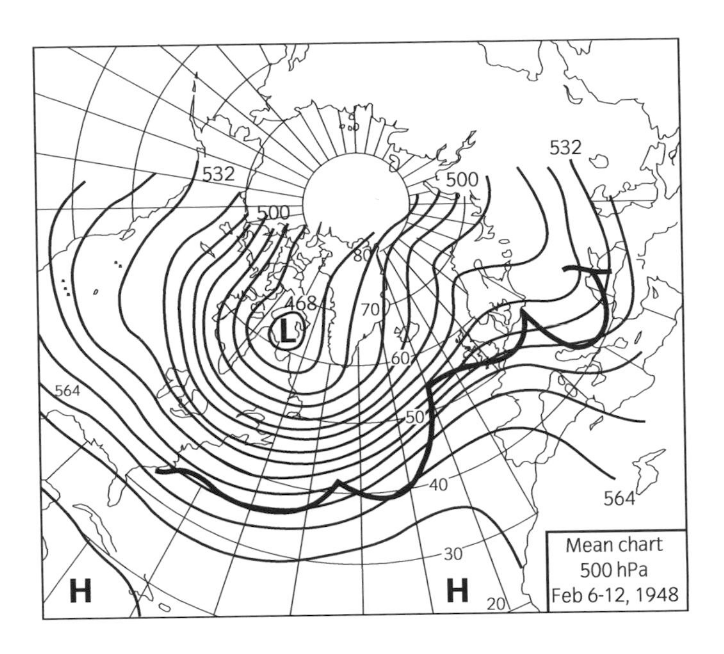

Figure 12.23 Contours of the 500 hPa surface averaged over seven days to reveal the underlying pattern of long waves. The surface fronts at an instant in the middle of the period comprise a series of cyclone waves associated with the long wave trough.

BOX 12.2 Variation of *f* with latitude, Rossby wave

Consider the rate of change of the Coriolis parameter f with y, the poleward distance along a meridian. According to Fig. 12.24 and the definition of radian measure (Box 4.4) a very small increment dy corresponds to a similarly small increment dϕ in latitude ϕ according to

$$dy = R\, d\phi$$

where R is the radius of the Earth. If follows that we can rewrite

$$\beta = \frac{\partial f}{\partial y} = \frac{1}{R}\frac{\partial f}{\partial \phi}$$

Substituting $2\,\Omega \sin\phi$ for f and differentiating we find

$$\beta = \frac{2\,\Omega\cos\phi}{R}$$

Substituting 7.27×10^{-5} rad s^{-1} for Ω, and 6,350 km for R, β is about 1.6×10^{-11} s^{-1} m^{-1} at latitude 45°.

A Rossby wave of wavelength L at this latitude will be stationary in a W'ly flow of 20 m s^{-1} if $L = 2\,\pi\,[U/\beta]^{1/2} = 2 \times \pi \times [20/(1.6 \times 10^{-11}]^{1/2} \approx 7 \times 10^6$ m $=$ 7,000 km. The length of the line of latitude 45° is $2\pi R\cos 45 = 2 \times \pi \times 6{,}350{,}000 \times 0.7071 \approx 28{,}200$ km, which would take almost exactly four of these stationary Rossby waves.

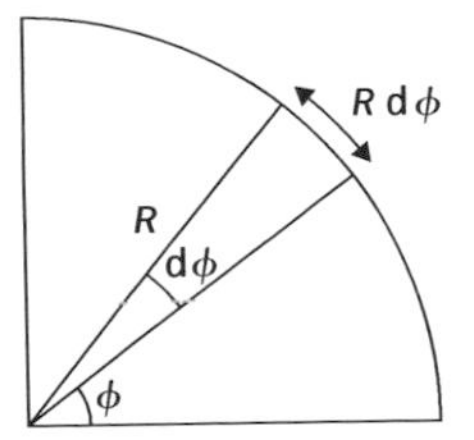

Figure 12.24 An Earth quadrant showing the meridional arc R dϕ subtending a little latitude range dϕ.

upwind side of the barrier, but substantial consequences remain. Since air is dynamically incompressible on this large scale, the vertical squeezing is compensated by horizontal divergence (Section 7.15), which in turn acts on the background planetary vorticity to produce anticyclonic curvature in the previously

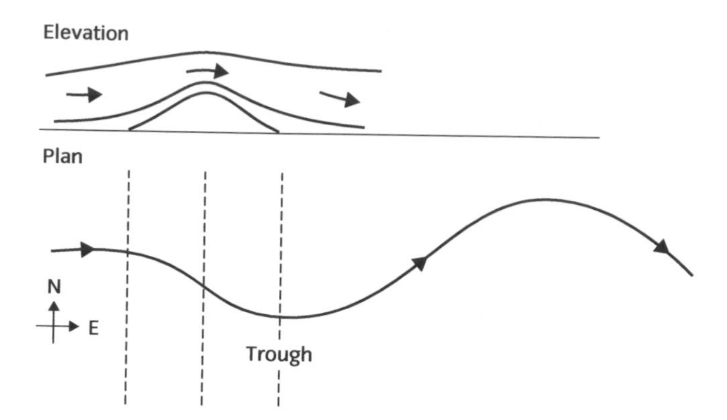

Figure 12.25 Formation of a large-scale trough in the lee of a meridionally extensive mountain range.

straight W'ly flow (curving S'wards in the N hemisphere). After the air has passed over the crest of the mountain ridges, the falling topography enforces large-scale horizontal convergence which in turn produces cyclonic curvature, tending to re-establish the initial absolute vorticity. However, the air is now at a somewhat lower latitude, where the planetary vorticity f is smaller, so that the air is left with residual cyclonic relative vorticity in the form of cyclonic curvature. There is consequently a *lee trough* on the downwind side of the mountain barrier (Fig. 12.25). Downwind of this the air moves polewards again, losing cyclonic curvature as it moves to regions with larger f, and beginning to curve anticyclonically as it passes the latitude of its original flow W of the barrier. The flow is now in exactly the situation considered at the beginning of the treatment of Rossby waves above, and will show long-wavelength meridonal undulations apparently anchored to the mountain barrier. In practice such anchorage appears only statistically, as a greater than average concentration of long-wave troughs just E of the Rockies.

Very significantly, all long waves seem to exert a controlling influence over the faster-moving baroclinic waves both by steering and amplitude control. For example, there is a tendency for baroclinic waves to be suppressed to the W of a long-wave trough, and enhanced to its E, which shows as a zone of repeated cyclogenesis. And once formed, the depressions tend to move as if steered by the contours of long-wave patterns in the middle troposphere.

12.6.4 Index cycle

On an even grander scale, the long-wave pattern around a complete hemisphere seems to *vacillate* (oscillate irregularly) between two extremes: one of maximum meridional waviness, and the other of minimum waviness (Fig. 12.26). The maximum is usually associated with one or more well-developed blocks equatorward of the nearly cut-off long-wave ridges, and the minimum corresponds to unusually straight zonal flow over extensive regions, with baroclinic waves developing and moving quickly E in the vigorous W'ly flow. The long wave set-up is quantified by the *zonal index* which is defined to be the zonally averaged pressure difference across an agreed meridional segment of middle latitudes at some particular level—for example between latitudes 35° and 55° at the 500 hPa level. The index corresponds to an average geostrophic wind in the zone and is sometimes expressed as such. Regardless of units used, it is small in contorted

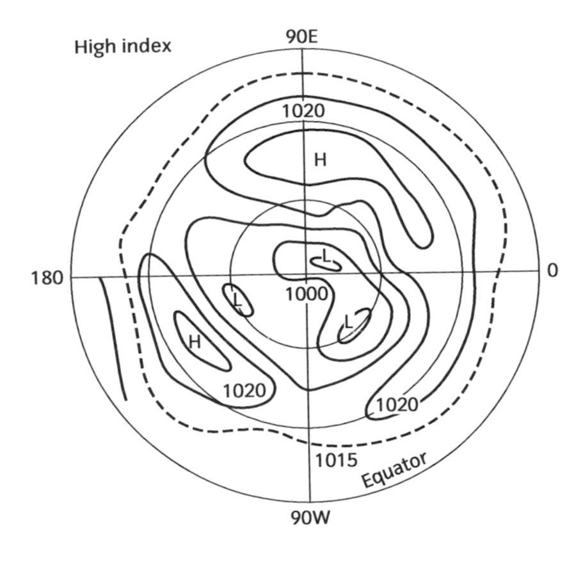

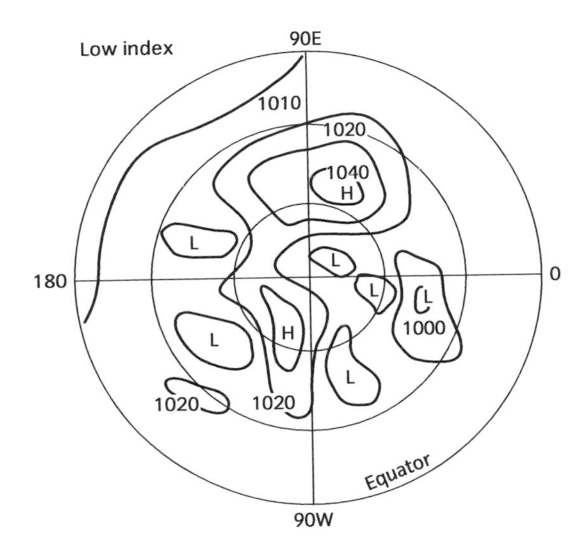

Figure 12.26 Extremes of index cycle as shown by two examples of seven day averages of sea level pressure in the N hemisphere. In the low-index example, the high pressures are centred over N America and Central Russia.

zonal flow, and large in straight zonal flow, and the vacillation from one extreme to the next (say from minimum to minimum) is called an *index cycle*. Such cycles have larger amplitudes in winter than in summer, and generally last between 3 and 8 weeks. They clearly correspond to significant events on the very largest possible scale of tropospheric flow—the circumpolar vortex and can extend their influence far above the tropopause by complex vertical transmission of wave energy. However, they are not yet susceptible to forecasting of even the most rudimentary kind. Since the periodicities of the zonal index often occupy a significant fraction of a season, they probably account for a considerable part of the year-to-year variations which are such a feature of short-term climatic variations (Section 14.4).

12.7 Polar lows and heat lows

It seems likely that there are several different types of low-pressure systems at work in mid-latitudes in addition to the extratropical cyclones described by the Norwegian cyclone model, though the latter is undoubtedly the most common and significant system. As an example let us consider one quite dynamic type, the *polar low* of the Eastern N Atlantic, and the relatively static and ephemeral *heat low*.

12.7.1 Polar lows

In the E Atlantic and W European regions there is a type of weak low-pressure system which occurs occasionally in winter and is notorious in Britain for its ability to produce disruptive snow falls. A classic example appears in Fig. 12.27, which shows the typical synoptic situation in which they occur. A fairly gentle N'ly flow, to the W of an old stagnant complex low over Scandinavia, is bringing very cold air off the Greenland ice cap and over Iceland to reach the British Isles from the NNW. Weak surface lows form in the vicinity of Iceland and move SE'wards with the air flow in the low troposphere, bringing substantial amounts of nimbostratus, cumulonimbus, and heavy falls of snow to the British Isles, before moving on into N France and weakening.

It was once thought that these systems were essentially conglomerations of fairly shallow cumulonimbus, but analysis *[55]* of the case of Fig. 12.27, amongst others, suggests that they are shallow baroclinic waves forming in strongly baroclinic N'ly flow. Although there are snow showers from cumulonimbus in the wake of the system, just as there are showers in the wake (i.e. to the W) of a normal depression, these seem to be secondary to the mass of nimbostratus associated with the polar low itself. A sophisticated radar, capable of measuring the horizontal motion of precipitation from the Doppler shift of reflected radar waves, indicated the presence of meso-scale convergence (and therefore ascent—Section 7.15) in the low troposphere, but failed to detect any updrafts strong enough to suggest the presence of cumulonimbus embedded in the nimbostratus. Isentropic analysis revealed a structure rather like that found in normal depressions, but confined to the lower half of the troposphere, limited to little more than 500 km in horizontal extent, and of course moving S'wards instead of E'wards. If there were fronts they were too weak to show on the radar and too small to show on the standard synoptic observations. Theoretical analysis *[54]* confirms that the shallow but strongly baroclinic zone between the cold air flow off the Greenland ice cap and the much warmer air to the W is capable of generating baroclinic waves with the observed dimensions and speeds of propagation, and development of rate, provided that wind speeds are not strong enough to produce destructive damping by turbulent friction (i.e. not more than 10 m s^{-1}). Analysis of more subtle effects, arising from heating by the relatively warm sea surface, suggests that development is inhibited unless the air flow is almost perpendicular to the sea-surface isotherms. Since local sea-surface isotherms are largely zonal, this may well be the reason why polar lows develop only in N'ly flows. Between them these restrictions ensure that polar lows develop in only a small fraction of N'ly air streams in the E Atlantic in winter.

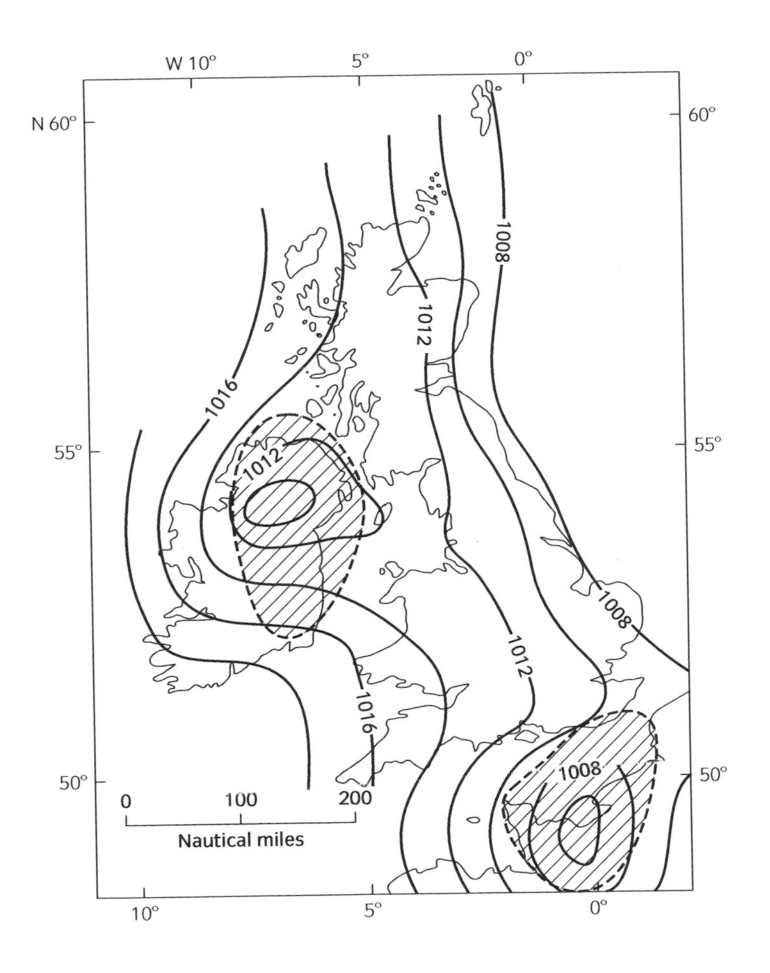

Figure 12.27 Two polar lows near the British Isles at 1800 Z on 8 December 1967. Sea level isobars show that the shallow lows were embedded in geostrophic N'lies. Areas of falling snow are hatched.

The two lows apparent in Fig. 12.27 produced substantial snow falls in the British Isles, including some exceeding 250 mm in average depth (about 25 mm equivalent rainfall) which almost totally disrupted traffic in parts of S England. Although some disruption arose because of that area's combination of high population density and low preparedness for snow, it is worth noting that two quite familiar features of snowfalls can make them especially disruptive of road traffic and other activities. One is the low density of fresh snow, which arises because the spiky arms of the snow crystals prevent close packing, and ensures that it lies initially to an average depth which is at least 10 times the equivalent depth of rainfall. The other feature is the tendency for recently fallen snow to drift in only quite moderate winds, moving from exposed surfaces and gathering in deep drifts where wind eddies form in the lee of walls and hedges. Narrow walled or hedged roads abound in lowland Britain and are often blocked in these conditions by a metre or two of snow in windy conditions while surrounding meadows have only centimetres. The disruption caused on the occasion of Fig. 12.27 served a useful purpose in attracting the attention of British meteorologists which led to a considerable increase in understanding of these weather systems. Forecasting of polar lows was almost impossible before the advent of regular polar-orbiting satellites and is still complicated by their rapid formation and short transit time to the reach the British Isles.

12.7.2 Heat lows

In warm summer weather in middle latitudes it is common to see weak puddles of low pressure developing daily over islands and peninsulas in the absence of any strong synoptic-scale pressure gradients. Sometimes they may be associated with showery activity in the afternoon and evening, and sometimes the skies may remain largely clear. These lows are a meso-scale, even synoptic scale, response to the rapid development of strong temperature contrasts between land and adjacent sea.

Initial stage

Figure 12.28 represents a typical mid-afternoon summer heat low over England. During the previous night the land had cooled below the temperature of the surrounding sea, but a few hours of morning sunshine have been enough to reverse the situation, thanks to the very small effective heat capacity of land compared with that of the sea (Section 10.4). The temperature of the land surface and its

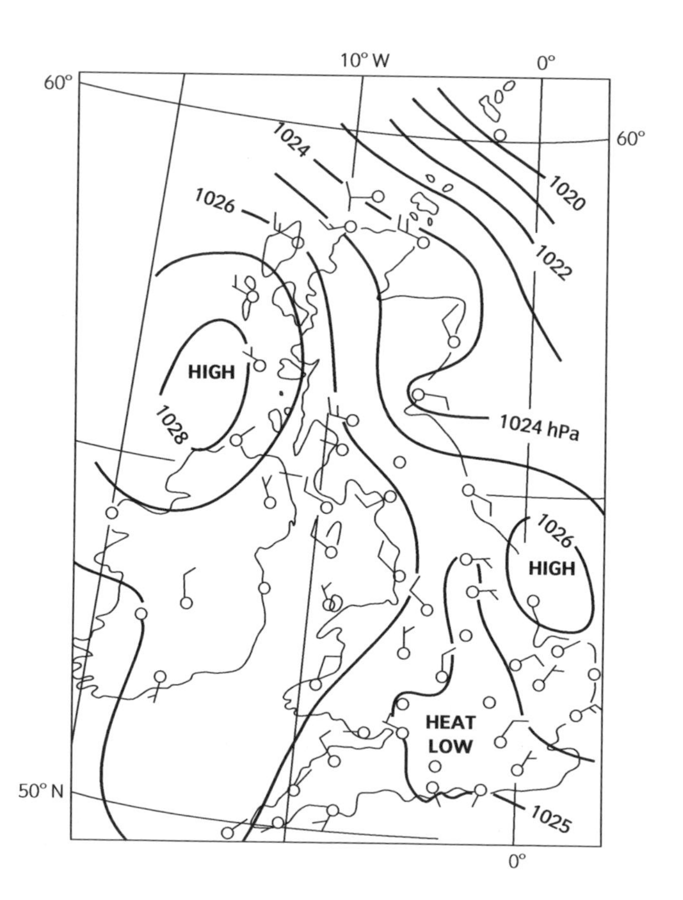

Figure 12.28 Sea level isobars showing a heat low over SE England at 1500 Z on 20 June 1960. Temperatures were just over 25 °C across the area of low pressure. Although not plotted here, a succession of earlier maps shows that the low developed as the land heated strongly in the preceding few hours.

NUMERICAL NOTE 12.2 Summer boundary layer inflation

(i) If the boundary layer average temperature is 22 °C ($\approx$ 295 K), and surface pressure is 1,000 hPa (10^5 Pa), then air density $\rho = p/R\,T = 10^5/287 \times 295 = 1.18 \approx 1.2$ kg m^{-3}, the mass of a 250 m air column is $250 \times 1.2 = 300$ kg m^{-2}, and the total atmospheric mass is

$$p_s/g = 10^5/9.8 \approx 10{,}200 \text{ kg m}^{-2}$$

If layer temperature rises by $\Delta T = 10$ °C, the layer depth $h = 250$ m will increase by Δh, where $\Delta T/T \approx \Delta\rho/\rho \approx \Delta h/h$ (from small changes Δ in $p = \rho\,R\,T$ at constant pressure) and the boundary layer inflation $\Delta h \approx 250 \times 10/295 \approx 8.5$ m. A more thorough approach uses the method of Section 9.6 and Box 9.6 to find $\Delta h = 8.8$ m.

(ii) If this increase Δh flows laterally off the inflated layer, the loss of an 8.8 m air column of density 1.2 kg m^{-3}, reduces surface pressure by $\Delta p = g\,\rho\,\Delta h = 9.8 \times 1.2 \times 8.8 \approx 103$ Pa ≈ 1 hPa.

boundary layer can rise 10 °C above the adjacent sea surface temperature. The land-based layer expands vertically (inflates) as it warms, raising the overlying air so that its isobaric surfaces dome slightly upwards. If the warming layer is 250 m deep and has average temperature 22 °C half way through its 10 °C warming and expansion, its depth will increase by about 8.5 m during the full expansion, raising the whole overlying atmosphere by this amount. Such inflation is discussed in Section 9.6 and calculated in NN 12.2.

Notice that the very slow upward motion of the top of the inflating layer (~ 1 mm s^{-1}) is not to be confused with the thermal updrafts embedded in the layer. The latter are strong (~ 1 m s^{-1}) localized and almost completely offset by intervening subsidence (Section 11.5), and are the means whereby sensible heat is pumped into the warming layer from the underlying surface (Section 10.7).

The inflated layer is now as sketched in Fig. 12.29a, which shows how the lifted air (virtually the whole atmosphere) will tend to slide outwards off the domed heated boundary layer. The air movement needed to accomplish this in say 3 hours is very small indeed for a small land area like Britain: Box 12.3 shows that radial speeds of about 1 cm s^{-1} will suffice for a land surface of radius 100 km. If we assume that the movement is complete when the isobars at the top of the heated boundary layer become horizontal again (Fig. 12.29b), then a mass of air equal to an 8.5 layer of boundary layer has been removed in the worked example, which corresponds to a loss of pressure of a little more than 1 hPa at the surface and throughout the heated layer (NN 12.2). The sea-level pressure over the warming land has now fallen by this amount below values over the surrounding sea, producing a weak heat low as shown in the mid-afternoon synoptic chart (Fig. 12.28).

Subsequent development

Even after the reduction of pressure by shedding of overlying air, the boundary layer is strongly baroclinic around the edges of the warmed land, since the isotherms dip down inland much more steeply than the isobars (Fig. 12.29b). The air in and immediately above this shallow baroclinic layer responds by developing a sea-breeze circulation as described in Section 10.11. The inflow of cool, denser sea air and outflow of warm land air tends to fill the heat

(a)

Vertical

Warmed layer

Land

Sea

Sea

(b)

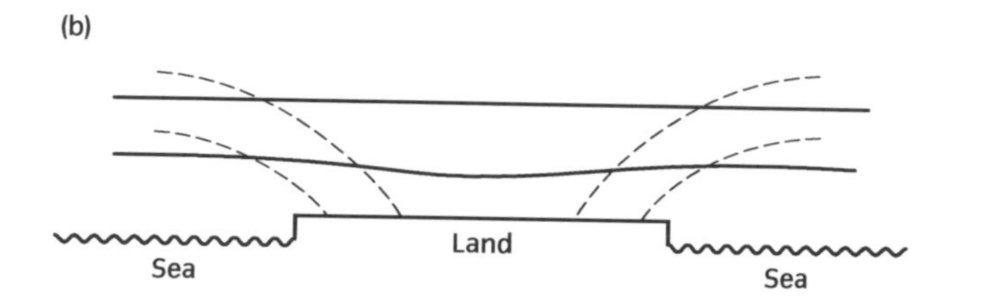

Figure 12.29 Idealized vertical sections showing the formation of a heat low. In (a) the isobars (solid lines) dome slightly upwards in and above the warmed layer (isotherms dashed), radially shedding air aloft. In (b) the shedding is complete, leaving isobars dished downwards in the warmed layer.

low, but the tendency is impeded by boundary layer friction, and by Coriolis deflection of the air flows, so that the low may be only slightly depleted before evening cooling begins to destroy it by reversing the heating by which it was created in the first place.

If the warming is not fully offset by subsequent nocturnal cooling, for example because of a change in the synoptic situation, then the residual shallow low may develop further with the next day's heating, and so on for several days. This applies more to land masses larger than Britain, such as France or Spain, where the scale is too large for the offsetting sea-breeze circulation to reach more than a small proportion of the total area in the course of a day. Coriolis deflection is crucial here; if the sea breeze persists for more than a few hours, it suffers Coriolis deflection (to the right in the N hemisphere) and tends towards quasi-geostrophic equilibrium with the pressure field and friction. The flow in the low troposphere begins to move cyclonically round the shallow low, forming a dynamic barrier to further filling, and confirming the feature as a consistent low-pressure system on the weather map.

If convection deepens the overland warming layer, the pressure deficit in the heat low will increase. However, this development will depend heavily on how convective stability is being influenced by other factors. If stability is being reduced, for example by synoptic-scale vertical stretching associated with the approach of a weak front, then convection may deepen quickly, producing outbreaks of heavy thundery showers. The associated cloud reduces the surface heating, but the deep, warm, cloudy layer may persist for some days as baroclinic adjustment encourages it to rise further in relation to its cooler surroundings. Such *thundery lows* often develop over France in the summer and give spells of humid, thundery weather in Britain as conglomerations of showers swing NW'wards in the weak circulation to the NE of the low.

If, however, convective stability is being maintained or increased, for example by subsidence associated with an adjacent anticyclone, then convection from the warmed land surface may remain confined in a fairly shallow layer, and the depth of the heat low will be limited by the temperature excess of the heated layer relative to adjacent cooler layers. Figure 13.7b shows this effect in sea-level pressures over North Africa, where a semi-permanent heat low over the W Sahara maintains a pressure deficit of about 20 hPa compared with the Azores High further W, in the presence of very high surface temperatures over land. Nothing comparable is apparent on the winter maps (Fig. 13.7a). Comparison of summer and winter over the Indian subcontinent shows that there are even more dramatic effects with the great monsoon there (Section 13.3).

BOX 12.3 Radial winds and heat lows

Consider a horizontal circular area $A = \pi R^2$ over which the surface pressure falls by Δp in time Δt. Assuming hydrostatic balance, the reduction in the mass of air resting on the area is $A\,\Delta p/g$, so that the average rate of fall of mass is

$$\frac{\pi R^2\,\Delta p}{\Delta t\, g} \qquad \text{B12.3a}$$

In the simplest model of a developing heat low this mass loss happens by air moving radially outwards at all levels above the warming and inflating boundary layer. Figure 12.30 shows a cylinder of air standing on A. If the radial air flow has speed V between heights z and $z + \mathrm{d}z$, the outward mass flux in this height range is

$$V\,\rho\,2\,\pi\,R\,\mathrm{d}z$$

Integrating through the full depth of the atmosphere above the relatively shallow boundary layer, and replacing $\rho\mathrm{d}z$ by $\mathrm{d}p/g$ through the hydrostatic approximation, we find that the total rate of export of mass from the cylinder standing on A is

$$\frac{2\pi R}{g}\int_0^{p_0} V\,\mathrm{d}p$$

where p_0 is the pressure at the base of the column. If V is uniform, or the appropriately weighted average radial speed, this expression becomes

$$\frac{2\pi R}{g} V\,p_0 \qquad \text{B12.3b}$$

The balance between the observed rate of pressure fall and the rate of mass export by radial flow is stated by equating expressions B12.3a and B12.3b. After rearrangement

$$V = \frac{R}{2p_o}\frac{\Delta p}{\Delta t}$$

If we take the case of a rather large and rapid pressure fall (3 hPa in 3 hr) we will find an upper limit to the required radial air flow speed. For an R value corresponding to a medium-sized island (100 km) we find V to be about 1.4 cm s^{-1}. Notice that this corresponds to a horizontal divergence ($\partial u/\partial x + \partial v/\partial y$ in Section 7.15) of a little over 10^{-7} s^{-1}.

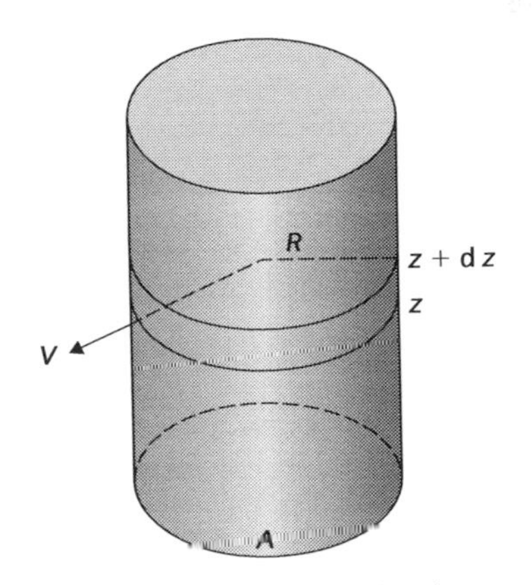

Figure 12.30 A vertical cylindrical column of air for estimating radial mass fluxes during the formation of a heat low (Box 12.3).

Checklist of key ideas

You should now be familiar with the following ideas.

1. The origins of our description and understanding of the extratropical cyclone.

2. Norwegian model of the ETC: birth in baroclinic zone, motion and life cycle, family series.

3. Mature ETC: patterns of temperature, wind, pressure, and weather in plan and cross-section, jet stream.

4. ETC passage: ridge, warm front, warm sector, cold front, cool showery aftermath; wind waves and storm surges.

5. Isentropic analysis and three-dimensional air flow in mature ETC: warm and cool conveyor belts producing cloud and rain, ascent along the cold front.

6. Types of anticyclone and their weather: ridge, block, continental (extratropical—Chapter 13).

7. Waves in the westerlies: ETCs as cyclone waves, long waves including Rossby waves, stationary waves by dynamics and topographical locking.

8. Polar lows as small baroclinic systems: formation and behaviour; heat lows: development, diurnal cycle of land and sea breezes, thundery low.

Problems

Outline answers to these problems can be found on the **Online Resource Centre**. Answers to odd numbered problems can be found under Student Resources, answers to even numbered problems under Lecturer Resources.

Note that the N hemisphere is assumed unless otherwise stated.

Level 1

12.1 In not more than one sentence in each case, cite observational evidence for the following large-scale properties of developing extratropical cyclones: broad-scale uplift; convergence in the low troposphere; divergence in the upper troposphere.

12.2 Whereabouts is the observer of the following series of events most likely to be in relation to the classical model of a family of depressions? Pressure rising and falling by about 20 hPa every couple of days; winds strongly W'ly, backing a little as cloud and rain move in during each fall of pressure, and veering in the subsequent clearances. Sketch the position on a model family.

12.3 A square-rigged ship is making N in strong W'ly winds just E of an advancing cold front. Unless the ship's course is changed, what is likely to be the situation of the ship after the passage of the surface front?

12.4 To the N of an old centre of low pressure a mass of cloud is observed advancing from the E, with the edge of its high cloud moving quickly N along the N–S edge. Use the thermal wind relation to decide whether the front is a warm or cold front.

12.5 Given that air at two levels, 2 km apart on a vertical sounding, has risen in parallel conveyor belts with a gradient of about 1 in 100, estimate the horizontal separation of their surface sources.

12.6 In the conveyor belt picture of air flows in fronts, a large region of heavy rainfall is likely to be associated with what aspects of the conveyor belt?

12.7 Severe cold wind damage to evergreen foliage was concentrated on the E sides of shrubs and trees during the severe winter of 1963 in the British Isles (and visible for years after). Where was the centre of the blocking high?

12.8 Outline the rationale for saying that the rapid onset of continental winters in middle and high latitudes is often associated with the radiative behaviour of snow.

Level 2

12.9 A low-pressure centre with sea-level pressure 980 hPa lies not far from a ridge with pressure 1012 hPa. Find the difference in sea levels which have adjusted hydrostatically to this pressure difference.

12.10 If air has risen to the 500 hPa level along a typical warm conveyor belt with a slope of about 1 in 75, estimate the geographical position of the region in which it was just leaving the atmospheric boundary layer.

12.11 Air enters the upper troposphere of an anticyclone (at the 400 hPa level) with a temperature of –30 °C and is saturated with respect to supercooled water. If it sinks to the 700 hPa level in 3 days, cooling by net radiation at an equivalent isobaric rate of 2 °C per day, find its temperature and relative humidity at the 700 hPa level. Use the tephigram in Fig. 5.7.

12.12 Using the method of Problems 7.15 and 16, find the average horizontal divergence which is needed in the lower troposphere of an anticyclone to accommodate a rate of subsidence of 1 cm s^{-1} in the middle troposphere there.

12.13 Using the expression for the speed of E'ward propagation of a Rossby wave, estimate the speed of a cyclone wave of wavelength 2000 km in middle latitudes. Although interesting and realistic, some of the assumptions underlying the pure Rossby wave do not strictly apply to a real cyclone wave. Which are these?

Level 3

12.14 Compare and contrast the isobaric and isentropic views of air flow in a mature depression, citing their observational strengths and weaknesses.

12.15 Compare and contrast the synoptic-scale structures of a depression and a blocking anticyclone in middle latitudes.

12.16 The term jet stream has unrealistic overtones if it conjures the picture of a fire-fighter's hose. Describe atmospheric jet streams realistically to emphasize their true drama, including an estimate of the mass flux in a typical jet core 100 km broad and 2 km deep.

12.17 In Problem 12.12, the subsidence associated with deflation by radiative cooling has been ignored. Assuming that the low troposphere as a whole is cooling at 2 °C per day in the anticyclone, estimate the associated rate of subsidence in the mid-troposphere. The original omission is therefore justified.

12.18 Draw up a list of all meteorological effects which involve the meridional variation of the Coriolis parameter, including a brief description of the involvement in each case.

12.19 What differences in regional climate would you expect if the Earth were entirely covered by ocean? (Note that climatologists might then be dolphins!)

12.20 Documentary videos of storm surges often confuse them with storm wind waves. Distinguish clearly between the two, and the different problems they present to coastal defences.

13 Large-scale weather systems in low latitudes

13.1 Introduction

Middle-and high-latitude weather and climate are strongly affected by the seasonal march of the Sun, even over the oceans. In the lower latitudes the contrast is more between land areas, whose seasons are dominated by large seasonal migrations of wet weather (including the enormous Asian monsoons), and the large ocean areas, where seasonal migrations of wet weather are much smaller. The seasonal changes are wet and dry rather than the warm and cold cycle of higher latitudes, though dry seasons tend to be warmer, being less cloudy. If the wet season follows the Sun, as happens in many regions, then the warm season may be out of step with the Sun. And between the tropics, the double passage of the Sun through the zenith at noon can encourage a double seasonality, which is often asymmetrical. In many regions there are strong diurnal rhythms of wind and rain which are modulated by the seasons.

In the subtropics, a term often loosely applied to the latitudes between the tropics and 30°, the rainy seasons are largely quenched by the persistent subtropical high-pressure systems, though some monsoon effects can occur over land. There can be a muted form of mid-latitude seasonality, as the noon Sun at latitude 30° ranges between 83.5° and 36.5° above the horizon (Fig. 8.10), though the absence of mid-latitude westerlies and the greater tendency to anticyclonic weather, gives it a very different feel. There is altogether a greater variety of seasonality in low latitudes than in high, where the range is largely between maritime and continental.

Though more than half of Earth's surface area is included in these lower latitudes (exactly half lies between latitudes 30 °N and S), the historical growth of meteorology in middle latitudes, the greater complexity of lower-latitude weather and climate, and the sparseness of data from the enormous tropical oceans, have combined to delay progress in understanding this major part of Earth's atmosphere. Fortunately the satellite observation network (Section 2.11) has redressed the observational imbalance in recent decades, and the global coverage required by modern numerical forecasting models has forced attention on the whole zone.

The climatology of low latitudes is outlined in Section 14.3, and the structure of the Hadley circulation and its relations with the Inter-tropical Convergence Zone and the Subtropical High-Pressure zone have been sketched in Sections 4.7 and 8.10. In the present chapter we focus on the three main types of synoptic-scale weather systems of low latitudes: the subtropical anticyclones, the monsoons and the family of tropical low-pressure systems (*tropical cyclones* if they contain cyclonic circulation).

13.2 Subtropical anticyclones

These archetypal anticyclones nearly girdle the Earth in the tropics and maintain the subtropical high-pressure zones which are a major feature of large-scale weather systems there (Section 4.7). They are very large, persistent, zonally elongated areas of high sea-level pressure, and are particularly well developed over the oceans, especially in winter. Figure 13.1 shows a typical actual situation in the North Atlantic, where the local subtropical high (STH) is known as the Azores High in Europe and the Bermuda High in North America, because it often straddles one or both of those popular holiday islands. Observations on Fig. 13.1 show that the weather is mostly fine, with few reports of precipitation, which is mostly very light when it happens, apart from showers which are usually confined to the W flanks of the high. Where skies are not clear, clouds are generally observed to be low cloud types, with stratocumulus the most common. There is very little cloud in the middle troposphere, as is particularly obvious when viewed from aircraft, where the sky is usually seen to be quite empty above whatever low cloud there may be (Fig. 13.2).

Radiosondes rising through such anticyclones show clearly why the middle troposphere is so free of cloud. Above the layer occupied by surface-based convection (Fig. 13.3), there is a relatively shallow layer which is either isothermal or contains a temperature inversion (temperature increasing with height). In either case the potential temperature increases sharply with height, and the layer acts as a convectively very stable interface separating the low troposphere from the potentially (or actually) warmer air aloft. This warm air has very low relative humidity (often below 20%), which prohibits cloud formation or maintenance. Even small liquid aerosol droplets often cannot escape evaporation (Section 6.8),

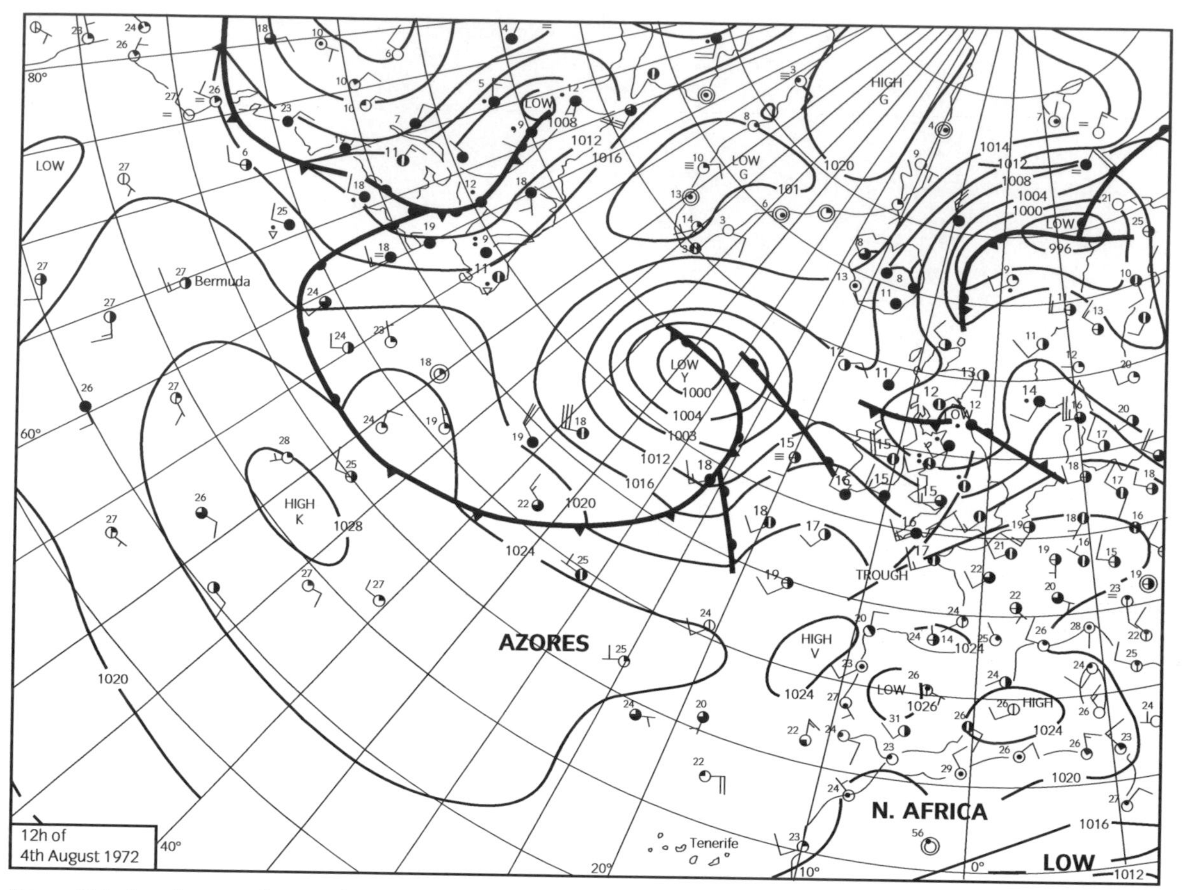

Figure 13.1 A surface map showing the Azores High in a well-developed state. Depressions and fronts are moving E'wards on its N flank, and a ridge of high pressure is extending N'wards at about 50 °W. Notice the lowering of sea-level pressures over NW Africa in association with the summer solar heating there.

with the result that the air in the middle troposphere is unusually clear. This contrasts sharply with the air below the temperature inversion (between it and the surface), which is often very hazy, especially over or downwind of the mostly arid continents. Viewed from mountains or low-flying aircraft, there is often a very sharp haze-top at the altitude of the base of the inversion. Figure 13.4 sketches the mechanics of this important and sometimes spectacular structure.

13.2.1 Subsidence

The high potential temperatures and low humidities of air in the middle troposphere of an anticyclone are maintained by gentle but persistent downward motion which is widespread across the system. As air sinks down from the high troposphere, it warms, and because there is virtually no cloud present (usually nothing more than a few wisps of cirrus or aircraft condensation trail), the warming tries to be dry-adiabatic. Because there is no cloud or precipitation to evaporate and rehumidify the sinking air, its specific humidity is conserved, and its relatively humidity falls as it sinks and warms.

Figure 13.2 A deck of stratocumulus, showing the globular patterning of its shallow layer. The sky above the deck is apparently cloudless.

If the descent of air was truly dry adiabatic, the warming and dehumidifying would be as shown in Fig. 13.5 and analysed in Box 13.1, and the subsided air would be even warmer and drier (lower relative humidity) than it is in fact. As shown in Fig. 13.3, temperatures in the subsiding air do not warm as much as the dry adiabatic process would imply: potential temperatures fall (Box 13.1) more or less steadily, which shows that the sinking air is losing heat diabatically as it sinks. This cooling is maintained by net radiative loss (the only mechanisms capable of reaching air isolated from convection by its own convective stability), in which weak direct solar warming is consistently exceeded by net terrestrial radiative cooling (Sections 8.3 and 8.4). Although it is usually unwise to assume that air flow in large-scale weather systems stays on a horizontal plan or a vertical elevation (Section 12.4), the picture of air sinking through the upper and middle troposphere and into the lower troposphere is fairly realistic in the case of large anticyclones. The 'potential' cooling of a sinking parcel (its potential temperature is falling though its temperature is rising) is only a few degrees per day, and the large-scale buoyancy effect this controls limits subsidence to less than 1 km per day (a few cm s^{-1}). However, the winds in the upper troposphere are often so weak, and the anticyclones so extensive, that individual air parcels stay within the system for the days it takes to sink into the low troposphere.

The stable layer in the low troposphere of an anticyclone (Fig. 13.3) is called a *subsidence inversion* because it is maintained by the subsidence of the warm, dry air which makes up the bulk of the air in the system. The inversion forms at the interface between the large-scale descent and the small-scale convection maintained by solar heating of the underlying surface, and is often quite strong. In practice it is often still called a 'subsidence inversion' even when it is not strong enough to maintain a true inversion—an increase of temperature with height. Though curious, the title is not dynamically misleading since its stability layer is still strong enough to act as an effective lid on the convection beneath.

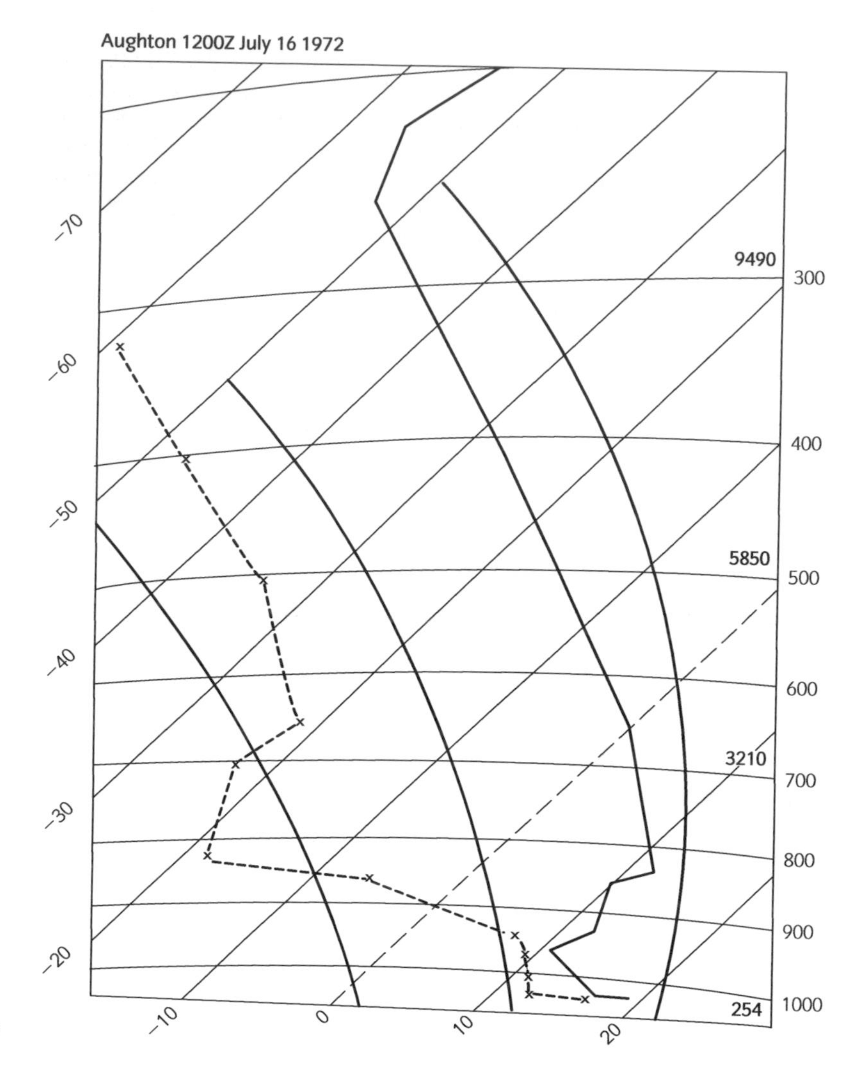

Figure 13.3 A radiosonde ascent through the interior of an anticyclone, showing a well-mixed convecting boundary layer surmounted by a subsidence inversion at about 950 hPa, and a deep layer of very dry subsiding air.

BOX 13.1 Drying by descent

Figure 13.5 presents idealized vertical temperature and humidity profiles in the middle and upper troposphere plotted on part of a tephigram (Section 5.8). The point P represents air which has reached the 300 hPa level (just over 9 km above sea level) by saturated adiabatic ascent, probably in a distant cumulonimbus. It has a temperature of about −38 °C (interpolating between the −30 and −40 °C isotherms) and a potential temperature of +60 °C, and the nearby dashed line shows that its saturated specific humidity is less than 0.75 g kg^{-1} (a full-sized tephigram or tables would show it to be 0.53 g kg^{-1}). We will suppose it is saturated, although in reality it is more likely to be slightly unsaturated. The curved line is the 20 °C saturated adiabat, along which the air might have risen in its distant ascent, which is nearly parallel to the dry adiabatic at 300 hPa—a measure of how little vapour there is left to condense at such low temperatures. In fact the saturated adiabat is asymptotic to a potential temperature of 61 °C (the equivalent potential temperature of the 20 °C saturated adiabat—Section 5.8).

If this air were now to sink dry adiabatically to the 500 hPa level (about 5.5 km above sea level) it would arrive at the point Q, found by moving down the dry adiabat through P to its intersection with 500 hPa. Its potential temperature is still 60 °C, and its vapour content is still what it was at P (0.53 g kg^{-1}) but its actual temperature is now about +1 °C (slightly above the heavy dashed 0 °C isotherm). At this much higher temperature (almost 40 °C higher) saturation would require over 6 g kg^{-1} of vapour (in fact 7.9 g kg^{-1}), which means that the relative humidity at Q is 6.7% (= 100 × 0.53/7.9). An equivalent statement is that the dew-point depression at Q would be about 32 °C, since the air at Q would have to be chilled from + 1 to −31 °C at constant pressure (500 hPa) to become saturated (at point R, the intersection with the saturated specific humidity line from P).

In the real atmosphere such subsidence is accompanied (indeed made possible) by considerable radiative cooling. If this was able to cool the sinking air along the 20 °C saturated adiabat from P (but by dry rather than saturated descent—the potential temperature being lowered by radiative cooling rather than by evaporation), the air would reach the 500 mbar level at S with a temperature of −9 °C. The specific humidity stays at the P value, as it did in the previous dry adiabatic descent, so that the relative humidity at S is about 15%, as you can confirm roughly from values on Fig. 13.5 (saturated specific humidity at S ≈ 3.5 g kg^{-1}, $RH = 100 \times 0.53/3.5 \approx 15\%$). This is still a very low value, compared with typical values of 60–100% in much of the rest of the troposphere.

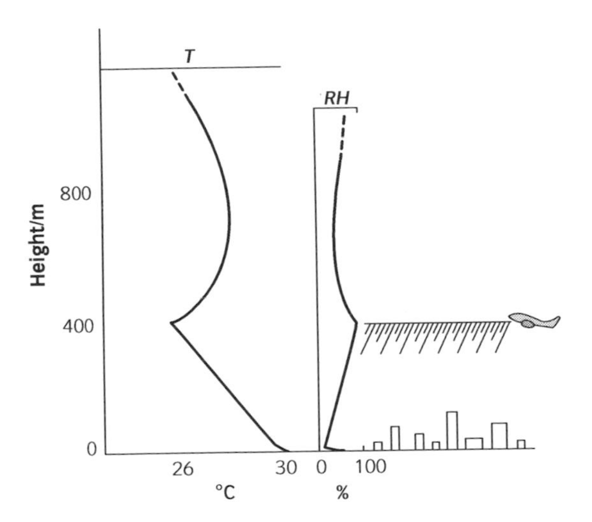

Figure 13.4 A schematic picture of the formation of a sharp haze top at the base of a strong subsidence inversion of an anticyclone. If the lifting condensation level of the surface-based convecting layer is a little below the inversion base, there will be stunted cumulus from cloud base to inversion base.

13.2.2 Cloud and smog

If the air in the convective boundary layer has typical humidity (much higher than above the subsidence inversion), the lifting condensation level (LCL—Sections 5.6 and 5.11) will often be a few hundred metres below the base of the subsidence inversion, and stunted cumulus will form at the top limit of convection from the surface. When convection is persistent, a layer of stratocumulus forms, which may be thick enough to reduce sunlight considerably, and even produce a little drizzle. Seen from above, however, the back scatter of sunlight makes the cloud layer look quite bright. Once formed, such a cloud layer tends to maintain itself by internal convection between its top surface, which is cooled by terrestrial

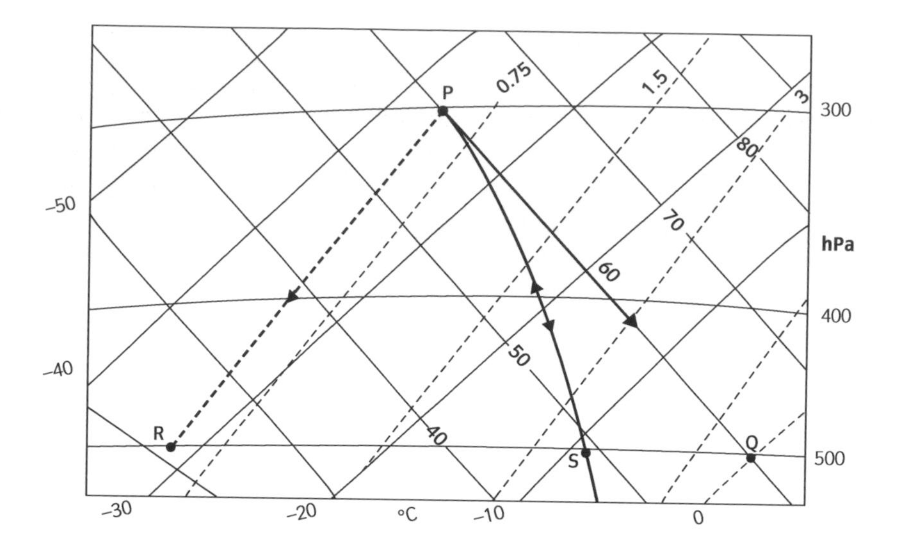

Figure 13.5 Drying by dry adiabatic and dry diabatic descent in the high troposphere (on part of a tephigram).

radiation more than it is warmed by sunlight, and its base, which is kept warm by the exchange of terrestrial radiation with the underlying warm surface, and sometimes by convection from the underlying surface. The consequent overturning can maintain the lumpy appearance of the cloud, even when there is little or no convection below cloud base (Fig. 13.2).

If the air in the convecting boundary layer is much drier that this, then the LCL will be so high that convecting air will still be unsaturated when its upward motion is stopped by the base of the subsidence inversion, and there will be no low cloud. The presence of the subsidence inversion may still be seen from the haze top which is often visible, even from the oblique viewpoint of a surface observer. In these conditions the solar warming of the surface may become intense in the middle of the day, warming the convective boundary layer as a whole and tending to eat into the base of the subsidence inversion. However, large-scale subsidence is often observed to be able to contain convection below the LCL for days on end, and is sometimes strong enough to lower the subsidence inversion to within a few hundred metres of the surface. In such circumstances, concentrations of industrial pollution are enhanced in the confined layer. The so-called Los Angeles smog is produced there and in many other cities under the influence of anticyclonic subsidence (Fig. 13.4). This type of smog is especially damaging in summer because the ultraviolet component of the strong sunlight encourages photochemical production of ozone and other biologically damaging substances in the confined boundary layer.

On other occasions the convective boundary layer may be several kilometres deep, and yet still not reach the lifting condensation level, because of the dryness of the air in the convecting boundary layer; this can happen over continental subtropical deserts such as the Sahara in North Africa. The systematic absence of substantial cloud means that rainfall in the subtropical highs is very low. In continental areas there is a belt of hot deserts in the subtropics of each hemisphere, including N Africa in the N hemisphere and Australia in the S hemisphere. Rainfall is sparse, intermittent and unreliable, coming in occasional deep convection during periods of weakness of the anticyclonic circulation. The rain when it comes may fall in torrential showers, running off the baked landscape in flash floods which rush down the gullies (*wadhis* to the Arabic nomads who have long eked a living in such harsh conditions in North Africa and Arabia) maintained by similar sporadic events in the past. Though the brief, inefficient moistening of the ground surface is enough to trigger the ecological miracle of the flowering desert, the water soon evaporates and the desert returns. Over the oceans the persistent excess of evaporation over precipitation under the subtropical highs keeps the surface waters significantly more saline than elsewhere.

13.2.3 Air flow

The anticyclonic circulation which is so pronounced in the lower troposphere, and which gives the whole type of system its title, can be related to the simplified vertical section in Fig. 13.6. Since air is persistently subsiding in the middle troposphere it must be persistently converging in the high troposphere and diverging in the low troposphere. The convergence aloft concentrates the vorticity there, making it more cyclonic (Section 7.16), but it seems that the initial vorticity is often very feeble, with the result that the circulation in the upper parts of a deep anticyclone is often only weakly cyclonic. As this air sinks into the low troposphere, however, it diverges strongly and its absolute vorticity is reduced so far that it becomes anticyclonic relative to the weather map. Although quite pronounced, this relative anticyclonic vorticity is limited by the fact that absolute vorticity cannot become negative on the large scale (Section 7.16), so that winds in anticyclones never reach the strengths which are quite common in cyclonic systems.

Near the surface, turbulent friction encourages divergence by allowing air to flow with a substantial component of motion outwards across the isobars towards lower pressures. On the W and polar flanks of the Subtropical High cells such diverging flow feeds air into the warm sectors of the mid-latitude depressions as they sweep E'wards, maintaining their warmth, and supplying air to the warm conveyor belts there (Section 12.4). On the equatorial flanks the diverging flow feeds the trade winds which angle SW'wards (NW'wards in the S hemisphere) down to the Inter-tropical Convergence Zone, fuelling the deep convection there (Section 4.7).

During the establishment of a particular subtropical high-pressure cell, surface pressure builds as convergence aloft exceeds divergence below, in a reversal of what happens during the formation of a depression, where divergence above exceeds convergence below and reduces surface pressures. However, during the quite long mature periods in which the central surface pressures of an anticyclone vary little, divergence and convergence must balance closely. No simple argument explains the continuing presence of the high surface pressure, despite the

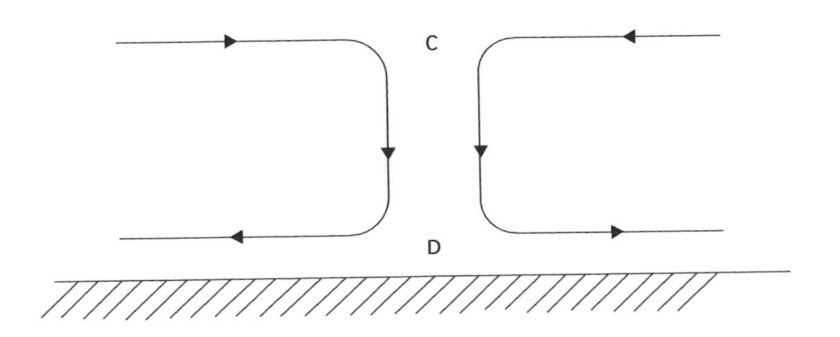

Figure 13.6 An outline of vertical and radial airflow in the troposphere of an anticyclone. Regions of synoptic scale horizontal convergence and divergence are labelled C and D respectively. Compare with a similar picture of a cyclonic system (Fig 7.33).

simplistic arguments in some descriptive texts. For example, high pressure is not a dynamic consequence of the downward motion of air in the main mass of air (like the pressure of a hose squirting water down onto a floor), since this would imply a gross failure of the hydrostatic approximation (Section 4.4), which actually is especially accurate in the quiet conditions prevailing in an anticyclone. It seems that dynamic constraints such as the Coriolis effect maintain the excess of atmospheric mass in the central parts of the anticyclone, once it has been established by the initial excess of convergence in the high troposphere.

The relative warmth of the subsiding air means that air pressure falls more slowly with increasing height than in adjacent cooler air masses, with the result that the upward doming of isobaric surfaces (or equivalently the concentration of high pressure on horizontal surfaces) increases with height, often into the upper troposphere. The high-pressure system is therefore warm and deep, unlike the shallow, cold highs which develop over continental interiors in winter (Section 12.5 and 14.3). Over the Sahara in summer, the subsiding warm core is superimposed on a shallow heat low at the very hot surface (Section 12.7), with the result that the depression of isobaric surface near the land surface gives way to upward doming in the middle and upper troposphere; a deep high therefore overlies some of the depressed sea-level pressures apparent over North Africa in Fig. 13.7b.

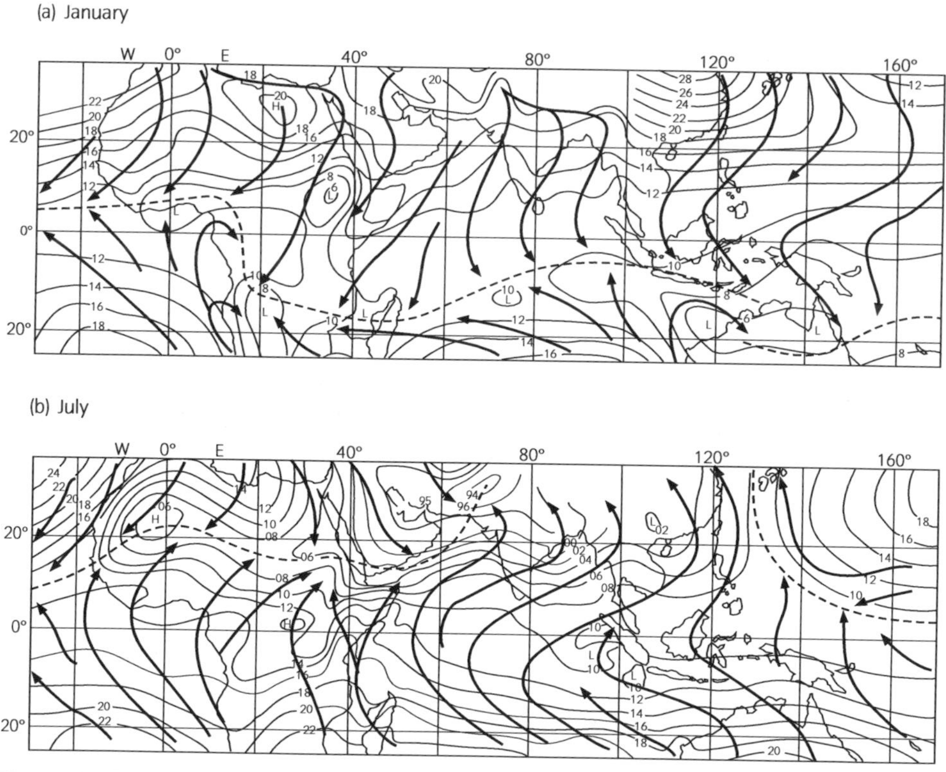

Figure 13.7 Mean pressure and airflow at the base of the troposphere in low latitudes in (a) January and (b) July. Isobars of mean sea-level pressure (from *[85]*) are labelled in hPa above 1,000 hPa. Arrowheaded thick lines are streamlines of flow near the surface (from *[76]*). Dashed lines mark zones of confluence (also from *[76]*), many of which correspond to the showery intertropical convergence zone described in Chs 4.7, 8.10, 13.3, and 14.3.

13.2.4 Global circulation

The subtropical anticyclones play a central role in the general circulation of the troposphere, being the descending branch of the Hadley circulation, and forming a quiet dynamic wall separating the troposphere of the vast intertropical region from higher latitudes. The barrier is made of a number of elongated cells, each tens of degrees of longitude in EW extent, between which there is substantial flow of warm air polewards and cool air equatorwards. And in the N hemisphere the barrier is disrupted permanently by the Himalayas and Tibetan plateau, and seasonally by the Asian monsoon. In the low troposphere the polar flanks of the cells feed warm air into the long waves and cyclone waves of middle latitudes, while the equatorial flanks feed the trade winds angling W'wards and equatorwards. In the high troposphere, the equatorial flanks receive air moving polewards in the upper part of the Hadley circulation, having ascended in the deep convection of the Inter-tropical Convergence Zone. The poleward flanks receive air which has ascended in middle latitude showers and fronts, some of which has had to make its way from the higher latitudes to which it had climbed in the warm conveyor belts. And all these inflows to the upper troposphere of the subtropical anticyclones feed the continuing subsidence which dominates the deep cores of these great quiet systems.

13.3 Monsoons

The great N'ward migration of cloud and rain over the Indian subcontinent which takes place each summer is known to virtually everyone on Earth through direct experience, school geography lessons, news reports, and locational fiction. It is often simply called the Indian monsoon, which correctly implies that there are monsoons in other regions, but tends to underplay the scale and interconnection of the huge seasonal events which affect Indo-China, Indonesia and China as well as India. Popular discussion also tends to focus on the wet summer, whereas this is only one part of an annual cycle, though very important because of the water it brings.

The word *monsoon* means season in Arabic, and dramatic seasonal change is a common thread which links a wide variety of climatic events occurring annually in low latitudes. Figure 13.7 shows how the air flow in the low troposphere shifts between January and July. In most longitudes there is some meridional migration of the convergence zones which are the local components of the Inter-tropical Convergence Zone (ITCZ—Section 4.7). Since these zones are dotted with cloud and rain maintained by the convergence of moisture-laden winds, their presence in any area is associated with the rainy season there. One of the largest meridional shifts occurs in the vicinity of the Indian subcontinent. In January the nearest convergence zone lies about 10° S of the equator (i.e. in the summer hemisphere), whereas in July, the nearest equivalent feature is the heat low and associated *monsoon trough* extending along the axis of the Ganges and Indus rivers, just S of the great zonal barrier of the Himalayas. There is another very large shift from N Australia to an ill-defined location in N China, and a substantial shift in E Africa. By contrast, the convergence zones over the Atlantic and Pacific Oceans migrate comparatively short distances, and the convergence zone in the E Pacific remains in the N hemisphere and close to the equator throughout the year.

13.3.1 Mechanisms

The much larger seasonal shifts of air flow over land as compared with ocean clearly must be related to the relatively very small effective heat capacities of land surfaces (Sections 10.4 and 10.5): as the zenith Sun moves from the tropic of Capricorn in late December to the tropic of Cancer in late June, so the zone of maximum thermal response by surface and atmosphere tries to march with it, and that march must be faster where the heat capacity is smaller, as it is over land. But the response is made complex and indirect by the role of water vapour and cloud: solar input to moist land surfaces is used largely to evaporate water (Section 8.6), which may then travel a long distance before giving up its latent heat in cloud formation. And the atmosphere tends to organize convergent flows of moistened air which gather latent heat from extensive ventilated areas before releasing it in narrow zones of cloudy uplift. Such convergence happens on a range of scales from the Hadley circulation (Fig. 8.21) downwards, all of which are too subtly sensitive to distributions of topography, land, and sea to be prescribed by a few mechanistic sound bites.

On the smallest scale, atmospheric response is localized in cumulonimbus, even the largest of which would be individually far too small to appear on maps like Fig. 13.7 (even if they were not smoothed by seasonal averaging). The ascent which constitutes the rising branch of the Hadley circulation actually takes place in a large number of very tall cumulonimbus (Section 11.5)—the *hot towers* whose narrow, vigorous updrafts are embedded in much larger volumes of relatively static or sinking air. And these are organized in a variety of types of large tropical weather system, whose observation remained fragmentary until the advent of the meteorological satellite, and whose mechanisms have yet to be outlined as clearly as have the large-scale weather systems of middle latitudes.

13.3.2 Rainy seasons

In terms of local climate, the result of all this is that regions within the range of the seasonal excursions of the ITCZ tend to have rainy seasons when it is close by, and dry seasons otherwise. In the simplest picture, places between the extremes of seasonal migration of the convergence zone would have two rainy seasons per year, but there are so many special factors at work in any particular location, especially overland, that the N'ward and S'ward migrations are often very different, and only one may be effective as a rainy season.

A rainy season is not simply a period of unceasing rain, or an unrelieved succession of showers: the rains come in bursts as one or other of the low-latitude weather systems develops or moves by, and a particular rainy season may be unusually wet or dry depending on the number, type, and intensity of systems occurring. Indeed the rather short length of many rainy seasons over any location, mean that year-to-year variability of rainfall is often higher than in middle latitudes. This is especially obvious in regions with short rainy seasons, but appears even in well-watered low-latitude regions (compare Table 13.1 and Fig. 14.1).

13.3.3 The Indian monsoon

Although the rains enter Myanmar (Burma) in April and May, they do not reach India until late May or early June. Then they often sweep across the S half of the

Table 13.1 Annual rainfall at Canton Island 2° 48′ S, 171° 43′ W

Year	Rainfall	Year	Rainfall
1957	1269 mm	1962	402 mm
1958	1597	1963	713
1959	759	1964	519
1960	492	1965	1433
1961	508	1966	1101
	mean 879		
	range 1597 to 402		
	standard deviation 414, i.e. 47%		

subcontinent in a spectacular *burst*, whose mechanism has intrigued generations of meteorologists, and which may owe more to rearrangements of flow in the high troposphere than to the prior establishment of the heat low in the NW of the subcontinent (Fig. 13.7) which used to be regarded as a prime cause.

The monsoon usually becomes established over the Indian subcontinent by late June, by which time there is a more or less continuous S'ly flow of warm, moist surface air towards, but not effectively into, the monsoon trough lying across the N of the subcontinent. (N'ly flow to this zone is blocked by the highlands of Tibet and Afghanistan.) The S'ly flow is in fact a continuation of the SE trades of the S hemisphere. As these cross the equator N'wards, the Coriolis deflection switches from leftward to rightward, and the flow veers to become SW'ly. This flow of already moist air picks up further large quantities of water vapour from the Arabian Sea (warmed to 28 or 29 °C by very strong sunshine before the onset of the monsoon) and reaches the W coast of India as the SW monsoon. Large quantities of rain fall as ascent is forced or triggered along the W edges of the great plateau, and substantial flows continue N'wards and merge with the S'ly flow from the Bay of Bengal. In the middle and lower middle troposphere there are important W'ly flows of air probably originating in Africa and even the Mediterranean.

Slow-moving synoptic-scale low-pressure systems, such as *subtropical cyclones* and *monsoon depressions* (the latter being one of the few types to bring rain to the monsoon trough) form in association with these flows and can produce large areas of rain and quite strong winds even though surface pressures are depressed by only a few hectopascals.

With the surface thermal equator established locally just a little N of latitude 25 °N, and substantial volumes of air rising in weather systems on its equatorial flank, the normal baroclinity is reversed through almost the full depth of the troposphere between this warmth and the cooler troposphere nearer the equator. The thermal wind relationship (Section 7.12) then ensures that the *tropical easterly jet stream* is a semi-permanent feature of that region in the summer months at latitudes about 15 °N (Fig. 12.7). The weather systems associated with the monsoon therefore form and act in the context of SW'ly flow at low levels and E'ly flow aloft, and it is not surprising that their movements are often sluggish and W'ward, and their mechanisms complex.

The SW monsoon begins to retreat in September and usually completes its withdrawal from the Indian subcontinent by the end of November. Well before

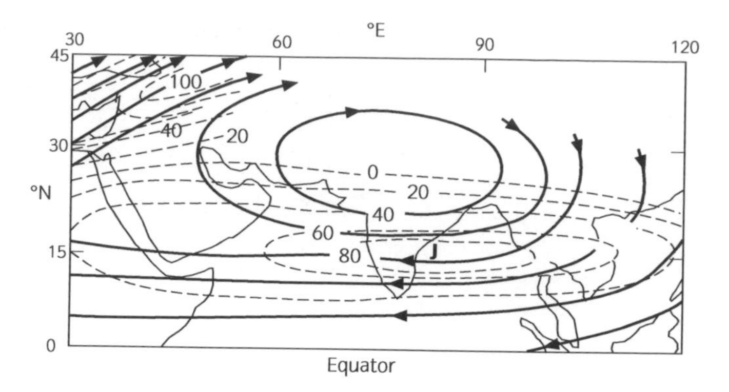

Figure 13.8 Streamlines (arrowed solid) and isotachs (dashed, labelled in knots) on the 200 hPa surface at 0300 Z on 25 July 1955, showing a strong E'ly jet stream over S Central India. The zero isotach marks the boundary between E'lies to the S and W'lies to the N, and a SW'ly jet stream lies SE of the Black Sea, in air flowing towards the N of the Tibetan massif.

the end of this period, the E'ly jet aloft is replaced by a narrow branch of the subtropical (W'ly) jet skirting the S flank of the Himalayas. In the N of the subcontinent the ensuing drier seasons are also much cooler because of the moderately high latitudes, and flows of cold air from neighbouring high ground. Weak low-pressure systems move SE along what was the axis of the monsoon trough, and later in the winter these may be followed by NW'ly surges of cold air which move right across the peninsula from the great reservoir which is building to the N and E of the Tibetan massif. But this barrier largely protects India from the fierce surges of the bitterly cold NE monsoon which affects China in winter (Section 14.2). Before the onset of winter proper, coastal areas of the Indian subcontinent may have been influenced by an important type of low-latitude weather system which we will consider in the next section—the tropical cyclone.

13.4 Tropical weather systems

13.4.1 Consequences of barotropy

The baroclinity of the tropical troposphere is weak and unstructured by comparison with higher latitudes. Solar heating of the surface and troposphere reaches a broad maximum between the tropics, and terrestrial radiation partly offsets this by an amount which leaves a radiative excess varying little with latitude (Fig. 8.9). The result is that the tropical troposphere (the whole troposphere between the tropics, including the equatorial zone) is nearly barotropic in comparison with the essentially baroclinic troposphere of higher latitudes. Horizontal temperature gradients at any level are usually very small, and differences of more than ~ 5 °C are comparatively rare away from the immediate influence of land and sea surfaces (strongest in the atmospheric boundary layer). For the same reason, seasonal changes in temperature are often so small over the great seas and coastal margins that the seasons in low latitudes are measured more by their contrasts in weather, especially rainfall, than they are by their contrasts in temperature.

Although the tropical troposphere is relatively barotropic, the small horizontal temperature gradients which do arise are associated with very significant patterns of cloud and associated weather. Tall cumulonimbus (hot towers) may be only a couple of degrees warmer than the mid-troposphere they penetrate, which may in turn be a couple of degrees warmer or cooler than its counterpart outside the conglomeration of cumulonimbus which constitutes the weather system. In addition, the weakness of the Coriolis effect at low latitudes means that most disturbances of the tropical troposphere are not even quasi-geostrophic in their mutual adjustment of wind and pressure fields, which means that the pressure field cannot be used to bolster the observed wind field in the way which is so useful at higher latitudes. (However, more than a few degrees of latitude away from the equator, the undisturbed environment is still geostrophic to the extent that, for example, the thermal-wind relationship holds at least qualitatively.) As a result, a vigorous synoptic-scale weather system producing ~100 mm of rainfall per day and winds of nearly gale force may be associated with a depression in sea-level pressure of only a few hectopascals, and with horizontal temperature differences aloft of only a very few degrees.

In operational meteorology such systems are usually known as *tropical depressions* unless surface winds reach gale force (Section 13.5). This sensitivity of linkage between behaviour and ambient conditions poses severe problems for observational, operational, and theoretical meteorology—problems which are compounded by the sparseness of observations enforced because so little of the tropical zone is land, and much of it is sparsely populated. Fortunately, the development of the meteorological satellite in recent decades, especially the geosynchronous type (Section 2.11), has improved things considerably. Analysis techniques have been developed to extract wind profiles from the movements of large clouds between consecutive pictures, and cloud and air temperatures from simultaneous measurements of terrestrial radiation at several wavelengths.

13.4.2 Tropical (or easterly) waves

In the zones polewards of the equatorial confluence zones, the trade winds maintain an E'ly component of flow in the low troposphere, and there is clear evidence from satellite pictures of these zones that much of the cloud and rain over the oceans occurs in large clusters of cloud, hundreds to thousands of kilometres apart, which remain observably coherent while moving W'wards at between 5 and 10 m s^{-1}—a speed which is usually a little slower than the air flow in the low troposphere. Overland, the picture is confused by topography and a strong diurnal rhythm, so that many clusters develop and decay without obvious motion. So although land-based weather systems are the more immediately relevant to people, the oceanic systems are the more obviously coherent and better understood. In addition, many weather systems which form over the tropical oceans (which cover about 75% of the Earth's surface between the tropics) significantly affect adjacent land areas, especially their E margins.

Many of the cloudy systems over the sea seem to move in association with weak troughs in sea-level pressure (Fig. 13.9) whose wave-like appearance on a weather map gives them their older name *easterly waves*. Note that the reversed pressure gradient compared with the mid-latitude norm of N'ward pressure lapse gives a superficial impression of a pressure ridge to natives of middle latitudes. In fact the trough represents a weak poleward extension of the convergence zone trough whose zonal axis lies ~10° of latitude equatorwards of the heart of the wave (along the foot of Fig. 13.9b).

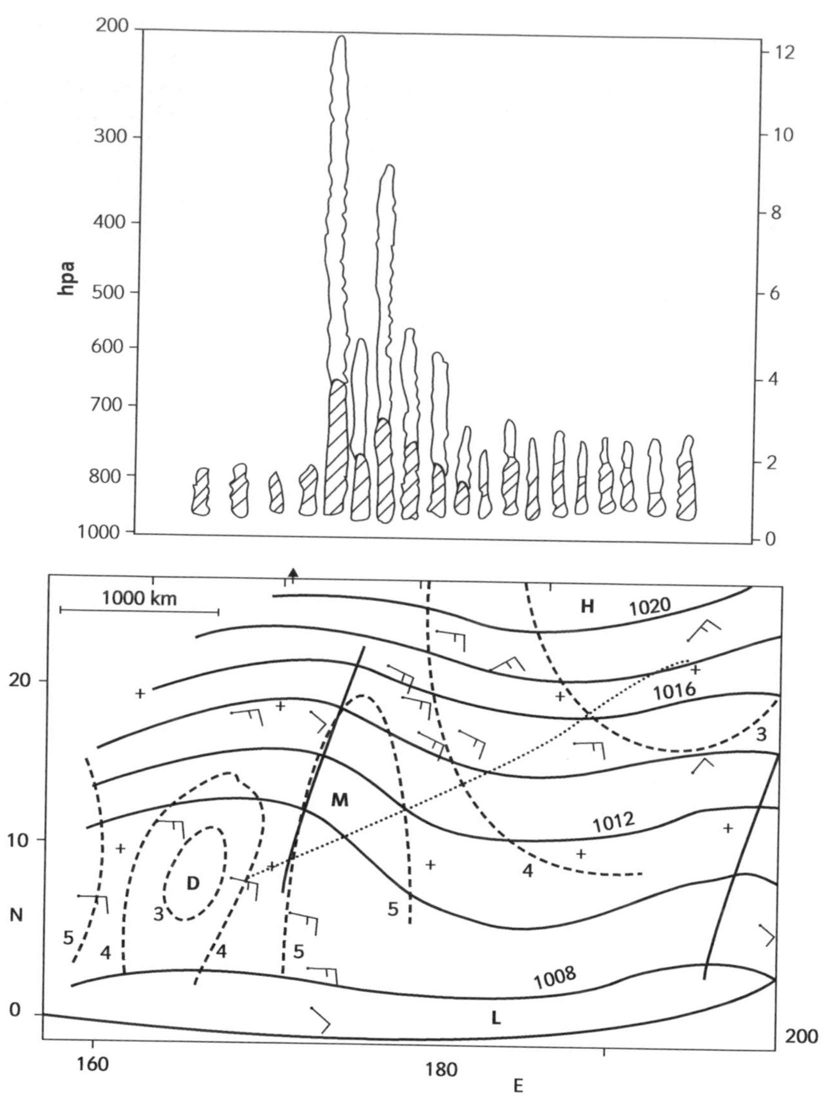

Figure 13.9 An Easterly Wave in the central Pacific in plan and vertical section. On the map, sea-level pressures and surface winds show a wave with trough line (bold near M) approaching the Marshall Islands, followed by another wave about 3000 km to the E. Synoptic surface winds are plotted as arrows with each full tail feather representing 10 kt. Column totals of water vapour as a proportion of atmospheric mass are plotted as dashed lines and labelled in parts per thousand, with dry and moist zones marked D and M. The dotted line was traversed by instrumented aircraft whose observations of cloud height are represented on the vertical section—average heights by hatched cloud towers and extremes by clear extensions. Horizontal extents of towers are purely diagrammatic.

In some cases there is marked asymmetry of cloud distribution—cumulonimbus and some altostratus being concentrated with rain and fairly strong winds E of (i.e. behind) the trough axis. This concentration is associated with gentle convergence in the low troposphere there and corresponding divergence to the W of the trough axis, as follows. The convergence implies quite substantial vertical stretching of the lower troposphere which serves to destabilize the layer (Section 11.3) and thereby encourage the growth of cumulonimbus. It also helps concentrate the fuel for cloudy convection—the water vapour entering the boundary layer from the sea surface. Stretching and convergence also deepens the layer with E'ly wind components, so that the W'lies are confined to the upper troposphere (Fig. 13.9). And it increases the absolute vorticity of the lower troposphere (Section 7.16). The latter shows up in the poleward excursion (which increases the local Coriolis parameter f) and cyclonic curvature of the air flow as it overtakes the wave crest. A nearly balancing slight divergence seems to be concentrated in the high troposphere, giving a much more asymmetrical vertical distribution of vorticity than is typical in middle latitude systems. To the W (i.e. ahead) of the wave axis each of these effects is reversed, maintaining relatively clear skies in the NE'ly flow.

In some cases the cloud is banded zonally across the trough axis, and in others again the cloud is distributed fairly homogeneously. There is conflicting evidence about whether or not all these cloudy systems are warmer or cooler on aggregate than their surroundings—whether the systems are *warm-* or *cold-cored*—but it is clear that they are not as clearly warm cored as their more energetic relatives (see below). The W'ward movement of such systems across the great oceans at intervals of a few days provides much of the rainfall in these zones, and in the coastal and island areas they affect. They have been studied relatively intensively in the N central Atlantic by American meteorologists interested in their effects on the Caribbean and surrounding regions, and their capacity to develop into *tropical storms* (next section) and further into *severe tropical cyclones*, known as *hurricanes* in the Caribbean, and *typhoons* and *cyclones* elsewhere. It is observed that a substantial number of the systems reaching the Caribbean originate in the lines of cumulonimbus which move W'wards across NW Africa in the summer. Others form in the Caribbean convergence zone and on fragments of fronts trailing down from mid-latitudes. The annual total observed in the Atlantic remains fairly constant at about 100, of which about 70 reach the Caribbean.

The subtropical cyclones mentioned in Section 13.3 are not limited to monsoon zones; they can be found in the trade wind zone, where they can show several closed isobars (at the standard 2 hPa interval) at sea level, even though they appear to originate in the upper troposphere. Unlike the warm cored tropical cyclone family, they are cold cored systems, though the temperature depressions are quite small and they can occasionally become slightly warm-cored through unusually large release of latent heat in the embedded clouds. There is evidence of banded cloud structure, and they are often individually very persistent.

13.5 Tropical storms and severe tropical cyclones *[56]*

Of the 70 tropical disturbances (mostly easterly waves) reaching the Caribbean each year, about 25 deepen to the point where there are several closed isobars on surface charts and strong cyclonic circulation, cloud, and rain over a substantial area. When winds reach gale force (17 m s^{-1} at 10 m—the boundary between Beaufort force 7 and 8) they are called *tropical storms*, and are equivalentin intensity to middle-latitude depressions, although they have no fronts. In summer and autumn, a variable number of these storms, averaging about eight in the Caribbean, intensify to the point where winds exceed hurricane force (Beaufort force 12–33 m s^{-1} at 10 m), whereupon they are known as *hurricanes* (in the Caribbean). As is well known by news reports and maritime fiction *[57]*, hurricanes are most intense in a broad ring surrounding a much calmer, very low-pressure *eye*. In the week or so in which they remain in their mature, violent state, they usually travel W'wards and then polewards. They are found in many tropical sea areas in addition to the central W Atlantic, Caribbean (Fig. 13.10), and E Pacific, being called variously severe tropical cyclones, severe cyclonic storms, tropical cyclones or cyclones in the Indian Ocean and around Australia, and *typhoons* in the rest of the Pacific Ocean. The nomenclature is frankly a mess, in sharp contrast to the regularity achieved by the WMO elsewhere. Regardless of regional title, they are all the same type of weather system, whose combination of intensity and extent makes it the most powerful type of synoptic scale weather disturbance observed in the Earth's atmosphere. In view of the confused nomenclature, in this book we will give them the general title *severe tropical cyclone*, since that identifies them clearly as extreme members of the tropical cyclone family, or hurricane or typhoon—their most popular regional titles.

Severe tropical cyclones form only over oceans where surface temperatures exceed about 26.5 °C over substantial surface areas at latitudes of at least 5°, but they occur in every such eligible zone except the South Atlantic. The annual total of about 80 severe tropical cyclones is small in comparison with totals of extratropical cyclones, and yet they continue to attract the attentions of meteorologists to an extent which is out of all proportion to their number. Two reasons dominate:

(i) They can be almost unbelievably destructive if they pass over vulnerable areas: buildings and crops suffer severely in sustained winds exceeding hurricane force (often greatly), and coastal areas risk inundation by the violent wind waves and large storm surges (Section 12.3 and below) which can be generated. Torrential rainfall often causes catastrophic flooding, and land and mudslides in hilly terrain.

(ii) They are unusual among tropical weather systems in having a very sharply defined inner structure which excites comment and investigation.

Although severe tropical cyclones have been studied intensively for many decades now, their formation from relatively very weak tropical disturbances is incompletely understood. Since they form in effectively barotropic conditions, there is no obvious source of available potential energy (Section 9.7) to compare with the baroclinity of mid-latitudes. There is some evidence that residual baroclinity

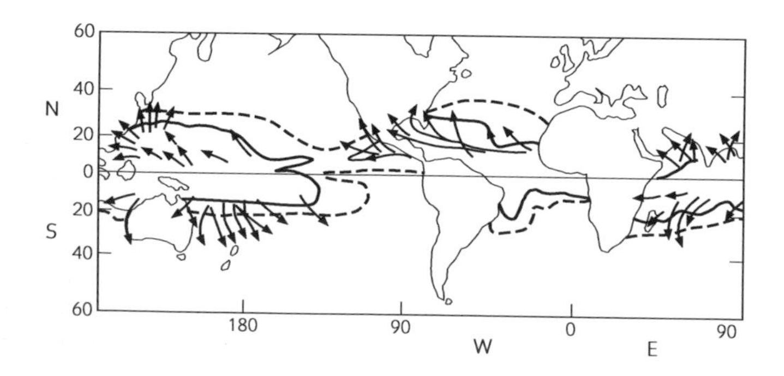

Figure 13.10 Representative tracks of severe tropical cyclones (arrowed thick lines) compared with mean sea-surface isotherms (25 °C dotted, 27 °C solid) for September in the N hemisphere and March in the S hemisphere. Note that the Mercator projection seriously underplays the fraction of the Earth's surface area affected by such storms.

imported from higher latitudes in the form of a weak upper trough may help trigger their dramatic development. However it happens, in a couple of days the central sea level pressure can fall by 20 hPa or more (normal enough in middle latitudes but unusual in the tropics), setting the stage for transition from tropical storm to severe tropical cyclone. This may still be prevented by premature landfall and consequent loss of the crucial supply of water vapour from the warm sea surface, or by entrainment of unusually cold air into the centre of the vortex in the high troposphere (which quenches the developing warm core of the storm); but if such factors do not intervene, the mature tropical cyclone develops rapidly by formation of a relatively localized ring of violent winds, updrafts, and rainfall around a central, relatively inactive eye with very low surface pressure.

13.5.1 Structure

Figure 13.11 is a typical satellite view of a mature severe tropical cyclone. The dense white ring extending out over 100 kilometres from the edge of the eye is a shield of cirrus fanning out into the high troposphere from the top of the ring of updraft surrounding the eye. The appearance of the shield and eye amid the more diffuse mass of high cloud is a sure sign that the severe tropical cyclone has tightened into full maturity. Penetration by special reconnaissance aircraft reveals and monitors the full power of the storm and helps with forecasting its trajectory and potential for damage.

Working inwards from the outer edges of the storm, the spiral arms of cloud stretching hundreds of kilometres outwards from the central structure, often mainly equatorwards and E'wards, are long curving lines of cumulonimbus forming in air spiralling into the storm in the lower troposphere (Figs. 13.12 and 13.14). The inner zone contains a dense mass of cumulonimbus with sheets of cloud spreading from them at all levels, circling faster and faster cyclonically as the centre is approached (Fig. 13.12), until the fastest flow is reached in the vicinity of a ring of giant cumulonimbus which often completely encircle the eye, and produce a particularly strong

Figure 13.11 Hurricane Felix over the S Caribbean about September 2, 2007, seen obliquely from the E by hand-held camera. The hurricane fills the whole panorama, but the wheel of violent weather about the eye is hidden under the much smaller annulus of solid cloud studded with overshooting updrafts. Two of the arms of deep cloud spiralling into the annulus are glimpsed in and above the left foreground. Maximum 1 minute winds at 10 m exceeded 160 mph, and the minimum sea-level pressure was 929 hPa. (See Plate 15).

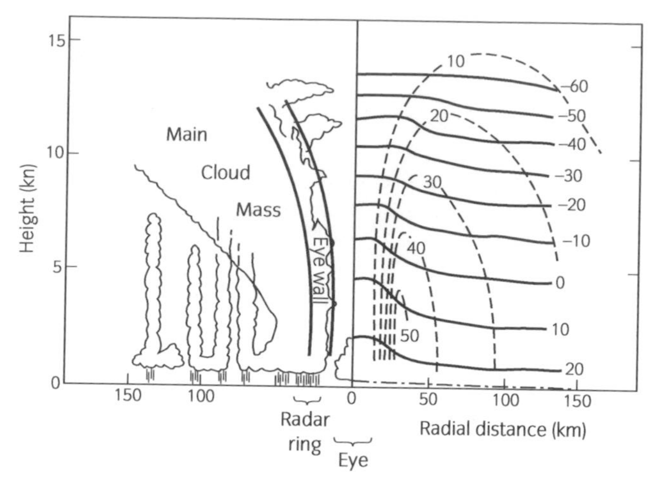

Figure 13.12 Vertical radial section through hurricane Helene on 26 September 1958. On the right-hand side the solid isotherms labelled in °C show the deep warm core of the storm. The dashed isotachs of tangential wind speed (in metres per second) show the powerful inverted vortex. The left-hand side is actually the right side again, but showing distributions of cloud, precipitation, the plan position of the strongest radar echo, and the boundaries of the eye wall as defined by the zone of maximum temperature gradient.

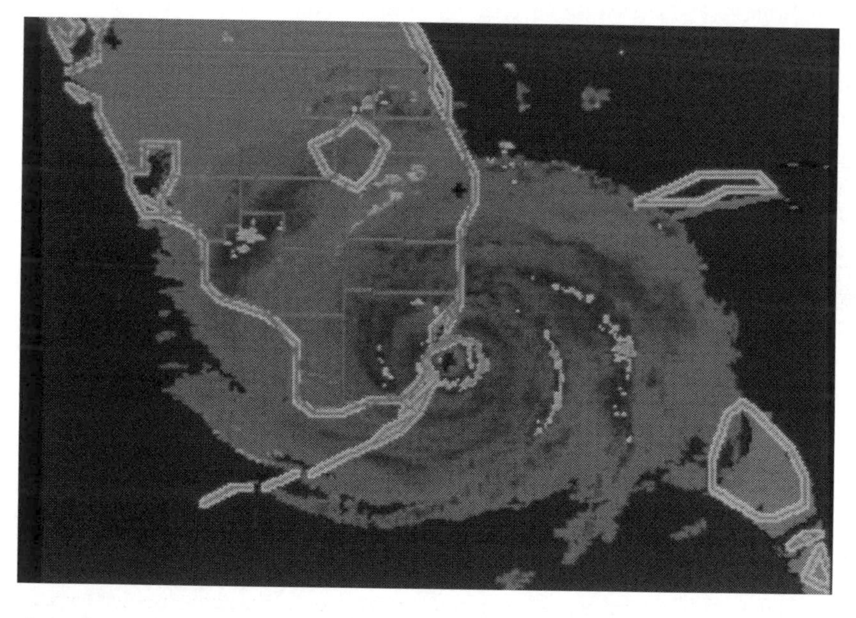

Figure 13.13 A PPI radar composite of Hurricane Andrew making landfall on SE Florida on August 4 1992. Echo is from precipitation in the central annulus of the storm, and shows spirals of enhanced echo leading into the solid narrow ring of intense echo (torrential rainfall) just outside the eye. The radius of the echo disk is over 100 km. (See Plate 16).

ring echo on weather radars (Fig. 13.13). The eye itself is often relatively cloud-free: a quiet amphitheatre usually only a few tens of kilometres across surrounded by the massive, rotating *eye wall* of encircling cumulonimbus (Fig. 13.15).

Temperatures rise sharply inwards at all levels below the high troposphere, reaching a warm plateau in the vicinity of the eye wall. Sea-level pressures fall towards the centre, slowly at first and then more and more rapidly to the minimum value which applies across the eye and is a useful measure of the severity

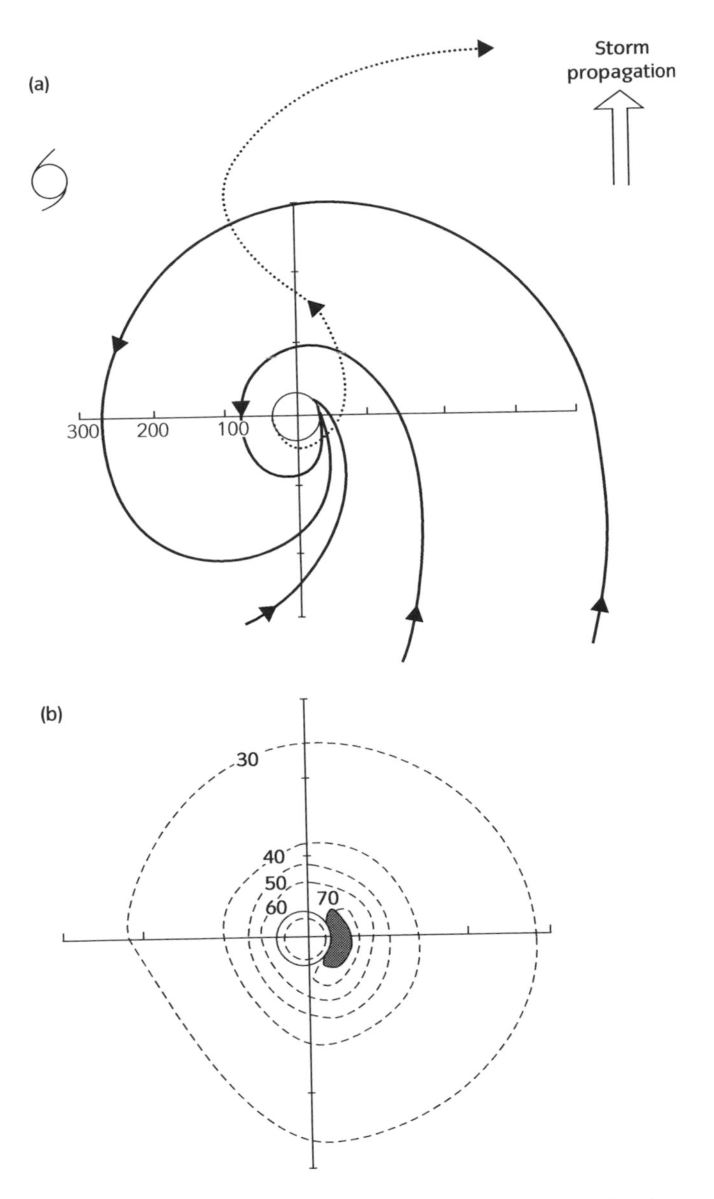

Figure 13.14 Winds in a very severe hurricane in the N hemisphere. (a) Solid streamlines show air spiralling inwards in the low troposphere, while the dotted streamline shows air spiralling outwards in the high troposphere—initially cyclonically like the low-level flow, but then anticyclonically. (b) The associated pattern of isotachs at the 10 m level is labelled in m s^{-1}. The black zone has winds in excess of 80 m s^{-1} (nearly 160 kt), which puts it in the top category (5) of the Saffir-Simpson scale.

of the storm. Most severe tropical cyclones have minimum sea-level pressures of 950 hPa or less. Such values are not very different from the minima in vigorous extratropical cyclones, although a few severe tropical cyclones have much lower minima, but the low pressures are concentrated in very much smaller areas, and are associated with pressure gradients which at their strongest (in the ring of maximum winds) can be ~1 mbar km^{-1}—more than an order of magnitude steeper than the largest synoptic-scale pressure gradients associated with extratropical cyclones. In fact the central zones of a tropical cyclone are structured on the meso-scale rather than on the synoptic scale, which made their detailed observation very sketchy before meteorological aircraft or satellites became available. On a conventional meteorological surface map, with sea-level isobars

Figure 13.15 Panoramic photograph of part of the eye wall of Hurricane Katrina (end of August 2005) taken from a meteorological observation aircraft crossing the eye in the mid troposphere. The floor of the eye on this occasion was carpeted with low cloud. Katrina was a category 5 storm with a very low minimum pressure of 902 hPa, and caused over 1,800 deaths in and around New Orleans.

drawn at 2 hPa intervals, the innermost 10 or so isobars are often not resolvable by synoptic observation, and are interpolated using a simple pattern of concentric circles spaced to match estimates of maximum winds. In fact the detail may be too fine for the thickness of the plotting lines.

13.5.2 Winds

The very high surface winds of the severe tropical cyclone are its most notorious feature, and exceed Beaufort force 12 (sometimes very considerably) in an annulus at least several tens of kilometres wide outside the eye. Facing in the direction of the storm's overall motion, highest winds occur to the right of the eye (in the N hemisphere), where the speeds of rotation and translation combine (Fig. 13.14), assuming velocities add as do the rim and top velocities of a wandering, spinning top. Sustained speeds as high as 75 m s^{-1} at the 10 m level have been recorded occasionally, though the chances of an official anemometer being in such mesoscale zones (and surviving flying debris) are small.

The effect of wind speeds on surfaces and structures increases roughly in proportion to the kinetic energy of the flow (i.e. to the square of the sustained wind speed—Eqn 10.25), though structures may have additional selective sensitivity to embedded gusts depending on their frequency—like the 'dancing bridge' in the Tacoma narrows in much lighter winds. Wind speeds above 50 m s^{-1} have enormous effects but last for several hours instead of the few minutes of a tornado. Over the sea they raise waves ~10 m high from trough to crest, and drive sheets of spray and foam over crests to the extent that the interface between air and water loses definition. Even modern well-found ships do not brave such conditions by choice; an accidental encounter between a United States battle fleet and a typhoon in the Philippine Sea in 1944 left 790 men dead, 3 ships sunk, many severely damaged, and 146 aircraft destroyed by the violent motion of their aircraft carriers.

As with extratopical cyclones, the zones of strongest winds produce storm surges (Section 12.3) in coastal waters, and sea levels can be raised or lowered by several metres for a few hours compared with the predicted astronomic tide, especially when amplified by shallow bays and estuaries. The actual response of the water is a very complex function of storm size, strength, and movement, as well as coast and sea-bed topography, but the net result in storms making perpendicular landfall on a coast is a substantial mound of water to the right of the storm centre (in the N hemisphere). In some situations resonance by surge reflection can lead to a dwindling series of resurgences, the first couple of which may be high enough to cause further trouble. If a high surge happens to coincide with a high astronomic tide there is obviously a serious risk of sea defences being breached in vulnerable areas, especially since walls, dunes, etc. are then unusually exposed to battering by large waves. Over 200,000 people were drowned in a few hours in 1970 when a severe tropical cyclone in the Bay of Bengal made landfall in the Ganges delta, where large numbers of people live within a few metres of sea level.

Over land, winds flatten crops, uproot trees, and severely damage or destroy weakly constructed buildings. Exposed livestock and people obviously risk death and injury from flying debris. Unfortunately in a way, severe tropical cyclones are sufficiently rare events at any particular location for societies not to be fully adapted and protected. So the emphasis is on reliable forecasting to allow time for window shuttering, housing loose items, and evacuating areas at greatest risk of flooding, etc. This touches on another aspect of their behaviour which is poorly understood. As the outline storm tracks in Fig. 13.10 show, many severe tropical cyclones travel W'wards initially in the E'lies in which they form, but on or after maturity tend to *recurve* sharply polewards. These irregularly curved paths are difficult to forecast accurately.

Flow mechanism

How are such winds produced and sustained for the considerable life (at least several days) of a mature severe tropical cyclone? The clustering of strong updrafts around the eye obviously requires a sustained inflow of air, which requires strong horizontal convergence, concentrated particularly in the lowest kilometre where friction encourages the flow to spiral in across the nearly circular isobars. In fact convergence must persist throughout the layer with significant cyclonic rotation, which is usually at least 10 km deep (Fig. 13.12). Throughout this layer, convergence continually concentrates pre-existing vorticity, which in the absence of any synoptic scale background is the planetary vorticity f. According to Box 13.2 and Fig. 13.16, conservation of angular momentum about a vertical axis in the centre of the storm implies that the tangential wind speed U reached after a ring of air has contracted from radius R_o (where air was stationary on the weather map) to a much smaller radius R, is given to a good approximation by

$$U = \frac{f R_o^2}{2R} \qquad 13.1$$

According to this, at latitude 20° a contraction from a radius of nearly 350 km would produce a wind speed of 100 m s^{-1} at radius 30 km (corresponding to a wind speed of about 50 m s^{-1} at height 10 m) which are typical values for

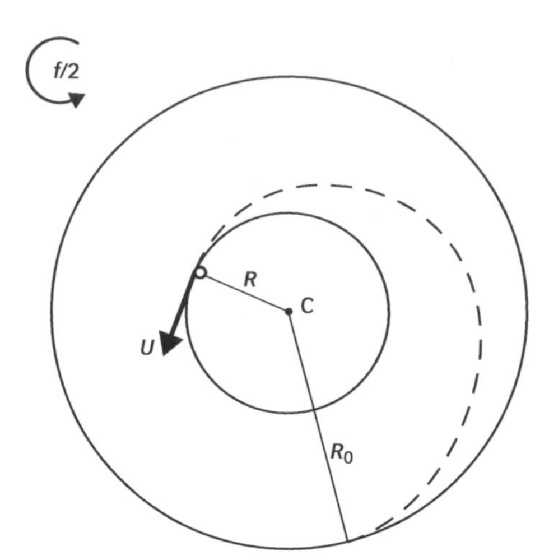

Figure 13.16 A diagram for portraying the conservation of angular momentum in air spiralling into the eye wall from an origin on the static rim of the storm (Box 13.2).

the ring of maximum winds. At latitude 5° the contraction would have to be from nearly 700 km for the same result (NN 13.1). The rarity of severe tropical cyclones at latitudes lower than about 5 °C suggests that the equatorial troposphere cannot sustain the extensive convergence needed to produce and sustain tropical cyclones.

It is observed that the effect of surface friction is to reduce the radial dependence of near surface winds from the $1/R$ predicted by the conservation of angular momentum to about $1/R^{0.6}$. However, the diverging flow at the very top of the troposphere (which spreads the canopy of anvil cloud seen by satellite) is above the reach of much of this friction, and tends to preserve the angular momentum of the ring of maximum winds more nearly as it diverges. The result is that, in zones outside the ring of maximum winds, wind speeds aloft are systematically much lower than at lower levels (Fig. 13.12). In fact when air spreads out to radii ~200 km it begins to rotate more slowly than the Earth's surface and hence turn anticyclonically on the weather map. The resulting anticyclonic curvature of upper-troposphere flow beyond the dense, central canopy makes the diverging high cloud in the outer parts conform to much the same spiral formation as the converging cyclonically curved lines of cumulonimbus in the lower parts of the troposphere, so that together they contribute to the characteristic spiral 'galactic arm' appearance which is used as the icon for a mature hurricane on large weather maps (Fig. 13.14).

Cyclostrophic balance

The reduction in tangential wind speeds in most of the upper parts of the central annulus of the storm (Fig. 13.12), described above, plays a vital role in enabling the storm to balance its wind and thermal structures. In the violent ring zone the horizontal accelerations arising from the ring rotation are so large that they swamp other accelerations relative to the map (unsteady linear accelerations, etc.) and therefore match the horizontal pressure gradient forces as closely as is normal in middle latitudes. However, since they far outweigh the weak Coriolis acceleration at these low latitudes (NN13.1), the resulting balance is *cyclostrophic* rather than geostrophic (Section 7.14). Assuming for simplicity

BOX 13.2 Convergence and conservation of angular momentum

Consider a narrow circular horizontal ring of air which begins at rest with radius R_o on the Earth's surface concentric with the tropical cyclone, and then shrinks to radius R while conserving angular momentum about the local vertical. Initially it has only the angular velocity of the surface itself, which is $f/2$, where f is the Coriolis parameter. Arbitrarily assuming unit mass for the ring, the initial angular momentum about the centre C of the circle is (Box 4.4, 7.14)

tangential speed $\times R_o =$

$$(\text{angular velocity} \times R_o) \times R_o = \frac{R_o^2 f}{2}$$

After shrinking to radius R, we suppose that the ring is turning with tangential speed U relative to the map. This contributes an additional angular velocity U/R to the planetary $f/2$, so that the total angular momentum is now $R^2\,(U/R + f/2)$. Equating initial and final expressions for angular momentum and rearranging, we find

$$U = \frac{f}{2R}\left[R_o^2 - R^2\right] \qquad \text{B13.2a}$$

When the first term in brackets is much larger than the second we have

$$U = \frac{f}{2R} R_o^2$$

which we can rearrange to find the initial radius needed to produce a chosen speed U at a chosen R

$$R_o^2 = \frac{2RU}{f} \qquad \text{B13.2b}$$

Notice that R_o increases with decreasing f (i.e. decreasing latitude): to produce a certain wind speed at a certain inner ring radius, greater convergence is necessary at lower latitudes.

NUMERICAL NOTE 13.1 Spinning up the winds of a severe tropical cyclone

- Take 50 m s^{-1} to be a typical wind speed in the violent ring (radius 30 km) of a severe tropical cyclone. At latitude 20°, the Coriolis parameter $f = 2\,\Omega \sin 20 = [2 \times 2\,\pi/(24 \times 60 \times 60)] \sin 20 \approx 5 \times 10^{-5}$ s^{-1} and the starting radius R_o needed to produce $U = 50$ m s^{-1} at $R = 30$ km by conservation of angular momentum is given by Eqn B13.2b, so that $R_o^2 = 2\,R\,U/f = 2 \times 30{,}000 \times 100/(5 \times 10^{-5}) = 6 \times 10^{10}$ m^2, and $R_o \approx 245$ km.

  At latitude 5°, $f = 1.3 \times 10^{-5}$ s^{-1} and the same calculation gives $R_o^2 \approx 4.6 \times 1011$, $R_o \approx 480$ km.
- Consider an air parcel moving at speed V cyclonically round a horizontal circle of radius R at latitude 20°. The hidden Coriolis acceleration is $f\,V$ centripetally (towards the centre of the circle) and the centripetal acceleration relative to the weather map is V^2/R, so that the ratio $CE/CO = V/(R\,f)$. At 20°, 50 m s^{-1} and on a circle of radius 30 km, $CE/CO = 50/(3 \times 10^4 \times 5 \times 10^{-5}) = 50/1.5 \approx 33$.

  The Coriolis acceleration is only about 3% of the centripetal acceleration on the weather map. The flow is essentially cyclostrophic (Section 7.14).
- In the violent ring, cyclostrophic balance ensures that the radial pressure gradient $\partial p/\partial R = \rho\,V^2/R \approx 1.2 \times 50 \times 50/(3 \times 10^4) = 0.1$ Pa m^{-1} = 1 hPa per kilometre.
- In the violent ring, the cyclostrophic thermal wind relation (Eqn 13.3) ensures that the radial temperature gradient $\partial T/\partial R = [T/(R\,g)]\,\partial V^2/\partial z = [2\,T/(R\,g)]\,V\,\partial V/\partial z$. Figure 12.11 shows that V falls from about 50 to 30 m s^{-1} as height increases from 4 to 8 km, in which $T \approx 270$ °C. Inserting this and $V = 40$ m s^{-1} and $R = 30$ km we have $\partial T/\partial R = [2 \times 270/(30{,}000 \times 10)] \times 40 \times 20/4{,}000 = 3.6 \times 10^{-4}$ °C m^{-1}, which is ≈ 0.4 °C per km. If temperature rose at this radial rate going horizontally inwards through a violent ring 20 km wide, the rise of air temperature would be about $0.4 \times 20 = 8$ °C.

that all the air at a radial distance R from the storm's centre is actually moving around that circle (real trajectories are complicated by radial motion and the translation of the storm), cyclostrophic balance between pressure gradient and centrifugal force is described by

$$\frac{V^2}{R} = \frac{1}{\rho}\frac{\partial p}{\partial R} \qquad 13.2$$

This balance reconciles the extremely strong horizontal pressure gradients and wind speeds found in the inner parts of the storm (NN 13.1), but it also implies a cyclostrophic equivalent of the thermal wind equation (Section 7.12 and Box 13.3):

$$\frac{\partial V^2}{\partial z} = \frac{g\,R}{T}\frac{\partial T_p}{\partial R} \qquad 13.3$$

relating the vertical gradient of the square of the wind speed to the radial isobaric temperature gradient. The directionality of this relation is the same as for the normal thermal wind: positive shear is associated with low temperatures to the left of the shear in the northern hemisphere.

In a tropical cyclone, this cyclostrophic thermal wind ensures that the distinctive warm core to the left of the cyclonic air flow (in the N hemisphere) is associated with negative shear (upward decrease of cyclonic winds) for cyclostrophically balanced flow—exactly what is maintained by the combination of surface friction and conservation of angular momentum. Inserting realistic wind speeds from Fig. 13.12 into Eqn 13.3 shows that the inward rise in temperature along a radius through the ring of maximum winds should be up to 8 °C in 10 km, in reasonable agreement with observation (NN13.1) at least more than a few kilometres above the surface. The warm core of the tropical cyclone is therefore dynamically confined by the deep vortex of high winds whose strength decreases upwards, in much the same way as adjacent warm and cold air in middle latitudes is confined by the dynamic barrier whose elevated core is the associated jet stream. In each case the warm air would spill out over the cold air, destroying the baroclinity in a few hours, if the dynamic barrier disappeared. However, in the

BOX 13.3 Thermal wind in cyclostrophic balance

The isobaric form of the equation for cyclostrophically balanced flow at speed V round a circle of radius R is

$$V^2 = R\,g\frac{\partial Z_p}{\partial R}$$

where $\partial Z_p/\partial R$ is the upward and outward slope of the isobaric surface which supplies the inward force to maintain the centripetal acceleration of the whirling air parcels. Proceeding as in Box 7.8 we apply this basic equation to two isobaric surfaces and subtract to find

$$V_2^2 - V_1^2 = R\,g\,\frac{\partial}{\partial R}\left[Z_{p2} - Z_{p1}\right]$$

where the right-hand side contains the radial thickness gradient. As in Box 7.8 we replace the thickness gradient by the gradient of the column mean temperature $\overline{T}$, which (overbar)

$$\frac{V_2^2 - V_1^2}{Z_{p2} - Z_{p1}} = \frac{R\,g}{T}\frac{\partial \overline{T}}{\partial R}$$

If we consider an extremely thin layer we get the differential form

$$\frac{\partial V^2}{\partial z} = \frac{R\,g}{T}\frac{\partial T_p}{\partial R}$$

The strongly negative radial temperature gradient in a hurricane therefore requires an upward reduction of wind speed.

case of the mid-latitude front, the dynamic balance involves the Earth's rotation directly, through geostrophic balance, whereas in the severe tropical cyclone, the cyclostrophic balance is built up from the weak Coriolis effect of low latitudes by powerful sustained convergence. Moreover, the direction of the balanced flow around the warm core of the storm requires a kind of deep, inverted jet stream with a circular jet core in the lower troposphere, with obvious relevance to the observed spectacular violence of surface winds.

13.5.3 Warm core

The very warm core of the severe tropical cyclone is a conspicuous feature, and one which distinguishes it from other members of the tropical cyclone family, which usually have only moderately warm cores, and tropical waves which are sometimes cool cored. (Remember that in all cases we must distinguish between the important overall core temperature and the warmer hot towers which they contain.) Though less spectacular than the winds, or the torrential rain and electrical activity which mark the inner parts of these storms, the warm core is quite noticeable without specialized observations, and many factual and resulting fictional accounts mention the oppressive warmth of the eye, where the distracting violence is temporarily reduced. The evolution and maintenance of the warm core is vital for the birth and prolonged maturity of the storm, but at first glance it is not easy to see how it is achieved. Severe tropical cyclones develop only over extensive warm sea surfaces, where deep cumulonimbus convection is endemic anyway, and the troposphere is consequently maintained in a nearly neutrally stable state for such convection (Section 5.11). How is it possible for a storm's population of cumulonimbus to achieve the observed substantial rise in core temperature in the absence of any hotter surface? To answer this question we must consider the mechanisms of hurricane development.

Figure 13.17 contains several temperature soundings in and around severe tropical cyclones. The average sounding for the hurricane season is conditionally unstable (Section 5.10) throughout most of the depth of the troposphere, so that when convection is triggered locally, a deep cumulonimbus will grow. However, if many showers break out in a smallish region, the local troposphere will become less encouraging of deep convection unless other processes intervene. There are at least two such processes.

(i) One process, which has been closely studied since it was proposed by Charney and Eliassen in 1964, involves positive feedback arising from frictional convergence of the boundary layer supplying warm moist air (convective fuel) to the bases of the cumulonimbus. Synoptic and meso-scale convergence concentrates and deepens this particularly moist layer with the result that more cumulonimbus convection is triggered and so on in an exponential growth which has realistic properties when investigated in detail. The process is called *conditional instability of the second kind* (CISK—*[35]*).

(ii) Another process becomes important when the central surface pressure has already fallen significantly (say by at least 20 hPa) below values outside the developing storm. The air spiralling in towards lower pressures in the boundary layer is then made potentially warmer through isothermal warming by contact with the uniformly warm sea surface. Figure 13.17 shows the advantage gained by having the warming take place at a pressure significantly lower than

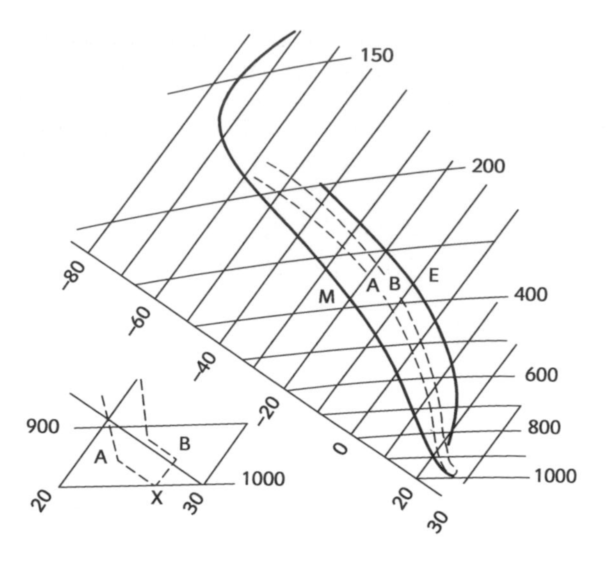

Figure 13.17 Temperature soundings in and around hurricanes, plotted on a tephigram. Curve M is the September Caribbean average. Curve A represents saturated adiabatic ascent from mean surface conditions, whereas B represents saturated adiabatic ascent after isothermal depressurization to 950 mbar (XB in inset). Curve E represents typical conditions in a hurricane eye. The difference between B and E in the upper troposphere is largely removed by allowing for liberation of latent heat of fusion in glaciating cloud.

surrounding sea-level values. Air rising in deep convection then follows a profile which is warmer at all levels, as if it had come from a significantly warmer sea surface at 1000 hPa. Cumulonimbus are thereby encouraged, and more heat is pumped into the active core of the storm, further reducing pressure by driving a

BOX 13.4 Low pressure and warm core

Suppose that a uniformly warm core of a severe tropical cyclone extends from the sea surface to height h, the top of the storm. If the mean scale height of the core is $H (= R\,T/g)$, then according to Eqn (4.8) the surface pressure p_s is related to the pressure p_h at the top of the core by

$$p_s = p_h\, e^{h/H} \qquad \text{B13.4a}$$

For a chosen h and p_h we can find the sensitivity of p_s to changes in H by differentiating Eqn B13.4a with respect to H

$$\frac{dp_s}{dH} = -\frac{p_s h}{H^2}$$

or

$$\frac{dp_s}{p_s} = -\frac{h}{H}\frac{dH}{H} = -\frac{h}{H}\frac{dT}{T} \qquad \text{B13.4b}$$

which shows how the air pressure at the base of an air column of *fixed height h* reduces as the column is filled with warmer air.

In a typical severe tropical cyclone, h is 13 km, and the column-average scale height is 8 km (corresponding to a column average temperature of 273 K), so that

$$\frac{dp_s}{p_s} = -1.63\,\frac{dT}{T} \qquad \text{B13.4c}$$

If the storm core is warmer than the environment of the core by $dT = 6$ °C throughout the column, then $dT/T = 6/273 \approx 0.022$, and the right-hand side of Eqn B513.4c is about – 0.036, showing that the sea-level pressure under a typical warm core is about 3.5% less than the sea-level pressure outside the storm core, but still within the storm. If the latter is 980 hPa, then the minimum pressure in the storm centre is about 35 hPa less than this, about 945 hPa.

Note that we have estimated the effect on surface pressure of filling the storm core with relatively warm air, without in any way explaining how that is achieved by the storm dynamics and thermodynamics.

net efflux of mass from the volume, further encouraging the cumulonimbus, and so on round the positive feedback loop. In mature storms the very low surface pressures ensure that the effect is large, equivalent to several °C, and contributing significantly to the efficiency of the tropical cyclone as a thermodynamic engine.

13.5.4 Pressure

Although vertical accelerations in some of the strong updrafts (and downdrafts) are just about significant in comparison with g, the resultant deviations from hydrostatic balance are localized and relatively small. Despite the storm's fury, hydrostatic equilibrium still prevails quite accurately on the meso and synoptic scales, so that atmospheric pressure at sea level is still a measure of the weight of overlying or flowing air. And since the weight of stratospheric air resting on the storm top is essentially uniform, the surface pressure beneath the storm is a measure of the radial distribution of atmospheric weight in the storm-filled layer (effectively the local troposphere). Being so much warmer, the storm core is significantly less dense than the surrounding air at all levels up to the storm top at 12–15 km above sea level (Fig. 13.12), so that atmospheric pressure at sea level in the core is considerably less than outside the core, and even less than beyond the edge of the whole storm. If the core is warmer than the rest of the storm by a typical 6 K in a column average 273 K, sea-level pressures there are about 35 hPa lower (Box 13.4) than further out—the pressure gradient being concentrated across the violent ring between the eye and the rest of the storm by the cyclostrophic balance outlined above. From the outer edge of the core to the edge of the storm (at least another 200 km further out) sea-level pressure may rise by further tens of hPa. This latter broad, gentle pressure gradient is typical of both tropical storms and severe tropical cyclones. It is the sharp-edged extra-low central pressure, the very warm core, and the violent inner ring which distinguish the severe tropical cyclone.

Surface pressure is essentially uniform across the eye, which is always free of deep convection, and sometimes completely free of cloud above low levels. Air is sinking into the eye from above the storm and warming to plug the storm centre with warm, unsaturated air. The descent is not dry adiabatic, since moist air is mixing in laterally from the turbulent eye wall to hold it close to the saturated ascent in the cumulonimbus ring, but the balance usually leaves it unsaturated and cloud-free.

13.5.5 The central zone

To summarize the spectacular energy and relatively fine structure of the inner zones of a mature severe tropical cyclone, imagine that you are observing the W'ward approach of the centre of a hurricane in the Caribbean, from the safe vantage point of a lighthouse. The NNW'ly surface winds (crossing the isobars at more than 20° because of surface friction) have been well above hurricane force (Beaufort force 12) for more than an hour, and the very low cloud base is only just visible through the driving, torrential rain and the blizzard of spray whipping off the very heavy seas. Lightning flickers frequently but indistinctly through the murk, but the thunder is inaudible above the roar of wind and sea. A tornado writhes briefly by, leaning towards the S in the strong wind shear, and marking a region of intense small-scale convergence below a particularly strong updraft. The barometer stands at 945 hPa and has fallen 20 hPa in the last hour. Suddenly the sky begins to lighten in the E, and the winds begin to ease. In five minutes they have fallen to gale force, the rain has largely stopped and through breaks in the low cloud you can see that the main mass of cloud is retreating to the W, giving glimpses of an enormous wall of dense cloud (the eye wall) down which some cloud fragments are falling like a waterfall in slow motion. A few blinks of Sun show the sea running high and confused, as wave trains spreading from beneath different parts of the eye wall cross and interfere, but the water surface is now largely free of spray and foam under the dying wind. The barometer has steadied at 944 hPa, but the still well in the lighthouse basement shows that the sea level is a metre above the expected tide and rising, in response to the storm surge in the local coastal waters, including an inverse barometer effect of the very low atmospheric pressure. Some exhausted birds are perching on the balcony of the lighthouse. The wind dies further and begins to change direction uncertainly. After nearly an hour of abatement it begins to freshen again, this time from the S, and soon the opposite eye wall is glimpsed approaching from the E and promising a virtual repeat of the earlier experience with winds and rain raging this time from the SSE.

Checklist of key ideas

You should now be familiar with the following ideas.

1. Very large area dominated by oceans, double seasonality with wet and dry seasons replacing warm and cold, monsoons over large land masses, and tropical weather systems.

2. Subtropical anticyclones ringing hemispheres around 30° latitude, seasonally variable over land, balance of convergence in the upper troposphere and divergence in the low troposphere; deep warm core filled with dry subsiding air; low-level subsidence inversion confining surface-based convecting layer, trapping pollution.

3. Monsoons as overland enhancements of seasonal migration of the intertropical convergence zone: example of particular geographical features in the Indian monsoon.

4. Barotropic troposphere and synoptic-scale convective weather systems: easterly waves and the tropical cyclone family: tropical depressions, tropical storms, and severe tropical cyclones (hurricanes and confused nomenclature).

5. Formation, structure, and mechanism of severe tropical cyclones, with meso-scale patterns of pressure, wind, and temperature around and in the eye, convergent spin-up of weak local rotation, cyclostrophic balance with inverted thermal wind shear, severe weather and storm surges.

Problems

Outline answers to these problems can be found on the **Online Resource Centre**. Answers to odd numbered problems can be found under Student Resources, answers to even numbered under Lecturer Resources.

Level 1

13.1 What family of low-latitude weather systems demonstrate that the Coriolis effect still matters there?

13.2 Sketch isobars and surface winds in a subtropical anticyclone in the S hemisphere.

13.3 Sketch isobars and surface winds in the undisturbed N flank of the ITCZ in the N hemisphere. Now add a weak depression (by a dotted distortion of the isobars) with an arrow for its motion relative to the surface.

13.4 Consider the ITCZ, Trade Winds, moderate tropical cyclones, severe tropical cyclones, subtropical anticyclones, and label them with one or more of the following descriptions of their air flow: NG (non-geostrophic), QG (quasi-geostrophic), G (geostrophic), CS (cyclostrophic).

13.5 Draw and label simple vertical sections to compare and contrast the warm cores of a subtropical anticyclone and a severe tropical cyclone.

13.6 Very simply: What is a monsoon? Why are they most marked overland? Compare and contrast with a heat low.

13.7 Compare and contrast air rising in an equatorial subthundershower and subsiding in an subtropical anticyclone by listing typical updraft speed, cloud, precipitation, turbulence, and humidity in each case.

13.8 Why do hurricanes weaken so quickly once they move inland? You should know of at least two reasons.

Level 2

13.9 Temporarily marooned on a small, low-lying island in the Caribbean hurricane season, you observe the following sequence: increased ocean swell from the E, clouding over from the SE, wind and rain strengthening from the NE (at this point you should go to the safest point on the island), veering to E before peaking at force 11 and veering to SE as it weakens. Sketch and justify your location relative to the passing hurricane. What was the safest point on the island?

13.10 Estimate the depth of a seasonal heat low, such as that over N Africa, associated with a layer of air 2 km deep which is on average 10 °C warmer than comparable air over adjacent seas.

13.11 Find the Coriolis parameter at latitudes 2, 5, 15, and 45° and in each case estimate the geostrophic wind speed associated with a pressure gradient of 3 hPa per 100 km in the low troposphere.

13.12 Use Eqn 5.18 and Fig. 5.7 to follow the thermodynamic path of air rising from a humid surface source with specific humidity 18 g kg^{-1}. It rises into the very high troposphere in a saturated adiabatic process and then sinks dry diabatically (warming by dry compression and cooling by net radiation) to reach the 800 hPa level with a temperature of 20 °C. What maximum potential temperature does it reach and where, and how long does it take to sink back to the 800 hPa level if it cools (loses potential temperature) at 2 °C per day? What important cycle are we trying to model thermodynamically?

13.13 Explain carefully how a subsidence inversion is maintained in the low troposphere of a subtropical anticyclone, including its effect on the depth and heat economy of the underlying surface-based convecting layer.

13.14 Find the angular velocity of the inner ring of air in the low troposphere of a hurricane. The air is moving at speed 50 m s^{-1} around a radius of 10 km (just outside the eye wall). Re-express this as a relative vorticity about the hurricane centre, and compare it with the planetary vorticity at latitudes 5 and 15° (Section 7.16) to see how much 'winding up' has been done by the hurricane. Briefly how has this happened?

Level 3

13.15 Consider air converging into the inner zone of a hurricane by flowing at speed V making an angle α with a circular isobar of radius R. If this happens through a height range h from the surface, and air density is ρ, write down an expression for the inward mass flux and equate it with the balancing upward mass flux through the 'lid' of this cylindrical layer. Show that the rate of uplift w for mass flux balance is $w = 2\ V h\ (\sin \alpha)/R$ and find the value of w when V is 50 m s^{-1}, h is 1 km, α is 30°, and R is 50 km. This is a very big value!

Notice I have cancelled ρ. If we had allowed for a realistic variation, in what direction would it have shifted the above value? I have also ignored the eye where there is no uplift. Put in a likely eye radius, find the direction and the % error in your first estimate.

13.16 The uplift in Problem 13.15 is unrealistically uniform. What is a more realistic model? If you are thinking of embedded violent cumulonimbus, that is sensible, but where are the cool downdrafts? Speculate freely but mechanistically that a hurricane may be convection organized so that there is no interference between rising and sinking air (just as a supercell severe local storm has reduced interference).

13.17 It seems that weather in low latitudes is dominated by organized convection. Discuss this idea with reference to the Hadley circulation, monsoons, and the whole range of tropical weather systems.

14 Climate and climate change

14.1 Introduction

Climatology is the study of average and typical weather, and therefore overlaps with meteorology in covering the conditions associated with the development and passage of weather systems, and the seasons and their distribution across the globe. However, its coverage stretches from there to much longer time periods, limited only by the availability of reliable data. Beginning with data from the instrumental period (mainly since 1850 CE), daily averages are averaged over a season or a year, annual averages are often compared with a 30-year average of annual averages, and departures from the average on each time scale are analysed to define prevailing weather, climate, and climate change. All this is done for each observing station, and then averaged to cover geographical regions from small countries to large continents, zones of latitude, the hemispheres, and the whole globe. Vast amounts of data are stored by the international meteorological network, and displayed and analysed to reveal the workings of weather and climatic processes and systems.

Traditional climate texts had largely completed a detailed description of world climate on this basis by the mid-twentieth century, just as people began to exploit the enormous potential of indirect evidence about climate from before the instrumental period, and to find that climate has varied significantly over all accessible time periods. In parallel with observational work, primitive climate modelling began by extending models developed for numerical weather forecasting. All this work has expanded greatly in recent decades, gaining in importance and urgency with the growing realization that *anthropogenic* climate change is beginning to dominate natural changes.

We begin with a brief survey of climate zones in the traditional manner, assuming meteorological processes and systems described earlier in the book.

14.1.1 The physical context of low- and mid-latitude climates

We have seen that the middle-latitude troposphere is chronically active, with opposing radiative imbalances at lower and higher latitudes (Section 8.7) maintaining a strong baroclinic zone whose deep W'ly flow is intrinsically unstable (Section 12.6), generating long waves, cyclone waves, and fronts. In addition in all latitudes there is a spectrum of smaller-scale disturbances from the meso-scale down to showery and smaller-scale convection. The seasonal march of the Sun produces large seasonal variations in solar and terrestrial radiative fluxes, and their net imbalance, driving seasonal rhythms of mean temperature, and diurnal temperature range, and associated weather, which are especially marked over land surfaces (Fig. 14.1), because of their small heat capacities (Section 10.5).

The climatic potential of the seasonal and meridional variations of solar elevation and length of daylight is apparent in Table 14.1 and Box 8.3, and is obviously greater in middle and high latitudes than it is in low latitudes.

(i) In high-latitude winters the Sun's rays fall very obliquely on the horizontal surface even at noon, spreading the solar beams, already attenuated by their long atmospheric paths, over large horizontal areas; and the hours of daylight are short. During the long nights, net emission of terrestrial radiation from the surface can produce strong cooling under clear skies. In high-latitude summers, by contrast, solar elevations reach low latitude values, and the days are very long.

(ii) At low latitudes the hours of daylight vary little between summer and winter, and the noon Sun is never very far from the zenith, reaching it in midsummer at the tropics, and passing through it twice annually in lower latitudes.

Two factors tend to offset meridional and seasonal variations of weather and climate in middle latitudes. The extratropical cyclones generated by the long- and cyclone-wave instability of the strong meridonal temperature gradients drive large heat fluxes polewards across middle latitudes both in the troposphere and in the ocean surface layers (e.g. the Gulf Stream and N Atlantic Drift—Section 8.10),

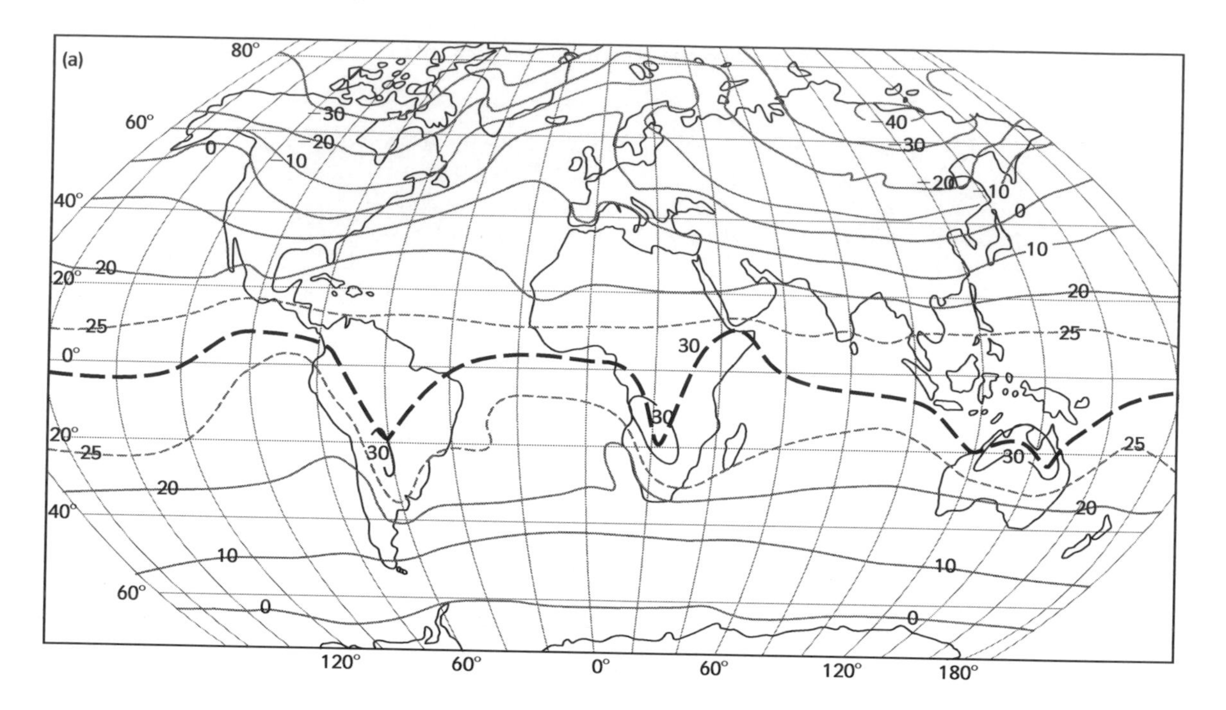

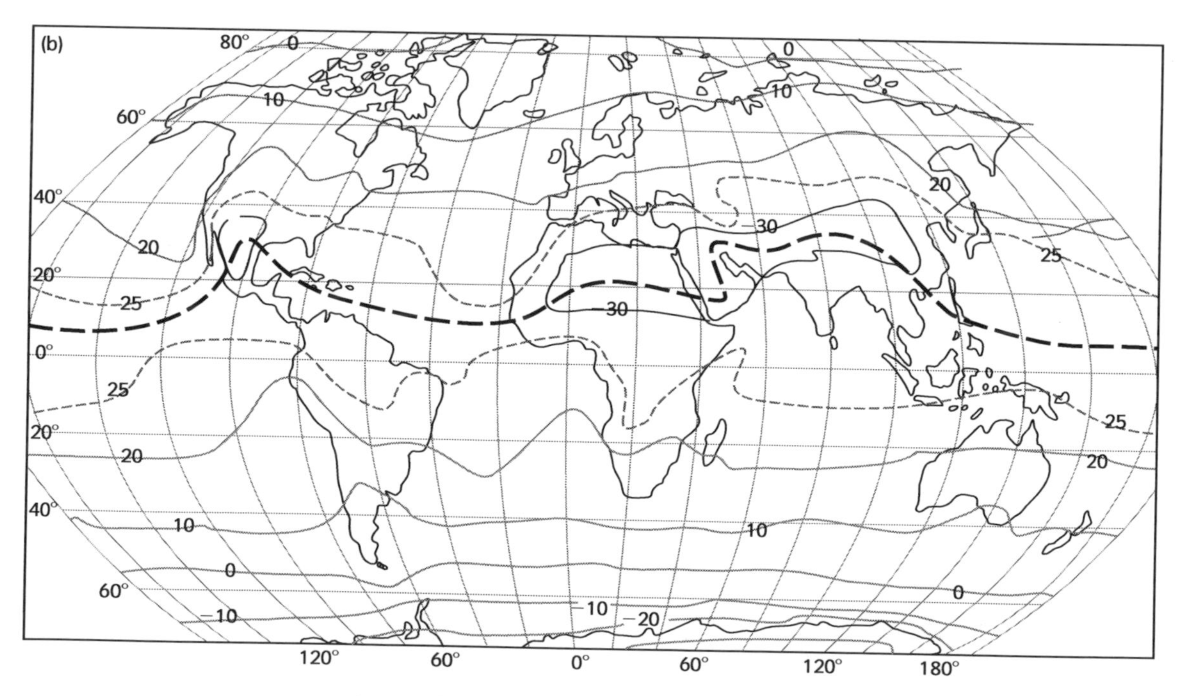

Figure 14.1a & b Mean surface temperatures in (a) January and (b) July. Isotherms are solid lines and dashed, and the line of maximum temperatures (which is not an isotherm) in each map is dashed bold. Note that the seasonal migration of temperature is much greater overland than over sea.

Table 14.1 Midwinter and midsummer noon solar elevations and day lengths at several latitudes

Latitude/°	Midwinter Sun Noon elev/°	Day length/hr	Midsummer Sun Noon elev/°	Day length/hr
60	5.5	2.7	53.5	18.5
40	26.5	9.2	73.5	14.8
20	46.5	10.8	86.5	13.2

and the very large heat capacities of the ocean surface layers smooth seasonal temperature variations over the oceans and western continental margin. Evaporation of water from oceans and moist land surfaces feeds large fluxes of latent heat, which limit daytime overland surface temperature maxima by evaporative cooling, and nocturnal minima by closing the atmospheric window with cloud. Since all these processes are concentrated by cyclonic activity over oceans and western continental margins, and reduced in continental interiors, they maintain a marked W–E zonation of climate from maritime to continental types (Fig. 14.1) which dominates the following discussion.

14.2 Mid-latitude climates

14.2.1 Western margins

Prevailing W'lies, and the weather systems they bear, strongly limit seasonal variations of temperature and rainfall (Figs. 14.1 and 14.3); they also raise annual average temperatures significantly above the zonal average, mainly by limiting winter cooling. This is especially so in the N hemisphere, because it contains a large proportion of all middle-latitude landmasses whose W sides would otherwise have much colder winters. Over the oceans, families of depressions are normal throughout the year, interspersed with irregular periods of blocking anticyclones, whose location and timing can considerably affect individual seasons. Contrasts between tropical and polar air masses are limited by their rapid accommodation to the local sea-surface temperature: for example, showers quickly warm the cool polar air flows in which they flourish, largely by pumping latent heat aloft. Summer heat lows (Section 12.7) may develop inland in slack air flow, giving thundery weather, but their warmth is tempered by cloudy shielding of sunlight. However, what might seem almost boring seasonal uniformity to natives of more extreme climatic zones, in fact contains a wealth of day-to-day variation which excites endless comment in native populations, and complex climatological commentary on subtle *climatic singularities [58]*, such as the extravagantly named European Monsoon—a tendency for cyclonic activity to increase in mid to late June, which has finally persuaded organizers of the Wimbledon (London) international tennis tournament to roof its centre court.

Temperature departures from the seasonal average arise from anticyclonic blocking (warming in summer, cooling in winter—Section 12.5), or vigorous cyclonic activity (cooling in summer, warming in winter), and can be comparable with the small seasonal variations, so that the warmest days of some winters can be warmer than the coolest days of summers. The prevailing W'lies maintain strong concentrations of rainfall on western coasts and hills, where there is a fine structure of relief rainfall and rainshadows, with significant overall reductions inland over distances of 10–100 km (Section 10.11). Where hills are modest, these E'ward reductions extend well inland, reflecting the ability of large weather systems to penetrate long distances before dying out. High N–S barriers like the Rockies and Andes can compress the transition from maritime to continental climates into much narrower corridors (Fig. 14.1).

Prevailing W'ly winds sweep inland, weakening near the surface with the increase of surface roughness (Section 10.9) in flowing from land to sea, fading patchily in the lee of N–S terrain barriers, and weakening progressively in depth over longer fetches (downwind distances) from the coast, as the driving cyclonic activity decreases inland. On W-facing coasts it is common to have daily runs of wind (Section 2.4) of more than 600 km at the 10 m standard level (an average 6.9 m s^{-1} or force 4—Box 2.2), encouraging reliable generation of electrical power by wind turbines.

Diurnal temperature variations at screen level (1.5 m) increase quickly inland, as the reduced thermal capacity of the overland fetch allows larger and larger response to the diurnal cycle in net radiation. This encourages a diurnal cycle in convection, and associated cloud and precipitation, which reverses seasonally: in winter, showery air masses (such as the polar maritime air which sweeps in behind cold fronts) quickly lose their showers as the air flows over progressively colder land; whereas in summer, convectively nearly stable flows are rapidly destabilized as they pass over progressively warmer land, giving sunny coasts and showers inland. The winter effect adds to the normal relief rain shadow inland from the W margins. Inland cooling sharply increases the incidence of frost and lying snow in winter.

Many of the climatic features of extensive W margins appear on smaller scales too, from the North Mediterranean in winter (where local cyclogenesis is encouraged by the strong baroclinity between the warm sea and the cold southern European land mass), to the local climates of E margins of the North American Great Lakes, and the Black and Caspian Seas.

14.2.2 Continental interiors

Just as W margins have maritime climates because they lie only a short fetch from the sea down the prevailing W'lies, so the rest of the continental areas have continental climates on account of their much larger fetches from the upwind ocean. Even E margins are largely continental in climate (Fig. 14.1), since cyclogenesis in the adjacent seas to the E often affects the margins only briefly before the young depressions move E'wards out of range.

For the reasons mentioned in the previous section, continental climates are typically much more extreme than maritime climates, meridionally, seasonally, and diurnally. For example, Fig. 14.1 shows that at latitude 50° the winter–summer swing in seasonal mean temperatures is from 5 to 15 °C on the W coasts of North America and Europe, whereas it is from about −10 to +20 °C in N–E America (longitude 80 °W) and −15 to +25 °C in Asiatic Russia (longitude 80 °E). The seasonal swings in continental interiors are largely associated with a transition from cold winter anticyclones to summer heat lows, but considerable differences between North America and Eurasia arise from their particular geographical situations, as follows.

North America

The smaller seasonal temperature range over the North American continent arises from its smaller size, a tendency for weak cyclogenesis in the lee of the Rockies (Section 12.6), and the presence of a large warm water surface to the south (in the Gulf of Mexico). The cold anticyclone in winter is a weak, unstable feature in comparison with its great Asiatic counterpart.

Much of the land N of latitude 60° has a near-Arctic climate, with widespread *permafrost* (permanently frozen ground under a shallow top layer which thaws and freezes seasonally). In winter, N'lies in the lee of the Rockies bring bitterly cold conditions from the Arctic continental source, which contrast sharply with conditions found in S'ly flows from the Gulf at other times. Strong outbreaks of the very cold air (*cold waves*) are associated with temperature falls of 10°C or more in a few hours. Even sharper contrasts occur when unusually strong W'lies force a very warm, dry *Chinook* flow down the E slopes of the Rockies (Section 10.11). Winter precipitation (snow) is sparse in the central interior, but increases sharply to the E of the Great Lakes, and Hudson Bay, before it freezes over in early winter and loses its localized maritime effect. Unlike its great Asian counterpart, the cold anticyclone of the North American winter is a statistical feature underlying considerable day-to-day variability.

In summer, S'ly flows of warm, humid air from the Gulf of Mexico are associated with very dramatic outbreaks of severe local storms along sharply defined cold fronts (Section 11.6). Less vigorous cyclonic activity draws vapour from the same source to rain on the East of the continent, but much summer precipitation falls in heavy showers not associated with any air mass contrast. Though still basically driven by the strong daytime solar heating of the land, these showers often occur at night, apparently in response to subtle changes in low-level air flow set off by cooling in the late evening. The balance between precipitation and evaporation in some of the great grain-growing regions is rather fine, leaving them vulnerable to quite small changes in global climate or agricultural practice, as in the great dust-bowl disaster of the 1930s.

Eurasia

The much greater zonal extent of the Eurasian continent (over twice that of North America, however the boundaries are defined) gives full play to the effects of oceanic remoteness, despite the lack of any real equivalent of the Rocky Mountain barrier to the W. A maximum winter–summer swing of seasonal mean temperatures from −40 to +15 °C is reached around longitude 130°E.

Winters are extremely harsh over a wide area. Permafrost reaches S of the Arctic circle from about 100 °E to about 180 °E (extreme eastern Siberia). Maritime influence from the Arctic Ocean is lost when polar pack ice extends to the N coast, and the barriers of the Tibetan and Himalayan massifs to the S prevent N'ward intrusion of warm air from the Indian Ocean. Snow settles permanently in late autumn over a wide area and remains until spring, raising the ground surface albedo and thereafter reflecting most of what little potential solar input there is (Section 10.3). Underneath the developing winter temperature inversion (Fig. 12.18), falling temperatures reduce the maximum vapour content and with it the local greenhouse effect, allowing the surface to cool even faster by terrestrial radiation to space (Section 8.5). Though the shallow layer of cold air produced in this way is channelled by valleys and dammed by ridges, it spills out over a wide area throughout the winter, producing intensely cold waves over N China and Korea. Precipitation is minimal over a wide area.

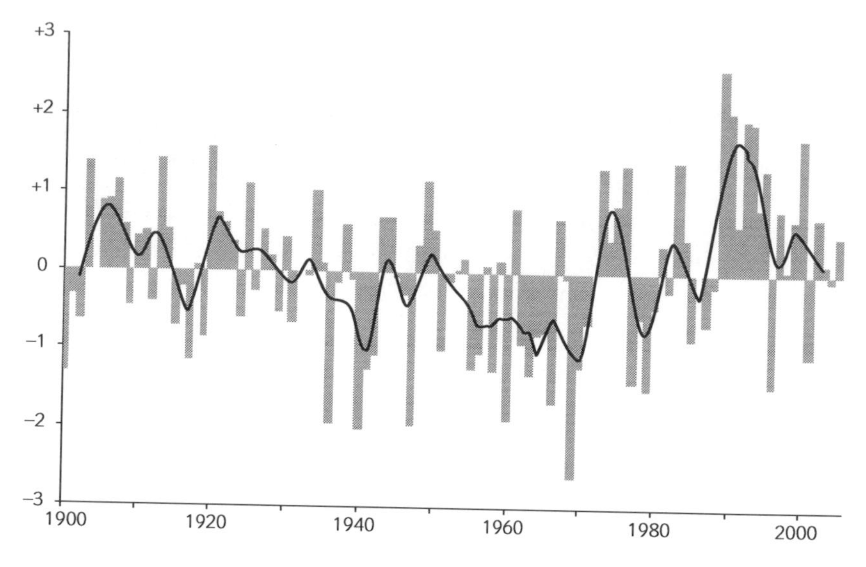

Figure 14.2 The North Atlantic Oscillation from 1900 to 2006. Annual average values are shown by a bar chart (expressed as a deviation from the 100-year average), and the solid black line is a 5-year running mean. Positive values occur when the N Atlantic airflow is more simply zonal; negative values occur when it is more meridionally wavy.

After a rapid transition into spring, summer conditions arrive with little of the ocean-induced lag apparent in maritime conditions. Temperature gradients are modest over a wide area (Fig. 14.1), and a reasonable rainfall total falls mainly as showers, though the N flanks of the Tibetan massif are semi desert. Parts of China, however, experience some maritime influence, with shallow depressions advancing NE in late spring, as part of the Asian monsoon, and occasional typhoons in late summer. In the autumn, the net radiation input quickly fades and reverses over a very wide area, and rain and snow alternate unpredictably (especially in the Eurasian west, as Napoleon and Hitler found to their cost), before winter proper sets in.

Gilbert Walker, a pioneer of modern climatology, discovered a statistical relationship with a time scale of several years, between annual average sea-level pressures in the Icelandic Low (in the high N Atlantic) and the Azores High (the E end of the N Atlantic subtropical high-pressure zone). As shown in Fig. 14.2, there is an irregular oscillation between years of low index (lower than average pressure differences between Iceland and the Azores) and years of high index. High-index years of this *North Atlantic Oscillation* (NAO) are associated with strong wet W'lies, and cool summers and warm winters. Low-index years have weaker W'lies, and winters in which depressions move E'wards along more S'ly tracks than usual, bringing cool, wet conditions to the N Mediterranean.

More recent work suggests that the NAO and other patterns of regional climate (such as ENSO—below) show enough statistical coherence to suggest the presence of global *teleconnections* which have yet to be understood mechanistically.

14.3 Low-latitude climates

14.3.1 Hadley circulation

The Hadley circulation (Fig. 8.21) supplies a framework for outlining significant features of low-latitude weather and climate. It covers the largely barotropic region between the tropics, and reaches polewards by another 10° of latitude—i.e. the area

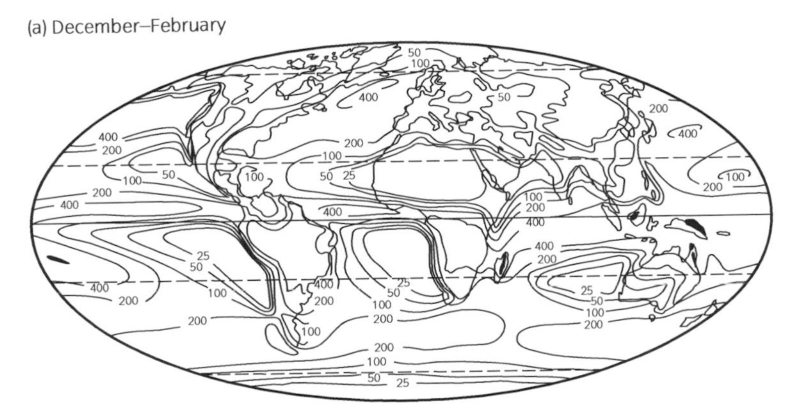

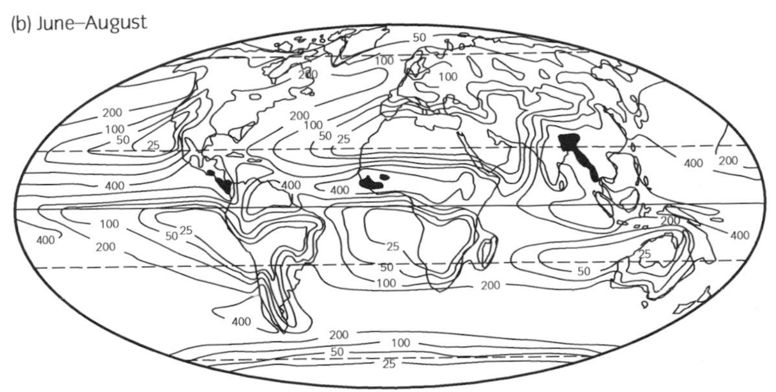

Figure 14.3 Average precipitation in millimetres in (a) December–February, and (b) June–August. To be consistent with a global annual average of about 800 mm, the global average in each quarter should be about 200 mm. Examine and compare the two maps to see the full extent of the permanent and seasonal features, and note that persistent patterns are roughly as expected from the Hadley Circulation.

in which the highest noon Sun of the year is never more than 10° from the zenith. Air mass contrasts are weak in comparison with higher latitudes, and there are no fronts, although thermal contrasts between land and sea can be significant, especially when sharpened by sea breezes or monsoon flows. However, there are very large meridional contrasts of water content, cloud, and precipitation between the dry subtropics and the wet equatorial zone, and comparable E–W contrasts, with large seasonal variations (Fig. 14.3), arising from distributions of land and sea.

Subtropical high-pressure zones (STH)

These include all of the hot land deserts of the Earth. Widespread tropospheric subsidence inhibits cloud and precipitation throughout the year, so that land is permanently arid and the scant water vapour input is advected from adjacent moist areas. The virtual absence of evaporative cooling during the day, together with strong nocturnal radiative cooling through the atmospheric window (enhanced by the absence of cloud and scarcity of vapour), can maintain a diurnal surface temperature range of 50 °C or more (Section 10.4). Over the oceans the STHs are marked by maximum sunshine, minimum rainfall, and maximum surface salinity, as evaporating fresh water leaves its salt behind.

On the equatorial flanks of the STHs there are brief, unreliable rainy seasons where the Inter-tropical Convergence Zone (ITCZ) reaches furthest poleward overland. In the North African Sahel, gross inter-annual rainfall variation and poorly understood longer-term variations have contributed to human misery on an awesome scale in the late twentieth century. In India, the dispositions of sea, land, and mountain barriers encourage a vast monsoonal invasion of cloud and rain in summer, which though it produces some of the world's highest rainfall totals on exposed hills, is still unreliable enough to cause large-scale human distress from time to time, especially near the N limits of its migration.

Trade winds (TW)

The Trade Winds are fed by air flowing W'wards and equatorwards from the bases of the STHs. As the air flow is gradually transformed by injection of water vapour from beneath (especially over the oceans), it becomes populated by cloudy weather systems which drift W'wards, providing much of the rainfall in those zones, especially when enhanced on E sides of islands and larger land masses. The N'ward flow across the equator in the Indian summer monsoon is turned E by the Coriolis effect, given rainfall enhancement on W coasts and slopes (Figs. 13.7b and 14.3). Over the warmest ocean surfaces at latitudes of at least 5° (i.e. with a large enough Coriolis parameter), some of these disturbances can develop into severe tropical cyclones in summer (Section 13.5). In addition to their notoriously strong winds, these storms can deliver very large (sometimes catastrophic) rainfalls on making landfall on islands and larger coasts. Indeed some recurve polewards out of low latitudes, still raining heavily enough to produce notable floods, in the NE USA for example.

Inter-tropical Convergence Zone (ITCZ)

The Trade Winds from each hemisphere's Hadley cell feed the ITCZ, supplying air and water vapour to the updraughts of organized populations of cumulonimbus. These billow up to the high equatorial tropopause, dropping torrential showers onto underlying surfaces, and maintaining most of the several thousand thunderstorms active in the Earth's atmosphere at any instant. In the weak overall air flow, the organization of such 'hot towers' is subtly sensitive to local geography and the diurnal cycle, giving each location its distinctive rainy season and diurnal patterns of rainfall, and producing an endless display of blossoming anvil clouds in satellite pictures (Fig. 1.3).

At sea, the slack and directionally variable air flow caused notorious delay, anxiety, and distress to sailors before the arrival of motorized ships. The expressive popular name *doldrum* for the zone and its effect on morale seems to be a conflation of dullness and tantrum—moods of sailors immortalized by *Coleridge's Ancient Mariner*.

14.3.2 ENSO

The zonal patterning which is obvious in Fig. 14.3 shows small but influential variations of uplift and subsidence which were detected statistically by Gilbert Walker in 1923, and confirmed and elaborated by subsequent observation and analysis (Fig. 14.4a). This *Walker Circulation* has been found to undergo an irregular reversal every few years, called the *Southern Oscillation* because it is linked to changes over a large part of the South Central Pacific (Fig. 14.4b). More recently the Southern Oscillation has been linked to the notorious *El Niño* (the child, meaning Christ child)—a Peruvian name for a hesitation in the Humboldt current which occurs there each December but every few years develops into

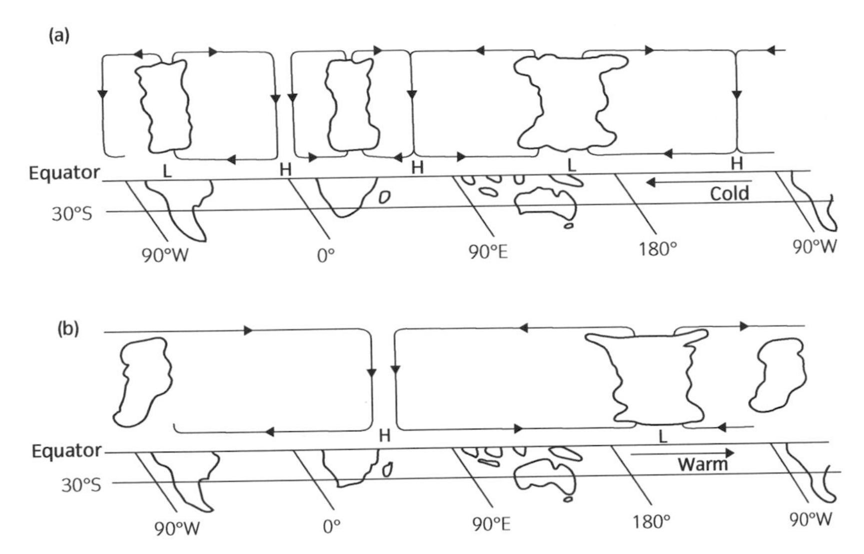

Figure 14.4a & b Vertical equatorial section of the Walker circulation in (a) normal and (b) ENSO modes, after Tourre (1985) in *[87]*. Canton Island is 3° S, 172° W. The term La Niña has recently be coined for an extreme form of the normal state.

a major breakdown. The resulting failure of the cold, fish-rich Humboldt Current for a season or longer, blights the local fishing industry and dependent communities. As a late afterthought, an unusually strong version of the normal cool situation has been called *La Niña*. The whole system has become known as *El Niño-Southern Oscillation* (ENSO).

Major ENSO events have occurred in 1925–26, 1982–83, and 1997–98, the last being particularly severe. Weak El Niños occurred annually from 1991 to 94, prompting speculation that it could be sensitive to global warming, and there have been El Niños since in 1997/8, with another close series in 2002/3, 2004/5, and 2006/7. There is some statistical evidence of a quasi-decadal cycle in ENSO activity, as shown by the two recent periods of frequent events.

In the normal Walker circulation, uplift and rainfall are enhanced over Indonesia and suppressed over the Humboldt current by a slight but persistent vertical circulation superimposed along the zonal central axis of the Hadley circulation in the Pacific. Subsidence along the equator can split the local ITCZ by producing a cloud-free corridor, clearly visible in satellite pictures, which can reach W'wards as far as Canton Island (central Pacific, 172 °W—Table 13.1). In the ENSO cycle, subsidence weakens over the failed Humboldt current, and uplift and rainfall are encouraged in a large zone centred on Canton Island.

Figure 14.4 shows that ENSO affects the vertical circulation along the full length of the equator, and there is statistical evidence of other related events in the Atlantic (a year or so later) and at higher latitudes as well. Schemes of *teleconnections* are proposed to explain such distant harmonies in tropospheric activity, many involving interactions between atmosphere and ocean. Though mechanisms are largely speculative so far, the statistical evidence points to a degree of global interconnectedness in both space and time, greatly extending Walker's pioneering work, and encouraging hope that forecasting of short-term climatic fluctuations may become possible when they are more fully understood. In particular they highlight the crucial role of air–sea interactions in imposing longer-term potentially predictable order on short-term atmospheric chaos.

14.4 Unsteady weather and climate

14.4.1 Seasonal variations

In most of the book so far we have tacitly assumed that time scales of atmospheric variability range from the brief wobbles associated with small-scale turbulence, to excursions usually lasting about a week associated with extratropical cyclones, etc., and that longer-term variations are limited to the relatively regular seasonal variations acting within their yearly framework. If this was the case, if there really were no longer-lasting variations in the atmospheric system, conditions in British summers (for example) would be very similar from year to year, since each season would usually contain so many randomly occurring transient variations that seasonal and annual averages of temperature, rainfall, etc. would be stable from year to year. Anecdotal experience and systematic data reveal a very different picture.

Figure 14.5 includes annual rainfalls for over 30 years from a station in NW England, showing quite large year-to-year variations. This and a wealth of other data (temperature, pressure, sunshine, etc.) from any site on the globe, show that particular seasons can vary considerably from year to year, even though the seasonal driving force (the annual cycle of the tilt of the N hemisphere towards and away from the Sun) is mathematically repetitive, and that these carry through to annual averages and totals.

In fact the assumptions that weather systems are short and random (i.e. unconnected in space or time) are incorrect for reasons implied above and listed below.

(i) Especially slow-moving extratropical cyclones *can* affect a given location for considerably more than a week, and hence significantly affect the average for that season and year. And *blocking anticyclones* (Section 12.5) can persist for weeks or even a month in a given region, replacing the normal sequence of E'ward-moving weather systems by a nearly rain-free lull which may be very

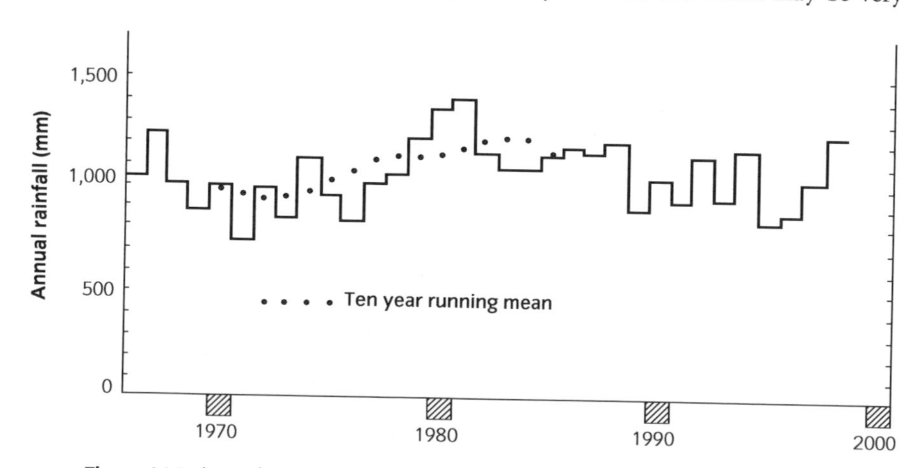

Figure 14.5 Annual rainfalls from 1966 to 1998 at a climatological station in NW England, on a 90 m ridge overlooking the Irish Sea. The visual average of 1,100 mm is typical of coasts in NW Europe and Scandinavia exposed to maritime W'lies, and is probably about 30% above the global, annual average.

considerably warmer or cooler than the seasonal average, depending on whether it is summer or winter.

(ii) Variability between consecutive summers, etc. can arise more subtly, as large, shadowy mechanisms inhibit or encourage extratropical cyclones and anticyclones over periods of weeks. Averaging mid-tropospheric air flow over at least 5 days reveals important slow-moving *long waves* in the mid-latitude westerlies (Section 12.6). As well as apparently guiding the development of individual extratropical cyclones, the patterns of these long waves develop and shift on time scales which are long enough to affect seasonal and annual averages very significantly. And even larger and longer-lived, but less obvious, teleconnections between weather patterns, such as NAO and ENSO (as above), are observed by careful data analysis across widely separated parts of the globe.

14.4.2 Annual variations and longer

Extending the argument of the previous subsection, if events such as blocks, lasting half a season at most, and teleconnected disturbances, lasting for a few years at most, really were the longest-lived disturbances of the atmosphere, we should expect averages over a sufficient number of consecutive years (at least 10 though 30 has become the norm) to be virtually static. As before, this is not observed.

Figure 14.5 shows that even after subjecting the data to a 10-year running mean (Fig. 14.5 and Table 14.2) there is clear evidence of coherent variation in time. It seems that the atmosphere is prone to disturbances even longer-lasting than anything associated with teleconnections like ENSO.

In fact continuing the above sequence by analysing climate over longer and longer time periods, we find that, across periods ranging from a few years to literally millions of years, there is no obvious upper limit to the time periods of significant, coherent climatic variation.

14.4.3 Central England Temperatures

In Fig. 14.6, temperatures in central England are extended backwards from the present to the beginning of the first sustained meteorological use of thermometers

Table 14.2 Running means of annual rainfalls in Fig. 14.5

The first 25 successive annual rainfall totals in Fig. 14.5 are as follows (in mm).

Year	1	2	3	4	5	6	7	8	9	10	11	12	
Total	1030	1250	1000	890	1000	720	995	850	1115	955	840	1005	mm
M10						*981*	*962*	*937*	*943*	*975*	*1012*	*1076*	*1117*

Year	13	14	15	16	17	18	19	20	21	22	23	24	25	
Total	1060	1210	1370	1360	1400	1160	1090	1085	1150	1190	1170	1200	885	mm
M10		*1148*	*1145*	*1158*	*1189*	*1208*	*1219*	*1218*	*1169*					

The arithmetic mean of yearly totals 1 to 10 is 981 mm, 2 to 11 is 962, 3 to 12 is 937 and so on, giving 10-*year running means* of rainfall in italics in the middle of each 10-year window. As shown on Fig. 14.5, this procedure smooths extreme year-to-year variations, and thereby highlights longer-lasting variations previously lost in the short period clutter. It can be performed on any temporal data sequence, using any averaging window which fits usefully within the sequence, as above.

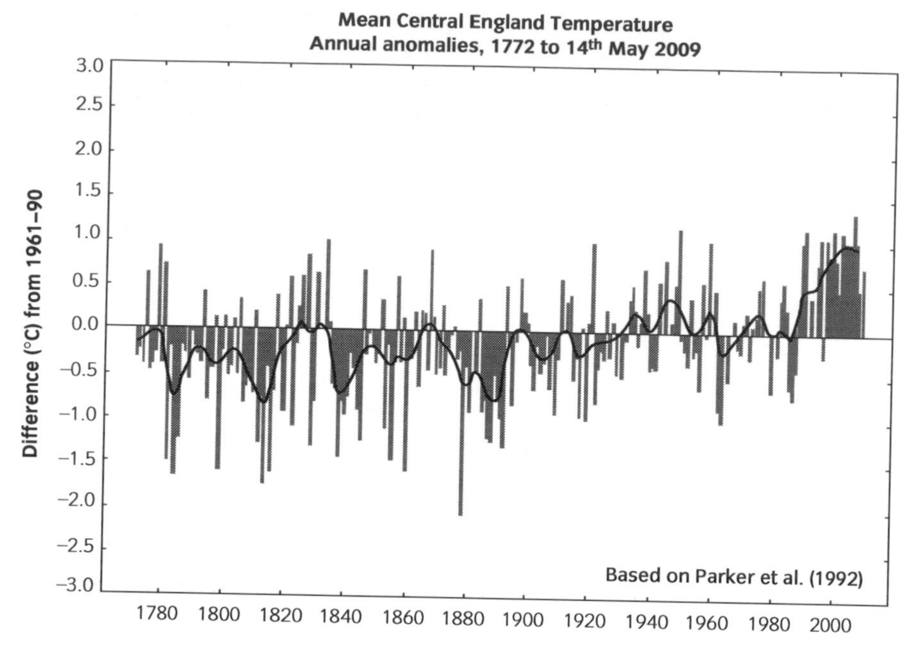

Figure 14.6 The Central England Temperature series of annual averages from 1772 to 2009 (i.e. omitting the first 113 years of the full CET, and averaging the core daily series). The smoothed line is equivalent to a ten-year running mean. The datum is for 1961–90 which is visually about 0.5 °C above the average to 1900, close to the average from 1920–1990, and sharply below the average and trend since 1990. Interannual variations are large (over 0.5 °C), and the smooth curve visually shows variability on scales of about 10 years, 50 years, and a 100-year upward trend since about 1900. This and other global series are subject to exhaustive statistical analysis for comparison with climatic mechanism.

in England (CE 1659). This Central England Temperature (CET) record continues to this day, and is the longest series of consistent air temperature measurements by thermometer anywhere in the world, predating the beginning of the international synoptic meteorological networks by nearly two centuries. Early data in the CET were particularly prone to variations in instrumental performance and exposure, and their preparation (by Gordon Manley *[60]*) required sustained, painstaking work in data discovery, collation, comparison, critical assessment, and judicious correction, much of it done before current interest in climate change had highlighted the urgent practical value of such data.

To smooth out fluctuations lasting only a few years, the data in Fig. 14.6 have been subjected to a ten-year *running mean* (as in Table 14.1), and yet considerable variations obviously survive. For example, there was a pronounced cold period in the British Isles centred around CE1700. Despite their seeming numerical insignificance, such variations in temperature and associated weather can strongly affect human society through agricultural production, etc. This cold period was in fact the culmination of two centuries of cooling (predating the start of the CET record) so widespread and systematic that it has become known as the *Little Ice Age*, even though it does not begin to compare in scale or intensity with true ice ages or their glacials (Section 14.5 and Fig. 14.9). Nevertheless in Scotland the harsher climate caused large-scale emigration in search of better conditions.

Many emigrated to NE Ireland, where conditions were slightly easier and good agricultural land was made available by forcible expulsion of native Irish, with social and political consequences apparent to this day *[61]*. On a lighter note, the short cool period centred around 1820 (now recognized as the last kick of the Little Ice Age, probably assisted by the Tambora eruption (Section 14.7)), coincided with the impressionable childhood of Charles Dickens, encouraging his later evocative descriptions of cold winters and Christmases in and around London, and filling Christmas cards to this day with scenes of snow which have been rare in the British lowlands for 150 years or more.

14.5 Climate change in history and prehistory

Meteorological measurements by instrument did not exist before the mid-seventeenth century, but interest in *climate change* (the name now given to all climatic variations lasting more than a few years) has become so strong in the last fifty years that an immense programme of scientific detective work is under way to outline and detail past weather and climate. In historical pre-instrumental times, old social and agricultural records are combed and analysed, and for these and the prehistoric aeons, a huge range of physical, chemical, geological, and biological methods is being developed and applied to assess climate throughout the lifetime of the Earth, as selectively mentioned below.

(i) In relatively recent times, analysis of the thickness of annual tree growth rings in cross-cut trunks of ancient trees can give useful clues about climate and climate change. For example, a Bristlecone pine in California has been analysed, its ring thicknesses giving a history of conditions affecting annual tree growth for every individual year since its birth in 2,833 BCE (Fig. 14.7). Analysis of pollen

Figure 14.7 A Bristlecone pine tree in the White Mountains of California. These dense resinous trees live for thousands of years, the oldest (the famous Methusaleh in the same region but hidden from vandals) being a little over 4,800 years old, so that the annual growth record in their trunk rings are highly prized by climatologists.

grains embedded in geologically dated soil and rock strata can show what flora were growing when the host stratum was the exposed land surface, and their otherwise known climatic tolerance used to indicate the prevailing climate (e.g. Fig. 14.8). Both these methods are limited to geologically recent periods.

(ii) Ice cores down through the current Greenland and Antarctic ice sheets contain solids and air trapped since the snow fell up to 300,000 years ago, whose analysis can illuminate the global climates of those times. For example, much more desert sand was deposited during the two *glacials* which occurred in this time period (see below), because the cool global climate was more arid; and the CO_2 content switch-backed in step with global temperatures as the oceans absorbed CO_2 when they were colder and expelled it when warmer (Fig. 14.8).

(iii) The ratio of oxygen-18 (^{18}O) to ^{16}O in calcite beds forming the ocean floor is high during *ice ages*, as the slightly more volatile $H_2{}^{16}O$ evaporates preferentially and moves polewards to be snowed into the great ice sheets and trapped, raising the proportion of $H_2{}^{18}O$ in the residual oceanic waters in which the marine organisms grow their shells. Analysis of the $^{18}O/^{16}O$ ratios in geologically dated deposits of these shells recovered from beneath the sea bed can reveal the presence of ice ages or warm epochs, and the glacial/interglacial episodes within an ice age.

(iv) A great many geological features are climatically revealing, from terrain sculpting by glaciers and ice sheets, to sandstones laid down in subtropical deserts

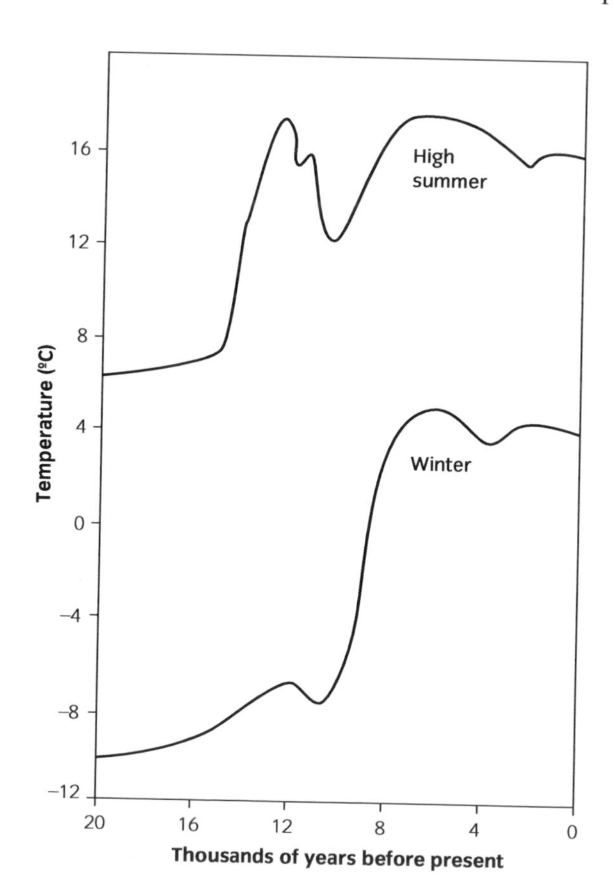

Figure 14.8 Surface temperatures in Central England in the last 20,000 years, estimated largely from pollen analyses. When allowance is made for systematic errors in carbon dating, the dating of the Younger Dryas is pushed back over 1,000 years.

(whose fossil dunes can even indicate prevailing winds), to bedding structure from ancient sea beds and coastlines, to coal seams formed by incompletely rotted vegetation whose formative climate can be assessed, and even to boulders dropped to the ancient sea bed from melting icebergs. Even when clues are inconclusive or ambiguous on their own, as is often the case, coherence revealed by crosschecking with other independent evidence encourages more confident interpretation, as in all the natural sciences.

14.5.1 Recent prehistory

Figure 14.8 includes summer and winter temperatures in central England over the last 20,000 years estimated by pollen analysis, and shows the rapid temperature rise associated with the ending of the most recent *glacial* epoch (see below and Box 14.1), one of whose ice sheets reached S to central England, and the onset of the current *interglacial* (the *Holocene*). The plotted temperatures are roughly what would have been logged by thermometers (had they been available) subjected to a running mean of several centuries.

The severe temporary cooling after the warm peak around 13,000 years BP (before the present) is known as the *Younger Dryas stadial*, named after the little tundra flower (*Dryas Octopetala*) whose buried pollen provided the first evidence of the sharp cooling which persisted for nearly 1500 years and re-established hundreds of metres of ice on the Western Highlands of Scotland, which had lost their full ice cap a few thousand years earlier. There is clear evidence of this ice resurgence around Loch Lomond, and many different geochemical and radiological indicators now confirm the event as a major interruption of the warming across NW Europe which preceded the final rapid establishment of the present warmer period (the Holocene interglacial) worldwide about 11,500 years BP. (Beware uncorrected ^{14}C dating which puts this at about 10,000 BP.)

Figure 14.9 takes us back further still, graphing temperatures in lowland Britain estimated from analyses of insect fauna over the last 120,000 years, and showing that the Holocene interglacial (with its Younger Dryas prelude) was preceded by a severe *glacial*—a cold period which lasted about 100,000 years, was at its coldest between 20,000 and 30,000 years ago, and was associated with widespread continental glaciation in middle and high latitudes in the N hemisphere, with vast ice caps kilometres deep, built and maintained by persistent accumulation of snow. The glacial's impact on the S hemisphere was less

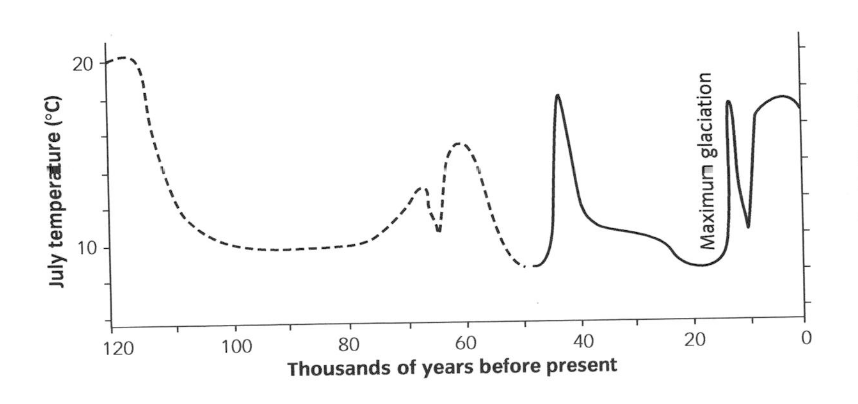

Figure 14.9 Surface temperatures in lowland Britain in the last 120,000 years estimated from evidence of insect life. Redrawn from *[61]*. Though lacking the gradual temperature descent of other more recent records, this one clearly shows the unsteady but deep glacial which ended only 10,000 years ago, and the previous interglacial over 100,000 years earlier.

spectacular, because most of Antarctica was fully glaciated already (as it still is, apart from the Antarctic peninsula), and there are relatively few other continental platforms for ice caps in high southern latitudes—though much of the S Andes became an ice cap finger reaching through middle latitudes.

The higher temperatures apparent nearly 120,000 years ago (Fig. 14.9) represent most of the previous short interglacial, much like our current Holocene, but apparently somewhat warmer, shorter, and climatically less stable.

Even longer records show that the cycle of Fig. 14.9 repeats back for several million years, with long glacials alternating with short interglacials in about 20 cycles since the beginning of the current *Cenozoic* or *Quaternary ice age*. The cycles are asymmetrical in time, with slow cooling and abrupt warming forming a characteristic saw-tooth pattern. And further back again, there is extensive evidence of a long prelude to the current ice age, with mostly gentle global cooling persisting over tens of millions of years, accentuated after the sudden appearance of the Antarctic ice cap just over 30 million years ago, until the fairly sudden appearance of ice caps on the continents ringing the Arctic Ocean about 3 million years ago began the current ice age proper.

14.5.2 The long view

Before the onset of the cooling to the present ice age, there was a warm period lasting about 200 million years in which there is no evidence of permanent continental ice near sea level anywhere on the globe, even at the poles. Further back still, there is clear evidence of several ice ages in the *Permian* and *Carboniferous* periods (350–250 million years ago—Fig. 3.15), a smaller event about 450 million years ago, and an ice epoch (the *Cryogenian*) lasting more than 150 million years and ending about 630 million years ago, in the *Precambrian*. At times this latter was so severe that continental ice caps appeared in low as well as in high latitudes, implying that most of the oceans must have been iced over and prompting the nickname *Snowball Earth [62]*. And before that there were extensive warm epochs interrupted by at least one ice age about two billion years ago.

A vast and growing mass of evidence therefore shows that, for most of Earth's history, the global climate seems to have been significantly warmer than now, with no permanent ice near sea level, and global sea levels standing about 100 metres higher than at present. In total, these *warm epochs* account for about 90% of Earth history; for most of the remaining 10%, the global climate has been distinctly colder than now, with extensive ice caps in high latitudes, and sea levels standing about 100 m lower than at present. And in the latest of these ice ages (the current one), global climate has been as warm as it is now for only 10% of the time. So even if previous ice ages followed the same glacial–interglacial rhythm as the present one (which is possible, though unlikely at times during Snowball periods), our present cool but not cold climate is a rare event, typical of no more than $\sim$ 1% of Earth history.

Earth's climate has been intrinsically unsteady throughout Earth history, and particularly so in the ice ages, including the last few million years. This realization is relatively recent, and has become established in parallel with the realization that human activity is now almost certainly warming the present interglacial as quickly as happened (naturally) during the final stages of the arrival of the present interglacial over 11,000 years ago.

BOX 14.1 Ice age terminology

Ice ages

In popular discussion the term 'ice-age' is widely used for what is technically called a *glacial*; for example the most recent glacial (ending between 15,000 and 11,000 years ago) of the current ice age is popularly called the 'last ice age'. The technical definition of an *ice age* is simply a climatic epoch in which there are permanent continental ice caps reaching down to near sea level, as there are now on Greenland and Antarctica. So we are still in an ice age, even though the N ice caps in particular are very much smaller than they were in the last and preceding glacials. In fact we are in the latest of a series of relatively short, mild *interglacials* which have punctuated but not interrupted the present ice age. A *stadial* is a cool interruption of an interglacial (like the Younger Dryas), while an *interstadial* is a warm interruption of a glacial. Ice ages are separated by long *warm epochs*, in which there is no permanent ice at or near sea level, even at the poles.

Ice caps

Ice caps are layers of self-compressed snow lying on continental platforms. They can be kilometres deep, and heavy enough to depress the slowly flexible continents on which they lie, weighing their rocky surfaces down to sea level, as is currently the case in parts of Antarctica and Greenland. Ice caps grow slowly as a glacial becomes established and reach dynamic equilibrium when coastal glaciers lose ice to the sea as quickly as it is replenished by inland snowfall. They grow slowly but shrink relatively quickly as an interglacial sets in. In interglacials, ice caps are confined to polar regions, but they can spread to upper mid-latitudes in the coldest parts of glacials, and apparently reached the equatorial zone several times in Snowball conditions.

Water substance trapped in ice caps has evaporated from oceans or moist land, lowering global sea levels by about 100 metres in the glacials of the present ice age, as compared with its interglacials. It is fresh water, though contaminated by dust, etc. deposited before burial under subsequent snow. Ice now 3.5 km below the present surface of the Antarctic ice cap at Vostok base (near the centre of the Antarctic ice cap) was deposited as snow about 450,000 years ago, so that a vertical ice core to that depth covers several glacial/interglacial cycles *[63]*.

Icebergs, sea ice, and ice shelves

Icebergs result when substantial chunks of ice *calve* (break off) from the coastal edges of ice caps, usually at the *snouts* (bottom ends) of glaciers. They float away in ocean currents, melting slowly and remaining a hazard to shipping until they melt completely.

Sea ice is very different, forming on the sea surface when water temperatures fall below about −1.8°C (the freezing point of normally salty seawater). It floats on the surface, thickening downwards as underlying seawater freezes onto its base, quickly at first and then more and more slowly as the thickening ice layer thermally insulates its base from the much colder

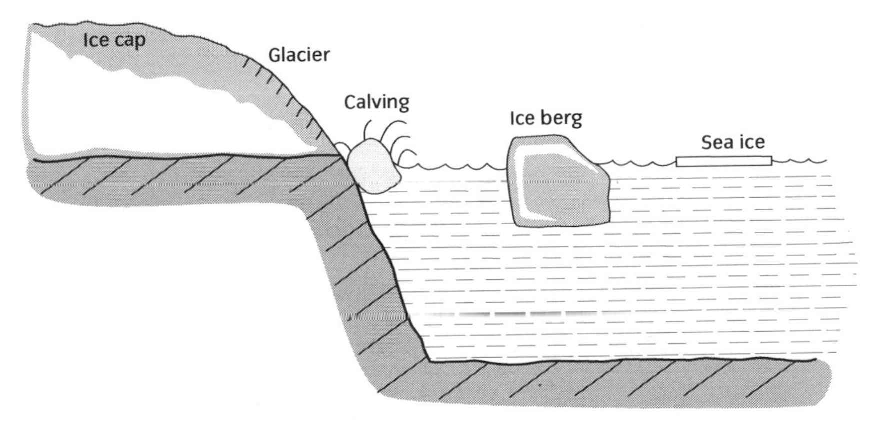

Figure 14.10 Icecaps, icebergs, and sea ice. Sea levels are unaffected by the formation or melting of floating ice, whether ice bergs or sea ice. But they are raised by the calving of icebergs, the melting and run-off of ice caps, and the expansion of water by warming.

surface air. In this way individual *ice floes* can thicken to a metre or so in a polar winter, forming *pack ice* when they collectively cover a substantial area. Sea ice can thicken further as ice floes raft over each other in converging sea currents, but seldom to more than 10 metres. There is a large seasonal rhythm of formation and melting of sea ice in the Arctic Ocean and round the edges of Antarctica, but global warming is currently reducing the winter ice cover in and around the Arctic Ocean, facilitating ocean transport there (including the fabled North West Passage from the NW Atlantic to the NE Pacific), disturbing wildlife, and further warming the polar zone as the reflective ice cover shrinks and is replaced by dark sunlight-absorbing water.

Around Antarctica very large *ice shelves* have grown partly by base-freezing and partly by surface snow deposition. When grounded in shallow water they can be more than 100 m thick. The ice shelf in the NW Weddell sea has recently shed a floating ice island which had an area of over 100 square kilometres before it began to break up.

Sea levels

Ice caps resting on continental platforms are built by abstraction of water from the world ocean. Even the smallish ice caps of the present interglacial represent the abstraction of about 70 m depth of water across the world ocean, and would raise sea levels by this amount if they were fully melted. Although it would take at the very least several centuries to supply the necessary heat and physically drain the cap areas, the warm conditions in which this might occur would significantly inflate the world ocean and take the total sea level rise to well over 100 m. This would return the oceans to what has been their state for 90% of Earth history, but the implications for the huge human populations now living in coastal margins would obviously be very great.

In the same way, sea levels during the long glacials of the present ice age have been about 100 m below the present state. Coastal outlines were very different then, with the upper slopes of many continental shelves dried out: for example, Britain was geographically part of Europe, the paleo-Thames was a tributary of the paleo-Rhine, and the fishing grounds of Dogger Bank in the North Sea were a tundra grazed by elk and mammoth, whose bones are trawled by fisher-folk still working there.

Note that the formation or melting of floating sea ice, and the melting of ice bergs, have no effect on sea levels, since by Archimedes Principle, floating ice displaces its own weight (and therefore volume) of sea water. When the berg is fully melted at around 0 °C, its meltwater exactly fills the volume of seawater displaced when it floated as ice. Icebergs do raise sea levels, but only when they fall into the sea from their parent land-supported glacier; there is no further effect as they melt.

14.6 External mechanisms of climate change

It is apparent from the brief summary of the previous section that climate varies on any time scale we choose to examine, only the very shortest of which overlap with those covered by conventional meteorology. This leaves a huge range of longer-lasting events demanding explanations many of which lie outside the meteorologically orthodox range. In recent years climatologists and other earth scientists have responded vigorously to this challenge, with the result that the field of climate change abounds in speculation and theory, much of it kept in a ferment by the continual arrival of new observational data. In this complex and fluid situation we consider a selective and largely descriptive outline of mechanisms under two headings—those working from outside the planet Earth (external) in the present section, and those working from inside it (internal) in Section 14.7.

14.6.1 The Sun

Since the Sun is the powerhouse of the atmospheric engine, significant variations in its output must affect climate, at least in principle (Chapters 8 and 9). Direct data is very limited because solar output has been measured with useful accuracy only since the early years of the twentieth century. Though variations in that period are observed to be numerically small ($< 0.5\%$), their effects could still be significant, given Earth's potential sensitivity to small radiative imbalances. Throughout the twentieth century, solar output seems to correlate positively with sunspot numbers, raising the possibility that sunspot activity, recorded since about CE 1600, might be a useful proxy for solar output. An eleven-year solar cycle in sunspots seems to have little climatic resonance, but it is suggestive that the Little Ice Age was most intense during the *Maunder sunspot minimum* (when there were hardly any sunspots), between CE 1650 and 1700. If this is more than coincidence, significant Sun-driven warming and cooling may be at work on moderately short climatic time scales, modifying the effects of other mechanisms driving rapid climate change, and complicating the current task of isolating and understanding anthropogenic effects. However, although the role of the Sun as an agent of climate change has been under active investigation for nearly a century, and has a passionate following among a small number of climatic experts, a substantial majority consider that its effects have probably been small for much of Earth's 4.5-billion-year history, including the present.

In Earth's first billion years it was quite another matter. The Sun too was young then, and studies of the life cycles of similar stars show that its radiative output would have been $\sim$ 30% below current levels. If all other things then had been as now, such a *young dark Sun* would have kept the Earth's surface so much colder, that the oceans would have been largely frozen (NN 8.2). And yet geological evidence suggests that oceans have been unfrozen from earliest times, except during Snowball Earth. The consensus view is that young Earth was kept warm enough because Earth's greenhouse (Section 8.5) was much stronger then, offsetting the effect of the cooler Sun, and maintaining the oceans as the fluid cradle and play room of life.

Towards the end of the Sun's life (after another 4 billion years or so, according to observations of the life cycles of similar stars) the Sun will have exhausted most of its hydrogen fuel, and entered its *red giant* phase—so-called because of its enormous growth in size and significant photospheric cooling, and reddening by Wien's law. In the process the Sun's luminosity will grow $\sim$ 1,000-fold, increasing the absolute temperature of the Earth's surface about six-fold (NN 8.2), radically changing its climate and terminating its biosphere. Human migration outwards in the solar system is not a permanent solution, as the Sun will quickly collapse to a dull (cold) white dwarf star with insignificant luminosity.

14.6.2 Earth's orbit and spin

The Earth moves in an elliptical orbit round the Sun (Box 14.2), and currently spins so that its equatorial plane is tilted 23.5° from its orbital plane. Since the magnitude and absolute orientation of that tilt are effectively constant on time scales of millennia, the N hemisphere (for example) is tilted towards the Sun and then away from it during the course of each complete orbit, maintaining the familiar annual migration of the height of the noonday Sun above the horizon. The S hemisphere experiences the same cycle displaced by six months. As shown in Fig. 8.10 and text, this arrangement maintains a large seasonal rhythm of solar input in middle and high latitudes in particular, which obviously dominates the seasonality of climate there. We will call this the 'tilt seasonality' mechanism to

BOX 14.2 Orbital geometry

Figure 14.11a shows Earth's slightly elliptical orbit, with the Sun at one focus (S) and the other focus void (V). A pencil point in the apex of a taut thread of length 2 l looped round pins driven into S and V trace the ellipse as it moves from *perihelion* (P), through *quadrature* Q, to *aphelion* (A), and back to P again through the other quadrature. Compared with the circle radius l which the same loop of thread would trace if S and V contracted to O, the ellipse is stretched along its *major axis* 2 a (PSOVA) and contracted along its *minor axis* 2 b (QOQ). If the distance SV is written as (2 a e), e is the *eccentricity* of the ellipse, and simple geometry shows that the separation of Sun and Earth increases and decreases as Earth orbits from P to Q, A, Q, and back to P, as listed in Table 14.3.

With Earth's current orbital eccentricity at 0.017, the ratio of the Sun–Earth separations at perihelion and aphelion $(1 - e)/(1 + e)$ is 0.967, and the ratio of the corresponding solar irradiances (0.967^2) is 0.935, showing a 6.5% decrease in irradiance between perihelion and aphelion. With an eccentricity of 0.060, the Sun–Earth separation ratio is 0.887, and the irradiance ratio is 0.787, giving a 21.3% decrease between perihelion and aphelion. With an eccentricity of 0.0005, Sun–Earth separations are 1.000, because the orbit is a circle to within the significant figures quoted in Table 1.2

Table 14.3 Orbital elements for current and 'recent' extreme eccentricities

eccentricity	PS	QS	AS	QS	PS again
e	$a\,(1 - e)$	a	$a\,(1 + e)$	a	$a\,(1 - e)$
$e = 0.017$	0.983 a	a	1.017 a	a	0.983 a
$e = 0.060$	0.940 a	a	1.060 a	a	0.940 a

distinguish it from a currently much smaller 'orbital seasonality' arising from the elliptical shape of Earth's orbit round the Sun. On much longer time scales, however, the ellipticity of Earth's orbit, and the magnitude and orientation of the relative tilt of its axis of spin are variable, with potentially important consequences for climate change.

(i) *Ellipticity* The Earth moves in an elliptical orbit with the Sun at one of its two foci (Fig. 14.11a and Box 14.2). In the current state of this orbit, the Earth is 3.3% nearer the Sun at its nearest point (*perihelion* in early January), than it is at its furthest point (*aphelion* six months later). Since the solar irradiance of Earth varies as the inverse square of the distance from the Sun (Section 8.1 and Box 14.2), the insolation just outside the Earth's atmosphere is 6.5% higher at perihelion than it is at aphelion, with solar inputs to Earth's surface varying in proportion. This maintains an 'orbital' seasonal cycle which is in step across both hemispheres, unlike the familiar 'tilt' seasons which are 6 months out of step in the two hemispheres. Though these orbital seasons are largely obscured in climatic effect by the much larger tilt seasons, they are readily measurable by radiometer. In the current state of Earth's orbit and spin (with perihelion in early January, etc.), orbital seasons tend to oppose Northern hemisphere tilt seasons and reinforce the Southern hemisphere ones, as in Table 14.4.

As all the planets orbit the Sun, their gravitational interactions continually distort their orbits, making them more or less elliptical in quasi-steady rhythms. The *eccentricity* (Box 14.2) of Earth's orbit has increased and decreased with rhythmic periods of about 100,000 and 400,000 years in astronomically recent times, and is now 0.017. Detailed astronomical calculation shows that the

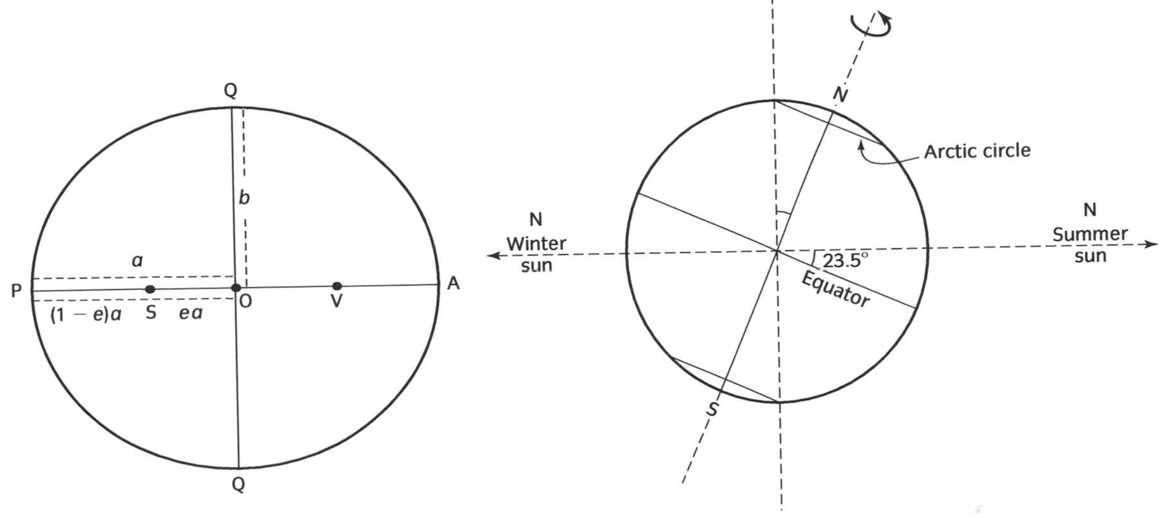

Figure 14.11a Earth's elliptic orbit round the Sun is here grossly exaggerated (ellipticity 0.44 instead of 0.017). P, Q, and A are the Earth positions at perihelion, quadrature, and aphelion respectively.

Figure 14.11b The spinning Earth in N hemisphere midsummer and midwinter, with the Arctic circle just fully illuminated and just completely shadowed.

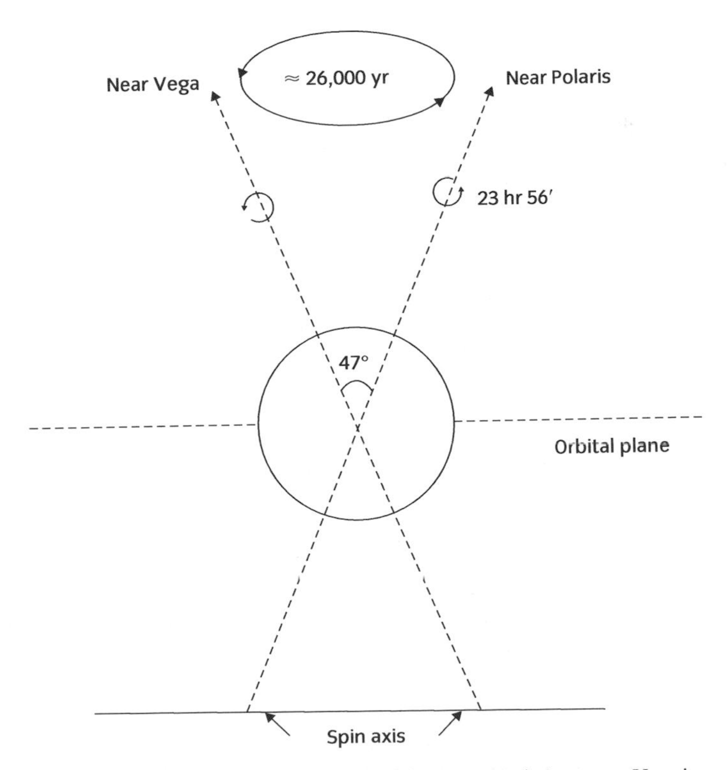

Figure 14.11c Precession of Earth's spin from near Polaris to near Vega in the N hemisphere. One cycle takes about 26,000 years.

Table 14.4 Current orbital and tilt summers (S) and winters (W)

	January	July	January
N hemisphere			
Orbital season	S	W	S
Tilt season	W	S	W
S hemisphere			
Orbital season	S	W	S
Tilt season	S	W	S

eccentricity of Earth's orbit has ranged between 0.0005 and 0.060 in the last ten million years, producing seasonal ranges of insolation varying from about 0% to 21% compared with the current 6.5%. Clearly the large values are associated with much stronger orbital seasonality than is current, adding and subtracting much more significantly from the tilt seasons. The relative timing of tilt and orbital seasons varies as outlined in (iii) below.

(ii) *Tilt magnitude* The Earth's equatorial plane (which is perpendicular to its axis of spin) currently makes an angle of 23.5° with the plane of its orbit around the Sun, with geometrical consequences shown in Fig. 14.11b and the seasonal rhythms in solar input outlined in Section 8.7 and Fig. 8.10. For example, at each hemisphere's midwinter, the Sun just fails to rise above the horizon at noon at latitude 66.5° ($= 90° - 23.5°$), and its pole is in the middle of its six-month polar night. In midsummer the Sun just fails to set at latitude 66.5° and the pole is in the middle of its six-month day. Simple spherical surface geometry (Box 14.3) shows that the area bounded by each of these *Arctic* and *Antarctic Circles*, with their extreme seasonal variations of insolation, is 8.3% of the area of its hemisphere, so that it is not surprising that they are climatically important for middle and high latitudes. Because of gravitational interactions within the solar system, the angle between the equatorial and orbital planes has varied regularly from 21.8° to 24.4° in astronomically recent times, completing a full cycle in about 41,000 years. Though these variations look small numerically, Box 14.3 shows that the area bounded by the Arctic and Antarctic Circles increases by over 20% as the tilt increases from 21.8° to 24.4°. In fact if the tilt varied by much more than this, the climatic consequences might well be so great as to disrupt the development of life. It is now known that Earth's spin axis is stabilized by its very large Moon, without which it would otherwise wander over a much wider range, as is the case with Mars and its tiny ineffective moons.

(iii) *Tilt orientation* Earth's axis of spin currently points near the star Polaris in the N hemisphere. However, from ancient times people have noticed that the centre of the apparent rotation of the night sky (where the spin axis points) moves slowly through the stars, and we now know that it completes a large circle in the night sky, returning to Polaris every 26,000 years. This happens because the Sun and Moon act gravitationally on the slight equatorial bulge of the spinning Earth, to make its axis of spin describe a cone about a fixed point in inertial space, much as an angled spinning top *precesses* about a vertical through its fixed base point, though for quite different dynamical reasons. Like the top, the angle of the cone described by the *precessing* spin axis is constant during precession, apart from the small variations mentioned in (ii).

As the axis precesses, the point on Earth's elliptical orbit corresponding to a particular spin orientation relative to the Sun (for example an equinox) moves along the orbit. Again the effect was first observed in ancient historic times through the creeping of the equinoxes through the annual cycle of the stars, and the set of related effects is now ascribed to this *precession of the equinoxes*. Modern measurement puts the advance at 20.4 minutes per year, or 1 day in 71 years. After 26,000 years ($\approx 71 \times 365$) the equinoxes will have worked their way round Earth's elliptical orbit, passing through the positions of perihelion and aphelion on the way, and putting the tilt and orbital seasons in and out of phase shown in Table 14.4. However, the apparent periodicity of 26,000 years assumes that Earth's elliptical orbit round the Sun is fixed, which is not the case.

The gravitational interactions within the solar system which vary the magnitude of Earth's orbital ellipticity, also vary its absolute orientation, causing its semi-major axis (for example) to sweep like a lighthouse beam within its orbital plane, making a full rotation in 112,000 years. Since this sweeping is in the opposite direction to the precession arising from spin axis precession, the time interval between successive coincidence of tilt midwinter and perihelion (for example) is reduced to 21,000 years (Fig. 14.12 and NN 14.1). For example, perihelion (orbital midsummer) currently occurs soon after N hemisphere tilt midwinter, when the hemisphere is fully tilted away from the Sun, so that orbit and tilt seasons are nearly in opposition (as in Table 14.4). But just over 10,000 years ago, perihelion occurred in N hemisphere midsummer, and aphelion in

NUMERICAL NOTE 14.1 Combining orbital and spin precessions

A point P describes the precession of an equinox (say N hemisphere spring) relative to Earth's elliptical orbit whose major and minor axes are initially supposed to be fixed in space. Modern astronomy shows that one full precession round the orbit would then take 26,000 years (26 kyr). In fact the axes of Earth's orbit are known to sweep in the opposite direction at a rate which completes a full turn in 112 kyr relative to non-rotating space. This means that one full precession relative to the sweeping perihelion will be completed in less that 26 kyr, since equinox and perihelion are moving in opposite directions round the orbit (Fig. 14.12). Calling the resultant period T kyr, in time T the elliptical orbit sweeps through an angle $\alpha = (T/112)\ 360$. For the two to coincide again after T, the equinoxes must go through an angle $360 - \alpha = (T/26)\ 360$ in the opposite direction. Putting these together

$$360 - (T/112)\ 360 = (T/26)\ 360$$

so that $1/T = 1/26 + 1/112$ and $T = 21.1$ kyr, as has been measured with increasing accuracy from early historic times, long before these astronomical kinematics were recognized.

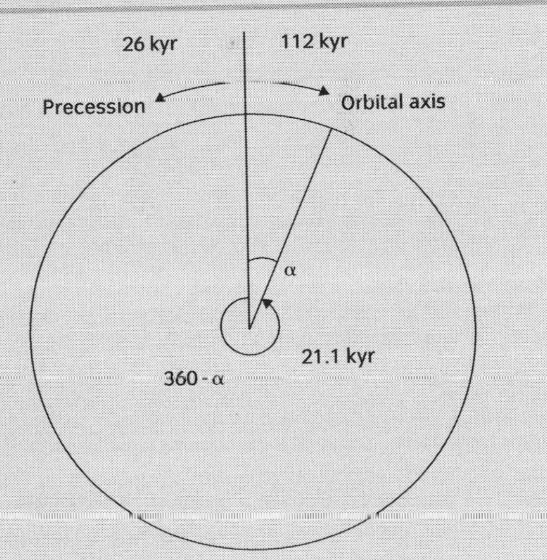

Figure 14.12 Combining spin axis precession with slower contrary precession of Earth's elliptical orbit, reduces the 26 kyr spin axis cycle to 21.1 kyr for the precession of the equinoxes.

midwinter, putting orbital and tilts in step in the N hemisphere and producing larger seasonal temperature swings there, instead of in the S hemisphere, as now. And the same thing will happen again about 10,000 years in the future. At a time of maximum eccentricity (and therefore maximum orbital seasonality) these movements in and out of step are very important climatically.

Climatic consequences

The effects on insolation of changes in Earth's orbit and spin were first investigated by the Scot James Croll in the mid-nineteenth century, when evidence of geologically recent glaciations was beginning to emerge from studies of glaciated landscapes. It was very thoroughly developed by the Serb Milutin Milankovitch in the mid-twentieth century. As more detailed and accurate evidence of glacials and interglacials in the current ice age accumulated, early scepticism about the climatic relevance of such *Croll–Milankovitch* mechanisms, has given way to a consensus that most of their rhythms can be seen in climatic data from the last three million years (the life of the current ice age). The confirmation has been statistical rather than mechanistic, but is judged to be statistically significant. Since the Croll–Milankovitch 'forcings' are not very strong, it seems likely that they act to trigger rather than drive climatic responses, which then develop according to internal mechanisms with positive feedback, such as the mobility of major meridional ocean currents like the Gulf Stream, and changes in polar albedos, especially in the N hemisphere, with its capacity for large changes in areas of ice cap and sea ice (Section 14.7, albedo).

It seems very likely that the timing of most of the glacial/interglacial fluctuations of the last three million years can be ascribed to Croll–Milankovitch triggering, although the sensitive framework for the current ice age was established by a much slower cooling, which began well before the sudden glaciation of Antarctica 34 million years ago—a cooling by slow internal mechanisms (see below), very different from Croll–Milankovitch. It is at least possible that Croll–Milankovitch triggering was at work in previous ice ages: the interplay between ellipticity, tilt, and precession being much as now in principle, though not in numerical detail, over the life of the Earth, and therefore likely to resonate with natural internal rhythms in each ice age, though such rhythms will differ considerably according to dispositions of land and sea and other internal factors.

14.6.3 Bolide impact

In recent decades, it has become clear that the Earth has been struck sporadically throughout its history by meteorites and comets large enough to cause intense, short-lived global climate change. The Chicxulub meteorite of 65 Myr BP has become particularly famous for its likely role in the *mass extinction* of living species, including the dinosaurs, but this is only one of $\sim$ 100 events of about this magnitude which have happened during the lifetime of the mature Earth (in its early days the bombardment was even more fierce, as the scarred face of the unweathered Moon still demonstrates). The amount of kinetic energy injected onto the Earth's surface by such impacts is truly astronomical. A typical large body of diameter 10 km, impacting at 30 km s^{-1}, has a kinetic energy of about 5×10^{23} J, whose nearly instantaneous conversion into heat is equivalent to the

NUMERICAL NOTE 14.2 Meteorite impact energy and heating potential

If the incoming body has radius R 5 km, and density ρ 2,500 kg m^{-3}, its mass M is $(4/3)\ \pi\ R^3\ \rho \approx 4\ R^3\ \rho \sim 10^{15}$ kg. If its impact speed V is 30 km s^{-1} its kinetic energy is $(1/2)\ M\ V^2 \approx 0.5 \times 10^{15} \times 10^9 = 5 \times 10^{23}$ J or about 10^8 Mt TNT equivalent, where 1 Mt TNT $= 4.2 \times 10^{15}$ J. If all this heat ΔQ were to be spread evenly throughout Earth's atmosphere mass $= 4\ \pi\ (6{,}350{,}000)^2 \times 10{,}000 \approx 5 \times 10^{18}$ kg and specific heat $C_p = 1{,}000$ J kg^{-1} K^{-1}, the rise in air global temperature would be $\Delta Q/(\text{mass}\ C_p) = 5 \times 10^{23}/(5 \times 10^{18} \times 10^3) = 100$ °C. In fact local heating is several thousand °C and global damage is done by impact dust, wildfire heat and smoke (text), and mega tsunamis.

simultaneous explosion of 100 million one megaton hydrogen bombs (NN 14.2). Devastation within a range of 100 km is total, with atmosphere, ocean, and sea bed blasted to incandescent vapour to a depth of tens of kilometres below sea level in a few seconds. In the next hour or so a hail of incandescent rock impacts the entire Earth's surface from sub-orbital trajectories, starting wildfires over all vegetated land areas, and quickly shrouding the globe in a dense pall of smoke which may persist for several years. This is everywhere so thick that solar input to the surface is essentially extinguished (planetary albedo rising to over 90%), and surface temperatures are reduced by 40 °C within a few weeks, killing a large fraction of the biosphere either directly (as on land), or by loss of photosynthesized food supply (as in the still warm oceans). Climatic recovery takes only a decade or so, but the biosphere is moved permanently into new evolutionary lines which take millions of years to develop and even longer to settle into mutual adjustment. In a very real sense, the fact that the writer and readers of this book are mammals rather than dinosaurs may well be testimony to a climatic event lasting only a few years 65 Myr ago. Some older impacts are believed to have had even bigger biological consequences *[64]*.

14.7 Internal mechanisms of climate change

14.7.1 Continental drift

Long-standing suspicion that the continents have been separating, moving, and reassembling on a time scale ∼ 100 Myr was suddenly confirmed in the mid-twentieth century when the discovery of physical mechanisms for sea-floor spreading and subduction led to the establishment of *plate tectonics* as the framework for describing the structure and behaviour of the surface layers of the Earth *[65]*. It was immediately clear that the relatively rapidly changing positions of continents and oceans must have been having a major impact on global and local climate throughout Earth's history.

A direct way in which this happens is by redistributing surface albedos across Earth's surface—land surfaces being considerably paler than oceans, ignoring transient Sun glitter. This effect must have been even greater before the rise of the land-based biosphere in the last 300 Myr, since well-watered land surfaces have been considerably darkened by vegetation (forests are nearly as dark as oceans—Table 10.1). For most of Earth history differences between land and sea albedos were considerably larger than now, and may have played a significant role in inducing ice ages when continents gathered in low latitudes, backscattering more of the strong solar input there and driving global cooling *[66]*. If this reached the stage of inducing deposition of permanent snow on high-latitude land surfaces, there would be strong positive feedback on a relatively short time scale (see **albedo** below).

There are many other and less obvious ways in which the changing distribution of land and sea can affect climate. The last 100 Myr will serve as an example—part of the extended aftermath of the breakup of the super continent Pangaea over 200 Myr ago. The situation about 40 million years ago (Fig. 14.13) shows several features believed to have been particularly important in supplying the preconditions for the current ice age, which began in earnest just 3.5 Myr ago, but has been in preparation for well over 50 Myr.

(i) The clustering of continents around the North pole, and the arrival of Antarctica over the South Pole has already occurred.

(ii) The opening of the Atlantic as a deep meridional ocean corridor from Pole to Pole is already well under way.

(iii) Australia is separating from Antarctica, and allowing cold circumpolar ocean currents to isolate Antarctica and begin its cooling and glaciation.

(iv) India is moving very rapidly N'wards to collide with southern Asia and raise the Himalayan massif.

(iv) The equatorial ocean corridor is closing, firstly to form the Mediterranean, and later (about 30 Myr after the map time) to separate the Atlantic and Pacific Oceans by forming the central American isthmus.

Figure 14.13 Global outline of continents and islands in about 40 Myr BP. Solid lines are coasts as they were then; dotted lines are where they would have been with today's lower sea levels. The high N Atlantic is distinctly narrower then than now (it was completely shut only 60 Myr earlier), a very (cartographically) distorted Australia restricts the circumpolar W'ly flow of the Southern Ocean round Antarctica, and there is largely unimpeded flow of E'ly ocean currents in the equatorial zone. The E Mediterranean closed later, and Panama closed shortly before the onset of the present ice age.

Discussion

By 100 Myr BP, a warm epoch had lasted for over 100 Myr. Global annual average surface temperatures were between 5 and 10 °C warmer than the current 15°C, the ocean deeps were much the same temperature as surface waters, and there were no polar ice caps, indeed no permanent ice anywhere except on high mountains. From about 70 Myr BP, the ocean deeps began their persistent cooling towards the present state in which all are filled with near freezing water at all latitudes.

Antarctica suddenly became largely ice covered 34 Myr ago, beginning the first phase of the current ice age and increasing the flow of cold water from its perimeter to the ocean deeps. The rise of the Tibetan massif as India began to collide with central Asia about 50 Myr ago seems to have been associated with global cooling *[67]*. It certainly has the capacity to lower atmospheric CO_2 levels (and hence the prevailing greenhouse effect) as Himalayan relief rainfall captures atmospheric CO_2 and produces rocky carbonates on exposed slopes, but the magnitude of this loss and associated cooling depends on the strength of opposing, mostly geological sources of atmospheric CO_2, which are complex and still unclear on geological time scales.

Incremental blocking of the equatorial ocean corridor and the warm, Trade-Wind-driven W'ward ocean current which previously threaded all low latitudes, was finally completed by the closing of the Central American Seaway less than 4 Myr BP (Fig. 14.13). This may have been the last geological input to global cooling, paradoxically by encouraging N'ward flow of warm water and water vapour which led to copious deposition of snow in high N latitudes cold enough for millions of years but starved of precipitation to form ice caps. The series of glacials and interglacials covering and partly uncovering the continental platforms (N America, Greenland, and N Eurasia) surrounding the Arctic Ocean began about 3.5 Myr ago as the Croll–Milankovitch rhythms triggered instabilities inherent in the atmosphere/hydrosphere/cryosphere system set up by this long and complex sequence of events.

14.7.2 Albedo

Earth's albedo in solar wavelengths depends mainly on scattering by tropospheric cloud, but also by tropospheric and stratospheric haze, and pale land surfaces, which will include polar snow and ice during ice ages. The distributions of cloud, snow, and ice are obviously sensitive to the atmospheric activity whose energy input they help to control, and will be active in the labyrinth of feedback loops which maintain the rich variety of climatic behaviour.

Tropospheric cloud encourages negative feedback to the extent that cloud cover increases with solar input, thereby offsetting some of that input by increasing backscatter to space. However, increasing cloud reduces *TR* output through the atmospheric window, increases the greenhouse effect, raises surface temperatures, and drives further cloud formation in a positive feedback loop. Snow and ice cover will also tend to operate in positive feedback loops, increasing surface solar absorption as they shrink, and decreasing it as they spread.

Venus, Earth, and Mars appear to have reached very different balances between these effects. Venus is at the hot and cloudy end of runaway positive feedback, with a huge greenhouse in its far infrared Venusian wavelengths and a very high albedo in solar wavelengths, leaving the planet with a low equivalent blackbody temperature, and a very high surface temperature under a massive CO_2 atmosphere (Section 8.5).

Mars is near the cold and cloudless end of runaway positive feedback, with a small greenhouse in infrared Martian wavelengths, virtually no cloud (apart from sporadic global dust storms), and seasonal polar caps of ice and frozen CO_2 under a thin atmosphere. Apparent evidence of water and ice flows in geologically recent times intriguingly suggests that the present Martian balance may not be permanent. Earth lies between these extremes with its considerable greenhouse in *TR* wavelengths locked in dynamic equilibrium with a smallish, mainly atmospheric albedo—an equilibrium whose balance shifts substantially between warm epochs and ice ages.

Ice cap albedos

For most of Earth's life there have been no permanent ice caps near sea level, and surface albedos have ranged narrowly from low ocean values to somewhat higher land values. However, much larger variations in surface albedo are associated with the variable distributions of snow and ice which are the defining feature of ice ages, including the present one, in which we know that the polar ice caps have responded to the Croll–Milankovitch forcings on time scales of about 10–100 kyr (Section 14.6).

Current interglacial

In the current interglacial, the Antarctic ice cap, and its adjacent ice shelves and sea ice occupy most of the area inside the Antarctic Circle, which encloses just over 4% of Earth's surface (Table 14.5). The identical area within the Arctic Circle contains most of the Greenland ice cap, and a large (though currently diminishing) area of permanent sea ice, together with seasonal snow and sea ice which largely completes the area coverage in the winter but disappears in summer. For simplicity we suppose that the Arctic and Antarctic Circles (*polar circles*) are filled with all of Earth's permanent snow and ice, covering 8.3% of the Earth's surface with permanent ice.

To gauge the importance of such ice cover for Earth's albedo, we must view Earth from the Sun, seeing it as an apparently flat circular disk, a seasonally variable part of which is covered by very reflective ice. Given the present 23.5° between Earth's equatorial and orbital planes, the percentage of the sunlit Earth disk covered by a chosen polar ice cap (the Arctic cap, for example) undergoes the following annual cycle (Fig. 14.14, Box 14.3, summarized in Table 14.5).

(i) midwinter: zero—completely (but only just) hidden from the Sun;

(ii) equinoxes: 1.4%—a straight-edged polar segment; and

(iii) midsummer: 6.3%—a distorted patch extending one third of the way from the disk's top to its apparent middle.

Last glacial maximum

Polar caps are much more extensive in glacial maxima, like the one at its largest about 20,000 years ago, with permanent ice reaching down to about latitude 50° in the N hemisphere. In the S hemisphere, ice cover was somewhat smaller, because of the absence of land platforms in sufficiently high latitudes to add to the Antarctic total, though the latter was considerably extended by permanent sea ice, and a long ice tongue covering the Andes to quite low latitudes. Again we deliberately oversimplify by saying that permanent polar ice reached down to latitude 50° in each hemisphere, covering 23.4% (Table 14.5) of Earth's spherical surface with permanent ice. Given Earth's present 23.5° tilt, the percentage of

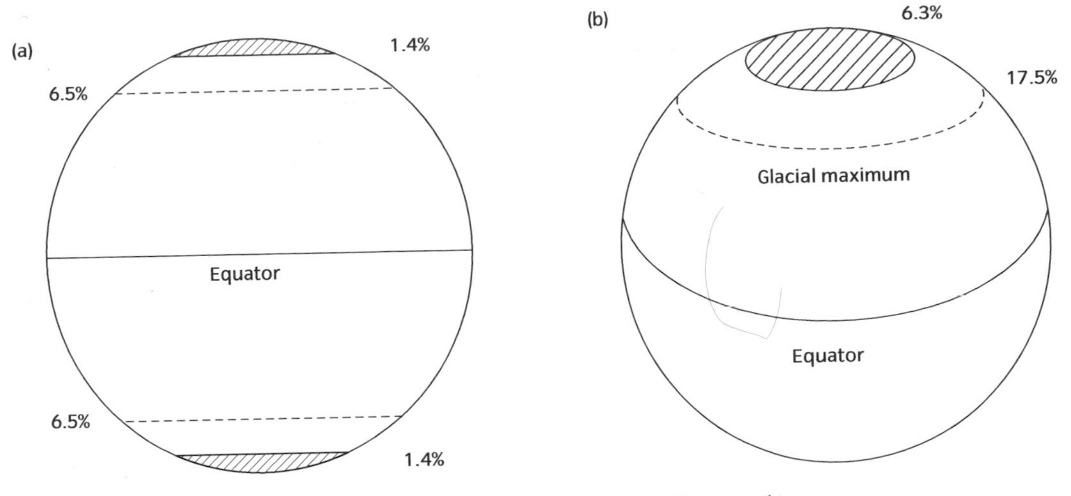

Figure 14.14 Polar ice caps as seen from the Sun in the equinoxes (a), midsummer (b), and midwinter (the lower half of (b)). The current Arctic/Antarctic circles are hatched, and the notional extent to latitude 50° in the last glacial is shown dashed in the equinoxes and midsummer. In midwinter it shows such a slim cusp at the Earth's rim that it would be difficult to draw it without exaggeration.

BOX 14.3 Polar and ice cap geometry

Cap area

Standard analysis shows that the surface area of a polar cap bounded by latitude ϕ is given by $A_c = 2\pi R^2 F_c$, where R is Earth's radius, and $F_c = (1 - \sin\phi)$ gives the cap area as a proportion of the surface area of the hemisphere ($2\pi R^2$).

The values for F_c in Table 14.4 show that 50% of Earth's surface lies above latitude 30°, 23.4% lies above latitude 50°, 13.4% lies above latitude 60°, and only 8.3% lies inside the Arctic or Antarctic Circles (polar circles). As Earth's spin axis has variably tilted in the last 10 Myr, the latitude of the polar circles has ranged from 65.6° to 68.3°, taking F_c from 0.089 to 0.071, and varying polar cap areas by over 20% between extremes.

Sun-facing disk areas

Equinox Seen from the Sun at an equinox (Fig. 14.14a), each polar cap appears as a polar segment with straight low latitude edge across the Earth disk. The segment area on the Earth disk is given by $A_{de} = \pi R^2 F_{de}$, where $F_{de} = 0.5\,[1 - \phi°/90° - (\sin 2\phi)/\pi]$ and ϕ is the bounding latitude of the cap and segment. The F_{de} value for latitude 66.5° shows that polar circles currently each bound a segment covering 1.4% of the solar disk, totalling 2.8% for the pair. This proportion is sensitive to the tilt of Earth's spin axis (90° – ϕ), increasing from 1.2% for polar circle latitude (PCL) 68.3° to 1.6% for PCL 65.6°, a range of nearly 30%.

Summer solstice At its summer solstice, a distorted version of a polar circle covers a proportion F_{dss} of the Earth disk (Fig. 14.14b), where $F_{dss} = \sin^3(90° - \phi)$. This has value 0.063 (i.e. 6.3%) for $\phi = 66.5°$, and is sensitive to the tilt of Earth's spin axis, increasing from 5.1% to 7.1% as PCL decreases from 68.3° to 65.6°, a range of over 30% for only 2.7° change in PCL.

Extended polar cap When an ice cap extends beyond the polar circle (e.g. to latitude 50°), the expressions for F_c and F_{de} apply and give the values tabulated above. The expression for F_{dss} (the fraction of Earth disk covered by the ice cap at its summer solstice) might seem to be $\sin 23.5 \sin^2 40 = 0.165$ or 16.5% (using the geometry of the tilted flat cap which perfectly models the tilted polar circle), but needs correction for the polar cap gaining sunlight by bulging and losing it by shadowing, so that the true proportion of the illuminated Earth disk is 17.3%. This small correction becomes rapidly larger for ice caps extending to latitudes lower than 50° at tilt 23.5°, as does the small fringe of cap illuminated at its winter solstice.

Note All geometrical results in this box follow from standard circular and spherical geometry, and are quoted here because they do not appear in texts known to the author. The detailed expression for F_{dss} for an ice cap extending beyond its polar circle is too complex to quote here.

Table 14.5 Sphere and disk geometry of one polar cap

Cap edge latitude ϕ	0	30	50	60	66.5	90°
F_c	1	0.5	0.234	0.134	0.083	0
F_{de}	0.5	0.195	0.065	0.029	0.014	0
F_{dss}			0.173	------	0.063	0
F_{dws}			0.001	------	0.000	0

the solar-facing disk covered by each large polar ice cap had an annual cycle as follows (Fig. 14.14, Box 14.3, and Table 14.6), being

(i) midwinter: ≈ 1%—thin crescent edge to summer half-disk;

(ii) equinoxes: 6.5%—a straight-edged polar segment; and

(iii) midsummer: 17.3%—distorted patch extending over half way to the apparent middle of the disk.

Climatic consequences

Table 14.6 includes simple estimates (NN 14.3) of average albedos of the sunlit Earth disk through an annual cycle in the current interglacial and in the last glacial maximum. In the glacial maximum, average disk albedo was about 36.7% compared with the 30.6% in the current interglacial, decreasing solar input to Earth by 9% of its current value (NN 14.3). These are very large variations in solar uptake by Earth, and must affect Earth's weather and climate significantly since they persist for ~ 10 kyr in interglacials and ~ 100 kyr in glacials. It is known from a range of evidence that in the last glacial maximum, the subtropical deserts covered a much larger range of latitude than now, and that the whole atmosphere was drier and dustier (leaving dust deposits in ice cap ice cores), as well as distinctly cooler in low latitudes and very much colder in middle latitudes.

Not only is the average solar input significantly reduced in glacial maxima, but its seasonal variation is significantly increased: the seasonal variation in albedo in the glacial maximum produces a 2.5% variation in solar input about the annual average, compared with 2.0% in the current interglacial. The seasonal effects in the extensive middle latitude edges of the great ice caps must have been very strong, with reduced poleward advection of heat and cloud allowing temperatures to fall steeply with the declining Sun. The vast interiors of the caps must have been much as Antarctica is now, though maybe quieter because of reduced atmospheric activity.

NUMERICAL NOTE 14.3 Albedo estimations for Table 14.4

Assume simplistically that average albedo inside the polar ice caps is 85%, whereas average albedo over the rest of the globe is 28%. At the equinoxes of the current interglacial, according to the relative disk areas of polar ice in Table 14.6 the global average albedo is [2.8 × 85 + 97.2 × 28]/100 = 29.6%. Using the same albedo values, the global albedo at a solstice is [6.3 × 85 + 93.7 × 28] = 31.6%, implying an observably realistic annual average of 30.6%.

Using the same values for the Earth at the last glacial maximum (20 kyr BP), the average albedo of the N hemisphere at the equinoxes and solstices are found in the same way to be 35.4 and 37.9% respectively, averaging to 36.7% and implying a reduction of solar input to the Earth from 69.4 units to 63.3 units compared with the current interglacial—a fall of about 9%.

The annual range in solar input between solstice minima and the equinox maxima is 2 units in 69.4 (about 3%) in the current interglacial, and 2.5 units in 63.4 (about 3.9%) in the last glacial maximum.

Table 14.6 Sunlit ice cover and global albedo

Current interglacial	Disk area/%		Disk albedo
solstices	1 @ 6.3	= 6.3	31.6%
equinoxes	2 @ 1.4	= 2.8	29.6
average			30.6
Last glacial maximum			
solstices	1 @ 17.3 + 1 @ 0.1	= 17.4	37.9
equinoxes	2 @ 6.5	= 13.0	35.4
average			36.7

Our simple model of the polar ice caps is unrealistically symmetrical between the hemispheres: the real S hemisphere having more ice cover in the interglacials (as now) and the N hemisphere having more in the glacials, as N America and N Eurasia become glaciated. Such asymmetry will tend to enhance response to the relative phases of the tilt and orbital seasons (Table 14.4). For example, the ending of a glacial may be encouraged when tilt and orbital summers coincide, especially if orbital ellipticity is large and summer insolation very high (Table 14.3). Summer after summer for several millennia (until tilt and orbit slide out of step again), the especially warm summers may reduce the N ice cap by more than it is rebuilt in ensuing winters, and the uncovered land and sea is able to absorb more sunlight and encourage the process. The very abrupt endings of glacials (presumably typified by the current interglacial between 16,000 and 11,000 yrs BP) suggests strong positive feedback at work.

The statistical significance of Croll–Milankovitch forcing is undeniable, but detailing mechanisms in a properly quantitative way will be an enormous task, involving robust understanding of a wide range of nonlinear responses in the Earth surface system.

14.7.3 Snowball Earth

The period from about 800 to 630 Myr BP contained several unusually intense and prolonged ice ages, and is sometimes known as the *Cryogenian* on this account *[62]*. In parts of these, there is considerable evidence that most if not all of the Earth was covered with permanent ice—a *snowball* condition. The totality of ice cover is still a matter of considerable debate, as are the mechanisms for the onset and ending of such extreme glaciations, which do not seem to have occurred in the rest of Earth history.

The development of a snowball state seems certain to involve a runaway albedo effect. However triggered, if the ice caps cover half Earth's surface (i.e. come down to 30° latitude and cover almost 40% of the equinox Earth disk seen from the Sun), Earth's albedo is increased very dramatically above the value associated with recent glacial maxima (to 50% compared with the 37% estimated above). At some point surface temperatures even in low latitudes will fall below freezing, and the remainder of the Earth will ice over, reaching a snowball condition once the ocean surfaces have cooled to radiative equilibrium temperatures. Atmospheric conditions would become very quiet as solar input dropped with rising albedo. If global albedo reached 80%, for example, corresponding to a nearly pure snowball, the global annual average heat input from the Sun would be reduced to less than 30% of its current value, drastically diminishing atmospheric activity and probably clearing the sky of what little cloud could survive the drastically lowered saturation limit of the very cold air.

Early evidence of snowball conditions was discounted on the grounds that a snowball condition once formed would be permanent, and Earth would still be a snowball. This objection disappeared when it was realized that CO_2 would no longer be continually washed by rainfall or snowfall from the cold, largely cloudless sky, allowing it to accumulate from volcanic eruptions and eventually produce a greenhouse sufficiently strong to warm and melt enough ice to initiate a runaway deglaciation towards a more normal ice age, or even a warm epoch.

The snowball condition continues to attract conflicting analyses and interpretations, which incorporate a wide range of evidence and mechanisms of climate change. The subsequent Cambrian explosion of complex life is inevitably suspected as being a delayed reaction to the end of the long hibernation of residual primitive life during the Cryogenian—a garden of Eden after nearly sterile frigidity.

14.7.4 Volcanism

Volcanic eruptions inject large volumes of CO_2 and reflective sulphate aerosol particles into the atmosphere, both of which can significantly modify the radiative equilibrium of the planet and hence its climate.

Sulphates

When eruptions are powerful enough, they can inject aerosol particles into the middle stratosphere where the absence of precipitation allows them to persist for years, covering the Earth with a thin reflecting veil which increases its albedo and reduces solar input. This happened on a very grand scale in April 1815 when Tambora (E Indonesia) produced by far the most powerful eruption anywhere on Earth in the last 2,000 years, causing great loss of life and destruction locally, and lowering global temperatures for several years. In N America and parts of

Europe, 1816 became known as 'the year without a summer', and crop failures and famine in many parts of the world added greatly to the original death toll. As in the case of many smaller eruptions before and since, the veils of aerosol in the stratosphere gave rise to unusually colourful sunrises and sunsets all round the world, and some may have inspired the artist Turner, who was then at the height of his powers and delighted in vivid impressions of sky and light.

The spectacular and closely studied eruption of Mt St Helens in 1980 produced relatively little stratospheric sulphate because the eruptive blast was much smaller and largely horizontal. Particulate emissions were largely confined to the troposphere where they were quickly washed out by cloud and rain, leaving little effect on global albedo.

CO_2

Volcanoes also produce large volumes of CO_2 which are a major source of atmospheric CO_2 on longer time scales (short time scales now being dominated by the biological cycle—Section 3.5), in which they are largely balanced by removal by solution in cloud, precipitation onto land and oceans, and subsequent involvement in geochemical and slow geobiological cycles. At present the average annual input of CO_2 by volcanoes is believed to be about 200 Mt yr^{-1} *[91]*. However, volcanism is unsteady on longer time scales, and there have been sustained periods of much higher rates of injection in the past. One of these is believed to have lasted for at least 100,000 years about 66 Myr BP, close to the time of the Chicxulub impact (Section 14.6). On its way to collision with Asia, India was then passing over a vigorous hot plume in the Earth's mantle which produced vast outpouring of flood basalts, whose severely eroded remains still form one of the biggest volcanic formations on Earth (the Deccan Traps of W India). Over 20 thousand Gt of CO_2 may have been released *[92]*—nearly 10 times the current mass of atmospheric CO_2. If the release lasted 100,000 years, the implied average rate of release is over 200 Mt yr^{-1}. Assuming that volcanic activity elsewhere remained at current average global levels, the Deccan eruptions may have doubled the global volcanic rate of CO_2 injection for 100,000 years, significantly raising its atmospheric concentration, increasing the global greenhouse, warming the atmosphere, acidifying the oceans, and cooperating with the Chicxulub impact in the mass extinction of living things.

Notice how volcanism connects the atmospheric engine with the independent and much more ponderous engine working inside the Earth, acting on surface conditions without being significantly affected in return. On account of their lack of feedback, volcanism and continental migration are more like external agents of climate change than the other internal processes listed here.

14.7.5 Nuclear winter

In 1971, Carl Sagan and others had to wait several months for a global dust storm on Mars to subside and reveal its surface to the first spacecraft to orbit another planet. Partly to pass the time they studied the dust cloud vertical temperature profile with the craft's infrared radiometer, noticed how the surface warmed and the cloud cooled as it subsided, and tried to model the behaviour for different amounts of raised dust. The models were later adapted to describe terrestrial volcanic dust, and the dust and smoke raised by a large asteroid impact, in each case estimating very strong cooling beneath a thick persistent cloud. By 1981 the methods had been applied to the smoke and dust raised by the detonations of nuclear weapons in a range of war scenarios imagined for the catastrophic breakdown of the cold war, then in a very fragile state.

A seminal study by Sagan and others *[68]* strongly suggested severe global cooling under a smoke layer which would fill much of the troposphere for a year or more. In addition, the injection of debris into the stratosphere would produce a fine layer of reflective dust, and cause widespread heavy depletion of stratospheric ozone layer by oxides of nitrogen—all this in addition to the obvious local destruction by fire and blast and short-lived radioactivity, and global biological damage caused by longer-lived radioactivity. The implied danger to humanity and the rest of the biosphere was so severe that the study became international, culminating in a conference in late 1983, including climatologists of the Soviet Union by satellite link. Subsequent studies with improved models have modified some of the conclusions in detail, but suggest that even a relatively 'small' war, exploding only 100 bombs (totalling perhaps 50 Mt TNT equivalent) could have a very significant global impact.

Copious tropospheric smoke from bombed cities acts like the wildfires triggered by asteroid impact on vegetated land. In either case, persistence is encouraged by the development of a large temperature inversion from the chilled base of the smoke layer to its top (as in a thick fog layer—Section 10.12), suppressing vertical motion, and leaving slow settling as the slow dominant removal process for cloud particles. Some of this was unfortunately tested during the torching of the Kuwait oil fields in the First Gulf War (1991), when surface temperatures under the extensive local smoke clouds were held down by more than 10 °C for weeks. Stratospheric dust operates as in the case of powerful volcanic eruptions, since bombs of more than about 0.5 Mt TNT inject their plumes well into the stratosphere. The damage to the ozone layer could be a very severe version of the ongoing ozone hole event (Box 3.3), with oxides of nitrogen replacing CFCs as the damaging agents.

The damage to the biosphere raises hugely complex issues being studied in the context of the 20 or so natural mass extinctions in the past *[64]*, including the extinction currently being driven by the vast range of environmental impacts by the peaceful but still very disruptive explosion of the human race. The strong suspicion remains that a large war, exploding several thousand nuclear bombs could cause a very rapid mass extinction of the biosphere comparable with the event 65 Myr BP.

14.7.6 Biosphere

The interactions between Earth's climate and life is a vast topic which will involve human study at the highest level and broadest extent for the foreseeable future, and is far beyond the scope of this book. Initially we tend to think of life as adapting to climatic conditions imposed by the physical and chemical conditions of the solar system and Earth. But the effects of the evolution of life on the development of the atmosphere, and the transformation of land surfaces as they became colonized in the last 300 Myr or so have been huge and far-reaching, as we see in the complete transformation of the carbon cycle from a slowish exchange dominated by geochemistry and geophysics to a frantically fast geobiological exchange (Section 3.5). The effect of this transformation on Earth's greenhouse through the activity of atmospheric CO_2 has been profound, and is now at the centre of our realization that life can strongly influence world climate, and that the human race in particular is fully capable of having an important effect on a very short time scale.

14.7.7 Ocean currents

Heat capacity

As mentioned in Sections 8.10, 10.5 and elsewhere, even quite shallow layers of water have large heat capacities which can store or transport very significant quantities of heat. On a global scale, the total heat capacity of the world oceans is over 1,200 times larger than the heat capacity of the whole atmosphere (NN 14.4), and to this extent dominates the temperature response of the combined ocean and atmosphere to any heat input or output which is thoroughly spread through them both. However, such thorough mixing can occur only on time scales of more than ~ 1,000 years, because of the relatively sluggish stirring of the ocean deeps. On shorter time scales, the effective heat capacity of the oceans is smaller because only a fraction of the ocean body is thermally involved. For example, the heat capacity of the top 100 m of the oceans, the part reached directly and indirectly by solar heating in the course of a year, is about 30 times larger than the total atmospheric heat capacity. This is still a large factor, and implies that successful weather or climate forecasts over periods of a season or more must include heat exchanges with the ocean, extending methods pioneered in recent decades based on analyses of routine satellite surveys of sea surface temperature.

NUMERICAL NOTE 14.4 Oceanic heat storage and transport

Since the world ocean would have a depth of 3 km if spread evenly over a smoothed Earth, the average ocean mass per horizontal square metre is depth × density = 3,000 × 1,000 = 3×10^6 kg m^{-2}, as compared with the atmospheric mass of about 10^4 kg m^{-2} (NN 4.4). This ratio of 300 in mass becomes a ratio of 1,260 in heat capacity when multiplied by the ratio of specific heat capacities water/air (4,200/1,000 = 4.2).

The top 100 m of the actual oceans (covering 70% of the Earth) has the mass and heat capacity of the top 70 m of the uniform world ocean, with a corresponding heat capacity of 1,260 × 70/3,000 ≈ 30 times the atmospheric value.

The Gulf Stream is about 100 km wide, 200 m deep, moving at 2 m s^{-1}, balanced by a diffuse 4 °C colder counter current with the same mass flux. Using Box 8.3, the net heat flux is mass flux × specific heat capacity × temperature difference = $(10^5 \times 200 \times 1{,}000 \times 2) \times 4{,}200 \times 4 = 6.72 \times 10^{14}$ W, i.e. 672 TW. A typical frontal air flow 100 km wide, 400 hPa deep (mass = 40,000/g), moving at 30 m s^{-1}, is balanced by counter-current 10 °C colder with the same mass flux. The net heat flux = $(10^5 \times 40{,}000/10 \times 30) \times 1{,}000 \times 10 = 1.2 \times 10^{14}$ W, i.e. 120 TW and 20% of the oceanic value.

Currents

Because of their much greater heat capacity, water currents can transport as much or more heat than their much faster atmospheric counterparts. For example, the Gulf Stream at about 2 m s^{-1} can maintain a net poleward heat flux of over 600 TW—several times more than the flow maintained by a mature polar frontal system with flow speeds of 30 m s^{-1} (NN 14.4). Estimating the hemispheric poleward heat fluxes averaged over a year is complicated by the number of such ocean currents, and the number, duration, and frequency of such frontal systems. Results suggest that ocean currents contribute about 25% of the total required to maintain the low and high latitudes in their average thermal state (Section 8.10). In addition the following example shows that the planet's heat economy can be strongly affected by very particular aspects of oceanic heat transport not covered by such general considerations.

The Younger Dryas stadial (Section 14.5) is a marked feature in Greenland, Britain, and Europe. By about 13,500 BP, Earth had warmed to near interglacial conditions, stabilizing at temperatures a little below the present as the great N American ice cap continued to melt and drain S into the Gulf of Mexico. In the later stages of this drainage, glacial meltwater broke into the N Atlantic through the St Lawrence estuary, spreading a sheet of low-density fresh water across the high N Atlantic. As it chilled and froze on the surface (as on a freshwater pond surface on a frosty night), permanent and seasonal sea ice returned across the high N Atlantic to an extent not seen since the ending of the glacial maximum several thousand years earlier, insulating the ocean from surface chilling by emission of terrestrial radiation and contact with cold air, and stopping the descent of chilled surface waters into the ocean deeps. These effects combined to divert the Gulf Stream (and its broad downstream tail—the North Atlantic Drift) to much lower latitudes. The stopping of heat flow from the surface of the NE Atlantic to the overflowing W'ly winds quickly reduced surface temperatures in the British Isles and NW Europe to glacial values for about a thousand years. Only when the flood of meltwater from N America had ceased did the ice disappear and the Gulf Stream swing polewards again, helping to establish the present Holocene interglacial in a few centuries of the most rapid global warming so far measured (with the possible exception of the present).

Deep ocean currents, like those fed by the subsidence of cold surface waters in the high N Atlantic, have long been known to be sensitive to changes in water density arising from very small changes in temperature and/or salinity. The quiet ocean interiors are so nearly frictionless that huge masses of water can be kept in motion by small persistent forces, carrying vast amounts of heat in the process. They are therefore very sensitive to density variations arising from small changes in surface temperature and salinity imposed by the atmosphere—a recipe for powerful feedback in the mechanisms of global and local climate change. The role of such sensitivity in the function of the oceanic conveyor belt (Sections 8.10 and 9.4) is being observed and modelled closely for this reason.

14.7.8 Greenhouse

The role of Earth's greenhouse (Sections 8.5 and 8.6) in climate and climate change is very great, especially in reacting and adjusting to changes imposed in other ways from within or outside the Earth system. The two largest components (water vapour and cloud) play vital roles in meteorology, climatology, and human life. Of these, water vapour content is linked to air temperature through the saturation limit, and exerts a major control on air temperature by liberating latent heat during saturated ascent (Section 5.7). As vapour content increases, the atmospheric window narrows, increasing the greenhouse effect by restricting the passage of terrestrial radiation to space. The resulting differences between the greenhouse effects in polar and equatorial atmospheres can be very substantial, even though each may be saturated and cloudless. Low cloud more than about a 100 m thick completely shuts the window, increasing the local greenhouse effect and restricting nocturnal surface cooling to an extent which is obvious and familiar to countless people who have no idea that it is an aspect of the greenhouse effect at work. And carbon dioxide is at the heart of many variations in the greenhouse, from earliest Earth history, when its much greater abundance kept Earth's surface above freezing in the weaker rays of the young dark Sun, through the huge spikes which probably ended each Snowball state, through the surges, plateaux and troughs which accompanied the global flowering of the biosphere since the Cambrian, through the sawtooth series of the current ice age, to the spike currently being injected by human industry as it burns the buried residue of that post-Cambrian flowering. All this activity is riddled with positive or negative feedback at levels which makes reliable evaluation and prediction of detailed consequences enormously more demanding of observation and theoretical modelling than can be covered in a descriptive outline like this.

14.8 Climate modelling

14.8.1 Historical

Earth's climate is maintained and changed as the atmosphere, oceans, and ice sheets adjust and readjust in response to changes imposed from outside the Earth and within, and those adjustments can range from near insensitivity, through simple proportionate response, to great sensitivity and instability.

The first Earth system to be modelled, in the general sense of the word, was the atmosphere, beginning with empirical rules of thumb suggested by data coming from the young synoptic observational network (1860 onwards), through the systematic inclusion of dynamical and physical and some chemical laws (1900 onwards), to the development of computer models extending the process of mathematical solution beyond the very narrow range of interactions capable of analytical solution (1950 onwards). The focus of this very successful enterprise was the mechanistic description of the behaviour of the atmosphere on time scales of seconds to weeks, and especially on successful forecasting on time scales of hours to a week. After rapid development up to about 1980, its progress is now more gentle, aided by the development of computer hardware and modelling, and by the uncovering of the extent and intensity of influential, slightly longer-term interactions with the oceans.

Oceanography grew out of the need for reliable charts of coasts, sea beds, currents, and tides for all who use the sea. Individual heroic expeditions in the nineteenth and twentieth centuries laid the foundations of modern understanding. Interactions with the atmosphere were obvious from the outset, but difficult to describe in useful detail (the Beaufort wind scale being a pioneering exception) apart from the great systems of wind-driven water currents. And the vast hidden world of the ocean depths is still only fitfully revealed by sporadic soundings. The profound cold of the deeps was obviously related to polar surface climate by descent of polar waters, and the shallow, warm, saline anticyclonic gyres just as obviously related to the atmospheric Subtropical Highs, but until recent decades, atmospheric and oceanographic scientists each tended to regard such interactions as inputs from another world.

Panoramic maps of sea surface temperatures by satellite in the 1970s encouraged the realization that interactions could be followed on a routine basis, and are now central to attempts to extend weather forecasting to the time scales of seasons, and to a blurring of the traditional distinction between weather and climate. Climate modelling in the modern sense began in the 1960s as an extension of computer weather forecasting into longer time periods, as if marine and other boundary conditions were static boundary conditions. Edward Lorenz's work *[2]* was seminal in several respects, revealing the unexpected presence of deterministic chaos (Section 1.6) in the atmosphere, and demonstrating the value of making repeated modelling runs to cover the range of behaviour arising from identical starting conditions. And there was gradual response to the need for flexible boundary conditions (such as sea surface temperatures) to take account of the important but slower dynamics and thermodynamics of the oceans.

With the recent deluge of data on past climate, and the realization that it has been dynamic on time scales from years to the age of the Earth, and the further realization that present and future climate change will be driven by an amalgamation of natural and anthropogenic factors, the stage was set by about 1980 for the explosive growth in the practice of computerized modelling of climate and climate change on many time scales and for many purposes, including the socially vital need to make reliable forecasts of climate change in the coming decades and centuries, and to identify and take steps to mitigate damaging features.

14.8.2 Climate models

According to a widely used introduction to climate modelling *[69]*, models can be considered under four headings, each of which relates to matter discussed in various parts of the present book. In ascending order of complexity, these model categories are as follows.

(1) Energy balance models which examine the behaviour of global or zonal boxes of the atmosphere in radiative equilibrium with the Sun and space. An absolutely basic example is the determination of Earth's equivalent blackbody temperature as the planet balances emission of terrestrial radiation and absorption of solar radiation (Section 8.2).

(2) Models with a single (vertical) dimension which accommodate an empirical representation of convection in a radiative balance with vertical structure. The treatments of idealized convection and the adiabatic lapse rates (Section 5.8) underlie the crucial convective aspects of these models.

(3) Models of intermediate complexity which often make use of zonally averaged data to illuminate the meridional structure of the climate, together with some vertical structure, and may include some representation of surface

processes (for example, the meridional transport of heat by ocean currents) and simple geochemistry. These build quantitatively on qualitative discussions in Chapters 3 and 8 and Section 14.7.7.

(4) General circulation models (GCMs) which build on the full three-dimensional representation of the atmosphere begun in Chapter 7 and developed in computerized weather forecasting models, incorporating representations of interactions with oceans, ice, the biosphere, and distributions of land and sea. These are the ultimate models in principle but are not necessarily the best models to use in many circumstances. Even if it were feasible, it will often not be an effective use of resources to try to develop and run a fully interactive representation of Earth's surface systems on the fine resolution of time and space now used in global weather forecasting. It will usually be better to limit the interactions between the atmosphere and land, sea, and ice to the time scales of the particular investigation. For example, the behaviour of the N Atlantic Ocean surface layers and sea ice on a time scale of ~10 yr (and their interactions with the atmosphere) may serve to model the equatorward displacement of the Gulf Stream and N Atlantic Drift by the melting of the North American ice cap and the Younger Dryas stadial (above). Whereas to model the long descent from the last warm period to the present ice age would require empirical representation of the movements of continents and ocean basins, the behaviour of the full depth of the oceans on a time scale of ~ 10 Myr, the development of the Antarctic ice cap and the removal of CO_2 by rainfall and weathering of the rising Himalayas.

At any particular stage in the long term, for example the Permian/Triassic (about 270 to 220 Myr BP) when all the continents were clustered in onc supercontinent (Pangaea) surrounded by a vast ocean (Panthalassa), and life was in its vigorous Palaeozoic stage, the dispositions of land and sea could be used as boundary conditions in which to run a fully detailed model atmosphere to see what its general circulation and climate might be like. This would need to be linked to a model for the circulation of the ocean, and some simple representation for the polar ice caps. If there was special interest in 251 Myr BP, and the end-Permian mass extinction of the biosphere (the greatest of all the mass extinction events known so far, which eliminated nearly all marine life *[64]*, then the huge emissions of CO_2 and SO_2, etc. from the Siberian Traps could be added, with representation of the strong global warming and the complex atmospheric and marine chemistry which proved so lethal to life.

14.9 Current and future climate change

The present (Holocene) interglacial was finally established about 11,000 years ago, and has since provided what we now realize is an unusually benign and stable climate for the agricultural revolution and the subsequent rise of global human society. But although climate has been much steadier than in the preceding glacial, and in the tumultuous transitions through the Little Dryas to the Holocene, the current interglacial has not been absolutely steady. Its opening millennia were globally wetter and at least 1 °C warmer than now (Fig. 14.15a), and there have been several markedly cool periods since, the last of which being the

Figure 14.15a In a mountainous region of SE Algeria, now too dry to support serious agriculture, cave drawings show that very different conditions prevailed about 5,500 years ago—6,000 years after the beginning of the Holocene interglacial. The lively cartoons show scenes of herding and hunting animals which have been extinct in that region for millennia, because of reduced rainfall. The arabic name for the area (Plateau of the Rivers) suggests that it remained wetter into very early historical times.

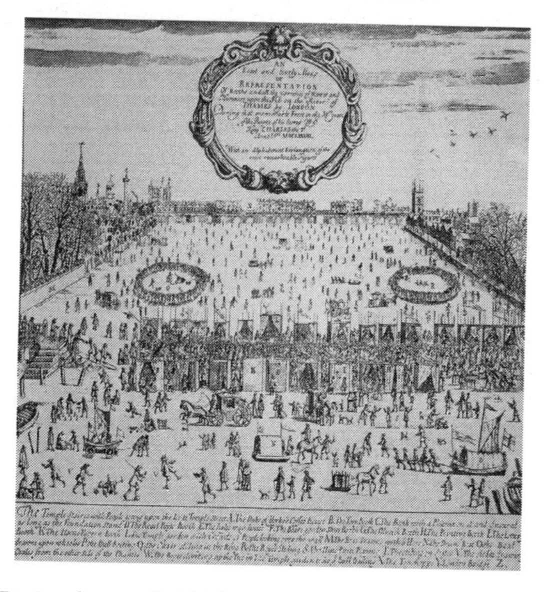

Figure 14.15b During the so-called Little Ice Age, the Thames in London froze right across in 24 winters between 1400 and 1814. This contemporary print shows the most famous (1683) of the resulting 'Frost Fairs', with an icy roadway lined by booths carrying thousands of Londoners across the river just upstream of the old London Bridge. The diarist John Evelyn records a darker side with meteorological perception 'by reason for the excessive coldness of the air hindering the ascent of the smoke, was so filled with the fuliginous steam of the sea-coal ...that one could hardly breath'. (See Plate 17).

Little Ice Age (Section 14.4) which bottomed around 1700 CE and had a last cold kick in the early 1800s, assisted by the Pinatubo eruption. Current understanding of these relatively small climatic variations is probably less secure than it is for the larger ones in the current ice age, since they are potentially attributable to many causal factors, but have had substantial effects on the biosphere and human living, and we must assume that they will continue for as long as the Holocene continues.

Our lack of understanding of current small amplitude climate change is unfortunate in principle, and very unfortunate in practice, because it seriously frustrates our ability to assess the importance of the really new addition to the list of factors driving climate change—the injection of CO_2 by developing human industry since about CE 1700 (Box 3.4). Though humanity has been affecting the atmospheric carbon cycle for several millennia, through deforestation and agriculture, results seem to have been buried or balanced by natural variations for most of that time, leaving atmospheric concentrations of CO_2 remarkably steady up to about CE 1700, when the effects of industrial production of CO_2 became apparent in the air bubbles trapped in contemporary ice cores from polar ice caps. Since then the smooth but well-marked increase out of the steady background has become an exponential eruption which will lead to a doubling of pre-industrial concentrations by CE 2050 whether or not we manage to begin to check the relentless increase in its industrial production.

Concern about the current industrial spike of CO_2 and other greenhouse gases, arises because of their involvement in the mechanics of Earth's greenhouse, variations of which have long been recognized as an important agent in climate change. The greenhouse role of CO_2 in particular has attracted attention since it was recognized by Arrhenius in the late nineteenth century; and the anthropogenic growth of CO_2 has been causing increasing concern since the 1950s (Box 3.4). There is now nearly universal expert agreement that anthropogenic CO_2 production is driving significant global warming by increasing Earth's greenhouse. As part of their continual monitoring of such anthropogenic climate change, climatologists in the Intergovernmental Panel on Climate Change (IPCC) of the World Meteorological Organization, make and remake detailed estimates of past, present, and future concentrations of greenhouse components and their greenhouse effectiveness *[70]*.

The implication that global industrialization, as it has developed in the last 300 years, may not be sustainable in its present form, challenges deeply ingrained assumptions about the human birthright, and excites disbelief among national and international commercial and political groups who see a promising future compromised. The ideal of robust, objective scientific argument between expert peers is not always appreciated by the general public, who until recently have been badly confused by inevitable controversy about major and minor technicalities, especially when highlighted by media which sometimes prize the entertainment value of public argument above a clear summary of the arguments themselves, including publicising the vital distinctions between consensus, majority, minority, and maverick expert views.

14.9.1 Radiative forcing

The conventional way of assessing the greenhouse effectiveness of atmospheric components (natural or anthropogenic), is to calculate their *radiative forcing* of the terrestrial energy balance. In essence, this takes the distribution and radiative behaviour of greenhouse components, and calculates the contribution they make to the atmosphere's opaqueness to terrestrial radiation (*TR*). Technically, radiative forcing is defined to be the reduction in net upward *TR* flux at the tropopause

Table 14.7 Change in radiative forcing from 1765 assuming the following IPCC scenarios for industrial and population growth *[33]*: (a) business as usual; (b) accelerated control policies (CO_2 emission half of 1985 value, stringent controls in developed countries, moderate controls elsewhere)

Period (CE)		CO_2	CH_4	N_2O	CFC_s	Total	(all W m^{-2})
1765–2000	(a)	1.85	0.69	0.12	0.29	2.95	—
	(b)	1.75	0.59	0.11	0.29	2.74	—
1765–2065	(a)	5.49	1.37	0.40	1.03	8.28	—
	(b)	2.77	0.52	0.24	0.68	4.22	—

(i.e. above most of the atmosphere which interacts with *TR*) produced by the components in question, expressed as a global average in W m^{-2}. For example, recent IPCC estimates suggest that by CE 2065, the total radiative forcing by CO_2 added by human industry since its modern phase began in 1765 will lie between about 2.8 and 5.5 W m^{-2} depending on whether or not production by combustion is curbed in the remaining decades. Even the lower value (2.8 W m^{-2}) is very significant, being over 1% of the current global average *TR* output from planet Earth (242 W m^{-2}—NN 8.2). The curiously precise starting date for industrial CO_2 production is bureaucratic in origin, and his little significance, since early industrial CO_2 production was very slow compared with the last 100 years.

It is important to realize that the calculation of radiative forcing deliberately assumes that almost everything about the atmosphere remains unchanged *except* for the calculated increase in opaqueness to *TR*. By definition, positive radiative forcing implies that the atmosphere is being warmed by net radiative gain, since the global average *SR* input is assumed to remain at 242 W m^{-2}, whereas *TR* output has fallen below this by the amount of the positive radiative forcing. Negative forcing implies net radiative loss and cooling. Of course the magnitude and distribution of actual resulting warming will depend sensitively on how the many layers of the troposphere, and underlying surfaces, adjust to the net energy input or output caused by the forcing, and these matters are complex and still incompletely understood (see below). Basically, radiative forcing provides the input for detailed models of climate change, but deliberately avoids second-guessing their output.

Table 14.7 presents recent estimates of radiative forcing by the main industrial and agricultural components for two overlapping periods, the first estimating the accumulated forcings at CE 2000, and the second projecting them into the middle of the current century, when atmospheric CO_2 is expected to have doubled its pre-industrial level. The results show very significant forcing by CO_2, with important contributions from several other greenhouse gases, all of which are more greenhouse effective than CO_2, molecule for molecule, but less effective in resultant effect because they are so much less abundant. For this reason it actually makes more sense to burn CH_4 from decaying rubbish dumps, converting it to CO_2 and H_2O before venting, and ideally producing useful heat or electrical power in the process. Some CFCs are greenhouse effective out of all proportion to their small but long-lived populations, in addition to being able to destroy stratospheric ozone (Section 3.3).

Calculations suggest that increases in total forcing since pre-industrial times could exceed 8 W m^{-2} by the middle of the present century, if human population and industry continue adding to Earth's greenhouse at current rates. At over 3% of the global average *TR* output of 242 W m^{-2}, this is a huge deviation from the pre-industrial balance.

14.9.2 Global response

Qualitative

Radiative forcing of Earth's atmosphere and surface will drive global warming until radiative equilibrium is re-established, but the degree and distribution of that warming is likely to be sensitive to interactions between and within the atmosphere, oceans, and surface ice, many of which are still quite poorly understood. The presence of positive feedback in such reactions will amplify the forcing effect and tend to make the Earth run away from the original state to something very different. By contrast, reactions with negative feedback will oppose the forcing, and tend to lead to a new balance not far from the original state. Overall radiative equilibrium of the planet enforces overall negative feedback (Section 8.2), by ensuring that *TR* output rises to check planetary warming, or falls to check planetary cooling. But within that overall constraint, powerful positive feedbacks can be at work, for example driving the planet Venus into its permanent hot condition, where it is maintained by negative feedback, and driving the cooling Earth to a Snowball state where it was maintained by negative feedback for millions of years until CO_2-driven warming reactivated the positive feedback in the opposite direction (Section 14.7).

The most important greenhouse gas (water vapour) must tend to drive positive feedback by increasing the greenhouse warming as air warms and its saturation limit rises. If the air remained saturated everywhere, the increase would be easily calculable, but saturation is always confined to patchy cloudy updrafts. And if global cloud cover increases with vapour content (as seems inevitable), the planetary albedo will increase, checking the warming by reducing solar input (negative feedback), while at the same time increasing the cloudy contribution to the greenhouse effect (positive feedback). Without much more detailed understanding, we cannot even be sure of the sign of the net effect of increasing cloud and therefore whether it will assist or oppose the effect of increasing water vapour. Many other factors involve positive feedback, including some operating on longer time scales. For example, reduction in seasonal snow and ice in higher latitudes increases solar absorption there, driving global and local warming (Section 14.7.2). And ocean warming exsolves dissolved CO_2 back into the atmosphere, further increasing the greenhouse effect and warming the ocean even more. Estimates suggest that a third or more of CO_2 produced by industry in recent centuries has gone into the oceans, some of it liable to exsolve as they warm, to produce lagged positive feedback—the most difficult of all response types to deal with until our understanding is more robust, and an example of a possible 'tipping point' in recent popular discussion.

And all these things are happening in parallel with continuing natural climate change, whose full understanding is still very incomplete, despite many recent advances. Given emerging evidence of very fast climate change both now and 11,000 years ago (during the ending of the Younger Dryas), humankind is facing a still poorly-defined but probably substantial threat to its future development and well-being, with political, social, and economic consequences that have become subjects of continual and sometimes heated debate in recent years.

Quantitative

Despite these and other complexities, we need to have some feeling for the realities of atmospheric response to anthropogenic radiative forcing, however rudimentary. Suppose that a positive radiative forcing is suddenly imposed on Earth's atmosphere. Consider the response in two general stages: initial warming and eventual result.

(i) *Initial warming* is potentially very fast because the atmosphere's heat capacity (including land surfaces) is so small. We see this every year in middle latitudes as the atmosphere responds to the very large radiative forcings imposed by the migration of the Sun in spring and autumn—Fig. 4.9 shows responses to short-term forcings of order tens of W m^{-2} (Fig. 8.11). We must also consider responses to much more persistent forcings.

If an extra radiative forcing of 8 W m^{-2} (as suggested by Table 14.7) is suddenly imposed, the atmosphere on its own would begin to warm at about 2°C per month (NN 14.5). As a global value this is unrealistically fast, because we have not included the heat capacity of ocean surface layers, which we know are responsible for the difference between maritime and continental climates. If we include the heat capacity of the top 100 m of the oceans, the part which is to be in fairly direct contact with the atmosphere, we multiply the heat capacity of the responding mass about 30-fold and reduce the initial warming rate to 0.07 °C per month (over 0.8 °C per year). Though still much faster than the very rapid warming at the end of the Younger Dryas, this is now slow enough to be controllable by at least some of the slower feedback reactions triggered by initial warming.

This raises an important point: the various responses of the system have intrinsic response times which ensure that the overall response can be very different depending on the speed of the imposed change, which our instantaneous forcing assumed to be infinitely fast. In fact by CE 2050 the 8 W m^{-2} forcing will have been reached by continuous increases spread over 350 years (though concentrated in the last 100 years). We see from the short response time for the atmosphere and surface ocean that they will in fact adjust quickly to incremental forcing, rather than let it build up before it triggers a seismic shift like the one assumed at the start of the present discussion.

NUMERICAL NOTE 14.5 Initial warming by radiative forcing

Sudden imposition of 8 W m^{-2} radiative forcing, reduces Earth's *TR* output to space by this amount, while *SR* input remains fixed, moving the previous radiative balance to a net gain dQ/dt of 8 W m^{-2}. Given an atmospheric column's heat capacity of $M\,C_p$ per unit area (10,000 × 1,000 = 10 MJ K^{-1} m^{-2}) we have $dQ/dt = M\ C_p\ dT/dt$, so that $dT/dt = dQ/dt/(M\ C_p) = 8/(10{,}000{,}000) = 8 \times 10^{-7}$ °C s^{-1} or 25.2 °C per year or 2.1 °C per month.

Adding the heat capacity of 100 m depth of water covering 70% of the Earth increases the total heat capacity of Earth's surface water and atmosphere from 10 to about 310 MJ K^{-1} m^{-2} (NN 14.4) and reduces its annual rate of warming to $25.2 \times 10/310 \approx 0.8$ °C.

Add 1 km depth of water (instead of 100 m) and the same recalculation raises the atmosphere + ocean heat capacity to 2,950 MJ m^{-2} and reduces the warming to 8.5 °C. in 100 years.

We can also see that the initial response will change depending on how long we take the 'initial' period to be. If we take it to be as long as 100 years, then we need to allow for the warming to reach much further into the ocean than the assumed top 100 m, since we know from the residence time for water in the oceans (Section 3.8) that the whole 3 km depth is reached by stirring in about 3,000 years. Since 100 m of ocean water are involved in seasonal temperature swings, we might guess that the response in a century could involve the top kilometre of ocean, decreasing the response to 8.5 °C per century (NN 14.5), which is less than an order of magnitude faster than the fastest warming rates now suggested observationally for the end of the Younger Dryas. In fact we need to incorporate the incremental involvement of greater and greater ocean depths with increasing time, and something similar will have to be done for all the other components of the response of atmosphere, oceans, and ice (for example, the melting of ice caps and the response of the carbon cycle to changes in land and sea temperatures), and the biosphere. It appears that even the initial response requires careful, detailed analysis of many factors.

(ii) *Eventual result* To raise some issues involved in assessing the ultimate change (the new equilibrium) accomplished by the assumed radiative forcing of 8 W m^{-2}, consider now a deliberately naive empirical adjustment dominated but not suppressed by net negative feedback.

Suppose that the Earth is in radiative equilibrium with *SR* input of 242 W m^{-2} balanced by an identical *TR* output from the top of the *TR*-opaque troposphere, which emits like a blackbody at temperature T_E, the equivalent blackbody temperature of the Earth (256 K). We are using the Initial model of Section 8.4 with realistic values.

We know that the average temperature of the base of the troposphere is 288 K, which corresponds to a blackbody emittance of 390 W m^{-2} (NN 8.3). This drives the upward *TR* cascade through the atmospheric greenhouse, maintaining the terrestrial output of 242 W m^{-2} from the high troposphere to space. The ratio of upward *TR* fluxes at the bottom and top of the troposphere is 390/242 (= 1.612), a measure of the 'gearing' of the greenhouse effect in this view.

If a positive radiative forcing of 8 W m^{-2} is now suddenly introduced, the radiative output from the top of the troposphere drops to 234 W m^{-2}, and the greenhouse gearing is increased to 390/234 (= 1.666). If this increased greenhouse gearing ratio is maintained while the whole troposphere warms by radiative forcing, we can use it to find the raised surface emittance which will eventually drive 242 W m^{-2} from the top of the troposphere, restoring radiative equilibrium with *SR* input (which we assume simplistically to be unchanged by any change in planetary albedo). That new surface emittance is 403.2 W m^{-2} (simply 1.666 × 242), which corresponds to a blackbody at about 290.4 K, according to Stefan's law.

This blindly empirical model suggests that the increased atmospheric greenhouse implied by a radiative forcing of 8 W m^{-2} could drive a global surface warming of about 2.5 degrees if the ratio of emittances is held at 1.666—a warming which would lift the planet significantly from its current interglacial state towards an ending of the current ice age. Of course the working assumption is simplistic, but it provides this discussion with a context in which to consider the positive feedbacks which might amplify such warming, and the negative feedbacks which might reduce it. Of these, the most difficult outcomes to foresee and manage are lagged reactions with strong positive feedback, so that warming remains moderate, apparently benign, until some 'tipping point' is passed, after which it becomes much more difficult to manage. As mentioned above, the oceanic branch of the CO_2 cycle may well have that capacity.

The major task of current modelling of climate change is to investigate and evaluate the range of possible reactions and interactions arising from the radiative forcing of Earth's greenhouse currently being imposed by large and increasing anthropogenic emissions. It is a daunting task which will require the concerted effort and skill of many scientists, but one which has been well begun in recent decades. Perhaps even more importantly, it requires the sympathy and readiness of the great mass of humankind to look beyond the comfort zones of the social, political, and economic norms established in the first stages of the ongoing global industrial revolution, now ending in some confusion.

14.9.3 Localized effects

Whatever the magnitude of the global climate response, some localized responses will be much larger, and others smaller. Current models of the changing general circulation during global warming tend to show maximum warming at high latitudes, as poleward heat fluxes in mid-latitude weather systems (Section 8.10) increase in the warmer, moister atmosphere—though equatorward diversion of the Gulf Stream and North Atlantic Drift could in principle offset this as it did in the prelude to the Holocene (Section 14.7). Such warming will melt polar ice caps and raise global sea levels, already rising through the thermal expansion of the warming seas. Magnitudes vary greatly with the climatological model used, and local coastal effects are sensitive to continuing local adjustments of land height: some land masses, like Scandinavia, are still rising so rapidly after losing the burden of their recent ice cap, that sea levels along their coasts will continue to fall for some time, even though the global sea level is rising.

Pronounced meridional variations of climate change arises also from the shifting of the great climate zones (Sections 14.2 and 14.3). There is clear evidence that during the most recent glacial, global cooling and marked cooling in middle and high latitudes (Fig. 14.9) were associated with substantial poleward movement and expansion of the subtropical arid zones associated with the descending branch of the Hadley circulation (Fig. 8.21). For example, ice cores from the Greenland ice cap show that desert dust was carried there in much larger quantities than is now the case. It is probable that further warming would reduce the subtropical arid zones still further, as it apparently did in the early Holocene (Fig. 14.14), which would help those currently struggling in marginal conditions in North Africa, for example. However, the sheer complexity of the atmospheric, oceanic, and ice cap systems, and their interactions, make such generalizations dangerous, unless and until substantiated by detailed modelling. And the potential presence of thresholds or tipping points arising from highly nonlinear processes warns against assuming that any smooth trend can be safely extrapolated. Again the Younger Dryas is an impressive example.

14.9.4 Social consequences

Human sensitivity to climate change can hardly be overemphasized. A general rise in sea level will threaten low-lying coasts with increased flooding, extending on a smaller scale the huge inundations which drowned the North Sea and Hudson

Figure 14.16a Two miles away from the spectacular waves of Fig 12.13, the waves and storm surge of the same violent N'ly gale caused extensive damage and flooding in Morecambe, and demolished part of this apparently solid sea wall.

Figure 14.16b About ten years later the vulnerable sea walls had been replaced by modern 'rock armour' which breaks the waves progressively rather than at a single wall.

Bay as the great ice caps melted after the last glaciation. Even quite small rises ($\sim$ 10 cm) will force vulnerable populations to improve their coastal defences to avoid increasing disruption by winter storms (Figs. 14.15 and 12.13). The human population is currently growing so rapidly that food production is already inadequate in many areas, and though modern agriculture has helped greatly in some regions, production limits imposed by land availability and social efficiency may well be reached over much of the planet within the next century. Imagine the possible effects of rapid, large-scale climatic change on this tense situation. Established agriculturally productive zones, like the North American wheat granary, could be seriously degraded, and social confusion might well seriously delay realizing the increased potential of other zones. Recent experience in marginal lands like the African Sahel shows how easily climatic deterioration can combine with social change to produce devastation: wet decades in the mid-twentieth century encouraged rapid growth in human and animal populations there, so that when sporadic long droughts returned, whole populations were forced to mortgage the longer-term future of their agriculture for the sake of short-term survival, leaving them increasingly vulnerable to further climatic variations.

Time is the crucial factor in all this. If climate change occurs over many human generations—for example 2,000 years for a major warming or descent towards another glaciation—then adjustment might hardly be noticed amidst all the other changes we bring on ourselves. If, however, there is substantial climate change in a period as short as one or two centuries, as seems to have happened in the final onset of the Holocene, then the adjustment of a crowded, anxious world will almost certainly be painful.

14.9.5 Epilogue

From earliest times people have looked at the sky and marvelled and wondered. The quotation in the flyleaf of this book has been sung in solemn acts of worship since before 500 BCE, when the Holocene had been established for about 9,000 years, and modern human society had been developing for nearly as long. Its celebration of the colour, power, and beauty of the heavens is timeless in its appeal, as is its very human drive to see a purpose and mechanism in the greater world they represent. Similar testimony comes from all cultures then and since. Such delight, curiosity, and yearning, when married with patient observation and classical Greek intellectual rigour, and cherished and developed by Islamic and then Christian cultures through the supposedly dark ages, led to the recognition of the organized simplicity of the movements of the planets round the Sun by Copernicus, Galileo, and Kepler, and thence via Newton to the elegant and comprehensive dynamical relationships sketched and applied in the heart of this book (Chapters 7, 8, and 9). Less than four centuries later, the natural sciences have developed into a huge, expanding, and intricate tapestry of concepts, observations, and techniques which allow each generation to find more and more to enjoy and explore in the world which it is their privilege to inherit and their responsibility to bequeath.

Because of its complexity, the science of the atmosphere has matured more slowly than was expected by Newton, Boyle, and other pioneers of modern science, but in the last 150 years it has come of age in various ways, some of which are sketched in this book. Out of the many stages of that progress I choose five which seem to me to have been especially important.

(1) The development of the electric telegraph by Morse and others around 1840 allowed synoptic observations to be collated and analysed in nearly real time, opening our eyes to all the larger scales of atmospheric structure and behaviour, and gradually to the grandeur and subtlety of their organization.

(2) The sustained application of the basics of fluid dynamics and thermodynamics to the understanding of atmospheric mechanism by the Norwegian school in the very early twentieth century lifted meteorology out of a period of overdependence on empirical observation, and set it on a path of sustained and rapid development which has not faltered since.

(3) The application from the 1950s onwards of electronic computers produced numerical solutions to the network of physical and dynamical equations lacing meteorology and climatology. This quickly freed modellers from the strait-jacket of analytical (algebraic) solutions, revealing a richness and creativity in atmospheric behaviour hitherto unrecognized by theoreticians, as well as underpinning most subsequent modelling and forecasting of meteorological and climatological behaviour.

(4) The development of satellite observations of the atmosphere and surface from about 1960, has confirmed the findings of generations of weather chart analysis, has added meso- and small-scale behaviour, displaying the whole to the human eye and brain in a way previously limited to individual observation of micro- and small-scale weather. The sheer awe and delight we experience in perusing such panoramas may well prove to be as creative as was, in the long run, the delight of the psalm singers in the clouds and stars which peopled their more limited heavens.

(5) The emergence of climatology in the last half century from its previous overshadowing by meteorology, and the associated explosive growth of evidence of past weather and climate, are currently revealing an astonishingly rich and vigorous climatic history. Underpinning all this is the vast time scale of the Earth, first sensed by geologists in the eighteenth century, and pursued and clarified in the teeth of scientific (end of Box 1.6) and antiscientific dogmatism.

The current growing awareness of past and present climate owes a huge debt of gratitude to the many people who pioneered work in this field before it became popular. I was privileged to meet one such person in my early years as a faculty member of the Department of Environmental Sciences at Lancaster University. Gordon Manley had retired as its first head by then, but returned occasionally to share his continuing enthusiasm for the Central England Temperature series (Section 14.4), and indeed for any aspect of British or world climate, with anyone who would listen. Despite his age, his delight was as fresh and articulate as that of the psalm singers of ancient times, and so I listened.

Our atmospheric home quite naturally excites our wonder and curiosity; to cherish it for our successors will require at least as much enthusiasm, patience, intellect, and wisdom as has been shown in the past—perhaps even more.

Checklist of key ideas

You should now be familiar with the following ideas.

1. The climatic potential of the geometry of Earth's shape, spin, and orbit.

2. Maritime and continental climates of middle latitudes as exemplified by North America and Eurasia, the North Atlantic Oscillations.

3. Low-latitude climates, the Hadley circulation, El Niño Southern Oscillation.

4. The unsteadiness of climate between years, decades, and centuries, and the Central England Temperature Series.

5. Climate change in history, prehistory, and geological epochs, the Holocene, the current ice age, ice age terminology, warm epochs, Snowball Earth.

6. External mechanisms of climate change: the Sun, Earth's orbit and spin, and their combinations in equinox precession, orbital and tilt seasons, Croll–Milankovitch triggering in the current ice age, bolides and their effects.

7. Internal mechanisms of climate change: continental and oceanic redistribution in the last 100 million years, albedo and glaciation, effects on global balance and seasons, onset and ending of a snowball event, volcanism, nuclear winter, oceans as heat reservoirs and heat transporters, possible role in the Younger Dryas, Greenhouse variations.

8. Climate modelling past and present, model types, models and time scales, radiative forcing as a way of gauging input and possible response, time scales of response, positive and negative feedback, initial and ultimate response, social implications and complications.

9. Landmarks on the road to modern meteorology and climatology.

Problems

Outline answers to these problems can be found on the **Online Resource Centre.** Answers to odd numbered problems can be found under Student Resources, answers to even numbered problems under Lecturer Resources.

Level 1

14.1–14.8 Write brief notes on any one of the following factors affecting climate: 1) continental and maritime influences in middle and high latitudes; 2) continental and maritime influences in low latitudes; 3) presence/absence of mountain barriers; 4) latitude; 5) teleconnections; 6) Hadley circulation; 7) low-latitude weather systems; 8) mid-latitude weather systems.

14.9 Briefly list the dominant climatic variations in two of the following time windows: 0–1 kyr BP; 0–150 kyr BP; 0–500 Myr BP.

14.10 Briefly list the dominant climatic variations in two of the following time windows: 0–15 kyr BP; 0–5 Myr BP; 0–4 1000 Myr BP.

14.11 Briefly define the terms ice age, glacial, interglacial, stadial, interstadial.

14.12 Describe and distinguish between ice caps, glaciers, icebergs, and sea ice.

14.13 Justify the statement that polar glaciation reduces summer temperatures more than it does winter temperatures.

14.14 Place in order of increasing difficulty (for human management), and briefly describe, the following types of feedback in the climate system: positive feedback, delayed positive feedback, negative feedback.

14.15 Why was there such initial scepticism about the suggestion that there might have been snowball conditions in the past? Briefly, how was this overcome?

14.16 At the time of final preparation of this text (June 2009), it was discovered that woolly mammoths had lingered in central England until 14,000 years BP, rather than 20,000 as previously thought. What might have killed them at this later date?

14.17 Defend the suggestion that for the conceivable future, no single model of our dynamic climate will be adequate.

Level 2

14.18 Consider the albedo of Earth's disc facing the Sun. Assuming the simple model of equally glaciated polar caps, what is the current seasonal rhythm of disc albedo? At times during the present ice age, orbital seasons have been much bigger. How will this affect the (tilt) seasonality of disc albedo, and what is the most effective combination of seasons for producing the warm summers necessary for ending a glacial? If we now allow the N hemisphere to be more glaciated than the S (as in recent glacials), what is the best combination for ending a glacial?

14.19 Estimate the rise in global sea level if all the Antarctic ice cap melts. By inspection the ice cap covers 66% of the area bounded by the Antarctic Circle, and the latter contains 4.15% of the Earth's surface area (Box 14.3). Assuming that the ice cap has uniform depth 2 km, find the volume of ice in the cap, the volume of water it would produce on melting (ice density is 0.92 water density), and the depth of this amount of water when spread across the 70% of the Earth covered by the world ocean.

14.20 Take the Chicxulub meteorite as described in NN 14.2, and assume that 10% of its KE at impact is used to heat and boil some of the sea water in the Gulf of Mexico. Calculate the mass of water which can be warmed from 25 to 100 °C and then boiled away and compare this with the 2.5×10^{18} kg of water in the Gulf. This is a much more realistic outcome than the one envisaged in NN 14.2, and probably occurs in the first few seconds after impact.

14.21 Distinguish clearly between 'tilt seasons' and 'orbital seasons' and qualitatively describe how they move in and out of step to produce a major component of the Croll–Milankovitch mechanics. What other aspects of Earth's spin and orbit contribute to Croll–Milankovitch?

Level 3

14.22 Photographs like Fig. 10.25 suggest that heat loss by open water in high latitudes is very much greater than heat loss by sea ice. Investigate this by considering the rate of heat loss through 30 cm of sea ice floating on water at 0 °C and overlain by air at −10 °C. The thermal conductivity of ice is 2 W m^{-1} K^{-1} (Box 8.3). Compare this with the rate of loss over open water according to Eqn 10.27, supposing that the 10 °C temperature difference is spread through a vertical metre of air and that the turbulence (visible in Fig. 10.25) is associated with a friction velocity of 0.3 m s^{-1}. In fact this models only one type of heat transport at work. What is the other, and how might it be affected by the presence or absence of ice?

14.23 Suppose that some tipping point in Earth's composite reaction to human activity was passed and led to a step of 2 W m^{-2} in radiative forcing, find the speed of thermal response of (a) the atmosphere alone, and (b) the atmosphere and the top 100 m of the oceans, supposing the warming was uniformly spread in each case and there was no further feedback. If the tipping actually took 10 years (still almost apocalyptically fast) how might the reaction change? Speculate broadly on how the real reaction to such forcing might differ.

14.24 Consider in qualitative detail the following systems from the point of view of their positive or negative feedback in global warming: global radiative equilibrium; cloud albedo; cloud greenhouse; water vapour; polar ice caps; CO_2 in the oceans.

Discussion topics

Even though many aspects of current climate change are uncertain, it is useful to speculate about many of them, but only if the specualtion is critical, balanced, and 'cool' (not driven by presupposiition, political or commercial agendas or overheated rhetoric), and is completely honest about the levels of uncertainty involved. A possible approach is to have two members of the class prepare contrasting briefing notes on the topic, and use these to frame a class debate. Examples could range from the technical (e.g. distinguishing between natural and anthropogenic climate change) to the party social (e.g. how to prepare people for socially significant but uncertain climate change).

Glossary

The Glossary lists terms which are sufficiently important, widespread, or peculiar to warrant separate summary definition. Many terms used in the definitions are themselves listed elsewhere in the Glossary to encourage you to refer to several entries to clarify any important definition. Except for very basic terms whose meaning is assumed, all technical terms can be traced through the Index to their definition in the main text, whether or not they appear in the Glossary.

absolute acceleration Acceleration measured relative to an unaccelerated (inertial) reference frame.

absolute temperature Temperature measured on the absolute or kelvin scale.

absolute vorticity Vorticity measured relative to a non-rotating reference frame.

absorptivity The proportion of incident electromagnetic radiant power absorbed by a surface, expressed as a fraction or percentage.

acceleration Rate of change of velocity with time.

accretion Growth of a body of water or ice by the collection of smaller particles or droplets.

adiabatic A thermal process is said to be adiabatic if it is *adiathermal* and *reversible*. In brief definitions reversibility is often ignored.

adiathermal In thermal isolation. Many short-term air parcel processes are nearly adiathermal.

advection Horizontal transport (of mass, heat, etc.) effected by horizontal exchange of air or water.

advection fog Fog formed by advection of warm air over a colder surface.

aerodynamic roughness length The height above the impermeable base of a rough surface at which wind speed falls to zero by extrapolating a log wind profile down through the overlying surface boundary layer.

aerological diagram *Thermodynamic diagram* for displaying the state of the atmosphere in depth, for example a *tephigram*.

aerosol Suspension of very small (maximum diameters of order 1 μm) particles or droplets in the atmosphere, especially concentrated in the low troposphere.

air The gaseous component of the turbosphere.

air frost Air temperature at or below 0°C.

air mass A synoptic-scale segment of troposphere whose temperature and humidity are usefully related to a geographical source region. Usefulness is often marginal in the presence of typically strong wind shears in the vertical.

air parcel A small body of air which, at least in principle, is coherent and identifiable throughout a useful period, for example a buoyant thermal in a cumulus cloud.

Aitken nuclei Smallest and most numerous of *condensation nuclei*.

albedo The proportion of incident electromagnetic radiation (usually solar) reflected or backscattered by a surface, expressed as a fraction or percentage.

alto Prefix to cloud type, meaning 'in the middle troposphere'; as in altocumulus.

altostratus Sheet cloud in the middle troposphere (typical of fronts).

anabatic flow or winds Air flow up sloping terrain.

anaerobic Without oxygen, as in some isolated water deeps.

aneroid Largely without air, as in the springy capsule of an aneroid barometer.

anemometer Instrument for measuring wind speed or run of wind.

angular momentum or moment of momentum The product of the tangential momentum of a body and its radial distance from the axis of rotation.

angular velocity The rate of turning of a body about an axis expressed in angle turned per unit time (usually radians per second).

anthropogenic Resulting from human activity.

anticyclone Synoptic-scale weather system with winds blowing anticyclonically (clockwise in the northern hemisphere) in the low troposphere.

anvil A fan of ice cloud at the top of a cumulonimbus.

aphelion Position or time of maximum separation of planet and Sun.

apparent *g* Gravitational acceleration measured relative to a fixed point on the Earth's surface.

Arctic sea smoke Steam fog produced when very cold air from polar land masses flows onto the sea.

atmospheric boundary layer or planetary boundary layer The part of the low troposphere (~ 500 m deep) most directly influenced by the underlying sea or land surface, including the laminar, surface, and Ekman boundary layers.

atmospheric waves Horizontal or vertical undulations of air flow on scales from the metric to the hemispheric, sometimes in short trains and sometimes solitary (e.g. see *cyclone* and *gravity waves*).

available potential energy That part of the atmospheric total potential energy which could be converted to kinetic energy by realistic redistribution of air parcels.

azimuth Wind direction specified by the angle in degrees from which the wind is blowing, counting clockwise from zero at true North.

backing Wind direction changing so that azimuth is decreasing.

baroclinic The condition of the atmosphere when large-scale isopycnic and isobaric surfaces are misaligned, usually in association with horizontal temperature gradients.

baroclinic instability Dynamic instability associated with baroclinic zones and their associated thermal winds, giving rise to cyclogenesis.

barograph A barometer with a graphical output.

barometer An instrument for measuring atmospheric pressure.

barotropic The condition of the atmosphere when it is not baroclinic.

Beaufort scale A scale of wind strength assessed by effects on surface features such as trees and sea state.

Bergeron–Findeisen A mechanism for the development of precipitation by preferred growth of ice crystals in clouds consisting largely of supercooled water.

biosphere The interacting network of organisms living on the Earth.

blackbody An ideal body which absorbs all incident electromagnetic radiation and emits it with maximum thermodynamic efficiency.

blackbody radiation Electromagnetic radiation emitted by an isothermal blackbody.

boundary layer A layer of fluid whose dynamic behaviour is directly influenced by an adjacent solid or liquid surface. Meteorological examples are the atmospheric boundary layer and the sheath of air enveloping a falling raindrop.

Bowen ratio The vertical flux of sensible heat expressed as a fraction or percentage of the latent heat flux at a particular site or across a geographical region.

Brunt–Väisälä frequency The frequency of vertical oscillation of an air parcel released after displacement from its equilibrium position in a convectively stable fluid layer.

buoyancy The net upward force experienced by a body immersed in a fluid of different density in the presence of a hydrostatic pressure gradient.

Buys-Ballot's law The law defining the relationship between the directions of horizontal pressure gradient and associated geostrophic wind.

cellular convection Convection organized in quasi-regular cells, as in stratocumulus or the cellular clustering of shower clouds.

Celsius scale The temperature scale in which the freezing and boiling points of pure water at standard atmospheric pressure are 0 and 100 units (°C) respectively.

centrifugal Away from the centre of a circle. Ignoring a centripetal acceleration incurs an apparent centrifugal force.

centripetal Towards the centre of a circle, as in centripetal acceleration.

chaos Apparently random behaviour of a deterministic system.

chinook A föhn wind near the Rocky Mountains.

cirrus A fibrous cloud of ice particles.

climate Weather conditions and their range typical of a region or site.

cloud A dense population of water droplets and/or ice crystals, each of which has diameter of order 10 μm.

cold front See *front*.

cloud seeding Encouragement of precipitation from clouds by freezing some supercooled cloud droplets and speeding the Bergeron–Findeisen mechanism.

collision and coalescence Precipitation development by accretion, in which larger particles or droplets grow as they fall through populations of smaller ones, collecting some by collision and subsequent coalescence.

condensation The growth of water or ice by net diffusion from contiguous vapour—the reverse of evaporation. In the case of ice growth, either process is often called sublimation.

condensation nuclei The components of an aerosol population which act as nuclei for the formation of individual cloud crystals or droplets in a realistically supersaturated vapour.

conditional instability Convective instability which arises when a previously unsaturated layer of air fills wholly or partly with cloud.

conduction of heat Heat transfer through spreading of enhanced thermal agitation by molecular impact.

continuity equation Formal description of the conservation of the mass of a body of moving air, relating the net mass flux into a fixed volume to the rate of accumulation of air mass there.

contour A height contour of an isobaric surface.

convection Usually the small-scale vertical exchange of air parcels, driven either by buoyancy (thermal convection—although density differences do not always depend entirely on temperature differences) or wind shear (mechanical convection). Thermal convection is usually assumed unless mechanical convection is specified. See *slope convection* for application to the large scale.

convective instability Tendency towards buoyant convection arising from disposition of atmospheric density.

convective stability Inhibition of buoyant convection arising from atmospheric density disposition.

convergence Negative divergence.

Coriolis acceleration Component of absolute acceleration arising from the motion of a body relative to the rotating terrestrial reference frame. In meteorology it usually means the component in the local horizontal arising from horizontal flow.

Coriolis parameter The factor $2\ \Omega \sin \phi$, where Ω is the Earth's angular velocity and ϕ is the angle of latitude.

covariance Average value of the product of deviations of matched pairs of values (of vertical and horizontal wind speeds, for example) from their respective mean values.

cryosphere Ice caps, pack ice, glaciers, permafrost—the frozen water substance at and below the Earth's surface.

cumulus The family of hill-shaped clouds ranging from small cumulus to large cumulonimbus. The term on its own usually denotes small cumulus or only slightly larger.

cumulus congestus A moderate or large cumulus cloud, growing conspicuously.

cumulonimbus A precipitating cumulus, often but not necessarily having an anvil.

cyclogenesis Cyclone formation, usually applied to extratropical cyclones.

cyclone Synoptic-scale weather system with winds in the low troposphere rotating about the local vertical in the same sense as the local terrestrial surface but faster (i.e. anticlockwise on a northern hemisphere map). Also a local name for a severe tropical cyclone in the Indian Ocean.

cyclone wave The wave-like deformation of flow in the middle and upper troposphere associated with an extratropical cyclone.

cyclonic shear Horizontal flow sheared in such a sense as to cause an embedded parcel to rotate cyclonically.

cyclostrophic flow Horizontal circular air motion in which centripetal acceleration on the weather map is effectively maintained by centripetal pressure gradient force, Coriolis and friction terms being negligible.

density Mass concentration expressed as mass per unit volume.

depression See *extratropical cyclone.*

detrainment Loss of mass from a rising thermal to surrounding air. The reverse of entrainment.

dew Water condensed onto a solid surface which has cooled to the dew-point of the contiguous air.

dew-point or dew-point temperature The temperature at which a given parcel of air becomes saturated during isobaric cooling with conservation of vapour content.

diabatic Not *adiabatic.*

diathermal Not *adiathermal.*

dimensions The dependence of a measurable quantity on the fundamental measurables—mass, length, time, and temperature. Thus the dimensions of velocity are length divided by time.

dimensionless number Any fundamental parameter (made up of a dimensionless assembly of observable factors) which usefully describes a regime of dynamic or thermodynamic behaviour, e.g. a Reynolds number.

diurnal With a daily cycle.

divergence (of velocity) The resultant rate of stretching of a fluid as given by the sum of the longitudinal gradients of the three components of flow velocity. In meteorology it usually refers in particular to the stretching components of air in the local horizontal.

drag coefficient The dimensionless ratio of the drag force on a body immersed in a fluid flow to the incident or adjacent fluid momentum flux.

drizzle Precipitation of water droplets with radii ~100 μm.

dry adiabat A line describing a dry adiabatic process on a thermodynamic diagram such as a tephigram.

dry adiabatic Describing an adiabatic movement or process of an air parcel in which there is neither evaporation nor condensation of the water substance—also called isentropic, because entropy is conserved.

dry adiabatic lapse rate Temperature lapse rate equal to the value found in dry adiabatic ascent or descent of an air parcel.

dry air Normal turbospheric air without water vapour.

dry-bulb thermometer Normal air *thermometer*, renamed to distinguish it from a wet-bulb thermometer.

dynamical coefficient of viscosity Tangential shearing stress arising from viscosity, divided by lateral shear (spatial velocity gradient). Also known as the Newtonian coefficient of the same.

dynamics Study of forces and associated motional deformation.

easterly Flowing or moving from the east.

easterly wave A synoptic-scale wave-like disturbance of the low tropospheric easterlies of low latitudes.

eddy Individual transient element of turbulence—often pictured as a translating and rotating air parcel.

eddy diffusion *Turbulent* mixing and spreading of heat, momentum, or material.

eddy viscosity Apparent viscosity arising from turbulence.

Ekman layer The topmost layer of the atmospheric boundary layer, between the free atmosphere and the surface boundary layer.

electromagnetic radiation Travelling waves of electric and magnetic disturbance, capable of passing through empty space. Of the full electromagnetic spectrum, solar and terrestrial radiation are the most important components in meteorology.

El Niño Long, temporary weakening of atmospheric subsidence of the E end of the Walker circulation associated with the failure of the cold southerly current there.

emissivity The actual emittance of a radiating surface expressed as a fraction of the blackbody value for the surface temperature.

emittance The flux of electromagnetic radiant energy emitted by unit area of a radiating surface.

energy The capacity to do work. See *kinetic*, *gravitational potential*, *radiant*, and *heat* energies for meteorologically important types.

ENSO El Niño Southern Oscillation: the *Walker circulation* cycle from *El Niño* to La Niña with associated *teleconnections*.

entrainment Incorporation of surrounding air by a rising thermal.

entrainment drag The effective drag on a thermal arising from continual sharing of momentum by entrainment of nearly static ambient air.

equation of motion Equation of relative acceleration with net force acting on an air parcel of unit mass—a form of Newton's second law of motion.

equation of state Equation relating pressure, density, and temperature of a parcel of an ideal gas or mixture of ideal gases.

equivalent potential temperature The potential temperature of an air parcel after all water vapour has been condensed by adiabatic decompression.

equinox A time when daylight and night are equally long.

evaporation Reverse of condensation.

exponential growth (or decay) Profile of growth (or decay) of a quantity whose gradient is directly proportional to the quantity itself.

extratropical cyclone A cyclone outside the tropics, distinguished from a tropical cyclone by greater scale, the presence of one or more fronts and the absence of great central intensity.

eye Relatively calm and cloud-free meso-scale centre of a severe *tropical cyclone*.

fall speed See *terminal velocity*.

fetch Distance upwind from point of observation to a significant location, as in 'standard synoptic anemometers should be sited to have an unobstructed fetch of at least 100 m'.

flux of energy, etc. The rate of flow of energy, etc. per unit time through a real or imaginary surface (normal to the flow unless otherwise specified).

flux density of energy, etc. The flux per unit area of real or imaginary surface.

fog Dense cloud in contact with a land or water surface, with density specified by visibility.

föhn A strong, dry, katabatic wind, produced by prior enforced ascent of air over high parts of the European Alps.

free atmosphere The atmosphere above the atmospheric boundary layer (i.e. above the gradient level).

freezing fog A fog of supercooled water droplets, freezing on impact with any solid surface.

freezing nuclei Those solid components of an aerosol population which act as nuclei for the freezing of supercooled cloud droplets.

frequency The number of events per unit time.

friction velocity A basic wind speed parameter defined by $\sqrt{T/\rho}$, where T is the turbulent horizontal wind stress in the surface boundary layer and ρ is the air density there.

front A swath of cloud and precipitation which is synoptic-scale in length and at least large-meso-scale in breadth, and is associated with a significant horizontal temperature gradient in the low troposphere of an extratropical cyclone. A front is called warm or cold depending on whether the warmer or colder air is advancing, and is called occluded when it connects the warm sector to a separated surface pressure minimum.

frontal zone A region of continually or seasonally preferred cyclogenesis. Also the meso-scale zone of maximum lateral horizontal temperature gradient associated with a front.

frost Deposition of ice on a land surface by diffusion and sublimation. When thick enough to produce marked whitening of vegetation (especially grass), it is called hoar frost.

gale A wind whose ten-minute average speed at height 10 equals at least 37 knots.

gas constant The constant in the equation of state for an ideal gas. In the universal form, a single universal gas constant applies to all gases. In the meteorological form, a different specific gas constant applies to each gas or uniform mixture.

general circulation Global air flow.

geostationary orbit An equatorial satellite orbit such that the satellite is stationary relative to the spinning Earth.

geostrophic wind A horizontal wind in which the Coriolis acceleration is exactly maintained by the horizontal pressure gradient force (or equivalently the Coriolis and pressure gradient forces exactly balance), i.e. when friction and acceleration relative to the Earth's surface are zero.

glacial Long cold spell in an *ice age*.

glaciation The process of conversion of supercooled water cloud into ice cloud.

gradient The spatial rate of change of an observable in the direction of maximum rate of increase (as in temperature gradient).

gradient level The lowest level in the troposphere which is so free of surface drag that actual and geostrophic winds are indistinguishable.

gradient wind A horizontal wind in which the combined Coriolis and centripetal accelerations relative to the Earth's surface, arising from cyclonic or anticyclonic flow, are exactly maintained by the horizontal pressure gradient force.

gravitational acceleration or apparent g The downward acceleration of a body falling freely *in vacuo*, as measured from a frame fixed to the Earth's surface. Equivalently it is the gravitational force per unit mass of the body, as measured from the same frame.

gravity waves Waves of fluid disturbance in which gravity is the predominant restoring force.

greenhouse effect The elevation of surface and low-troposphere temperatures which is maintained by the atmosphere's transparency to solar radiation and opaqueness to terrestrial radiation.

gust A short positive departure from the ten-minute average wind speed.

Hadley circulation The background vertical and meridional circulation of air in low latitudes, consisting of two opposing Hadley cells, each having air rising in the intertropical convergence zone and sinking in a subtropical anticyclone.

hail Millimetric or larger precipitation particle of ice, formed by accretion of ice crystals and rapidly freezing supercooled water droplets.

heat The energy of a material which is stored in the form of sensible heat (kinetic energy of rotational, vibrational and translational thermal motion of constituent atoms and molecules) and/or latent heat.

heat capacity The amount of heat required to raise the temperature of a body by one degree kelvin. In air or any other highly compressible fluid, its value differs considerably depending on whether heating occurs at constant volume or constant pressure, the latter being the larger.

heat source, sink Locations where heat enters or leaves a system.

high-pressure zones (highs) Regions of raised atmospheric pressure at mean sea level. Also known as anticyclones.

hoar frost See *frost*.

Holocene Current *interglacial* (last 11,000 years).

humidity The vapour content of air. See *relative* and *specific humidity*.

humidity mixing ratio The mass of vapour as a fraction of the mass of dry air with which it is mixed in a moist air parcel. Numerically indistinguishable from specific humidity in all but the most humid air.

hurricane A severe *tropical cyclone*.

hydrologic cycle The network of pathways of the water substance through the oceans, land surfaces, and atmosphere.

hydrosphere The shell of water which nearly envelopes the Earth in the form of oceans and inland seas. Sometimes includes the *cryosphere*, groundwater, and atmospheric water substance.

hydrostatic equation The formal expression of pure hydrostatic equilibrium.

hydrostatic equilibrium Balance between gravitational downward pull on an air parcel and upward pressure gradient force arising from vertical pressure lapse. Such equilibrium applies very accurately to all but the most violently disturbed parts of the atmosphere.

hygrograph A recording hygrometer.

hygrometer An instrument measuring humidity, often by measuring relative humidity.

hygroscopic Tending to attract and condense ambient water vapour.

ice age Climate epoch cold enough to have permanent seasonal ice at sea level, usually in polar regions.

ideal gas A gas which behaves as if its molecules were infinitely small, interacting only by perfectly elastic collision at the instant of collision, and therefore obeying the equation of state for an ideal gas.

ideal gas constant See *gas constant*.

inertial reference frame A reference frame with zero absolute acceleration.

infrared radiation Electromagnetic radiation lying beyond the red end of the visible spectrum, but not far beyond.

insolation Solar irradiance of a surface, sometimes totalled over a finite time period such as a day.

instability The condition of a body or system which responds to a specified disturbance by increasing the disturbance until an irreversible change has taken place. Sometimes used in context to mean convective instability.

interglacial Short warm spell *in* an *ice age*.

internal energy The sensible heat capacity of an ideal gas as assessed from its specific heat at constant volume.

interstadial Warm interruption of a *glacial*.

Inter-tropical Convergence Zone (ITCZ) The zone of persistent convergence of air flow in the low troposphere in very low latitudes.

inversion An increase of atmospheric temperature with height (an inversion of the normal tropospheric lapse).

ionosphere The region of the upper atmosphere (usually reckoned above 50 km height) which is chronically in a significantly ionized state.

irradiance The energy flux density of electromagnetic radiation impinging on a real or imaginary surface.

isentropic See *dry adiabatic*.

isobar An isopleth of atmospheric pressure.

isobaric At constant pressure.

isopleth A line or surface with uniform value of a specified property.

isopycnal or isopycnic An isopleth of air density.

isotach An isopleth of wind speed.

isotherm An isopleth of air temperature.

isotropic Independent of direction; thus microscale turbulence is nearly isotropic while larger scales are vertically squashed.

jet stream A relatively narrow and shallow stream of fast-flowing air, usually in the high troposphere (see *polar front* and subtropical jet streams).

joule The SI unit of energy.

katabatic winds Air flow down sloping terrain.

kelvin The unit or scale of absolute temperature.

kinetic energy The energy which a body has in respect of its motion.

knot A wind speed of one nautical mile per hour.

lag Sluggishness of a measuring instrument's response to an imposed change, usually measured by the time to make a 67% response.

laminar flow Smooth, viscosity dominated flow.

lapse rate See *temperature lapse rate*.

latent heat The heat which is given out when gases and liquids liquefy or solidify, and which is absorbed when solids or liquids melt or evaporate.

lateral acceleration Acceleration across the direction of flow, e.g. centripetal acceleration in cyclonic or anticyclonic flow.

lee trough A synoptic-scale low pressure system formed or maintained in the lee of a long ridge of high ground.

lifting condensation level The level at which air would become saturated if lifted dry-adiabatically conserving its vapour content (often referring to initial conditions at screen level).

lightning The large electric sparks produced in and around thunderclouds.

linear acceleration Acceleration in the direction of flow.

Little ice age Cool climatic period in N Atlantic and NW Eurasia from about 1400 to 1800 CE.

log wind profile The nearly linear profile of wind speed in the surface boundary layer when plotted against log height.

long waves Near hemispheric scale waves in mid-latitude tropospheric westerlies, the large context in which *cyclone waves* form.

luminosity Total radiant power emission by Sun or star.

mass The quantity of material in a body as assessed by its reluctance to accelerate when acted on by a force when free of all other constraints, or its weight in a gravitational field.

mechanical convection Vertical exchange of air parcels accomplished by turbulence driven by vertical wind shear.

mechanical energy Sum of a body's *kinetic energy* and its gravitational *potential energy*.

melting band The horizontal band of enhanced radar echo produced by snow melting at the 0°C level in extensive precipitating clouds.

meridional Along a line of meridian.

meso-scale Spatial scales intermediate between small and synoptic scales of weather systems.

mesosphere The region of the upper atmosphere lying between the stratosphere and the thermosphere.

microphysics of clouds Physical processes active on the scale of individual cloud and precipitation droplets and particles.

millibar (mbar) The former meteorological unit of pressure; 100 pascals.

mole A mass of pure material whose mass in grams is numerically equal to the material's atomic or molecular weight, and therefore contains Avogadro's number of atoms or molecules. The letter e is often dropped.

momentum The product of the mass and velocity of a body—a vector quantity.

monsoon Pronounced seasonal variation of air flow in the lower troposphere in tropical and subtropical regions—for example, the Indian monsoon. Sometimes used for similar but weaker behaviour in higher latitudes.

neutral stability The state of a system which accepts a specified disturbance without further response. Often used to mean neutral convective stability: the state of an atmospheric layer which accepts imposed vertical displacement of air parcels without positive or negative response (i.e. which is on the boundary between convective instability and stability).

newton SI unit of force—a force which will *accelerate* a mass of 1 kg at one metre per second per second.

nimbostratus An extensive layer of cloud precipitating rain and/or snow.

non-inertial reference frame A reference frame with non-zero absolute acceleration.

normalize To divide a quantity by a more fundamental quantity of the same dimensions to produce a non-dimensional ratio.

Normand's theorem A particular thermodynamic relation between temperature, thermodynamic wet-bulb temperature, and dew-point temperature, expressed in terms of a construction on a tephigram or other thermodynamic diagram.

northerly Flowing or moving from the North.

number density Abundance measure in numbers per unit volume.

occluded front See *front*.

order of magnitude The typical magnitude of a quantity to the nearest integral power of 10.

partial pressure The pressure which would be exerted by a particular component of a gaseous mixture if all other components were removed without other change.

pascal SI unit of pressure, one *Newton* per square metre.

perihelion The time in its annual orbit when Earth is nearest the Sun.

period The interval between consecutive similar stages in a repeating cycle of events.

photochemistry Chemical processes encouraged by incident photons of electromagnetic radiation (solar radiation in meteorology).

photodissociation Molecular dissociation caused by absorption of photons of electromagnetic radiation (solar radiation in meteorology).

photoionization Ionization caused by absorption of photons of electromagnetic radiation (solar radiation in meteorology).

photosynthesis Synthesis of water and carbon dioxide to form oxygen and sugars in the presence of sunlight and photosynthetic organisms.

planetary boundary layer See *atmospheric boundary layer*.

Poisson's equation Equation relating initial and final absolute temperatures and pressures of an ideal gas undergoing dry adiabatic compression or decompression.

polar front The middle-latitude frontal zone separating air flowing from tropical and polar source regions—a favoured site of cyclogenesis.

potential energy The energy of a body (or system of bodies) in respect of its position in a field of force. Used in meteorology to mean gravitational potential energy.

potential temperature The temperature of an air parcel after dry adiabatic compression or decompression from its actual pressure to 1000 mbar.

power Energy input or output per unit time.

precipitable water content Amount of *water vapour* in the atmosphere expressed as a depth of *rainfall*.

precipitation Ice particles or water droplets large enough (~ 100 μm or larger) to fall at least 100 m below cloud base before evaporating. See *drizzle*, *hail*, *rain*, and *snow*.

pressure The apparently continuous and isotropic force exerted on unit area of any real or imaginary surface because of bombardment by molecules of contiguous fluid.

pressure gradient force The net force on an air parcel arising from its location in a pressure gradient.

pressure tendency The rate of change of pressure at a fixed location, usually at the Earth's surface.

psychrometer An instrument measuring vapour content by means of wet- and dry-bulb thermometry.

radar Radar set detecting and ranging reflection of *electromagnetic radiation*, usually in the centimetric wavelength range.

radian The angle subtended at its centre by an arc of a circle equal to its radius.

radiation See *electromagnetic radiation*.

radiant flux Flow of electromagnetic radiant energy per unit time.

radiant flux density Radiant flux through unit area of real or imaginary surface, usually perpendicular to the rays.

radiation fog Fog produced by net radiative cooling of the underlying surface.

radiometer An instrument to measure fluxes of electromagnetic radiation.

radiosonde A balloon-borne package of thermometer, barometer, hygrometer, and radar reflector, for sensing the troposphere and low stratosphere during free ascent.

rain Precipitation in the form of millimetric-sized water droplets (as distinct from drizzle).

rainfall Amount of rain fallen on a surface expressed as a depth of water on a horizontal surface.

rain gauge An instrument to measure totals or rates of rainfall, including drizzle and melted hail and snow.

reference frame A set of moving or stationary spatial axes used as a basis from which to measure body motion and dynamics.

reference process See *thermodynamic reference process*.

reflectivity The proportion of incident radiation reflected by a surface, expressed as a fraction or percentage.

relative acceleration Acceleration measured relative to a specified reference frame, which may itself have absolute acceleration.

relative humidity The vapour content of air (measured as vapour density or pressure) as a percentage of the vapour content needed to saturate air at the same temperature.

relative vorticity Vorticity about the local vertical relative to the local tangential plane of the Earth.

residence time The average time spent by a particle in one particular component of a system of components in dynamic equilibrium (for example, by a water molecule in the oceans between precipitation and the next evaporation).

reversible Of a thermodynamic process which is always so close to equilibrium that it can be reversed by a minute change in the set-up.

Reynolds number The dimensionless ratio of fluid acceleration and accelerations induced by viscosity typical of a particular flow regime.

Richardson number The dimensionless ratio of velocity shears and buoyancy forces typical of a particular flow regime.

ridge An elongated zone of high pressure.

Rossby number The dimensionless ratio of relative accelerations and Coriolis accelerations typical of a particular regime of synoptic-scale flow.

run of wind The length of air flow registered by an anemometer in a certain time period (often an hour or a day).

saturated adiabat A line on a tephigram, or other thermodynamic diagram, corresponding to a saturated adiabatic process.

saturated adiabatic lapse rate *temperature lapse rate* associated with *saturated adiabatic* ascent or descent of an air parcel.

saturated adiabatic process An adiabatic process in which air is kept in an exactly saturated state by evaporation or condensation of water substance.

saturation The state of water vapour which is in dynamic equilibrium with a contiguous plane surface of pure water or ice through balanced fluxes of evaporating and condensing water molecules.

saturation vapour pressure *Partial pressure* of a *saturated* vapour.

scalar A measurable quantity having magnitude but no direction.

scale analysis Analysis of the rough magnitudes of individual terms in dynamic and thermodynamic equations, to determine their relative significance.

scale height The vertical height interval in which a certain atmospheric property (such as pressure) decreases to a certain proportion (such as 1/10) of its value at the base of the interval.

scattering The process of chaotic deflection of electromagnetic radiation by impact on rough surfaces.

screen See *Stevenson screen.*

seeding of clouds See *cloud seeding.*

severe local storm A very severe cumulonimbus.

shear Cross-flow gradient or difference of the direction and/or speed of fluid flow (speed if direction is unspecified).

shear stress The tangential force per unit area of sheared fluid which is associated with the shear through molecular or turbulent exchange across the flow.

slope convection Synoptic-scale ascent and descent associated with extratropical cyclones and anticyclones.

smog Polluted fog.

snow Precipitation in the form of well-developed hexagonal ice crystals, either singly or in conglomerate flakes.

Snowball Earth A glacial so severe that Earth is mostly covered by permanent ice.

solar constant The average value of normal solar irradiance just outside the Earth's atmosphere. The term can be used for other planets.

solarimeter An instrument to measure solar irradiance.

solar radiation Electromagnetic radiation from the Sun, especially in the wavelength range 0.3–3 μm which contains nearly all the total irradiance.

sound Longitudinal pressure waves audible to the human ear.

specific heat capacity The quantity of heat needed to raise the temperature of unit mass of a particular substance by one degree kelvin.

specific humidity The mass of water vapour as a proportion of the total mass of moist air of which it forms a part.

specific mass The mass of a certain material as a proportion of the total mass of mixture in which it is mixed and of which it forms a part.

spectrum The distribution of variance or power across wavelength or frequency.

speed Rate of movement relative to a specified reference frame, irrespective of direction. The magnitude of velocity.

squall A sudden onset of strong, gusty winds.

squall line or line squall A linear organization of vigorous cumulonimbus in the United States.

stability The condition of a body or system which responds to a specified disturbance by opposition and suppression. Often used in meteorology to refer to convective stability in particular.

stadial Cold interruption of an *interglacial.*

state Thermodynamic condition of a material (usually air in meteorology) often specified by *temperature*, *pressure*, and quantity.

station pressure Atmospheric pressure measured at station level before correction to sea level.

steam fog Fog formed by evaporation from a relatively warm water surface into cooler air.

Stevenson screen A white, wooden ventilated box used to allow thermometers to measure air temperature without radiative influence.

storm A general term applied to any type of weather system associated with strong surface winds.

stratified Horizontally layered.

stratopause Temperature maximum between *stratosphere* and *mesosphere.*

stratosphere The relatively quiet atmospheric layer between the troposphere and mesosphere.

streamline A line which is instantaneously parallel to fluid flow.

sublimation See *Sublime.*

sublime Pass from solid to *vapour* states or vice versa, without passing through the liquid state.

subsidence Gentle sinking of air, as in *anticyclones.*

subsidence inversion The temperature inversion in the low troposphere of the subsiding core of an anticyclone.

subtropical high pressure Zones of systematically raised surface pressure between latitudes 20° and 40°.

subtropical jet stream A westerly jet stream centred near the 200 hPa level at the polar extremity of each *Hadley* cell.

superadiabatic lapse rate A rate of fall of temperature with height which exceeds the dry adiabatic lapse rate.

supergeostrophic wind Wind with speed larger than the local geostrophic value.

supersaturation Greater vapour pressure or density than is required for saturation at the prevailing temperature.

synoptic observations Observations made simultaneously at the established synoptic grid of positions.

synoptic scale The minimum horizontal spatial scale of weather system which is well defined by the synoptic observation network.

teleconnections Observed synchronous or sequenced climatic anomalies in remote locations.

temperature The concentration of sensible heat in a body, measured according to an arbitrary scale such as the Celsius and Kelvin scales.

temperature lapse rate The rate of fall of temperature with increasing height. See *dry* and *saturated adiabatic lapse rates*.

tephigram A thermodynamic diagram having temperature and potential temperature as perpendicular axes.

terminal velocity or fall speed The speed at which a particular body's weight is balanced by its drag as it falls through a particular fluid.

terrestrial radiation Electromagnetic radiation emitted by materials at terrestrial surface and atmospheric temperatures.

thermal A buoyant atmospheric eddy, especially those of size ~ 100 m and larger.

thermal conductivity The ratio of heat-flux density to temperature gradient for any particular heat-conducting material.

thermal convection Buoyant convection driven by temperature differences.

thermal diffusivity Thermal conductivity divided by the product of density and specific heat capacity of a heat-conducting material.

thermal wind The vertical shear of geostrophic wind.

thermocline A region below a water surface in which temperature increases upwards.

thermodynamic diagram A diagram whose axes are thermodynamic parameters chosen so that area represents energy.

thermodynamic reference process An *adiabatic* (*dry* or *saturated*) series of thermodynamic states of an air parcel used for comparison with complex atmospheric reality.

thermodynamic wet-bulb temperature The lowest temperature to which an air parcel can be cooled by adiabatic evaporation of water into it.

thermograph A recording thermometer.

thermometer Instrument measuring its temperature against a chosen scale.

thermosphere Top layer of the *turbosphere* and beyond.

thickness The vertical depth of a slab of air bounded by chosen isobaric surfaces.

thunder The audible noise made by lightning.

tornado Intense, cloud-cored vortex extending from the base of a severe local storm to the surface.

torque The turning moment of a force.

total energy Sum of kinetic, *gravitational potential* and *internal energies* of a mass of air.

total potential energy The sum of the internal and gravitational potential energies of an atmospheric column.

trade winds The conspicuously reliable winds blowing obliquely in the low troposphere from a subtropical high to the intertropical convergence zone.

trajectory The path traced by a moving air parcel.

transmissivity The proportion of a flux of electromagnetic radiation which is transmitted through the atmosphere (or part of it).

tropical cyclone A cyclonic disturbance in tropical regions. When surface winds exceed 33 m s^1 it is called a hurricane in the Atlantic, a typhoon in the Pacific, and a cyclone in the Indian Ocean.

tropical storm Tropical cyclone with maximum wind force 8 to 11.

tropopause Boundary between *troposphere* and *stratosphere*, often a well-marked temperature minimum.

troposphere The layer of air continually stirred by weather systems (bounded by the surface and the tropopause).

trough An elongated zone of low atmospheric pressure at a horizontal surface.

turbosphere The atmospheric layer in which diffusive equilibrium is swamped by convective equilibrium (from the surface to about 90 km above sea level).

turbulence Apparently chaotic fluid motion.

typhoon See *tropical cyclone*.

ultraviolet radiation Electromagnetic radiation between the violet end of the visible spectrum and X-radiation.

universal gas constant The constant linking pressure, volume, and absolute temperature of a mole of ideal gas.

unstable The condition of a system which amplifies a particular type of imposed disturbance, the type of instability depending on the type of disturbance, e.g. convective, dynamic, and baroclinic.

upper-air observations Synoptic observations made by radiosonde or equivalent technique.

vacillation Irregular fluctuation between minimum and maximum sinuosity of middle tropospheric flow in middle latitudes.

vapour The gaseous form of a substance. Often used as an abbreviation for water vapour.

vapour pressure The partial pressure of water vapour.

vector A vector quantity has both magnitude and direction.

veering Wind direction changing so that azimuth is increasing.

velocity Rate of movement relative to a specified reference frame, including changes of direction. Velocity is constant only for steady motion in a straight line. See *speed*.

viscosity The friction in gases and liquids arising from molecular exchange and impact. Sometimes called molecular viscosity to distinguish it from eddy viscosity.

vortex Roughly circular flow. Atmospheric vortices appear on scales from centimetric (highly transient) to hemispheric (semi-permanent).

vorticity The measure of fluid rotation. In synoptic-scale meteorology the term usually means the relative vorticity about the local vertical.

Walker circulation Zonal vertical atmospheric circulation across the Pacific along or near the equator.

warm front See *front*.

warm sector The wedge of relatively warm troposphere between a warm front and the following cold front.

water substance Water in any of its physical states—liquid, ice, or vapour.

watt One joule per second—the SI unit of power.

wavelength The shortest distance between adjacent wave crests (or any other similar parts) in a train of waves. In a solitary wave it is taken to be the length of the perceptible distortion.

waves Regular or nearly regular distortions of physical properties which repeat regularly (or nearly so) in space and time.

weather radar *Radar* detecting and ranging reflections from rain, hail, or snow.

weight *Gravitational* force on a body.

westerly Flowing or moving from the West.

wet-bulb depression The depression of wet-bulb temperature below dry-bulb temperature.

wet-bulb potential temperature The wet-bulb temperature after an air parcel is taken saturated adiabatically to 1000 mbar. Equivalently it is the temperature at which the saturated adiabat of the parcel crosses 1000 mbar on a tephigram.

wet-bulb temperature The temperature registered by a wet-bulb thermometer.

wet-bulb thermometer A thermometer which registers the temperature reached by free evaporation from a saturated wick.

wind Air motion relative to the Earth's surface, usually its horizontal component.

wind shear Often the vertical shear of horizontal wind. See *shear*.

wind strength or force Wind assessed by its effect on sea or land surfaces according to the Beaufort scale.

wind stress The horizontal stress exerted on a real or imaginary surface because of adjacent wind shear.

wind vane A device to measure wind direction.

work Product of force and distance moved by its point of application in the direction of the force.

Younger Dryas *Stadial* just before the full onset of the *Holocene*.

zonal Along a line of latitude.

zonal index A pressure index of sinuosity of large-scale tropospheric flow in middle latitudes.

Bibliography, references, and other sources

Textbooks

The following is a selection of books which I have found useful and interesting in ways mentioned below. They were used mainly in writing and teaching during the preparation of the earlier edition, but several were in action again during the preparation of the present edition—being old familiars on my shelves. Some may now be in more recent editions than I have listed. You should find some of them around—they were the big books in their day—and anyway you will sense from my thumbnail comments what I would look for in a textbook in any generation. If you can find a copy of Hess, hang onto it or persuade your library to do so: it is a model of clarity and unpretentious rigour on the theoretical side.

To make full use of my text you will need to refer to books on basic mathematics and physics, because meteorology depends heavily on those disciplines. Which is best for you is a personal matter, but something comprehensive is best for mathematics (like Kreyszig below), so that you can look for useful details as well as principles, and something slimmer but physically clear is best for physics. For the bottom line in honest clarity (not necessarily simplicity) Feynman is often worth a look and always entertaining—mostly Volume I, but a little on fluids in Volume II.

The opinions are of course personal. In every case the best thing is to look and see whether it suits your background and needs.

The order is alphabetical.

Atkinson, B.W. (Ed.) (1981) *Dynamical Meteorology: An Introductory Selection*. Methuen, London.

Although the wide range of authors is responsible for a variety of levels of treatment, the overall coverage is excellent, with good discussions and just enough formalism to show the way to more thorough treatments. Graphics are adequate.

Feynman, R.S., Leighton, R.B., and Sands, M. (1964) *The Feynman Lectures on Physics (Volumes I, II, and III)*. Addison-Wesley Publishing Company, Reading, Mass.

Feynman's conversational and yet very thorough approach to the physical nature of things has become a classic. I often find him picking away at puzzling things which others are content to leave embalmed in an equation.

Goldstein, M. and Goldstein, I.F. (1995) *The Refrigerator and the Universe*. Harvard University Press, Cambridge, Mass.

A really good down-to-earth introduction to thermodynamics to go with the typical functional university text. In the end it goes much further than you need for meteorology, but the grounding is solid, interesting and clear.

Grant, I.S. and Phillips, W.R. (2001) *The Elements of Physics*. Oxford University Press, Oxford.

Good, clear, comprehensive text, uncluttered by a surfeit of pedagogical devices, again as per my general comments.

Hess, S.L. (1959) *Introduction to Theoretical Meteorology*. Holt, Rinehart and Winston, New York.

An old text, famous for its uncluttered clarity of treatment of thermodynamics, radiation, and dynamics (without vectors!) Minimal graphics and little explicit comparison with observation detract but not critically.

Holton, J.R. (1972) *An Introduction to Dynamic Meteorology*. Academic Press, New York.

Coverage as implied by the title but pitched at people with considerable experience in applied mathematics. Though the treatment is quite formal throughout, the associated discussion always encourages physical understanding. The treatment follows the conventional bias towards synoptic-scale dynamics.

Kreyszig, E. (1988) *Advanced Engineering Mathematics*. John Wiley and Sons, New York.

Very compendious and clear. Do not be put off by the thickness and detail; as in my general comments above, it is what you need (with some patience) unless you are going to re-invent the wheel.

Lamb, H.H. (1982) *Climate, History and the Modern World*. Methuen, London.

A brief outline of the ideas and methods of climatology followed by a fascinating survey of climate, mainly since the last glaciation, with special emphasis on social effects.

Lockwood, J.G. (1979) *Causes of Climate*. Edward Arnold, London.

A compact, semi-quantitative survey of climate and climatic change mostly over the last 100,000 years.

Ludlam, F.H. (1980) *Clouds and Storms: The Behaviour and Effect of Water in the Atmosphere*. Pennsylvania State University Press, University Park and London.

A very large and thorough treatment of many aspects of precipitating and non-precipitating clouds (but not thunderstorm electricity). Unusual for combining cloud physics and small-, meso-, and synoptic-scale meteorology, this book is posthumous testimony to the physical insight and elegant treatment typical of this remarkable scientist, whom I was privileged to have as PhD supervisor. Though not a teaching text, it touches usefully on a wide range of observable meteorological phenomena.

McIntosh, D.H. and Thom, A.S. (1969) *Essentials of Meteorology*. Wykeham Publications (Taylor & Francis), London.

A small, concise book covering a wide range of meteorological essentials in surprising depth. The treatment of the tephigram is particularly useful. Dynamics are treated using vector notation. Minimal graphics.

Mason, B.J. (1962) *Clouds, Rain and Rainmaking*. Cambridge University Press.

A classic introduction to the physics of clouds, precipitation, and thunderstorm electricity.

Neiburger, M., Edinger, J.G. and Bonner, W.D. (1982) *Understanding our Atmospheric Environment*. W.H. Freeman, San Francisco.

A substantial introductory text, mostly about meteorology but with just enough about pollution and other human interaction to justify the title. The treatment is mostly qualitative but is nevertheless meteorological (rather than physical geographical) in style. Good graphics.

Oke, T.R. (1978) *Boundary Layer Climates*. Methuen, London.

A good introduction to boundary-layer meteorology and climatology, with a substantial section on anthropogenic effects. The treatment is mainly semi-quantitative but there are many useful tables of values.

Palmen, E. and Newton, C.W. (1969) *Atmospheric Circulation Systems*. Academic Press, New York.

A classic survey of understanding of synoptic- and larger-scale weather systems, with some treatment of organized vigorous small-scale convection. Although the treatment is not at the introductory level, the discussion is full and rewards patient study.

Riehl, H. (1978) *Introduction to the Atmosphere*. McGraw-Hill Kogakusha Ltd, Tokyo.

Probably the best purely qualitative introduction to meteorology, with good photographs in the later editions.

Rogers, R.R. (1976) *A Short Course in Cloud Physics*. Pergamon Press, Oxford.

A thorough treatment of the whole range of meteorological cloud physics with the curious exception of thunderstorm electricity.

Scorer, R.S. (1978) *Environmental Aerodynamics*. Ellis Horwood, Chichester.

An unusual and stimulating book, beginning with a formal vector treatment of fluid dynamics which is well beyond introductory level, but then proceeding to a discussion of many smaller-scale aspects of atmospheric behaviour with useful emphasis on physical reality as well as theoretical nicety. It includes related topics such as chimney-plume dispersion and bird soaring.

Simpson, R.H. and Riehl, H. (1981) *The Hurricane and its Impact*. Basil Blackwell, Oxford.

A largely qualitative description of these fearsome storms with considerable emphasis on their destructive power.

Stanley, S.M. (1993) *Exploring Earth and Life through Time*. W. H. Freeman, New York.

A descriptive survey like this will help you sense the awesome and beautiful context of Earth's history in which to set any approach to climate and climate change.

Turco, R.P. (1997) *Earth under Siege*. Oxford University Press, Oxford.

This covers a wide range of physical and chemical aspects of the Earth, with special reference to anthropogenic effects. The coverage of geochemical cycles is especially useful to fill out my brief outline.

Wallace, J.M. and Hobbs, P.V. (1977) *Atmospheric Science: An Introductory Survey*. Academic Press, New York.

A modern text on fairly traditional lines, with very solid thermodynamics and radiation, but a much lighter treatment of dynamics. The coverage of cloud, precipitation, and electrical processes is particularly detailed. Diagrams and photographs abound.

Wayne, R.P. (1985) *Chemistry of Atmospheres*. Clarendon Press, Oxford.

A very full account of the chemistry of the terrestrial and other planetary atmospheres.

Wickham, P.G. (1970) *The Practice of Weather Forecasting*. HMSO, London.

An excellent summary of synoptic-scale weather analysis and manual forecasting with lots of maps and soundings from actual cases.

Journals

Most meteorological journals are pitched well beyond the introductory level, though they can be skimmed with profit by looking at diagrams and conclusions before trying to understand the physical meaning of the more formal sections. The popular journal of the Royal Meteorological Society (*Weather*) is an important exception. Though the level of articles is quite variable, there is a good range from the practical to the theoretical in each monthly issue, with a useful policy of public education.

Royal Meteorological Society, 104 Oxford Road, Reading, Berkshire, RG1 7LJ.

The American Meteorological Society publishes a monthly *Bulletin* for its members and others which contains technical reviews and updates which are often informally and accessibly expressed.

45 Beacon Street, Boston, Ma 02108, USA.

Scientific American publishes review articles on climatological and other atmospheric matters (amongst others) in a clear and graphic style by world authorities in the subjects.

Tephigrams

British Meteorological Office blank tephigrams (Metform 2810) from Her Majesty's Stationery Office can be ordered through The Government Bookshop, PO Box 569, London SE1 9NH. Other materials are listed and may be available through the Marketing Services section of the Meteorological Office, London Rd, Bracknell, Berks RG12 2SZ, UK.

References (numbered in the text)

1 Hobbs, P.V. (2000). *Introduction to Atmospheric Chemistry.* Cambridge University Press, Cambridge.

2 Lorenz, E. (1993). *The Essence of Chaos.* UCL Press, London.

3 Prigogine, I. (1980). *From Being to Becoming.* W.H. Freeman, San Francisco.

4 Lucretius, trans. R. Latham (1966). *The Nature of the Universe.* Penguin, Harmondsworth.

5 Met. Office (1969). *Handbook of Meteorological Instruments.* HMSO, London.

6 Met. Office (1982). *Observer's Handbook.* HMSO, London.

7 Green, J.S.A. (1956). *Weather*, **22**:4, 128–31.

8 Kreyszig, E. (1988). *Advanced Engineering Mathematics.* John Wiley and Sons, New York.

9 Met. Office (1964). *Hygrometric Tables, Part II Screen Values, Part III Aspirated Values.* HMSO, London.

10 Tritton, D.J. (1988). *Physical Fluid Dynamics, 2nd Edn.* Clarendon Press, Oxford.

11 Ward, R.C. and Robinson, M. (1989). *Principles of Hydrology, 3rd Edn.* McGraw-Hill, New York.

12 Met. Office (1982). *Cloud Types for Observers.* HMSO, Norwich. See [59].

13 Wickham, P.G. (1970). *The Practice of Weather Forecasting.* HMSO, London.

14 Hargreaves, J.K. (1979). *The Upper Atmosphere and Solar–Terrestrial Relations.* Van Nostrand Reinhold, Wokingham.

15 Turco, R.P. (1997). *Earth Under Siege.* Oxford University Press, Oxford.

16 Graedel, T.E. and Crutzen, P.J. (1993). *Atmospheric Change: An Earth System Perspective.* W.H. Freeman, San Francisco.

17 Wallace, J.M. and Hobbs, P.V. (1977) *Atmospheric Science: An Introductory Survey.* Academic Press, New York.

18 Press, F. and Siever, R. *Earth.* W. H. Freeman, San Francisco.

19 Lovelock, J. (1988). *The Ages of Gaia.* Oxford University Press, Oxford.

20 Zemansky, M.W. (1957). *Heat and Thermodynamics.* McGraw-Hill, New York.

21 Barry, R.G. (1981). *Mountain Weather and Climate.* Methuen, London.

22 Blundell, S.J. and Blundell, K.M. (2006). *Concepts in Thermal Physics.* Oxford University Press, Oxford.

23 Uman, M.A. (1984). *Lightning.* Dover Publications, New York.

24 Kibble, T.W.B. and Berkshire, F.H. (2004). *Classical Mechanics.* Imperial College Press, London.

25 Goldstein, M. and Goldstein, I.F. (1995) *The Refrigerator and the Universe.* Harvard University Press, Cambridge, Mass.

26 Hess, S.L. (1959) *Introduction to Theoretical Meteorology.* Holt, Rinehart and Winston, New York.

27 Ludlam, F.H. and McIlveen, J.F.R. (1969). *Meteorol. Mag.* **98**, 233–246.

28 Rogers, R.R. (1976) *A Short Course in Cloud Physics.* Pergamon Press, Oxford.

29 Ludlam, F.H. (1980) *Clouds and Storms: The Behaviour and Effect of Water in the Atmosphere.* Pennsylvania State University Press, University Park and London.

30 Wikipedia. *Lightning.*

31 Eady, E.T. (1949) *Tellus*, 1:3, 33–52.

32 National Weather Service (1979). *NWS Observing Handbook no. 1* (1979). US Department of Commerce, NOAA, Facsimile Products, Washington DC.

33 Houghton, J. (2002). *Physics of Atmospheres, 3rd Edn.* Cambridge University Press, Cambridge.

34 Goldstein, H. (1959). *Classical Mechanics.* Addison-Wesley, Reading, Mass.

35 Holton, J.R. (1972) *An Introduction to Dynamic Meteorology.* Academic Press, New York.

36 Scorer, R.S. (1997). *Dynamics of Meteorology and Climate.* John Wiley and Sons, Chichester. A greatly extended version of Scorer, R.S. (1978) *Environmental Aerodynamics.* Ellis Horwood, Chichester.

37 London, J. and Sasamori, T. (1971) *Space Res.*, **XI**, 639–49.

38 Monteith, J.L. (1989). *Principles of Environmental Physics.* Edward Arnold, London.

39 Pickard, G.L. and Pond, S. (1983). *Introductory Dynamical Oceanography.* Butterworth-Heinemann, Oxford.

40 Feynman, R.S., Leighton, R.B., and Sands, M. (1964) *The Feynman Lectures on Physics (Volumes I, II, and III).* Addison-Wesley Publishing Company, Reading, Mass.

41 Van Mieghem, J. (1973). *Atmospheric Energetics.* Clarendon Press, Oxford.

42 Oke, T.R. (1987). *Boundary Layer Climates, 2nd Edn.* Routledge, London.

43 Stewart, I. (1997). *Does God Play Dice?: The Mathematics of Chaos.* Penguin, London.

44 Lettau, H.H. and Davidson, B. (Eds). (1957). *Exploring the Atmosphere's First Mile, Vols I and II.* Pergamon Press, London.

45 Solomon, S. (2001). *The Coldest March.* Yale University Press, New Haven.

46 Aanenson, C.J.M. (1962). *Gales in Yorkshire in February 1962.* Met. Office Geophysical Memoir, HMSO, London.

47 The Tornado and Storm Research Organization (TORRO). www.torro.org.uk

48 Kuettner, J.P. (1971). *Tellus*, **23**, 404–26.

49 Agee, E.M., Chen, T.E., and Dowell, K.E. (1973). *Bull. Amer. Met. Soc.*, **54**:10, 1004–12.

50 Defoe, D. (1704, 2005). *The Storm.* Penguin, London.

51 Brown, S. (1962.) *World of the Wind.* Bobbs-Merrill, Indianapolis.

52 Minnaert, M. (1954). *The Nature of Light and Colour in the Open Air.* Dover, New York.

53 Carlson, T.N. (1980). *Mon. Ewa. Rev.*, **108** Oct, 1498–509.

54 Mansfield, D.A. (1974). *Quart. J. Roy. Met. Soc.*, **100**:427, 541–54.

55 Harrold, T.W. and Browning, K.A. (1969). *Quart. J. Roy. Met. Soc.*, **95**:406, 710–23.

56 Simpson, R.H. and Riehl, H. (1981) *The Hurricane and its Impact.* Basil Blackwell, Oxford.

57 Hughes, R. (1957). *In Hazard.* Amongst others, Penguin, Harmonsdworth.

58 Manley, G. (1952). *Climate and the British Scene.* Collins, London.

59 World Meteorological Organization (1975) *International Cloud Atlas.* WMO, Geneva.

60 Manley, G. (1974). *Quart. J. Met. Soc.*, **100**:425, 389–485.

61 Lamb, H.H. (1982). *Climate, History and the Modern World.* Methuen, London.

62 Walker, G. (2006). *Snowball Earth.* Bloomsbury, London.

63 National Oceanic and Atmospheric Administration, US Department of Commerce. www.noaa.gov.us.

64 Benton, M.J. (2008). *When Life Nearly Died.* Thames and Hudson, London.

65 Van Andel, T.H. (1994). *New Views on an Old Planet.* Cambridge University Press, Cambridge.

66 Budyko, M.I. (1974). *Climate and Life.* Academic Press, New York.

67 Raymo, M.E. (1994). *Ann. Rev. Earth Planet. Sci.*, **22**, 353–83.

68 Turco, R.P., Toon, A.B., Ackerman, T.P., Pollack, J.B. and Sagan, C. (1983). *Science*, **222**:4630, 1283–92.

69 K McGuffie, K. and A Henderson-Sellers, A. (2005). *A Climate Modelling Primer, 3rd Edn.* John Wiley and Sons, Chichester.

70 Intergovernmental Panel on Climate Change (IPCC). WMO, UNEP. 4th Assessment report (AR4), 2007. AR5 due 2014. www.ipcc.ch

71 Munn, R.E. (1966). *Descriptive Micrometeorology.* Academic Press, New York.

72 Herman, J.R. and Goldberg, R.A. (1978, 1985) *Sun, Weather and Climate.* Dover Publications, New York.

73 Simpson, J.E., Mansfield, D.A. and Milford, J.R. (1977). *Quart. J. Roy. Soc.*, **103**, 47–76.

74 Saunders, P.M. (1961). *J. Meteorol.*, **18**, 451–67.

75 Ludlam, F.H. (1966). *Tellus*, **XVIII**:4, 688–98.

76 Palmen, E. and Newton, C.W. (1969) *Atmospheric Circulation Systems.* Academic Press, New York.

77 Collinge, V. et al. (1990). *Weather*, **45**, 10.

78 Browning, K.A., (1971). *Weather*, **27**, 320–40.

79 Matejka, T.J., Houze, R.A. and Hobbs, P.V. (1980). *Quart. J. Roy. Met. Soc.*, **106**, 29–56.

80 Petterssen, S. (1956). *Weather Analysis and Forecasting. Vol 1, Motion and Motion Systems*. McGraw-Hill, New York.

81 Godske, C.L., Bergeron, T., Bjerknes, J. and Bundgaard, R.C. (1957). *Dynamic Meteorology and Weather Forecasting*. American Meteorological Society, Boston, Mass. and Carnegie Institute, Washington, DC.

82 Lyndulph, P.E. (1977) *Climates of the Soviet Union*, Vol. 7 of Landsberg, H.E. (Ed.) *World Survey of Climatology*, Elsevier Scientific, Amsterdam.

83 Flohn, H. (1969). *Climate and Weather*. Weidenfeld and Nicolson, London.

84 Stevenson, C.M. (1967). *Weather*, **23**, 156–60.

85 Ramage, C.S. (1971). *Monsoon Meteorology*. Academic Press, New York.

86 Simpson, R.H. and Riehl, H. (1981) *The Hurricane and its Impact*. Basil Blackwell, Oxford.

87 Barry, R.G. and Chorley, R.J. (1989). *Atmosphere, Weather and Climate*. Methuen, London.

88 Lancaster University, Hazelrigg Weather Station. Met. Office Climatological Station Number 7236. www.es.lancs.ac.uk/hazelrigg/

89 Historical Central England Temperature Data (HadCET). Met. Office. badc.nerc.ac.uk/data/cet/

90 Stanley, S.M. (1993) *Exploring Earth and Life through Time*. W. H. Freeman, New York.

91 Gerlach, T.M. (1991) Transactions of the Amer. Geophys. Union, 72, no. 4, 249, 254–6.

92 McLean, D.M. (1985) *Cretaceous Research*, 6, 235–9.

Sources of information

- Most major national meteorological organizations operate useful and interesting websites. The National Weather Service of the USA is particularly impressive, but the UK Met. Office and Meteo France and others are full of information, provided you ignore the cartoon maps produced for casual consumption. There are useful web pages on meteorological topics for those with more than a casual interest. More detailed information is often freely available through the NWS, but less so from the Met. Office unless you can demonstrate an educational need. Tephigrams and SkewTlogP diagrams and technical manuals abound (and in the World Meteorological Organization, Geneva) but are often very functional and hard to reach.
- Most meteorological journals are pitched well beyond introductory level, though they can be skimmed with profit by looking at diagrams and conclusions before trying to follow the more formal sections. The popular journal of the Royal Meteorological Society (*Weather*) is an important exception. The technical level is accessible to the non-specialist, without being slipshod, and there is a reasonable range of types of topic from the observational to the theoretical in each monthly issue, with a good policy on public education.

  www.rmets.org/index.php

  The American Meteorological Society publishes a monthly *Bulletin* for its members and others which contains technical reviews and updates which are often informally and accessibly expressed.

  www.ametsoc.org/pubs/

  The international weekly journals *Science* and *Nature* publish occasionally in the fields of meteorology, climate, and climate change, and these are especially useful when summarized or commented on for non-specialists in the topics.

  www.sciencemag.org/ www.nature.com/nature/journal/

  At a more popular level, *Scientific American* and the *New Scientist* publish articles in our fields, sometimes by scientific reporters, and sometimes by leaders in their fields, often with informative diagrams.
- The quality of media reports on the science of weather and climate ranges randomly from the useful to the useless, even in the most respected newspapers and channels, because of the very variable competence of science reporters, the pressures of page space and air time, and an almost willful failure to understand the nature of scientific debate.

The web and its search engines have revolutionized the whole basis of searching for and checking on information about weather and climate. Obviously you have to be critical, but that is good practice anyway. Most pages I have consulted in Wikipedia have been useful. Bloggers are often obviously in the grip of conspiracy paranoia, and therefore harmless. More subtle ones could be a problem, but remember that Alfred Wegener (the prophet of continental drift) was dismissed as an eccentric loner for much of his life and for 20 years after his untimely death in 1930.

Index

H

I

J

K

L

W

Z

Made in the USA
Lexington, KY
21 August 2014